STONE AGE INSTITUTE PUBLICATION SERIES

Series Editors Kathy Schick and Nicholas Toth

Stone Age Institute

Gosport, Indiana

and

Indiana University,

Bloomington, Indiana

Number 1.

THE OLDOWAN: Case Studies into the Earliest Stone Age

Nicholas Toth and Kathy Schick, editors

Number 2.

BREATHING LIFE INTO FOSSILS:

Taphonomic Studies in Honor of C.K. (Bob) Brain

Travis Rayne Pickering, Kathy Schick and Nicholas Toth, editors

Number 3.

THE CUTTING EDGE:

New Approaches to the Archaeology of Human Origins

Kathy Schick and Nicholas Toth, editors

Number 4.

THE HUMAN BRAIN EVOLVING:

Paleoneurological Studies in Honor of Ralph L. Holloway

Douglas Broadfield, Michael Yuan, Kathy Schick and Nicholas Toth, editors

---

The Stone Age Institute Press Publication Series

Comment by Series Editors

The Stone Age Institute Press has been established to publish critical research into the archaeology of human origins. The Stone Age Institute is a federally-approved non-profit organization whose mission is investigating and understanding the origins and development of human technology and culture throughout the course of human evolution, and to disseminate information to the general public regarding science education, in particular about our evolutionary origins. The ultimate goal of these undertakings is to provide a better understanding of the human species and our place in nature.

An unbroken line of technology and culture extends from our present day back in time at least two-and-a-half million years. The earliest stone tool-makers were upright-walking, small-brained ape-men, and from these primordial origins, the human lineage embarked upon a pathway that is extraordinary and unique in the history of life. This evolutionary pathway has shaped our bodies, our brains, and our way of life, and has had an ever-accelerating impact on the earth and its other organisms. Well over 99 percent of the time span of this technological evolution occurred in the Stone Age, so it is no exaggeration to say that modern humans have evolved as Stone Age creatures.

The Stone Age Institute's goal is to provide a haven for this research, providing support for scholars from around the world conducting cutting-edge human origins research. Our research team and associates include professional research scientists in long-term positions, postdoctoral research fellows, and visiting scholars. Fieldwork, collecting primary data from important archaeological sites, as well as laboratory studies and experimental archaeological research are all critical components of our operation. Other vital functions of the Institute include organizing workshops - convening top scientists to discuss and work together on special topics - and organizing and sponsoring conferences, lectures, and internet websites to disseminate information to a broader, general public.

An equally critical component of our operations is the publication of research results in journal articles and books. *The Human Brain Evolving: Paleoneurological Studies in Honor of Ralph L. Holloway* is the fourth volume in a new publication series initiated by the Stone Age Institute Press. This volume emanates from a Stone Age Institute conference in honor of the lifework of Ralph Holloway and brain evolution. This meeting convened researchers at Indiana University and the Stone Age Institute for two days in April of 2007 to present a wide variety of research pertaining to the evolution of the human brain. The Stone Age Institute Press is devoted to the publication of seminal research in the form of high-quality, data-rich volumes, including information and interpretations from original research as well as critical syntheses of research results.

For more information on the Stone Age Institute, including its research programs, lectures, conferences, and publications, we invite you to visit our web site at *www.stoneageinstitute.org*. To explore our most recent science education project on evolution and "Big History," visit the web site *www.fromthebigbang.com*.

Kathy Schick and Nicholas Toth
Co-Directors, Stone Age Institute
Series Editors, Stone Age Institute Press Publication Series

STONE AGE INSTITUTE PUBLICATION SERIES
NUMBER 4
Series Editors Kathy Schick and Nicholas Toth

THE HUMAN BRAIN EVOLVING:

Paleoneurological Studies in Honor of Ralph L. Holloway

Editors

Douglas Broadfield
Florida Atlantic University

Michael Yuan
Columbia University

Kathy Schick
Stone Age Institute & Indiana University

Nicholas Toth
Stone Age Institute & Indiana University

Stone Age Institute Press · www.stoneageinstitute.org
1392 W. Dittemore Road · Gosport, IN 47433

FRONT COVER CAPTIONS

Center: Portrait of Ralph L. Holloway.
Upper left: A modern human brain.
Upper right: Ralph measuring landmarks on an endocast ca. 1976.
Lower right: Homo habilis cranium KNM-ER-1813 from Koobi Fora, Kenya (photo by Holloway).
Lower left: Ralph with an endocast of the Flores "hobbit" cranium.

Published by the Stone Age Institute.
ISBN-10: 0-9792276-3-1
ISBN-13: 978-0-9792276-3-9

Dedication

To
Michael Sheng-Tien Yuan
(1959-2008)

Michael Sheng-Tien Yuan, D.D.S., M.A., M.S., Ph.D. was born in Taiwan (ROC) on November 17, 1959. In an earlier life before he chose orthodontics and anthropology as his careers, Michael was an internationally recognized actor, earning acclaim for his role in the Taiwanese motion picture *Jade Love* (1984), which received a number of Golden Horse Awards and was featured at film festivals around the globe. In 1988 after completing his dental degree at National Taiwan University and two years of compulsory military service, Michael moved to New York to work on his Master's in orthodontics from Columbia University. After completing his orthodontics degree Michael entered the Ph.D. program in Anthropology at Columbia. Shortly after entering the anthropology program Michael had his last brush with the acting bug when he was offered the lead in Ang Lee's first American film, *The Wedding Banquet* (1993). Michael turned down the offer, choosing instead to focus on his new career.

Michael was Ralph Holloway's student from the moment he entered the anthropology program. He worked closely with Ralph and Doug Broadfield on a number of endocast projects, bringing his ample artistic skills to the science of paleoneurology. Though his dissertation work was on dental development, Michael is most remembered in anthropology for his contributions to our understanding of human brain evolution. After completing his Ph.D. in 2000, Michael moved back to Columbia's College of Dental Medicine as an Assistant Professor of Clinical Dental Medicine. As with everything he did Michael threw himself entirely into his new profession, anatomist. He earned a third dental degree in 2003.

In his short tenure at Columbia, Michael became one of the most beloved instructors in not only the College of Medical Dentistry, but also Columbia's College of Physicians and Surgeons, earning three Teacher of the Year awards. He was promoted to Associate Professor in 2008. After Michael became ill he continued to teach from his hospital bed, sending the medical and dental students details of the various procedures he underwent, and relating them to the current anatomical region the students were learning. True to the way he lived, the dedication on his last teaching award summarizes our feelings for Michael: …in appreciation for his "wisdom, gentleness, and ability to … accept life's challenges and use them to grow."

Contributors

John M. Allman (cebus@caltech.edu) Division of Biology, California Institute of Technology, Pasadena, California, USA

Nicole Barger (nbarger@ucsd.edu) Department of Anthropology, University of California at San Diego, USA

Doug Broadfield (broadfie@fau.edu) Department of Anthropology, Florida Atlantic University, Boca Raton, Florida, USA

Emiliano Bruner (emiliano.bruner@cenieh.es) Centro Nacional de Investigación sobre la Evolución Humana, Burgos, Spain

Daniel P. Buxhoeveden (buxhoeve@gwm.sc.edu) Department of Psychology, University of South Carolina, Columbia, South Carolina, USA

Joseph M. Erwin (agingapes@gmail.com)Foundation for Comparative and Conservation Biology, Needmore, PA

Dominique Grimaud-Hervé (dgherve@mnhn.fr) Département de Préhistoire, Muséum national d'Histoire naturelle, Paris, France

Patrick R. Hof (patrick.hof@mssm.edu) Department of Neuroscience, Mount Sinai School of Medicine, New York, NY

Ralph L. Holloway (rlh2@columbia.edu) Department of Anthropology, Columbia University, New York, New York, USA

Karin Isler (kisler@aim.uzh.ch) Anthropological Institute & Museum, University of Zürich-Irchel, Zürich

Jason A. Kaufman (kaufman@caltech.edu) Division of Biology, California Institute of Technology, Pasadena, California, USA

James E. King (kingj@U.Arizona.edu) *Emeritus*, Department of Psychology, University of Arizona, Tucson, Arizona, USA

Bruce Lahn (blahn@bsd.uchicago.edu) Department of Human Genetics, The University of Chicago, Chicago, Illinois, USA

David Lordkipanidze Georgian National Museum, Departement of Geology and Paleontology, Academy of Sciences of Georgia, Tbilissi, Georgia

Alan Mann (mann@Princeton.edu), Department of Anthropology, Princeton University, Princeton, New Jersey, USA

Robert D. Martin (rdmartin@fieldmuseum.org) Department of Anthropology, The Field Museum, Chicago, Illinois, USA

Janet Monge (jmonge@sas.upenn.edu), Department of Anthropology, University of Pennsylvania, Philadelphia, Pennsylvania, USA

Francine "Penny" Patterson The Gorilla Foundation, Woodside, California, USA

Mary Ann Raghanti (mraghant@kent.edu) Department of Anthropology and School of Biomedical Sciences, Kent State University, Kent, OH

James K. Rilling (jrillin@emory.edu) Department of Anthropology and Center for Behavioral

Neuroscience, Emory University, Department of Psychiatry and Behavioral Sciences, Emory University School of Medicine and Division of Psychobiology, Yerkes National Primate Research Center, Atlanta, Georgia, USA

Duane M. Rumbaugh (drumbaug@aol.com) *Emeritus*, Language Research Center and Department of Psychology, Georgia State University, Atlanta, Georgia, USA

E. Sue Savage-Rumbaugh (suerumbaugh@gmail.com) Great Ape Trust, Des Moines, Iowa, USA

Natalie Schenker (nmschenker@ucsd.edu) Department of Anthropology, University of California at San Diego and Department of Anthropology, The George Washington University, Washington, D.C., USA

Kathy Schick (kaschick@indiana.edu) Stone Age Institute and Indiana University, Bloomington, Indiana, USA

P. Tom Schoenemann (toms@indiana.edu) Anthropology Department, Indiana University, Bloomington, Indiana and Stone Age Institute, USA

Katerina Semendeferi (ksemende@ucsd.edu) Department of Anthropology and Graduate Program in Neurosciences, University of California at San Diego, USA

Chet C. Sherwood (sherwood@gwu.edu) Department of Anthropology, The George Washington University, Washington, D.C.

Francys Subiaul (subiaul@aol.com) The George Washington University, Mind, Brain & Evolution Center, Department of Speech & Hearing Science, Washington, D.C., USA

Jared Taglialatela (jtaglial@kennesaw.edu, jtaglia@emory.edu) Department of Biology, Kennesaw State University and Yerkes National Primate Research Center, Emory University, Atlanta, Georgia, USA

Nicholas Toth (toth@indiana.edu) Stone Age Institute and Indiana University, Bloomington, Indiana, USA

J. Michael Tyszka (jmt@caltech.edu) Brain Imaging Center, California Institute of Technology, Pasadena, California, USA

Eric J. Vallender (eric_vallender@hms.harvard.edu) New England Primate Research Center, Harvard Medical School, Southborough, Massachusetts, USA.

Anne H. Weaver, Ph.D. (oneweaver@msn.com) Anthropology Department, University of New Mexico, Albuquerque, New Mexico, USA

Michael Yuan (deceased) Columbia University, New York, New York, USA

THE HUMAN BRAIN EVOLVING: PALEONEUROLOGICAL STUDIES IN HONOR OF RALPH L. HOLLOWAY

EDITED BY

DOUGLAS BROADFIELD, MICHAEL YUAN, KATHY SCHICK AND NICHOLAS TOTH

STONE AGE INSTITUTE PRESS, 2010

CONTENTS

Preface

For those who work in the area of human brain evolution Charles Duell's, Commissioner of the United States Patent Office, 1899 mythical quip that "Everything that can be invented has been invented," seems abundantly appropriate to the nearly fifty year long career of Ralph L. Holloway, Jr. As each of us has struggled to introduce the field to a supposedly new idea, we each at some point have run headlong into a similar idea proposed by Ralph some ten, twenty, or forty years earlier. While this may seem dejecting, few researchers in the field have ever been made to feel this way by Ralph. Instead we and the field in general have been propelled forward by Ralph's boundless curiosity.

On April 27 – 28, 2007 researchers from Europe and the United States gathered at the Stone Age Institute and Indiana University to celebrate and pay tribute to Ralph Holloway's unparalleled contributions to the field of human brain evolution. Reflecting the diversity of Ralph's research, the symposium was an eclectic mix of the top minds in the field who over a two period synthesized ideas from the fossil record, state-of-the-art imaging, neuroscience, behavior and genetics, culminating in this festschrift, *The Human Brain Evolving: Paleoneurological Papers in Honor of Ralph L. Holloway*.

The content of the symposium ranged from the latest unpublished findings, modern revisions of previously proposed hypotheses of brain evolution, and new syntheses of current information. The papers presented in this volume represent in part many of the ideas presented at the symposium as well as reflections on the diverse discussions of the topics covered and the presentation of new data collected in response to conversations held at the symposium, making *The Human Brain Evolving: Papers in Honor of Ralph L. Holloway* unique in that a collection of such diverse topics in the field of human brain evolution have never been presented before in one place.

The symposium leading to this volume was the culmination of discussions that began back in the mid-1990s between one of our beloved colleagues to whom this book is dedicated, Michael Yuan, and Doug Broadfield. At that time Ralph discussed his retirement as eminent. Originally, Ralph had planned to retire from Columbia University around 2007, making the timing of the symposium apropos. The fruition of the symposium is due in large part to Kathy Schick and Nick Toth, co-directors of the Stone Age Institute, who not only threw the entire assets of the institute behind the planning and execution of the symposium, but also sacrificed their personal time and energy to put together what for many was a flawless and memorable weekend. In addition, the symposium was only possible through the generous support of not only the Stone Age Institute, but also the College of Arts and Sciences of Indian University, the Office of the Provost of Indiana University, the Indiana University Foundation, Carol Travis-Henikoff, Anthony Hess and Richard Foley. The conversations generated by the various talk were often sparked by the discussants, Bill Kimbel and Leslie Aiello, who managed to shed new light on the topics present through deft synthesis of the data and hypotheses. Finally, we are indebted to Mila Norman and Blaire Hensley-Marschand of the Stone Age Institute who worked behind the scenes to make sure no detail in pulling off the symposium was missed, and to Amy Sutkowski and Lawrence Buchanan for their help in the layout and design of this volume.

Often when a symposium is planned the organizers attempt to draw together a slate of experts in the field that often have a relationship with the honoree. In many ways a symposium of this nature is a reflective look back on the esteemed career of a beloved colleague by friends and former students. In determining the list of attendees it was necessary to pare a list of nearly one hundred individuals down to approximately twenty-five. While this unenviable task is often relegated to the organizers, we also solicited Ralph's input. The result was a menagerie of experts less chosen for their close personal relationship with Ralph and more simply because Ralph was fascinated by an individual's research. As a consequence of satisfying Ralph's unrivaled curiosity, the symposium was not just a meeting of minds, but also a meeting of firsts. It was one of the first symposia that brought together individuals from what before then were widely disparate fields such as paleoanthropology and molecular biology. It was also ironically the first time Ralph had ever met some of the participants in person. That someone would participate in a symposium for an individual they had never met is a genuine reflection of the admiration and respect scientists the world over have for Ralph Holloway.

Doug Broadfield
October 2010

Introduction

Douglas Broadfield, Kathy Schick and Nicholas Toth

The human brain is arguably the most important product of our evolution. How our early ancestors moved about the landscape looks to the question, when did hominins separate from the ancestor we share with chimpanzees, but it does not answer the question, when did our ancestors begin to think like us? It is assumed that the answer to this question is as simple as looking at the fossil record and recording the time when brain size increased above that of other apes. While early researchers such as Keith, Smith, and Broca among others assumed that human cognition is the result of brain size, more recent studies have done much to elucidate our understanding of human brain evolution.

For over a century researchers from all areas of science have weighed in on the evolution of the human brain usually drawing from their own knowledge in medicine, neuroscience, behavior or paleontology. In many ways the study of human brain evolution was viewed more as a hobby rather than a research career path. The result was a confusing oversimplified picture of brain evolution. However, this armchair approach to the subject all changed with Ralph Holloway's 1967 publication *The Evolution of the Human Brain* in the journal *General Systems*, albeit he did not know quite yet what to make of endocasts. In less than ten years, though, Holloway (with credit also going to Harry Jerison) had established a new field of study, *paleoneurology*, or the study of nervous system evolution.

From April 27-28, 2007 researchers from North America and Europe came together at Indiana University and the Stone Age Institute for *The Human Brain Evolving: Papers in Honor of Ralph L. Holloway* to celebrate the achievements of Ralph Holloway as well as to present the current status of the fledgling field of paleoneurology and to discuss its future. The result of the conference as demonstrated in the works presented here is that Holloway's seminal work has been responsible for turning paleoneurology into a dynamic field that cuts across all disciplines. The most noticeable characteristic of the works presented here is that they are not limited to a single line of study, say endocasts. Instead paleoneurology has evolved into a diverse field that draws data from every available technique, including genetics, behavior, the fossil record and imagining modalities. As a result, the chapters of this volume are loosely arranged into seven different topical themes.

The first set of papers looks at the theoretical concept of brain evolution. **Chapter 1** by Ralph Holloway is a retrospective originally published in *Annual Review of Anthropology* that sincerely captures the evolution of the field. What is most revealing in this chapter is that Holloway like others before him had difficulty early in his career accepting the validity of fossil endocasts as analytical tools. The reader will also learn about the principal themes of brain evolution first developed by him that still dominate the field today. **Chapter 2** by Robert Martin and Karin Isler is the result of 30 years of research on the Maternal Energy Hypothesis, an idea Martin first introduced in 1981 that examines the developmental strategies that go into the evolution of large brains. The Maternal Energy Hypothesis looks at mammalian brain evolution in general, but the concepts of the hypothesis are directly applicable to human brain evolution. The last chapter in this section, **Chapter 3** by Tom Schoenemann, takes a closer look at a specific problem in human brain evolution, the importance of brain size. Here Schoenemann argues that absolute increases in brain size observed in the fossil record likely resulted in concomitant alterations in cognition, spurring the development of features such as language.

The next section of papers looks at evidence that focus on the hard evidence for brain evolution, endocasts. **Chapter 4** by Ralph Holloway presents a new interpretation of the controversial LB1, *Homo floresiensis*, endocast. Here it is suggested that LB1 displays characters in the endocast that are similar to modern microcephalics while also possessing features that may be autapomorphic. The bottom line, though, according to Holloway is that the jury is still out on the status of LB1 until more specimens can be found. In **Chapter 5**, Dominique Grimaud-Hervé and David Lordkipanidze take an in-depth look at a couple of possible ancestors to *H. floresiensis* in the Dmanisi fossils D 2280 and D 2282. Like LB1 these have been controversial for their taxonomic designations based on a unique suite of features absent in contemporaneous fossils. Based on the affinities observed in these endocasts to early members of *H. erectus* the authors conclude that D 2282 may be *H. ergaster* while D2282 is likely *H. erectus*. **Chapter 6** by Emiliano Bruner examines the evolution of the parietal cortex in later human groups, primarily Neanderthals and early modern humans. Through a morphometric analysis of endocasts, Bruner proposes that the parietal lobes of some late hominins increased allometrically with brain size, but that in modern humans the parietal lobes increased non-allometrically or are larger than what would be predicted for a hominin with our brain size. The final paper in this group, **Chapter 7**, by Anne Weaver takes a look at one of the most overlooked brain regions, the cerebellum. The cerebellum, an area dedicated to motor coordination and related tasks, is assumed to have changed little during the course of primate brain evolution. However, approximately 30,000 years ago the relative size of the cerebellum compared to the size of the cerebrum changed – the cerebellum became relatively large next to the cerebrum when compared to our most immediate ancestors. Weaver suggests that while the evolutionary reasons for the change are uncertain, the changes appear to occur around the time significant mutations in regulatory genes such as microcephalin and ASPM appear.

Anne Weaver's analysis of cerebellar evolution in *H. sapiens* is a beautiful introduction to the chapter that follows. **Chapter 8** on *Study of Human Brain Evolution at the Genetic Level* stands alone as a unique section on the genes important to brain development by two of the leading researchers in the field, Eric Vallender and Bruce Lahn. Several years ago Lahn's lab at the University of Chicago published several seminal papers on the importance of understanding evolution, including brain evolution, through the action of genes. Two of those genes ASPM and microcephalin according to Vallender and Lahn, may figure prominently in brain evolution, primarily the evolution of modern human brain size. In addition, Vallender and Lahn cautiously remind the reader that work in this area of paleoneurology is only in its infancy and that much of the heavy lifting is yet to be done.

The next section of the book focuses on a collection of papers that examine brain evolution through examination of the neurological and neurocytoarchitectural evidence derived through a variety of techniques as well as development. In **Chapter 9**, Katerina Semendeferi and her team look at the evidence for human brain reorganization by examining the brains of our closest living primate relatives, the apes. One of the most important outcomes of their studies is the conclusion that the human brain is the result of mosaic evolution, increasing disproportionately not only in more derived regions, but also in areas assumed to be conservative in function and development.

One research method that is just beginning to impact the field is the application of diffusion tensor imagining. Previously the only methods for deriving brain pathways and activity were invasive experiments and certain imaging techniques. Among the imaging techniques that can be used to view brain activity in real time are the two, Positron Emission Tomography (PET) and Functional Magnetic Resonance (fMRI), that can generally provide insight into the connectivity and functionality of various brain pathways. However, the biggest impediment to using these modalities to study brain evolution is figuring out how to get your subject, say a chimpanzee, to lie still in the machine and cooperatively follow your commands without moving too much. Today a new technique known as Diffusion Tensor Imaging (DTI) is permitting researchers the first opportunity to access brain regions once off limits with other techniques. **Chapter 10** by Jim Rilling and **Chapter 11** by Jason Kaufman et al. provide insight into the value and application of this revolutionary approach. Rilling uses *in vivo* images of macaque, chimpanzee and human brains to demonstrate the difference observed between these species with regard to brain activity and lateralization. Complementing Rilling, Kaufman et al. apply DTI to a preserved gorilla brain to reconstruct various fiber tract pathways. This methodological paper for the first time looks at the capacity to delve into the histology of the brain in valuable specimens that are often unavailable for sectioning, using DTI. It also provides important information on the current shortcomings of imaging techniques (see also Chapter 4), and the need for more research on the methodology of imaging techniques.

Chapter 12 examines an often-overlooked aspect of brain organization, minicolumns. Here Dan Buxhoeveden revives Pasko Rakic's seminal work in the 1970s that found that the brain is organized into minicolumns that are a product of how the brain develops its various neurocytarchitectural layers. Buxhoeveden hypothesizes that the development of minicolumns directly affects features such as brain size and organization. This echoes to a certain degree Vallender and Lahn's paper discussing the effects of genes on brain development and evolution, suggesting that the development of additional minicolumns that may affect brain size as well as the variability among minicolumn size and distribution are likely the result of mutational events. Along these specific lines, in **Chapter 13** Raghanti et al. examine the effects of neurotrasmitter systems in cognition. However, unlike minicolumns where some of the genetics that likely affect their development have been discerned, the evolutionary consequences of the genes behind neurotransmitters are far from being well understood. Few researchers have attempted to link neuromodularity to human brain evolution, but as Raghanti et al. demonstrate the differences in the function of the three neurotransmitters studied are so acute between humans and chimpanzees that they not only provide an explanation for cognitive differences between us and our closest living relative, but also a potential explanation for certain human neuropathologies such as schizophrenia.

In **Chapter 14**, Doug Broadfield investigates sex differences in the corpus callosum. In 1982 Holloway and Kitty de Lacoste published a seminal paper on sex differences in the corpus callosum of humans, finding that the corpus callosum of females was proportionally larger than males. This paper set off a firestorm, and for the second time in his career Holloway earned credit for establishing a new field in neuroscience, this being sex differences and the brain. As with other questions examined here it is understood that modern humans display a particular feature apparently unique to the species. As with the question of when did the brain enlarge, Holloway's work on the corpus callosum raised the issue of when sex differences appear in hominins. In this study Broadfield looks at the corpus callosum of chimpanzees to determine if sex differences are an ancient feature of brain evolution or a recent phenomenon. In this case it appears that human sex differences in the corpus callosum are unique, albeit apparently built on a trend that stretches back to the last common ancestor we shared with chimpanzees.

The next two papers in the volume may appear somewhat out of place among studies of neurotransmitters and endocasts, but as the reader with gather from these two papers, understanding development is imperative to understanding brain evolution. The first of these papers (**Chapter 15**) is by Janet Monge and Alan Mann, two researchers that have spent much of their careers studying issues of development. Here Monge and Mann provide new evidence that provides valuable insight into modern human development as well as early hominin development. By knowing how our early ancestors developed we can have a better idea of how brain growth proceeded as well as obtain reliable measurements of adult brain size from juvenile specimens. In the second paper related to dental development our late colleague Michael Yuan presents data that supplements and in some ways challenges the work of Mann and Monge. In **Chapter 16**, Yuan looks at perikymata counts in Asian populations to determine if the use of perikymata in determining age in modern human populations needs to be reexamined.

The last section of this volume looks at brain evolution through behavior. While there are countless studies on animal and primate cognition, most do not examine animal cognition through the lens of human brain evolution. Two papers look at human brain evolution through two behaviors that are central to human cognition, learning and language. **Chapter 17** by Francys Subiaul examines human brain evolution through imitation learning. Imitation is an important skill for many animals, especially primates. We learn most of the skills we acquire early in development through imitation. By looking at macaques, apes and humans Subiaul concludes that human imitation did not evolve singularly, but instead arose as a result of mosaic evolution acting on multiple brain regions over the course of millions of years through selective pressures arising from ecological, technological and sociological means. **Chapter 18** by Duane Rumbaugh, Sue Savage-Rumbaugh, James King, and Jared Tagialatela takes a look back at the Rumbaughs' work on language acquisition in chimpanzees and bonobos. They conclude that based on the capacity for apes to acquire language along with other expressions of intelligence they display human brain evolution did not proceed with the development of human brain structure *de novo*, but instead occurred via to co-option and exaptation of structures that exist with the ape brain. In this volume's final chapter (**Chapter 19**), Nicholas Toth and Kathy Schick review the prehistoric archaeological record and examine how it correlates with probable brain reorganization and speciation in the course of human evolution. They document the appearance of novel behaviors over the past several million years of evolution and assess their cognitive complexity.

Almost fifty years ago Ralph Holloway exploded out of Berkeley to take on a field that was transitioning from the assumption that brain evolution was a simple matter of evolving big brains to the realization that human brain evolution is an immensely complex problem. Relying on little more than infinite curiosity and an innate ability to synthesize presumptively disparate data, Ralph Holloway changed the face of Anthropology. Ralph's contributions are too numerous to count. He lent his knowledge of neuroscience to the fossil records to make endocasts relevant not just for what they reveal about brain size, but also for the features they possess and what they can tell us about the mosaic evolution of the brain. He also charted the course for others by telling anyone interested where they should be looking for answers (e.g., inside the brain, in behavior, through sex differences, through modern variation). The result of this lifetime of work is a vibrant field, producing this volume on a variety of topics all related to brain evolution. To paraphrase Isaac Newton, if we the editors, authors, and others in the field see a little further, it is because we stand on the shoulders of giants. Ralph Holloway is certainly one of those giants.

CHAPTER 1

THE HUMAN BRAIN EVOLVING: A PERSONAL RETROSPECTIVE

RALPH L. HOLLOWAY

ABSTRACT

Minor controversies notwithstanding, the evolution of the human brain has been an intermingled composite of allometric and non-allometric increases of brain volume and reorganizational events such as the reduction of primary visual cortex and a relative increase in both posterior association and (most probably) prefrontal cortex, as well as increased cerebral asymmetries, including Broca's and Wernicke's regions, with some of these changes already occurring in australopithecine times. As outlined in Holloway (1967), positive feedback ("amplification-deviation") has been a major mechanism in size increases. Exactly how this mélange of organs evolved will require many more paleontological discoveries with relatively intact crania, an unraveling of the genetic bases for both brain structures and their relationship to behaviors, and a far more complete picture of how the brain varies between male and female, and different populations throughout the world. After all, the human brain is still evolving, but for how long is quite uncertain.

INTRODUCTION

One of my goals is trying to understand how humankind evolved, and in particular, why we have become the most dangerous species on the planet. I attribute this quandary of the species to its brain, and the capacity thereby to create by means of arbitrary symbols, systems of patterned insanity, that is, delusional systems that nevertheless sustain us. This belief follows from my definition of human culture:

> as that biosocial evolutionarily-derived and socially-sustained ability, possessed only by human beings as members of societies, which organize experiences in a blend of both arbitrary and iconic symbol representations. These representations can be imposed by any level or unit of human social structure, including the individual. (Holloway 1981a; see also Holloway, 1967, 1969a, 1996).

The key element here is "imposed" meaning forced upon or done against resistance.

I recognize that this not a view shared by most people, and I could well be wrong about the patterned insanity I regard as part of human behavior (particularly religion and politics, despite what few eufunctions may attend, at least as far as I understand human history). Because the human brain is the most important constructor of experience and reality, it would be important to know how it came to its present state. Some knowledge of comparative neuroscience, the relationships between individual variation and behavior, molecular neurogenetics, and paleoneurology, or the study of the only truly direct evidence, the endocasts of our fossil ancestors, is necessary. Endocasts, i.e., the casts made of the internal table of bone of the cranium, are rather impoverished objects (the cerebrum is covered by three meningeal tissues) to achieve such an understanding, but these are all we have of the direct evolutionary history of our brains and should not be ignored. Most of my professional career has involved the study of these objects.

To cover all the evidence for human brain evolution would be an impossible task in this retrospective essay. Fortunately, a fine review of human brain evolution has been published by Annual Review of Anthropology (Schoenemann 2006), as well as by Rilling (2006), Buxhoeveden & Casanova 2002, and Preuss et al, 2003, and these articles save me the task of restating all the evidence (see also Grimaud-Hervé 1997, Holloway et al., 2004a, Weaver 2005), and allow me to be more personal in my reflections.

Biographical

Getting Out of Drexel, New Mexico, Los Angeles

My early college education started at Drexel Institute of Technology in Philadelphia, where I was enrolled in the cooperative program of metallurgical engineering. The cooperative program in the early 1950s meant half the year was spent in classes, and the other half was spent in industry, meaning some job appropriate to one's major. I was lucky enough to work at Armco Stainless Steel Co. in Baltimore, and while I never did succeed in inventing transparent stainless steel (from my boyhood science fiction fantasies), I was allowed to experiment with extreme temperatures on various alloys of stainless. Three and half years later, I had my first choice of an elective course, which could be either public speaking for engineers or reading (again) Huckleberry Finn and Tom Sawyer. I chose the former.

Family matters took me to the University of New Mexico in Albuquerque, and I was admitted on probation since my Drexel grades in calculus left something to be desired. I was thirsting for knowledge, and took a course in Anthropology and a course in Geology. These courses affected me profoundly, and I decided to become an anthropologist. My father rebelled, and to shorten this tale, I became a geologist, since it would be more likely that I could be employed in the latter pursuit than the former. Indeed, upon graduating in 1959 with experience as a roughneck in southwest Texan oil fields, and working in a geophysics lab, I was unable to get a job in geology, there being a major recession at that time. I ended up in Burbank, California working on heat resistant metals with Lockheed Aircraft. I remember going to night school and taking a course taught by Dr. Jack Prost at the University of California, Los Angeles, and a course on metal fatigue, just to keep the schizoid quality of my life in motion. A year later, I was admitted to the PhD program in Anthropology at the University of California, Berkeley; I departed Los Angeles, and gratefully moved to the Bay Area.

Getting out of Berkeley

My first mentor at Berkeley was Professor Sherwood Washburn, who was extremely kind to me in offering graduate student research support. Washburn insisted on my taking various anatomy courses until I suggested to him that I wished to take a course (then taught by Marian Diamond) in neuroanatomy. He was appalled and told me that he would no longer be my mentor if I studied neuroanatomy. I was flabbergasted: how could anyone understand how humankind evolved without understanding how the brain evolved? His response was that I would become too specialized and would not be a physical anthropologist, an argument I found entirely unconvincing. However, if one looks at the textbooks in physical anthropology of the 1950s through the present, one will find it rare to see more than one page devoted to the brain, and what will be discussed is only the size of the organ. The recent text by Sanford et al. (2008) is an exception because one author, John Allen, is a neurobiologist who has also studied the lunate sulcus (Allen et al, 2006).

Washburn (and Irvin DeVore) had just come back from field studies in Ambolselie Game Park studying baboon behavior, and I think he wanted me to do the same. At the time, I thought primate studies were interesting, but I could not fathom using baboons as a theoretical model for understanding human evolution because I regarded each species a terminal end product of their own line of evolutionary development. Despite the warning, I took the neuroanatomy course and worked eventually with Diamond on the effects of environmental complexity on the cortex in rats. In 1966, I wrote the first paper on the effects of environmental complexity training on dendritic branching, using Golgi-Cox methods (Holloway 1966c).

My next mentor was Professor Theodore McCown, who was completely open and supportive regarding my burning interests in the brain. In 1964, I completed my dissertation, after much hassle with Washburn regarding a doctoral dissertation topic, and he was not a member of my committee. My dissertation was of the library variety, a review of quantitative relations in the primate brain [Holloway 1964; the first part of which was published in Brain Research (Holloway 1968), but the second half was mysteriously lost between the editors in Holland and Switzerland…]. I regarded endocast studies as possibly useless, and this gave me a burning desire to do empirical research and not armchair anthropology. Ironically enough, considering my experiences in geology and engineering, 1964 was a banner year for entering the job market, and I received several offers, most notably from Columbia University and Cornell. My first wife's folks were from New York, I took the Columbia position. My father had died prior to this, so this triumph was unknown to him.

Early Columbia University

My position at Columbia was mostly as a service to sociocultural anthropology, and I taught at both undergraduate and graduate levels. At that time, we were fully committed to the four-field approach, an approach now completely rejected by the cultural anthropologists at Columbia, the majority of which appear strongly postmodern, post colonialist, feminist, and political. I suppose in the earlier days, had I been more aggressive about constructing a biological anthropology program at Columbia, my stay would have been a more pleasant experience, but I was quickly isolated and marginalized at Columbia, and remain so. Instead, I tried to stay true to scholarship and research, and not politics. Fortunately, I was (and am) saved by my mighty tenure.

Harry Shapiro from the American Museum of Natural History was an Adjunct, and he and I shared the responsibilities of educating graduate students in the department. I tried to continue my research on the effects of environmental complexity on dendritic branching; both my children referred to me as the "man who draws spiders", as dendritic branching was done in my darkened office, tracing the dendrites against a sheet of paper attached to the wall, while manipulating the depth of focus on the microscope, there being no joy sticks or computers in those days. My hope was to do research on the quantitative histology of the cerebral cortex of different primates including humans, but no lab facilities were available. I approached my chairman, Morton Fried at the time, and asked for his interceding with the Biology Department, in the hope that they might provide some space and histological help. The answer was brutal: Cyrus Levinthal and Eric Kandel responded to Fried somewhat as follows: "if we do not know what is happening in the brain of Aplysia, the sea-slug, how could we possibly learn anything from the primate brain? No." Kandel, of course, went on to win a Nobel Prize for his research. Admittedly, this was a hard lesson for me regarding the hubris of molecular biologists, but I survived it. My early papers in those days were attempts at synthesizing hominid brain evolution (Holloway 1966a, b; 1967; 1968; 1969a,b; 1972a; 1973a), and were of the armchair variety, although I still regard certain papers [1967, 1969a,b; 1975b; 1976; these two latter papers suggested that throwing with force and accuracy selected aspects of brain evolution, well before Calvin's (1983) book, which took this idea much further] as some of my best attempts, despite their speculative hue.

On to Paleoneurology

Indeed, the above experience led me to seek a semester's leave, and I and my family went off to South Africa to look at australopithecines and endocasts under the guidance of Professor Phillip V. Tobias. This was in 1969, and I guess my encounters with the New York police during the 1968 student demonstrations (I experienced testicular trauma at the end of the police blackjacks…) were a sympathetic note to Tobias and the apartheid policies in South Africa that he was fighting. In any event, the experience settled my career, and I became a dedicated paleoneurologist. Ironically, my dissertation had explicitly found endocasts to be useless, particularly when I found that descriptions of Sinanthropus were more primitive than Homo erectus from Java, despite being later in time.

I met Professor Raymond Dart there, who had so kindly sent me all of his reprints when I was at Berkeley, and I became convinced that the Taung endocast needed independent study, despite the detailed work of George Schepers (Schepers 1946). My main concern at the time was finding an accurate volumes for the hominids (Holloway 1970a,b; 1973b), and trying to find an objective method(s) for deciding whether the cortex was reorganized as Dart had previously claimed (Dart 1925, 1926, 1956). This meant trying to determine the exact location of the infamous lunate sulcus, which is almost always the anterior boundary of primary visual cortex, or area 17 of Brodmann. Was it in a typical ape anterior position as Keith (1931) figured it, or was it indeed in a posterior, more human-like position as Dart had originally claimed? Little did I realize how contentious this question would turn out to be (30+ years!), as I acquired my long-standing opponent, Dean Falk. My estimate of the Taung endocast volume came out to 404 ml, double the volume of the 202 ml hemi-endocast I had constructed under the scrutiny of both Tobias and his fabulous assistant, Alun Hughes (Holloway 1970a). This value was quite less than the 525 ml previously reported, and I was pleased that both Alun and Philip did not find fault with my reconstruction. I particularly enjoyed working on the SK 1587 endocast from Swartkrans (Holloway 1972b) at the Transvaal Museum. Of course, nothing is static in paleoneurology, and the Taung endocast volume has been recently deflated (i.e., 382 ml) by Falk & Clarke (2007) in a paper filled with questionable methods, the most grievous being that they never bothered to define a midsagittal plane, an absolute requisite when trying to mirror-image a half-portion of an endocast (R. Holloway, manuscript in preparation). Falk et al. (2000) proposed some minor deflations of other australopithecine endocast volumes, and replies will ensue in the future.

Apparently, my skills were growing, and I believe Tobias let Louis Leakey know I could be trusted with the fossils. And so in 1971-1972, my family and I spent a sabbatical year in Kenya and South Africa working on australopithecines, habilines, and H. erectus. (So many anecdotes, so little space, but I shall always remember Louis' kindness to me and my family when he was in such considerable pain.)

I returned to Kenya a couple more times to work mostly on the habilines, and my visit in the late 1970s, in particular, allowed me to make an undistorted endocast of the famous KNM-ER 1470 cranium My observations on Broca's area were recorded in Richard Leakey's books (where I had determined that these were of a Homo-like form and found a cranial capacity of 752 ml.) My method scared the dickens out of Richard, because I filled the latex coated interior of the cranium with plaster of Paris to avoid any distortion while it was still in the cranial portion. When Richard saw this, he asked how in the hell I would get it out, and I told him to come by next day. He did, and lo and behold, there sat the perfect endocast, and there sat the undistorted cranium! (Given the existing breaks in the cranium, simply dissolving the glue joints and extricating the endocast without any damage to the fossil itself was an easy task.)

I believe it was during a 1978 visit, perhaps earlier, that Richard approached me in the Center's lab, and asked if Dean Falk could take some impressions ("peels") from the cranium, and I said yes, but did not know that she would later publish her observations (Falk 1983a) without either acknowledging my agreement, or even mentioning my findings, which were mentioned in Leakey's books (Leakey and Lewin 1978, Leakey 1981). At the time, I was supposed to be preparıng a full description to be included in Bernard Wood's monographic treatment of the Kenyan discoveries (Wood 1991). My results (Holloway 1983d), in very abbreviated form, were published in the journal Human Neurobiology, of which Doreen Kimura was a founder, but which did not survive very long as a journal. In the latter part of 1972 I went briefly to Indonesia to make endocasts from the newly discovered Homo erectus crania (Sangiran 4, 10, 12, and 17) in Dr. Teuku Jacob's lab at Jogjakarta. The hospitality was splendid, but the weather abominable.

The Armchair Stuff, Compulsive Collecting of Data, and More Controversies

Meanwhile, throughout the late 1970s and early 1980s, my interests broadened to more theoretical approaches to human brain evolution (albeit my 1967 paper in General Systems was a major beginning), and are reflected in my paper published in 1979, where I tried to synthesize brain size, brain reorganization, structural and regulatory genes, and allometry in the volume edited by Hahn, Jensen and Dudek (Holloway 1979, in Hahn et al. 1979) (see Figure 1). At this time I was in the midst of conceptual battles with my colleague Harry Jerison, who appeared, at least to me, to have little regard for the concept of reorganization (Holloway 1969b, 1974; see also Holloway 1966a for a critique of the extra neuron model Harry had offered in 1963). I was honored to give the James Arthur Lecture on the Evolution of the Human Brain (Holloway 1975b), in which I suggested, as I had in my earlier (Holloway 1967) paper, that selection for social behavioral complexity was what had driven the evolution of the hominid brain. (I would have been wiser to have called it "Machiavellian Intelligence," or the evolution of the "social brain", the current popular jargon which ignores earlier publications). The paper on relative encephalization quotients (EQ) measures (Holloway & Post 1982) was an important contribution also. My 1969 paper, "Culture: A Human Domain," was an attempt to describe what humans did as quite different from what other primates were doing, although if I were to rewrite that paper, I would find many more areas of behavioral continuity between our symbolically mediated behavior, and theirs. At that time, I thought the basis of human language, the use of arbitrary symbols, had aspects of cognitive processes that were shared by stone toll making.

A recent paper by Stout et al. (2008) using fMRI on Nick Toth while he was making stone tools indicates a possible connection between language sites in Broca's and Wernicke's regions of the cerebral cortex and stone tool making, something I had suggested in the above paper on the possible similarities, cognitively, between language and tool-making behavior.

I had, by 1978, made close to 200 latex rubber endocasts of modern humans and apes and monkeys, and compulsively collected thousands of data points on a comparative collection of these endocasts, including fossil hominids, apes, and modern humans, using a stereoplotter that was suggested to me by Dr. Alan Walker. This gadget measured the dorsal surface of the endocast every ten degrees in two planes, and took the distance from a homologous central point to the endocast surface, thus avoiding problems with allometric corrections. These results (Holloway 1981c) indicated that the region of greatest shape difference was in the posterior parietal region, which I thought was a buttress to my belief that relative expansion of the posterior parietal lobe had occurred early in hominid evolution, and was indicated on the Taung endocast, as well as the AL 162-28 specimen from Hadar, Ethiopia (Holloway 1983a, Holloway & Kimbel 1986). This was also a time in which I published my observations on the Spy Neandertal endocasts, the Omo endocast, and the Solo endocasts (Holloway 1980a,b; 1981b,d,e; 1983b; 1985b). More recently, I have been making endocasts of modern Homo sapiens, from sectioned crania in the bone lab at Columbia, and from the Von Lauschan collection at the American Museum of Natural History, adding roughly another 75 specimens to the growing sample size of the 15-20 that I did much earlier. Included among these latter specimens are 5-6 microcephalic endocasts (thanks go to Milford Wolpoff, who lent the crania to me) and a couple of extreme cranial deformation examples. These have all been done using "Dentsply Aquasil LV" dental impression material, which has, hopefully, a much longer shelf life than the earlier latex endocasts, many of which have deteriorated. Thanks to the efforts of Dr. Janet Monge and Dr. Tom Schoenemann, these endocasts (not the more recent human ones) have been scanned.

The Lunate Sulcus

Dart (1925,1926,1956) had believed that the Taung child's endocast showed definite signs of reorganization toward a more human-like condition on the basis of his belief that the lunate sulcus, which defines the anterior boundary of primary visual striate cortex, Brodmann's (1909) area 17, was visible on the Taung natural endocast. The cortex anterior to the lunate sulcus would be the parietal and temporal lobe association cortex, where higher cognitive functions occur. I trumpeted the concept of reorganization in my dissertation and early papers (e.g., Holloway 1966b, 1967,1979) and, indeed, still believe the concept to be of value as an additional set of

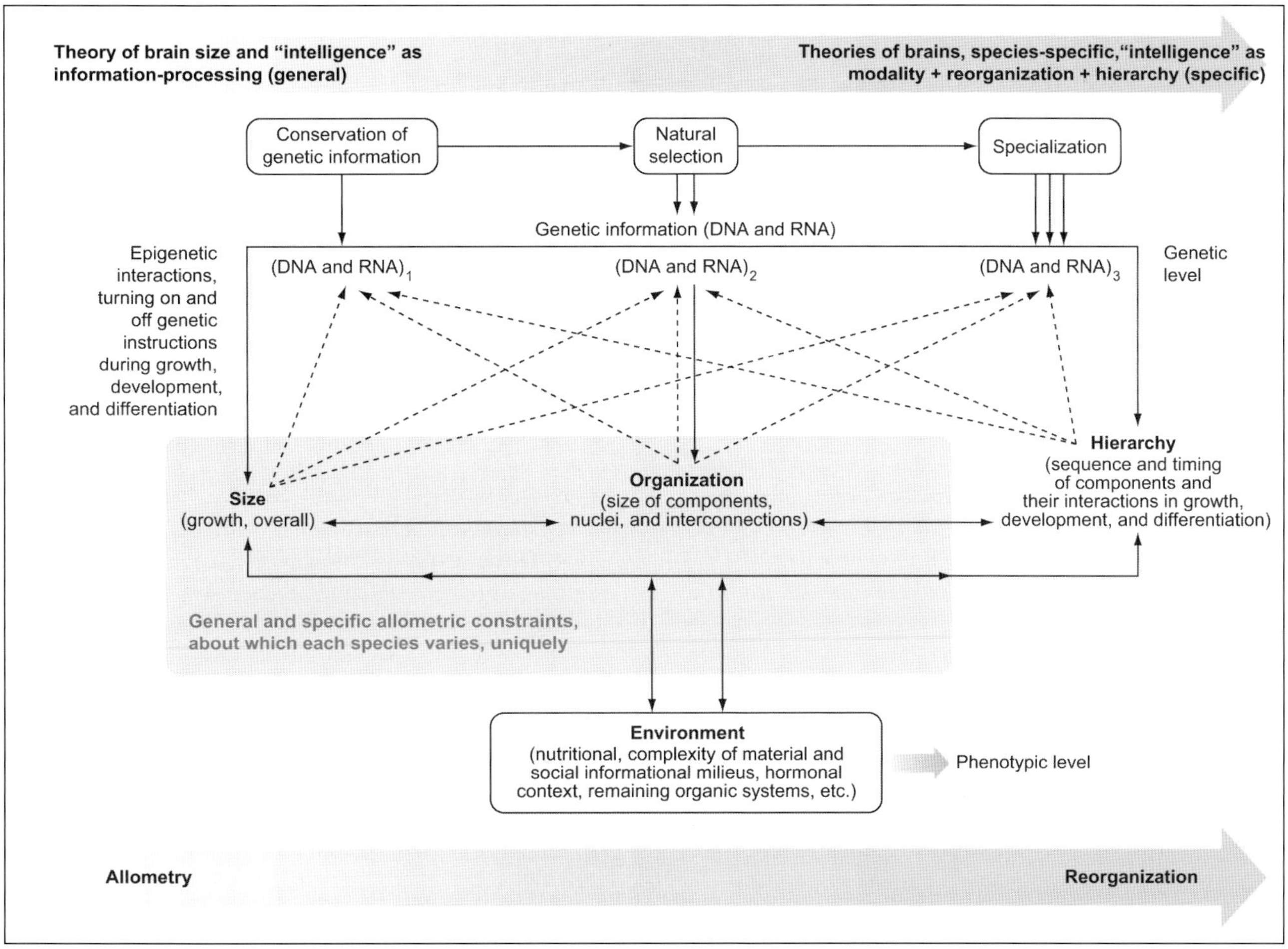

Figure 1 The brain is a composite of size, organization, and hierarchy, which is acted on at the phenotype level by natural selection throughout the life of the organism. Mathematical formulations and prediction tests are so far applied only to the shaded box containing size and organization. This model conceives of natural selection variously acting on three subsets of genetic information (DNA+RNA123), which also interact with each other and the developing and differentiating organism in epigenetic fashion. Allometrists and brain mass theorists are almost totally working within the framework of the left-hand side of the diagram.

quantitative changes that are not directly caused by brain size increase alone. How the brain is organized as well as its size is of great importance. (I came to this conclusion before 1964 when I made a seminar presentation in one of Washburn's classes, demonstrating that some human microcephalics with brain sizes that some gorillas might deride as diminutive were nevertheless able to talk. That meant to me that something in their brains was organized differently than in the great apes.) Most biological anthropologists ignore organization and cathect on brain size, which is a bit unscholarly. Dart, after all, had studied under Grafton Elliot Smith (see Smith 1904), the major claimant and champion of the lunate sulcus, and Dart himself wrote his dissertation on the evolution of the turtle brain, which of course has no lunate sulcus. This history was covered (Holloway 1985a; see also Holloway 1988a,b; Holloway et al. 2001a,b; 2003; 2004b) because Dean Falk had previously restudied the Taung endocast and decided that the lunate was represented by a small dimple placed well anterior of the lambdoid suture, and even more anteriorly than would be found in almost all chimpanzees, gorillas and orangutans all without any measurements based on a comparative sample (Falk 1980;1983a,b,c; 1985; 1989). It was, however, the question of a possible lunate sulcus in the Hadar AL 162-28 A. afarensis that received the most unwelcome confrontation with Falk. She (Falk 1983b) incorrectly oriented the Hadar specimen so that the cerebellar hemispheres protruded beyond the cerebral cortices. Further, the depression along the lambdoid suture region, which she regarded as the lunate sulcus, was placed in an anterior, ape-like position, which simply reflected her own bias. She had apparently accepted my earlier (Holloway 1983c) identification of the posterior end of the interparietal sulcus (IP), which usually abuts the lunate sulcus. I was reluctant to accept the depression as a true lunate sulcus because I had found many of my Pan endocasts had a distinct "sulcus" just immediately anterior to the lambdoid suture, which I name the "prelambdoid pseudosulcus", and which is actually caused by the posterior and inferior lip of the parietal bone. Clark et al. (1936) showed this artifact very clearly when they rubbed off

Table 1 Changes in the reorganization of the hominid brain based on endocasts (After Holloway et al. 2004a)

Brain changes (reorganization)	Taxa	Time (mya)	Endocast evidence
Reduction of primary visual striate cortex, area 17, and relative increase in posterior parietal cortex	*A. afarensis*	3.5 to 3.0	AL 162–28 endocast
	A. africanus	3.0 to 2.0	Taung child, Stw 505 endocast
	A. robustus	~2.0	SK 1585 endocast
Reorganization of frontal lobe (third inferior frontal convolution, Broca's area, widening prefrontal)	*Homo rudolfensis*	2.0 to 1.8	KNM-ER 1470 endocast
	Homo habilis		Indonesian endocasts
	Homo erectus		
Cerebral asymmetries, left occipital, right-frontal petalias	*H. rudolfensis*	2.0 to 1.8	KNM-ER 1470 endocast
	H. habilis, H. erectus		Indonesian endocasts
Refinements in cortical organization to a modern *Homo* pattern	*H. erectus* to present?	1.5 to 0.10	*Homo* endocasts (*erectus, neanderthalensis, sapiens*)

Table 2 Major cortical regions involved in early hominid evolution (with major emphasis on the evolution of social behavior and adapting to expanding environments) (After Holloway et al. 2004a)

Cortical regions	Brodmann's areas	Functions
Primary visual striate cortex	17	Primary visual
Posterior parietal and anterior occipital (peri- and parastriate cortex)	18, 19	Secondary and tertiary visual integration with area 17
Posterior parietal, superior lobule	5, 7	Secondary somatosensory
Posterior parietal, inferior lobule (mostly right side. Left side processes symbolic-analytical)	39	Angular gyrus perception of spatial relations among objects, face recognition
Posterior parietal, inferior lobule (mostly right side. See above)	40	Supramarginal gyrus spatial ability
Posterior superior temporal cortex	22	Wernicke's area, posterior superior temporal gyrus, comprehension of language
Posterior inferior temporal	37	Polymodal integration, vision, auditory input. Perception and memory of objects' qualities
Lateral prefrontal cortex (including mirror neurons)	44, 45, 47 (also 8, 9, 10, 13, 46)	Broca's area (Broca's Cap), motor control of vocalization, language
		Complex cognitive functioning memory, inhibiton of impulse, foresight, etc.

Table 3 Major size changes in human brain evolution (After Holloway et al. 2004a) aAllometric means related to body size increase or decrease, whereas nonallometric refers to brain size increase without a concomitant body-size increase.

Brain changes	Taxa	Time (mya)	Evidence
Small increase, allometric[a]	*A. afarensis* to *A. africanus*	3.0 to 2.5	Brain size increases from 400 ml to 450 ml, 500+ ml
Major increase, rapid, both allometric and nonallometric	*A. africanus* to *Homo habilis*	2.5 to 1.8	KNM-1470, 752 ml (Ca 300 ml)
Small allometric increase in brain size to 800 ml–1000 ml (Assumes *habilis* was KNM 1470-like)	*Homo habilis* to *Homo erectus*	1.8 to 0.5	*Homo erectus* brain endocasts and postcranial bones, e.g., KNM-ER 17000
Gradual and modest size increase to archaic *Homo sapiens* mostly nonallometric	*Homo erectus* to *Homo sapiens neanderthalensis*	0.5 to 0.10	Archaic *Homo* and neandertal endocasts 1200 to 1700+ ml
Small reduction in brain size among modern *Homo sapiens*, which was allometric	*Homo s. sapiens*	0.015 to present	Modern endocranial capacities

the charcoal soot from the endocast surface and compared the endocast to the actual brain. Later, Bill Kimbel and I (Holloway & Kimbel 1986) tried to set the matter straight by pointing out Falk's error of orientation and the fact that the distance between her purported lunate sulcus and the occipital pole was only 15 mm, roughly half the distance that occurs normally in chimpanzee brains of roughly the same volume, i.e., 385-400 ml. Measuring the distance between the occipital pole (the most posterior point on the occipital lobe) and the lunate sulcus on ~80 chimpanzee hemispheres, suggested that the Hadar A. afarensis AL 162-28 specimen was almost 3 standard deviations outside of the mean chimp value, which varied between 25 and 30+ mm. (Holloway et al. 2001a,b, 2003, 2004b).

This brouhaha was part of a larger theoretical issue, i.e., whether an increase in brain size must necessarily precede any organizational shift in brain components, or a reduced primary visual cortex relative to the size of the brain. Jerison (1990), Falk (1983b), and Armstrong et al. (1991) appeared to take the position that the brain did not reorganize until there was an increase in brain size, and I was taking the position, as had Dart before me, that reorganization took place prior to the increase in brain volume. I believed then and remain convinced today that the earliest hominids, i.e., Australopithecus africanus , A. afarensis and A. garhi, had brains definitely different than any ape's, despite their small size, and that natural selection had worked on more complex social behaviors (Holloway 1967, 1975b), as would be expected if the relative reduction in the primary visual cortex (PVC) signaled a relative increase in parietal association cortex.

Hopefully, the newer A. africanus brain endocast of Stw505 (from Sterkfontein, South Africa), with its clear-cut posterior location of a lunate sulcus (Holloway et al. 2004b) will convince most skeptics that, indeed, the australopithecine hominids had reorganized brains despite their overlapping in size with ape brains. Whether biological anthropology textbooks will recognize this possibility is another matter. As near as I can determine, many of the textbooks in biological anthropology only discuss brain volume in hominids (Sanford et al. 2008 being an exception).

In 1990, I had the honor of being a participant in the Fifth Interdisciplinary Fyssen Symposium, in which I presented a paper "Toward a synthetic theory of human brain evolution", eventually published in 1995 (Changeux and Chavaillon, 1995). This was the first time I tried to present a framework where brain size increases were interspersed with reorganizational changes. The point here was to suggest that different selection pressures occurred at different times regarding both size and organization. Falk, in her usual sarcastic manner characterized the paper as the same old stuff (Falk, 1997) even though this was truly a newer synthesis. If she has disagreed with my premises and outlines, I would have been pleased and would have regarded such as a positive step in our skirmishes, but instead it was just an opportunity to denigrate and ignore my ideas without ever providing counter-evidence, or discussing what was wrong with the data presented.

Tables 1, 2, and 3 (updated from Holloway et al. 2004a) present my recent synopsis of the evidence I presented then.

Another major brouhaha with Falk and her colleagues emerged after White & Falk (1999) asserted that the Omo L338y-6 australopithecine from Ethiopia had an occipital-marginal sinus drainage pattern that allied the specimen to robust australopithecines. Having studied and described the original specimen (Holloway 1981b), and not a cast of a cast, I was amazed to see this publication and hear these claims. I examined my original endocast reconstruction and, as I clearly remembered, could find no trace of such a sinus. Tim White and his colleagues were kind enough to secure a new mold of the posterior section of the newly cleaned Omo specimen and serially sectioned it through the purported region claimed by White & Falk. There was absolutely no sign of a marginal sinus on this specimen (Holloway et al. 2002). The presentation of these findings at an American Association of Physical Anthropology meeting caused an extremely emotionally charged encounter between me and David DeGusta on the one side and Falk and White on the other, each armed with their own endocast copies. (Fortunately, at 430 ml, the endocasts could not do much damage even if thrown, despite being made of plaster.)

One last example might be of interest: in Braindance, (Falk 2004, pp. 165-66) discusses her "radiator hypothesis" (Falk 1990) as a proven hypothesis and then provides a partial quote from my critique which appeared in Brain and Behavioral Sciences (Holloway 1990a), focusing on my belief that her hypothesis had the structure of a simple just-so story and was unduly speculative. What Falk then left out were my eight points regarding the lack of any empirical demonstration that show an increase in blood cooling associated with cranial capacity increase, upon which the fossil evidence is simply mute. Nor did she ever bother to respond in any detail to Kimbel's (1984) paper and (1990) critique. My paper (Holloway 1980c) on a re-analysis of the Pakkenberg & Voigt (1964) data on Danish brain weights which showed very clearly on p. 113 that body size alone could not explain the difference in male/female brain weights, a result she also finds in her 2004 edition of Braindance. Our work (de Lacoste-Utamsing & Holloway 1982; Holloway 1990b; Holloway et al. 1993) on the corpus callosum was not mentioned in her discussions of sexual dimorphism, nor our work on cerebral asymmetries (Holloway & de Lacoste-Lareymondie 1982).

On a more positive note, I was honored in 2007 with a two-day conference ("The Human Brain Evolving: Papers in Honor of Ralph L. Holloway") held on my behalf in Bloomington, Indiana, where 20+ colleagues came together to give papers on various aspects of brain evolution. This conference was sponsored by the Stone Age

Institute and Indiana University under the leadership of Drs. Nick Toth and Kathy Schick and also organized by two former students, Drs. M.S. Yuan and D.C. Broadfield. I take this as a validation of my research.

A Brief Aside on What Constitutes Evidence for Hominid Brain Evolution

This little battle, however, brings forth an interesting question about how valuable paleoneurology and comparative neuroanatomy are in discussing hominid evolution. As I have tried to point out in several places (e.g., Holloway et al. 2004a), the only direct evidence for hominid brain evolution is paleoneurology, the study of endocasts, despite the paucity of that information. Perhaps, in the future, molecular neurogenetics might be able to provide more details regarding what elements of the brain (neurotransmitters to gross neuroanatomy, i.e., gyri, sulci, fiber tracts, overall size; see, for example, Sherwood et al. 2003 regarding Broca's homologue in chimpanzees) have changed during hominid evolution. At the moment, however, such data are not available, and comparative neuroanatomy remains the study of extant (not extinct) animal brains, each of which have undergone their own separate evolutionary path development to their present condition, whatever that may be.

Give these questions some serious thought: Is today's chimpanzee brain (against which we do so many comparisons, whether in terms of size or structure) the same as that of the last common ancestor of hominids and chimps? Has the chimpanzee brain evolved during the past 5-7 million years? If so, are our comparisons with the present day chimpanzee on target? Should the same questions be asked of other areas of comparative primate comparisons, e.g., dentition, locomotion, behavior? The incomplete brain endocast of Proconsul africanus, of roughly 12 mya, appears to show an anthropoid pattern of having the lunate sulcus in an anterior position (which all extant anthropoids share) (Radinsky 1974, 1975, 1979). So perhaps with this characteristic, the derived condition (lunate sulcus in a posterior position, indeed an autapomorphy) for Homo is a reasonable conclusion that can be translated into functional (i.e., behavioral) terms, such as what we know about the role of posterior parietal association cortex in perception of objects, their position, recognition of faces, social behavior, and aspects of language reception. Herein lies the great value of comparative neuroanatomy: It is the essential link between neurobiological and behavioral variation writ both large and small. Still, where are the studies that link what we know of species-specific behavioral patterns and neuroanatomy in the primates? Where is the research that explains, neurologically, the behavioral differences between chimpanzee, gorilla, and orangutan. Even trying to describe the behavioral differences between Pan troglodytes and Homo sapiens is difficult, despite clear-cut differences in brain anatomy that have been described. I ask these questions not to detract from comparative studies, but simply in the hope of sharpening our analytic

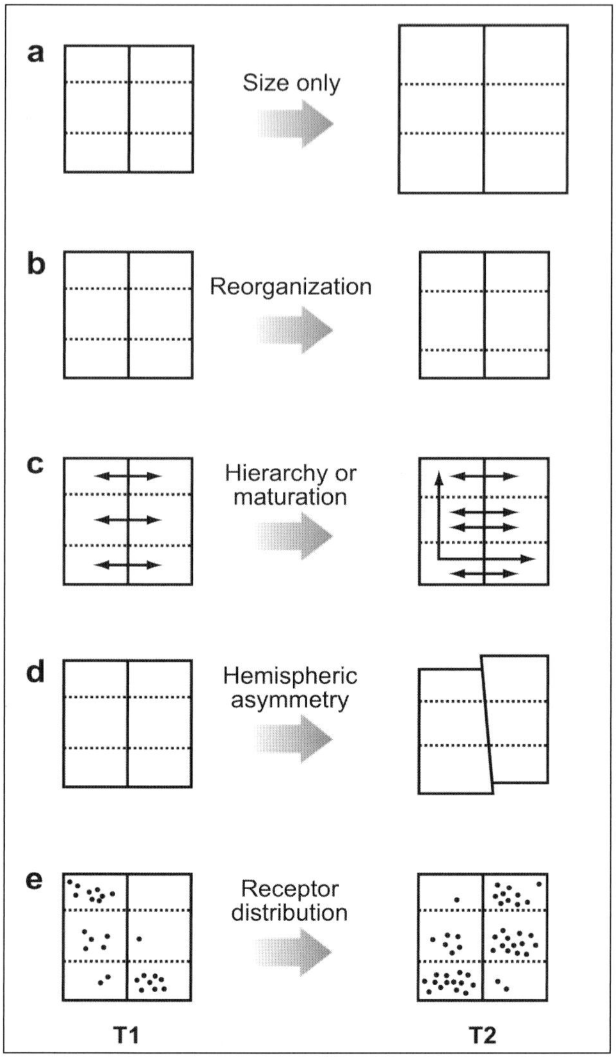

Figure 2 Figure 2 shows several different scenarios where it is possible to reorganize the brain without any apparent increase in size, from T1 (time 1) to T2 (some time after an arbitrary interval of evolution). The horizontal dashed lines represent the central sulcus and lunate sulcus, respectively, with the frontal lobe facing upward. The vertical line divides the two cerebral hemispheres. Thus in part (a), Time 1 to Time 2 involves an increase in size without changing any parts of the brain. In part (b), the lunate sulcus moves posteriorly, but brain size remains constant from T1 to T2. In part (c), different fiber tracts mature at different times and differentially increase or decrease. In part (d), the two hemispheres are asymmetric (left-occipital and right-frontal width petalias), but overall brain size remains constant. In part (e), brain size is constant, but neuroreceptors are differently distributed between T1 and T2 (Praire and mountain voles, and oxytocin receptors come to mind.).Needless to say, some of these scenarios cannot be detected on endocasts (parts c and e, and sometimes b). These are a few alternative ways to reorganize a brain without increasing its size.

abilities and to caution against the wholesale use of extant species' morphology in trying to understand human brain evolution. So much of the primate behavior I have read and the speculation that follows regarding hominid evolution seem to be based on the premise that the chimpanzee has had no further evolution since our split with Pan-like hominoids roughly 7 MYA.

Enter the "Hobbit", Homo floresiensis: an ongoing tango

In the above context, a parallel problem exists with respect to comparing modern-day human pathology with ancient hominid discoveries. The recent controversy over the "hobbit" Homo floresiensis, whether or not it is a true new species of hominid (see Brown et al. 2004 for original claim and description) or a case of pathology, has not been settled (e.g, Henneberg & Thorne, 2004; Hershkovitz et al. 2007; see also Richards 2006). I have spent more than two years studying the endocast of this creature and am still sitting on the fence as to whether or not it is a case of microcephaly or some other pathology, or a new species (Holloway et al. 2006). As cogent as the arguments of Jacob et al. (2006) and Martin et al. (2006a,b) might appear, I agree with the depictions in Falk et al. (2005) of the virtual endocast compared to modern Homo sapiens, Homo erectus, chimpanzee, and microcephalic H. sapiens [an unfortunate choice of one extremely small microcephalic (278 ml)], and the observation that there are no microcephalic brains yet published that show the suite of features found on the "hobbit" endocast, although the example by Martin's et al. (2006b) of the Indian microcephalic comes very close. What I see are : (a) extreme platycephaly, (b) extremely thin and protuberant gyri recti of the prefrontal lobe, (c) appearance of a smallish prefrontal lobe and temporal lobes as seen on the undistorted left side, (d) unusually spread cerebellar lobes, and (e) a peculiar triangular-shaped occipital sinus. These observations leave me sitting on the proverbial fence regarding a new species or pathology argument. The point here is that modern pathology (e.g., primary microcephaly) may not match what appears to be a possible pathology 13-18 thousand years ago. The full spectrum of microcephaly and other pathological conditions affecting the brain has not been available to study or illustrated in recent articles. My consultations with several pediatric neurologists suggest that they see it (the "hobbit") as pathological, but it does not match what they've seen in cases of true primary microcephaly. The original "virtual endocast" published by Falk et al. (2005) shows that they selected the damaged and inferiorly deflected right temporal lobe as a model for their "virtual endocast" when it is the left temporal lobe that was intact, and which, incidentally, appears rather small in comparison to the total size of the endocast. Their 417 ml volume is more likely to be 400 ml. In any event, this tango will not end until more of these creatures are discovered and described.

Having been kindly provided with an endocast made from the stereolith of the LB1 cranium by Peter Brown, I have never once been asked to referee any papers on the LB1 endocast. At the time of these writings Dean Falk and I are among a small number of practicing paleoneurologists (actually, so are Emiliano Bruner and Dominique Grimaud-Hervé, and Anne Weaver) to have worked on these endocasts. Promises made by Mike Morwood to receive the CT scan data so that I could make an independent study of the endocast have never materialized, and I strongly argue that independent study is sorely needed. But this tango is a common occurrence in paleoanthropology, where access to fossil specimens tends to be rigorously guarded (e.g., Atapuerca, Dmanisi, etc.)

Brain Variation and Tottering on the Edge of Political Incorrectness

The 1980s became a period of intensive data collecting. One of the first steps was requesting from Pakkenberg in Denmark the data from their study of Danish autopsies (Pakkenberg & Voigt 1964). These authors kindly sent me the data which I reanalyzed in 1980 (Holloway 1980c) because I was interested in exploring ranges of variation within a species of derived neuronal statistics such as extra numbers of neurons, EQs (encephalization quotients), and relative brain size. I was intrigued by techniques such as partial correlations, and was getting interested in possible sex differences in loss of brain weight with age and EQs, and indeed was able to show that the difference between male and female brain weights could not be fully explained by differences in body size. I was surprised to find that in males, the brain correlated significantly with stature, but the same effect did not hold for females.

At this time I had a brilliant graduate student, Christine de Lacoste-Lareymondie who was doing her dissertation on the distribution of fibers in the human corpus callosum. I remember approving and encouraging this project but insisted that she had to find out as much as possible about the variation of the corpus callosum, including variation by sex. From a small sample she had collected, Kitty discovered that females appeared to have larger corpus callosa relative to brain size than did males and that the splenium in particular seemed more bulbous in females than in males. We thought this was a very intriguing find and sent a manuscript to Science. Science then asked for the data because, indeed, our sample was very small. They accepted the paper (de Lacoste-Utamsing & Holloway 1982), and this created a minor cottage industry for the next couple of decades as to whether or not the corpus callosum was indeed relatively larger in females. Many people argued that it was equal, but seldom used our methods or seemed to understand we were talking about a relative size (Holloway et al. 1993). Blistering commentary depicting us as sexist and worse came [e.g., Fausto-Sterling's (1985) "Myths of Gender"; Bishop and Wahlsten (1997)]. We also were unaware that Bean (1906) had earlier made a similar finding, and his being a well-known racist provided these authors with the necessary guilt by association, which social scientists so savor. Finally, thanks to the sophisticated analytic paper by Richard Smith (2005) in Current Anthropology, a case to legitimate ratio data was proven. In those days, sex differences in the brain were really politically incorrect, particularly as a vast sea of feminist literature was being produced. Today, sex differences in the human brain are commonly accepted (e.g., Gur et al. 2002; Kimura 2003; Haier et al. 2005; Narr et al. 2007). These experiences were not pleasant, however, and I found myself sort of a pariah in one realm and a hero in another, and it had a lasting effect on my quest for truth, replication, and letting data trump emotional biases. I am afraid the same principles apply to possible ethnic ("racial") differences in the brain, because without knowing how the brain varies in the human species, it is impossible to understand fully how this organ evolved. Furthermore, given the sensitivity of the brain to environmental insult from conception on, sound knowledge of such variation, whether in overall size, maturation schedules, neuroreceptor sites, etc., is required to determine the most efficient therapeutic measures to take to ensure proper nutrition and other nurturance for the developing brain. A full understanding of the respective roles and interplay of nature and nurture particularly with respect to worldwide distributions on intelligence tests scores is impossible without knowledge of how the human brain varies and why it does so. It would be nice if human variation could be celebrated as our most precious evolutionary heritage and hope instead of prohibiting the study of our variation.

In the late 1970s and early 1980s I collected autopsy data from the Pathology department at Columbia's College of Physicians and Surgeons (now CUMS). I was interested in age, sex, and ethnic effects on brain size changes through time as might be found in cross-sectional data. Roughly 2000 cases were collected, without personal identifications, and all cases of brain pathology were culled out of the data set. The results, unpublished, were roughly the same as found in Ho et al. (1980, 1981) work on a sample from Milwaukee, which indicated that African American brains were statistically significantly lower in weight than were European American brains, that is, of course referring to the mean values. Ho et al. (1980) concluded that cultural effects were the reason behind the difference. Interestingly, Lieberman (2005) in his review of Rushton's (2000, 2002) claims regarding ethnic (racial) differences in brain sizes and behaviors ignored this work by Ho et al. Needless to say, Tobias's oft-cited paper on brain weight collecting methods (Tobias 1970) was cited to claim that autopsy data on brain weights are useless. Unfortunately, however problematic such data are, one tends to forget that autopsies are not done discriminately. Once the body is on the morgue slab, the autopsy is conducted in exactly the same fashion irrespective of the cadaver's race, and thus comparisons of such data collected by the same pathologist or medical examiner are surely valid, depending on which variables are being compared. Comparing data collected by different examiners may of course be difficult, and perhaps statistical meta-analyses would be in order. To my knowledge, none exists.

Simply put, this research area remains an intensely political and near-suicidal enterprise. (Indeed, one colleague suggested I should incinerate the data; another suggested this kind of study had led to his relatives perishing in the Nazi concentration camps.) The continuing gap in African American and European-descent test scores on various cognitive tests (particularly IQ) throughout the US and the world (Lynn & Vanhanen 2006) is a source of tremendous concern and acrimonious debate. Indeed, Jon Marks claimed he "outed" me as a "racist" (Marks 2000; see Holloway 2000 for reply) in his biological section of the American Anthropologist Newsletter, because I had the temerity to defend Arthur Jensen against Loring Brace's assertion that Jensen was a bigot. I had read a lot of this literature (e.g., Jensen 1998) including Jensen's infamous 1969 piece in the Harvard Law Review, and did not find him a racist. I remain appalled at our discipline which regards him as such and which invented the appellation "Jensenism" to tar and feather him. I remain interested in the possibility that different populations have variation both in their brains and behavior, but the issue is so politically incorrect that one cannot even approach such a study with anything but trepidation. (For example, the Annual Review article by Freedman & DeBoer, 1979, was declared by socio-cultural students at Columbia as racist, and therefore not to be read!) If one disbelieves there are populational differences in the weight and/or structure of the brain, they should examine the papers by Klekamp and his colleagues, particularly regarding the finding that the primary visual striate cortex of Australian aborigines is significantly larger than in brains from people of European descent (Klekamp et al. 1994). This paper is, to my knowledge, the only paper published since the 1930's demonstrating a real difference in brain morphology between modern populations (the last compilation of some of these earlier studies on brain morphology differences between different populations can be found in C.J. Connolly's 1950 book, External Morphology of the Primate Brain, which is a sort of bible for most people working in paleoneurology. See also Kochetkova 1978.) Of course, there is Gould's Mismeasure of Man, another bible of sorts, but which should be read along with Michael's (1988) Current Anthropology paper which found that Morton's rankings were correct and which Gould ignored in his later editions of the same book. There is certainly no evidence that Paul Broca used his elbow on the scales when measuring brains of peoples of European descent!

Additional autopsy data sets await my attention, including some 5,000 cases from Hong Kong collected by my colleague Philip Beh, and ~7,500 cases from Singapore, the latter of multiple ethnicities. I hope to get to these data sets when I retire.

Postscript and Acknowledgments

On November 27, 2007, the science section of the New York Times ran a profile of me and my work (Balter 2007). While grateful that I could make it within the New York Times, I wish more had been said of my other interests in brain research. Thus far, neither my colleagues in the Anthropology department nor the Columbia University Administration have acknowledged the article or the previously mentioned conference.

I am very grateful to the many colleagues who mentioned the honor, and to the many students I have encountered over the decades that have truly rewarded me with their intelligence, wit, and support as well as the temerity to disagree. In particular, I mention Michael Yuan, Douglas Broadfield, Chet Sherwood, Francys Subiaul, Sam Marquez, Lynn Copes, and Jill Shapiro, who read earlier versions of this paper and who offered many useful corrections, as well as former students Christine de Lacoste-Azziz, Peter Heilbroner, Jeffrey Schwartz, Este Armstrong, Joan Witkin, Jason Kaufman, and Peter Post. My special thanks go to Nick Toth and Kathy Schick for their interest in my work and their friendship, and for hosting the conference at the Stone Age Institute. I am grateful to Carole Traviz-Henikof for her role in that honor. The encouragement and friendship of the late Clark Howell are sorely missed. My colleagues Janet Monge, Alan Mann, Jason Lewis, Robert D. Martin, Alan Walker, Dominique Grimaud-Hervé, Emiliano Bruner, James Rilling, Tom Schoenenmann, Patrick Gannon, Daniel Buxhoeveden, John Allen, Katerina Semendeferi, Milford Wolpoff, John Hawks, Anne Weaver, and Carol MacLoed deserve special mention. To Peter Brown goes a special thanks for allowing me to study the LB1 endocast, regardless of whether I agree with him! I would not have been able to make any contributions to paleoneurology without the cooperation and hospitality I received from the Leakey family in Nairobi, Kenya, and the staff at the Center. I owe a similar debt to the late Raymond Dart, Phillip V. Tobias, Bob Brain, the late Alun Hughes, the late Teuku Jacob and Ralph von Koenigswald, A. Leguebe, Roger Saban, Yves Coppens, Ian Tattersall, Eric Delson, Gary Sawyer of the AMNH, and Theya Molleson of the BMNH. I continue to enjoy the collegiality and support of Tim White, Bill Kimbel, Yoel Rak, Gen Suwa, Berhane Asfaw, W. Henry Gilbert, Scott Simpson, and all their colleagues in Ethiopia, and I look forward to continuing studies on more hominid endocasts from there. To Chuck McAlexander and Dr. Graham Kavanagh go special thanks for their support. Lastly, my wife, Dr. Daisy Dwyer, has given me so much and put up with it all.

References

Allen JS, Bruss J, and Damasio H. 2006. Looking for the lunate sulcus: a magnetic resonance imaging study in modern humans. Anat. Rec. (A) 288A:867-76.

Armstrong E, Zilles K, Kurtis M, and Schleicher, A. 1991. Cortical folding, the lunate sulcus and the evolution of the human brain. J. Hum. Evol. 20:341-48.

Balter M. 2007. In study of human brain evolution, zeal and bitter debate. Profile of Scientist at work, Ralph Holloway. New York Times, Nov. 27:F2

Bean RB. 1906. Some racial peculiarities of the Negro brain. Amer. J. Anat. 5:353-432.

Bishop, K.M., and D. Wahlsten. 1997. Sex differences in the human corpus callosum: myth or reality? Neurosci. Biobehav. Rev. 21:581-601.

Brodmann K. 1909. Verleichende Lokalisationslehre der Grosshirnrinde in ihren Prinzipien dargestellt auf Grund des Zellenbaues. Leipzig: Barth.

Brown P, Sutikna T, Morwood MJ, Soejono RP, Jatmiko W, Saptomo E, Due RA. 2004. A new small-bodied hominin from the late Pleistocene of Flores, Indonesia. Nature 441: 624-28.

Buxhoeveden DP, Casanova MF. 2002. The minicolumn and evolution of the brain. Brain Behav. Evol. 60:125-51.

Calvin W. 1983. The Throwing Madonna: Essays on the Brain. New York: McGraw-Hill.

Changeux JP, Chavaillon J, eds. 1995. Origins of the Human Brain. Oxford: Clarendon Press.

Clark WE Le Gros, Cooper DM, Zuckerman S. 1936. The endocranial cast of the chimpanzee. J. R. Anthropol. Inst. 66:249-68.

Connolly CJ. 1950. External Morphology of the Primate Brain. Springfield, IL: Thomas.

Dart RA. 1925. Australopithecus africanus: the man-ape of South Africa. Nature 115:195-99.

Dart RA. 1926. Taung and its significance. Nat. Hist. 26:315327.

Dart RA. 1956. The relationships of brain size and brain pattern to human status. S. Afr. J. Med. Sci. 21:23-45.

de LaCoste-Utamsing C, Holloway RL. 1982. Sexual dimorphism in the human corpus callosum. Science 216:1431-2.

Falk D. 1980. A reanalysis of the South African Australopithecine natural endocasts. Am. J. Phys. Anthropol. 53:525-39.

Falk D. 1983a. Cerebral cortices of East African early hominids. Science 222: 1072-74.

Falk D. 1983b. Hadar AL 162-28 endocast as evidence that brain enlargement preceded cortical reorganization in hominid evolution. Nature 313:45-47.

Falk D. 1983c. The Taung endocast: A reply to Holloway. Am. J. Phys. Anthropol. 60:479-90.

Falk D. 1985. Apples, oranges, and the lunate sulcus. Am. J. Phys. Anthropol. 67:313-15.

Falk D. 1989. Ape-like endocast of "ape-man" Taung. Am. J. Phys. Anthropol. 80:335-39.

Falk D. 1990. Brain evolution in Homo: the 'radiator' theory. Behav. Brain Sci. 13:333-81.

Falk D. 1997. Book review: Origins of the Human Brain. Am. J. Hum. Biol. 9:766-67.

Falk D. 2004. BrainDance: Revised and Expanded Edition: New Discoveries about Human Origins and Brain Evolution. Gainseville, FL: Univ. Press Fla.

Falk D, Clarke R. 2007. New reconstruction of the Taung endocast. Am. J. Phys. Anthropol. 134(4):529-34.

Falk D, Hildebolt C, Smith K, Morwood MJ, Sutikna T, Brown P, Jatmiko T, Saptomo EW, Brunsden B, Prior F. 2005 The brain of LB1 Homo floresiensis. Science 308: 242-45.

Falk D, Redmond Jr. JC, Guyer J, Conroy GC, Recheis W, Weber GW, Seidler H. 2000. Early hominid brain evolution: a new look at old endocasts. J. Hum. Evol. 38:695-717.

Fausto-Sterling A. 1985. Myths of Gender. New York: Basic Books.

Freedman DG, DeBoer MM. 1979. Biological and cultural differences in early child development. Annu. Rev. Anthropol. 8: 579-600.

Grimaud-Hervé D. 1997. L'évolution de l'encéphale chez Homo erectus et Homo sapiens. Paris: CNRS Ed.

Gur RC, Gunning-Dixon F, Bilker WB, and Gur RE. 2002. Sex differences in temporo- limbic and frontal brain volumes of healthy adults. Cereb. Cortex 12:998-1003.

Hahn ME, Jensen G, Dudek BC. 1979. Development and Evolution of Brain Size. New York: Academic.

Haier RJ, Jung RE, Yeo RA, Head K, and Alkire MT. 2005. The neuroanatomy of general intelligence: sex matters. Neuroimage 25: 320-27.

Henneberg M, Thorne A. 2004. Flores may be pathological Homo sapiens. Before Farming 1:2-4.

Hershkovitz I, Kornreich L, Laron Z. 2007. Comparative skeletal features between Homo floresiensis and patients with primary growth hormone insensitivity (Laron Syndrome). Am. J. Phys. Anthropol. 134:198-208.

Ho K-C, Roessmann U, Straumfjord JV, Monroe G. 1980. Arch. Pathol. Lab. Med. 104:640-45.

Ho K-c, Roessmann U, Hause L, and Monroe G. 1981. Newborn brain weight in relation to maturity, sex, and race. Ann. Neurol. 10:243-46.

Holloway RL. 1964. Some quantitative relations of the primate brain. PhD thesis, Univ. Calif., Berkeley. 174pp.

Holloway RL. 1966a. Cranial capacity and neuron number: a critique and proposal. Am. J. Phys. Anthropol. 25:305-14.

Holloway RL. 1966b. Cranial capacity, neural reorganization, and hominid evolution: a search for more suitable parameters. Am. Anthropol. 68:103-21.

Holloway RL. 1966c. Dendritic branching: some preliminary results of training and complexity in rat visual cortex. Brain Res. 2: 393-96.

Holloway RL. 1967. The evolution of the human brain: some notes toward a synthesis between neural structure and the evolution of complex behavior. Gen. Systems 12:3-19.

Holloway RL. 1968. The evolution of the primate brain: some aspects of quantitative relations. Brain Res. 7:121-72.

Holloway RL. 1969a. Culture: a human domain. Curr. Anthropol. 10(4):395-412.

Holloway RL. 1969b. Some questions on parameters of neural evolution in primates. Ann. NY Acad. Sci. 167(1):332-40.

Holloway RL. 1970a. Australopithecine endocast (Taung specimen, 1924): a new volume determination. Science 168:966-68.

Holloway RL. 1970b. New endocranial values for the australopithecines. Nature 227:199-200.

Holloway RL. 1972a. Australopithecine endocasts, brain evolution in the Hominoidea and a model of hominid evolution. In The Functional and Evolutionary Biology of Primates, ed. R. Tuttle, pp. 185-204. Chicago: Aldine/Atherton Press.

Holloway RL. 1972b. New Australopithecine endocast, SK 1585, from Swartkrans, South Africa. Am. J. Phys. Anthropol. 37:173-86.

Holloway RL. 1973a. Endocranial volumes of the early African hominids and the role of the brain in human mosaic evolution. J. Hum. Evol. 2:449-59.

Holloway RL. 1973b. New endocranial values for the East African early hominids. Nature 243:97-99.

Holloway RL. 1974. On the meaning of brain size. Science 184: 677-679.

Holloway RL. 1975a. Early hominid endocasts: volumes, morphology, and significance. In Primate Functional Morphology and Evolution, ed. R Tuttle, pp. 393-416. The Hague: Mouton.

Holloway RL. 1975b. The Role of Human Social Behavior in the Evolution of the Brain. The 43rd James Arthur Lecture on the Evolution of the Human Brain. New York: Am. Mus. Nat. Hist. 45 pp.

Holloway RL. 1976. Paleoneurological evidence for language origins. In Origins and Evolution of Language and Speech, ed. SR Harnad, HD Steklis, J Lancaster, 280:330-48. New York: Ann. NY Acad Sci.

Holloway RL. 1979. Brain size, allometry, and reorganization: toward a synthesis. In Development and Evolution of Brain Size: Behavioral Implications, ed. ME Hahn, C Jensen, BC Dudek, pp. 59-88. New York: Academic.

Holloway RL. 1980a. Indonesian "Solo" (Ngandong) endocranial reconstructions: some preliminary observations and comparisons with Neandertal and Homo erectus groups. Am. J. Phys. Anthropol. 53:285-95.

Holloway RL. 1980b. The O.H. 7 (Olduvai Gorge, Tanzania) hominid partial brain endocast revisited. Am. J. Phys. Anthropol. 53:267-74.

Holloway RL. 1980c. Within-species brain-body weight variability: a reexamination of the Danish data and other primate species. Am. J. Phys. Anthropol. 53:109-21.

Holloway RL. 1981a. Culture, symbols, and human brain evolution: a synthesis. Dialect. Anthropol. 5:287-303.

Holloway RL. 1981b. The endocast of the Omo juvenile L338y-6 hominid specimen. Am. J. Phys. Anthropol. 54:109-18.

Holloway RL. 1981c. Exploring the dorsal surface of hominoid brain endocasts by stereoplotter and discriminant analysis. Philos. Trans. R. Soc. London B Biol .Sci. 292:155-66.

Holloway RL. 1981d. The Indonesian Homo erectus brain endocasts revisited. Am. J. Phys. Anthropol. 55:503-21.

Holloway RL. 1981e. Volumetric and asymmetry determinations on recent hominid endocasts: Spy I and II, Djebel Ihroud I, and the Sale Homo erectus specimens, with some notes on Neandertal brain size. Am. J. Phys. Anthropol. 55:385-93.

Holloway RL. 1983a. Cerebral brain endocast pattern of Australopithecus afarensis hominid. Nature 303: 420-22.

Holloway, R.L. 1983b. Homo erectus brain endocasts: volumetric and morphological observations with some comments on cerebral asymmetries. L'Homo erectus et la place de l'homme de Tautavel parmi les hominidés fossiles. Congr. Int. Paléontol. Hum., 1er, Nice. 16-21 Oct. 1982, pp. 355-66.

Holloway RL. 1983c. Human brain evolution: a search for units, models, and synthesis. Can. J. Anthropol. 3:215-32.

Holloway RL. 1983d. Human paleontological evidence relevant to language behavior. Hum. Neurobiol. 2:105-14.

Holloway RL. 1985a. The past, present, and future significance of the lunate sulcus in early hominid evolution. In Hominid Evolution: Past, Present, and Future, ed. PV Tobias, pp. 47-62. New York: Liss.

Holloway RL. 1985b. The poor brain of Homo sapiens neandethalensis: see what you please. In Ancestors: The Hard Evidence, ed. E Delson, pp. 319-24. New York: Liss.

Holloway RL. 1988a. "Robust" Australopithecine brain endocasts: some preliminary observations. In The Evolutionary History of the "Robust" Australopithecines, ed. F Grine. Pp. 97-106. New York: Aldine de Gruyter.

Holloway RL. 1988b. Some additional morphological and metrical observations on Pan brain casts and their relevance to the Taung endocast. Am. J. Phys. Anthropol. 77:27-33.

Holloway RL. 1990a. Comments on Falk's radiator hypothesis. Behav. Brain Sci. 13:360.

Holloway RL. 1990b. Sexual dimorphism in the human corpus callosum: its evolutionary and clinical implications. In From Apes to Angels: Essays in Anthropology in Honor of Phillip V. Tobias, ed. GH Sperber, pp. 221-28. New York: Wiley-Liss.

Holloway RL. 1995. Toward a synthetic theory of human brain evolution. In Origins of the Human Brain, ed. J-P Changeux, J Chavaillon, pp. 42-54. Oxford: Clarendon.

Holloway RL. 1996. Evolution of the human brain. In Handbook of Human Symbolic Evolution, ed. A Lock, C Peters, pp. 74-116. New York: Oxford Univ. Press.

Holloway RL. 2000. Holloway denies label. Anthropol. News 41(5):4.

Holloway RL, Anderson PJ, Defidine R, Harper C. 1993. Sexual dimorphism of the human corpus callosum from three independent samples: relative size of the corpus callosum. Am. J. Phys. Anthropol. 92: 481-98.

Holloway RL, Broadfield DC, Yuan MS. 2001a. The parietal lobe in early hominid evolution: newer evidence from chimpanzee brains. In Humanity From African Naissance to Coming Millennia, ed. PV Tobias, MA Raath, J Moggi-Cecchi, GA Doyle, pp. 365-71. Florence, Italy: Florence Univ. Press.

Holloway RL, Broadfield DC, Yuan MS. 2001b. Revisiting australopithecine visual striate cortex: newer data from chimpanzee and human brains suggest it could have been reduced during australopithecine times. In Evolutionary Anatomy of the Primate Cerebral Cortex, ed. D Falk, K Gibson, pp. 177-86. New York: Cambridge Univ. Press.

Holloway RL, Broadfield DC, Yuan MS. 2003. Morphology and histology of the chimpanzee primary visual striate cortex indicate that brain reorganization predated brain expansion in early hominid evolution. Anat. Rec. 273A:594-602.

Holloway, RL, Broadfield, DC, Yuan, MS. 2004. Brain Endocasts: The Paleoneurological Evidence: Vol. 3, The Human Fossil Record. ed. JH Schwartz, I Tattersall. New York: Wiley-Liss.

Holloway RL, Brown P, Schoenemann PT, Monge J. 2006. The brain endocast of Homo floresiensis: microcephaly and other issues. Am. J. Phys. Anthropol. S42:105.

Holloway RL, Clarke RJ, Tobias PV. 2004. Posterior lunate sulcus in Australopithecus africanus: was Dart right? Comptes Rendus Palevol. 3:1-7.

Holloway RL, de Lacoste-Lareymondie MC. 1982. Brain endocast asymmetry in pongids and hominids: some preliminary findings on the paleontology of cerebral dominance. Am. J. Phys. Anthropol. 58:101-10.

Holloway RL, Kimbel WH. 1986. Endocast morphology of Hadar hominid AL 162-28. Nature 321:536.

Holloway RL, Post DG. 1982. The relativity of relative brain measures and hominid mosaic evolution. In Primate Brain Evolution: Methods and Concepts. ed. E Armstrong, D Falk, pp. 57-76. New York: Plenum.

Holloway RL, Yuan MS, Broadfield DC, Degusta D, Richards GD, Silvers A, Shapiro JS, White TD. 2002. Missing Omo L338y-6 occipital-marginal sinus drainage pattern: ground sectioning, computer tomography scanning and the original fossil fail to show it. Anat. Rec. 266:249-257.

Jacob T, Indriati E, Soejono RP, Hsu K, Frayer DW, Eckhardt RB, Kuperavage AJ, Thorne A, Henneberg M. 2006. Pygmoid australomelanesian Homo sapiens skeletal remains from Liang Bua, Flores: population affinities and pathological abnormalities. Proc. Nat. Acad. Sci. USA 103:13421-26.

Jensen AR. 1998. The g Factor: The Science of Mental Ability. Westport, CT: Praeger.

Jerison HJ. 1973. Evolution of the Brain and Intelligence. New York: Academic.

Jerison, HJ. 1990. Fossil evidence on the evolution of the neocortex. In Cerebral Cortex, Vol. 8A. ed. EG Jones, A Peters, pp.285-310. New York: Plenum.

Keith A. 1931. New Discoveries Relating to the Antiquity of Man. New York: Norton.

Kimbel WH. 1984. Variation in the pattern of cranial venous sinuses and hominid phylogeny. Am. Phys. Anthropol. 63:243-263.

Kimbel WH. 1990. Comments on Falk's radiator hypothesis. Behav. Brain Sci. 13:359.

Kimura D. 2003. Sex and Cognition. Cambridge, MA: Bradford Books.

Klekamp J, Reidel A, Harper C, Kretschmann HJ.1994. Morphometric study on the postnatal growth of the visual cortex of Australian Aborigines and Caucasians. J. Brain Res. 35 :531-48.

Kochetkova VI. 1978. Paleoneurology. Washington, DC: Winston.

Leakey RE. 1981. The Making of Mankind. New York: Dutton.

Leakey RE, Lewin R. 1978. People of the Lake: Mankind and Its Beginnings. Garden City, NY: Anchor.

Lieberman, L. 2001. How "Caucasoids" got such big brains and why they shrank. Curr. Anthropol. 42(1) 69-95.

Lynn R. and Vanhanen T. 2006. IQ & Global Inequality. Augusta, GA: Washington Summit.

Marks J. 2000. Scientific racism in your face. Anthropol. News 41(2):60.

Martin RD, Maclarnon AM, Phillips JL, Dobyns WB. 2006a. Flores hominid : new species or microcephalic dwarf? Anat. Rec. A Discov. Mol. Cell Evol. Biol. 288:1123-45.

Martin RD, Maclarnon AM, Phillips JL, Dussubieux L, Williams PR, Dobyns WB. 2006b. Comment on The brain of LB1 Homo floresiensis. Science 312:999b.

Michael J. 1988. A new look at Morton's craniological research. Curr. Anthrop. 29(2) : 349-54.

Narr KL, Woods RP, Thompson PM, Szezko P, Robinson D, et al. 2007. Relationships between IQ and regional cortical gray matter thickness in healthy adults. Cereb. Cortex 17:2163-71.

Pakkenberg H, Voigt J. 1964. Brain weight of Danes. Acta Anat. 56:297-307.

Preuss TM, Cáceres M, Laucher J, Zapala MA, Redmond JC, et al. 2003. Using genomics to identify human brain specializations. Am. J. Phys. Anthropol. Suppl. 36:171.

Radinsky LB. 1974. The fossil evidence of anthropoid brain evolution. Am. J. Phys. Anthropol. 41:15-28.

Radinsky LB. 1975. Primate brain evolution. Am. Sci. 63:656-63.

Radinsky LB. 1979. The Fossil Record of Primate Brain Evolution. New York: Am. Mus. Nat. Hist.

Richards GD. 2006. Genetic, physiologic and ecogeographic factors contributing to variation in Homo sapiens: Homo floresiensis reconsidered. J. Evol. Biol. 19:1744-67.

Rilling, JK. 2006. Human and nonhuman brains: are they allometrically scaled versions of the same design? Evol. Anthrop. 15(2):65-77.

Rushton JP. 2000. Race, Evolution, and Behavior: A Life-History Perspective. Port Huron, MI: Charles Darwin Res. Inst. 3rd ed.

Rushton JP. 2002. Race, brain size, and IQ. Gen. Psychol. 37:28-33.

Schepers GWH. 1946. The endocranial casts of the South African ape men. In The South African Fossil Ape Men: the Australopithecinae, Transvaal Mus. Mem. 2, ed. R Broom, GHW Schepers. Pretoria: Transvaal Mus.

Schoenemann PT. 2006. Evolution of the size and functional areas of the human brain. Annu. Rev. Anthropol. 35(1):379-406.

Sherwood CC, Broadfield DC, Hof PR, Holloway RL. 2003. Variability in Broca's area homologue in great apes: implications for language evolution. Anat. Rec. 271A:276- 85.

Smith GE. 1904. The morphology of the occipital region of the cerebral hemispheres in man and apes. Anat. Anz. 24:436-51.

Smith RJ. 2005. Relative size versus controlling for size. Curr. Anthropol. 46(2):249-73.

Stanford C, Allen JS, Anton SC. 2008. Exploring Biological Anthropology: The Essentials. Upper Saddle River, NJ: Pearson Prentice Hall.

Stout D, Toth N, Schick K, Chaminade T. 2008. Neural correlates of Early Stone Age toolmaking: technology, language and cognition in human evolution. Phil. Trans. R. Soc. London B Biol. Ser. B:1-11.

Tobias PV. 1970. Brain size, grey matter and race-fact or fiction? Am. J. Phys. Anthrop. 32:3-26.

Weaver AH. 2005. Reciprocal evolution of the cerebellum and neocortex in fossil humans. Proc. Natl. Acad. Sci. USA 102:3576-80.

White DD, Falk D. 1999. A quantitative and qualitative reanalysis of the endocast from the juvenile Paranthropus specimen L338y-6 form Omo, Ethiopia. Am. J. Phys. Anthropol. 110:399-406.

Wood BA. 1991. Hominid Cranial Remains. Oxford: Clarendon Press.

CHAPTER 2

The Maternal Energy Hypothesis of Brain Evolution: An Update

Robert D. Martin and Karin Isler

Abstract

Bivariate scaling analyses can reveal interesting correlations between individual biological variables, but inference of actual causal links in complex networks requires multiple tests to satisfy the criterion of isolation. Mammalian brain tissue has high energy demands, so energy supply is inevitably a key issue in evolution of the primate brain, especially for large-brained hominids. Various hypotheses have proposed a direct link between brain size and metabolic turnover in adults, but the author's Maternal Energy Hypothesis (MEH) instead focuses on energy supplied by the mother during brain development up to weaning. This hypothesis is supported by various empirical findings, but it has also been challenged, particularly on the basis that these findings do not survive tests conducted to eliminate effects of phylogenetic inertia. New comparative analyses of brain size in mammals with improved datasets have, however, confirmed links to both basal metabolic rate and gestation period, complying with core predictions of MEH. The evidence now available in support of MEH is reviewed and some implications for brain evolution are explored. A widely recognized general trend towards increase in average relative brain size during mammalian evolution has recently been challenged by a study of brain size in bats that inferred, exclusively through analysis of data from extant species, that brain size has actually undergone reduction in numerous lineages. It is shown that the statistical test used to test for directionality of evolution was inappropriate. A review of fossil evidence for brain evolution in primates, cetaceans and carnivores confirms the generally accepted trend towards increased average brain size through the Tertiary. Progressive increase in mammalian relative brain size over time is at least partially attributable to increases in the level and efficiency of maternal investment. Energetic aspects, including those invoked in the MEH, are of special importance for outstandingly large-brained mammals such as hominids.

Introduction

Analysis of non-linear scaling relationships between individual variables and body size (allometric analysis) is now a standard tool in biology. The basic approach is bivariate analysis in which the *X*-axis is usually some measure of body size (e.g. body mass) and the *Y*-variable is a parameter of interest (e.g. brain mass). The standard allometric scaling formula is a power function $Y = kX^{\alpha}$, in which α is the scaling exponent and k is the scaling coefficient. Scaling relationships can be examined both within species (growth allometry; intraspecific scaling among adults) and between species (interspecific allometry). In the following text, attention will be directed exclusively at interspecific allometric relationships in which paired X and Y values represent means for individual species. A classic example of such an allometric relationship is provided by the scaling of basal metabolic rate (BMR) to body mass across placental mammal species, for which an empirically determined scaling exponent value of 0.75 is now widely (although not universally) accepted. Logarithmic conversion of the two variables transforms the scaling formula into a linear relationship with the equation $\log Y = \alpha * \log X + \log k$, such that the values of α and k can be determined by fitting a best-fit line. Basic concepts, methods and issues in allometric analysis have been extensively reviewed elsewhere (e.g. Gould, 1966, 1975; Harvey and Mace,

1982; Schmidt-Nielsen, 1984; Martin, 1989; Martin and Barbour, 1989; Reiss, 1989; Harvey and Pagel, 1991; Martin et al., 2005; Bonner, 2006).

Despite the relative simplicity of the standard bivariate approach to allometric scaling, it has been progressively recognized that allometric analysis is beset with complex problems. Three such problems involve statistical issues. The first of these, choice of an appropriate best-fit line, has long been recognized and is covered by an extensive literature. The least-squares regression is most widely used to determine a best-fit line in allometric analysis, but it entails two basic assumptions that are unlikely to be met with interspecific datasets: (1) The X-variable is measured without error; (2) The Y-variable is clearly dependent on the X-variable. For this reason, various authors have preferred to use alternative approaches that avoid these assumptions, such as the major axis or reduced major axis. However, the basic model underlying all of these parametric line-fitting techniques (least-squares regression, reduced major axis and major axis) is the bivariate normal distribution, yet interspecific datasets commonly do not conform to such a distribution. For this reason, a non-parametric, iterative method was developed as an alternative for fitting a line to bivariate data in allometric analyses (Isler et al., 2002).

A second widespread problem that has regrettably received far less attention is the potential existence of structural heterogeneity in datasets. Quite often, individual subsets in a sample of species show different scaling relationships, commonly showing similar values for the allometric exponent (α) but dissimilar values for the allometric coefficient (k). Separate scaling relationships for such subsets can be referred to as grades, and vertical separation of best-fit lines for those subsets in a bivariate plot can be said to involve grade distinctions or shifts. An illustrative example is provided by scaling of BMR in marsupials and placentals. The best-fit line for marsupials has essentially the same slope as that for placentals ($\alpha \approx 0.75$ in both cases), but the value of the allometric coefficient is lower. In other words, marsupials generally tend to have a lower BMR value at any given body mass than placentals. On average, for any given body mass the basal energy consumption of a marsupial will be about 30-35% less than that of a placental (MacMillen and Nelson, 1969; Dawson and Hulbert, 1970; Martin, 1990). Numerous examples of such grade distinctions are known, but there is no widely recognized method for their objective detection in any given dataset. As a rule, grade distinctions are identified in practice because the investigator decides to conduct separate analyses for selected subsets of data (e.g. for taxonomic groups suspected on *a priori* grounds to be potentially divergent with respect to the variable investigated). In primates, for instance, it is well known that there are several fundamental differences between strepsirrhines (lemurs and lorisiforms) and haplorhines (tarsiers and higher primates). It is therefore advisable to check for grade distinctions between these two groups in any analysis of allometric scaling in primates.

In fact, the non-parametric line-fitting method reported by Isler et al. (2002) has an incidental benefit in providing a direct indication of the existence of clear-cut grades in a given dataset (Martin et al., 2005). This property was explored with respect to distinct grade distinctions in the scaling of gestation period in placental mammals and of neonatal body mass in primates. With the former, placental mammals with well-developed precocial offspring generally have distinctly longer gestation periods relative to adult body mass than those with poorly developed altricial offspring. With the latter, among primates, individual neonatal body mass relative to maternal mass is distinctly greater in haplorhines than in strepsirrhines. However, the signal yielded by the non-parametric line-fitting method is weak even in cases where such clearly marked grades are present, underlining the difficulty facing objective detection of grade distinctions within a dataset. Failure to recognize the existence of grades within a datatset can lead to erroneous interpretations, as a single best-fit line determined for an entire dataset will usually indicate substantially different values for α and k compared to those inferred for the individual grades. In the case of gestation periods in placental mammals, for example, the empirical value for α indicated by a line fitted to the entire sample is ≈ 0.25, whereas separate best-fit lines for altricial and precocial mammals yield an α value of ≈ 0.15. It is reasonable to regard the latter as biologically meaningful and the former as an artefact arising from grade confusion.

A third, relatively recently recognized statistical problem involved in interspecific allometric studies is a potential biasing influence exerted by phylogenetic relatedness. As noted in a seminal paper by Felsenstein (1985), data points for individual species may not be statistically independent because of their differential degrees of relatedness within the phylogenetic tree. In principle, phylogenetic inertia might distort empirically determined scaling relationships. A possible remedy for this potential problem is calculation and analysis of differences ("independent contrasts") between values for sister taxa in the phylogenetic tree (Harvey and Pagel, 1991; Purvis and Rambaut, 1995). This method is now widely used, but it has a number of drawbacks (Martin et al., 2005). In particular, the method of calculation leads to marked exaggeration of effects of error terms in the data (Ricklefs and Starck, 1996). Ironically, because closely related species tend to have very similar body mass and hence similar values for any correlated variables, a comprehensive sample with many sister taxa will generate contrast values in which error terms are particularly prevalent. As there is no means of distinguishing between measurement error and "error" due to adaptive biological deviation from an idealized scaling relationship, the implications for analyses conducted using independent contrasts are difficult to decipher. However, one practical conclusion that can be drawn is that adequate monitoring of data quality to reduce observational errors to a minimum is absolutely crucial for any analysis using independent contrasts.

Quite apart from these three largely statistical problems, particular caution is required in any attempt to infer causality from any correlations that emerge from scaling analyses. It cannot be emphasized enough that correlation should not be simply equated with causality. Moreover, the value of the correlation coefficient (*r)* for a bivariate relationship is not a reliable guide to the likelihood of a direct causal link between the variables concerned. In any biological context, one very good reason for this is that networks of variables are commonly involved, such that analysis of just two variables in isolation may well yield a statistically strong correlation in the absence of any underlying causal link. An exquisite example of invalid extension from correlation to causality is provided by a report that the frequency of citation of authors declines across the alphabetical sequence of surnames (Tregenza, 1997). The title of that report ("Darwin a better name than Wallace?") playfully reflected the inference that the observed significant negative correlation reflected some causal connection. However, it was subsequently pointed out that the frequency of surnames beginning with any given letter also declines across the alphabetical sequence. Once this confounding factor is taken into account, the apparent correlation between the alphabetical sequence of surnames and citation frequency becomes non-significant (Shevlin and Davies, 1997). The authors of that rectification emphasized the importance of compliance with the criterion of isolation (i.e. excluding all potential confounding variables) when attempting to proceed from observed correlation to inference of a likely causal relationship. One useful approach in tackling networks of biological variables is analysis using partial correlations, which can theoretically permit identification of a persistent correlation between any two variables after excluding the effects of all others. However, the success of such an approach depends upon reliable identification of all variables that should be considered in the analysis.

The Maternal Energy Hypothesis

Formulation of the Maternal Energy Hypothesis (MEH) with respect to the relationship between brain size and body size in placental mammals (Martin, 1981, 1983) was initially prompted by two complementary sets of findings: (1) The scaling relationship between brain and body size in placental mammals is comparable to that for basal metabolic rate (BMR). (2) There are convincing indications of a link between brain size and gestation. Hofman (1983a) reached similar conclusions from these same lines of evidence. The brain is unusual compared to most other bodily organs in that most of its growth is achieved relatively early in ontogeny. In all mammals, a large part of brain development is completed by weaning, so it is clearly heavily dependent on resources provided by the mother. Accordingly, the MEH postulates that the size of the brain in an adult may be linked not to that individual's own BMR but to that of its mother (Figure 1).

It was long held that the empirical exponent value for the scaling relationship between brain size and body size is ≈0.67 (von Bonin, 1937; Jerison, 1973; Gould, 1975). An exponent value of 2/3 was interpreted as indicating some kind of connection between brain size and body surfaces, fitting the interpretation that brain size is linked to information flow to and from surface effectors and/or receptors. Interestingly, it had also been argued in earlier studies that the exponent value for scaling of basal metabolic rate to body size is ≈0.67 (Rubner, 1883). This was similarly interpreted as reflecting a relationship to body surface area. However, analysis of larger, improved datasets revealed that the value of the scaling exponent for BMR is actually ≈0.75, although small-bodied mammals are a special case (Brody and Procter, 1932; Brody, 1945; Kleiber, 1932, 1947, 1961; Hemmingsen, 1960; Schmidt-Nielsen, 1984; McNab, 1986, 1988). In comparable fashion, analysis of expanded datasets for placental mammal species eventually revealed that the exponent for brain:body size scaling actually has an empirical value of ≈0.75, similar to that for basal metabolic rate (Bauchot, 1978). For instance, Martin (1981) reported the following scaling formula derived by fitting a major axis to data for 309 placental mammal species:

$$\log_{10} E = 0.76 * \log_{10} P + 1.77 \ (r = 0.96)$$
(where E = brain mass in mg and P = body mass in g)

Broadly similar findings were reported from a series of other studies (e.g. Eisenberg, 1981; Armstrong, 1982, 1983, 1985, 1990; Hofman, 1982, 1983a,b). The sample analysed by Martin (1981) was subsequently expanded to 477 species, yielding a closely similar result (Martin, 1998):

$$\log_{10} E = 0.77 * \log_{10} P + 1.66 \ (r = 0.98)$$
(where E = brain mass in mg and P = body mass in g)

For comparability with other studies, this formula can be converted into the following form using natural logarithms and g instead of mg for brain mass:

$$\log_e E_M = 0.77 * \log_e P - 3.08$$
(where E = brain mass in g and P = body mass in g)

Most recently, a greatly enlarged dataset including 1129 placental mammal species from all 18 extant orders (Isler and van Schaik, in review) has almost tripled the available sample size. Analysis of this expanded dataset, taking the reduced major axis as a best-fit line (Figure 2), yields a result very close to those reported by Martin (1981, 1998):

$$\log_e E = 0.77 * \log_e P - 3.03 \ (r = 0.98)$$
(where E = brain mass in g and P = body mass in g)

It is hence established beyond reasonable doubt that the empirical value of the scaling exponent for the relationship between brain mass and body mass across placental mammals, taking raw data for individual species, exceeds 0.67 and is ≈0.75. However, it could be argued

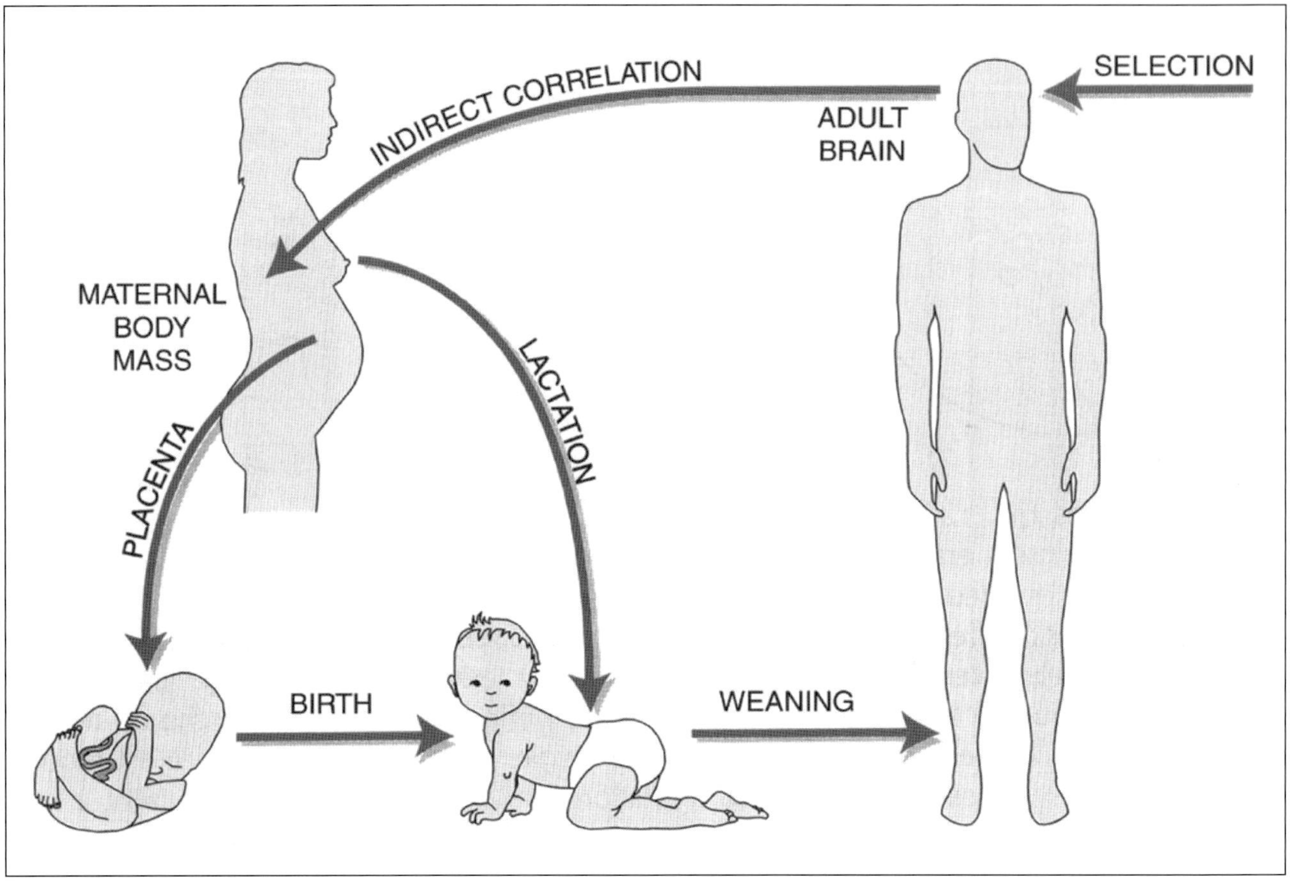

Figure 1 Schematic illustration of the Maternal Energy Hypothesis (MEH). Maternal resources provide for brain development prenatally via the placenta throughout gestation and postnatally through lactation up until the time of weaning. The eventual size of the adult brain is then determined by limited post-weaning growth. Correlations between brain size in an adult individual and other variables such as basal metabolic rate may hence be indirect, reflecting the body size of the mother rather than the body size of the adult itself. In addition to the mother's metabolic capacity, the eventual size of the adult brain can be influenced by variables such as gestation period and the duration of lactation.

that this result is biased by over-representation of particular orders of mammals. Bats (n = 315) and rodents (n = 340) together make up over half of the sample of 1129 placental mammal species, while at the other extreme 8 orders are represented by only 1-4 species (Dermoptera, Hyracoidea, Macroscelidea, Pholidota, Proboscidea, Scandentia, Sirenia, Tubulidentata). One simple pragmatic approach to offset this problem is to take overall average logarithmic values for brain and body mass for individual orders of mammals. This approach in fact yields a very similar result, with a slight reduction in the value of the allometric exponent value to 0.75 and a small improvement in the correlation coefficient (Figure 3):

$$\log_e E = 0.75 * \log_e P - 3.08 \ (r = 0.99)$$
(where E = brain mass in g and P = body mass in g)

In a more systematic approach designed to counteract the potential problem posed by species-rich taxa, Martin and Harvey (1985) presented logarithmic averages for brain and body weights calculated through successively higher taxonomic levels, ranging from genera up to orders. They obtained a scaling exponent value of 0.72 (95% confidence limits: 0.68-0.77). A scaling exponent value of ≈0.75 is therefore not attributable to a bias arising from the influence of species-rich orders.

Similarity in the empirically determined exponent values for scaling of BMR and adult brain size across placental mammals indicates a broadly isometric relationship between these two variables, i.e. simple proportionality regardless of body size (Mink et al. 1981; Hofman, 1983b; Martin, 1998; Fig. 4). Of course, such similarity in scaling could be merely coincidental. Moreover, it is well known that the exponent value for scaling of brain mass to body mass in mammals changes with taxonomic level of analysis (Martin and Harvey, 1985), and it is difficult to decide on the appropriate value for comparisons (Martin, 1990). Demonstration of a causal relationship requires extensive additional testing to ensure compliance with the criterion of isolation. Moreover, even if the existence of a connection between adult brain size and BMR is convincingly established, different interpretations are possible. One immediate possibility is that there is some direct connection between adult brain size and BMR. In this vein, Armstrong (1982, 1983) suggested that the size of the brain may be con-

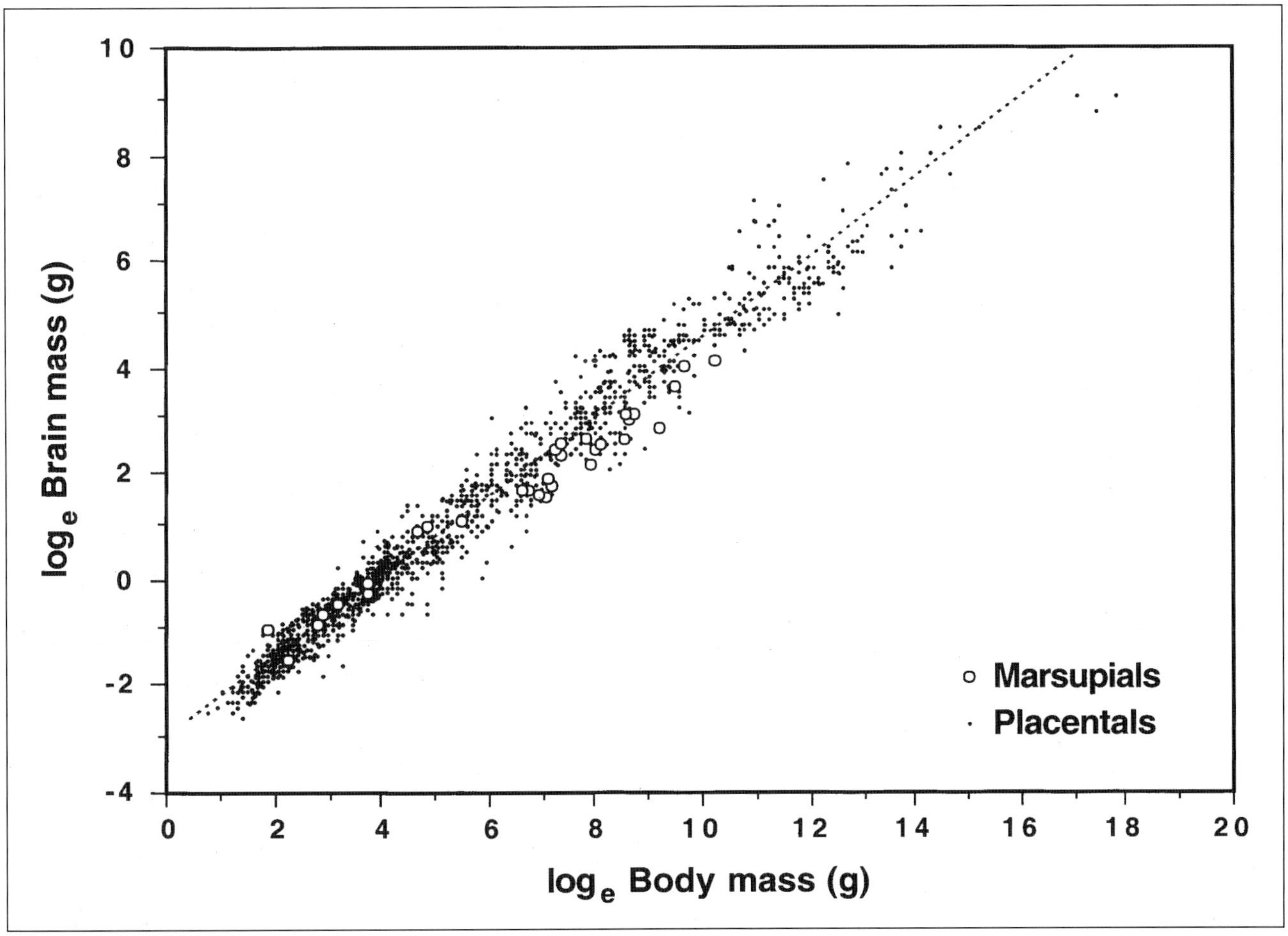

Figure 2 *Scaling of brain mass (g) to body mass (g) for 1129 placental mammal species including representatives from all 18 extant orders. The scaling formula indicated by the best-fit line (reduced major axis) is: loge E = 0.77 . loge P –3.03 (r = 0.98). For comparison, data for 31 marsupial species (not included in calculation of the best-fit line) have been superimposed on the graph. (Dataset for placental mammals compiled by Karin Isler and Carel van Schaik; dataset for marsupials from Haight and Nelson, 1987.)*

strained by the size of systems delivering oxygen and glucose and the rate of oxygen turnover, while Hofman (1983b) noted the need for compatibility between the energy demands of the brain and production and transport of oxygen by the body as a whole. In support of her interpretation, Armstrong explicitly cited the broadly isometric relationship between adult brain size and BMR shown in Figure 4.

However, postulation of a direct link between brain size and BMR in the adult conflicts with a number of other findings. First of all, for mammals generally the range of variation in relative brain size greatly exceeds the range of variation in BMR relative to body size (Martin, 1998). Overall, brain size shows a 25-fold range of variation relative to body size, whereas relative variation in BMR shows only a 4-fold range. There is hence considerable variation in adult brain size that cannot be explained by a direct relationship with BMR. There is also conflict with an observed grade shift towards higher values in the distribution of relative brain sizes among primates compared to other placental mammals. This grade distinction is not matched by a corresponding shift in the distribution of BMR values relative to body size (Mink et al., 1981; Leonard and Robertson, 1992; Leonard et al., 2003, 2007; Martin, 1998). Hence, the larger average brain mass of primates (Martin and Harvey, 1985; confirmed here in Fig. 3) is not explicable on the basis of a higher average BMR level. Armstrong (1982) in fact acknowledged that primates have larger brains than expected from their BMR values in comparison to other mammals, and a clear grade shift towards larger brains in primates is seen in a plot of residual values for adult brain mass against residual values for BMR, both determined relative to body mass (Armstrong, 1983). Armstrong proposed that primates allocate a larger proportion of available energy to the brain, but did not explain how primates can seemingly escape a constraint that supposedly limits brain size in other mammals. This same point applies even more emphatically to the human brain. Humans have an exceptionally large brain (the largest relative to body size recorded among placental mammals), yet the human BMR value relative to body size is quite close to the average condition for placental mammals generally. Finally, the absence of a direct connection between adult brain size and BMR is also indicated by data for marsupials. As already noted, BMR

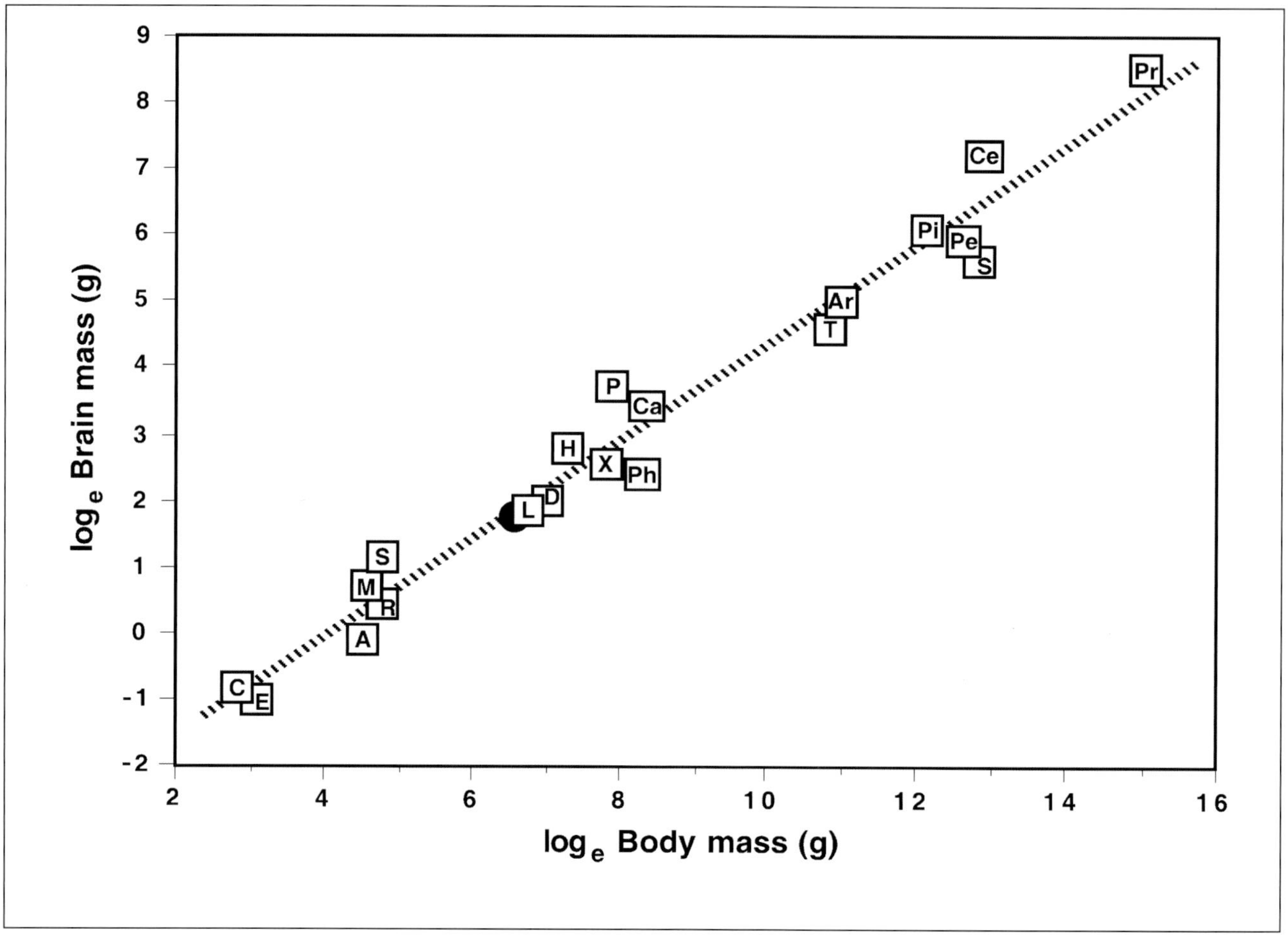

*Figure 3 Scaling of brain mass (g) to body mass (g) for placental mammals using the dataset shown in Fig. 2 but taking mean values for 20 ordinal groupings. Key: A = Afrosoricida; Ar = Artiodactyla (part of Certartiodactyla); C = Chiroptera; Ca = Carnivora; Ce = Cetacea (part of Certartiodactyla); D = Dermoptera; E = Eulipotyphla; H = Hyracoidea; L = Lagomorpha; M = Macroscelidea; P = Primates; Pe = Perissodactyla; Ph = Pholidota; Pi = Pinnipedia (part of Carnivora); Pr = Proboscidea; R = Rodentia; S = Scandentia; Si = Sirenia; T = Tubulidentata; X = Xenarthra. The scaling formula indicated by the best-fit line (reduced major axis) is: loge E = 0.75 * loge P –3.08 (r = 0.99). Note that the point for primates (P) is the highest relative to the best-fit line. (Dataset for placental mammals compiled by Karin Isler and Carel van Schaik.)*

relative to body mass in marsupials is approximately 30-35% below the average condition for placental mammals. Other things being equal, therefore, the existence of a direct link between adult brain size and BMR would surely predict distinctly smaller average relative brain size in marsupials than in placentals. However, there is complete overlap between individual values for marsupials and placentals in a plot of brain mass against body mass (Fig. 2), and the average condition for marsupials lies almost directly on the best-fit line determined for ordinal average values of placental mammals (Fig. 3). There is nonetheless an intriguing differentiation among marsupials in that small-bodied species tend to have relatively large brains compared to the average condition for placentals, whereas large-bodied species tend to lie below the best-fit line for placental mammals (Fig. 2). This may indicate that, compared to placentals, marsupials experience increasing constraints on brain development with increasing body size. Overall, however, it cannot be argued that the lower average BMR level in marsupials is associated with uniformly smaller brain size than in placentals. Indeed, small-bodied marsupials have quite large brains compared to placentals of comparable body size. Clearly, marsupials must have adaptations that permit them to develop quite large brains despite their generally lower BMR level.

An alternative approach that might potentially avoid the problems posed by inference of a direct link between adult brain size and BMR is the notion that there is a trade-off between the size of the brain and the size of other organs with high energy demands in the adult body. A prominent example of this is the Expensive Tissue Hypothesis (ETH) proposed by Aiello and Wheeler (1995; see also Aiello et al. 2001), which specifically invokes a trade-off in the adult individual between brain size and gut size. In principle, such a trade-off could explain why some species can have larger brains than others with equivalent energy resources in the adult con-

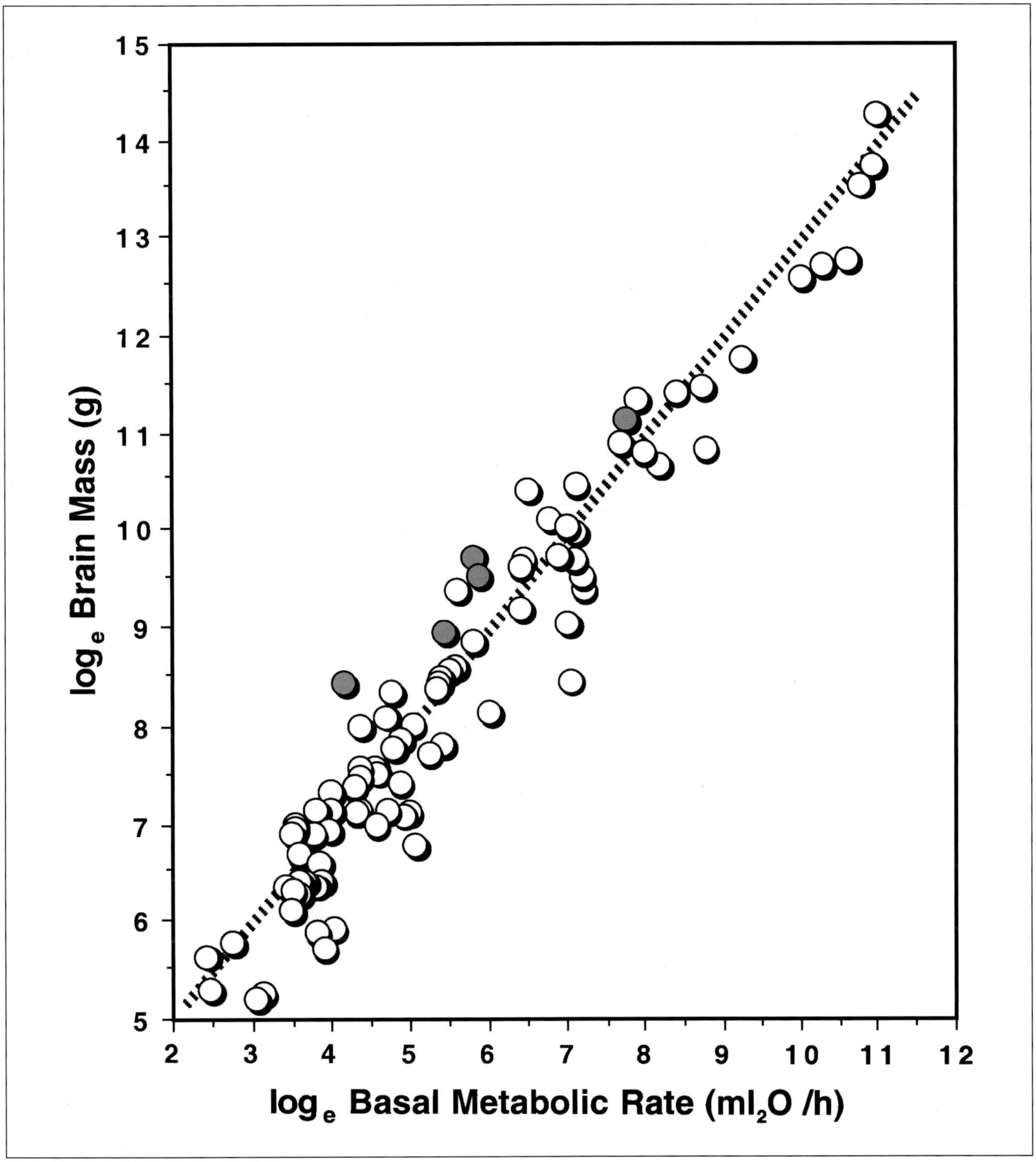

Figure 4 *Scaling of brain mass (g) to basal metabolic rate (ml 02/h) for 88 placental mammal species. Grey symbols indicate primates in the sample (n = 5); white symbols indicate other mammals. The reduced major axis yields an empirical value of 1.01 for the scaling exponent, confirming the isometric relationship (dirtly proportionality) indicated by the line.*

dition. One prediction of the ETH is that there should be a negative relationship between residual values for brain mass and gut size relative to body mass. Aiello and Wheeler (1995) tested this prediction with a sample of primates and reported that there is, indeed, the expected negative relationship. However, as with any result from a bivariate comparison, alternative explanations are possible. For instance, efficient digestion of leaves requires a resident population of symbiotic bacteria in the gut, either in the stomach or in the caecum, so folivorous (leaf-eating) mammals would be expected to have a relatively large gut. There are also indications that folivorous mammals have lower BMR than fruit-eating mammals (Clutton-Brock and Harvey, 1980; McNab,

1980, 1986), so the MEH would predict that leaf-eaters should have relatively small brains compared to fruit-eaters (frugivores) because low maternal BMR would limit fetal brain growth. Hence the reported negative relationship between residuals for brain size and gut size in primates is compatible with the MEH as well as with the ETH. Clearly, further testing is necessary to assess the relative merits of the two hypotheses. It should also be noted that primates do not have systematically smaller gut sizes than other mammals to compensate for their generally larger brains, as would be predicted from the ETH (Snodgrass et al., 2007). Leonard et al. (2003, 2007) have proposed instead a trade-off between brain mass and muscle mass in primates, notably in humans. In light of results from various comparative studies, Barton (2006) has suggested that the concept of trade-offs against brain size should be considered in relation to energetically expensive tissues generally rather than exclusively in relatioAs in other cases involving scaling relationships among primates, one possibility is to extend comparisons to other mammal groups. A suitable test case is provided by bats. Eisenberg and Wilson (1978) identified a marked grade distinction in the scaling of brain size to body size in bats, with frugivorous species having larger brains than insectivorous (arthropod-eating) species. This finding was replicated for a much larger sample of bat species by Jones and MacLarnon (2004). However, arthropod-eating mammals generally have smaller guts relative to body size than fruit-eating species, so frugivores would be expected to have relatively larger guts as well as relatively larger brains. Accordingly, it was pointed out that a direct trade-off of the kind predicted by the ETH would not be expected in this case (Martin, 1996). Jones and MacLarnon (2004) subsequently confirmed this expectation, showing that relative brain size in bats shows a positive rather than negative correlation with relative gut size. It should be noted, incidentally, that insectivorous bats typically have lower BMR relative to body mass than frugivorous bats. Hence, the difference in relative brain size between these two dietary categories could be explained by the MEH.

A test of the ETH was also conducted using data for 21 bird species (Isler and van Schaik, 2006a). Taking residuals calculated from raw values for gut mass and brain mass relative to body size, non-significant negative correlations were found for both individual species and family-level averages ($p = 0.53$ and $p = 0.43$, respectively). A significant negative correlation was found with the contrast values for species ($p < 0.03$), but this result was not confirmed by analysis of a larger dataset with intestine lengths for 192 bird species. By contrast, Isler and van Schaik (2006a) found a significant negative correlation between brain mass and pectoral muscle mass, interpreted as indicating a trade-off between brain size and locomotor costs in birds.

The second stimulus that led to formulation of the MEH was evidence for a connection between gestation period and brain size in mammals. Such a link was first clearly indicated by the seminal finding of Sacher and Staffeldt (1974) that there is a tighter correlation for the relationship between neonatal brain mass and duration of gestation than for the relationship between neonatal body mass and gestation period. Their reported result is replicated in Figure 5 by an analysis conducted with a similar dataset for 92 placental mammal species. The value of the coefficient of determination (r^2) for the relationship between neonatal brain mass and gestation period is 0.84, whereas that for the relationship between neonatal body mass and gestation period is only 0.72. In other words, only 16% of variation in neonatal brain mass cannot be attributed to variation in gestation period, whereas 28% of variation in neonatal body mass is attributable to factors other than gestation. Taken in isolation, this difference is suggestive but not compelling. However, partial correlations from a 4-way analysis of adult body mass, gestation period, neonatal body mass and neonatal brain mass reveal an even clearer distinction. The partial correlation between gestation period and neonatal brain mass is 0.71, whereas that between gestation period and neonatal body mass is only 0.12. By contrast, the partial correlation between neonatal brain mass and adult body mass is 0.176, whereas that between neonatal body mass and adult body mass is 0.75. Hence, neonatal brain mass seems to be associated primarily with gestation period, whereas neonatal body mass is linked more particularly to adult body mass. In light of their original finding of a tighter correlation between neonatal brain mass and gestation period, Sacher and Staffeldt (1974) suggested that the brain might serve as a pacemaker for mammalian development. However, this is only one possible interpretation of the observed correlation, and it is noteworthy that this finding is entirely compatible with the MEH.

Another key observation is that there is a clear grade distinction between primates and other placental mammals with respect to the relationship between neonatal brain and body mass. It has already been noted that a plot of ordinal averages for the scaling of brain mass to body mass in adults indicates larger relative brain size in primates compared to other mammals (Figure 3). However, there is considerable overlap between individual primate species and other mammal species in the adult condition. By contrast, there is very little overlap between primates and other mammals in a plot of brain mass against body mass for neonates. In primates, brain mass at birth is approximately twice as large as in other placental mammals, relative to neonatal body mass (Sacher, 1982; Martin, 1983, 1996). On the one hand, this reveals that the larger average brain sizes relative to body size found in adult primates can be traced to a marked difference in prenatal development, squarely placing the emphasis on the maternal contribution. On the other hand, the weakening of the distinction between primates and other mammals by the time that the adult condition is attained indicates that factors intervening after birth can influence the ultimate outcome. It is important to recognize that the distinctly larger brain sizes of primates at birth,

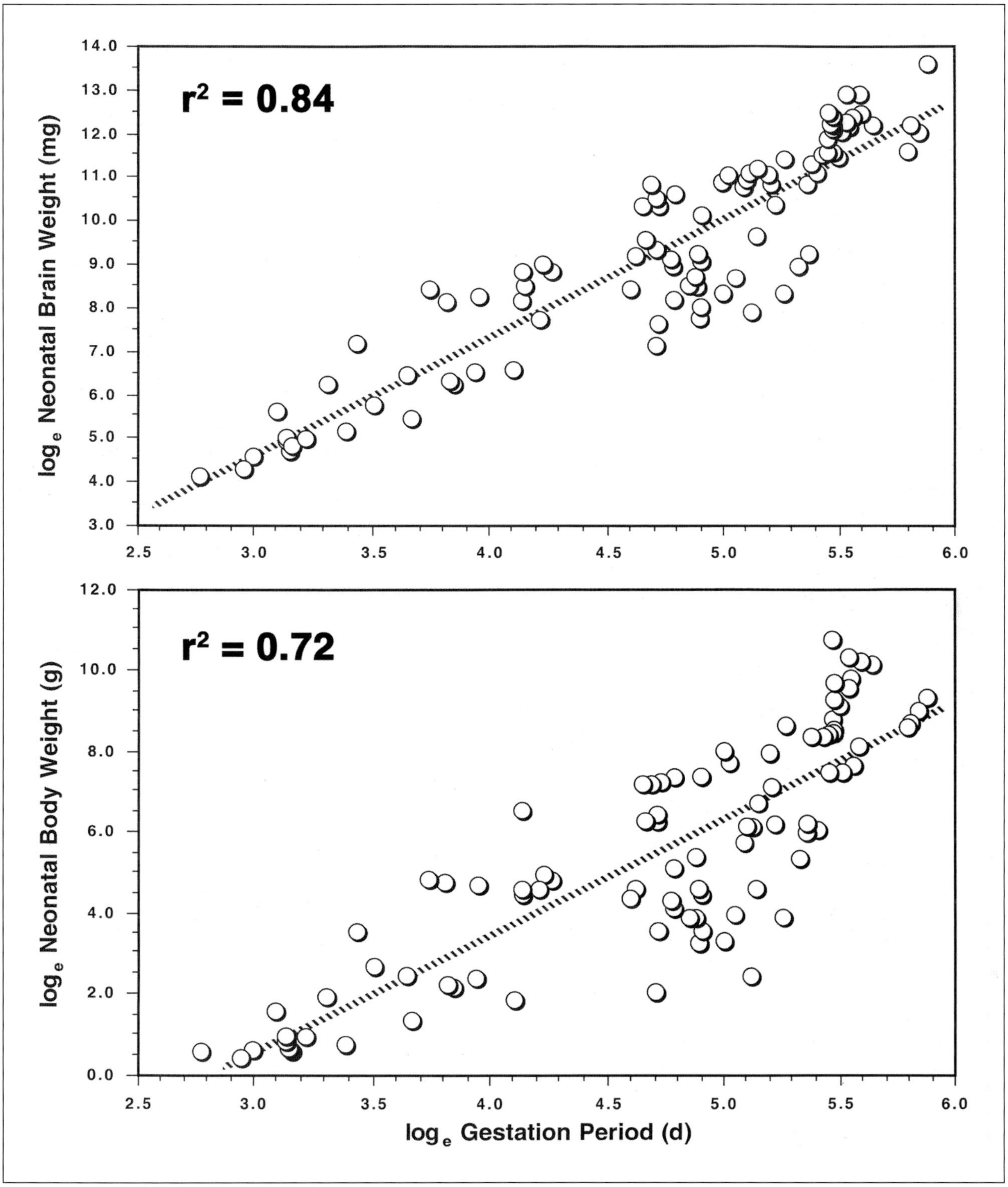

Figure 5 *Plots of neonatal brain (mg) and body mass (g) against gestation period (d) for 92 placental mammal species. Best-fit lines are least-squares regressions (provided for visual orientation only). The wider scatter of points around the line in the plot for neonatal body mass against gestation period is reflected by the lower value for the coefficient of determination (r2).*

relative to neonatal body mass, could be attained in two different ways. One possibility is that development of the fetal brain in primates is comparable to that in other mammals and that development of the rest of the body is restricted. Alternatively, it is possible that primate mothers actually invest more resources in development of the infant brain. These alternatives can be tested by plotting neonatal brain and body mass separately against adult body mass (Figure 6). As can be seen, primates overlap completely with other mammals with respect to the overall size of the neonate relative to adult body mass, but there is a clear grade distinction with respect to the size of the neonatal brain relative to adult body mass. Hence, the evidence shows that, in comparison to other mammals, primate mothers do actually invest more resources in the development of the fetal brain. It should be emphasized that brains of fetal primates are uniformly larger (relative to body mass) than those of other mammals throughout development, showing that increased maternal investment is consistently maintained during pregnancy in primates (Sacher, 1982; Martin, 1983).

Predictions of the MEH are also supported by combined analysis using partial correlations of adult brain size, adult body size, BMR and gestation period for the sample of 51 placental mammal species mentioned above (Martin, 1996, 1998). This analysis revealed persistent positive associations linking BMR to both body mass and brain mass, and linking gestation period to brain mass. Brain mass also showed a persistent positive association with body mass. However, the positive correlation between gestation period and BMR originally seen with the raw values was eliminated and replaced by a negative partial correlation. These results have now been confirmed with a much larger dataset for 320 placental mammal species. The results hence confirm that BMR and gestation period are both correlated with brain weight after eliminating the effect of body size (partial correlation coefficients: BMR—brain weight 0.214, gestation period—brain weight 0.307). At the same time, the negative partial correlation between gestation period and BMR (-0.233) indicates that relatively large brain sizes in mammals may be attributable either to longer gestation periods or to elevated BMR but not to both factors in combination. It should also be noted that, in a study restricted to primate genera, Little (1989) used path analysis to infer that gestation period and estimated metabolic rate are both connected to adult brain size.

In a study specifically focussing on bats, Jones and MacLarnon (2004) took data for 313 species to conduct a comparative test of three hypotheses concerning the rôle of energetics in the evolution of larger brains: (1) direct metabolic constraint; (2) ETH; (3) MEH. Their analyses provided virtually no support for the proposed link with basal metabolic rate invoked by any of the three hypotheses, but they did show that independent effects of gestation length and body mass can account for 95.9% of the variance in brain mass in bats. These authors hence demonstrated that the duration of maternal investment in bats plays an important part in the attainment of adult brain mass. They aptly noted that their results underline the crucial need to test the general applicability of any evolutionary hypothesis developed for a single clade in isolation by examining other clades with different evolutionary backgrounds. It should be noted, incidentally, that some bats are highly unusual with respect to the relationships between hibernation, BMR and reproductive parameters, so this may explain why no overall association between BMR and brain size was found in this case.

The maternal energy hypothesis focusses on the part played by maternal resources in brain development and the likelihood that they place constraints on the ultimate size of the brain in adulthood. However, selection to meet particular functional requirements will also exert an influence on brain size. Ideally, the concept of maternal energy constraints and that of selection favouring particular brain functions should be combined in a single model. A possible candidate is provided by the 2-phase hypothesis of brain size evolution proposed by Aboitiz (1996). This hypothesis proposes that brain size is influenced by both "passive" growth (general adjustment to body size) and "active" growth (adaptation in response to particular behavioural needs). The MEH can at least partially account for passive adjustment of overall brain size to body size, while selection of individual brain components to serve particular brain functions would eventually translate into increased brain size. Recognition of maternal investment as a key feature of "passive" brain growth permits refinement of the Aboitiz model in that an increase in overall brain size can result from an increase in maternal metabolic turnover or from an increase in the duration of gestation. Another implication of the model is that the quest for links between particular behavioural developments and increased brain size should focus particularly on associations between behaviour and particular parts of the brain rather than on overall brain size. For primates, visual components of the brain are of particular interest (Barton, 2006).

Additional support for the MEH emerges when the contributions of gestation period and BMR are combined (Martin, 1998). When brain size residuals were examined in relation to summed residuals for basal metabolic rate and gestation period, there was a marked improvement in the correlation coefficient compared to the values found with residuals for basal metabolic rate or gestation period in isolation ($r = 0.38$ for BMR alone; $r = 0.38$ for gestation period alone; $r = 0.55$ for BMR and gestation combined). This suggests that BMR and gestation period together account for ≈30% of variation in relative brain size between species. It should be noted that the MEH actually predicts that maternal BMR and gestation period would primarily influence neonatal brain size and that other factors (e.g. maternal investment through lactation) can intervene in the interval between birth and attainment of adult brain size. Various other lines of evidence support the inference that maternal resources are of particular importance for the evolution of the mam-

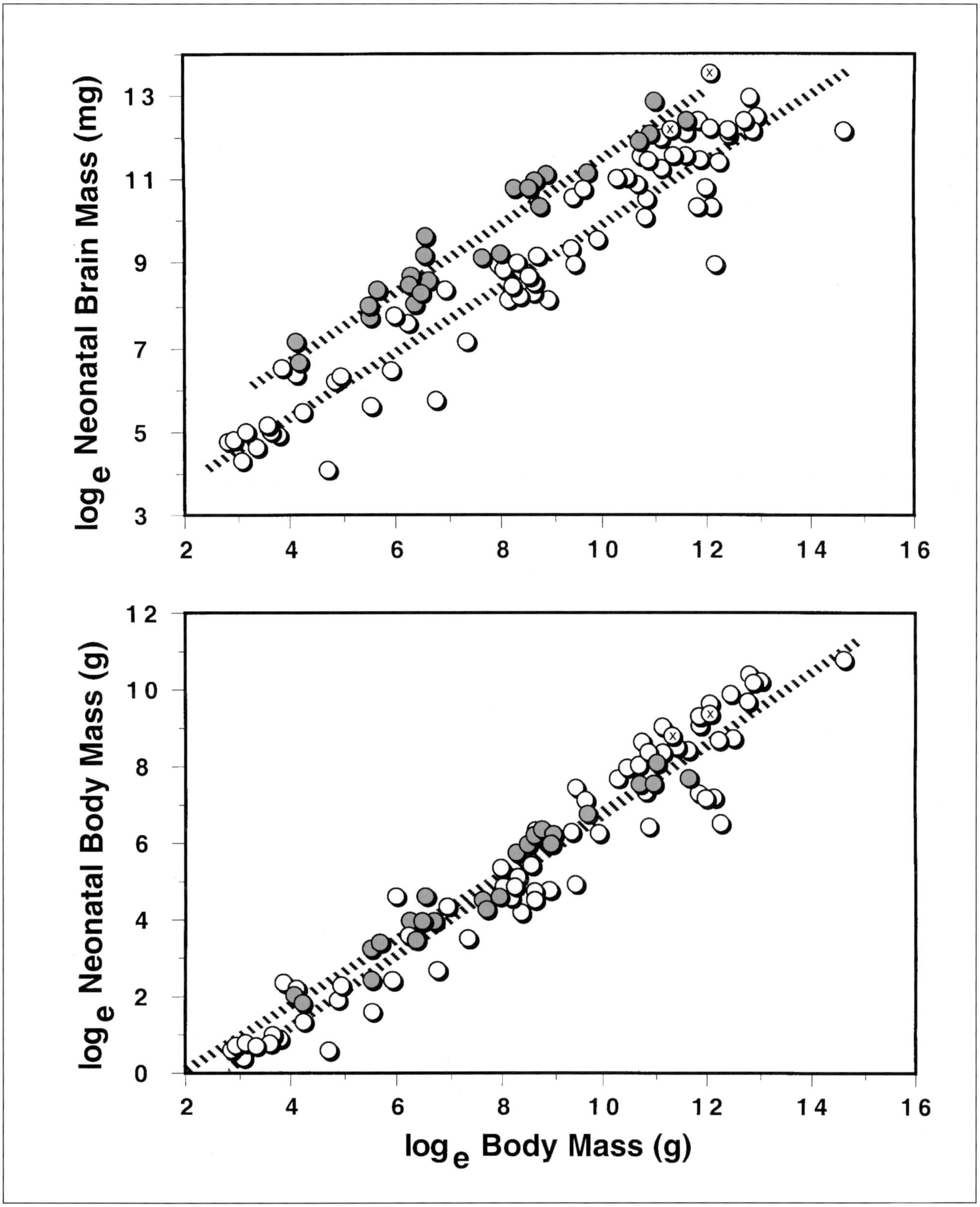

Figure 6 *Plots of neonatal brain mass (g) and neonatal body mass (g) against adult body mass (g) for a sample of 92 placental mammal species. Primates (shaded symbols) show an upward grade shift relative to other mammals (unshaded symbols) for neonatal brain mass, as indicated by the separate least-squares regression lines (provided for visual orientation only). By contrast, for neonatal body mass least-squares regression lines indicate no difference between primates and other mammals.*

malian brain. Experimental work on genomic imprinting, for example, has shown that maternally expressed genes specifically favour brain development (Keverne et al., 1996a). Furthermore, these maternally expressed genes favour higher centres of the brain (the "executive brain": neocortex + striatum), while paternally expressed genes promote more basal brain regions (the "emotional brain": hypothalamus + septum) instead (Keverne et al., 1996b).

Challenges to the Maternal Energy Hypothesis

Although several lines of evidence can hence be cited in support of the MEH, it has been subject to various challenges. One such challenge came from a test conducted by McNab & Eisenberg (1989) to investigate the proposed connection between brain size and BMR. These authors correctly noted that the MEH explicitly predicts that there should be a positive correlation between the residual values for brain size and BMR, both calculated relative to body size. They reported that analysis of data for 174 mammal species (including monotremes and marsupials) indicated that the relationship between relative brain size and relative BMR was not statistically significant ($p = 0.08$), although the trend was indeed positive as predicted. As noted by Martin (1998), however, the analysis conducted by McNab & Eisenberg (1989) was flawed because a parametric test was used to determine statistical significance. Such a test requires normality of distribution in the values compared, but it was applied after normally distributed logarithmic residual values had been converted to quotients with a strongly skewed distribution. A non-parametric test (Spearman rank correlation) applied to the derived quotient values revealed that the relationship is, in fact, statistically significant ($r_S = 0.17$; $p = 0.025$). As an alternative approach, a parametric test (Pearson correlation) was applied to the logarithmic residual values, also yielding a statistically significant result ($r = 0.16$; $p = 0.040$). Hence, the residual values for BMR and brain size reported by McNab & Eisenberg (1989) are actually significantly correlated. Despite this significance, however, the correlation is surprisingly weak in view of the other findings reported above. In fact, Martin (1998) reported a much stronger positive correlation from an analysis of 51 placental mammal species ($r = 0.38$; $p = 0.005$). The reason for this discrepancy has now emerged with the discovery that the dataset used by McNab and Eisenberg (1989) was itself seriously flawed. Data for brain sizes in rodents, taken from Mace et al. (1981), were systematically distorted because of the inadvertent addition of 0.59 g to the brain mass of every species (Isler and van Schaik, 2006b). Because rodents contributed disproportionately to the sample analysed by McNab and Eisenberg (≈45% of species included), the inaccurate values dramatically affected the results reported. Following correction of that error, a significant correlation between relative BMR and relative brain size was in fact found (Isler and van Schaik, 2006b). This amendment is particularly noteworthy because Aiello & Wheeler, (1995) specifically cited the doubly flawed paper by McNab and Eisenberg (1989) in their original presentation of the ETH. They stated (p. 211) that their "conclusions are derived from the general observation that there is no significant correlation between relative basal metabolic rate and relative brain size in humans and other encephalized mammals." That statement has now been invalidated.

A quite different challenge to the MEH arises from the claim that the results may have been biased by phylogenetic inertia (Pagel & Harvey 1990; Barton, 1999). From initial studies that attempted to offset effects of phylogenetic relatedness by conducting data analysis at the family level, it was reported that there was no significant relationship between BMR and adult brain size for mammals generally (Pagel and Harvey, 1988a), although a significant relationship between gestation period and neonatal brain size did remain (Pagel and Harvey, 1988b). Subsequently, Harvey and Pagel (1991) indicated that the exponent value for scaling of brain mass to body mass in placental mammals is reduced from 0.75 to 0.69 following contrasts analysis. Using a maximum likelihood approach, Pagel (1999) later reported exponent values of 0.59 for mammals and 0.48 for primates. Moreover, Barton (1999) reported that for primates no significant correlation between adult brain size and BMR or gestation period remains after application of the independent contrasts method (see also Barton, 2006). Yet Martin (1998) had reported in the meantime that an analysis of data for 51 placental mammal species had revealed that a highly significant correlation between adult brain size and BMR is found even after calculation of residual values determined from independent contrasts relative to body mass contrasts ($r = 0.465$; $p = 0.001$). However, the correlation between residuals for brain size and gestation period, although remaining positive, was found to be non-significant ($r = 0.203$; $p = 0.116$). Curiously, these conclusions are the opposite of those reported by Pagel and Harvey from analyses at the family level. Those authors found a significant correlation between relative brain size and gestation (Pagel and Harvey, 1988b) but not between relative brain size and BMR (Pagel and Harvey, 1988a). It has already been noted that there is a major pitfall in calculation of independent contrasts arising from magnification of the effects of error terms (Ricklefs and Starck, 1996; Martin et al., 2005). Because of this, assurance of data quality in a representative dataset is absolutely crucial. A recent analysis of a large, carefully monitored dataset for 347 mammal species has now demonstrated that there is in fact a significant correlation between BMR and adult brain mass after controlling for the effects of both body size and phylogenetic inertia (Isler and van Schaik, 2006b). The same finding has since been confirmed for primates taken in isolation (Isler et al., in press).

*Table 1. Partial correlation coefficients (r) between adult brain mass and maternal energy investment per offspring (MEI = (gestation period * BMR)/litter size), partialling out adult body mass and neonatal body*

	All species (N=229)		**Precocials (N=72)**		**Altricials (N=147)**	
	r	p	r	p	r	p
Raw data	0.362	<0.0001	0.637	<0.0001	0.198	0.017
Independent contrasts	0.177	0.008	0.401	0.0007	0.108	0.205

A further challenge of the MEH was raised by Pagel and Harvey (1988b), who suggested that multiple litters should be taken into account in the attempt to allow for the maternal energy input that every single offspring receives. Isler and van Schaik (in review) have found that relatively large-brained mammalian mothers produce fewer, but individually heavier offspring than small-brained mothers. This is in accordance with the MEH, which would additionally predict that, for a given neonate mass, a large-brained mother should invest more energy in a single offspring than a small-brained mother.

In other words, it is predicted that, after partialling out body mass and neonate mass, adult brain mass should still be positively correlated with maternal energy investment per offspring. To test this, we defined maternal energy investment (MEI) per offspring as gestation length multiplied by BMR and divided by litter size. Independent contrasts were calculated with PDAP:PDTree (Garland et al., 1992) in Mesquite (Maddison & Maddison, 2007), using the supertree of Bininda-Emonds et al. (2007). Precocial and altricial mammals were analysed separately, excluding bats. Species were defined as precocial if the young open their eyes at birth or shortly thereafter.

Partial correlation coefficients from this analysis are given in Table 1. In analyses of raw logarithmic species means, MEI and brain mass are positively and significantly correlated in all three groups (all mammals, precocials and altricials), whereas the correlation in altricials is no longer significant if independent contrasts are analysed. This might be explained by the fact that in altricial mammals a large proportion of brain growth is accomplished after birth, up to the age at weaning. To test whether the MEH also applies to the weaning period, we would need better data on weaning mass than presently available. However, our analyses thus far fully support the MEH. We conclude that large-brained precocial mothers indeed invest more energy in every single offspring, and, apart from producing heavier neonates in the first place, also invest more energy per unit neonate mass than relatively small-brained mothers, because the larger brain is so costly to grow.

Yet another potential challenge to the MEH is posed by the recent claim that the value of the scaling exponent for the relationship between BMR and body mass in placental mammals is not 0.75 but 0.67 (White and Seymour, 2003). This finding is puzzling because application of contrasts analysis had in fact confirmed the exponent value of 0.75 for scaling of BMR to body mass in placental mammals (Harvey and Pagel, 1991). In fact, White and Seymour (2003) reached their conclusion after excluding certain groups of mammals because of potential high energy turnover associated with digestion and after transforming the BMR data from the raw values that are usually used in analyses. As there are also problems with the published version of the dataset (confirmed by Michel Genoud, pers. comm.) further analyses are required to assess the validity of the results reported

Fossil Evidence for Mammalian Brain Evolution

An empirical finding that has emerged for all mammalian taxa for which an adequate fossil record is available is that average relative brain size tends to increase over time. Marsh (1874) originally noted that Eocene mammals from Wyoming generally have small brains in comparison to their modern counterparts. Indeed, he noted that in some "the brain cavity was hardly more capacious than in the higher reptiles". For example, in the primitive ungulate *Uintatherium* brain size was approximately one eighth of that in a modern rhinoceros of comparable body size. Progressive increase in the size of the brain was also reported in the evolution of horses. Subsequently, Edinger (1929, 1948) cast doubt on the existence of general trend towards increasing brain size, but she misinterpreted the fact that (other things being equal) a simple ratio of brain size to body size declines with increasing body size because of the negatively allometric scaling of brain size. Jerison (1970, 1973) subsequently demonstrated that, if relative brain size is calculated with due attention to allometric scaling, there is in fact a general trend towards increase in the average value over time. However, the spread of values also increases over time, with a few species showing little or no increase in relative brain size. Accordingly, some extant placental mammals (notably certain insectivores) have relative brain sizes that are little different from those of early placentals, but most have distinctly larger brains than any early fossil relatives. It is quite possible that brain size reduction has sometimes occurred in individual lineages, but this seems to be a relatively rare occurrence.

Average relative brain size clearly increased through the Tertiary as a general rule within primates (Martin, 1990). Relative brain size in Eocene primates is generally below that in extant prosimians, although there is some overlap in values between omomyiforms (often interpreted as relatives of tarsiers) and certain strepsirrhines (lemurs + lorisiforms). However, relative brain size in omomyiforms is below that found in extant tarsiers. In the earliest higher primates for which brain size is documented (early Oligocene *Aegyptopithecus*, *Catopithecus* and *Parapithecus*), relative brain size is comparable to that in extant prosimians but below the range of values for extant monkeys and apes. It has also been shown that in the Miocene New World monkey *Chilecebus* relative brain size is smaller than in extant platyrrhines (Sears et al., 2008). A similar trend towards increase in relative brain size over time has also been clearly demonstrated through analysis of an impressive dataset for toothed cetaceans by Marino et al. (2004). Eocene archaeocetes have very small brains compared to more recent toothed cetaceans, and an overall trend through the Tertiary is seen, leading up to the notably large brains of modern dolphins and their relatives.

It is equally well established that relative brain size increased over time during the evolution of mammalian carnivores. Initial data provided by Jerison (1970, 1973) showed that early Tertiary carnivore relatives (creodonts) and archaic ungulates (condylarths) had relatively small brains compared to their modern counterparts. Radinsky (1977, 1978) subsequently provided additional evidence showing that relative brain size was smaller in creodonts and the earliest known carnivores during the early Tertiary. In an analysis of encephalization quotients (EQ values) within the carnivoran suborder Caniformia, Finarelli and Flynn (2007) showed that taxa early in the evolutionary history of the group possessed significantly lower median values than extant taxa. A pronounced upward shift in median values was found at the Miocene-Pliocene transition. A gradual increase in variance around median relative brain size was also found. Reconstructions of ancestral EQ values revealed that increased encephalization took place in parallel across all major caniform clades, with the possible exception of skunks. A subsequent study focused specifically on brain size in Canidae in order to reveal underlying trends that might be masked in a more wide-ranging investigation (Finarelli, 2008). A shift towards higher encephalization in crown Caninae was found relative to a basal grade of encephalization in Hesperocyoninae, Borophaginae and *Leptocyon*. However, at this level of analysis no associated change in variance was found.

Widespread acknowledgment of a general trend towards increasing brain size during mammalian evolution was recently challenged in specific relation to bats by Safi et al. (2005). These authors concluded that brain size actually decreased over time in numerous bat lineages. However, their analysis was entirely based on analysis of relative brain size in extant bats, with no reference whatsoever to the fossil record. The results reported by Safi et al. are entirely dependent on their application of a statistical test that supposedly identifies directionality in the data (i.e. a trend towards increasing or decreasing brain size) in relation to a phylogenetic tree. The outcome of the test was that no statistically significant directionality was detectable. Once this inference has been made, it necessarily follows that increase in brain size in some lineages over time must be balanced by decrease in brain size in other lineages, such that the overall average remains unchanged over time (i.e. lacks directionality). It is hence only to be expected that a decrease in brain size was inferred in approximately half of bat species in the phylogenetic reconstruction presented by Safi et al. (2005). In fact, the test of directionality applied by Safi et al. (2005), using the software CONTINUOUS (Pagel, 1997, 1999; now included in BayesTraits) tests whether a directional change parameter should be included in the model of evolution of the trait under consideration. This parameter effectively measures the regression of trait values across species against total path length from the root of the tree to the tips. The CONTINUOUS manual states that it detects any general trends towards a dominant direction of evolutionary change (i.e. whether species have got bigger, smaller, faster, longer, etc.). The test can only be used with trees that have some variation in total path length from the root to tip species. In consequence, the test cannot be applied to the commonly used trees with branch lengths estimated as time elapsed on the basis of molecular data, because all extant taxa must exhibit the same distance from the root. Making matters worse, Safi et al. (2005) applied the test to a tree in which all branches between bifurcations are of equal length (=1). In their tree, therefore, species from species-rich taxa groups necessarily exhibit a longer pathway from root to tip than other species. Thus, they in fact tested whether speciose taxonomic groups differ in relative brain mass from taxonomic groups with fewer species, and did not find any indication of this. Any conclusions about directionality of brain size evolution drawn from this test are thus invalid.

In the absence of any attempt to verify their results by comparison with the fossil record, the findings reported by Safi et al. (2005) must in any case be treated with great scepticism. The potential dangers of reconstructing changes in size over time without reference to the fossil record are aptly illustrated by an analysis of body size of the mammalian order Carnivora by Finarelli and Flynn (2006). Among caniform carnivorans (Canidae, Ursidae, Pinnipedia and Musteloidea), many subgroups are now represented predominantly by large- or small-bodied species and the distribution of body sizes among extant species across the phylogeny in fact suggests a pattern of decreasing body size from an ancestral value of 10-50 kg. However, estimated body sizes for fossil representatives of a given caniform taxon often lie

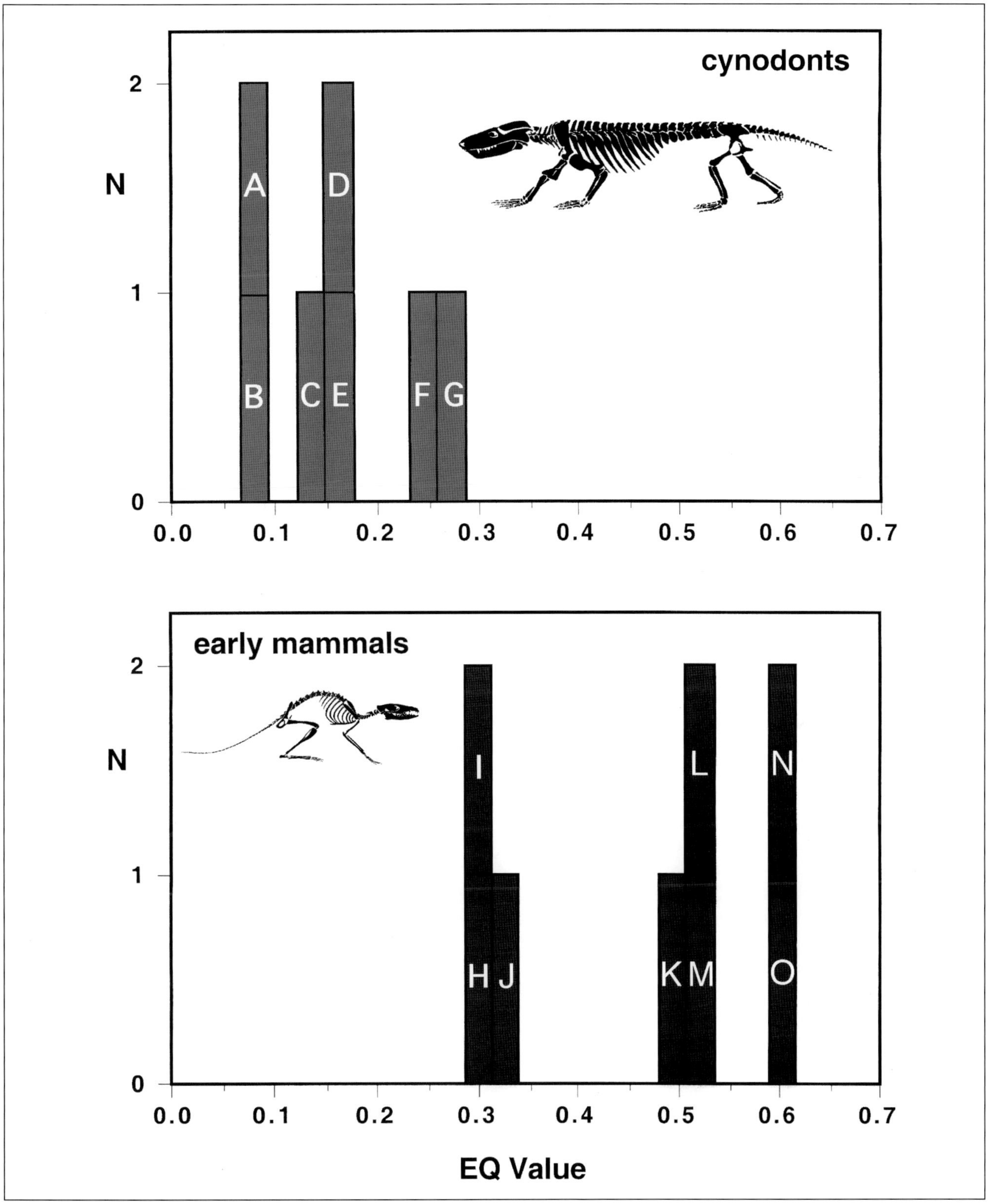

Figure 7 *Histogram showing relative brain sizes in cynodont therapsids (n = 7) and early mammals (n = 8), as indicated by encephalization quotient (EQ) values. EQ values were calculated using the formula determined for 309 modern placentals by Martin (1981). An EQ value of 1 indicates the average condition for modern placentals. Data on brain and body size derived from Jerison (1973), Crompton and Jenkins (1978), Quiroga (1980, 1984), Kielan-Jaworowska (1983, 1984), Krause and Kielan-Jaworowska (1993), Kielan-Jaworowska and Lancaster (2004), Macrini et al. (2007). Key to cynodonts: A =* Thrinaxodon*; B =* Exaeretodon*; C =* Probelesodon*; D =* Probainognathus*; E =* Diademodon*; F =* Therioherpeton*; G =* Massetognathus. *Key to early mammals: H =* Vincelestes *(placental); I =* Kennalestes *(placental); J =* Triconodon*; K =* Asioryctes *(placental); L =* Chulsanbaatar *(multituberculate); M =* Ptilodus *(multituberculate); N =* Zalambdalestes *(placental); O =* Kryptobaatar *(multituberculate).*

well outside the observed ranges for extant members, so the modern distribution of body sizes is not representative of the evolutionary history of the group. When 367 fossil taxa were included with 149 extant species for a combined analysis designed to reconstruct ancestral body sizes, a small-bodied ancestor (1-5 kg) was indicated both for Caniformia and for the monophyletic subclade Arctoidea (Ursidae, Pinnipedia and Musteloidea). As was aptly noted by Finarelli and Flynn (2006): "Evolutionary trends can reduce the accuracy of character state reconstructions, especially for methods assuming Brownian motion as the model for character change. This is because an estimated root value under such a model will always be some form of weighted average of observed values for terminal taxa (Schluter et al., 1997), and if a trend moves the range of observed character state values beyond the ancestral condition, it will be difficult, if not impossible, to accurately reconstruct the condition at the ancestral node (Garland et al., 1999; Oakley and Cunningham, 2000)."

There are therefore good reasons to question the findings reported by Safi et al. with respect to the evolution of relative brain size in bats. To test whether brain size evolution in bats was directional or not, the most obvious approach would be to seek data on brain size in fossil bats at different times in the Tertiary. However, the fossil record for bats is relatively poor, so it might prove very difficult to conduct an adequate test of the presence or absence of directionality in brain size evolution. Given the compelling evidence from diverse mammalian fossils for a general trend towards increasing brain size in mammalian evolution (Jerison, 1973), it seems highly unlikely that the conclusions drawn regarding bats will survive proper testing.

The existence of a general trend towards increase in relative brain size over the course of mammalian evolution is of particular interest in the context of the MEH. Given that resources provided by the mother throughout gestation and lactation seem to be of particular importance for the development of the brain, the emergence and subsequent refinement of pregnancy and suckling are presumably connected to evolutionary changes in brain size over time. It is, for example, to be expected that the origin of lactation in early mammals and its presence in the common ancestor of monotremes, marsupials and placentals some 200 million years (Ma) ago might have been accompanied by an increase in relative brain size. Most extant reptiles show no parental behaviour, so development of the offspring prior to independent feeding relies entirely on the resources provided in the egg when laid by the mother. Moreover, interspecific scaling of brain size to body size follows a different trajectory in reptiles compared to mammals. The size of the egg is related to the mother's metabolic capacity, while the size of the brain in the hatchling is related to the egg's metabolic capacity. The outcome is a lower exponent of $\approx$0.56 for brain:body scaling in reptiles (Martin, 1981), which imposes a handicap that increases with increasing body size. In a further crucial development at a later stage of mammalian evolution, egg-laying was replaced by internal development of the offspring (vivipary) in the common ancestor of marsupials and placentals at least 135 Ma ago. Retention of the developing egg within the mother's body at once permitted continuous provision of maternal resources to the developing offspring, as opposed to reliance on a one-off provision of resources in an externally deposited egg. Overall, these considerations lead to the expectation that relative brain size might have increased in the earliest mammals and would probably have increased even further with the origin of vivipary.

The earliest mammals arose from the cynodonts, advanced mammal-like reptiles (therapsids), close to the Triassic/Jurassic boundary about 200 Ma ago. Estimates of brain size and body size are now available for 7 cynodonts (*Diademodon*, *Exaeretodon*, *Massetognathus*, *Probainognathus*, *Probelesodon*, *Therioherpeton* and *Thrinaxodon*), providing an adequate basis for comparison with early mammals. Unfortunately, very little is known about relative brain size in Jurassic mammals, so comparison of cynodonts with the earliest known mammals is not yet possible. However, data are available for the Late Jurassic *Triconodon*, an Early Cretaceous placental (*Vincelestes*) 3 Late Cretaceous placentals (*Asioryctes*, *Kennalestes*, *Zalambdalestes*), 2 Late Cretaceous multituberculates (*Chulsanbaatar*, *Kryptobaatar*) and the Palaeocene multituberculate *Ptilodus*. As can be seen from Figure 7, the cynodonts uniformly have relatively smaller brains than the early mammals. The early mammals, in turn, have relative brain sizes that consistently lie below the average condition for modern mammals (indicated by an EQ value of 1). However, the values for the early mammals do overlap with the lower end of the range for modern mammals. This evidence confirms the expectation that relative brain size should be increased in early mammals relative to advanced mammal-like reptiles, but should lie below the average condition for extant mammals. Although no information on relative brain size is available as yet for the earliest mammals, there is an indirect indication that expansion of brain size relative to the reptilian level was probably under way quite early in mammalian evolution. It is generally accepted that the multituberculates diverged very early in mammalian evolution, and some authors have indeed linked them to monotremes rather than to the lineage leading to marsupials and placentals. Yet the values for multituberculates in Figure 7 are comparable to those for early placentals.

Following the origin of vivipary in the lineage leading to the ancestry of marsupials and placentals, refinements in intrauterine development would have permitted a further increase in provision of maternal resources. Placentation is generally poorly developed in marsupials and gestation periods are very short, although provision of maternal resources through suckling is enhanced by an extended pouch life. Marsupials generally have only

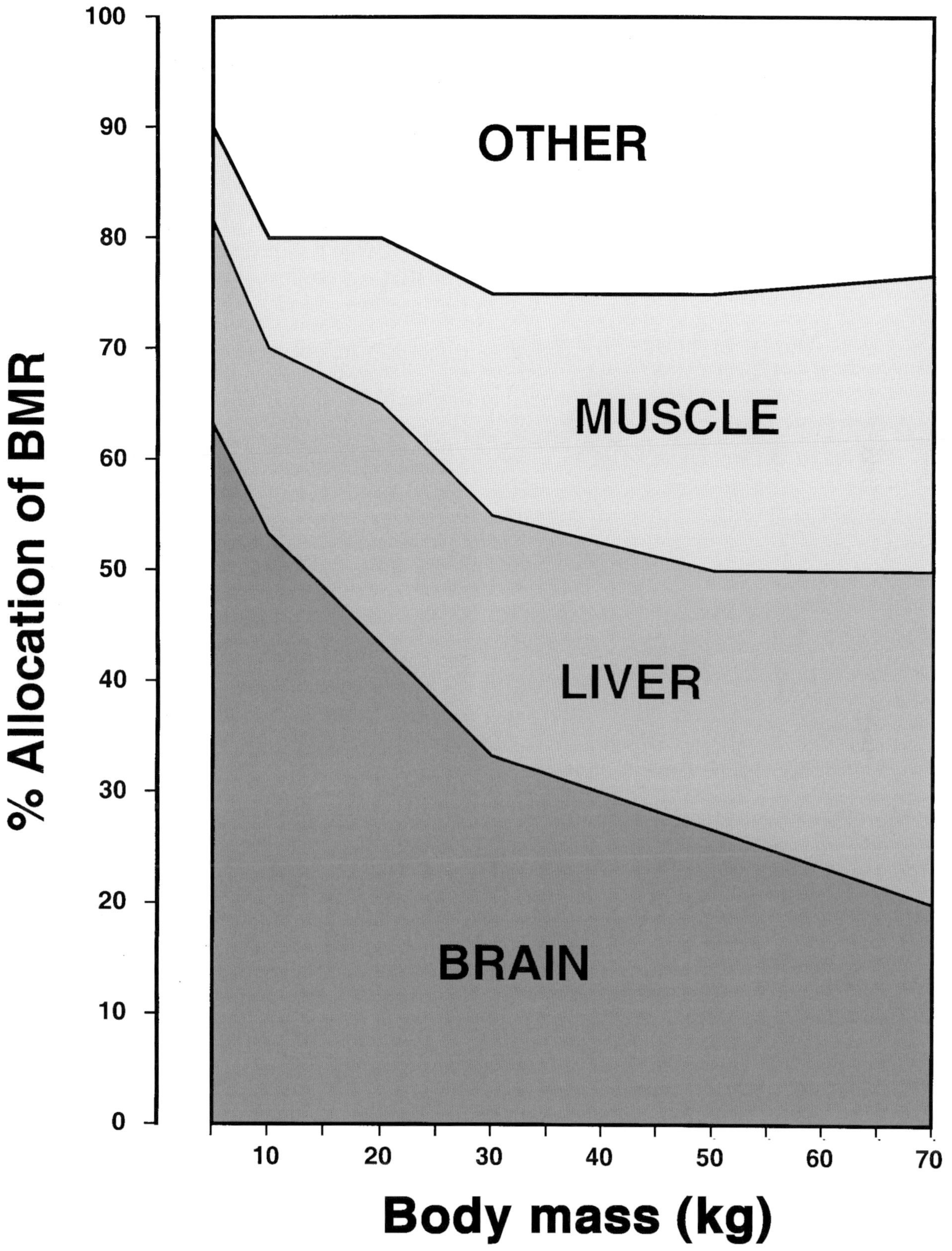

Figure 8 Proportional allocation of BMR to brain, liver, muscle and other tissues in humans at different body mass, ranging from birth to adulthood. (Data from Holliday, 1986.)

small to moderately developed brains, as would be expected from these constraints. Among placental mammals, however, many lineages have developed extended gestation periods, which are associated with precocial offspring (well developed and usually singletons). This contrasts with the condition in mammals that produce litters of poorly developed altricial offspring, which have markedly shorter gestation periods (Martin et al., 2005). Comparative evidence indicates that the altricial condition is primitive for placental mammals. Refinements in placentation doubtless occurred in parallel in many lineages during the course of evolution of placental mammals, and this would have provided a basis for increased provision of maternal resources. However, maternal resources are also provided during lactation, so this provides an additional avenue for maternal investment in the development of the offspring's brain.

A Note on Implications For Hominid Brain Evolution

Because the modern human brain is the largest, relative to body size, among mammals generally, the problem posed by energy demands is particularly acute. Indeed, this problem is most marked early in postnatal life in comparison to adults. Data for allocation of BMR to different tissues in humans at different body sizes (Holliday, 1986) show that allocation of energy to the brain is predominant early in life (Fig. 8). Leonard et al. (2003, 2007) have shown that dietary quality (rather than BMR) may be a key factor in ensuring an adequate supply of energy to the brain in adult humans. In a plot of residuals for diet quality and brain size (both calculated relative to body mass), humans are clear outliers in having an unusually high dietary quality in comparison to other primates. As noted by Leonard et al. (2003, 2007), the relatively small gastrointestinal tract of humans is consistent with adaptation for a high-quality diet and may, in itself, have no direct connection with brain size. Mounting energetic requirements accompanying increasing brain size over the course of human evolution must clearly be considered as a fundamental issue (Martin, 1983). One thought-provoking attempt to take this into account has explored implications of carnivory for increased brain size in human evolution (Vasey and Walker, 2001).

Acknowledgments

This paper is a modified and expanded version of a presentation given by the first author at the Symposium in Honor of Ralph L. Holloway "The Human Brain Evolving" held at Indiana University, Bloomington on April 26th-29th, 2007, jointly sponsored by the Stone Age Institute and Indiana University. The content of the paper benefited greatly from interactions with Michel Genoud, Ann MacLarnon and Carel van Schaik. Thanks go to Ann MacLarnon and Ben Rudder for assistance in compilation of the original dataset used for various allometric analyses. We are also grateful to Edna Davion for logistical support and comments on the manuscript.

References

Aboitiz, F., 1996. Does bigger mean better? Evolutionary determinants of brain size and structure. Brain Behavior and Evolution 47, 225-245.

Aiello, L.C., Bates, N., Joffe, T.H., 2001. In defense of the expensive tissue hypothesis: ontogeny, maternal care and organ size. In: Falk, D., Gibson, K.R. (Eds.), Evolutionary anatomy of the primate cerebral cortex. Cambridge University Press, Cambridge, pp. 57-78.

Aiello, L.C., Wheeler, P., 1995. The expensive tissue hypothesis: the brain and the digestive system in human and primate evolution. Current Anthropology 36, 199-221.

Armstrong, E., 1982. A look at relative brain size in mammals. Neuroscience Letters 34, 101-104.

Armstrong, E., 1983. Relative brain size and metabolism in mammals. Science 220, 1302-1304.

Armstrong, E., 1985. Relative brain size in monkeys and prosimians. American Journal of Physical Anthropology 66, 263-273.

Armstrong, E., 1990. Brains, bodies and metabolism. Brain, Behavior and Evolution 36, 166-176.

Barton, R.A., 1999. The evolutionary ecology of the primate brain. In: Lee, P.C. (Ed.), Comparative primate socioecology. Cambridge University Press, Cambridge, pp. 167-203.

Barton, R.A., 2006. Primate brain evolution: Integrating comparative, neurophysiological, and ethological data. Evolutionary Anthropology 15, 224-236.

Bauchot, R., 1978. Encephalization in vertebrates: A new mode of calculation for allometry coefficients and isoponderal indices. Brain, Behavior and Evolution 15, 1-18.

Bininda-Emonds, O.R.P., Cardillo, M., Jones, K.E., MacPhee, R.D.E., Beck, R.M.D., Grenyer, R., Price, S.A., Vos, R.A., Gittleman, J.L., Purvis, A., 2007 The delayed rise of present-day mammals. Nature 446, 507-512.

Bonner, J.T., 2006. Why size matters: From bacteria to blue whales. Princeton University Press, Princeton.

Brody, S., 1945. Bioenergetics and growth. Rheinhold, New York.

Brody, S., Procter, T.C., 1932. Relations between basal metabolism and mature body weight in different species of mammals and birds. Research Bulletin of the Missouri Agriculture Research Station 166, 89-101.

Clutton-Brock, T.H., Harvey, P.H., 1980. Primates, brains and ecology. Journal of the Zoological Society of London 190, 309-323.

Crompton, A.W., Jenkins, F.A., 1978. Mesozoic mammals. In: Cooke, H.B.S., Maglio, V. (Eds.), Evolution of African mammals. Harvard University Press, Cambridge, pp. 46-55.

Dawson, T.J., Hulbert, A.J., 1970. Standard metabolism, body temperature and surface areas of Australian marsupials. American Journal of Physiology 218, 1233-1238.

Edinger, T., 1929. Die fossilen Gehirne. Ergebn Anat Entwickl-Gesch 28, 1-249.

Edinger, T., 1948. Evolution of the horse brain. Mem Geol Soc Am 25, 1-777.

Eisenberg, J.F., 1981. The mammalian radiations: An analysis of trends in evolution, adaptation and behaviour. Athlone Press, London.

Eisenberg, J.F., Wilson, D.E., 1978. Relative brain size and feeding strategies in the Chiroptera. Evolution 32, 740-751.

Felsenstein, J., 1985. Phylogenies and the comparative method. The American Naturalist 125, 1-15.

Finarelli, J.A., 2008. Testing hypotheses of the evolution of encephalization in the Canidae (Carnivora, Mammalia). Palaeobiology 34, 35-45.

Finarelli, J.A., Flynn, J.J., 2006. Ancestral state reconstruction of body size in the Caniformia (Carnivora, Mammalia): the effects of incorporating data from the fossil record. Society of Systematic Biologists 55, 301-313.

Finarelli, J.A., Flynn, J.J., 2007. The evolution of encephalization in caniform carnivorans. Evolution 61, 1758-1772.

Garland, T., Harvey, P.H., Ives, A.R., 1992. Procedures for the analysis of comparative data using phylogenetically independent contrasts. Society of Systematic Biologists 41, 18-32.

Garland, T., Midford, P.E., Ives, A.R., 1999. An introduction to phylogenetically based statistical methods, with a new method for confidence intervals on ancestral values. American Zoologist 39, 374-388.

Gould, S.J., 1966. Allometry and size in ontogeny and phylogeny. Biological Reviews 41, 587-640.

Gould, S.J., 1975. Allometry in primates, with emphasis on scaling and the evolution of the brain. Contributions to Primatology 5, 244-292.

Haight, J.R., Nelson, J.E., 1987. A brain that doesn't fit its skull: A comparative study of the brain and endocranium of the koala, *Phascolarctos cinereus* (Marsupialia: Phascolarctidae). In: Archer M (Ed.), Possums and opossums: Studies in evolution. Surrey Beatty & Sons, Sydney, Australia, pp. 331-352.

Harvey, P.H., Mace, G.M., 1982. Comparison between taxa and adaptive trends: Problems of methodology. In: King's College Sociobiology Group (Eds.), Current problems in sociobiology. Cambridge University Press, Cambridge, pp. 343-361.

Harvey, P.H., Pagel, M.D., 1991. The comparative method in evolutionary biology. Oxford University Press, Oxford.

Hemmingsen, A.M., 1960. Energy metabolism as related to body size and respiratory surfaces, and its evolution. Reports of the Steno Memorial Hospital and Nordinsk Insulin Laboratorium 9, 1-110.

Hofman, M.A., 1982. Encephalization in mammals in relation to the size of the cerebral cortex. Brain, Behavior and Evolution 20, 84-96.

Hofman, M.A., 1983a. Evolution of the brain in neonatal and adult placental mammals: A theoretical approach. Journal of Theoretical Biology 105, 317-322.

Hofman, M.A., 1983b. Energy metabolism, brain size and longevity in mammals. Quarterly Review of Biology 58, 495-512.

Holliday, M., 1986. Body composition and energy needs during growth. In: Falkner, F., Tanner, J.M. (Eds.), Human growth: A comprehensive treatise. Vol. 2: Postnatal growth biology. Plenum Press, New York, pp. 101-117.

Isler, K., Barbour, A.D., Martin, R.D., 2002. Line-fitting by rotation: A nonparametric method for bivariate allometric analysis. Biometrical Journal 44, 289-304.

Isler, K., Kirk, E.C., Miller, J.M.A., Albrecht, G.A., Gelvin, B.R., Martin, R.D., in press. Endocranial volumes of primate species: Scaling analyses using a comprehensive and reliable dataset. Journal of Human Evolution.

Isler, K., van Schaik, C.P., 2006a. Costs of encephalization: the energy trade-off hypothesis tested on birds. Journal of Human Evolution 51, 228-243.

Isler, K., van Schaik, C.P., 2006b. Metabolic costs of brain size evolution. Biology Letters 2, 557-560.

Isler, K., van Schaik, C.P., in review. The Expensive Brain: a framework for explaining evolutionary changes in brain size.

Jerison, H.J., 1970. Brain evolution: new light on old principles. Science 170, 1224-1225.

Jerison, H.J., 1973. Evolution of the brain and intelligence. Academic Press, New York.

Jones, K.E., MacLarnon, A.M., 2004. Affording larger brains: Testing hypotheses of mammalian brain evolution on bats. American Naturalist 164, E20-E31.

Keverne, E.B., Fundele, R., Narasimha, M., Barton, S.C., Surani, M.A., 1996a. Genomic imprinting and the differential roles of parental genomes in brain development. Developmental Brain Research 92, 91-100.

Keverne, E.B., Martel, F.L., Nevison, C.M., 1996b. Primate brain evolution: genetic and functional considerations. Proceedings of the Royal Society of London B 263, 689-696.

Kielan-Jaworowska, Z., 1983. Multituberculate endocranial casts. Palaeovertebrata 13, 1–12.

Kielan-Jaworowska, Z., 1984. Evolution of the therian mammals in the Late Cretaceous of Asia. Part VI. Endocranial casts of eutherian mammals. Palaeontologia Polonica 46, 157–171.

Kielan-Jaworowska, Z., Lancaster, T.E., 2004. A new reconstruction of multituberculate endocranial casts and encephalization quotient of *Kryptobaatar*. Acta Palaeontologia Polonica 49, 177-188.

Kleiber, M., 1932. Body size and metabolism. Hilgardia 6, 315-353.

Kleiber, M., 1947. Body size and metabolic rate. Physiological Reviews 27, 511-541.

Kleiber, M., 1961. The fire of life: An introduction to animal energetics. John Wiley, New York.

Krause, D.W., Kielan-Jaworowska, Z., 1993. The endocranial cast and encephalization quotient of *Ptilodus* (Multituberculata, Mammalia). Palaeovertebrata 22, 99-112.

Leonard, W.R., Robertson, M.L., 1992. Nutritional requirements and human evolution: A bioenergetics model. American Journal of Human Biology 4, 179-195.

Leonard ,W.R., Robertson, M.L., Snodgrass, J.J., Kuzawa, C.W., 2003. Metabolic correlates of hominid brain evolution. Comparative Biochemistry and Physiology A 136, 5-15.

Leonard, W.R., Snodgrass, J.J., Robertson, M.L., 2007. Effects of brain evolution on human nutrition and metabolism. Annual Review of Nutrition 27, 311-327.

Little, B.B., 1989. Gestation length, metabolic rate, and body and brain weights in primates: epigenetic effects. American Journal of Physical Anthropology 80, 213-218.

Mace, G.M., Harvey, P.H., Clutton-Brock, T.H., 1981. Brain size and ecology in small mammals. Journal of Zoology London 193, 333-354.

MacMillen, R.E., Nelson, J.E., 1969. Bioenergetics and body size in dasyurid marsupials. American Journal of Physiology 217, 1246-1251.

Macrini, T.E., Rougier, G.W., Rowe, T.B., 2007. Description of a cranial endocast from the fossil mammal *Vincelestes neuquenianus* (Theriiformes) and its relevance to the evolution of endocranial characters in therians. The Anatomical Record 290, 875-892.

Maddison, W.P., Maddison, D.R., 2007. Mesquite: A modular system for evolutionary analysis. Version 2.01. http://mesquiteproject.org.

Marino, L., McShea, D.W., Uhen, M.D., 2004. Origin and evolution of large brains in toothed whales. The Anatomical Record 281A, 1247-1255.

Marsh, O.C., 1874. Small size of the brain in Tertiary mammals. American Journal of Science and Arts 8, 66-67.

Martin, R.D., 1981. Relative brain size and metabolic rate in terrestrial vertebrates. Nature 293, 57-60.

Martin, R.D., 1983. Human brain evolution in an ecological context (52nd James Arthur Lecture on the Evolution of the Human Brain). American Museum of Natural History, New York.

Martin, R.D., 1989. Size, shape and evolution. In: Keynes, M. (Ed.), Evolutionary studies—A centenary celebration of the life of Julian Huxley. Eugenics Society, London, pp. 96-141.

Martin, R.D., 1990. Primate origins and evolution: a phylogenetic reconstruction. Princeton University Press, New Jersey.

Martin, R.D., 1996. Scaling of the mammalian brain: the maternal energy hypothesis. News in Physiological Sciences 11, 149-156.

Martin, R.D., 1998. Comparative aspects of human brain evolution: Scaling, energy costs and confounding variables. In: Jablonski, N.G., Aiello, L.C., (Eds.), The origin and diversification of language. California Academy of Sciences, San Francisco, pp. 35-68.

Martin, R.D., 2007. The evolution of human reproduction: A primatological perspective. Yearbook of Physical Anthropology 50, 59-84.

Martin, R.D., Barbour, A.D., 1989. Aspects of line-fitting in bivariate allometric analyses. Folia Primatol 53, 65-81.

Martin, R.D., Genoud, M., Hemelrijk, C.K., 2005. Problems of allometric scaling analysis: Examples from mammalian reproductive biology. The Journal of Experimental Biology 208, 1731-1747.

Martin, R.D., Harvey, P.H., 1985. Brain size allometry: Ontogeny and phylogeny. In: Jungers, W.L., (Ed.), Size and scaling in primate biology. Plenum Press, New York, pp. 147-173.

McNab, B.K., 1980. Food habits, energetics and the population biology of mammals. American Naturalist 116, 106-124.

McNab, B.K., 1986. The influence of food habits on the energetics of eutherian mammals. Ecological Monographs 56, 1-19.

McNab, B.K., 1988. Complications inherent in scaling the basal rate of metabolism in mammals. The Quarterly Review of Biology 63, 25-54.

McNab, B.K., Eisenberg, J.F., 1989. Brain size and its relation to the rate of metabolism in mammals. American Naturalist 133, 157-167.

Mink, J.W., Blumenshine, R.J., Adams, D.B., 1981. Ratio of central nervous system to body metabolism in vertebrates: its constancy and functional basis. American Journal of Physiology 241, R203-R212.

Oakley, T.H., Cunningham, C.W., 2000. Independent contrasts succeed where ancestor reconstruction fails in a known bacteriophage phylogeny. Evolution 54:397-405.

Pagel, M.D., 1997. Inferring evolutionary processes from phylogenies. Zoologica Scripta 26, 331-348.

Pagel, M., 1999. Inferring the historical patterns of biological evolution. Nature 401, 877-884.

Pagel, M.D., Harvey, P.H., 1988a. The taxon-level problem in the evolution of mammalian brain size: Facts and artifacts. American Naturalist 132, 344-359.

Pagel, M.D., Harvey, P.H., 1988b. How mammals produce large-brained offspring. Evolution 42, 948-957.

Pagel, M.D., Harvey, P.H., 1990. Diversity in the brain size of newborn mammals: allometry, energetics or life history tactics? Bioscience 40, 116-122.

Purvis, A., Rambaut, A., 1995. Comparative analysis by independent contrasts (CAIC): an Apple Macintosh application for analysing comparative data. CABIOS 11, 247-251.

Quiroga, J.C., 1980. The brain of the mammal-like reptile Probainognathus jenseni (Therapsida, Cynodontia). A correlative paleoneurological approach to the cortex at the reptile-mammal transition. Journal für Hirnforschung 21, 299-336.

Quiroga, J.C., 1984. The endocranial cast of the mammal-like reptile Therioherpeton cargnini (Therapsida, Cynodontia) from the Middle Triassic of Brazil. Journal für Hirnforschung 25, 285-290.

Radinsky, L.B., 1977. Brains of early carnivores. Paleobiology 3, 333-349.

Radinsky, L.B., 1978. Evolution of brain size in carnivores and ungulates. American Naturalist 112, 815-831.

Reiss, M.J., 1989. The allometry of growth and reproduction. Cambridge University Press, Cambridge.

Ricklefs, R.E., Starck, J.M., 1996. Applications of phylogenetically independent contrasts: A mixed progress report. Oikos 77, 167-172.

Rubner, M., 1883. Über den Einfluss der Körpergrösse auf Stoff- und Kraftwechsel. Z Biol 19, 535-562.

Sacher, G.A., 1982. The role of brain maturation in the evolution of the primates. In: Armstrong, E., Falk, D. (Eds.), Primate brain evolution. Plenum Press, New York, pp. 97-112.

Sacher, G.A., Staffeldt, E., 1974. Relation of gestation time to brain weight for placental mammals. American Naturalist 108, 593-615.

Safi, K., Seid, M.A., Dechmann, D.K.N., 2005. Bigger is not always better: when brains get smaller. Biology Letters 1, 283–286.

Schluter, D., Price, T., Mooers, A.O., Ludwig, T., 1997. Likelihood of ancestor states in adaptive radiation. Evolution 51, 1699-1711.

Schmidt-Nielsen, K., 1984. Scaling: Why is animal size so important? Cambridge University Press, Cambridge.

Sears, K.E., Finarelli, J.A., Flynn, J.J., Wyss, A.R., 2008 Estimating body mass in New World "monkeys" (Platyrrhini, Primates), with a consideration of the Miocene platyrrhine, *Chilecebus carrascoensis*. American Museum Novitates 3617, 1-29.

Shevlin, M., Davies, M.N.O., 1997. Alphabetical listing and citation rates. Nature 388, 14.

Snodgrass, J.J., Leonard, W.R., Robertson, M.L., 2007. Primate bioenergetics: an evolutionary perspective. In: Ravosa, M.J., Dagosto, M. (Eds.), Primate origins: adaptations and evolution. Springer, New York. pp. 703-737.

Tregenza, T., 1997. Darwin a better name than Wallace? Nature 385, 480.

Vasey, N., Walker, A.C., 2001. Neonate body size and hominid carnivory. In: Stanford, C., Bunn, H. (Eds.), Meat-eating and human evolution. Oxford University Press, New York, pp. 332-349.

von Bonin, G., 1937. Brain-weight and body-weight of mammals. The Journal of General Physiology 16, 379-389.

White, C.R., Seymour, R.S., 2003. Mammalian basal metabolic rate is proportional to body mass(2/3). The Proceedings of the National Academy of Sciences USA 100, 4046-4049.

CHAPTER 3

The Meaning of Brain Size: The Evolution of Conceptual Complexity

P. Tom Schoenemann

Abstract

A complete understanding of exactly how to interpret changes in brain size during human evolution remains a major unresolved question. A common misconception is that absolute brain size is not behaviorally relevant, and that only relative brain size (controlling for body size via, e.g., encephalization quotients) has any evolutionary importance. It is argued that this is unlikely to be a valid interpretation of brain size, and that absolute brain size itself is behaviorally relevant, both theoretically and empirically. It is argued that - whatever else brain size increases brought - they likely resulted in fundamental increases in the complexity of conceptual understanding. This, in turn, likely played a central role in spurring language evolution.

Introduction

The increase in size of the human brain over human evolution is one of the most extensively and clearly documented changes of any species so far documented in the fossil record. Cranial capacity estimates have been made for over 150 separate hominid specimens covering over 3 million years of evolution (Holloway et al., 2004). The increase in cranial capacity indicates a ~3-fold increase in brain volume during this period. Because of the tremendous costs of increasing brain tissue, this increase cannot reasonably be explained as anything other than adaptive (Smith, 1990). The specific costs include the fact that brain tissue is one of the most metabolically expensive tissues in the human body (Hofman, 1983), larger brains take longer to mature (Harvey and Clutton-Brock, 1985), and there is a conflict between the biomechanical efficiencies of narrow hips in bipeds and the need for a large birth canal for increasingly larger brained infants (Lovejoy, 1975). Thus we cannot explain the increase in the human brain without accepting that there must have been some substantial benefit. Presumably this benefit had to do with behavioral abilities, but exactly what was selected for is not clear (Schoenemann, 2006).

Brain Size and Body Size

One question that has received considerable discussion regarding the increase in brain size is exactly how to account for body size increases. It has long been known that body size correlates with brain size across mammals, and this has led to a variety of measures of 'relative brain size' that take body size into account. The mos t commonly used of these measures is Jerison's encephalization quotient (EQ), which is simply a ratio of a species brain size divided by the average brain size of a mammal with the same body size (Jerison, 1973). The average brain size of mammals at different body sizes is estimated empirically. Modern humans have EQ's of between 5 and 7, depending on the mammalian sample used to estimate the average mammal brain/body relationship (Jerison, 1973; Martin, 1981).

Calculating EQ is straightforward, but interpreting species differences with respect to what it means behaviorally is completely unclear. There is an unfortunate tendency in the human paleontology literature for EQ to be treated as if it were something akin to IQ. The assumption seems to be that brain size variation that is explained (in the statistical sense) by body size differences therefore has no behavioral implications. For example, Kappelman (1996), in a paper assessing the possibility of estimating body mass from eye orbit dimensions and thereby allowing EQ estimates of individual fossil specimens, suggests that "...the long period of quite consistent EQs through the nearly 2 million years of premodern *Homo* would predict a pattern of "behavioral sameness", which should stand in marked contrast to the behaviors associated with modern humans and their relatively higher EQ." (p. 271). Similarly, Wynn (2002) writes that "Although the brain size of Nariokotome was larger than earlier hominids, so was his body size; there was only a small increase in relative brain size (compared to, say, *Homo habilis*)... It is not clear from the cranial capacity that a significant increase in braininess accompanied this adaptive shift [in the species niche]." (p. 399). And Wood and Collard (1999) write that "Although there are substantial differences in the mean absolute brain size of the australopiths on the one hand and the *Homo* species on the other, some of these differences are almost certainly not meaningful when differences in the body size proxy are taken into account." (p. 69). These researchers seem to believe that relative brain size (EQ or some similar measure) is the most valid criteria for judging behavioral abilities among species of hominids.

Exactly why this assumption is made regarding the interpretation of EQ is usually not explicit. It may be due to a mistaken belief that if brain scales with body, this likely indicates some sort of developmental constraint between them (Schoenemann, 2006). Under this conceptualization, one reason this might be true is that larger bodies might require larger brains to run them with the same level of sophistication. The extra brain mass that is associated with larger bodies (in the statistical sense) therefore isn't available for additional or more complex behavioral functions, because it is completely devoted to simply maintaining the basic functional requirements of the additional body body mass. Kappelman seems to suggest this when he states: "It appears to be the case that many early studies of the tempo and mode of hominid brain evolution focused on brain size only because most workers either assumed that there was no appreciable variation in body mass beyond that seen in modern humans, or that too few data existed to test the question." (p. 268).

The problem with this perspective is that brain size could be associated with body size for reasons other than some sort of developmental constraint, or that a larger body somehow needs a larger brain to run it. It is clear from analyses of brain/body relationships in mammals (and other groups of animals) that species vary tremendously on how much they invest in brain tissue: even at a given body size, the largest mammal brains can be more than 10 times greater than the smallest ones (Finlay et al., 2001; Schoenemann, 1997). This belies the view that body size imposes a tight constraint (developmental or otherwise) on brain size.

A better explanation for the association between brain/body size may be that brain size is constrained – but not determined – by the metabolic resources that are available to a species (Martin, 1981). These metabolic resources are in turn constrained by body size. Because of competition among and within species for survival, species will tend toward the higher end of the possible brain sizes that are supportable given metabolic constraints placed by their body sizes (Schoenemann 2006). This will lead to an association between brain size and body size, but not because they are developmentally or functionally linked. It would also explain the wide range of brain sizes at a given body size in mammals. For some species niches (e.g., those occupied by many primates, and humans in particular), the behavioral benefits of large brains may be more important than they are for other species, and as a consequence, those species would be expected to devote a greater proportion of their metabolic resources to growing and maintaining brains (versus other body components). As a result, brain sizes would tend to vary with body size across mammals (and within other major groups of animals), but with a large range of variation due to the myriad of possible adaptive niches (varying in their cognitive demands) that species find themselves in. Under this model, absolute brain size would actually be expected to be *more* relevant to behavior than relative brain size. Relative brain size would still be important, in that it would index the extent to which a species invests in (or the extent to which a species niche values) brain-related functions. However, under this conceptualization it would be a mistake to assume that species of significantly different body masses are likely equivalent in their behavioral abilities solely because they have the same relative brain size.

At a purely theoretical level, furthermore, there are reasons to believe that absolute brain size is more behaviorally important than relative brain size. First, species with equivalent EQ's but different body sizes (and hence, different absolute brain sizes) do not have (or lack) equivalent numbers of extra neurons (neurons in excess of – or less than, if they have EQ's less than 1 – those predicted by brain/body scaling). Jerison (1973) devised a way to estimate the number of these "extra neurons," based on empirical estimates of the relationship between neuron density and brain size. While Holloway (1974) has cautioned against the uncritical use of such estimates, it is nevertheless clear that, e.g., a large-bodied species with an EQ of 2.0 will have many more extra neurons than a small-bodied species with the identical EQ. For example, using Martin's (1981) body/brain scaling formula for mammals, cotton-top tamarins (*Saguinus oedipus*) have and EQ of ~1.8, which is a bit higher than the EQ of ~1.7 found for common chimpanzees (*Pan troglodytes*). However, in absolute terms, cotton-top tamarins have only ~4 g greater total brain mass than predicted for a mammal of their body mass, whereas chimpanzees have ~167 g extra (which alone is 17 times the size of an entire tamarin brain). From a basic circuit-design/information-processing perspective, it is hard to believe that these species would nevertheless have essentially the same cognitive abilities simply because they have very similar EQ's. To argue otherwise is to believe that larger bodied species need more neurons to accomplish the same sort of cognitive processing, solely because they have bigger bodies. This is analogous to suggesting that radios in dump trucks should be expected to require many more electrical circuits than radios in small cars, solely because trucks are so much bigger. Since brain circuits appear to be very flexible, in that the processing of various cognitive functions can be fairly rapidly shifted to different regions if need be (e.g., 5 days of artificial blindness in normal sighted people learning braille appears to lead to tactile information being processed in the primary visual cortex – which no longer has visual information to process, Merabet et al., 2008), it is hard to see why larger bodies would need more neurons to accomplish the same cognitive functions. Barring some compelling empirical reasons to believe otherwise, our starting assumption should always be that greater numbers of neurons should translate into the potential for more sophisticated cognitive processing.

It is also important to recognize that the evolutionary costs associated with brain size appear to be a function of absolute brain size, *not* relative brain size. The extra metabolic costs of larger brains, for example, are a function of the total mass of neural tissue, not a function of the relative size of this tissue with respect to body size. Using the above species comparison again, chimpanzees have much greater additional metabolic costs for their brains than do tamarins, even though they have about the same EQ. These larger metabolic costs may not require a disproportionately larger share than in smaller bodied animals, since larger bodied species have greater total metabolic resources to draw upon. However, larger bodies also have greater overall metabolic demands generally. If relative brain size is the proper index of behavioral ability, then, everything else being equal, species should evolve towards smaller body sizes to save the metabolic costs (of both larger bodies and larger brains) while maintaining the same behavioral abilities.

Similarly, maturation time is much more strongly a function of absolute brain size than of EQ. Using nonhuman primate data from Harvey and Clutton-Brock (1985), age at menarche correlates with log brain weight at r=.83 (p<.000001), whereas it correlates with EQ only at r=.59 (p<.00001, EQ estimated using Martin's 1981 equation). Again, two species differing in absolute brain size but with exactly the same EQ would nevertheless likely differ substantially in their average maturation time. Everything else being equal, shorter maturation time is an evolutionary advantage because it translates into more descendents per unit time. If relative brain size really is the proper index of behavioral ability, then species would again be expected to evolve smaller bodies (and hence smaller brains) to reduce maturation time while maintaining the behavioral advantages of the same EQ.

Thus, appropriately smaller body size would be an advantage if relative brain size indexes behavioral ability, because it would allow the species to maintain relative brain size while decreasing the costs of larger brains. It is true that there are independent costs and benefits to body size changes (so reducing body size might have other costs). If behavior is really indexed by EQ, then we have to come up with an independent explanation for the increase in body size during hominin evolution (Kappelman, 1996; Wood and Collard, 1999). However this is complicated by the fact that the increase in body size occurs during a period in which hominins are becoming increasingly independent of their environment through the use of stone tool technology. Increasing behavioral flexibility is generally thought to be the primary adaptation of hominins. If relative brain size really is the appropriate measure of behavioral ability, we should expect, on this account, *decreasing* body size in hominins over time rather then increasing body size, since it would result in lower evolutionary costs while maintaining behavioral ability. However, if instead *absolute* brain size is the better index of behavioral ability, then larger body size in hominins could simply reflect the need to have greater total metabolic resources to help pay for their increasingly large brains.

Ultimately, the question of whether absolute brain size or relative brain size is a better index of behavioral dimensions is an empirical one. Studies of this issue are complicated by the problem of fairly assessing behavioral differences between species. Species differ in both exactly what motivates them (e.g., types of food) as well as in the types of sensory information they focus on (e.g., visual vs. olfactory). If a species fails at a particular task, it might be because it is cognitively limited, but it might also just be because the task favors a sensory modality that isn't the species strength, and/or the species is not properly motivated (Essock-Vitale and Seyfarth 1986; Striedter 2005)? Humans compared to dogs are particularly biased towards the visual domain and happen to particularly favor sugar. A visual task that rewards performance with candy would therefore not be a fair assessment of the inherent cognitive abilities of dogs. This said, there must be *some* reason why species vary in brain size, and if we find some behavioral task that does in fact correlate with aspects of brain size (either relative or absolute), it is useful starting point for hypotheses about exactly why brains differ the way they do across species. Differences in sensory emphases and types of motivation across species are unlikely to result in a purely spurious association with aspects of brain size.

Their are some behavioral associations with relative brain size that have been found. Several studies show significant associations between relative brain size and aspects of diet. Among primates, fruit-eaters tend to have larger relative brain size than leaf-eaters (e.g., Milton, 1988). Bats that subsist on fruits, flowers, meat, fish, or blood tend to have larger relative brain size than bats who are insect-eaters (Eisenberg and Wilson, 1978; Hutcheon et al., 2002). Striedter (2005) suggests the diet-related findings may be explained under a "clever foraging" hypothesis: "...highly encephalized species [those with larger relative brain sizes] tend to forage (or hunt) strategically, taking into account the habits of their food (or prey), while less encephalized species tend to graze (or hunt) opportunistically." (p. 119). Thus, it isn't necessarily the case that hunters are more encephalized than non-hunters. The key seems to be more in the difficulty of finding one's food. For example, it is likely easier for bats to find flying insects than to find fruit, because flying insects are everywhere but fruit is distributed patchily across both time and geographic space (Milton 1988). In some cases relative brain size is more closely associated with some behavioral dimension. While there are some insect-eating bats that have larger brains in absolute size than many fruit-eating ones, the fruit-eaters almost universally have larger relative brain sizes than insect-eating ones (Striedter, 2005).

However, these dietary associations are complicated by the fact that the direction of causality is not clear. Because larger brains place increased metabolic loads on species, it is entirely possible that larger brained species must eat higher quality, more nutrient dense foods to pay these costs (Aiello and Wheeler, 1995). In other words, do the cognitive demands of different types of dietary specializations cause increased relative brain size, or does increased relative brain size occur for other reasons, and simply require certain kinds of foods as a result? The primary metabolic fuel for the brain is glucose, which happens to be found in high quantities in fruits. This could explain part of the tendency of primates with larger relative brain sizes to focus on fruits vs. leaves (though it does not explain, e.g., smaller relative brain size found among insect-eating bats, since insects are fairly nutrient-dense).

Another issue concerns the extent to which broad dietary classes, such as "fruit-eating," are too general as descriptions of behavior to be of much use. There are many types of fruit, and many types of fruit eaters. Fruit-eating would seem to encapsulate a very different level or categorization of behavior than does "problem solving," "behavioral flexibility," or even "3-dimensional spatial recognition." There are likely significant differences in the cognitive demands of various kinds of fruit-eating adaptive niches. Collapsing them all together into a single category leads to such a general level of description as to be helpful only for very coarse levels of understanding. Such correlations are likely of limited value for understanding human brain evolution.

What evidence is there for behavioral correlations with absolute brain size? It turns out that for a large number of studies, absolute brain size is either as good as, or an even better predictor of behavior than is relative brain size (Schoenemann, 2006). Although many studies of brain/behavior associations across primates usually emphasize measures of relative brain size, absolute brain size is invariably also associated with the behavioral dimensions assessed. For example, in Dunbar's (1992) study of 38 primate species, mean group size correlated r=.87 (p<.001) with neocortex ratio (neocortex vs. the rest of the brain), but it also correlated r=.74 (p<.001) with the absolute size of the neocortex by itself. Reader and Laland (2002) similarly showed that the frequency of observations of social learning in 32 primate species correlated r=.69 (p<.00001) with 'executive brain ratio' (ratio of [neocortex + striatum] to brainstem, but that it also correlated r=.58 (p<.0005) with absolute 'executive brain' size (neocortex + striatum) alone. They also found correlations between the frequencies of innovation observed in primate species and 'executive brain ratio' of r=.58 (p<.0005), but also correlations with absolute 'executive brain' size alone of r=.49 (p<.005). Thus, for these aspects of behavior, relative brain size is only marginally more highly correlated than is absolute brain volumes, suggesting that absolute brain size alone is indexing important behavioral variation.

For some particularly interesting behavioral tasks, relative brain size is actually worse than absolute brain size in predicting ability. One example is the speed at which an individual learns that you want it to discriminate between two objects ("learning sets"), and is essentially a measure of how fast they 'learn to learn'. This type of learning task works as follows. First, the subject is repeatedly given the choice of selecting one object out of a pair, with only one of these choices being rewarded in some way (the pair of objects stays the same during this time). When the subject demonstrates that they have learned which object gets them the reward (by consistently selecting this object in subsequent trials), a new pair of objects is then presented, again with only one of them consistently earning a reward. New pairs of objects are introduced as soon as the subject demonstrates they have learned which object of a particular pair is being rewarded. If the subject learns the basic idea behind this task (i.e., that one of a pair will always be rewarded), they learn which object of subsequent pairs is rewarded with fewer and fewer trials. The speed at which the subject gets better at this type of task (learns to learn) can be indexed by assessing the % correct on the second presentation of each set of objects (the first presentation of a given pair can only be guessed at). As the subject learns, their likelihood of selecting the correct object on the second presentation increases. Human children learn this type of task after only a few learning sets (pairs of objects), whereas it takes rats over a 1000 learning sets to approach only ~60% correct on the second trial (50% is guessing randomly, Passingham 1982).

It turns out that this experimental behavioral measure is much more strongly correlated with absolute brain size than it is with EQ. Figure 1 shows the relationships between log learning set slopes (calculated from the second-trial % correct plots) against either log brain size or log EQ (data from Riddell and Corl, 1977). Though the number of species is small (N=11), it is clear that brain size is much more strongly associated with learning set slope than is EQ (though the relationship for both is nonlinear even for log transformed data). Though

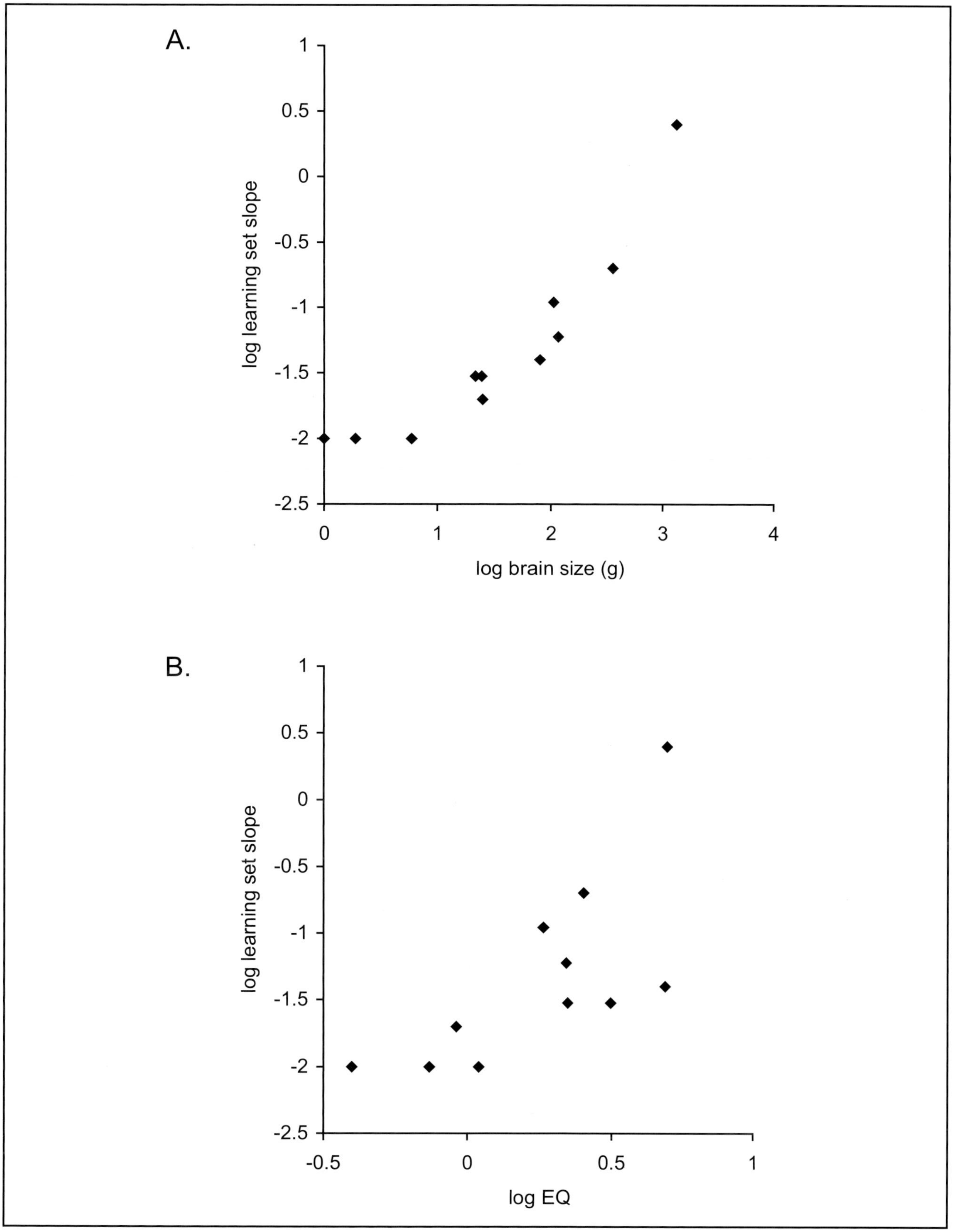

Figure 1: Associations between learning set slope and A. log brain mass (g), B. log EQ. Learning set slope is a measure of how fast a species learns which of two items is associated with a reward. EQ calculated using Martin's (1981) formula. Data from Riddell and Corl (1977). The relationships are nonlinear: [log learning set slope] = .237[log brain size (g)]2 - 2.075, r=.98, p<.001; [log learning set slope] = 2.556[log EQ]2 - 1.761, r=.61, p<.05

it is theoretically possible that some of the differences between species may be due to insufficient motivation and/or differences in sensory emphasis (the learning set studies collected by Riddell and Corl 1977 used visual discrimination), however these differences would then have to independently correlate perfectly with brain size. This is of course possible, but unlikely.

More recently, Rumbaugh and colleagues have devised an ingenious method for controlling for possible cross-species learning confounds (Rumbaugh, 1997; Rumbaugh et al., 1996). Their technique involves training two sets of subjects on a discrimination task, but training them to different levels of accuracy (67% correct vs. 84% correct). The subjects are then tested with the conditions reversed, such that the object that was initially not rewarded now is not, and the object that wasn't rewarded now *is*. The score for a species (which Rumbaugh and colleagues refer to as the "Transfer Index") is the difference between the two groups on their percentage correct for these new, reversed-reward trials. Thus, the measure is insensitive to the total number of trials needed to get to some level of accuracy. Instead, it measures how different levels of learning in a species (however long it takes to be achieved) affect subsequent learning. As such, it is much less sensitive to problems of motivation and/or differences in sensory abilities. For some species, such as the talapoin monkey (*Cercopithecus talapoin*), learning the task to 84% accuracy results in relatively *poorer* performance when the rewards are reversed than if the task was learned only to 67% accuracy. In other words, the better this species learns a to favor a particular object in a discrimination task, the harder it is for it to switch. By contrast, for species such as the gorilla (*Gorilla gorilla*), learning the task to 84% accuracy results in *better* performance on subsequent trials where the rewards are reversed. Gorilla's seem to learn the general idea behind the task, such that they are flexibly able to apply the *idea* of descrimination (as a concept) to a series of tasks, rather than simply learn a series of object discriminations, each essentially disconnected from the rest.

What is particularly interesting about this work is that Transfer Index correlates r=.82 (N=13, p<.001) with the absolute amount of brain a species has in excess of that predicted by their body weight ("extra brain volume", Rumbaugh, 1997; Rumbaugh et al., 1996). It does not correlate significantly with EQ, however. Talapoin monkeys have an EQ of 2.9, whereas gorillas have an EQ of only 1.2 (Schoenemann 1997), for example, yet talapoin monkeys have the lowest transfer index score while gorillas have one of the best scores (trailing only a group of language-trained apes).

Furthermore, a recent exhaustive meta-analysis of the literature by Deaner et al. (Deaner, 2006) show that some species consistently tend to do better across a wide range of behavioral tasks, and that this cannot easily be explained by methodological confounds. Furthermore, Deaner et al. (2007) show that the absolute brain volume correlates most strongly with the relative rankings of general behavioral ability revealed by their meta-analysis. Various measures of relative brain size (such as EQ) were much worse.

Thus, it is quite clear that absolute brain size is strongly associated with important and interesting behavioral dimensions across species. It is important to note that studies of more broad behavioral domains indexed by the size of the social group (Dunbar, 1992) and the tendency towards social learning and innovation (Reader and Laland, 2002) show the highest correlations with EQ, whereas controlled laboratory studies focusing on 'learning to learn' show the lowest correlations. It is therefore not legitimate to ignore or discount changes in absolute brain size during human evolution when assessing behavioral evolution.

This said, it is important to note some caveats. First, not every cognitive domain is necessarily associated with larger brain size. Echolocating bats seem to be able to accomplish extraordinary behavioral feats of sound processing without requiring large brains (or large EQ's). Second, although between-species associations between brain size and behavioral ability presumably require non-zero brain/behavior associations within species, these can be very small while nevertheless remaining highly evolutionarily significant (Schoenemann et al., 2000). Finally, we don't want to forget that there can also be meaningful localized associations between brain anatomy and behavior that likely have played critical roles in human evolution. These constitute part of what Holloway refers to by functional reorganization (Holloway, 1995).

Localized Brain/Behavior Functions

The brain is of course not an undifferentiated mass of neurons, but does have a significant degree of localization of function. This localization appears to be quite flexible, however, as has been revealed by studies of changes in localization of function in individuals who lose a limb or some form of sensory input (e.g., permanent or even temporary blindness as discussed above, see also Ramachandran, 2004). Studies of cortical maps in species with various specialized behavioral adaptations show predictable changes in the relative proportions of particular areas of their cortex (Krubitzer, 1995). Star nosed moles (*Condylura cristata*) have very little need for visual information, as they live most of their life underground, and predictably they have very small visual cortices. About half the cortex of the echolocating ghost bat (*Macroderma gigas*) us devoted to processing sound information (Krubitzer, 1995). This pattern holds even within the human brain: it has long been recognized that the size of various regions of both the primary motor and primary somatosensory areas are proportional to the degree of elaboration of function for a given part of the body. This is usually depicted graphically with a 'homunculus', in which the size of different parts of the homunculus are drawn approximately proportional to the relative size of the corresponding portions of the cortex devoted to those areas.

There are also a few studies suggesting that variation within humans in the size of specific areas of the cortex predicts behavioral abilities mediated by those areas. We have shown, for example, that size of a proxy measure of the prefrontal (i.e., all cortex anterior to the corpus callosum) correlates with performance on the Stroop test of the ability to focus on key stimuli in the face of distractors (Schoenemann et al., 2000). This is consistent with the finding that children with attention-deficit/hyperactivity disorder (ADHD) have smaller superior prefrontal volumes than healthy controls (Hill et al., 2003). More recently, we have found an association between the size of areas of the corpus callosum and behavioral domains in an MRI study of health human females. This work was inspired by Ralph Holloway's many studies of sex differences in the corpus callosum, and his hypotheses regarding its possible evolutionary explanation (de Lacoste-Utamsing and Holloway, 1982; Holloway, 1990; Holloway et al., 1993; Holloway and de Lacoste, 1986; Holloway and Heilbroner, 1992). Specifically, Holloway has suggested that social communication might have been particularly strongly selected for in females, and that this is likely to require more cross-talk between the cerebral hemispheres, but that visuospatial abilities (which males tend to be better at on average) might be better processed in one dominant hemisphere (Holloway et al., 1993). If this is correct, we might expect to find that the mid and anterior portions of the corpus callosum, which connect temporal and frontal lobe areas thought to be important to social domains, would be larger in individuals who are particularly socially adroit. Conversely, the splenium (posterior portion) of the corpus callosum, which connects visual cortical areas and portions of the parietal lobes known to be involved in spatial processing, might be expected to be smaller in individuals who are particularly good at spatial tasks. In a sample of 36 female sibling pairs, we found patterns consistent with this: smaller splenia were associated with better performance on a mental rotation task, whereas larger anterior and mid portions of the corpus callosum were associated with a greater degree of social interaction (specifically: the number of people the subject reported talking to for more than 5 minutes in the last week). Figure 2 illustrates these relationships (unpublished data). These data suggest that localized variation even within species may be associated with behavioral differences.

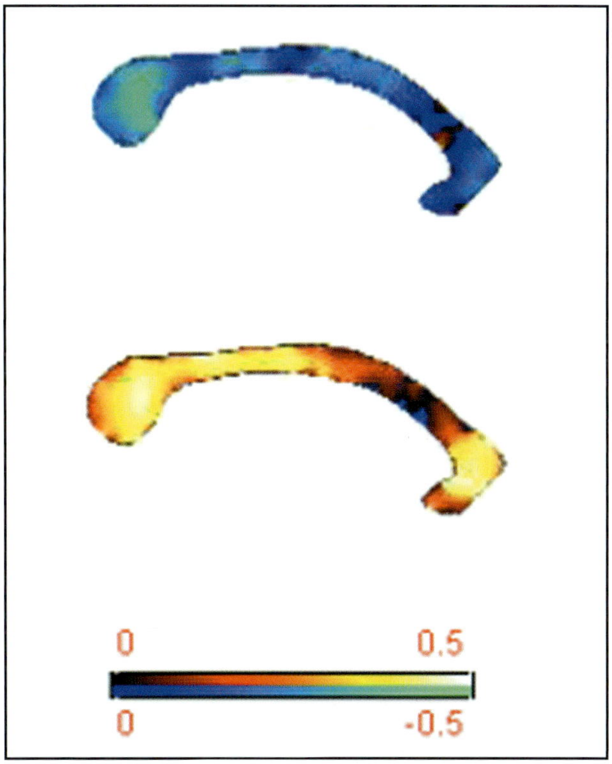

*Figure 2: Associations between size variation in cross-sectional areas of the corpus callosum and **A.** mental rotation ability, **B.** degree of social interaction. Black-to-yellow indicate increasingly larger positive correlations; blue-to-green indicate increasingly larger negative correlations. Mental rotation ability is generally negatively associated with localized corpus callosum size, whereas degree of social interaction is generally positively associated. Mental rotation ability was tested using a computerized version of the Vandenberg and Kuse (1978) test. Degree of social interaction was indexed by reported number of individuals talked to in the last week for more than 5 minutes. Localized anatomical variation was quantified using non-rigid deformation techniques (see e.g., Avants et al. 2006). These relationships are correlations of sibling differences in anatomy with sibling differences in behavior, thereby controlling for possible between-family confounds, such as socio-economic status, that might lead to artifactual correlations between anatomy and behavior (Schoenemann 2006).*

Thus, both overall absolute brain size, as well as at least some localized neuroanatomical variation, appear to be associated with behavior. Both of these need to be recognized in any complete understanding of human evolution. It should be noted here also that the absolute size of localized regions are likely to be behaviorally relevant *independent* of their relative size compared to other regions, or the rest of the brain. Partly, there is no reason to believe *a priori* that a given circuit is likely to work less well if there are more circuits in other regions (subserving other behaviors) than if there are fewer circuits in those other regions. However it is also difficult to square the supposed critical importance of relative area size with the fact that evolutionary costs of neural tissue are a function of absolute size, not relative size. If relative size of a circuit was generally the most behaviorally relevant measure, then species would have evolve very small brains, but with just the right proportions, thereby saving the metabolic and maturational costs but maintaining the behavioral benefits.

Functional Localization as a Consequence of Increasing Brain Size

Comparative studies of brain size differences across species have highlighted an important change that appears to go hand-in-hand with brain size increase. As brain size increases, different areas of the cortex become less directly connected with each other. This appears to be a related to the fact that the number of connections between neurons (or between cortical columns) has to increase much faster than the increase in neurons, if each is to remain equally well connected (meaning: directly connected) to all others (Ringo, 1991). It is structurally much easier for smaller brains to have more direct connections between more areas than it is for larger brains. This fact is reflected in the proportions of white vs. gray matter in different sized brains. White matter consists primarily in connective axons between relatively distant areas, whereas gray matter consists primarily of neuron cell bodies and dendrites. If equal connectivity is to be maintained between neurons, we should expect white matter to increase much faster than gray matter. Empirically, larger brains do in fact have proportionately more white matter than smaller brains, but not enough to maintain equal connectivity among all regions (Ringo 1991).

This has an important general functional consequence, because it means that as brains increase in size, areas are increasingly able to carry out processing independent of other regions. This leads inevitably to functional specialization, in which different areas process different kinds of information in different ways. Empirically, larger brained species have been shown to have larger numbers of distinct cortical areas (Northcutt and Kaas, 1995). Rodents, for example, have only 5–8 visual areas whereas primates have 20–30 (Northcutt and Kaas, 1995). Changizi and Shimojo (2005) showed there is in fact a predictable relationship between the number of distinct, identifiable cortical areas across mammals and a species brain size.

This increase in functional specialization has important behavioral consequences. First, new specialized areas allow for more sophisticate processing of particular types of information. This is an important component of the story, but I wish to focus here on a more general consequence, that operates at a higher hierarchical level of brain function. Specifically, the greater the independence of different areas, the greater will be the sophistication of processing of information overall. Such independence makes parallel processing increasingly possible, and this has significant consequences because it leads to greater sophistication in behavioral response.

Furthermore, 'actions' (outputs of various kinds from different areas) can increasingly be separated from inputs. One can imagine a continuum of types of neural circuits, with simple reflex loops involving a single synapse (e.g., startle reflexes that close the eyelids when fast-moving objects approach the eyes) to complex deliberative circuits involving many subunits (functional areas) processing many different kinds of information both serially and in parallel. The later type of circuits are by definition not fast, but they are smart, flexible, and adaptive – the epitome of 'thoughtful' processing. Increasing numbers of increasingly separate functional areas inevitably leads to a wholly different kind of behavioral repertoire, that we generally associate with intelligence. Note also that this is a consequence of increasing absolute brain size, not of increasing relative brain size (except, of course, insofar as those two are conflated, as occurred during much of human evolution).

Conceptual Complexity and Brain Size

The concept of intelligence is notoriously difficult to define to everyone's satisfaction. However, whether or not one agrees that intelligence refers to the degree of complexity of information processing, the increase in the number of quasi-independent processing areas occurring as a consequence of increasing brain size would at least have lead to an increase in the complexity, subtlety, and sophistication of our conceptual universe (Schoenemann, 2005). By "conceptual universe" I mean the totality of all our conceptual understanding, whether it is closely grounded in direct sensory experience (e.g., [hot (temperature)] [water], [ball], etc.), or is more abstract ([evolutionary fitness], [contingent], [love], etc.). Conceptual complexity may be thought of as proportional to the number of independent dimensions the brain can meaningfully distinguish. "Meaningful" distinctions, in this view, would be differences the brain can detect in patterns of stimuli that, in turn, make a difference in how the brain can respond. "Dimensions" are aspects of reality that the species is sensitive to (e.g., chemicals in the air, liquids, or solids, electromagnetic radiation, air pressure vibration, levels of energy, physical pressure, time) as well as internally generated states (emotions, logic, etc.). To see why we should expect brain size to be relevant to this aspect of our cognition, it is necessary to think about how concepts are instantiated in our brain.

Concepts appear to be networks of activation between different areas of the brain, which more or less specialize in particular types of processing of particular types of information. Functional imaging studies support this contention by showing, for example, that passively reading action words that refer to different body parts activates the same cortical areas as does movement of the implied body part (Pulvermüller, 2005). Thus, the concept [kick], brought to mind by the word "kick", activates the areas involved in actually kicking. Similarly, imagining (but not actually seeing) an object often activates the primary visual areas that are active when the object is seen (Damasio et al., 1993; Kosslyn et al., 1993; Kosslyn and Thompson, 2003). Behavioral studies on correlations between different word meanings suggest that the organization of features associated with different word meanings plays a critical role in the organization of semantic memory (McRae et al., 1997). These studies indicate also that information does not flow exclusively in one direction, from primary sensory areas on to secondary sensory and association areas. The activation of primary sensory areas can occur as a result of internally generated activity in other areas of the brain.

Furthermore, even relatively simple conceptual awareness is typically the result of the combination of processing from a variety of cortical areas. For example, our experience of taste is actually the result of the interaction of olfactory (smell) and gustatory (taste) inputs (e.g., banana 'taste' is actually a smell). Auditory perception of simple phonemes is partly a function of concurrent visual input of a speaker's face (McGurk and MacDonald, 1976). Thus, conceptual awareness requires the integration of processing from different areas, and this integration is made possible by neural connections between areas.

It stands to reason that the more processing areas a brain has, the greater the degree of complexity of the possible interactions between these areas. Since conceptual awareness involves activating neural networks connecting different areas, and since larger brains have larger numbers of quasi-independent specialized processing areas, larger brains can potentially create a greater diversity of concepts and a richer and more subtle conceptual understanding. Deacon's (1997) thesis regarding the evolution of symbolic understanding incorporates this idea of conceptual awareness requiring the integration of different neural areas, but he argues that language required an additional step not found in other animals (potentially explaining why other animals don't have language). For Deacon, the key is the ability of conceptual networks to interact more directly with each other, rather than being tied to their grounding in basic sensory information. This would ultimately allow the brain to think entirely conceptually – essentially to form concepts about concepts. Whether or not humans are truly unique in this regard is debatable, but it certainly is clear that human brains have a much greater potential for creating a much greater diversity of conceptual networks.

One simple way to illustrate how simple brain size increases might lead to massive increases in conceptual complexity is to note how fast the logically possible ways of combining different processing areas together increases as a function of the total number of areas. Assume for argument's sake that, as a (gross) simplification, a single concept involves the interaction of a unique subset of n processing areas. The total number of concepts would then be the total number of unique subsets of n areas, which can be shown to be 2^n. This means that the total number of concepts would *double* with the addition of each new area. To put this into context, consider that Changizi and Shimodo's (2005) equation estimating the number of distinct areas as a function of brain size predicts that chimpanzee-sized brains would have ~100 areas, whereas a human-sized brain would have ~150 areas. There are 2^{50} times as many unique subsets of 150 areas as there are of 100 areas.

This simple calculation should not be taken as a straightforward estimate of the degree of difference in conceptual complexity between chimpanzees and humans, of course. For one thing, not every unique combination of processing areas leads necessarily to a unique concept. The concept of [baby] involves such ideas as soft skin and hair (a tactile sensation), small physical size (a visual and/or tactile-pressure sensation), various cries and other sounds (acoustic sensations), and so forth. One would not want to argue that the concept of [baby] necessarily *requires* concurrent activation of all these areas. A species lacking some particular sensory processing area relevant to the human conceptual understanding of [baby] would not necessarily lack the concept of [baby] completely, even though it would clearly be different in some potentially important way. Similarly, the activation of, say, one less area than is typical for the complete concept of [baby] in humans does not constitute a completely unique concept. It would, however, likely be subtly different. The nature of conceptual networks is that activation of a portion of the network usually leads to the activation of the entire network.

Another complication is that, as alluded to above, there appear to be real differences in the complexity of internal processing of particular areas in different species. The pattern for the human somatosensory cortex, in which regions corresponding to parts of the body for which we have more sensitivity are expanded, is a manifestation of a more general pattern across species. Racoons (*Procyon lotor*), for example, have distinct cortical gyri for individual digits on their hands (and relatively large somatosensory cortices generally compared to carnivores), which correspond to their highly developed manual dexterity(Krubitzer, 1995). This indicates that the same area in different species can differ substantially in the complexity of information processing that can be accomplished in given cortical areas. However, the complexity of processing is at least loosely indexed by the size of a given area, and this must translate into a difference in the subtlety and sophistication of conceptual understanding for which that area participates in creating. All of this suggests that the degree of complexity of conceptual understanding can reasonably be considered a function of brain size.

Prefrontal Cortex and Conceptual Understanding

Having many different areas processing many different kinds of information in many different ways is not – by itself – sufficient to produce useful thinking or conceptual understanding. What is needed is a way to organize and prioritize processing from different areas in a meaningful way. It appears that a variety of areas in the prefrontal cortex are specialized for just this sort of processing. The prefrontal cortex also appears to play a general oversight role with respect to processing in other areas of the brain, and in planning generally (Damasio, 1985). Drugs that are used to moderate the symptoms of Attention Deficit/Hyperactivity Disorder (ADHD), such as ritalin, act by making the prefrontal cortex *more* active, for example [they are in fact stimulants, but they are highly specific with respect to what they stimulate cite????]. The prefrontal cortex is also active when learning a new task (Cabeza and Nyberg, 2000), when making any free choice that isn't tightly constrained by the context (Frith et al., 1991; Lau et al., 2004), as well as when experiencing surprising events (Fletcher et al., 2001). With respect to the question of conceptual awareness, areas in the prefrontal appear to be centrally involved in conceptual/semantic information processing (Gabrieli et al., 1998; Gaillard et al., 2000; Kerns et al., 2004; Luke et al., 2002; Maguire and Frith, 2004; Thompson-Schill et al., 1997; Thompson-Schill et al., 1998).

If larger brains tend to have increased numbers of cortical areas, and more cortical areas lead to greater possible complexity of conceptual understanding, and the prefrontal cortex plays a key role in organizing the interactions between these areas, we might expect there to be a biased elaboration in the prefrontal cortex with increasing brain size. Furthermore, since the increase in possible interactions between areas increases geometrically with the increase in areas, we might expect the prefrontal to increase much faster than the rest of the brain (i.e., positive allometry). The evidence in fact supports this prediction. Semendeferi et al. (2002) found positive allometry for the entire frontal (of which the prefrontal is a subset) with respect to the rest of the brain. Our own study found statistically significant positive allometry for a proxy measure of the prefrontal itself (i.e., total cerebrum anterior to the corpus callosum, Schoenemann et al., 2005). This is also true of the cytoarchitectural data from Brodmann (1912), though the data just misses statistical significance (the slope of the regression line predicting log prefrontal cortical area from log non-prefrontal cortical area is 1.13, with 95% confidence intervals ranging from 0.99 to 1.28; N=11, excluding humans). A cytoarchitectural study of area 10, a subset of the prefrontal cortex, in apes shows particularly strong positive allometry (slope of the regression predicting area 10 volume from total brain volume is 1.64, with 95% confidence intervals ranging from 1.16 to 2.12; N=5) (see also Holloway, 2002; data from Semendeferi et al., 2001). Area 10 is particularly important for planning in general (see references in Semendeferi et al., 2001), and is also specifically implicated in semantic processing (word meanings, e.g., Gabrieli et al., 1998; Luke et al., 2002). By contrast, area 13 of the prefrontal, which is more closely associated with aspects of social cognition and not semantic and/or conceptual information processing, does not show evidence of positive allometry in apes, but instead appears to be isometric (slope of the regression predicting area 13 volume from the volume of the rest of the brain is 1.01; N=5, NS). Thus, the prefrontal itself, as well as at least one relevant subdivision – area 10, appears to increase in size faster than the rest of the brain, which is exactly what we would predict given its oversight role organizing activity in posterior areas of the brain.

There is also some evidence that the human prefrontal is particularly enlarged, above that predicted by these positively allometric relationships (i.e., that it got bigger even faster than one would predict from primate brain scaling relationships). Brodmann's (1912) data suggest this (Deacon, 1997), as do studies estimating cortical folding in anterior vs. posterior regions (Armstrong et al., 1991; Rilling and Insel, 1999). Studies estimating the relative increase in size of the prefrontal from studies morphing other species brains in to human brains also support this contention (Avants et al., 2006; Van Essen, 2005; Zilles, 2005). Uylings and Van Eden (1990) do not show increased prefrontal in humans, but their measure of prefrontal is based on thalamic projection patterns, which show much more overlap in smaller brained species, thereby confounding the analysis. The human frontal lobe as a whole (which includes areas in addition to the prefrontal) – although significantly bigger in absolute terms than any other primate – is slightly smaller than primate trends predict (though not statistically significantly so, see figure 2 of Semendeferi et al., 2002). Our proxy of the prefrontal, on the other hand, suggests that humans do have significantly more prefrontal than primate trends predict (figure 4 of Schoenemann et al., 2005). This seems to suggest that, as one looks at increasingly anterior portions of the cerebrum, that humans are disproportionately enlarged beyond even what the positive allometry in primates predicts (see discussion in Schoenemann, 2006).

As the prefrontal cortex plays an oversight role organizing activity in posterior regions, it is not surprising that it has extensive connections to many areas of the brain (Deacon, 1997). Given that larger brains tend to have greater numbers of cortical areas, we might also expect there to be a particular bias with respect to estimates of connectivity to and from the prefrontal. One way to estimate this is through comparing white matter volumes in this region among primates, since white matter contains mostly long-distance axonal connections. Our own study found that the white matter regions of the human prefrontal showed the greatest degree of disproportion compared to primate scaling trends, in fact accounting for most of the disproportion of the prefrontal as a whole (Schoenemann et al., 2005). Schenker et al.'s data (2005) also suggest a disproportionate increase in white matter of the entire frontal (Schoenemann, 2006).

These studies are consistent with the idea that increasing brain size led to dramatic increases in conceptual complexity, and that this required disproportionate increases in the size of the prefrontal over other areas.

Conceptual Complexity and the Evolution of Language

Given that language presupposes a rich conceptual awareness in order to make communication (and/or conceptual ('symbolic') thinking) a useful exercise in the first place, the likely importance brain size increase has for understanding language evolution is straightforward. The connection between increasing conceptual complexity and language evolution has in various ways been pointed out repeatedly, particularly with respect to the role brain size likely has played in this equation (Deacon, 1997; Gibson, 1988; Gibson, 2002; Gibson and Jessee, 1999; Gibson et al., 2001; Schoenemann, 1999; Schoenemann, 2005). The relevance of brain size evolution to language evolution also has a long history, going back at least to Darwin himself (Darwin, 1882; Dunbar, 1996; Nadeau, 1991; Wang, 1991; Washburn, 1960).

Although it has been claimed that the use of natural language syntax and grammar are unique to humans, and that other animals cannot learn them (e.g., Pinker, 1994), descriptions of "universal grammar" (i.e., those grammatical features common to all languages) appear to simply reflect general descriptions of our conceptual universe rather than a series of specific rules (Schoenemann, 1999; Schoenemann, 2005). It is true that individual grammatical rules found in individual languages are often quite specific, to the extent that linguists do not understand how they could be learned without highly specific innate cognitive structures (an argument from "personal incredulity", Dawkins, 1986). However, these are invariably restricted to specific languages, and as such cannot be considered "universal" without special pleading. The features that are universal turn out to be general things like hierarchical structure, rules specifying argument structure (e.g.,: who did what to whom; the specific ways this is accomplished however vary across languages), a noun-verb distinction (which also varies across languages), and so forth (see Pinker and Bloom, 1990).

Because these features appear to reflect our conceptual understanding, they raise the question of whether the rules of syntax and grammar that are supposed to be unique to human language are actually simply cultural manifestations of our underlying conceptual understanding of the world (Schoenemann, 1999; Schoenemann, 2005). Pinker and Jackendoff (2005) seem to suggest this when they state that "...the only reason language needs to be recursive is because its function is to express recursive thoughts. If there were not any recursive thoughts, the means of expression would not need recursion either." (p. 230). Though they themselves believe that recursion in language is not "...a straightforward externalization of a single [internal] recursive system..." (p. 231), this does not rule out recursion's emergence from these apparent conceptual precursors. In fact, a great deal of work, particularly in computational modeling, has suggested that a structure can emerge (in the cultural evolutionary sense) simply from repeated attempts at communication among individuals (Kirby, 2000; Kirby and Christiansen, 2003). Whether this ultimately explains all grammar and syntax found in human languages is an open question, but enough has been shown so far as to make claims that it can't possibly do so obviously premature.

Regardless of one's position on this question, however, language evolution was clearly built on a rich conceptual structure that predated language itself. This in turn appears to owe its existence to the dramatic increase in brain size that occurred during our evolutionary history.

Acknowledgments

Like most of the contributors to this volume and the conference that preceded it, I owe a great deal of gratitude to Ralph Holloway, not only for inspiring me to see that the evolution of the brain was both a possible and rewarding intellectual pursuit, but also for being a supportive friend and colleague during my career. I can only hope to produce a fraction of the good work he has accomplished over his career to date. I don't expect he will agree with all I have written, but I have learned over the years that he can respect even those he disagrees with. I hope that is true this time! I also want to thank Doug Broadfield, Michael Yuan, Nick Toth, and Kathy Schick, for organizing the The Human Brain Evolving conference, as well as this volume. It is with great sorrow that we have had to endure Michael Yuan's passing. Hopefully this volume will serve as a tribute to him as well as Ralph. Lastly, I wish to thank Vince Sarich, William Wang, Craig Martell, Thomas Lee, Morten Christiansen, Jinyun Ke, James Minett, Ching Pong Au, Feng Wang, John Allen, Karen Schmidt, and Reina Wong for discussing these ideas with me over the years, and hopefully helping me think more clearly about them. The errors that remain are of course my own.

References

Aiello, L.C. and Wheeler, P., 1995. The expensive tissue hypothesis: The brain and the digestive system in human and primate evolution. Current Anthropology 36, 199-221.

Armstrong, E., Zilles, K., Curtis, M. and Schleicher, A., 1991. Cortical folding, the lunate sulcus and the evolution of the human brain. Journal of Human Evolution 20, 341-348.

Avants, B.B., Schoenemann, P.T. and Gee, J.C., 2006. Lagrangian frame diffeomorphic image registration: Morphometric comparison of human and chimpanzee cortex. Medical Image Analysis 10, 397-412.

Brodmann, K., 1912. Neue Ergebnisse über die vergleichende histologische localisation der grosshirnrinde mit besonderer berücksichtigung des stirnhirns. Anatomischer Anzeiger supplement 41, 157-216.

Cabeza, R. and Nyberg, L., 2000. Neural bases of learning and memory: functional neuroimaging evidence. Curr Opin Neurol 13, 415-421.

Changizi, M.A. and Shimojo, S., 2005. Parcellation and area-area connectivity as a function of neocortex size. Brain Behav Evol 66, 88-98.

Damasio, A.R., 1985. The frontal lobes. In: Heilman, K. and Valenstein, E. (Eds.), Clinical neuropsychology. Oxford: Oxford University Press. pp. 339-375.

Damasio, H., Grabowski, T.J., Damasio, A., Tranel, D., Boles-Ponto, L., Watkins, G.L. and Hichwa, R,D., 1993. Visual recall with eyes closed and covered activates early visual cortices. Society for Neuroscience Abstracts 19, 1603.

Darwin, C., 1882. The Descent of Man and Selection in Relation to Sex, 2nd Edition. London: John Murray.

Dawkins, R., 1986. The Blind Watchmaker. New York: W.W. Norton & Company, Inc.

de Lacoste-Utamsing, C. and Holloway, R.L., 1982. Sexual dimorphism in the human corpus callosum. Science 216, 1431-1432.

Deacon, T.W., 1997. The symbolic species: the co-evolution of language and the brain. New York: W.W. Norton.

Deaner, R.O., 2006. Do some taxa have better domain-general cognition than others? A meta-analysis of nonhuman primate studies. Evolutionary Psychology 4, 149-196.

Deaner, R.O., Isler, K., Burkart, J. and van Schaik, C., 2007. Overall brain size, and not encephalization quotient, best predicts cognitive ability across non-human primates. Brain, behavior and evolution 70, 115-124.

Dunbar, R., 1996. Grooming, Gossip and the Evolution of Language. London: Faber and Faber.

Dunbar R.I.M., 1992. Neocortex size as a constraint on group size in primates. Journal of Human Evolution 22, 469-493.

Eisenberg, J.F. and Wilson, D., 1978. Relative brain size and feeding strategies in the Chiroptera. Evolution 32, 740-751.

Essock-Vitale, S. and Seyfarth, R.M., 1986. Intelligence and social cognition. In: Smuts, B.B., Cheney, D.L., Seyfarth, R.M., Wrangham, R.W. and Struhsaker, T.T., (Eds.), Primate Societies. Chicago: University of Chicago Press. pp. 452-461.

Finlay, B.L., Darlington, R.B. and Nicastro, N., 2001. Developmental structure in brain evolution. Behavioral and Brain Sciences 24, 263-308.

Fletcher, P.C., Anderson, J.M., Shanks, D.R., Honey, R., Carpenter, T.A., Donovan, T., Papadakis, N. and Bullmore, E.T., 2001. Responses of human frontal cortex to surprising events are predicted by formal associative learning theory. Nat Neurosci 4, 1043-1048.

Frith, C.D., Friston, K., Liddle, P.F. and Frackowiak, R.S.J., 1991. Willed action and the prefrontal cortex in man: A study with PET. Proceedings of the Royal Society of London, Series B 244, 241-246.

Gabrieli, J.D., Poldrack, R.A. and Desmond, J.E., 1998. The role of left prefrontal cortex in language and memory. Proceedings of the National Academy of Sciences USA 95, 906-913.

Gaillard, W.D., Hertz-Pannier, L., Mott, S.H., Barnett, A.S., LeBihan, D. and Theodore, W.H., 2000. Functional anatomy of cognitive development: fMRI of verbal fluency in children and adults. Neurology 54, 180-185.

Gibson, K.R., 1988. Brain size and the evolution of language. In: Landsberg, M.E., (Ed.), The Genesis of Language: A Different Judgement of Evidence. Berlin: Mouton de Gruyter. pp. 149-172.

Gibson, K.R., 2002. Evolution of human intelligence: the roles of brain size and mental construction. Brain Behav Evol 59, 10-20.

Gibson, K.R. and Jessee, S., 1999. Language evolution and expansions of multiple neurological processing areas. In: King, B.J., (Ed.), The Origins of Language: What Non-human Primates Can Tell Us. Santa Fe, New Mexico: School of American Research Press. pp. 189-227.

Gibson, K.R., Rumbaugh, D. and Beran, M., 2001. Bigger is better: primate brain size in relationship to cognition. In: Falk, D. and Gibson, K., (Eds.), Evolutionary Anatomy of the Primate Cerebral Cortex. Cambridge: Cambridge University Press. pp. 79-97.

Harvey, P.H. and Clutton-Brock, T.H., 1985. Life history variation in primates. Evolution 39, 559-581.

Hill, D.E., Yeo, R.A., Campbell, R.A., Hart, B., Vigil, J. and Brooks, W., 2003. Magnetic resonance imaging correlates of attention-deficit/hyperactivity disorder in children. Neuropsychology 17, 496-506.

Hofman, M.A., 1983. Energy metabolism, brain size, and longevity in mammals. Quarterly Review of Biology 58, 495-512.

Holloway, R.L., 1974. Review: On the Meaning of Brain Size. Science 184, 677.

Holloway, R.L., 1990. Sexual dimorphism in the human corpus callosum: Its evolutionary and clinical implications. In: Sperber, G.H., (Ed.), From Apes to Angels: Essays in Anthropology in Honor of Phillip V Tobias. New York: Wiley-Liss. pp. 221-228.

Holloway, R.L., 1995. Toward a synthetic theory of human brain evolution. In: Changeux, J.-P. and Chavaillon, J., (Eds.), Origins of the Human Brain. Oxford: Clarendon Press. pp. 42-54.

Holloway, R.L., 2002. Brief communication: how much larger is the relative volume of area 10 of the prefrontal cortex in humans? American Journal of Physical Anthropology 118, 399-401.

Holloway, R.L., Anderson, P.J., Defendini, R. and Harper, C., 1993. Sexual dimorphism of the human corpus callosum from three independent samples: Relative size of the corpus callosum. American Journal of Physical Anthropology 92, 481-498.

Holloway, R.L., Broadfield, D.C. and Yuan, M.S., 2004. The Human Fossil Record, Volume 3. Brain Endocasts - The Paleoneurological Evidence. Schwartz JH, and Tattersall I, editors. Hoboken: John Wiley & Sons.

Holloway, R.L. and de Lacoste, M.C., 1986. Sexual dimorphism in the human corpus callosum: An extension and replication study. Human Neurobiology 5, 87-91.

Holloway, R.L. and Heilbroner, P., 1992. Corpus callosum in sexually dimorphic and nondimorphic primates. American Journal of Physical Anthropology 87, 349-357.

Hutcheon, J.M., Kirsch, J.A. and Garland Jr., T., 2002. A comparative analysis of brain size in relation to foraging ecology and phylogeny in the Chiroptera. Brain, behavior and evolution 60, 165-180.

Jerison, H.J., 1973. Evolution of the Brain and Intelligence. New York: Academic Press.

Kappelman, J., 1996. The evolution of body mass and relative brain size in fossil hominids. Journal of Human Evolution 30, 243-276.

Kerns, J.G., Cohen, J.D., Stenger, V,A. and Carter, C.S., 2004. Prefrontal cortex guides context-appropriate responding during language production. Neuron 43, 283-291.

Kirby, S., 2000. Syntax without natural selection: How compositionality emerges from vocabulary in a population of learners. In: Knight, C., Studdert-Kennedy, M. and Hurford, J.R., (Eds.), The Evolutionary Emergence of Language. Cambridge: Cambridge University Press. pp. 303-323.

Kirby, S. and Christiansen, M.H., 2003. From language learning to language evolution. In: Christiansen, M.H. and Kirby, S., (Eds.), Language Evolution. Oxford: Oxford University Press. pp. 272-294.

Kosslyn, S.M., Alpert, N.M., Thompson, W.L., Maljkovic, V., Weise, S.B., Chabris, C.F., Hamilton, S.E., Rauch, S.L. and Buonanno, F.S., 1993. Visual mental imagery activates topographically organized visual cortex: PET investigations. Journal of Cognitive Neuroscience 5, 263-287.

Kosslyn, S.M. and Thompson, W.L., 2003. When is early visual cortex activated during visual mental imagery? Psychol Bull 129, 723-746.

Krubitzer, L., 1995. The organization of neocortex in mammals: are species differences really so different? Trends in Neurosciences 18, 408-417.

Lau, H.C., Rogers, R.D., Ramnani, N. and Passingham, R.E., 2004. Willed action and attention to the selection of action. NeuroImage 21, 1407-1415.

Lovejoy, C.O., 1975. Biomechanical perspectives on the lower limb of early hominids. In: Tuttle, R.H., (Ed.) Primate Functional Morphology and Evolution. The Hague: Mouton. pp. 291-326.

Luke, K.K., Liu, H.L., Wai, Y.Y., Wan, Y.L. and Tan, L.H., 2002. Functional anatomy of syntactic and semantic processing in language comprehension. Hum Brain Mapp 16, 133-145.

Maguire, E.A. and Frith, C.D., 2004. The brain network associated with acquiring semantic knowledge. Neuroimage 22, 171-178.

Martin, R.D., 1981. Relative brain size and basal metabolic rate in terrestrial vertebrates. Nature 293, 57-60.

McGurk, H. and MacDonald, J., 1976. Hearing lips and seeing voices. Nature 264, 746-748.

McRae, K., de Sa, V.R. and Seidenberg, M.S., 1997. On the nature and scope of featural representations of word meaning. Journal of Experimental Psychology: General 126, 99-130.

Merabet, L.B., Hamilton, R,, Schlaug, G., Swisher, J.D., Kiriakopoulos, E.T., Pitskel, N.B., Kauffman, T. and Pascual-Leone, A., 2008. Rapid and reversible recruitment of early visual cortex for touch. PLoS ONE 3, e3046.

Milton, K., 1988. Foraging behavior and the evolution of primate intelligence. In: Byrne, R.W. and Whiten, A., (Eds.), Machiavellian Intelligence. Oxford: Clarendon Press. pp. 285-305.

Nadeau, R., 1991. Minds, Machines and Human Consciousness. Chicago: Contemporary Books.

Northcutt, R.G. and Kaas, J.H., 1995. The emergence and evolution of mammalian neocortex. Trends Neurosci 18, 373-379.

Passingham, R.E., 1982. The Human Primate. San Francisco: W.H. Freeman and Company Limited, 600 Market Street, S.F. CA 94104. 390 p.

Pinker, S., 1994. The Language Instinct: How the Mind Creates Language. New York: Harper Collins Publishers, Inc.

Pinker, S. and Bloom, P., 1990. Natural language and natural selection. Behavioral and Brain Sciences 13, 707-784.

Pinker, S. and Jackendoff, R., 2005. The faculty of language: what's special about it? Cognition 95, 201-236.

Pulvermüller, F., 2005. Brain mechanisms linking language and action. Nature Reviews Neuroscience 6, 576-582.

Ramachandran, V.S., 2004. A Brief Tour of Human Consciousness. New York: Pearson Education.

Reader, S.M. and Laland, K.N., 2002. Social intelligence, innovation, and enhanced brain size in primates. Proceedings of the National Academy of Sciences USA 99, 4436-4441.

Riddell, W.I. and Corl, K.G., 1977. Comparative investigation of the relationship between cerebral indices and learning abilities. Brain, Behavior and Evolution 14, 385-398.

Rilling, J.K. and Insel, T.R., 1999. The primate neocortex in comparative perspective using magnetic resonance imaging. Journal of Human Evolution 37, 191-223.

Ringo, J.L., 1991. Neuronal interconnection as a function of brain size. Brain, Behavior and Evolution 38, 1-6.

Rumbaugh, D.M., 1997. Competence, cortex, and primate models: a comparative primate perspective. In: Krasnegnor, N.A., Lyon, G.R. and Goldman-Rakic, P.S., (Eds.), Development of the Prefrontal Cortex: Evolution, Neurobiology, and Behavior. Baltimore, MD.: Paul H. Brookes. pp. 117-139.

Rumbaugh, D.M., Savage-Rumbaugh, E.S. and Wasburn, D.A., 1996. Toward a new outlook on primate learning and behavior: complex learning and emergent processes in comparative perspective. Japanese Psychological Research 38, 113-125.

Schenker, N.M., Desgouttes, A.M. and Semendeferi, K., 2005. Neural connectivity and cortical substrates of cognition in hominoids. Journal of Human Evolution 49, 547-569.

Schoenemann, P.T., 1997. An MRI Study of the Relationship Between Human Neuroanatomy and Behavioral Ability [Dissertation]. Berkeley: University of California, Berkeley.

Schoenemann, P.T., 1999. Syntax as an emergent characteristic of the evolution of semantic complexity. Minds and Machines 9, 309-346.

Schoenemann, P.T., 2005. Conceptual complexity and the brain: Understanding language origins. In: Wang, WS.-Y. and Minett, J.W., (Eds.), Language Acquisition, Change and Emergence: Essays in Evolutionary Linguistics. Hong Kong: City University of Hong Kong Press. pp. 47-94.

Schoenemann, P.T., 2006. Evolution of the Size and Functional Areas of the Human Brain. Annual Review of Anthropology 35, 379-406.

Schoenemann, P.T., Budinger, T.F., Sarich, V.M. and Wang, W.S., 2000. Brain size does not predict general cognitive ability within families. Proceedings of the National Academy of Sciences USA 97, 4932-4937.

Schoenemann, P.T., Sheehan, M.J. and Glotzer, L.D., 2005. Prefrontal white matter volume is disproportionately larger in humans than in other primates. Nat Neurosci 8, 242-252.

Semendeferi, K., Armstrong, E., Schleicher, A., Zilles, K. and Van Hoesen, G.W., 2001. Prefrontal cortex in humans and apes: a comparative study of area 10. American Journal of Physical Anthropology 114, 224-241.

Semendeferi, K., Lu, A., Schenker, N. and Damasio, H., 2002. Humans and great apes share a large frontal cortex. Nat Neurosci 5, 272-276.

Smith, B.H., 1990. The cost of a large brain. Behavioral and Brain Sciences 13, 365-366.

Striedter, G.F., 2005. Principles of Brain Evolution. Sunderland, MA.: Sinauer Associates.

Thompson-Schill, S.L., D'Esposito, M., Aguirre, G.K. and Farah, M.J., 1997. Role of left inferior prefrontal cortex in retrieval of semantic knowledge: a reevaluation. Proceedings of the National Academy of Sciences USA 94, 14792-14797.

Thompson-Schill, S.L., Swick, D., Farah, M.J., D'Esposito, M., Kan, I.P. and Knight, R.T., 1998. Verb generation in patients with focal frontal lesions: a neuropsychological test of neuroimaging findings. Proceedings of the National Academy of Sciences USA 95, 15855-15860.

Uylings, H.B.M. and Van Eden, C.G., 1990. Qualitative and quantitative comparison of the prefrontal cortex in rat and in primates, including humans. In: Uylings, H.B.M., Van Eden, C.G., De Bruin, J.P.C., Corner, M.A. and Feenstra, M.G.P, (Eds.), Progress in Brain Research, Vol 85. New York: Elsevier Science Publishers. pp. 31-62.

Van Essen, D.C., 2005. Surface-based comparisons of macaque and human cortical organization. In: Dehaene, S., Duhamel, J.-R., Hauser, M.D. and Rizzolatti, G., (Eds.), From Monkey Brain to Human Brain. Cambridge, Massachusetts: MIT Press. pp. 3-19.

Vandenberg, S.G. and Kuse, A.R., 1978. Mental rotations, a group test of three-dimensional spatial visualization. Perceptual and Motor Skills 47, 599-604.

Wang, W.S.-Y., 1991. Explorations in language evolution. Explorations in Language. Taipei, Taiwan: Pyramid Press. pp. 105-131.

Washburn, S.L., 1960. Tools and evolution. Scientific American 203, 63-75.

Wood, B. and Collard, M., 1999. The human genus. Science 284, 65-71.

Wynn, T., 2002. Archaeology and cognitive function. Behavioral and Brain Sciences 25, 389-438.

Zilles, K., 2005. Evolution of the human brain and comparative cyto- and receptor architecture. In: Dehaene, S., Duhamel, J.-R., Hauser, M.D. and Rizzolatti, G. (Eds.), From Monkey Brain to Human Brain. Cambridge, Massachusetts: MIT Press. pp. 41-56.

CHAPTER 4

Human Brain Endocasts, Taung, and the LB1 Hobbit Brain

Ralph L. Holloway

This Chapter is dedicated to the memory of Dr. Michael S. Yuan, who conceived this Festschrift, and whose untimely death in 2008 robbed many of us of a cherished friend, colleague, and scientist.

Introduction, Lines Of Evidence

There are really four approaches toward understanding how the human brain evolved:

(1) comparative neuroanatomy, which can compare the brain structures, size differences, pathways, and neurochemistry between living, extant species, each of which is the terminal end product of their own line of evolutionary development. This approach is indispensable for understanding the relationships between behavioral and neural variations. As such this approach is ***indirect***, because the animals compared are not evolutionary stages, but end products, each with a separate evolutionary history of variable time depth. For example, anthropologists and neuroscientists are often comparing present day chimpanzee brains with modern monkey and human brains, and the implicit assumption is that the chimpanzee of today is the same as the chimpanzee ancestor which split from the hominid line some 5-7 million years ago. Since we do not have a fossil record for the chimpanzee we cannot be sure that it hasn't undergone significant evolutionary change since 5-7 million years ago (see, for example, the papers of Rilling, 2006; Schoenemann, 2006; Semendeferi, this volume; Holloway, 2009).

(2) The second approach is paleoneurology, or the study of brain endocasts made from the crania of fossil animals, in particular, hominins from about 3-4 million years ago to the present. This is, at present, the only ***direct*** evidence available for human brain evolution (see Holloway et al., 2004 and associated references).

(3) A newer approach, however, is developing which offers great promise for understanding human brain evolution, and that is the nascent field of molecular neurogenetics, which in time may be able to unravel the actual genetic/target tissue/behavioral changes that took place in the past.

(4) The fourth method is simple speculation, often derided as "just-so stories" that abound in both the scientific and lay press. Examples include cooling of the brain/radiator hypothesis; singing Neandertals, increased senses of humor to whet the selective and receptive appetite of females, throwing, tool-making, the need for high protein sources, the necessity for cooking meat, working memory, etc, etc. All of these, none of these, or some of these could be true, but behavior simply does not fossilize, and testing these ideas against either comparative, or paleoneurological evidence is extremely difficult if not practically impossible. Nevertheless, insofar as such speculations challenge us with more focused attention to variables involved, testable hypotheses can emerge, leading to further testing. Of course the fossil record is limited, but who is to say that we perceive all of what presently is available in all of its details and associations.

Paleoneurology: The Direct Evidence

What sorts of data can one retrieve from the paleoneurological approach? First, one needs to remember how poor the data is, as each endocast is simply a cast of the interior of a skull, and therefore not a brain, but rather an impression left on the internal table of bone

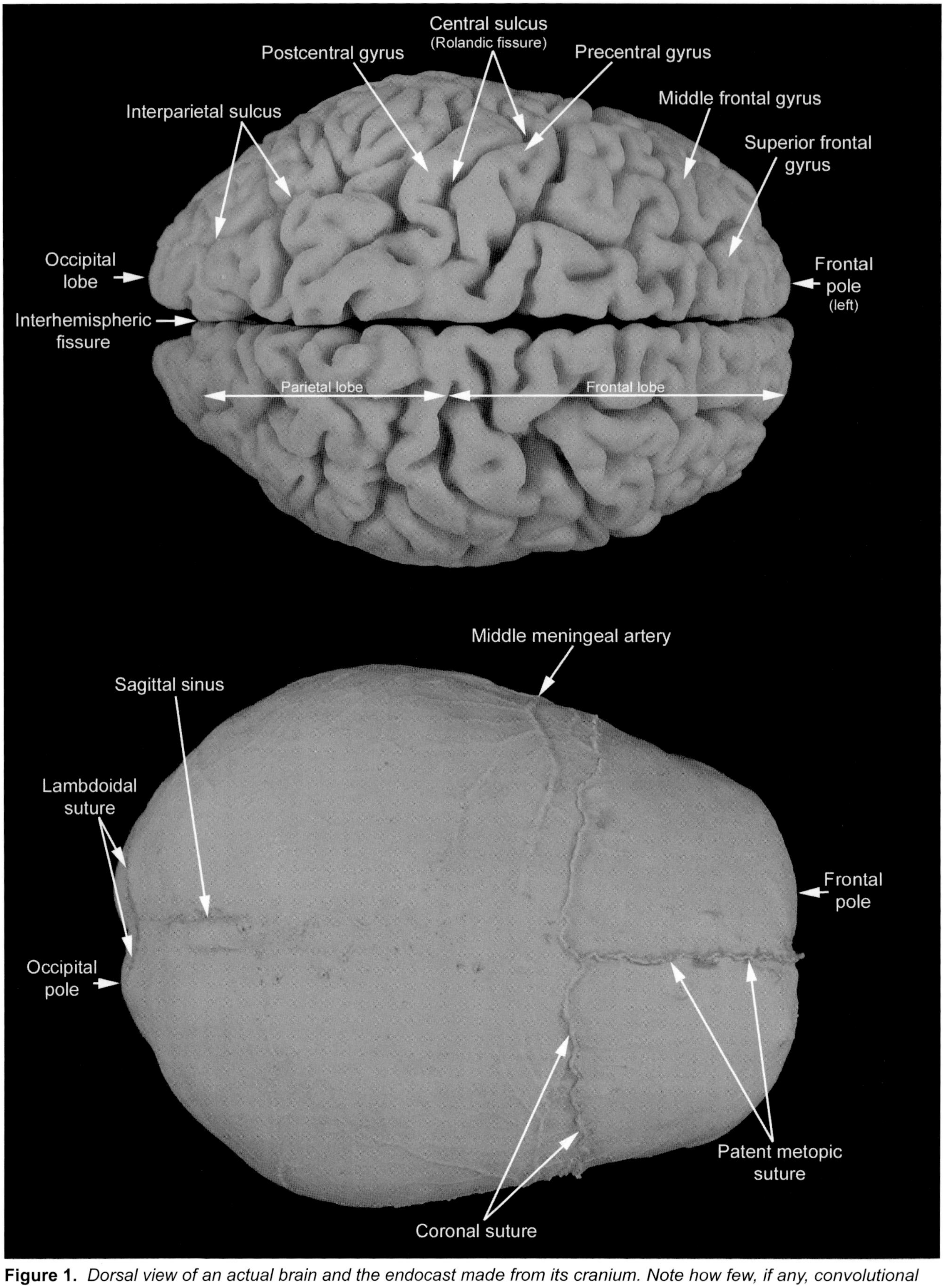

Figure 1. *Dorsal view of an actual brain and the endocast made from its cranium. Note how few, if any, convolutional details are retained on the endocast.*

meaning the outer layer of the dura mater, while the once living brain was covered by two additional menigeal tissues, arachnoid tissue (containing cerebral spinal fluid) and the pia mater, which invests itself on the cortical surface. All three meninges "conspire", so to speak, against the complete and faithful impressions of the cerebral and cerebellar convolutions on the internal table of the skull.

Four different types of important data can be retrieved from endocasts:

(1) First, and probably most important, is the overall size (volume) of the endocast, which provides a close approximation to actual brain weight of the once throbbing living brain. The accuracy obtained will depend on the completeness of the cranial remains, the amount of distortion, and if based on CT scans, the density profiles used in defining edges and contrasts.

(2) Depending on how well some of the gyri and sulci are impressed through the dura mater, once can get some idea of the extent and form of the various cerebral lobes, i.e., prefrontal, frontal, parietal, temporal, and occipital, however controversial the interpretations. In general, the central sulcus which divides frontal and parietal lobes is seldom visible on hominin endocasts, except possibly in its most superior portion. The precentral sulcus is more frequently visible, but never in its full extent. Similar difficulties exists in defining the boundaries of parietal, temporal and occipital lobes since no cytoarchitectonic data is present, and one must use classical landmarks of neuroanatomical lobar divisions, which in addition to being arbitrary, are not plainly visible on endocasts. On the other hand, one can make out, albeit dimly, possible sulci and gyri such as superior and middle temporal gyri, lunate sulcus and inferior occipital sinus, retrocalcarine sulcus, supramarginal and angular gyri on the parietal lobe (although never completely), and while the Broca's cap regions of the prefrontal lobe may show a morphology similar to that found in modern humans from *Homo erectus* on, it is rarely the case that the *pars opercularis, p. triangularis,* and *p. orbitalis* can be readily visualized.

(3) In humans, and those fossil hominins attributed to the genus *Homo*, one usually finds that the cerebral hemispheres are asymmetrical, and close correlations with handedness have been shown between the petalias (projections of either occipital and/or frontal lobes more to one side than the other.) Indeed, some of the fossils provide evidence of cerebral asymmetries from about 2 million years ago to the present. The prefrontal regions, which include Broca's caps of the third inferior frontal convolution, including *pars opercularis, triangularis*, and *orbitalis*, are sometimes asymmetrical as in modern humans. These asymmetries suggest, through the assumption of homologous structure and function, that the basic human-like organization of the brain was established early in hominin evolution. In other words, if we know from studying modern humans through neuroanatomy, surgical procedures, PET, MRI, fMRI, and dissection, that handedness involves cerebral asymmetry and specialization, and that Broca's regions are involved in an important manner with language production, and we find these same appearances on say, Neandertal or *Homo erectus* endocasts, can we reasonably speculate that these hominids also had cerebral dominance and language behaviors similar, if not identical to our own? As to how old such asymmetries are is clouded by the more recent findings of asymmetries in chimpanzees. Asymmetries are of course found in many different animals.

(4) One can take numerous measurements on the endocasts, and using multivariate statistical procedures, attempt to show more objectively how endocast shape patterns vary between different hominin groups. These biometrical approaches are becoming standard in endocast descriptions as seen in the many articles by Bruner (this volume), Bruner and Holloway 2010, Grimaud-Hervé (this volume), and Wu et al (2010).. These approaches are needed to test preliminary qualitative observations regarding endocast morphologies and taxonomic differences, as well as evolutionary behavioral explanations. These are difficult as almost every single measurement taken is allometrically related to overall volume.

Mosaic Brain Evolution In Hominins

If both the data for brain size increases through time and key reorganizational features of the brain' surface, such as petalias, asymmetries in Broca's regions, a reduction in primary visual striate cortex, or area 17 of Brodmann are accurate, it is clear that the human brain underwent an evolutionary trajectory that intercalated size increases with organization changes, and that the evolution of the human brain was a complex mosaic affair, involving more than simple brain size increase and encephalization.

Even the australopithecines, the earliest hominins of 2-4 million years ago, show that despite their ape-sized brain volumes, they had a cortex reorganized toward a more human pattern, as evidenced by the appearance of the lunate sulcus in a more posterior position. However, their frontal lobes do not show Broca's regions similar to what we find in *Homo*. Indeed, the earliest *Homo* is the famous KNM-ER 1470 specimen, which at a volume of 752 ml, shows clear-cult petalial asymmetries of the human pattern (left occipital/right frontal width) and Broca's regions that are human-like in external morphology. This occurs at 1.8 to 2.0 million years ago. The size increases through evolutionary time are both allometrical and non-allometrical, the former being related to increased body sizes. But by 1.5 million years ago, as shown by the Nariokatome child skeleton from Turkana, Kenya, *Homo erectus* had a modern human body size, but a brain that appeared to vary between 750 and 900 ml. Any increase in brain size thereafter would be basically, non-allometric, i.e., unrelated to body size increase. Thus, the paleoneurological evidence is suggesting that selection pressures were surely varied over the course of hominin evolution. Selection for increased body size

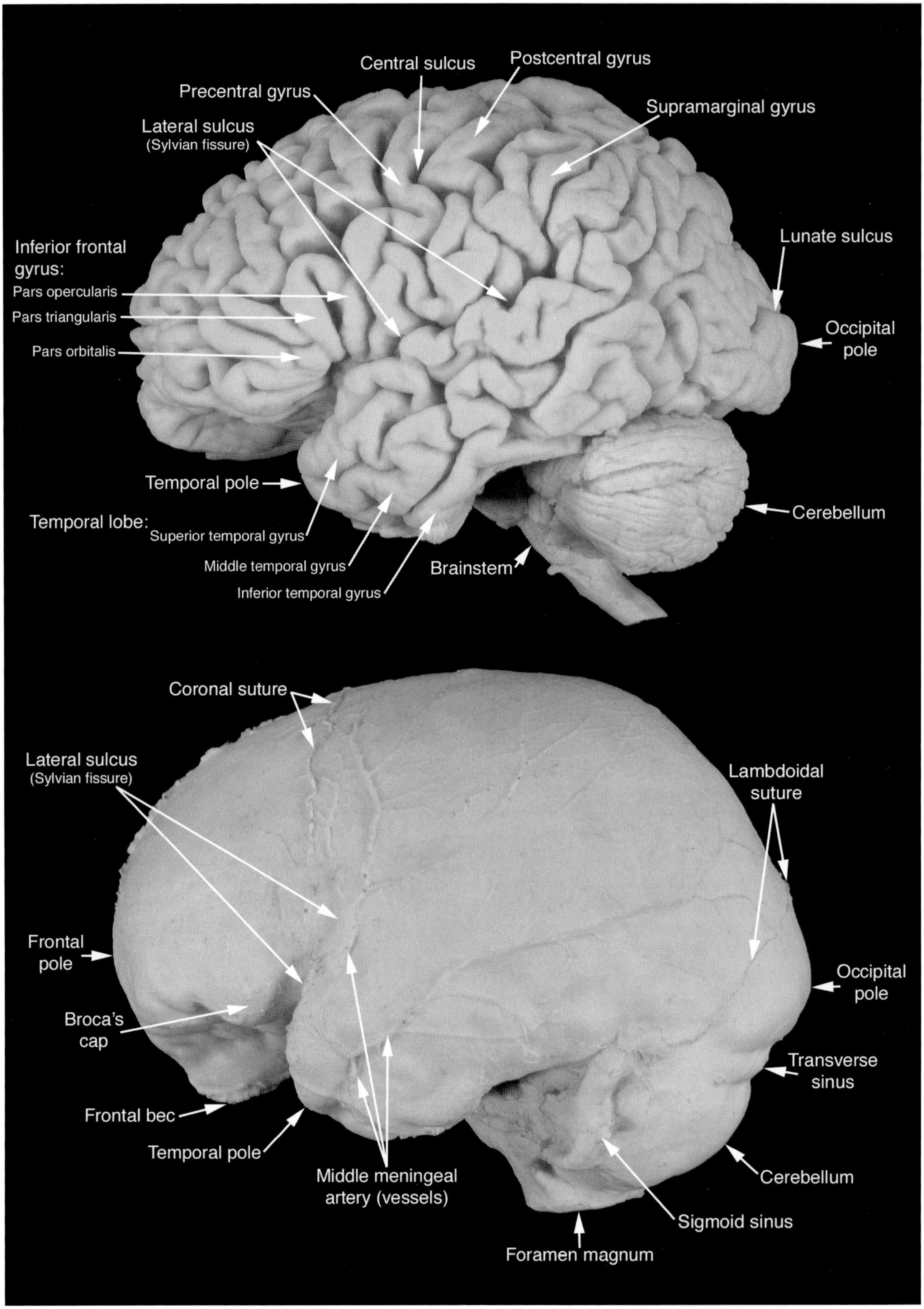

Figure 2. *The left lateral surface of the brain and endocast*

could have provided part of the increase in brain size in early hominin evolution, i.e., from *Australopithecus* to early *Homo*. Thereafter, and throughout the course of hominin evolution selection pressures for brain size increase, without attending body size increases were occurring, although it is possible that the Neandertals could be an exception, in that their body sizes (lean body mass) and adaptations to cold climates may have necessitated larger bodies, with a concomitant increase in brain volume, as their average slightly surpasses our own volume. (See Bruner, this volume, for biometric analyses showing differential size changes in frontal and parietal lobes; see also Tables 1, 2, & 3 in Chapter 1.)

Some Ongoing Issues In Paleoneurology

Seduction by Laser Scanning, or the Demotion of the Taung Endocast's Size

Falk and Clarke (2007) have a recent technical paper in which, by using laser scanning of a replica of the Taung endocast they attempted to place a mirror image of the right side on the left, and came up with a resulting 382 cc endocast volume, considerably less than my previously water displacement 404 cc version (Holloway 1970).

Such a technique of course requires the assumption of perfect symmetry of the two cerebral sides, and the careful definition of a true midline. Neither of these two requisites appears in their paper. Instead, a mirror image of the right side is imposed onto the missing left side with delineating and describing a midline. What is more, their figure of the dorsal surface of the Taung endocast shows that the left and right sides are asymmetrical! (See Fig. 1) As pointed out in the Holloway 1970 paper, a slice of Taung endocast 1 mm in thickness would result in only 7 cc of volume, and it doubtful the definition of the midline in that paper was off by more than a mm.

The main advantages of laser (and CT) scanning of crania, or endocasts as in the above case, is that the methods are non-invasive, and depending on the skills of the investigators, can correct more easily for distortions and missing portions, than one can with plasticene. Still, the results depend heavily on the techniques used and the skill and understanding of brain and cranial anatomy on the part of the investigators. The history of the various volume estimates of Taung, STS 71 (Conroy et al. 1998, 2000; Holloway 1970, 1972, 1973, 1983, 1999; Holloway et al. 2004), and Stw505 as well as the Hobbit LB1, *H. floresiensis* (see below), suggests that my previous volumes were correct after all (see Holloway et al. 2004).

The Hobbit Brain

The major argument existing today is whether or not the brain of the hobbit (based on a single cranium, LB1) is that of a pathological microcephalic (primary, secondary, or yet unknown), or a true non-pathological species

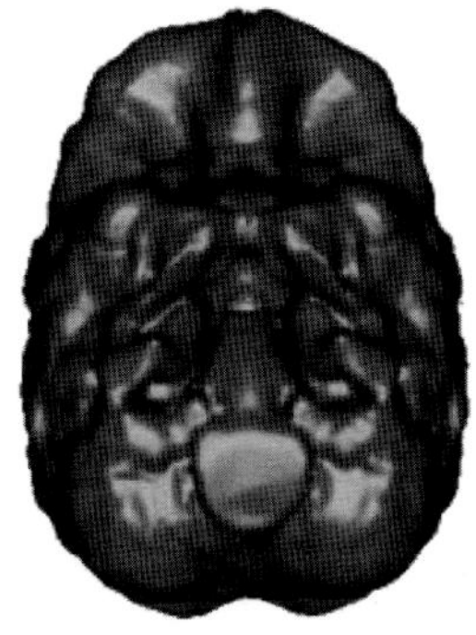

Fig.3 Basal and dorsal views of the Falk and Clarke (2007) mirror image technique applied to Taung. Note that the left and right sides are not symmetrical. (After Falk and Clarke, 2007)

that evolved through some unknown process of island dwarfing. There is a third possibility which is that the brain of the hobbit shows pathology unlike the pathological appearance that one sees in cases of primary or secondary microcephaly, and that these are not recognized because the full range of variation in the broader condition known as "microcephaly" hasn't been thoroughly studied. Indeed, finding a large sample of these individuals either illustrated or measured hasn't yet happened. Falk and her colleagues believe the brain is simply that of a new species (Falk et al. 2005, 2007, 2009), probably derived from some earlier *Homo erectus* ancestor (indeed *Australopithecus* is now also in the running) that underwent dwarfing, but was otherwise normal, and not pathological. Martin and his colleagues (Martin et al. 2006), and Henneberg & Thorne (2004) regard the hobbit as severely pathological, the pathology involving not just the brain, but the entire skeleton. A recent entire issue of the *Journal of Human Evolution* has been devoted to showing that this hominin is not pathological.

My own opinion (Holloway et al. 2006) is that more fence straddling is prudent, even if I am only hanging by an ankle. I do agree with Falk and her colleagues that the LB1 endocast does not look like any cases of primary microcephaly found in modern populations that have yet been published or presented, and this includes the Indian microcephalic presented by Martin. I come to this conclusion after having made endocasts of several microcephalics (primary and secondary) from the Pathology collection of the University of Michigan, through the kindness of Dr. Milford Wolpoff, and from Gary Sawyer and Dr. Ian Tattersall from the American Museum of Natural History, two from the Museum of Comparative Zoology, courtesy of Dr. Dan Lieberman, and one intriguing Indian microcephalic from India that Dr. Robert Martin and Dr. Susan MacLarnon sent me. Finally, I have one endocast sent to me by Dr. Dominique Grimaud-Hervé from Paris of a case of Seckel's ("Bird-headed Dwarf") microcephaly. Some of these were shown to the audience in Alaska during my AAPA April,

2006 presentation. Aside from the two cases of Seckel's Syndrome and the Indian microcephalic, all of the primary microcephalic endocasts I have seen and studied have relatively enlarged cerebellar lobes compared to their diminutive cerebral cortices. The secondary cases, with larger brain volumes do not show cerebellar protrusion, nor do the overwhelming majority of 198 cases of ape endocasts (Holloway et al. 2010) thus ruling out protruding cerebellar lobes as a derived feature in LB1 (Falk et al. 2009). None, however, show the extreme degree of platycephaly (flattening of the brain) that occurs in LB1. Furthermore, none show the extremely protuberant and narrow prefrontal gyri (recti) which are so striking on LB1, and which I regard as a possible pathology, perhaps akin to microgyria, in which 4 rather than the normal 6 layers of the cortex develop. Additionally, no microcephalic I have seen shows the peculiar spreading of the cerebellar lobes that one sees on the hobbit brain cast, or the peculiar trigonal-shaped eminence on the dorsal surface of the brain stem, which cannot be a blood sinus feature.

Discussions with neurologists, pediatric and otherwise appear to confirm that this trigonal structure doesn't appear in modern human brains, so one is tempted to regard this as an autapomorphy. Much more study is required here (see Figures 4-6, and Table 1 of possible hypotheses). It is best to remember that the full range of variability in external appearances of microcephalic brains (particularly secondary forms) has not been studied, and the possible pathologies I am suggesting remain a possibility in LB1, however unlikely these are viewed at present.

This particular hominin provides also a window on aspects of scientific cooperation in studying these remains. To date, only Falk and her colleagues have provided a study of the endocast using CT scanning where descriptions of exact procedures and smoothing techniques are not available. I have personally tried, since the first description of this find to obtain the CT scan data for an independent assessment of the endocast, and am delighted to have recently obtained the original CT scan data from Dr. Michael Morwood. While this older, now redundant CT scan that has been replaced by a more recent micro-CT scan, the scan data show very clearly that the original endocast required considerable reconstruction, and that the endocast I received from Dr. Peter Brown, and those I made from the stereolith, match exactly the older scan data.

These data show that the right temporal lobe was severely displaced laterally and inferiorly, and that the left temporal lobe was not distorted, and appears relatively small, at least to my eye, which is not in agreement with Falk et al. (2009) claim regarding it as a derived character state in *Homo floresiensis*. Aside from the prominent, but very thin *gyri recti* of the prefrontal lobe, this part of the frontal lobe also appears relatively small to me (see Fig. 7).

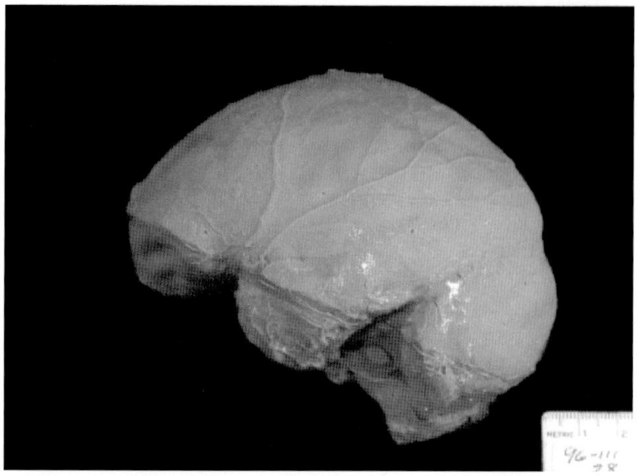

Figure 4. Lateral view of a typical case of primary microcephaly. Notice, in particular, the relatively large appearance of the cerebellar lobes relative to the cerebral cortex. (Figures 4-6 are from pictures taken at the 2006 American Association of Physical Anthropologists annual meeting)

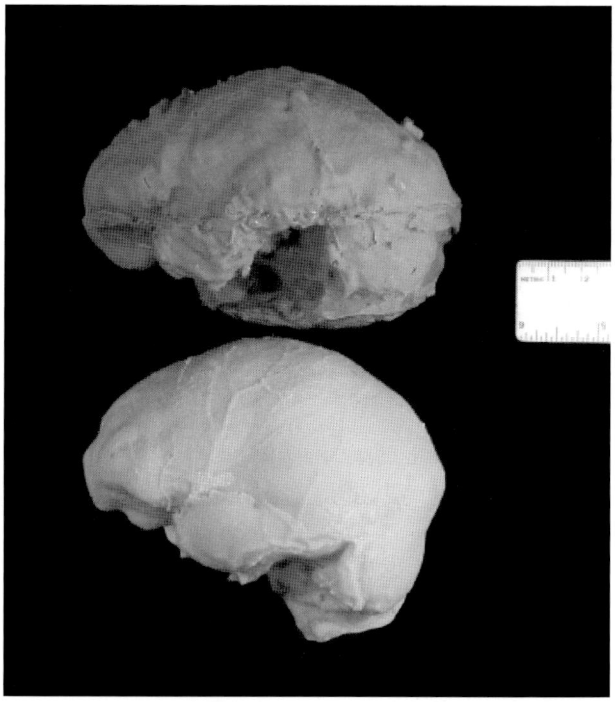

Figure 5. The top endocast is the left lateral view of the LB1 endocast (as resulting from original CT scan data), and the bottom picture is that of an Indian microcephalic with the same brain size. In this case the cerebellar lobes of the microcephalic appear almost normal in relative size, but the height of the endocast is quite different from that of LB1.

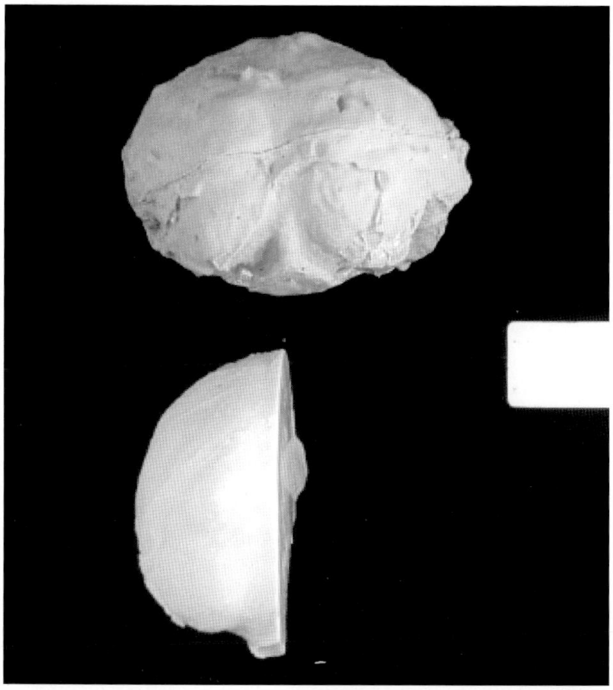

Figure 6. Occipital views of both endocasts. Note that the platycephaly so evident on the LB1 (top) endocast is not present on the microcephalic endocast. Notice also the trigonal eminence between the cerebellar lobes of LB1.

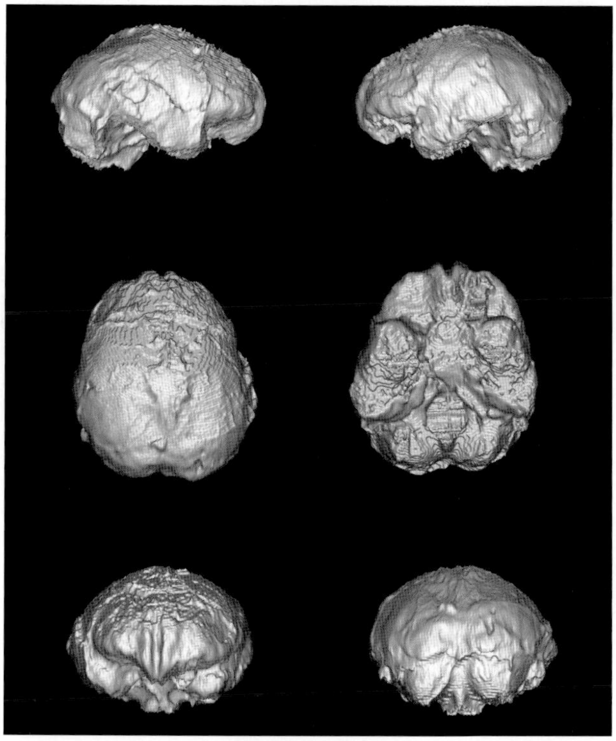

Figure 7. Six standard views of the LB1 endocast segmented using ITK-SNAP, Version 2.0, from the CT scan data, which I received from Michael Morwood and his Indonesian colleagues. These are un-smoothed, and show in particular, the damage to the right temporal lobe.

Table 1 below provides a humorous summary of the possible interpretations of the LB1 hobbit.

Table 1.

IDENTITY CRISIS
• If it's Monday, LB1 must be: **a microcephalic**
• If it's Tuesday, LB1 must be: **a new species, *Homo floresiensis***
• If it's Wednesday, LB1 must be: **a sex slave...**
• If it's Thursday, LB1 must be: **all of the above...**
• If it's Friday, LB1 must be: **an alien from another galaxy**
• If it's Saturday, LB1 must be: **a derived form of australopithecine**
• If it's Sunday, LB1 must be: **an endemic dwarfed from derived from *Homo erectus***
• (Fortunately the "intelligent designer" only had 7 days)

The Political Correctness Angle or the 800-pound gorilla in the room...

Political correctness within biological anthropology, at least as far as the nervous system is concerned, involves the notion that the human species may very well vary from the top of the head down to the toes, but not in the brain, or if the latter is true, the variations are without any behavioral importance. To realize otherwise might lead down the slippery slope of racist history and racism. We know that this is very unlikely, that human groups do have variability in terms of brain size, although we know very little if anything about whether biological populations differ in how their brains are organized (However, see Klekamp et al. 1994 regarding Australian Aboriginal striate cortex volumes). It would surely be an amazing instance of genetic conservatism if all of the thousands of regulatory and structural genes related to the brain and its growth and development were the same in every population. We know considerably more about sex differences in the brain, and to suppose that these differences did not arise through evolutionary selection pressures for aspects of social behavior, or have no genetic basis is just silly in my opinion. It is strange to read so many accounts of how we became smarter and smarter during our evolution when our brains became larger, but such variation in modern human groups has absolutely no behavioral significance today. It is equally strange to talk about how the human sexes are complementary to each other in terms of child care, learning, and subsistence, but then insist that no hard-wired differences in the brain exist between the sexes, when dozens of articles in neuroscience journals indicate otherwise. How else could they have arisen? It would be more accurate to say that the differences in brain size among human populations today, while perhaps statistically significant, are a rather small difference compared to the 1000 ml increase that occurred during hominid evolution over the last 2-3 million years. As to whether or not there are significant behavioral differences, such as IQ, or other cognitive tests, raises many difficult methodological and moral issues, which combined with an almost species-specific bent toward PC discourages most, if not all investigations into modern human brain variation, despite excellent studies showing maturational, white fiber matter differences throughout the brain (see for example, Rushton and Ankney, 2007, 2009, for a review that might receive rebuke from many social scientists, but yet remains disproven). Perhaps, in the future, as molecular neurogenetics becomes more advanced, we might know more about how the human brain varies, and how such variation relates to behavior, within a cultural context (in particular, nutrition and diet, disease and parasite vectors), and how and why such differences, however minor, evolved. In my opinion, without a fuller knowledge of how the human species' brain varies, it is extremely difficult, if not impossible, to know how the human brain really evolved. I believe we can only benefit, both medically and scientifically, from knowing more about how we vary as a species.

As I have said to my classes many times, human variation is one of the best things we possess as a species, and should be treasured and celebrated, not feared. (Holloway 2008).

Acknowledgements

Needless to say, I am very grateful to the Editors of this volume, and to all of the authors who have honored me with their chapters, friendship, and collegiality. I hope our collaboration toward understanding how the human brain evolved will continue well into the future!

References

Bruner E, Holloway RL. 2010. A bivariate approach to the widening of the frontal lobes in the human genus. *J.Hum. Evol.* 58:138-146.

Conroy GC, Weber GW, Seidler H, Tobias PV, Kane A, Brunsden B. 1998. Endocranial capacity in an early hominid cranium from Sterkfontein, South Africa. *Science* 280:1730-31.

Conroy GC, Falk D, Guyer J, Weber GW, Seidler H, Recheis W. 2000. Endocranial capacity in Sts 71 (*Australopithecus africanus*) by three-dimensional computed tomography. *Anat. Rec.* 258:391-96.

Falk D, Hildebolt C, Smith K, Morwood MJ, Sutikna T, Brown P, Jatmiko T, Saptomo EW, Brunsden B, Prior F. 2005 The brain of LB1 *Homo floresiensis*. *Science* 308: 242-45.

Falk D, Clarke R. 2007. New reconstruction of the Taung endocast. *Am. J. Phys. Anthropol.* 134(4):529-34.

Falk, D, Hildebolt C, Smith K, Morwood MJ, Sutikna T, Jatmiko, Saptomo EW, Imhof H, Seidler H, Prior F. 2007. Brain shape in human microcephalics and *Homo floresiensis*. *Proc. Natl. Acad. Sci.* USA104: 2513-18.

Falk D, Hildebolt C, Smith K, Morwood MJ, Sutikna T, Jatmiko, Saptomo EW, Prior F. 2009. LB1's virtual endocast, microcephaly, and hominin brain evolution. *J. Hum. Evol.* 57:597-607.

Henneberg M, Thorne A. 2004. Flores may be pathological *Homo sapiens*. *Before Farming* 1:2-4.

Holloway RL. 1970. New endocranial values for the australopithecines. *Nature* 227:199-200.

Holloway RL. 1972. Australopithecine endocasts, brain evolution in the Hominoidea and a model of hominid evolution. In *The Functional and Evolutionary Biology of Primates,* ed. R. Tuttle, pp. 185-204. Chicago: Aldine/Atherton Press.

Holloway RL. 1973. Endocranial volumes of the early African hominids and the role of the brain in human mosaic evolution. *J. Hum. Evol.* 2:449-59.

Holloway RL. 1983. Human brain evolution: a search for units, models, and synthesis. *Can. J. Anthropol.* 3:215-32.

Holloway RL. 1999. Hominid brain volume. *Science* 283:34

Holloway RL. 2008. The Human Brain Evolving: A Retrospective. *Annu. Rev. Anthropol.* 37:1-19.

Holloway RL. 2009. Brain Fossils: Endocasts. In *Encyclopedia of Neuroscience*: Vol. 2, ed. LR Squire, pp. 253-61. Oxford: Academic.

Holloway, RL, Broadfield, DC, Yuan, MS. 2004. *Brain Endocasts: The Paleoneurological Evidence:* Vol. 3, *The Human Fossil Record*. ed. JH Schwartz, I Tattersall. New York: Wiley-Liss.

Holloway RL, Brown P, Schoenemann PT, Monge J. 2006. The brain endocast of *Homo floresiensis*: microcephaly and other issues. *Am. J. Phys. Anthropol.* S42:105.

Holloway RL, Schoenemann T, Monge J. 2010. The Hobbit brain: some questions about its "derived" features. *Am. J. Phys. Anthropol.* S50:130.

Klekamp J, Reidel A, Harper C, Kretschmann HJ.1994. Morphometric study on the postnatal growth of the visual cortex of Australian Aborigines and Caucasians. *J. Brain Res*. 35 :531-48.

Martin RD, Maclarnon AM, Phillips JL, Dobyns WB. 2006a. Flores hominid : new species or microcephalic dwarf? *Anat. Rec. A Discov. Mol. Cell Evol. Biol.* 288:1123-45.

Preuss TM, Coleman GQ. 2002. Human-specific organization of primary visual cortex: alternating compartments of dense Cat-301 and calbindin immunoreactivity in layer 4A. *Cereb. Cortex* 12:671-91.

Rilling, JK. 2006. Human and nonhuman brains: are they allometrically scaled versions of the same design? *Evol. Anthrop*. 15(2):65-77.

Rushton JP, Ankney CD. 2007. The evolution of brain size and intelligence. In *Evolutionary Cognitive Neuroscience, ed.* SM Platek, JP Keenan, TK Shackelford, pp. 121-61. Cambridge, MA: MIT Press.

Rushton JP, Ankney CD. 2009. Whole brain size and general mental ability: a review. *Int. J. Neurosci.* 119:691-732.

Schoenemann PT. 2006. Evolution of the size and functional areas of the human brain. *Annu. Rev. Anthropol.* 35(1):379-406.

Wu X, Schepartz LA, Liu W. 2010. A new *Homo erectus* (Zhoukoudian V) brain endocast from China. *Proc. R. Soc. B.* 277:337–344.

CHAPTER 5

The Fossil Hominid Brains of Dmanisi: D 2280 and D 2282

Dominique Grimaud-Hervé and David Lordkipanidze

Abstract

Since the discovery of the first mandible in 1991 at the site of Dmanisi, many other human fossil remains have been found at the site in association with archaeological artefacts and faunal remains dated between 1.81 and 1.77 My. Dmanisi is probably the oldest site outside of Africa, and was most likely an important migration route into Europe and Asia from Africa. Analysis of the two first hominid endocasts of D 2280 and D 2282 has been done. These endocats were compared with contemporaneous African fossil hominids (*Homo habilis*, *Homo rudolfensis*, *Homo ergaster*) and continental and insular Asiatic ones (*Homo erectus*). Dmanisi's endocasts are similar in size with the earliest specimens of *Homo*, while cerebral form, vascular middle meningeal pattern, cerebral morphology are more similar to Asiatic *Homo erectus*. Based on these similarities they can be assigned to the same taxon as early representatives of the genus *Homo* (*Homo ergaster* or even *Homo habilis*), suggesting Dmanisi played an essential role in the earliest settlement of eastern Asia.

Keywords

Dmanisi, Brain, Human evolution, Cognitive capacities, Eurasia

Introduction

Discoveries of human fossil remains at Dmanisi are now well known, yielding five skulls, four mandibles and many postcranial elements and associated archaeological assemblages (Gabunia 1992, Gabounia & Vekua 1995, Gabounia et al., 1999, 2000, 2002; Vekua et al., 2002; Jashashvili, 2002; Rightmire et al. 2005; Lordkipanidze et al., 2006, 2007; Lumley et al. 2006). The faunal remains mostly consist of Villafranchian species assigned to the Late Villanian and Early Biharian, living in a mosaic environment of open steppe and gallery forests (Gabunia et al. 2000, 2001).

Most of the human remains have been discovered in layers V and VI, dated between 1.85 and 1.77 My (Gabunia et al., 2000, 2001; Lumley et al., 2002). Dmanisi is an important site for the understanding of early human migrations, since it is the oldest site outside of Africa, and it is in a geographically strategic position at the intersection of Africa, Europe, and Asia.

Skulls D 2280 and D 2282, an adult male and subadult female respectively (Lumley & Lordkipanitze, 2006 ; Lumley *et al.*, 2006), were scanned in superior view with a General Electric High Speed HAS scanner at the CHNO des Quinze-vingts in Paris under the care of Pr Cabanis with following acquisition parameters for both Georgian hominids: scanner energy 120 kV, 100 mA, 1.0 mm-thick slices; 25 cm field of view and 0.488 pixel size with a pixel matrix of 512*512 for D 2280; 23 cm field of view and 0.449 pixel size with a pixel matrix of 512*512 for D 2282. The sections were used to create three-dimensional computer models of both specimens using Mimics 8.1 software (Materialise N.V.). Stereolithographic reproduction from scanner data has been done.

Material

Fossil hominids'endocasts (Muséum national d'Histoire naturelle, Paris; Columbia University, New York)

OH 7, OH 13, OH 16, OH 24
KNM-ER 1813, KNM-ER 1470, KNM-ER 3733, KNM-ER 3883, KNM-WT 15000
Trinil 2, Sangiran 2, Sangiran 10, Sangiran 12, Sangiran 17, Sangiran 38
Ckn.D 1.PA.17. (Sin.II), Ckn.E 1.PA.16. (Sin.III), Ckn.L 1.PA.98. (Sin.X), Ckn.L 2.PA.99. (Sin.XI), Ckn.L 3.PA.100. (Sin.XII)

The above fossil material has been chosen to compare Dmanisi's hominids with contemporaneous fossil hominids from Africa: Olduvai (Tanzania) attributed to *Homo habilis*, East and West Rudolf (Kenya) attributed to *Homo habilis*, *Homo rudolfensis* and *Homo ergaster* and from insular Asia with more recent fossil from Sangiran and Trinil (Java) and from continental Asia in Zhoukoudian Lower Cave (China). Asiatic hominids are attributed to *Homo erectus*. The obtained results are also compared to the actual extinct human sample.

Endocasts from Olduvai (Tanzania) are not well preserved and those from East and West Turkana (Kenya) are in poor quality, so our data have been completed with results from Tobias (1987, 1991), Begun & Walker (1993), Holloway (1978, 1983), Holloway et al. (2004) and Saban (1984).

Modern population (Muséum national d'Histoire naturelle de Paris)

n = 103 from Europe, continental and insular Asia, Africa, America and Oceania

595, 713,723, 726, 727, 728, 729, 730, 731; 732-3, 733, 748, 749, 754, 755, 764-1, 784, 788, 789, 794-3, 798-2, 800, 808, 1294-2, 1489, 1490, 1865, 3635, 3662, 3663, 3664, 3665, 3666, 3667, 3668, 3669, 3670, 3671, 3672, 3673, 3674, 3675, 3676, 3677, 3678, 3679, 3680, 3681, 3682, 3683, 3684, 3685, 3686, 3688, 3689, 3690, 3691, 3692, 3693, 3694, 3695, 3696, 3697, 3698, 3699, 3700, 3702, 3703, 3704, 3705, 3706, 3707, 3708, 3709, 3775, 3827, 3828, 4362, 4815, 5720, 5733, 9843, 9844, 9852, 9853, 9854, 10109, 10111, 10112, 10113, 10114, 12033, 19246, 21413, 24636, 24940, 24942, 25027, 25536, 25620, 27429, 30189, 30195.

Methods

Morphological description of encephalic relief and vascular imprints (venous sinuses and middle meningeal system) is realized on Dmanisi's endocasts. Comparison is done with human fossils from Africa and Asia. Cranial capacity has been estimated directly by immersion of endocasts in water, repeated three times, and the results averaged.

A traditional metrics study (linear and angular measurements) (Table 1) was done in order to compare absolute and relative values between both Dmanisi's endocasts and African and Asiatic fossil hominids as well as a large extinct modern human reference sample. Principal components analysis was performed to synthesize information contained in 11 variables, selected in relation to the preservation of the Dmanisi endocasts. These measurements included width (WBE), average hemispheric length (LME), occipito-cerebellar projection (DOCE), both height (HGQE) and (HBRE), angular data (XBE, XLSE, XLIE) and sagittal chord of each cerebral lobe (CFR, CPA, COC). These 11 variables have been used on 47 specimens with 21 fossil hominids and 26 actual extinct human.

In the 3D geometric morphometrics study, particular care has been required to choose maximum common landmarks preserved on Dmanisi's endocasts and on the fossil human comparison sample from Africa and Asia. 3D coordinates of 28 anatomical landmarks were digitized on each endocast with a Microscribe 3DX digitizing arm (Table 2). Selection of 6 sagittal landmarks along interhemispheric fissure and 22 (11 X 2) parasagittal landmarks was done for this comparative analysis to determine the morphometrical affinities of Dmanisi's hominids. Landmarks coordinates have been fitted by Generalized Procrustes Analysis (O'Higgins 2000, Rohlf & Marcus 1993).

Preservation and brief description of encephalic and vascular imprints of Dmanisi's endocasts

The cranial cavity of D 2280 is perfectly preserved, producing a high quality endocast where all the relief and depressions are clearly visible. Unfortunately D 2282 is deformed with the inferior part of right hemisphere pushed inside and exhibits an altered internal surface. Many irregularities corresponding to sediment deposits disturbs full observation of the endocranial surface. Encephalic relief is visible only on the left side of the endocranium.

Both Dmanisi's endocasts have been described previously (Grimaud-Hervé *et al.* 2006). Most important morphological characters can be reminded here (Fig.1, Fig.2).

Vascularization

Concerning dura mater sinuses, superior sagittal sinus, visible on posterior part of D 2280, runs into the left lateral sinus and is well defined with noticeable relief approximately 8mm in diameter. On D 2282 the superior sagittal sinus is not noticeable on the sagittal or left lateral side. It is, though, discernable on the right with relief of 7.5mm. From this it is assumed that the superior sagittal sinus runs with the right lateral sinus on D 2282. Asymmetry is observed on these two fossil hominids, but without a preferential side being obvious. No spheno-parietal sinus is observed. Three sagittal arach-

Table 1. Metric variables used in traditional multivariate analysis 2D

Left lateral view
Maximal length (LME), average of right (LMDE) and left (LMGE) hemispheric length, measured from the most anterior point (endoglabella) to the most posterior (endo-opisthocranion) of the endocast
Maximum height of maximum hemispheric length (HGQE), average of right (HGQDE) and right (HGQGE) heights
Maximum endobregma height of maximum hemispheric length (HBRE), average of right (HBRDE) and right (HBRGE) heights
Frontal chord (CFR), between the most anterior point of the frontal lobe and central fissure at the midsagittal plane
Parietal chord (CPA), between central fissure and perpendicular scissure at the midsagittal plane
Occipital chord (COC) between perpendicular scissures and most depressive point of Herophile torcular at the midsagittal plane
Occipito-cerebellar projection (DOCE) measured by occipital projection perpendicular from Herophile torcular
Bregmatic angle (XBE) between maximal length and chord between most anterior point of the frontal pole and endobregma, average between right (XBDE) and left (XBGE) angle
Angle comprised between chord from perpendicular fissures at the midsagittal plane to endo-opisthocranion and chord from endo-opisthocranion to Herophile torcular (XLTE), average between right (XLTDE) and left (XLTGE) sides of the brain
Angle comprised between chord from perpendicular fissures at the midsagittal plane to endo-opisthocranion and maximal length of hemisphere (XLSE), average between right (XLSDE) and left (XLSGE) sides of the brain
Angle comprised between chord from endo-opisthocranion to Herophile torcular (XLIE), average between right (XLIDE) and left (XLIGE) sides of the brain
Upper view
Maximum width of the endocast (WBE), subdivided in right (WMDE) and left (WMGE) width
Maximum width on parietal lobes (WBE)
Maximum width on triangular part of third frontal gyrus (WCBE)
Right frontal surface (FRD)
Left frontal surface (FRG)
Right parieto-temporal surface (PTD)
Left parieto-temporal surface (PTG)
Right occipital surface (OCD)
Left occipital surface (OCG)
Right Hemispheric surface (HD) = FRD + PTD + OCD
Left hemispheric surface (HG) = FRG + PTG + OCG
Brain surface = HD + HG

Table 2. Landmarks points digitalized in geometrical morphometrics 3D

Sagittal points	
1	Base of encephalic rostrum between both left and right first frontal convolution in midsagittal plane
11	Intersection between left postcentral sulcus and interhemispheric fissure
14	Posterior interhemispheric point (= most depressed point of Herophile torcular)
15	Intersection between left and right perpendicular scissures and interhemispheric fissure
16	Intersection between precentral scissures and interhemispheric fissure
17	Middle point of frontal arch
Left parasagittal points	
2	External edge of left encephalic rostrum
3	Orbital part of left third frontal convolution
4	Point of maximal curvature of triangular part of left third frontal convolution
5	Upper point of left sylvian valley (between opercular part of third left frontal convolution and left temporal lobe)
6	Most anterior point of left temporal pole
7	Left Euryon (corresponding to maximal endocranial width)
8	Point of maximal curvature of left supramarginal gyrus
9	Anterior point of left interparietal sulcus, means base of left first parietal convolution
10	Middle point of anterior edge of left first parietal convolution
12	Upper point between left temporal and left cerebellar lobes (= upper point of left temporo-cerebellar excavation)
13	Point of maximal curvature of left occipital pole
Right parasagittal points	
18	External edge of right encephalic rostrum
19	Orbital part of right third frontal convolution
20	Point of maximal curvature of triangular part of right third frontal convolution
21	Upper point of right sylvian valley (between opercular part of right third frontal convolution and right temporal lobe
22	Most anterior point of right temporal pole
23	Right Euryon (corresponding to maximal endocranial width)
24	Point of maximal curvature of right supramarginal gyrus
25	Anterior point of right interparietal sulcus, means base of right first parietal
26	Middle point of anterior edge of right first parietal convolution
27	Upper point between right temporal and right cerebellar lobes (= upper point of right temporo-cerebellar excavation)
28	Point of maximal curvature of right occipital pole

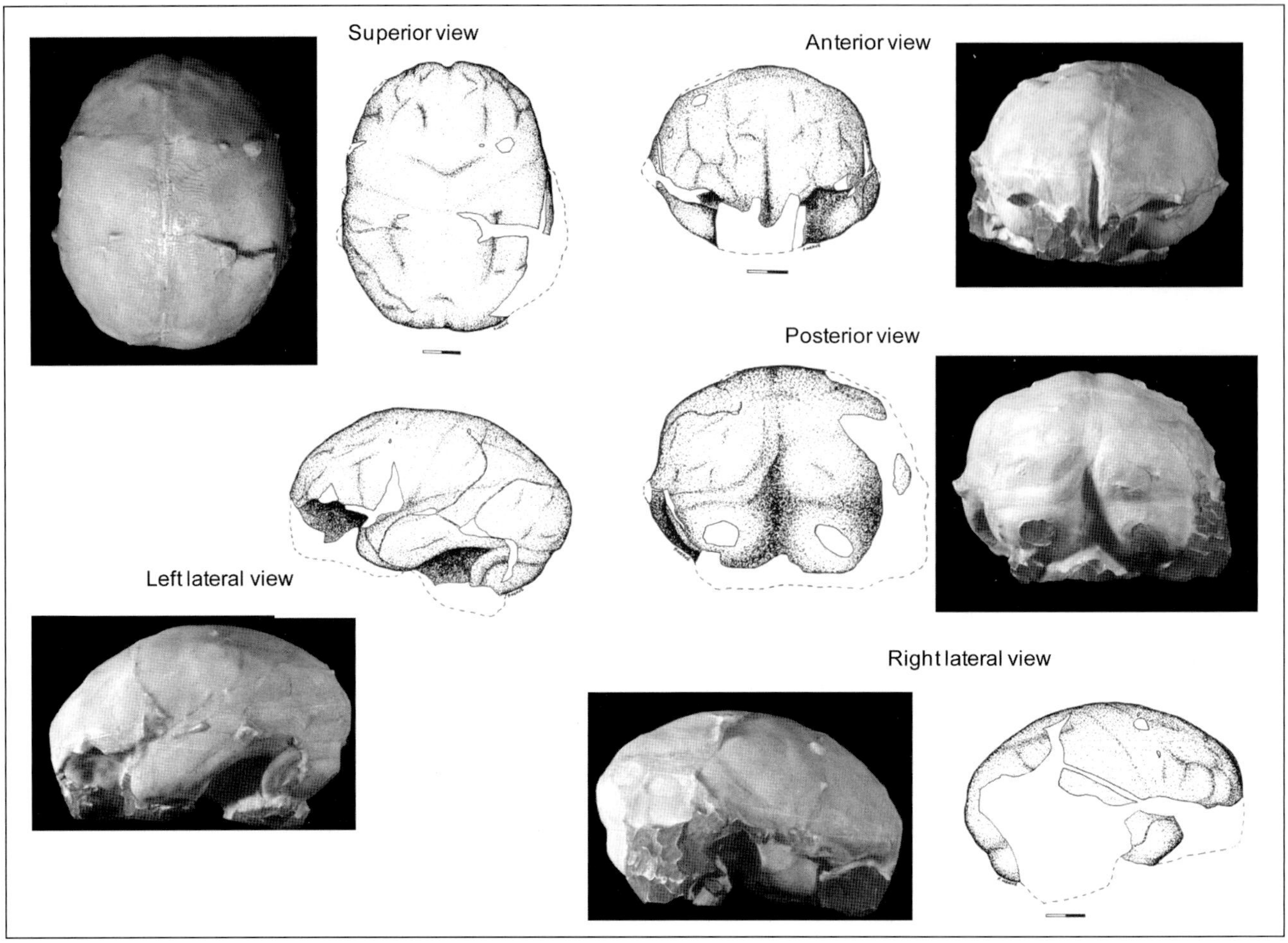

Figure 1. D 2280 – Endocast in left and right lateral, superior, anterior and posterior views

noid granulations are noted in superior part of precentral convolutions on D 2282, one on the right and two on the left hemisphere.

The middle meningeal pattern is very poorly represented. On left hemisphere of D 2280 it is best preserved with individualization of two branches on the second temporal convolution. The anterior meningeal artery appears reduced and disappears at the third frontal convolution. The posterior ramus is more developed, and is subdivided in both directions into an oblique and lambdoidal branches. Neither ramifications nor anastomoses are observed on the Dmanisi's endocasts. On D 2282, just the superior middle meningeal branches are apparent on left hemisphere. The middle meningeal pattern exhibits plesiomorphies, but are poorly patterned with the absence of a spheno-parietal sinus in both Dmanisi's endocasts.

Encephalic relief

The longitudinal cerebral fissure is wide, resulting in a significant separation between the frontal lobes in both Dmanisi's endocasts. The longitudinal cerebral fissure displays a significant separation between the hemispheres until posterior hemispheric. The junction with perpendicular fissure constitutes a depression forward of endolambda. On the left hemisphere of D 2280 the lateral fissure is weakly impressed and inclined up and backward, becoming straightened at its extremity; it is also observed on the left side of D 2282. The lateral fissure is situated in the prolongation of the lateral valley, which is very wide, separating Broca's cap from the temporal pole. The junction between the central fissures is situated behind endobregma (35mm on D2280; too deformed on D 2282). Based on Holloway (1982) there is a left frontal petalia on D 2280 and a right right frontal petalia on D 2282, confirming the asymmetry observed on vascular pattern.

The precentral sulcus is weakly impressed, but detectable about 15mm in front of the central one, which is the breadth of precentral gyrus on both Dmanisi's hominids. A long and narrow encephalic rostrum is clearly individualized on D 2280 (this region is not preserved on D 2282) with the right and left first frontal convolution widely separated as noted before. The breadths of these convolutions appears equivalent. The anterior ramus of the central fissure is clearly impressed with individualized relief of the orbital part of the third frontal gyrus equal on both sides of D 2280. The left Broca's area shows the central fissure on D 2280 as well as the contralateral side in D 2282.

The postcentral sulcus is situated nearly 20mm behind the central one on D 2280. The postcentral gyrus is just little more developed than precentral. There was a smaller difference observed between these two struc-

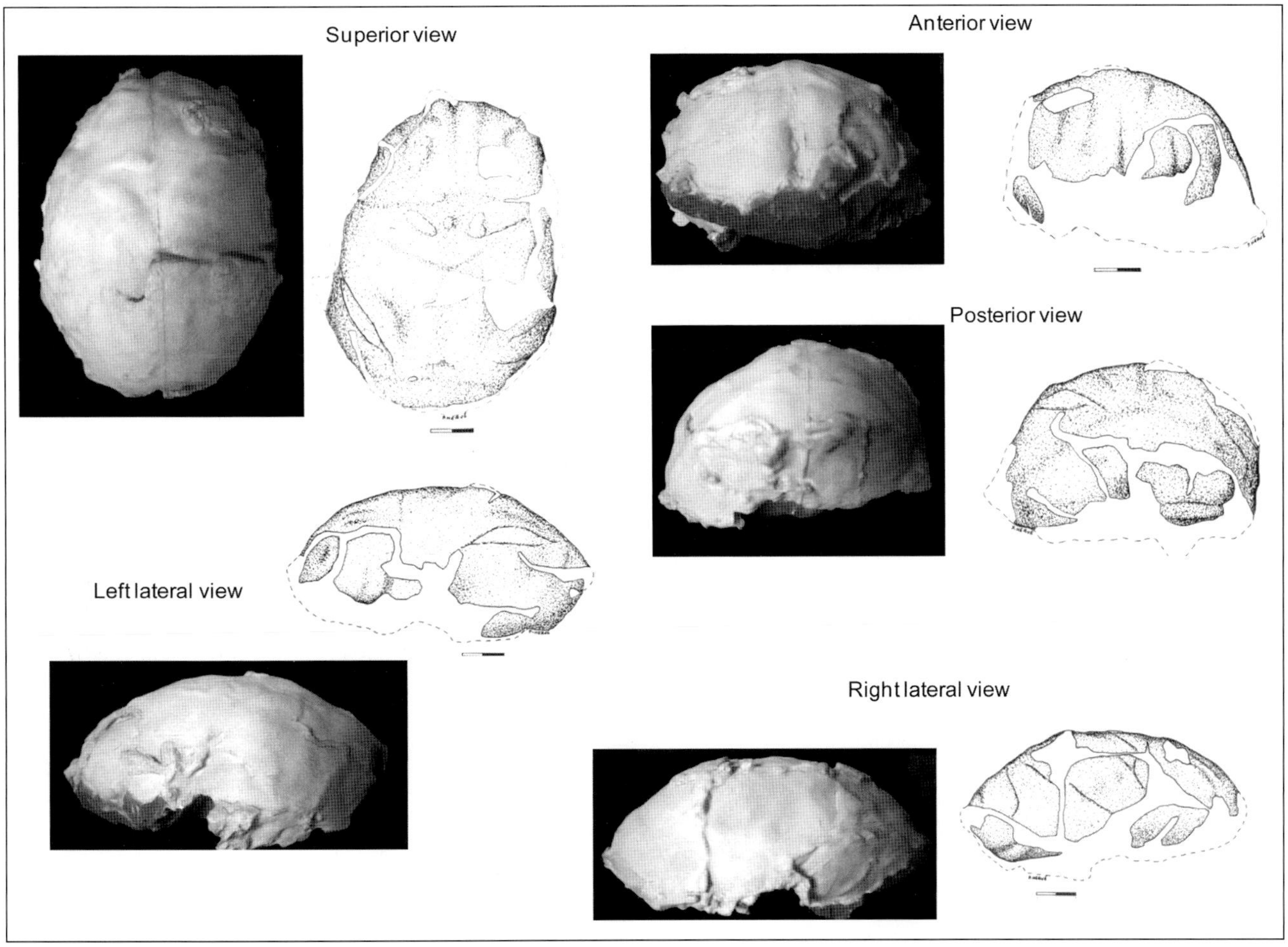

Figure 2. D 2282 – Endocast in left and right lateral, superior, anterior and posterior views

tures in D 2282. The breadth of first parietal convolution decreases posteriorly (28 to 20mm on D 2280, 20 to 15mm on D 2282). The supramarginal lobule is well delimited on D 2280, more than it is in D 2282. The angular lobule surface is smaller with weaker relief on D 2280 contrary to what is observed in D 2282. The temporal lobes converge towards the brain with the temporal pole set back behind Broca's area on D 2280. The temporal poles are not preserved in D 2282. In addition, the occipital region in this hominid is too altered to provide any data. In D 2280, *sulcus lunatus* is in the posterior position commonly observed on genus *Homo*.

Left occipital lobe is in a set back position compared to the right on D 2280 with the reverse being the case in D 2282. Instead the occipital lobes of D 2282 are extensions of the temporal and parietal lobes without a clear boundary. The cerebellar lobes are situated under the occipital ones. These occipital and cerebellar posterior positions are the expression of weak cerebral rolling and opening of the basi-cranial angle, which are primitive conditions commonly observed in contemporaneous fossils hominids.

Morphological comparison

Vascular imprints

Any fossil endocast studied shows a spheno-parietal sinus which is scarce in the modern human sample. The superior sagittal sinus asymmetry corresponds with greater development of one hemisphere (Delmas & Chifflet 1950). When this posterior cerebral region is preserved, two patterns appear: In the first one, the superior sagittal sinus goes into the left transverse sinus which is generally larger than the right, like on OH 24 (Tobias, 1991), Sangiran 2 and 10 (Grimaud-Hervé and Saban 1996) and D 2280. In the second pattern the sagittal sinus goes into the right transverse sinus on D 2282, OH 13 and OH 16 (Tobias, 1991), ER 3883 and WT 15000 (Begun & Walker, 1993), Trinil 2, Sangiran 12 and 17, hominids from locus E and L of Zhoukoudian Lower Cave, and 80% of the modern sample (Grimaud-Hervé 1997, 2004). This latter group shows a more developed left hemisphere of significant length.

No relationship has been established between the meningeal system and the venous sinuses (Paturet 1964). The results of both parts are going to be treated independently. For the middle meningeal pattern, two groups appear. In the first one with the posterior branch is very

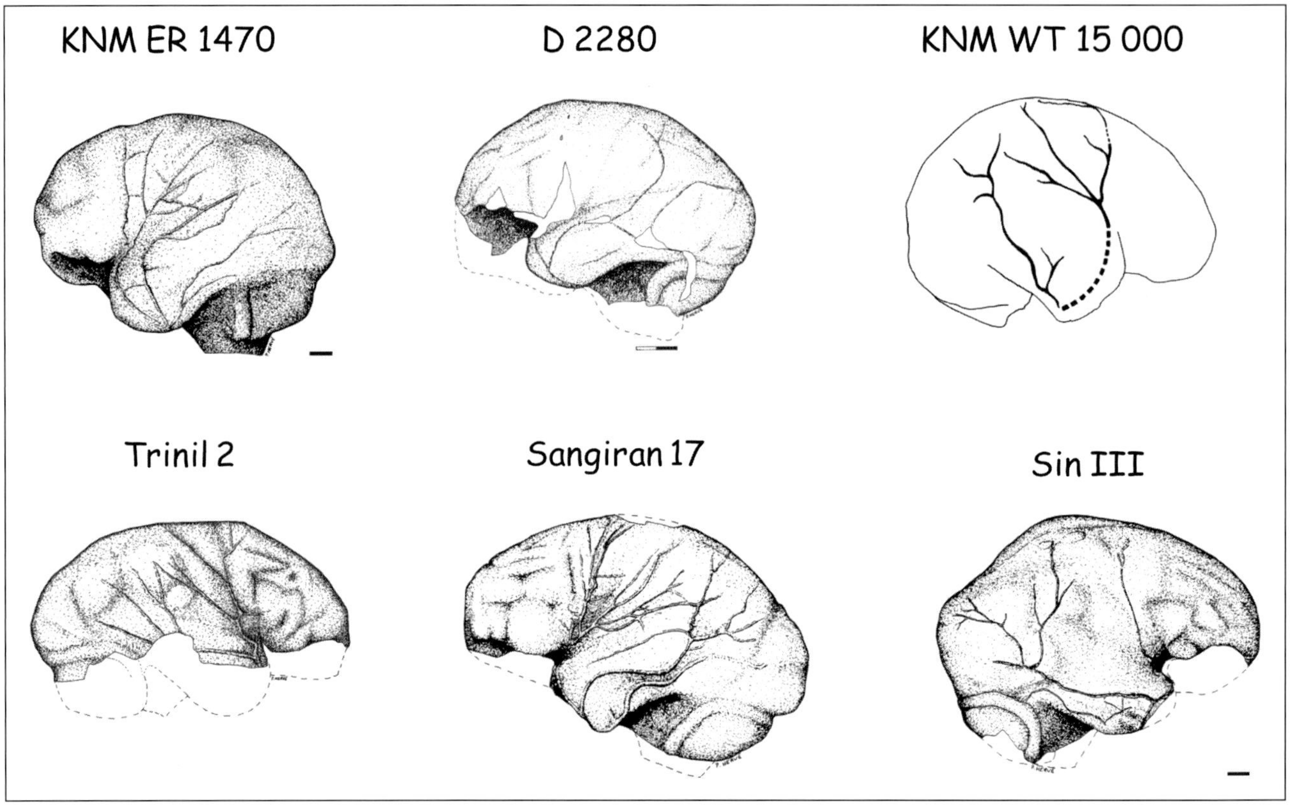

Figure 3. Middle meningeal pattern

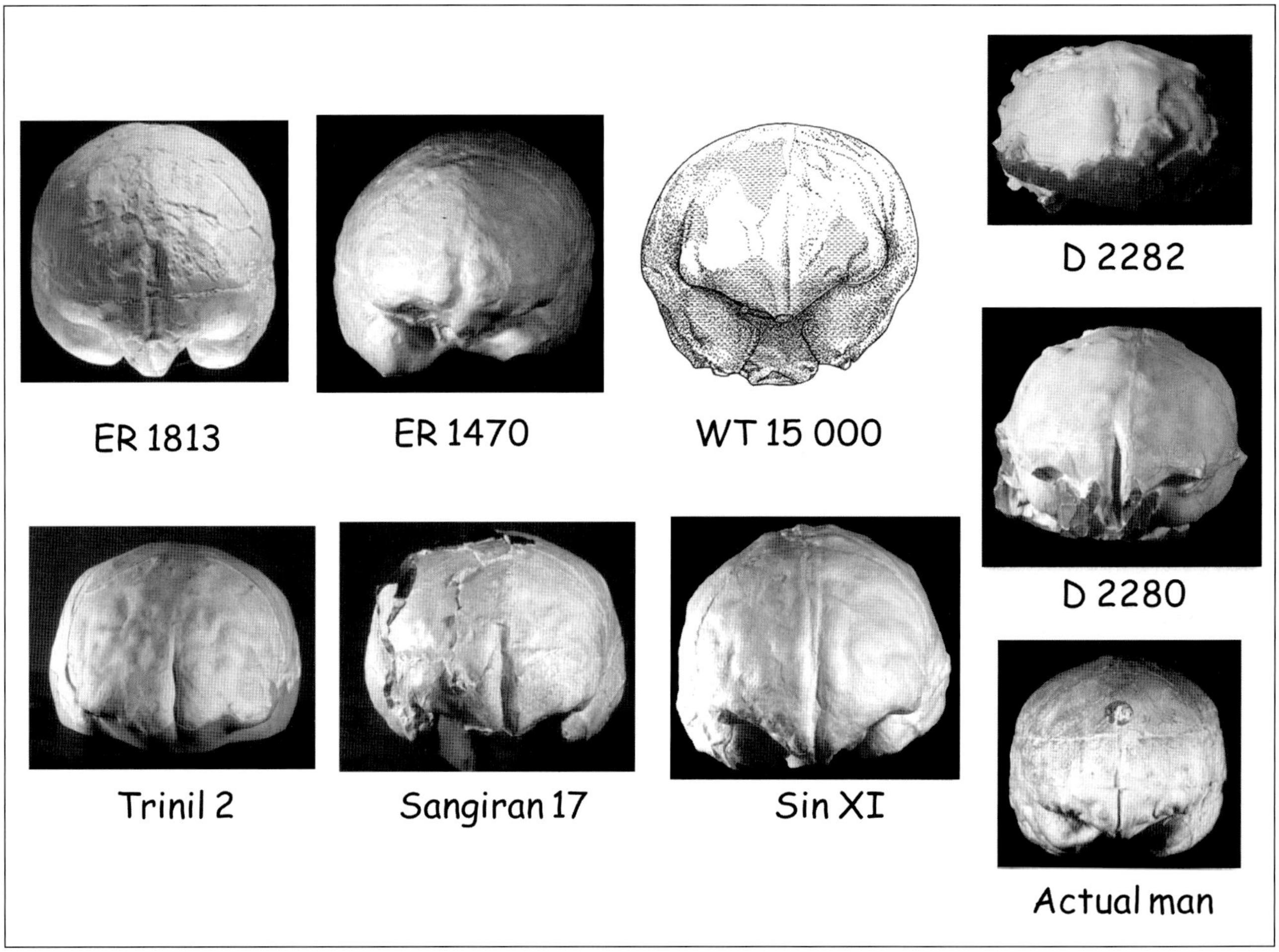

Figure 4. Endocranial transversal shape in anterior view

developed as exemplified in OH 7, D 2280 and Zhoukoudian Lower Cave hominids. In the second with the posterior branch is equivalent or reduced in prominence compared with the anterior middle meningeal branch such as is observed in ER 1813, ER 1470, ER 3883, WT 15000, Trinil and Sangiran hominids, and a great majority of the modern human sample (94%) (Fig.3). Although the differences between this sample and others are not significant, it appears that the morphology of the posterior branch of the middle meningeal artery is a plesiomorphic character.

Encephalic imprints

The encephalic rostrum, which is considered a plesiomorphic feature, has been observed on all studied fossil hominids where this region is preserved. It corresponds to the extension of the left and right first frontal convolutions next to the orbital roofs, which are separated by a wide, deep longitudinal cerebral fissure, particularly on the Dmanisi and Asiatic endocasts. Nevertheless, the outline varies according to region such that it is regularly concave on all African hominids, whereas it consists of two parts separated by an angle on Asiatic hominids and Dmanisi. This morphological character is very well individualized on that second group. A clear regression of the encephalic rostrum combined with a narrower interorbital space is observed on the modern sample. This may indicate a reduction in the importance of olfaction in the modern sample versus earlier fossils.

In the anterior view the same distinct group differences are observed. African *Homo ergaster*'s endocranial outline, in particular, is narrow and high, and is regularly convex from left to right in the orbital part of third frontal convolutions without an interruption in the transversal cerebral curvature. The outline of D 2280 is closer Asiatic *Homo erectus* with a wider and lower endocranial transversal shape. It is less marked in its convexity, and the thrd frontal convolution is interrupted in the medial region of the frontal lobes. The parietal lobes are interrupted by a vast depression corresponding to the middle frontal sulcus as it runs into the interparietal. The outline of this feature shows a slight twist similar to that observed on all fossil hominids of this group (Fig.4).

The frontal and parietal areas are smoothly limited on the African specimens compared with the more marked relief in Dmanisi, particularly D 2280 as well as the Asiatic hominids. The lateral valley is very broad in OH 24, ER 1813 and ER 1470 as well as *Homo ergaster*, while it is less broad in Asiatic *Homo erectus* and D 2280, meaning there has been greater development of the frontal and temporal lobes in these latter groups. In moderns the two areas so little if any separation in moderns, indicating a closer affinity of the Dmanisi and Asian fossils to moderns than African fossils. However, all of the fossil hominids show the frontal lobes converging anteriorly, while they remain parallel in modern human sample.

The breadths of the pre- and postcentral convolutions tend to correspond to particular evolutionary stages. It is notable that the postcentral gyrus is nearly always broader than the precentral in the endocasts sampled here. In the African hominids the postcentral convolution is equal to precentral one (OH7, OH13, OH16, OH24, KNM ER 1470, KNM ER 3883, KNM WT 15000), or slightly wider (KNM ER 3733). In Asiatic *Homo erectus* these two convolutions have either equivalent breadths, or the precentral gyrus is more developed, this phenomenon is more accentuated in the more recent Javanese fossil hominids from Ngandong and Sambungmacan. This indicates that there is an increase in motor areas between the early and later fossils. With regard to this morphological feature the Dmanisi endocasts are similar to KNM ER 3733. Thus, this derived feature is found on the two first representative hominids inside and outside of Africa.

In the lateral view it is noteworthy to note that the posterior encephalic shape differences between the African sample and the Asiatic hominids and D 2280. In the African sample the occipital lobes bulge only slightly beyond the cerebellum posteriorly, and are not delineated anteriorly from the parietal lobe. In the second group the occipital lobes are clearly projecting backwards, their anterior part situated above cerebellar lobes. This position corresponds to a rotation of the cerebellum under the cerebral mass, and the role of basicranial flexion in the position of the cerebrum through time.

In the superior view the D 2280 (D 2282 is too damaged) endocranial outline is very different from African *Homo ergaster*. In *H. ergaster* the anterior frontal region is strong in relation with the lateral cerebral convergence, showing slight cerebral development at this level. The D 2280 outline is globular, and is closer from the Asiatic *Homo erectus* shape with a noticeable widening of cerebral region from the lateral valley to the orbital part of third frontal convolutions.

Results of Metrical Study

Cranial capacity

The average of three tests corresponding to immersion of the endocast in water is presented in Table 1. These measurements present different values from those observed n the Dmanisi hominid endocranial capacities as measured using mustard seed (Gabounia *et al.* 2000, Lumley *et al.* 2006).

The cranial capacity values (Table 3) show a regular increase from more the ancient African and Dmanisi hominids to Asiatic *Homo erectus* and modern humans (Fig.5). The very small cranial capacity of D 2282 appears to be the result of damage. The measurement presented here is approximately 100ml less than previous measurements, placing this individual within the range of African *Homo habilis*, while D 2280 is towards to lower end of Asiatic *Homo erectus*. The encephalization quotient, recommended by some researchers (Armstrong

Table 3. Cranial capacity in ml

Fossil hominids	Direct method	References
D 2280	790	775 (Gabounia et al., 2000) 770 (Lumley M.A. de et al., 2005)
D 2282	(645)	650 (Gabounia et al., 2000) 625 (Lumley M.A. de et al., 2005)
D 2700		645 (Vekua et al. 2002), 600 (Lee, 2005)
D 3444		625 – 650 (Lordkipanidze et al. 2006)
OH 7	-	674 (Tobias 1975, 1991)
OH 13	618	673 (Tobias 1975, 1991)
OH 16	-	638 (Tobias1975, 1991)
OH 24	556	594 (Tobias 1975, 1991)
KNM-ER 1813	500	510 (Holloway, 1978)
KNM-ER 1470	760	752 (Holloway, 1978)
KNM-ER 3733	715	848 (Holloway, 1983)
KNM-ER 3883	785	804 (Holloway, 1983)
KNM-WT 15000	885	880 (Begun et Walker, 1993)
OH 12	656	
OH 9	1118	1067 (Holloway, 1975)
Bouri		995 (Asfaw et al., 2002)
Buia		800 (Abbate et al., 1998)
Sale		930-960 (Jaeger,1975), 880 (Holloway,1981a)
Trinil 2	930	943 (Holloway, 1975)
Sangiran 2	840	815 (Holloway, 1975)
Sangiran 10	840	855 (Holloway, 1978)
Sangiran 12	-	1059 (Holloway, 1978)
Sangiran 17	960	1004 (Holloway, 1978)
Ckn.D 1.PA.17 - Sin II	995	
Ckn.E 1.PA.16 - Sin III	915	915 (Weidenreich, 1943)
Ckn.L 1.PA.98 - Sin X	1245	1225 (Weidenreich, 1943)
Ckn.L 2.PA.99 - Sin XI	1020	1015 (Weidenreich, 1943)
Ckn.L 3.PA.100 - Sin XII	1020	1030 (Weidenreich, 1943)
Homo sapiens (n=103)	$\mu = 1520$	Min=1190; Max=1940; VarCoef=11

1985, Bauchot et Stephan 1969, Hartwig-Scherer 1993, Holloway and Post 1982b, Jerison 1975, Mac Henry 1976, Martin 1990, 1995, 1996, Rightmire 1986, 2004, Ruff *et al.* 1997, Rosenberg *et al.* 2006, Tobias 2006), has not been calculated due to lack of data on stature.

Univariate dimensions (Table 4)

Univariate distributions are reported for both Dmanisi's endocasts, each hominid fossil group and for modern humans. Concerning maximal hemispheric lengths, it is interesting to note important differences between both ancient African hominids from East Turkana, ER 1470 and 1813. KNM ER 1813 is closer to the Olduvai specimens OH 16 and OH 24 than KNM ER 1470. However, both Dmanisi endocasts are similar to African *Homo ergaster* with distributions between 138.5 and 154 mm. As noted before (Grimaud-Hervé et al. 2006), Asiatic *Homo erectus* values are large when compared with Zhoukoudian Lower Cave hominids, which are near the modern human range. From *Homo habilis* to *Homo sapiens* the overall increase is 35.2%.

Maximal endocranial breadth (WME) is situated in the postero-inferior position on the second temporal gyrus in African and Asiatic fossils hominids, and is also found in the same poistion in the Dmanisi specimens, while it is positioned around the first temporal convolution in *Homo sapiens*. This endocranial measurement is scarcely joined with the endobiparietal maximal breadth (WBE), which is positioned around the supra marginal gyrus. WME and WBE are reduced in ER 1813 compared to ER 1470 (respectively 18.5mm and 25 mm). With the Dmanisi specimens D 2280 is closer to ER 1470 as well as smaller Asiatic hominids such as Ckn.D 1.PA.17 or Sangiran 2. D 2282 is closer to the Olduvai hominid values. The overall increase from *Homo habilis* to *Homo sapiens* for this measurement is between 38.7% and 33.8%.

Table 4. Univariate dimensions

	LME	WME	WBE	WPPE	WCBE	WPFE	HBBE	HGQE	HBRE	CFR	CPA	COC	XBE	DOCE	XLTE	XLSE	XLIE
D2280	145		105		88	98	95	57.75	57.25	101	50	48	46	11.75	151.75	74.75	77.25
D2282	140	100	96		80	84	75	42.25	40.25	94	44	45	38.75		114.5	56	58
H.habilis																	
N	4	5	5	5	2	2	2	4	4	3	3	2	1	3	1	1	1
Average	127.4	103.1	93.6	94.8	78.0	83.5	90.0	45.0	40.8	64.3	47.3	38.0	45.5	11.1	145.0	67.5	77.5
Var Coef	3.6	4.0	5.7	7.7	7.3	9.3	1.6	4.8	9.4	27.9	12.9	11.2		81.2			
ER1470	137.5	114.5	111	109	84	92	103	52.25	45.5	97	48	44	50.5	6.75	126.5	62	64.5
H.ergaster																	
N	3	3	3	2	2	1	3	3	3	3	3	3	3	3	3	3	3
Average	147.2	119.2	115.5	112.0	87.0	70.0	96.9	52.2	47.7	91.7	51.0	52.3	41.4	3.8	157.8	75.5	81.9
Var Coef	5.4	1.2	1.1	3.8	5.0		5.2	5.3	5.1	6.7	9.0	9.6	12.4	27.2	6.1	5.4	11.0
H.er.Java																	
N	4	5	5	5	4	4	3	5	5	4	5	5	4	5	5	5	5
Average	154.4	124.0	114.0	111.8	90.4	92.5	103.3	58.0	54.7	109.8	43.8	52.8	45,7)	11.5	124.9	60.8	64.1
Var Coef	4.1	4.2	4.6	4.8	4.9	11.5	3.0	3.6	1.8	2.0	14.2	6.7	4.4	19.9	6.1	6.3	9.2
H.er.China																	
N	5	5	5	5	3	3	5	5	5	5	5	4	5	4	4	4	4
Average	165.7	124.0	112.6	111.8	95.3	100.3	112.0	61.4	57.8	119.6	47.4	59.5	43.0	14.6	120.9	58.0	62.9
Var Coef	3.6	3.8	4.7	5.6	3.0	4.0	4.4	5.0	5.0	4.2	8.3	10.7	5.4	25.2	6.5	4.6	8.5
Extinct humans																	
Nb Spec.	105.0	105.0	105.0	104.0	105.0	105.0	104.0	104.0	104.0	104.0	105.0	105.0	27.0	103.0	26.0	26.0	26.0
Average	170.5	136.5	132.1	130.4	105.1	116.6	121.2	66.6	58.8	122.5	73.3	59.1	51.9	9.6	136.3	68.2	68.1
Var Coef	4.9	5.7	5.5	5.8	7.1	6.9	5.0	12.2	14.1	5.3	11.8	10.7	7.4	43.7	6.8	7.4	6.6
Maximum	187.3	152.0	150.0	147.5	127.0	138.0	136.0	88.8	85.0	137.5	88.0	77.0	59.8	20.0	151.3	75.3	76.0
Minimum	148.5	116.0	114.0	108.0	93.0	100.0	105.0	52.0	44.0	44.0	56.0	58.0	44.0	1.3	116.5	56.5	60.0

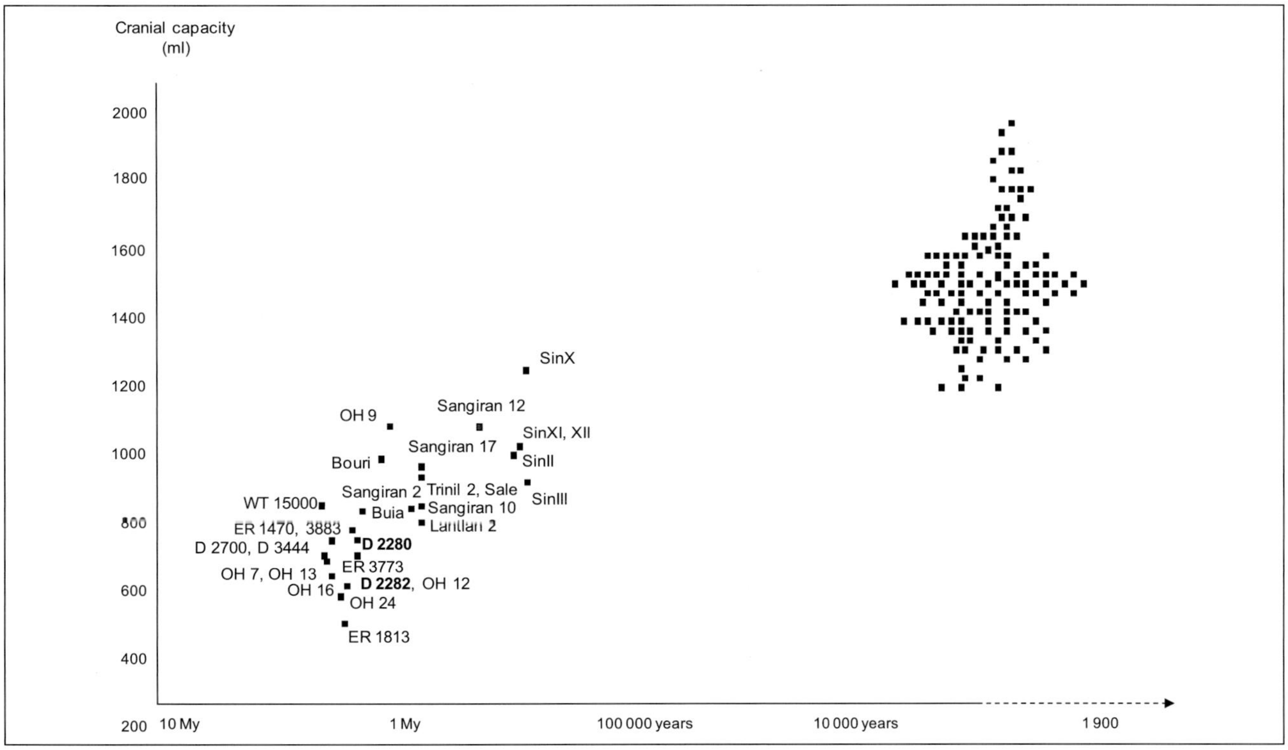

Figure 5. Cranial capacity

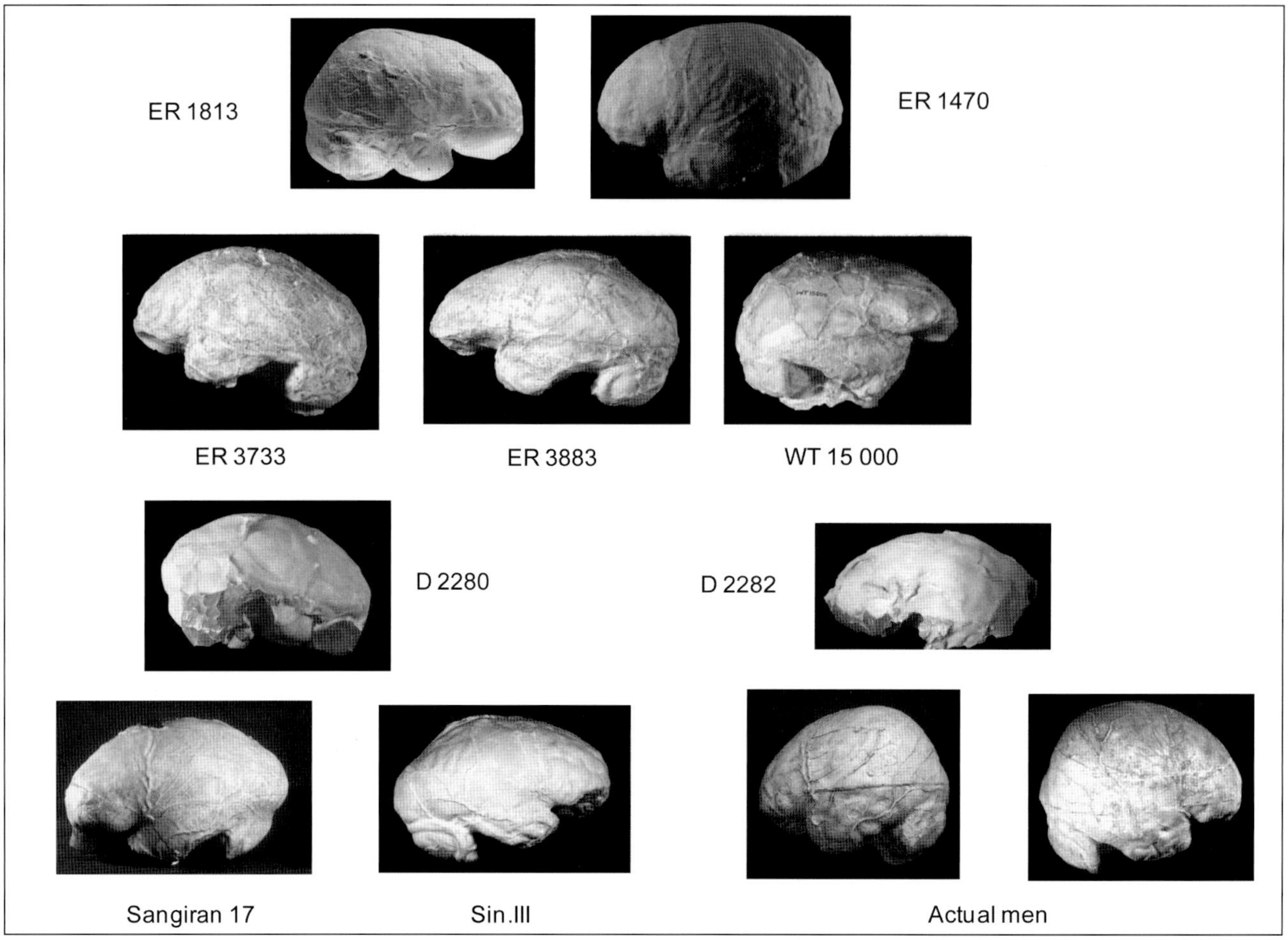

Figure 6. Occipito-cerebellar projection

With regard to the opercular part of third frontal convolutions the difference between ER 1813 and ER 1470 breadths (WPFE) is again similar to (78 and 92 mm) those of Dmanisi hominids (98 et 84 mm). ER 1813 and D 2282 are closer to OH 24 (82mm). D 2280, though, is closer Asiatic *Homo erectus* with regard to this feature, displaying an increase in the posterior frontal region not observed in African specimens. Growth between *Homo habilis* and extinct *H. sapiens* reaches more than 40%. The maximal breadth on the triangular part (WCBE) is close to the opercular part in African hominids, but is situated lower, becoming more central, in *Homo sapiens*. An important difference is noticed between ER 1813 and 1470 when compared to the Dmanisi hominids. In this measurement D 2280 is near *Homo ergaster* and Asiatic *Homo erectus* Sangiran-Trinil averages. Zhoukoudian Lower Cave are larger in comparison. An increase of 33% is noted between the more ancient fossil hominids of this study and *Homo sapiens*.

Total height (endobregma-endobasion = HBBE) measurement requires the preservation cerebral regions that are rarely preserved. D 2280 is similar to *Homo ergaster*'s, but smaller than Asiatic *Homo erectus* (Javanese = 103.3mm, Chinese = 112mm). There is an increase of about 35% between *Homo habilis* and *sapiens*.

Partial height corresponding to the upper part of the brain (HGQE) is used when the basal part of the brain is not preserved. This maximal point above endoglabella-endoopisthocranion is situated between endobregma and the central fissure. In general, this measurement is positioned more anteriorly in fossil hominids, but is positioned more posteriorly in more recent hominid groups. In Dmanisi D 2282 does not provide reliable data with regard to this measurement, while D 2280 does. D 2280 is closer to WT 15000 and Javanese *Homo erectus*, while the Zhoukoudian hominids (61.4mm) are closer to *Homo sapiens* (77.3mm). A major increase in height is observed in upper portions of the hemispheres (approximately 50%) from *Homo habilis* to *Hom sapiens*.

Concerning the occipito-cerebellar projection, there is a very clear distinction between the fossil hominids in this sample (Fig.6). In general, the regular sagittal curve of African specimens results in less overlap with the occipital obes when compared with Asian *H. erectus* specimens, which have protruding lobes. The occipital poles of African hominids from East and West Turkana only project slightly (average = 3.8mm). There is greater occipital projection in the nine Asian specimens value (average = 12.8mm). The occipital projection in D 2280 (D 2282 is too damaged) is similar to other *Homo erectus* (average = 11.8mm). Finally, the average in *Homo sapiens* is 9.65mm, but with greater variation near 50%.

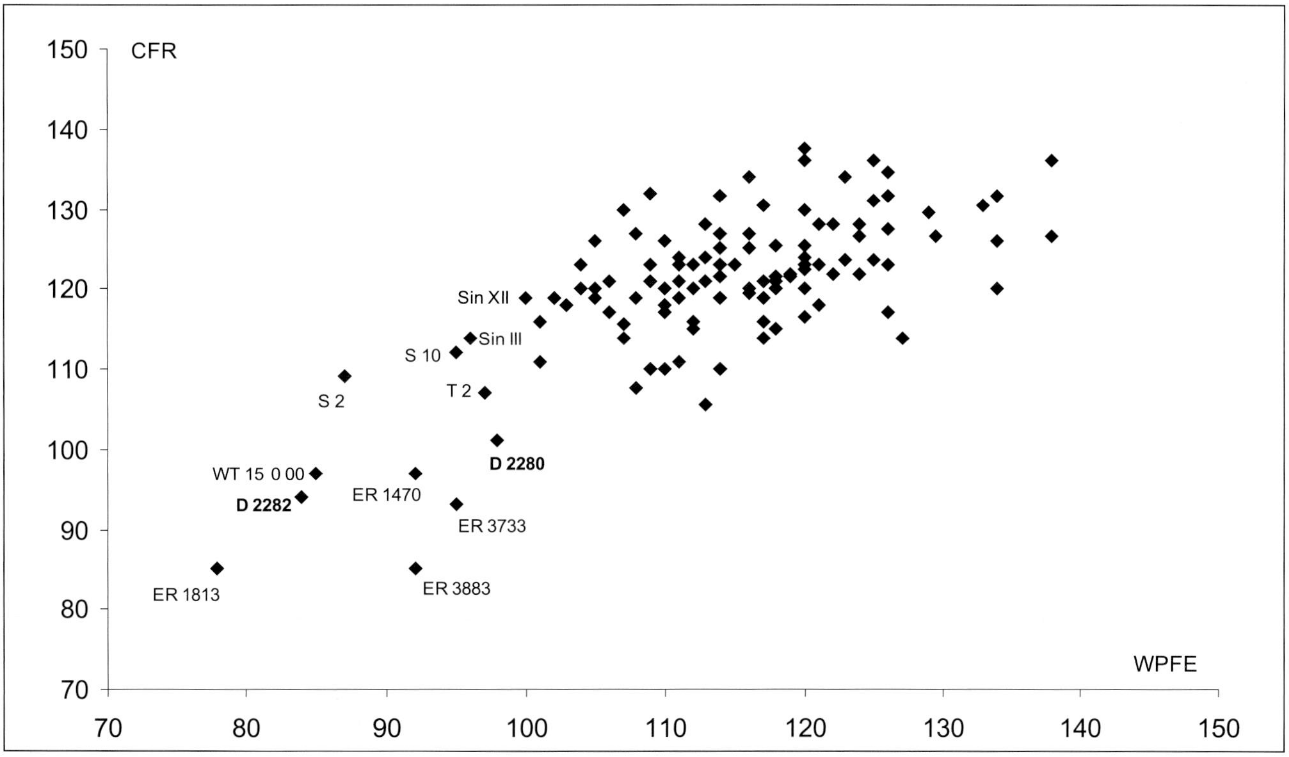

Figure 7. Ratio of frontal chord (CFR) on F3 opercular part (WPFE) of the brain

Thus, Dmanisi D 2280 is clearly different from the African configuration, and is closer to that of *Homo erectus*, confirming similar configurations of this posterior region in Georgian and Asian endocasts.

In conclusion, the Dmanisi values of maximal hemispheric length are more similar to those of *Homo ergaster* than other specimens. Yet, with concern to maximal breadths D 2282 is closer to *Homo habilis*, while D 2280 is closer to *Homo ergaster* and gracile Asian *Homo erectus*. D 2280 exhibits significant frontal widening similar to Asiatic *Homo erectus*, a feature not observed in African specimens. This increase occurs more anteriorly in the triangular parts of third frontal convolutions. Finally, D 2280 is more similar to *Homo ergaster* with regard to total endocranial height, but is closer to Asiatic *Homo erectus* if only the upper region of the hemisphere is considered. Its occipito-cerebellar projection emphasizes the same pattern as Asiatic hominids. Thus, there is a significant distinction between African specimens one hand and the Dmanisi specimens, which tend to associated more with the Asiatic hominids, in particular with regard to the protrusion of the occipital lobes.

An analysis of the sagittal chord of each cerebral lobe (frontal = CFR, parietal = CPA and occipital = COC) displays an increase of 49% in the frontal and occipital lobes and 56% for the parietal lobes. The parietal sagittal chord is equal or superior to the occipital one in African specimens, *Homo sapiens* and the Dmanisi hominids. In Asiatic *Homo erectus* the occipital sagittal chord is always more developed which can be considered as an autapomorphy of this group.

With regard to the total sagittal curvature, the observed ratios are quite stable with a more marked increase in the parietal region between African specimens and *Homo sapiens*. Asiatic *Homo erectus* displays a primitive character with less development in the medial sagittal portion of the brain. This is possibly due to the longer occipital representation observed in this group. This is particularity specific to Asiatic *Homo erectus*, and, thus, can be interpret as an autapomorphy.

The frontal sagittal chord compared to the maximum frontal breadth (Fig.7) is over 100 in fossil hominids and *Homo sapiens*, meaning that the breadth is more developed than length. This index is nearly the same between *Homo ergaster* and *H. sapiens*, while all Asiatic *Homo erectus* and Dmanisi possess a higher index, indicating reduced transversal development of the posterior frontal region. This new result confirms the morphological distinction of the fossil samples. The parietal sagittal chord compared to the maximal parietal lobe width is not significant. Instead the values merely distinguish fossil hominids from *Homo sapiens*, indicating an increase in parietal lobe development in this later group.

Bivariate dimensions (Table 5)

The ratio between maximal endocranial length and breadth shows the same repartitions between D 2282 and D 2280, which are distributed between African and Asiatic specimens. There is an apparent increase in length from ancient to more recent *Homo erectus*, but then this measurement remains stable throughout *Homo sapiens*. The breadth values are more similar between the groups, albeit the relative position of maximum breadth on the endocast varies (Fig.8).

	WME/ LME	WBE/ LME	HBBE/ LME	HBBE/ WBE	HGQE/ LME	HGQE/ WBE	CFR/ WPFE	XBE	XLTE	XLSE	XLIE
D2280			65.5	90.5	39.8	55.0	103.1	46	151.75	74.75	77.25
D2282	71.4		53.6	78.1	30.2	44.0	111.9	38.75	114.5	56	58
H.habilis											
N	4	4	2	2	4	4	1	1	1	1	1
Average	80.4	72.5	71.4	101.8	35.3	49.0	109.0	45.5	145.0	67.5	77.5
Var Coef	5.5	6.7	1.4	5.6	3.9	9.5					
ER1470	83.3	80.7	74.9	92.8	38.0	47.1	105.4	50.5	126.5	62.0	64.5
H.ergaster											
N	3	3	3	3	3	3	3	3	3	3	3
Average	81.2	78.7	65.9	83.9	35.5	45.2	101.5	41.4	157.8	75.5	81.9
Var Coef	6.4	6.6	4.8	5.9	6.1	5.7	11.1	12.4	6.1	5.4	11.0
H.er.Java											
N	4	4	3	3	4	5	4	4	5	5	5
Average	79.8	73.0	67.2	92.6	37.5	50.8	119.9	45.7	124.9	60.8	64.1
Var Coef	1.3	1.2	2.8	2.7	2.9	4.4	12.4	4.4	6.1	6.3	9.2
H.er.China											
N	5	5	5	5	5	5	3	5	4	4	4
Average	74.9	67.9	67.6	99.5	37.1	54.6	116.3	43.0	120.9	58.0	62.9
Var Coef	2.9	2.7	2.6	3.9	2.5	4.7	1.8	5.4	6.5	4.6	8.5
Extinct humans											
Nb Spec.	105	105	104	104	104	104	105	27	26	26	26
Average	80.2	77.7	71.3	71.3	39.2	50.4	105.4	51.9	136.3	68.2	68.1
Var Coef	7.5	701	6.2	6.2	13.3	10.8	6.8	7.4	6.8	7.4	6.6
Maximum	92.8	92.2	85.5	85.5	54.5	66.2	121.5	59.8	151.3	75.3	76.0
Minimum	66.1	64	61.5	61.5	29.1	40	85.6	44	116.5	56.5	60

Table 5. Bivariate dimensions

The relationship between maximal endocranial length and height was determined using the total height of the endocast when available or the upper part of the hemisphere when the basal portion was missing. The results place Dmanisi in an intermediate position between African *Homo ergaster* and Asiatic *Homo erectus*. With regard to breadth the Dmanisi fossis lump with the African and Asian fossils. Bregmatic angle (between maximal length and the intersection with endobregma: XBE) is similar compared with the other fossils, while this measure is slightly wider in *Homo sapiens*, implying the development of frontal rounding and recurving inferiorly through time towards *H. sapiens*. Also, it appears that from this sample there is an increase in frontal convexity through time

The occipital angle was measured in lateral view (between endolambda – endoopisthocranion and endoopisthocranion-superior edge of transverse sinus = XLTE). This angle is opened in ER 1813 (145°), similar to that of *Homo ergaster* (157.8°) and D 2280 (151.8°). However, values from Asiatic *Homo erectus* are lower with the angle being more closed between the superior squama and nuchal plane of occipital bone, a feature unique to this group.

The inferior angle (XLIE) is more opened than the superior one (XLSE) on all studied fossil hominids. This supports other morphological results where the weak superior angle accentuates the position of the occipital lobes relative to the extension of parietal and temporal lobes posteriorly. In *Homo sapiens*, these two angles are nearly equal implying an increase of the superior part of the lobe in relation to the occipital lobes, which are in a lower and more inferior position in comparison with the parietal lobes.

In conclusion, the index shows a symmetrical increase in the length, width and height of the brain through time. The short height of the superior part of the brain identifies fossil hominids as platyencephalic. In addition, the endocranial angles analysis shows a steady widening of the bregmatic angle compared to the occipital angle in African sample, albeit the reverse is true in Asiatic and Dmanisi hominids. All of the fossils have an inferior angle that is more open than the superior one, a condition not observed in *Homo sapiens*.

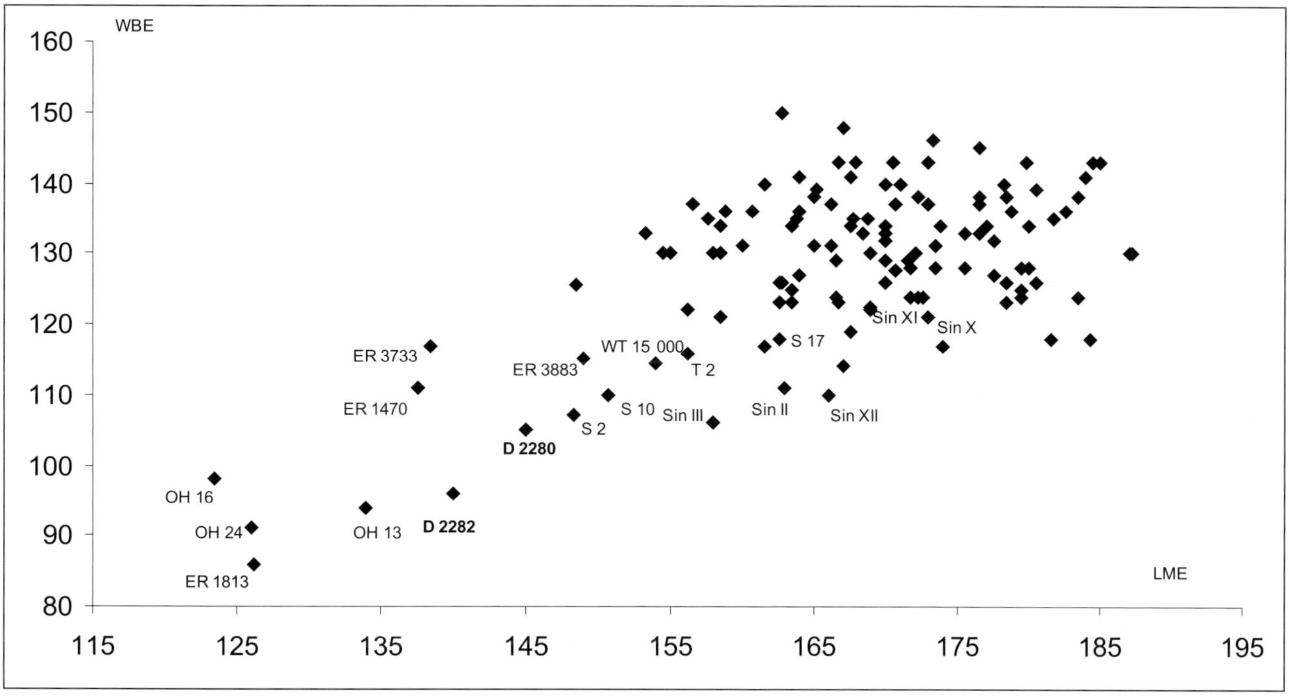

Figure 8. Ratio of the biparietal width (WBE) on the maximal length average (LME)

	SET	RH	LH
D 2280	(306,3)	(162,8)	143,5
Homo habilis (n)	260,8 (3)	131,5 (3)	129,3 (3)
ER 1470	313	153	160
Homo ergaster (n)	310,5 (n=2)	157,7 (n=2)	158,3 (n=3)
Homo erectus Java (n)	342.1 (n=2)	175,4 (n=2)	166,7 (n=3)
Homo erectus China (n)	372,3 (n=3)	187,8 (n=3)	184,5 (n=4)
100 *H.sap.* (Var Coef)	494,6 (7,7)	248,4 (8,3)	246,2 (7,6)

Table 6. Tab. 6 : Total endocranial surface (SET), right (RH) and left (LH) hemispheric areas

Endocranial surfaces

D 2282 is too damaged with regard to this assessment; however, the preservation of D 2280 is more amenable to such a description. Data on the frontal lobes of D 2280 position it between those from African *Homo ergaster* on one hand and Asiatic *Homo erectus* on the other. The left parieto-temporal area as well as the occipital ones are within the lower limit of variation observed in African *Homo ergaster*. An important difference is observed between the right and left parieto-temporal areas of D 2280, which displays deformation in the inferior part of the right lobe, meaning right hemispheric and total brain measurements from this specimen are only descriptive.

Asymmetry

Endocranial asymmetry was observed on the fossil and modern samples. It is assumed that each hemisphere is devoted to particular tasks with the left one being more oriented towards learning and analyzing, for example, articulate language organization, and the right one being more specialized towards emotion and relational aspects. However, we must keep in mind that the two cerebral hemispheres complement each other in any task execution as emphasized in Schmidt-Nielsen (1998), Bruner (2003), Stout et al. (2000, 2007, 2008), Sherwood et al. (2003, 2008), and Holloway et al. (2004).

Asymmetry in endocasts is often attributed to a particular petalial pattern (Holloway, 1982a). For example leftward asymmetry or right-handedness is assumed from right frontal (RF) - left occipital (LO) petalia. This petalial pattern is normally attributed to species within the genus *Homo* (LeMay, 1976; Holloway et al., 1982b; Gilissen 2001).

In our sample, any geographic or chronologic association appears in the endocranial outline analysis among the studied human fossils. D 2280 shows left frontal and right occipital petalia similar to ER 1813, Sangiran 2,

	RFr	LFr	RPr	LPr	ROc	LOc
D 2280	58,7	58,5	(84)	67	20,1	18
Homo habilis (n)	47.2 (3)	46.5 (3)	64.7 (3)	64.7 (3)	19.7 (3)	21 (3)
ER 1470	56	55	76	81	21	24
H. ergaster (n)	54.5 (3)	54.8 (3)	80 (3)	78 (2)	23,2 (3)	23,7 (3)
H. erectus Java (n)	65,7 (3)	58,8 (4)	81,2 (3)	79,8 (3)	28,5 (4)	28,1 (5)
H. erectus China (n)	72,0 (3)	69,8 (4)	83,5 (3)	80,7 (4)	32,4 (4)	34,0 (5)
100 *H. sapiens* (Var Coef)	86.1 (12,1)	85.5 (11)	131.9 (10,2)	128.4 (8,8)	30.4 (18,2)	32.8 (19,2)

Table 7. Right and left frontal (RFr, LFr), Temporo-parietal (RTPa, LTPa) and occipital (Roc, Loc) lobes areas

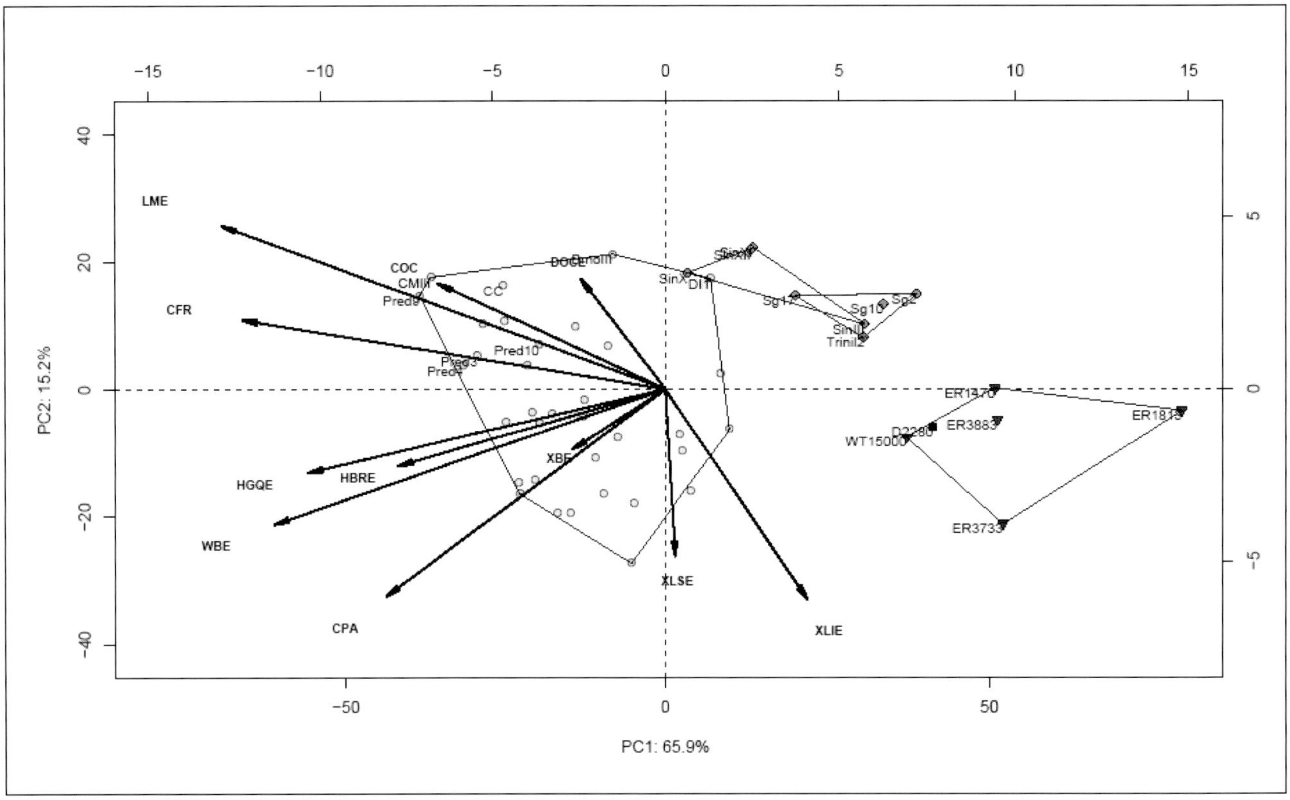

Figure 9. PCA 2D (11 variables and 47 specimens)

10, 17 and Ckn.L.2.PA.99. A right frontal – Left occipital petalia is observed in OH 13, OH 16, ER 1470, ER 3733, ER 3883, WT 15000, CKn.E.PA.16, Ckn.L.1.PA.98 and Ckn.L.3 PA.100 (Tables 6 & 7).

Results obtained from the *Homo sapiens* sample (n=100) are consistent with Gilissen's conclusion that a left occipital – right frontal petalia is common in modern humans (77%). It is a well-established fact now that the pattern of combined left-occipital and right-frontal petalias became more common in human evolution through time as asymmetrical lateral protrusions of the cerebral hemispheres became more common. Generally, brain asymmetries correspond with handedness. For example, left hemispheric dominance generally indicates right manual dexterity (Galaburda et al., 1978). However, recent studies do not support the association between this petalial torque pattern and right-handedness (Good et al., 2001; Hervé et al., 2006).

The conclusion of this analysis is that there is variability in observed asymmetry moving from early *Homo* through moderns, a feature confirmed by Holloway & De La Coste-Lareymondie (1982b), who stated that "this pattern is not consistently present in nonhuman primates or in hominid fossil brain endocasts until *Homo erectus*". Hemispheric asymmetry corresponding to lateralization phenomenon is essential in tracing evolutionary changes in brain organization and cognition. Comparison of the fossil hominid sample through metrical as well as morphological analyses demonstrates potential correlations with the behavioural stages of human evolution (Holloway, 1981b, 1982; Holloway et al., 2004; Chieze, 1983; Saban, 1984; Gilissen, 2001; Sherwood et al., 2008).

	LMDE	LMGE	HGQDE	HGQGE	HBRDE	HBRGE	WMDE	WMGE	XBDE	XBGE	DOCDE	DOCGE	XLTDE	XLTGE
D2280	146	144	58	57.5	57	57.5	66	57	46	46	11	12.5	151	152.5
D2282	140	140	41.5	43	39.5	41	48	52	39	38.5				114.5
H.habilis														
N	4	4	4	4	4	4	4	5	1	1	3	3	1	1
Average	127.8	127.1	45.8	44.3	41.6	39.9	52.8	49.6	48.0	43.0	11.5	10.7	148.0	142.0
Var Coef	3.6	3.6	5.7	3.9	11.1	7.7	7.2	3.4			87.3	75.2		
ER1470	137	138	52	52.5	45	46	52	59	50	51	5.5	8	129	124
H.ergaster														
N	3	3	1	3	1	3	3		1	3	1	3	1	3
Average	147.7	146.7	54.0	52.5	46.0	49.3	61.0		45.5	41.7	3.0	4.0	148.5	159.8
Var Coef	6.1	6.2		17.3		18.4				21.8		25.0		5.7
H.er.Java														
N	4	4	4	5	4	5	5	6	3	4	4	4	4	5
Average	154.8	154.1	57.5	58.4	55.1	54.2	62.6	61.4	44.0	46.4	12.6	10.3	125.0	124.4
Var Coef	4.3	3.9	4.1	3.1	1.1	2.4	6.2	2.5	2.3	4.8	18.4	21.3	8.5	7.7
H.er.China														
N	5	5	5	5	5	5	5	6	5	5	4	4	4	4
Average	165.4	166.0	59.8	63.0	56.2	59.4	61.7	62.3	42.5	43.4	14.8	14.4	119.3	122.5
Var Coef	3.4	3.9	5.1	4.9	5.8	4.2	2.4	6.1	6.3	5.0	16.3	34.2	3.4	9.7
Extinct humans														
Nb Spec.	105	105	104	104	104	104	105	105	27	27	103	103	26	26
Average	170.1	170.8	66.6	66.6	58.8	58.9	68.2	68.5	51.9	51.8	8.8	10.5	139.8	132.8
Var Coef	4.9	5.0	12.5	12.4	14.1	13.1	6.5	7.5	8.1	7.3	55.0	41.5	6.4	12.3
Maximum	187	189	90	87.5	85	76.5	79	88	61.5	60	21	21	161.5	156.5
Minimum	147	150	50	52	44	44	56	58	44	43	0	1	117.5	99.5

Table 8 : Asymmetry

Univariate metrics analysis

On all of the fossil and modern human samples (Table 8) the difference between left and right hemispheric length is always less than 4mm. There is not apparent trend for asymmetry to be directed more towards one side versus the other in African or Asiatic fossil hominids. A slight asymmetry is observed in D 2280. In modern humans (n=109), though, there is a slight trend toward eftward asymmetry (54.1%) (170.8mm, VarCoef=5mm) (13.8% equivalent and 32.1% right with 170.1mm, VarCoef=4.9).

Right maximal hemispheric width is slightly larger on all African fossil hominids, D 2280 and Javanese *Homo erectus* from Trinil and Sangiran. However, no real trend appears in Zhoukoudian Lower Cave sample or in moden humans (n=105; left width=68.5mm, VarCoef=7.5; right width=68.2, VarCoef=6.5)

Concerning the superior hemisphere height at endobregma or endovertex level, all African fossil hominids from Olduvai and the Turkana region, Asiatic *Homo erectus* from Sangiran and Trinil, both Georgian fossil hominids, and the modern humans sample do not show any particular trend. Only Zhoukoudian Lower Cave hominids exhibit a more elevated height (towards the left). However, the meaning of this last this result should be considered with caution, since the result is drawn from only five individuals.

No asymmetric trend has been observed concerning bregmatic angle opening on left or right hemisphere in fossil or modern samples. The right occipital angle opening is greater in KNM ER 1813 and 1470 as well as in modern humans (139.8° VarCoef=6.4, left=132.8°, VarCoef=12.3). However, the other endocasts studied here are distributed equitably between left and right asyemmetries. Similar results were observed for the superior and inferior parts of the occipital angle.

No particular trend appears concerning the occipito-cerebellar distance in modern humans, which instead shows tremendous variability. In the fossil hominid sample, the repartition is equivalent between both groups, confirming either a left or right petalia distribution without any chronologic or geographic association. No asymmetry appears in the well preserved frontal and occipital lobes surfaces of D 2280.

Cerebral relief and vascular imprints

The study of the variation of encephalic relief and vascular imprints provides useful data on asymmetries through time. The endocranial surfaces of Olduvai and Turkana (East and West) are incompletely preserved. According to Begun and Walker (1993) relief of the left frontal is more developed in KNM ER 1813, but less so in KNM WT 15000 and KNM ER 1470. KNM ER 3733 and 3883 are too damaged for comparison. All Asiatic *Homo erectus* specimens from Sangiran, Trinil (Java) and from Zhoukoudian Lower Cave (China) have less relief in left frontal lobes. In moderns 72% exhibit left frontal relief against 16% for with more developed relief on the right frontal lobe (12% are equal).

D 2280 preserves some relief, displaying equivalent left and right frontal relief, except for the *pars triangularis* of third frontal convolution which seems bulge more on the right hemisphere. Both of these surfaces are clearly delimited from *pars orbitalis* and *pars opercularis*.

The opercular part of second parietal gyrus produces an eminence on both hemispheres of D 2280 similar to that seen in modern humans. This area is a little more developed in the left parietal lobes of Asiatic *Homo erectus*. The supramarginal gyrus and angular gyrus are more accentuated with clearer limits on left temporo-parietal lobes of Asiatic *Homo erectus*, 45% of modern humans (37% on the left, 18% equivalent), and in D 2280. It is more developed on right hemisphere in OH13; however, the endocranial surface preservation of other fossils from Olduvai do not allow for any comparison. No difference was noted by Begun and Walker (1993) for this cortical region for WT 15000.

In modern human right-handers it is common to observe a larger right frontal lobe associated with a left one showing a more developed third frontal convolution (in particular the orbital and triangular parts) and left dominance of the supramarginal gyrus, which is included in Wernicke's area (LeMay, 1976). However, according to Gannon et al. (1998) and Sherwood et al. (2003) great apes exhibit humanlike asymmetry in Broca's area homologue and planum temporale, which is more localized to the supramarginal gyrus. Thus, there is debate about whether the humanlike asymmetry patterns expressed in non-human primates autapomorphic or plesiomorphic characters that could be expected to be resent in our early ancestors (Sherwood et al., 2008).

With concern to vascularization no relationship has been established between the meningeal system and the venous sinuses (Paturet 1964). No major trend has been observed between fossil hominids, which correlates with the results for frontal and occipital petalias. This is unlike modern humans, which exhibit a sagittal sinus going to the right transverse sinus on 77% (21% on the left and 2% indetermined), and have a more developed left hemisphere. The poor preservation on the right surface of D 2280 doesn't allow for any comparison between the middle meningeal pattern in both hemispheres. This vascular system seems more developed on the left side of the brain of Javanese *Homo erectus*, but is symmetrical in Chinese hominids. This left predominance reaches 43% (13% right). in modern humans, while 45% are symmetrical. Based on the fossil hominid sample vascular asymmetry seems to be considered autapomorphic character of modern humans, albeit we are uncertain what the correlation may be to cognitive abilities. In conclusion, there is no general trend with regard to vascularization in fossil himinids. The results show a rightward dominance for the frontal and temporo-parietal lobes, and a leftward dominance for the occipital lobes is associated with more developed cerebral relief on left side of the brain.

Principal components analysis

Correlation matrices suggest a strong relationship between the above data, except DOCE which corresponds to occipito-cerebellar projection (measures in lateral and superior views) (Table 9). The factorial weights table of each variable on each principal component demonstrates that all retained dimensions are correlated with the first principal component. It is the size effect that implies that small hominid endocasts are placed in positive values and larger ones in negative values. The result is that fossil hominids from Africa and Asia are in the right part of the graph, while more recent humans from the upper Palaeolithic to moderns are in the left part (Table10).

Principal component analysis applied to the endocast coordinates shows a polarized axis where the two first components reach 81% (Fig. 9). The first one (65.9% of total variance) is related to anterior cerebral region development, while increasing of the frontal and parietal chords is associated with elevation of the central region of the brain between the precentral and postcentral sulci (corresponding to HGQE, HBRE and XBE) on one hand and supramarginal gyrus development (corresponding to WBE) on the other. So the first axis clearly separates fossil hominids into two groups, African and Asiatic ones with short, low and narrow general dimensions and short frontal and parietal sagittal chords from modern humans.

The second component (15.2% of total variance) shows closing of the inferior part of the occipital angle, which is related to the occipito-cerebellar projection (DOCE), position of cerebral lobes and posterior cerebral rotation. In fossil hominids, the more open inferior occipital angle is observed in African endocasts with D 2280, which shows the most plesiomorphic pattern for this feature. The Asiatic sample, Javanese and Chinese, exhibits a slight narowing of the angle, but a higher occipito-cerebellar projection. D 2280 could be an Asiatic variation, since its absolute value for occipito-cerebellar projection is clearly outside that of the African sample. DOCE, which is only slightly smaller in PC1, is very large in the second component. However, this relative position of Dmanisi is the result of 11 variables. Its XLIE value is closer to African specimens. Endocasts situated on the negative axis show small a occipito-cerebellar projection (African hominids) than those ones with positive values in the superior where there is a notable backwards projection (Asiatic hominids).

The first axis of variation separates fossil hominids from modern humans. Three groups are clearly delimited on the graph. Endocasts in the inferior right quarter are the smallest in the three general dimensions (length, width, heights). The frontal angle is more narrow in this group, illustrating the low value of general cerebral bending. This group exhibits most opened angle in the inferior part of the occipital region, emphasizing a plesiomorphic configuration. The Asiatic sample, subdivided into Javanese as more archaic, and Chinese as more modern, demonstrates closeness in the inferior occipital angle and projection of occipito-cerebellar region. The pattern of increasing values from the African group to modern humans with concern to general dimensions, opening of frontal angle and closeness of the inferior occipital is clearly emphasized on this analysis.

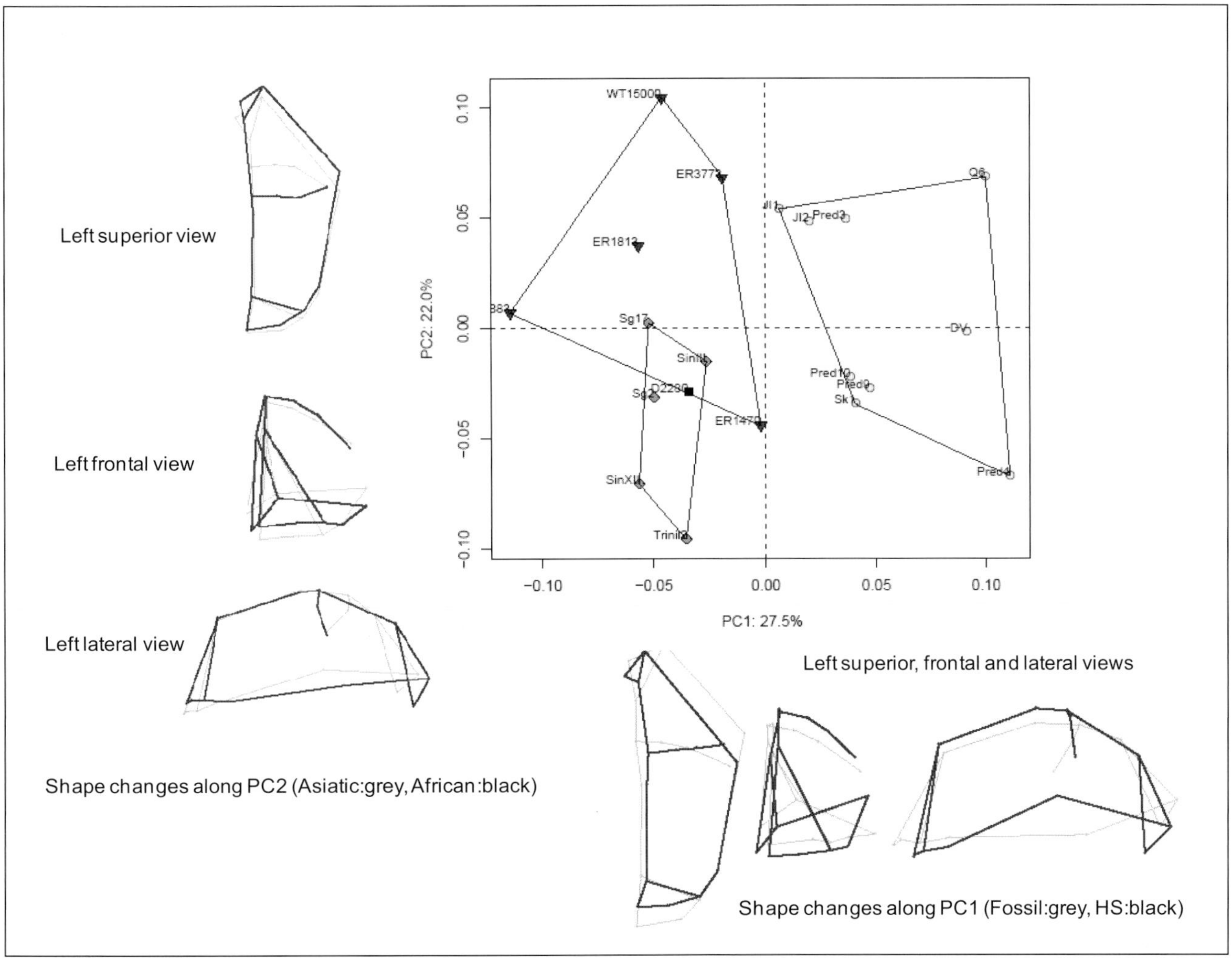

Figure 10. PCA 3D with shape changes along PC1 and PC2

	DOCE	HGQE	HBRE	LME	WBE	CFR	CPA	COC	XBE	XLSE	XLIE
DOCE	1										
HGQE	0,32	1,00									
HBRE	0,36	0,92	1,00								
LME	0,63	0,68	0,55	1,00							
WBE	0,17	0,83	0,68	0,68	1,00						
CFR	0,53	0,87	0,78	0,85	0,74	1,00					
CPA	0,12	0,65	0,56	0,53	0,66	0,47	1,00				
COC	0,53	0,63	0,49	0,73	0,58	0,66	0,25	1,00			
XBE	0,01	0,73	0,60	0,23	0,60	0,48	0,48	0,40	1,00		
XLSE	-0,65	0,12	0,10	-0,27	0,20	-0,16	0,29	-0,19	0,19	1,00	
XLIE	-0,81	-0,31	-0,22	-0,66	-0,20	-0,51	-0,06	-0,59	-0,17	0,68	1

Table 9. Correlation matrix (11 variables and 47 specimens)

Variable	CP 1	CP 2	CP 3	CP 4	CP 5
LME	-0.47	0.36	-0.41	-0.11	0.32
WBE	-0.41	-0.29	0.02	-0.60	0.04
HGQE	-0.38	-0.18	0.31	0.19	-0.16
HBRE	-0.28	-0.17	0.39	0.42	-0.06
DOCE	-0.09	0.24	-0.05	0.21	-0.13
CFR	-0.45	0.15	0.28	0.15	0.47
CPA	-0.30	-0.46	-0.66	-0.41	-0.22
COC	-0.24	0.23	0.12	0.33	-0.55
XBE	-0.10	-0.13	0.17	0.03	-0.36
XLSE	0.01	-0.37	0.04	-0.20	0.08
XLIE	0.15	-0.47	0.11	-0.15	0.38
Sdt.Deviat	27.41	13.17	8.66	7.03	5.92
Pr Variance	0.66	0.15	0.07	0.04	0.03
Cum Prop	0.66	0.81	0.88	0.92	0.95

Table 10. Factorial weights of variables on principal components

Morphometric affinities of Dmanisi brains

Particular care has been taken to choose the most landmarks preserved on the Dmanisi endocasts and other fossils. Procrustes superposition was peformed on 14 sagittal and left parasagittal landmarks digitalized in three dimensions (x,y,z) on 20 fossils specimens (Tale 2). This selection covers the majority of the left hemispheric surface, and is the result of the preservation state of D 2280. The D 2280 endocast was compared with African and Asiatic fossil hominids, and to upper Palaeolithic *H. sapiens* by principal components analysis. The two first components account for 49.5% of the total variance. The first component (27.5%) clearly separates two groups. All of the modern humans are on the positive side of the axis (Fig.10), while almost all of the ancient fossil hominids are placed on the negative side, including the Dmanisi specimen. Negative values correspond with elongated and low endocast shapes due to the lower position of the medial part of the brain, maximal endocranial width in a lower and more posterior position, and, finally, frontal and occipital extremities placed in prolongation of the maximal length. All these morphological features are considered as plesiomorphic.

On the positive side of the axis the shortness of the brain is underlined by the lower position of the anterior frontal part, which is rounding downward and backward, and the occipital and cerebellar lobes, which round downward and frontward. As a result these two cerebral regions are brought closer. Clear elevation of the posterior frontal and anterior parietal regions is noticed in more rounded endocasts. These phenomena are associated with the more anterior and superior position of the maximal endocranial width at the base of the parietal lobes. Thus, on the graph all modern humans are clearly situated on the positive side, and all the fossil hominids, African and Asiatic, are mixed together on the negative side of the axis.

Individualization in the two groups appears on the second component (22%). Asiatic hominids are placed in the negative part of axis, Africans in the positive part. Dmanisi hominids are intermediate between both fossil plots, while modern humans are intermixed with the fossils. The outline of the endocranial shape emphasizes a longer hemisphere with similar elevation in African hominids compared to Asiatic ones. Maximal endocranial width is situated superiorly in the second group. In superior and lateral views the different outlines of the first parietal convolution between African and Asiatic hominids is apparent. In African endocasts this is regularly convex between the sagittal plane and the maximal endocranial width contrary to Asiatic hominids which show an interruption in the curvature with a depression towards the level of the interparietal sulcus. Dmanisi joins the Asiatic sample in possessing a large and depressed sulcus. Dmanisi's position on the graph is intermediate between African and Asiatic groups, perhaps indicating the acquisition of derived characters similar to those observed in Asiatic *Homo erectus*.

Function

Clear regression of the encephalic rostrum, associated with a narrower interorbital space, is observed between the hominid fossil sample and modern humans. This difference can perhaps be functionally interpreted as an indication of a decreasing emphasis on olfaction. According to Sherwood et al. (2008), brain size enlargement in human evolution might have led to a greater degree of functional neocorticalization with this structure taking on more direct influence of other brain regions, allowing for greater voluntary control over actions, contributing to human-specific behavioural abilities, such as the modality and stimulus dependence of language. Broca's area, located in the third frontal gyrus, is a key component of the cortical circuitry in language production. It is more developed on the left hemisphere in 95% of humans as demonstrated by functional imaging or cortical stimulation studies (Parrot, 1981; Habib et al., 2000; Sherwood et al., 2003). For Arbib (2005) the role of Broca's area is more important that expected. Recent brain imaging data suggest important non-linguistic functions relevant to language development in the inferior frontal gyrus, revisiting the role of Broca's area in language, which has surely played as crucial a role in the evolution of human speech as gestural communication has in nonhuman primates. According to Lieberman (2007), "the starting points for human speech and language were perhaps walking and running". Earlier humans would be an intermediate stage in the evolution of language, indicating that this process was gradual and not an abrupt phenomenon. Of course, some form of speech, or different form of communication must have been in place in archaic hominids, allowing for culture and knowledge transmission through generations. Nevertheless, since the frontal lobes are involved in functions such as abstraction, planning and articulate language, their expansion and development are of great importance.

Expansion of the parietal lobes in modern humans (Grimaud-Hervé, 1997; Bruner 2004) is associated with enlargement of the temporal lobes. Both have key functions with regard to language comprehension, verbal memory and face recognition. Parietal cortex expansion is also important in human evolution, affecting visuospatial and sensory integration, multimodal processing and social communication (Holloway 1995). Important in the evolution of manual dexterity, the extension of posterior parietal cortex is presumed to have aided changes related to object manipulation, motor planning and, therefore, stone tool production (Stout & Chaminade, 2007). But we have to keep in mind that many asymmetries are expressed in nonhuman primates, and that plesiomorphic characters represent the substrate for a pre-adaptation to hemispheric specialization in human evolution. In this context, Aboititz et al. (2006) have emphasized the existence of a cortical sensory-motor auditory-vocal circuit, which was probably present in monkeys, and which served as the precursor for the cortical language circuits in the human brain (Broca's and Wernicke's areas). This idea is supported by recent neuroimaging studies in the monkey (Semenferedi and Damasio, 2000; Gil-da-Costa et al., 2006). Of course, study of the fossil record is limited to endocranial geometry and morphological description, which correspond only to macroscopic pattern. Unfortunately, any information that could be available on the architecture of the cerebral tissues and their associated functions is lost in the fossil record.

Using techniques as MRI (Magnetic Resonance Imaging) or FDG-PET (FluoroDeoxyGlucose Positron Emission Tomography), descriptions of activated cerebral areas during particular duties is possible, and allows one to establish the relationships between function and brain structure. Functional imaging research on modern humans cannot directly infer the cognitive capacities of extinct *Homo* species, but does permit speculation with regard to the development of evolutionary significant behaviours. The results of experimental toolmaking (Oldowan and Acheulean) by expert subjects (Schmidt-Nielsen, 1998; Stout et al., 2000, 2008) emphasizes the importance of visuomotor coordination, postural deportment, proprioception and hierarchical action organization. Increased activation of the ventral premotor and inferior parietal elements of the parietofrontal praxis circuits in both hemispheres and of the right hemisphere homologue of Broca's area suggest coevolutionary hypotheses linking the emergence of language, toolmaking, functional lateralization and association cortex expansion in human evolution (Falk, 1992, 2005; Gibson and Ingold, 1993; Holloway 1981b; Holloway et al., 2004; Tobias, 1991).

Actions involved in the toolmaking task are reflected in the activation of the left inferior parietal lobe, while knowledge of tools and tool-use are reflected in activation of the left posterior temporal cortex. According to Stout et al. (2008), results of functional imaging research with modern humans cannot directly reveal the cognitive capacities or neural organization of extinct hominin species but can clarify the relative demands of specific, evolutionarily significant behaviours. As expected in this study, expertise was associated with increased inferior parietal lobe activation during Oldowan toolmaking, but contrary to this expectation, this activation was strongly bilateral. Regions adjoining human anterior interparietal sulcus are also involved in the storage of visuospatial properties associated with tool manipulation. Thus, bilateral activation revealed in Stout et al. (2008) shows that expert performance is supported by an enhanced knowledge of the action properties of the tool and the body system, rather than semantic knowledge about appropriate patterns of tool use. The authors conclude that the task of Oldowan toolmaking is inherently bimanual with distinct but complementary roles for the two hands which confirm that hypotheses linking language capacities and tool-use typically focused on left hemisphere have to be discussed. The right hemisphere seems to play an important role in language processing (Bookheimer 2002), and contributes to elements of perception and action on larger spatio-temporal scales. Particular tasks require cortical association structures (Gilissen, 2001; Aboitiz and Garcia 1997; Amuntz et al., 1999; Wu et al., 2006). The archaeological record of technological change in understanding human cognitive evolution has to be reassessed, since it may be likely that ancient hominids and modern humans could have been capable of utilizing similar tool making techniques. For example, archaic tools, which are not indicative of their cognitive abilitis, can be found with modern humans (Carbonell et al., 1995; Shea, 1997; Lévêque et al., 1993; d'Errico et al., 1998; Mellars, 1996; Roebroeks and van Kolfscoten, 1995). These neurological results obtained from magnetic resonance imaging of expert archaeologists perhaps provide an explanation for the absence of asymmetrical frontal and temporo-parietal lobes in African fossil hominids associated with Oldowan culture. In addition, a slight left predominance has been found in Asiatic fossil hominids, which is related to Wernicke's area of auditory comprehension.

Conclusion

Results of Dmanisi's brain study seem to link D 2282 with the African fossil hominid sample (cranial capacity, univariate and bivariate dimensions). These measurements place D 2282 closer to *Homo habilis*, suggesting it possesses many plesiomorphic morphological features commonly observed in *H. habilis* as well as *Homo ergaster*. A lateralization phenomenon is observed in the orientation of longitudinal superior sinus on African and Asiatic hominids, but no chronological or geographical groups have been noticed. This asymmetry is observed primarily in the most ancient fossil hominids of our studied sample. This sinuses lateralization corresponds with the observed petalial patterns noted on all of the specimens here.

The middle meningeal system of Dmanisi's hominids is very poor with scarce ramifications and no anastomoses unlike those observed on most ancient African hominids. Asiatic *Homo erectus* meningeal system is more developed with more ramified branches. The Dmanisi endocranial morphological features as compared to African fossil hominids from Olduvai, East and West Turkana, and to Asiatic ones from Trinil and Sangiran in Java and Zhoukoudian in China, allows one to emphasize some of the similarities which can be interpreted as plesiomorphies. This is the case of the systematic encephalic rostrum present on all specimens, which corresponds to the first and second convolutions that invaginate between the orbital roof and are determined by the size of the interorbital space. This morphological character is related to olfaction, and is well developed in all early fossil hominids.

Frontal and parietal relief are scarcely individualized in the African sample, and are more marked with less diffuse limits in D 2280 and Asiatic *Homo erectus*. The Sylvian valley, which is particularly broad in the African sample, is narrower on Dmanisi and Asiatic hominids with clear development of frontal and temporal lobes which lie more closely together because of closing of Sylvian valley. In most ancient Asiatic *Homo erectus* the precentral gyrus breadth is often wider (or equivalent to the postcentral). This trend is accentuated on more recent Asiatic *Homo erectus* such as Ngandong for example. Most ancient African hominids exhibit equality of the breadth of the pre- and post-central convolutions. ER 3733 (on which this character can be observed) exhibits a slightly broader post central gyrus. The same observation is also true of both Dmanisi's hominids as well as modern humans.

Convergence of results obtained from morphological, univariate and bivariate analysis, and morphometrical 2 and 3D analyses contribute to a synthetic aproach concerning the phyletic position of the Dmanisi hominids. Size factors distinguished African hominids and Dmanisi from Asiatic sample. Those excluding size emphasize the morphological features that are traditionally difficult to quantify. These show an unquestionable closeness between D 2280 (D 2282 is too damaged to be integrated into this analysis) and Asiatic hominids, which are distinct from the African sample as well as modern humans.

The primary results, based on the small fossil sample, possibly suggests that the first African representatives from the genus *Homo* (*Homo habilis* or *Homo ergaster*) could have evolve into archaic *Homo sapiens*, which could be the ancestor of anatomically modern *Homo sapiens* in Africa. In addition, these African fossils may also provide a link to the Dmanisi fossils, which share strong affinities with Asian *Homo erectus*, as well. One of these fossils hominids (D 2280) exhibits derived characters similar to those observed in Asiatic *Homo erectus*, implying that the Dmanisi fossils lie near the origin of the Asian fossils.

Acknowledgements

I am very grateful to Doug Broadfield, Michael Yuan, Kathy Schick and Nicolas Toth for inviting me to participate to the Symposium "The Human Brain Evolving: Papers in Honor of Ralph L. Holloway" at The Stone Age Institute and Indiana University in Bloomington. Particular thanks are due to Chet Sherwood, Doug Broadfield and Ralph Holloway for their scientific remarks and editing. I am sincerely grateful to Caroline Simonis-Sueur and Miguel Caparros who had the patience to help me concerning multivariate analysis, and Antoine Balzeau for his technical help and all of them for comments on an early draft of this paper.

References

Abbate E, Albianelli A, Azzarolo A, Benvenuti M, Tesfamariam B, Bruni P, Clarke RJ, Ficcarelli G, Macchiarelli R, Napoleone G, Papini M., Took L, Sagr, M, Tecle TM, Torre D, Villa I. 1998. A one-million-year old *Homo* cranium from Danakil (Afar) Depression of Ethiopia. Nature, 393:458-460.

Aboitiz F, Garcia GL. 1997. The evolutionary origin of language areas in the human brain. A neuroanatomical perspective. Brain Res Rev, 25:381-396.

Aboitiz F, Garcia RR, Bosman C, Brunetti E. 2006. Cortical memory mechanisms ans language origins. Brain and language. 98:40-56.

Amunts K, Schleicher A, Bürge U, Mohlberg H, Uylings HBM, Zilles K. 1999. Broca's region revisited: cytoarchitecture and inter subject variability. J Comp Neurol, 413:319-341.

Arbib MA. 2005. From monkey-like action recognition to human language : an evolutionary framework for neurolinguistics. Behavior and Brain Sciences, 28:105-167.

Armstrong E. 1985. Allometric considerations of the adult Mammalian brain, with special emphasis on Primates. In:"*Size and scalling in primate biology*", W.L. Jungers Eds, Plenum Press New-York, 115-146

Asfaw B, Gilbert WH, Beyene Y, Hart WK, Renne PR, WoldeGabriel G, Vrba E, White T. 2002. Remains of *Homo erectus* from Bouri, Middle Awash, Ethiopia. Nature, 416:317-320.

Bauchot R, Stephan H. 1969. Encéphalisation et niveau évolutif chez les Simiens. Mammalia, 33(2):225-275.

Begun D, Walker A. 1993. The Endocast. In:" *The Nariokotome* Homo erectus *skeleton*", A. Walker and R. Leakey Eds, 327-358.

Bruner E. 2003. Fossils traces of the human thought : paleoneurology and the evolution of the genus *Homo*. Rivista di Antropologia, Journ Anthropol Sc, 81:29-56.

Bruner E. 2004. Geometric morphometrics and paleoneurology : Brain shape evolution in the genus *Homo*. JHE, 47:279-303.

Caparros M. 1997. *Homo sapiens* archaïques : un ou plusieurs taxons (espèces) ? Analyse cladistique et morphométrique. Doctorat du M. N.H.N., Paris, 263p.

Carbonell E, Bermudez de Castro JM, Arsuaga JL, Diez JC, Rosas A, Cuenca-Bescos G, Sala R, Mosquera M., Rodriguez XP. 1995. Lower Pleistocene Hominids and Artifacts from Atapuerca-TD6 (Spain). Science, 269:826-832.

Chieze C. 1983. *L'asymétrie cérébrale au cours de l'évolution*. Mémoire de D.E.A., Géologie du Quaternaire, Paléontologie Humaine, Préhistoire, M.N.H.N., Paris VI, 31p.

Delmas A, Chifflet J. 1950. Le pressoir d'Hérophile. C R Ass Anat, Louvain, 123-131.

Errico d' F, Zilhao J, Julien M, Baffier D, Pelegrin J. 1998. Neanderthal Acculturation in Western Europe ?. A Critical Review of the Evidence and its Interpretation. Current Anthropology, 39:1-23.

Falk D. 1992, 2005. *Brain dance : new discoveries about human origins and brain evolution*. New York: Henry Holt and Company.

Gabounia L. 1992. Der menschliche Unterkiefer von Dmanisi (Georgien, Kaukasus). Romisch-Germanischen Zentralmuseums Mainz, 39:185-210.

Gabounia L, Vekua A. 1995. A Plio-Pleistocene hominid from Dmanisi, East Georgia, Caucasus, Nature, 373:509-512.

Gabounia L, Vekua A, Lordkipanidze D. 1999. A hominid metatarsal from Dmanisi (Eastern Georgia), Anthropologie, XXXVII(1/2):163-166.

Gabounia L, Vekua A, Lordkipanidze D, Swisher CCIII, Ferring R, Justus A, Nioradze M, Tvalcherelidze M, Anton SC, Bosinski G, Joris O, Lumley MA, Majsuradze G, Mouskhelishvili A. 2000. Earliest Pleistocene cranial remains from Dmanisi, Republic of Georgia: taxonomy, geological setting, and age. Science, 288:1019-1025.

Gabounia L, Anton SC, Lordkipanidze D, Vekua A. Justus A, Swisher CCIII, 2001. Dmanisi and dispersal. Evol. Anthropol., 10:158-170.

Gabounia L, Lumley MA, Vekua A, Lordkipanidze D, Lumley H. 2002. Découverte d'un nouvel hominidé à Dmanisi (Transcaucasie, Géorgie), C.R.Palevol, 1:243-253.

Galaburda AM, LeMay M, Kemper TL, Geschwind N. 1978. Right-left asymmetries in the brain. Science, 199:852-856.

Gannon PJ, Holloway RL, Broadfield DC, Braun AR. 1998. Asymmetry of chimpanzee planum temporale: humanlike pattern of Wernicke's brain language area homolog. Science. 279:220-222.

Gibson K, Ingold T. (Eds) 1993. *Tools, language and cognition in Human evolution*. Cambridge, Cambridge University press.

Gil-da-Costa R, Martin M, Lopes MA, Munoz M, Fritz JB, Braun AR. 2006. Species-specific calls activate homologs of Broca's and Wernicke's areas in the macaque. Nature Neuroscience, 8:1064-1070.

Gilissen E. 2001. Structural symmetries and asymmetries in human and chimpanzee brains. In: "*Evolutionary anatomy of the primate cerebral cortex*", Falk and Gibson Eds, 187-215.

Good CD, Johnrude, Ashburner J, Henson RN, Friston KJ, Frackowiak R.S. 2001. Cerebral asymmetry and the effects of sex and handedness on brain structure : a voxel based morphometric analysis of 465 normal adult brains. NeuroImage, 14:685-700.

Grimaud-Hervé D. 1997. L'évolution de l'encéphale chez l'*Homo erectus* et l'*Homo sapiens*. Les cahiers de Paléoanthropologie. CNRS Eds, 405 p.

Grimaud-Hervé D. 2004. Endocranial vasculature, In "*The human fossil record* ». Vol.3 : Brain endocasts, the paleoneurological evidence, Holloway, RH, Broadfield DC and Yuan MS, Wiley Press, New York, 273-282.

Grimaud-Hervé D, Saban R 1996. Les empreintes vasculaires observées sur les moulages endocrâniens d'hominidés fossiles et actuels. Anthropologie, Brno, XXXIV/1:27-34

Grimaud-Hervé D, Lordkipanidze D, Lumley MA, Gabounia L. 2006. Etude préliminaire des endocrânes de Dmanisi D 2280 et D 2282. L'anthropologie, 110:732-765.

Habib M, Joanette Y, Roch Lecours A. 2000. Le cerveau humain et les origines du langage. Médecine Sciences, 2:171-180.

Hartwig-Scherer S. 1993. Body-weight prediction in early fossil hominids: towards a taxon –"independent" approach. AJPA, 92: 7-36.

Hervé PY, Crivello F, Perchey G, Mazoyer B, Tzourio-Mazoyer N. 2006. Handedness ans cerebral antomical asymmetries in young adult males. NeuroImage, 29:1066-1079.

Holloway RL. 1975. Early hominid endocasts : volumes, morphology, and significance for hominid evolution. In: Primate functional, Morphology and Evolution. RH Tuttle, Mouton, 393-425.

Holloway RL.1978. Problems of brain endocast interpretation and African hominid evolution. In (C Jolly Ed) Early hominids of Africa, 379-401. London: Duckworth.

Holloway RL. 1981a. The Indonesian *Homo erectus* brain endocasts revisited. AJPA, 55:503-521.

Holloway RL. 1981b. Culture, symbols and human brain evolution : a synthesis. Dialectical Anthropology, 5:287-303.

Holloway RL. 1982. *Homo erectus* brain endocasts volumetric and morphological observations with some comments or cerebral asymetrics. 1er Cong. Int. Pal. Hum., Nice, 355-369.

Holloway RL, 1983. Human brain evolution: a search for units, models and synthesis. Canad. J. Anthropol., 3:215-230.

Holloway RL. 1995. Toward a synthetic theory of human brain evolution. In: Changeux JP, Chavaillon J (Eds), Origins of the human brain. Clarendon Press, Oxford, 42-54.

Holloway RL, Post DG. 1982a. The relativity of relative brain measures and hominid mosaic evolution. In Primate Brain Evolution: methods and concepts (Ed Armstrong E, Falk D) 57-76, New York : Plenum publishing Co.

Holloway RL, De La Coste-Lareymondie MC. 1982b. Brain endocast asymetry in Pongids and Hominids : some preliminary finding on the paleontology of cerebral dominance. AJPA, 58:101-110.

Holloway RH, Broadfield DC, Yuan MS. 2004. "The human fossil record ». Vol. 3: Brain endocasts, the paleoneurological evidence, Wiley Press, New York, 315p.

Jaeger JJ. 1975. Découverte d'un crâne d'hominidé dans le Pléistocène moyen du Maroc. In *« Problèmes actuels de Paléontologie–Evolution des vertébrés »*. Col.Int. CNRS : 218:897-902.

Jashashvili T. 2002. Fossiles humains post-crâniens de Dmanisi, Humérus et clavicule. DEA. du MNHN, Paris, 90p.

Jerison HJ. 1975. Fossil evidence of the evolution of the human brain. Annual Review of Anthropology, 4:27-58.

Lee SH. 2005. Brief communication : Is variation in the cranial capacity of the Dmanisi sample too high to be from a single species ? AJPA, 127: 63-266.

LeMay M. 1976. Morphological cerebral asymmetries of modern man, fossil man, and nonhuman primates. Ann NY Acad Sci, 280:349-366.

Levêque F, Backer AM, Guibaud M. 1993. Context of a Late Neandertal. Implications of Multidisciplinary Research for the Transition to Upper Palaeolithic Adaptations at Saint-Cesaire, Charente-Maritime, France, Prehistory Press, Monographs in World Archaeology n°16.

Lieberman DE. 2007. Homing in on early *Homo*. Nature, 449:291-292.

Lordkipanidze D, Vekua A, Ferring R , Rightmire GP, Agusti J, Kiladze G, Mouskhelishvili A, Nioradze M, Ponce de Leon M, Tappen M, Zollikofer C. 2005. The earliest toothless hominin skull. Nature, 717-718.

Lordkipanidze D, Vekua A, Ferring R, Rightmire GP, Zollikofer C, Ponce de Leon M, Agusti J, Kiladze G, Mouskhelishvili A, Nioradze M, Tappen M, 2006. A fourth hominid skull from Dmanisi, Georgia. Anat. Rec., 288A:1146-1157.

Lordkipanidze D, Jashashvili T, Vekua A, Ponce de Leon MS, Zollikofer CPE, Rightmire GP, Pontzer H, Ferring R, Oms O, Tappen M, Bukhsianidze M, Agusti J, Kahlke R, Kiladze G, Martinez-Navarro B, Mouskhelishvili A, Nioradze M, Rook L, 2007. Postcranial evidence from early *Homo* from Dmanisi, Georgia. Nature, 449:305-310.

Lumley H, Lordkipanidze D, Féraud G, Garcia T, Perrenoud C, Falguères C, Gagnepain J, Saos T, Voinchet P. 2002. Datation par la méthode ^{40}Ar/^{39}Ar de la couche de cendres volcaniques (couche VI) de Dmanisi (Géorgie) qui a livré des restes d'hominidés fossiles de 1.81Ma. C R Palevol, 1:181-189.

Lumley MA, Lordkipanidze D. 2006. L'Homme de Dmanisi (*Homo georgicus*), il y a 1 810 000 ans. C R Palevol, 5:273-281.

Lumley MA, Gabounia L, Vekua A, Lordkipanidze D. 2006. Les restes humains du Pliocène final et du début du Pléistocène inférieur de Dmanisi, Géorgie (1991-2000). I – Les crânes (D2280, D2282, D2700). L'Anthropologie, 110:1-110.

Mac Henry H.M. 1976. Early body weight and encephalization. AJPA, 45 : 77-84.

Martin RD. 1990. Primates Origins and Evolution : A phylogenetic reconstruction. Princeton University Press.

Martin RD. 1995. La taille du cerveau et l'évolution humaine. Pour la Science, 210:60-67.

Martin RD. 1996. Scaling of the mammalian brain: the maternal energy hypothesis. News Physiol Sci., 11:149-156.

Mellars P. 1996. The Neandertal Legacy. An Archaeological Perspective from Western Europe. Princeton : Princeton University Press.

O'Higgins P. (2000) The study of morphological variation in the hominids fossil record : biology, landmarks and geometry. Journal of Anatomy, 197:103-120.

Parrot JL. 1981. Paul Broca et la physiologie du langage. Rev Pal Déc, 9 (90):15-26.

Paturet G.1964. Traité d'Anatomie humaine. Système nerveux. IV, 1183 p., Masson.

Rightmire GP. 1986. Body size and encephalization in *Homo erectus*, In « Fossil Man – New facts, new ideas ». Anthropos, 23: 39-150.

Rightmire GP. 2004. Brain size and encephalization in Early to Mid-Pleistocene *Homo*. AJPA, 124: 109-123.

Rightmire GP, Lordkipanidze D, Vekua A. 2005. Anatomical descriptions, comparative studies and evolutionary significance of the hominin skulls from Dmanisi, Republic of Georgia. JHE, 1-27.

Rightmire GP, Lordkipanidze D. 2007. The role of Dmanisi in early hominin dispersal. AAPA abstracts, 199.

Roebroeks W & Van Kolfscoten T. 1995. The earliest occupation of Europe, Proceedings of the European Science Foundation Workshop at Tautavel (France, 1993), University of Leiden.

Rohlf FJ, Marcus LF. 1993. A revolution in morphometrics. TRENDS in Ecology and Evolution, 8(4):129-132.

Rosenberg KR, Zune L, Ruff CB. 2006. Body size, body proportions, and encephalization in a Middle Pleistocene archaic human from northern China. PNAS, 103 (10):3552-3556.

Ruff B, Trinkaus E, Holliday TW. 1997. Body mass and encephalization in Pleistocene *Homo*. Nature, 387:173-176.

Saban R. 1984. Anatomie et Evolution des veines méningées chez les hommes fossiles. Comité des Travaux historiques et scientifiques, Mémoire des sciences de la terre, 11, ENSB- CTHS, 289p.

Schmidt-Nielsen K. 1998. How much structure is enough? In "Principles of Animal Design. The optimization of Symmorphosis debate", Eds ER.Weibel, CRTaylor, LBolis, 11-12, Cambridge University Press

Semenferedi K, Damasio H. 2000. The brain and its main anatomical suvdivisions in living usinf magnetic resonance imaging. JHE, 38:317-332.

Sherwood CC, Broadfield DC, Holloway RL, Gannon PJ, Hof PR. 2003. Variability of Broca's area homologue in African great apes: implications for language evolution. Anatomical Record, A271A:276-285.

Sherwood CC, Subiaul F, Zawidzki TW. 2008. A natural history of the human mind: tracing evolutionary changes in brain and cognition. J Anat, 212:426-454.

Stout D, Toth N, Schock K. 2000: Stone tool-making and brain activation: position emission tomography (PET) studies. J of Archaeol Sc, 27:1215-1223.

Stout D, Chaminade T. 2007. The evolutionary neuroscience of tool making. Neuropsychologia, 45:1091-1100.

Stout D, Toth N, Schick K, Chaminade T. 2008. Neural correlates of early stone age toolmaking: technology, language and cognition in human evolution. Phil.Trans.R.Soc.B., doi:10.1098/rstb.2008.0001, published on line, 11p.

Tobias PV. 1975. Brain evolution in the *Hominidae*. In : Primate functional morphology and evolution, RH Tuttle Eds, Mouton, 353-392.

Tobias PV. 1987. The brain of *Homo habilis*: a new level of organization in cerebral evolution. JHE, 16: 741-761.

Tobias PV, 1991. Olduvai Gorge. The skulls, endocasts and teeth of *Homo habilis*. Cambridge Press University, 2 tomes, 921p.

Tobias PV. 2006. Longevity, death and encephalization among Plio-Pleistocene hominins. International Congress Series, 1296:1615.

Vekua A, Lordkipanidze D, Rightmire G, Agusti J, Ferring R, Majsuradze G, Mouskhelishvili A, Nioradze M, Ponce de Leon M, Tappen M, Tvalcherelidze M, Zollikofer C. 2002. A new skull of early *Homo* from Dmanisi, Georgia, Science, 297:85-89.

Weidenreich F. 1943. The skull of *Sinanthropus pekinensis*. A comparative study on a primitive hominid skull. Paleontologica Sinica, 10, 298 p.

Wu X, Schepartz.E, Falk D, Liu W. 2006. Endocranial cast of Hexian *Homo erectus* from South China. A.J.P.A., on line.

CHAPTER 6

The Evolution of the Parietal Cortical Areas in the Human Genus: Between Structure and Cognition

Emiliano Bruner

Abstract

Recently, the renewed interest for concepts such as morphological integration, functional craniology, and the analysis of covariance patterns has spawned a change in paleoneurology that has to be interpreted as the study of the evolutionary variations in the relationships between brain and braincase. The parietal lobes have been hypothesised to have undergone important morphological changes in early hominid evolution. Nonetheless, the role of these areas within the evolution of the human genus has been rather neglected because of their alleged lack of association with "higher" cognitive functions. Some allometric constraints could have affected the changes in the parietal surfaces in non-modern humans. On the other hand, modern humans display a non-allometric change in the shape of these areas which are bulging at their midsagittal profile. Such changes raise questions on the relationship between structural rearrangements and cognitive adaptations. Focusing on the parietal surface, it seems that the upper lobule and the intraparietal sulcus might be directly involved in the evolution of the modern human brain morphology. This is particularly intriguing considering the many cortico-cortical reciprocal connections between these areas and the prefrontal ones. Most of all, they are directly involved in integrating inner and outer information to reproduce a subjective "virtual inner reality" necessary not only to organise movements, but also to make decisions, to perform thought experiments, and to handle the interaction between the self and such imagined space physically and conceptually. Whether or not the origin of the modern human lineage coincided with the origin of a modern human brain is still to be investigated.

Functional Craniology and Endocranial Morphogenesis

Morphogenesis is a complex process in which a poligenic and pleiotropic network linking genes and characters is expressed within a given functional and structural framework. Structure (both in terms of biomechanical and developmental constraints) and function generate the unique combination of forces and physical relationships in which a specific genetic background produces a given phenotype. Following the principles of functional craniology (Moss and Young, 1960), morphogenesis is the result of two components, namely growth (changes in size) and development (changes in shape). The correct balance between these two components leads to the normal phenotype, while an improper redirection of the growth forces leads to subpathological or pathological dysmorphologies associated with osteoblast/osteoclast induction or timing and rate of sutural activity (Moss, 1959). Neurocranial vault morphogenesis is mainly related to brain growth for the size changes and to the connective meningeal tensors for the shape variation. The principal connective tensors are the falx cerebri and the tentorium cerebelli, exerting forces on five main points: crista galli, small wings of the sphenoid, and petrous pyramids.

Of course, such simplification is useful to produce hypotheses to investigate these processes, but subtle variables can play an important role in the general structural management. For example, although brain growth and pressure are the principal forces leading to the modelling of the vault, strains are integrated by responses and inductions through the dura mater and the vascular system by mechanical transduction or by biochemical signalling (Henderson et al., 2004). Figure 1 synthesises the main components and relationships within the brain versus braincase matrix.

The embryological context is also rather heterogeneous, with some components originating from the mesoderm and others from the neural crests, some through direct ossification, others with cartilaginous precursors (Jiang et al., 2002; Morriss-Key & Wilkie, 2005). For example, both frontals and parietals are dermal bones, but the former is ectoderm-derived while the latter has mesodermic contributions. Such differences make morphogenesis a polyphasic process with many possible steps in which small changes could exert large phenotypic variations during evolution.

The reticulated system of forces, functions, constraints, and genes, underlying the actual morphological variability through ontogeny and phylogeny has convinced biologists to move from the study of single traits and isolated characters to the analysis of the patterns of covariance, i.e. *morphological integration* (Olson and Miller, 1958; Cheverud, 1996). Phenotype is hence interpreted not as a sum of features, but as a combination of relationships between features. This is clearly true for the human skull (Bookstein et al., 2003; Bruner, 2007) as well as for the relationship between the neurocranium and brain (Richtsmeier et al.; 2006; Bruner and Ripani, 2008).

Paleoneurology, as the study of the nervous system in extinct species (Holloway, 1978; Falk, 1987; Bruner, 2003a), deals exactly with this last issue: the interpretation of the endocranial morphology as the result of the integration between its structural (developmental, biomechanical) and functional (neural, cognitive) components.

Neurocranial Shape Variation and Parietal Lobes

Despite the never-ending struggle on the lunate sulcus debate (see Holloway et al., 2003), we can currently state that if any differences between Australopithecinae and the other apes did occur, it was at the posterior parietal boundary. Holloway very early recognised that such differences, because associated with the visuo-spatial integration and recognition of the outer reality, could provide a relevant rearrangement of the ecological and social organisation of the early Hominids (see Holloway, 1995). And, through a pioneering stereoplotting surface analysis, the parietal areas were hypothesised to be a crucial source of morphological variation in both Hominoids and Hominids (Holloway, 1981).

The parietal lobes have been generally described as "associative cortex" by virtue of the many connections (neural and anatomical) with the other districts. Excluding the postcentral gyrus, mainly involved in the

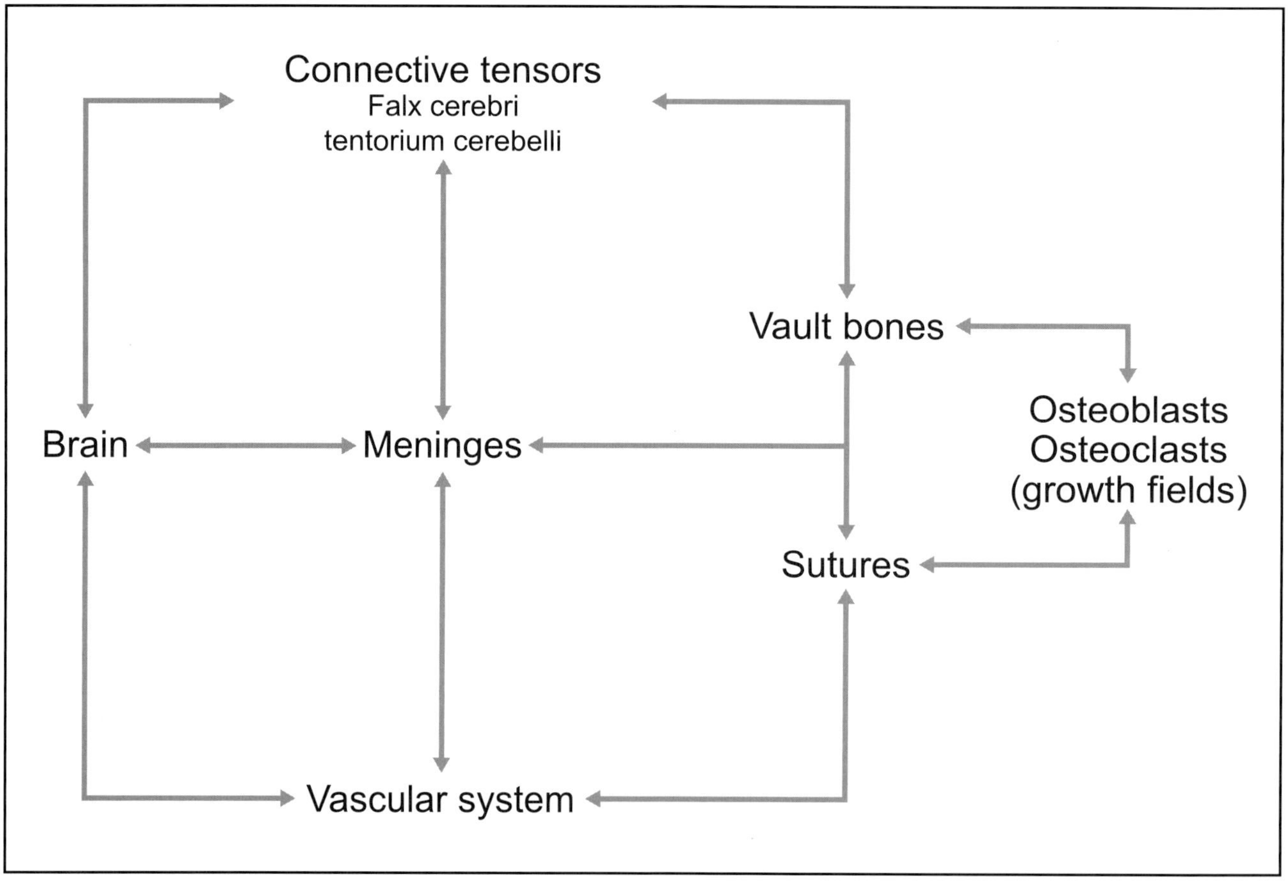

Figure 1. Paleoneurology deals with the morphogenetic relationship between neurocranium and brain, associated with functional and structural responses between hard and soft tissues resulting from the interaction between genetic programs (cellular differentiation and activation) and developmental forces (strains, biomechanics).

somato-sensorial system, the posterior parietal areas are basically divided into upper and lower parts, separated by the intraparietal sulcus (Eidelberg and Galaburda, 1984). The upper lobule almost gradually fades into the occipital one, both in terms of gross anatomy and cytoarchitecture. The lower lobule is part of the Wernicke area, including the over-studied angular and supramarginal gyri. The intraparietal sulcus is a rather peculiar structure, providing a large part of the parietal surface deepened into the cortical volumes, showing different cytoarchitectonic patterns, and supporting heterogeneous neural functions (e.g., Bisley and Goldberg, 2003; Choi et al., 2006). Its displays at least five different morphological patterns, showing generally (about 75% of the cases) a continuity with the postcentral sulcus (Ebeling and Steinmetz, 1995).

Considering the evolution of the human skull, its globularity has always been described as the main traits and trends associated with the encephalisation process (Lieberman et al., 2002). Actually, quantifying such variations and analysing the midsagittal cranial shape in the human genus, modern humans stand apart from the other taxa mostly by virtue of their fronto-parietal bulging (Bruner et al., 2004; Fig. 2).

Moving from the ectocranium to the endocranium, some evidence comes from simple traditional metrics. Using the main endocranial diameters (hemispheric length, frontal and maximum widths, and vault mid-height) to perform a factor analysis on a sample of 21 endocasts from the human genus (see Bruner et al., 2003 for details), the first vector is easily recognised as an allometric component, with all positive loadings, accounting for almost 90% of the total variation (Fig. 3). This is to be expected, considering the simple metrics involved and the large correlation between these diameters. It is nonetheless worth noting that the first factor is almost parallel to the hemispheric length vector. Therefore, we can assume that hemispheric length in humans is a good (and easy, and quick) proxy for encephalisation, or at least for cranial capacity. Hence, we can use these diameters both in a traditional approach as a size index, and in geometric superimpositions as a relevant baseline. Us-

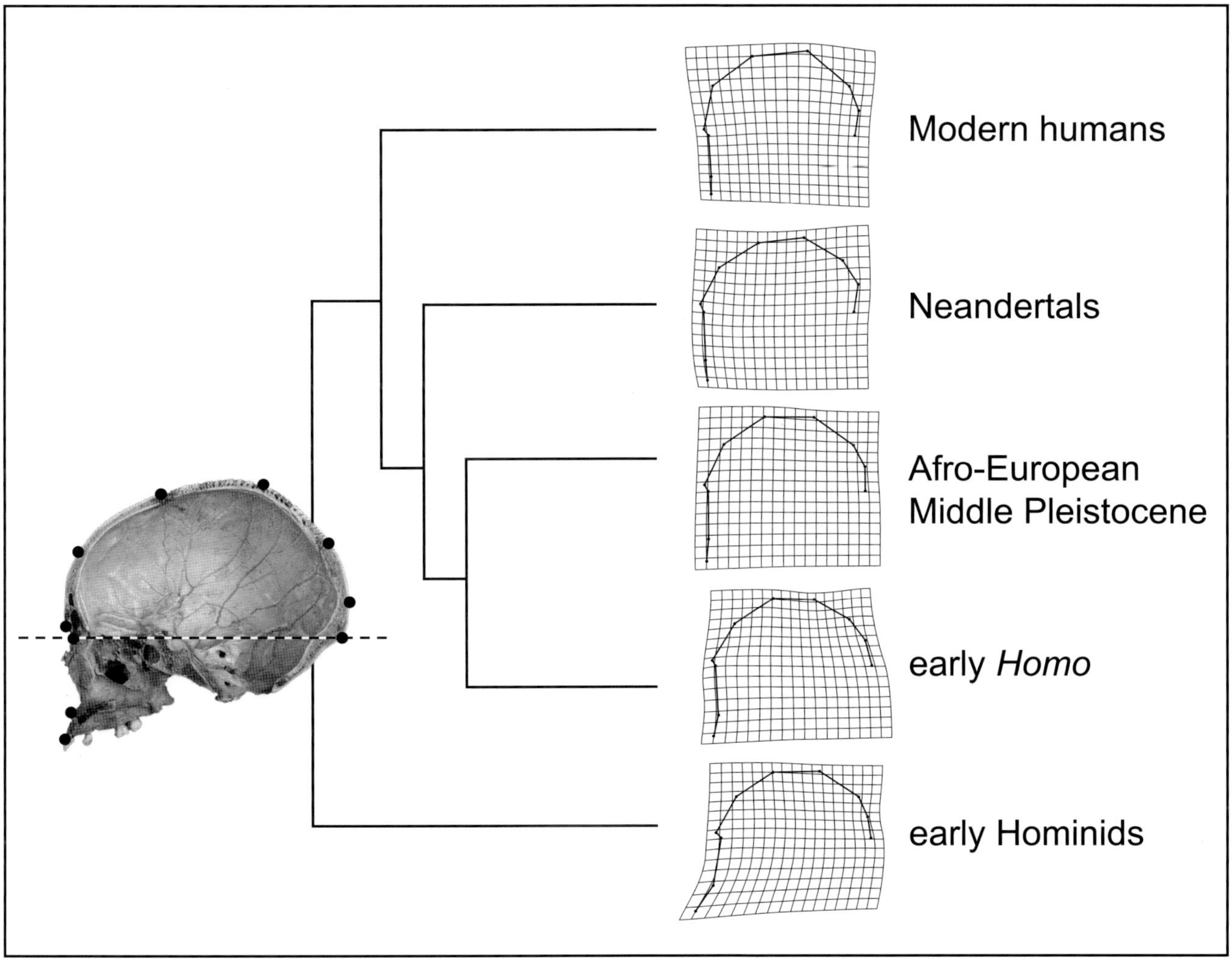

Figure 2. The ectocranial midsagittal profile largely characterises the major extinct human taxa. Apart from a general trend towards reduction of the facial block, Neandertals show a specific projection of the midface, while modern humans show a definite bulging of the parietal profile. Here, average midsagittal configurations from the main Hominid groups are compared using a nasion-inion baseline, and the thin-plate spline deformation grids. The degree of facial reduction and the parietal bulging set modern humans apart from the rest of the human morphotypes (after Bruner et al., 2004).

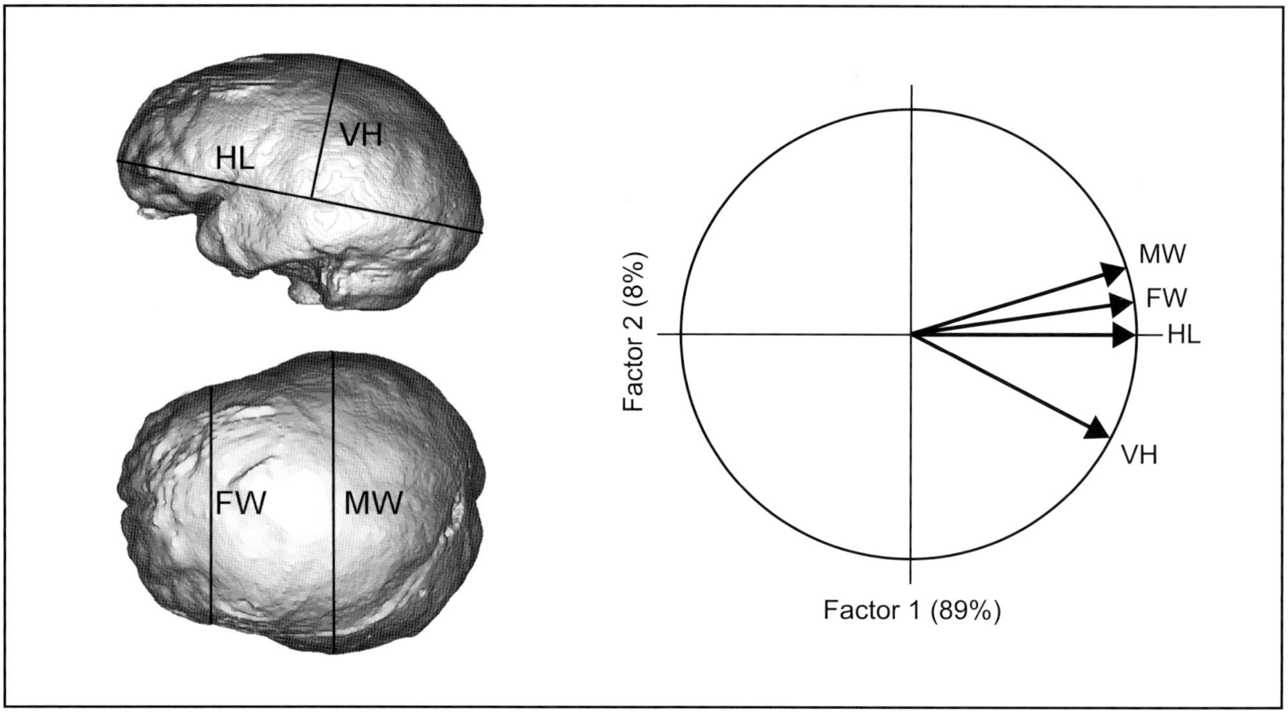

Figure 3. The main endocranial diameters can be used to perform a factor analysis in the human genus (see Bruner et al., 2003). HL: hemispheric length; VH: midvault height; MW: maximum endocranial width; FW: frontal width (at the Broca's cap). The first component is largely allometric, while the second is associated with inverse relation between height and width. Interestingly, the hemispheric length vector is parallel to the first component, i.e., among the main raw endocranial diameters the hemispheric length is a good linear proxy for brain size. The diameters are shown on the digital reconstruction of the endocast of Saccopastore 1 (Bruner and Manzi, 2008).

ing simple endocranial diameters like these, it has been described how modern humans display largest parietal length and height when compared with the *Homo* allometric trajectories (Bruner, 2004). Interestingly, encephalisation in the human genus is associated with a relative shortening of the parietal chord, with the exception of modern humans showing a discrete morphological change because of a definite enlargement of the parietal diameters (Bruner et al. 2003).

Similar results are supported and further detailed using geometrical endocranial models. Figure 4 shows the comparison between a mean modern human lateral endocranial configuration and the mean Neandertal figure, using the fronto-occipital baseline, visualised through thin-plate spline deformation grids (Bookstein, 1991), and mapping of the Euclidean distance differences on two representative specimens (Bruner, 2008a). The registration according to the same hemispheric length shows the main differences at the parietal outline, the deformation grids suggest that the main spatial changes are represented by the parietal bulging, and the Euclidean distance matrix evidences an absolute enlargement of the parieto-cerebellar diameters.

Using the Procrustes superimposition (i.e., translating, scaling, and rotating the geometric models to minimise the residual coordinate differences; Bookstein, 1991), a three dimensional comparison between average endocranial shapes in archaic humans, Neandertals, and *Homo sapiens*, shows very scanty differences between the formers, and a marked morphological change in the latter, associated with the parietal midsagittal enlargement (Bruner et al., 2003). Again, the same results were confirmed by a two-dimensional analysis of the lateral endocranial profile performed through multivariate statistics, mean shapes, and phenetic distances, suggesting a morphological gap between the modern and non-modern variations (Bruner, 2004).

These analyses were computed using homologous landmarks of the brain, which of course are difficult to recognise on the endocasts, requiring experience and a lot of caution. Nonetheless, the same results can be obtained using simple geometrical references not associated with given anatomical structures. Figure 5 shows a Procrustes comparison of the lateral profile between the Salé (archaic *Homo*, Africa, about 400 ka) and the Combe Capelle (modern human, Europe, about 25 ka) endocasts. The endocast from Salé was supposed to be a good example of basic *Homo* endocranial morphology because of the absence of specific derived traits, including the marked projection of the occipital lobes displayed by the Asian *Homo erectus* (Bruner, 2003b, 2004). Nonetheless, there is a certain disagreement on this point, and in other studies, this endocast has been hypothesised to be largely comparable with those from the Asian groups (Holloway et al., 2004). Unfortunately, the endocast shows damage exactly at the parieto-oc-

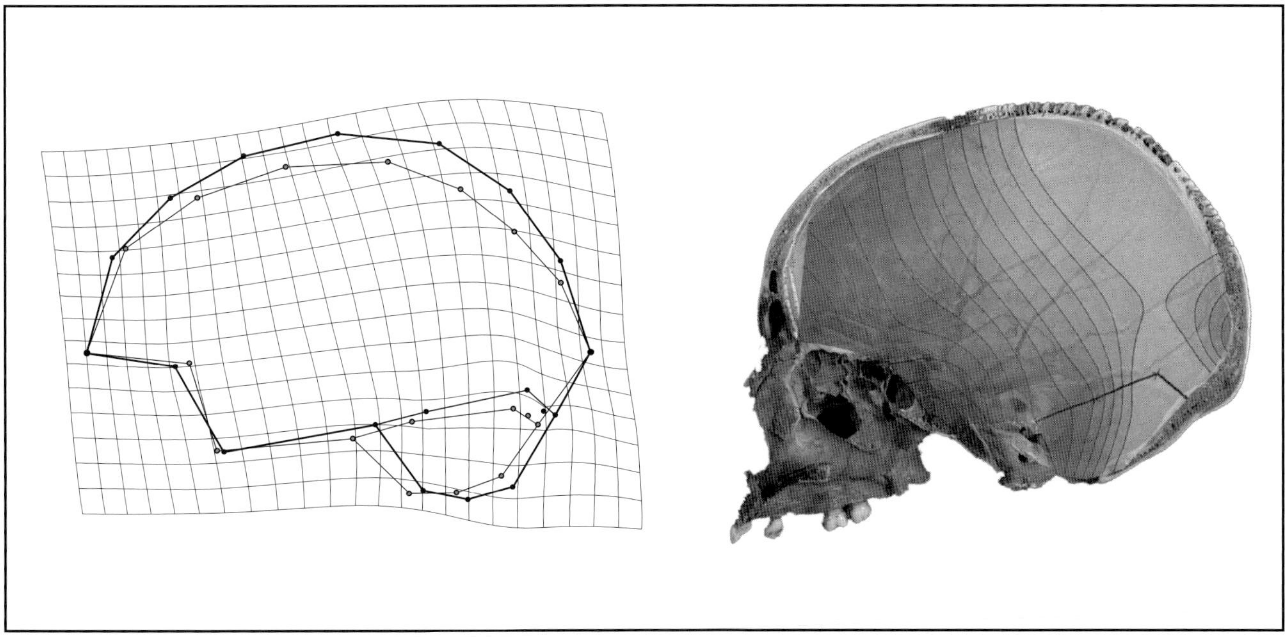

Figure 4. Geometric comparison of the lateral endocranial profile in modern humans (thick links) and Neandertals (thin links), through fronto-occipital superimposition and thin-plate spline deformation grids (left) and mapping of Euclidean distance matrix analysis (right)(see Bruner, 2008a for details). The baseline comparison is computed on mean shapes, while EDMA data are from two representative complete specimens: La Ferrassie 1 for Neandertals, and one Mesolithic Italian fossil for modern humans. The main differences can be clearly detected at the upper parietal areas, both in terms of shape (grid deformation and superimposed profile describing the parietal bulging) and form (EDMA map; dark grey: shorter diameters in the modern specimen; light grey: longer diameters in the modern specimen).

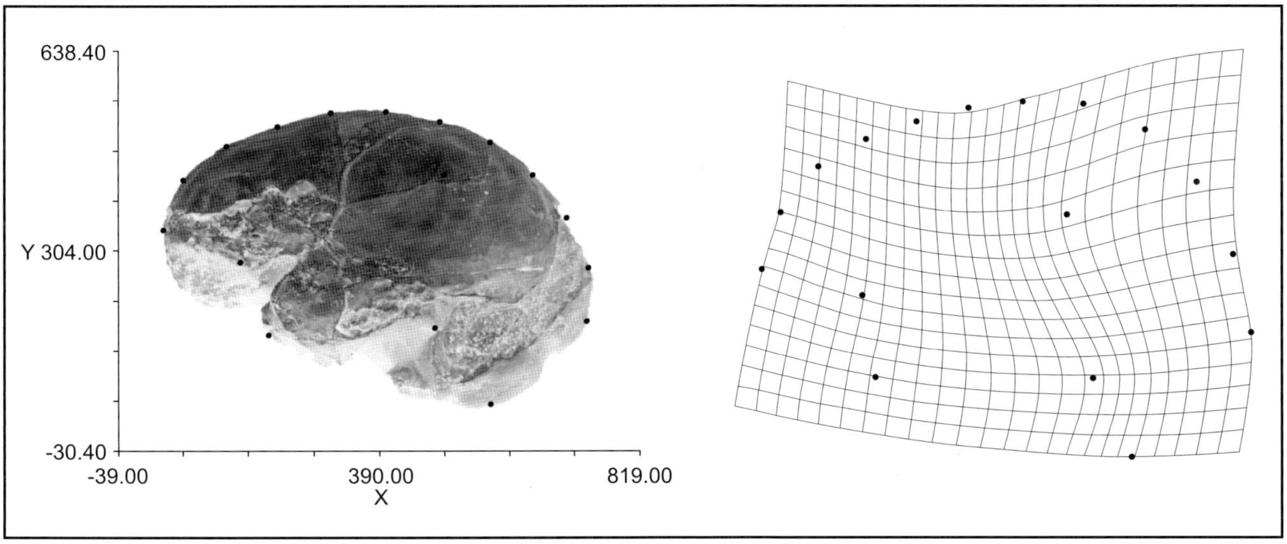

Figure 5. Using sliding landmarks to delineate the upper (fronto-occipital) endocranial profile by using a Procrustes superimposition and minimisation of the bending energy, the result is similar to the previous comparisons. The lateral endocranial configuration from the 400 ka Moroccan specimen from Salé (left) is superimposed onto the configuration of the 25 ka European specimen from Combe Capelle, showing again bulging of the upper parietal surface associated with convolution of the brain morphology. In this comparison it can be also recognised a certain lengthening of the temporal lobe. Although these areas have been hypothesised to have undergone a relative enlargement in modern humans, in this case it may be related just to a specimen-specific morphology, being not always detected in other similar comparisons between modern and non-modern endocasts. Superimposition and deformation grids are computed by using tpsSplin 1.20 (Rohlf, 2004). Both endocasts are from the University La Sapienza, Roma.

cipital boundary, hampering a robust assessment of the missed morphology. Furthermore, it must be always taken into account that paleoneurology necessarily relies on different endocranial collections, with comparisons made upon casts from different authors, different materials, and different historical periods. Interestingly this specimen also shows a certain lateral bulging of the parietal surfaces.

After lateral photography, the endocranial profile of the two specimens was modelled using some main anatomical references, and 10 sliding-landmarks between the frontal and the occipital poles (see Zelditch et al., 2004 for further details on the geometric morphometric tools). Again, after Procrustes superimposition and thin-plate spline interpolation the parietal bulging is easily recognised as the main morphological change of the endocranial geometry.

Sliding landmarks can be also used to perform a principal component analysis of the fronto-parieto-occipital profile, from the anterior insertion of the crista galli to the internal occipital protuberance (Fig. 6). The first component explains 55% of the total variance, being associated with parietal bulging and occipital flattening, characterising the modern human hemispheres. The second component separates Neandertals and archaic humans mainly because of the occipital projection of the latter.

Of course, because the brain versus neurocranium is a unique structural and functional system, changes in a given region could be associated with differences in other related districts. Accordingly, the bulging of the upper parietal areas described in modern humans can be the result of at least three different processes: 1) a change in the upper parietal neural mass; 2) a change in other neural areas influencing the position and topology of the upper parietal surface (e.g., the lower parietal structures); 3) a change of the skull organisation (e.g., the cranial base) involving rearrangement and redistribution of the endocranial volumes.

Some information to better evaluate this framework can be provided by comparing directly the parietal components by using again a landmark-based approach (Fig. 7). Superimposing the lateral parietal morphology from Salè and from a modern human endocast using the hemispheric length as a baseline, the lateral sulcus shows a similar position and orientation, and the lower parietal

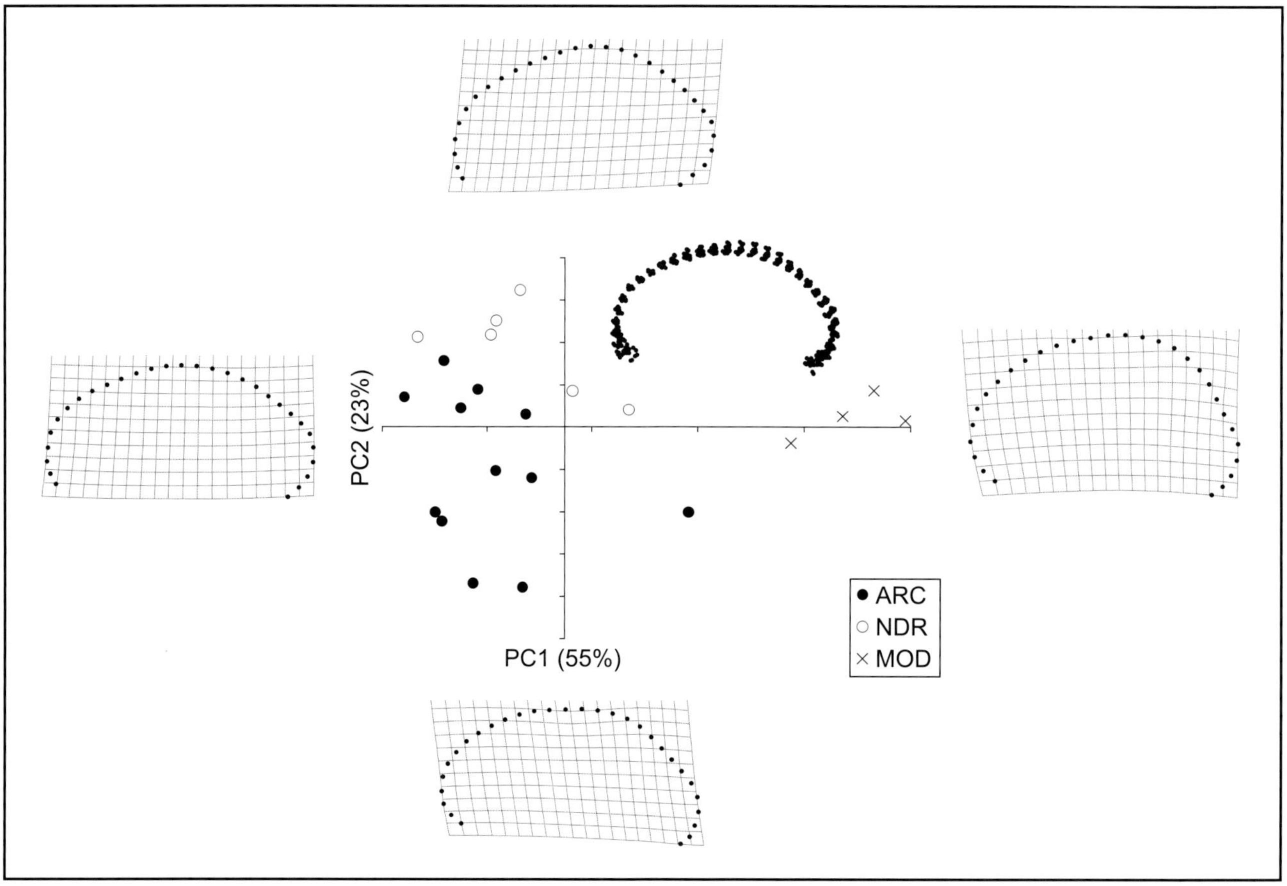

Figure 6. A Principal Component Analysis was performed using sliding landmarks and Procrustes registration to outline the fronto-parieto-occipital profile (from crista galli to internal occipital protuberance), by using tpsRelw 1.45 (Rohlf, 2007). The first component separates modern (MOD) from non-modern specimens because of the parietal bulging of the former. The second component separates (to a lesser degree) archaic humans (ARC) from Neandertals (NDR), because of the occipital projection of the former. ARC: Salé, Arago, Trinil 2, Zhoukoudien 3, Zhoukoudien 12, Sambungmacan 3; NDR: Saccopastore 1, Guattari, La Chapelle-aux-Saints; MOD: Combe Capelle, Vatte di Zambana (both hemispheres).

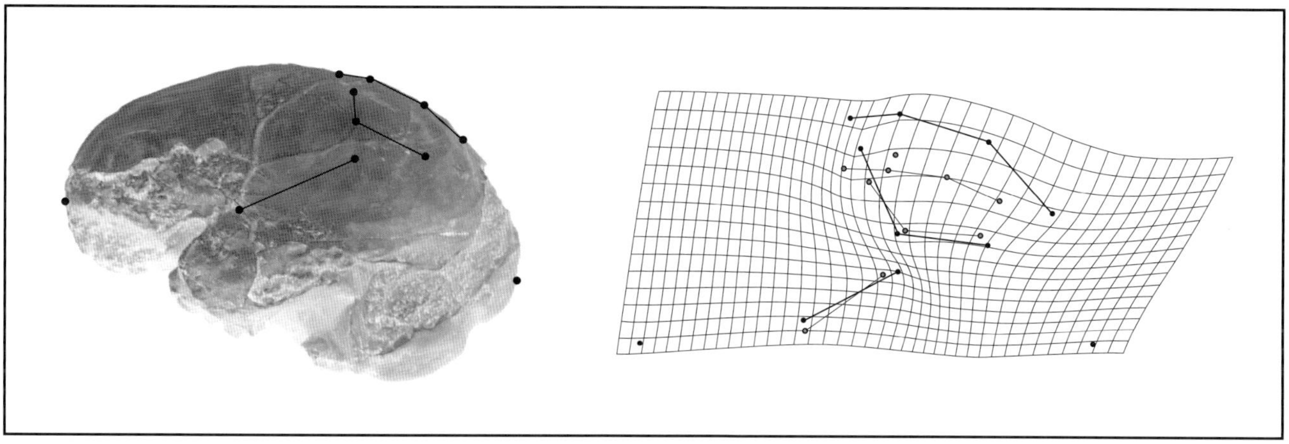

Figure 7. The parietal areas from Salé (Middle Pleistocene, North Africa) and Vatte di Zambana (Mesolithic, Italy) are compared through fronto-occipital registration and thin-plate spline deformation grids, by using Morpheus et al. (Slice, 2000). Apart from the frontal and occipital poles, the configuration includes the lower fronto-parietal boundary and the posterior edge of the lateral sulcus, the supramarginal and angular gyri, the anterior edge of the intraparietal groove, the anterior and posterior midsagittal boundaries of the parietal lobes, the midagittal projection of the postcentral sulcus, and the midpoint on the upper lobule midsagittal profile. Each configuration is the mean of five independent resampling procedures. According to the hemispheric length (wireframes) and minimum deformation (grids), the differences are clearly localised at the upper parietal volumes, enlarged in the modern specimen. The intra-parietal area seems to delineate the lower border of such expansion.

areas do not display relative enlargements. The thin-plate spline deformation grids (which are superimposition-independent and account for the minimum spatial deformation required for the geometrical fitting of the two systems of coordinates) further confirm this evidence. Once more, changes localised at the upper parietal lobule seem to be the more striking features of modern human endocranial morphology. The intra-parietal area seems to separate an area of relative expansion (upper lobule) from an area of relative compression (lower lobule), at least in lateral view.

A final indication comes from the endocranial traces of the middle meningeal vessels, as record of fossilised physiological and morphogenetic processes. The patterns of these vascular imprints show interesting differences within the human genus in its complexity, position, and general organisation (Grimaud-Hervé, 1997). Although the endocranial angiogenesis has an active role in neurocranial growth and development (Henderson et al., 2004), the vascular organisation is largely influenced by the neurocranial structural and functional environment (O'Laughlin, 1996). Using a fractal analysis, the degree of reticulation of the meningeal vessels has been demonstrated to be similar in Neandertals and archaic humans, but definitely higher in *Homo sapiens* (Bruner et al., 2005). The increasing reticulation of the middle meningeal vessels concerns the whole endocranial surface, mostly through its anterior branches, but it is particularly stressed at the parietal surface (e.g., Saban, 1982). The evolution and morphogenesis of these vessels has been largely ignored (Falk, 1993; Bruner and Sherkat, 2008). The more complex branching pattern and larger number of anastomoses detected in modern humans through endocranial imprints may be related to a more reticulated vascular system (associated with cognitive or metabolic functions), or to a larger number of traces left on the endocranial wall (associated with some structure/pressure differences). In both cases, they once more suggest that in modern humans some factors have induced changes in the brain versus braincase relationship at the parietal surfaces.

Of course, the fossil record is far from being a robust statistical sample, and there are some interesting exceptions. One of these is the European Middle Pleistocene parietal from Arago (Fig. 8a) showing no midsagittal bulging but a rather large parietal surface and branched vascular traces. Other reticulated middle meningeal traces can be described for the Neandertals from Biache-Saint-Vaast (Saban, 1979) and for some fragmented parietals from Krapina (Bruner et al., 2006).

The Evolution of the Parietal Areas in the Human Genus

According to the shape differences in the endocranial profile of the genus *Homo* and the patterns of morphological covariation associated with the human extinct variability, it has been hypothesised that some structural constraints could have characterised the evolution of the parietal areas (Bruner, 2004). Considering the non-modern variation, as the brain gets larger the longitudinal and vertical diameters of the parietal areas do not keep pace with the frontal and occipital changes. This negative allometry of the parietal profile leads to a morphological compression and flattening of the parietal areas along the encephalisation trajectories. Such relative

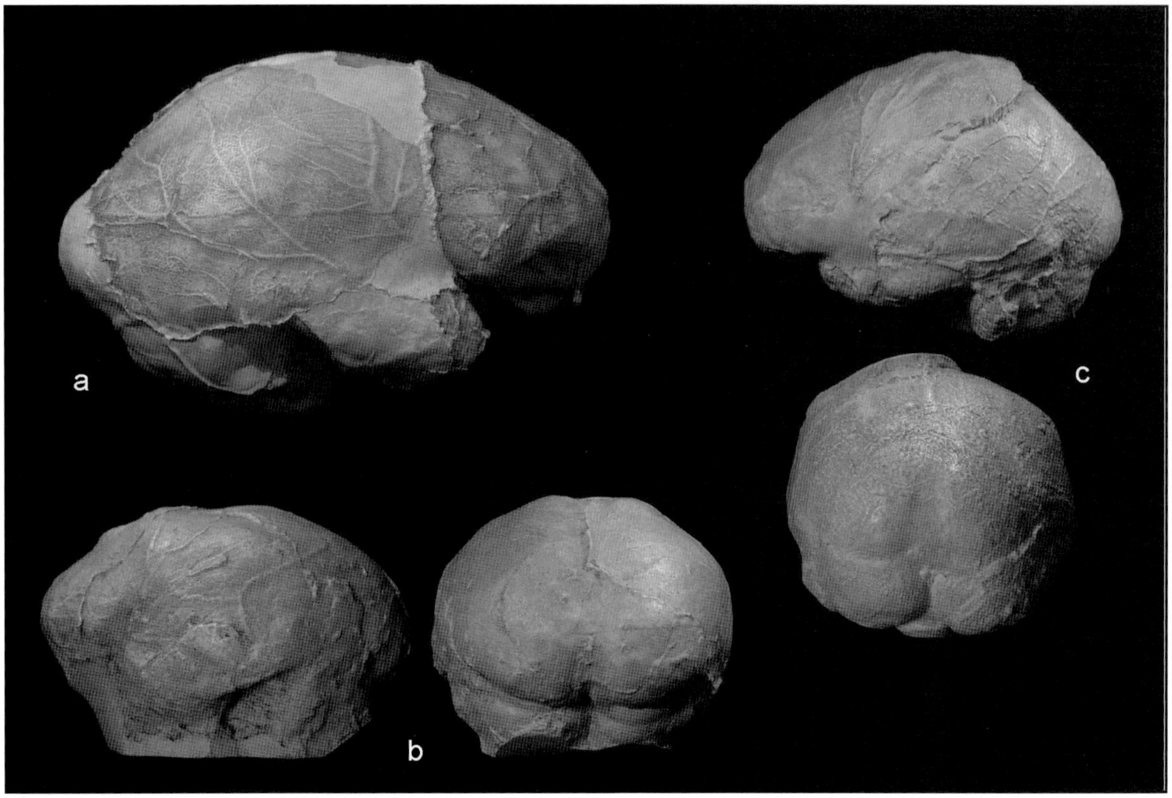

Figure 8. The reconstructed endocast from Arago at the Istituto Italiano di Paleontologia Umana, Roma (a). The parietal surface is rather large and bossed, with reticulated traces of the middle meningeal vessels. The endocast from Jebel Irhoud (b) and Skhul V (c) at the Institut de Paleontologie Humaine, Paris, in left and posterior views. The first shows a general archaic appearance, but with a certain lateral bulging of the upper parietal areas (most evident on the left side) as described for Neandertals. The second unfortunately is damaged at the anterior parietal boundaries, but the overall parietal morphology is closer to the modern human figure.

shortening and flattening of the upper parietal structures could have been induced by two factors, mostly based on the tight causal relationship between the brain and vault morphogenesis, in which the former largely determines the latter (Moss and Young, 1960; Enlow, 1990). Firstly, the position of the parietal areas between the frontal and occipital ones may suggest that, while the anterior and posterior volumes are able to arrange their topology according to the allometric changes of the hard (the cranial base) and soft (the subcortical structures) tissues, the interposed volumes are forced to vary accordingly. Secondly, being the vault shape largely associated with the strains of the meningeal tensions, it may be hypothesised that allometric and structural constraints may be related to the relationship between cortical volumes and the falx cerebri. For example, such a relationship can be easily influenced by a classical spatial interaction between structures growing at the power of three (the brain) and structures growing at the power of two (the falx).

Neandertals were the most encephalised non-modern human group, showing a sort of vault upward "bending" possibly related to this allometric pattern. Interestingly, Neandertals often display supernumerary ossicles at the parieto-occipital boundary, revealing a sort of "morphological instability" of those areas (Sergi, 1934, 1948). Such hypostotic traits, even when sub-pathological, suggest a lack of morphogenetic balance during ontogeny (Manzi et al., 1996), revealing some possible evolutionary limits, i.e we can assume that at the parieto-occipital boundary Neandertals could have been characterised by a loss of balance between size (growth) and shape (developments) changes during ontongeny. This is not particularly surprising, the basic organisation of their neurocranial system having evolved at the end of the Pliocene for brains of 600-700 cubic centimetres.

Considering this hypothesis about the non-modern endocranial variation of the human genus, the modern configuration can be interpreted in two ways. First, the parietal rearrangement in *Homo sapiens* could have been a structural solution to the allometric endocranial constraints. Of course, such a solution could have revealed some interesting cognitive involvements. Alternatively, the cognitive changes associated with the upper parietal areas could have been the selective force leading to the morphological changes, which secondarily could have led to the structural solution to trespass the allometric constraints.

The Neandertal lineage displayed a "Neandertal brain" from 100-120 ka, as suggested by the morphology of the Saccopastore (Bruner and Manzi, 2008) and possibly Krapina (Bruner et al., 2007) endocasts. Nonetheless, they reached a "classic" morphology around 50-60 ka.

However, the modern endocranial organisation is supposed to have evolved at least around 100 ka. The skull from Jebel Irhoud (Morocco, about 150 ka), although showing a plesiomorph neurocranial morphology, displays a modern-like overall profile (Bruner et al., 2004). The endocast (Fig. 8b - see Holloway, 1981) shows a non-modern morphology (Bruner et al., 2003). More interestingly, the parietal morphology displays a Neandertal-like lateral expansion leading to the *en bombe* profile in posterior view (Bruner, 2003b; but see Grimaud-Hervé, 2005). The endocast from Skhul V (Near East, around 120 ka), supposed to be a full anatomically modern human, shows the modern-like parietal bulging but not so stressed like in the Upper Pleistocene European fossil record. Unfortunately, some damage at the mid-parietal surface hampers a reliable assessment of the endocranial upper morphology (Fig. 8c).

The first modern humans shared the Mousterian lithic assemblage with the Neandertals. Also, the cognitive evidence of higher level processing capability ("enhanced working memory"; see Wynn and Coolidge, 2003, 2004, 2006; Coolidge and Wynn, 2005) are definitely recognisable much after the first appearance of the modern fossil record. All this incomplete evidence lead us to question whether or not the origin of the modern human lineage coincided with the origin of the modern human brain. This issue is particularly intriguing, and it will represent the most interesting topic in paleoneurology in the next years.

Of course, even if these morphological changes are actually related to the enlargement of the upper parietal cortical areas, the exact nature of such differences must be further investigated, being possibly related to an increased number of neurons, or increased number of connections, or even increased glial component (for example, to support metabolism). On the other hand, this anatomical change can be surely investigated in terms of functional craniology and morphological integration, including considerations on the overall cranial architecture. For example, the modern neurocranial globularity was hypothesised to be a consequence of changes at the temporal and frontal poles (Lieberman et al., 2002). Now, in the evolution of the human genus, the frontal lobes display only some allometric variations (Semendeferi et al., 1997: Rilling, 2006), mainly related to lateral enlargement (Bruner, 2004), and without any relevant changes of the midsagittal profile (Bookstein et al., 1999). Concerning the temporal lobes, although they could show some derived traits mostly related to the lateral morphology (Bastir et al., 2008), their changes are mostly associated with structural and functional constraints related to the biomechanical association of the middle fossa with the underlying mandibular structures (Bastir et al., 2004a, 2004b). Therefore, both the frontal and temporal areas seem hardly related to the neurocranial globularity described for the modern human populations. On the other hand, the parietal enlargement should be carefully considered when the general geometric convolution of the modern brain is acknowledged, associated with forward shifting of the cerebellar and temporal lobes, cranial base flexion, and closure of the interposed spaces (Sylvian valley at the lesser wings of the sphenoid and temporal valley at the petrous pyramids).

Interestingly, the occipital and parietal bones have been hypothesised to be part of a single integrated unit with modern humans and Neandertals being the extremes of a continuous structural trajectory, characterised respectively by bulging occipital and flat parietals, and bulging parietals and flat occipital (Gunz & Harvati, 2007). This information raises two relevant questions: 1) whether the modern human transition has been discrete or more gradual; 2) whether the modern human transition has been based on an actual morphological reorganisation or simply on the variation of pre-existing relationships. Of course, these questions can be only investigated after increasing the fossil record from North Africa, East Africa, and Levant, associated with the second half of the Middle Pleistocene.

Clearly, neontological studies are also needed to move further on these topics, being the current knowledge on the endocranial morphogenesis rather fragmented and heterogeneous.

A large amount of MRI brain studies suggest that the temporal lobes are the only areas in modern humans showing a definite exceeding volume when the allometric pattern of the non-human primates is taken into account (Rilling & Insel, 1999; Semendeferi & Damasio, 2000; Rilling & Seligman, 2002; Rilling, 2006). As already mentioned, a forward shifting of the anterior temporal areas in modern humans has been also described relatively to other extinct human species (Bastir et al., 2008). Unfortunately, mostly because of the blurred boundaries between the parietal and occipital lobes, no volumetric comparisons are currently available for the parietal areas alone. Clearly, it must be assessed whether or not the parietal volume in the modern human brain fits the expected allometric value for primates. I suspect that even these areas could show a positive departure from the allometric trajectory of the primates brain organisation, in its volumetric component or considering the surface/volume relationship.

Other information comes directly from the neurogenetic process of the modern cortical areas. During the brain ontogeny the upper parietal areas reach maturation very early compared with other structures (Gogtay et al., 2004). Considering the common statement that early maturing structures are the most primitive, or the upper parietal cortical areas are not so derived, or the statement is quite misleading! On the other hand, there is evidence that single gene changes can promote/demote the growth of large cortical surfaces (like in polymicrogyria; see Rakic, 2004), suggesting that discrete neural evolutionary steps are at least possible. Finally, there are very interesting approaches remarking the role of neurons as biomechanical tensors in shaping the brain morphology directing the growing forces during ontogeny (Van Es-

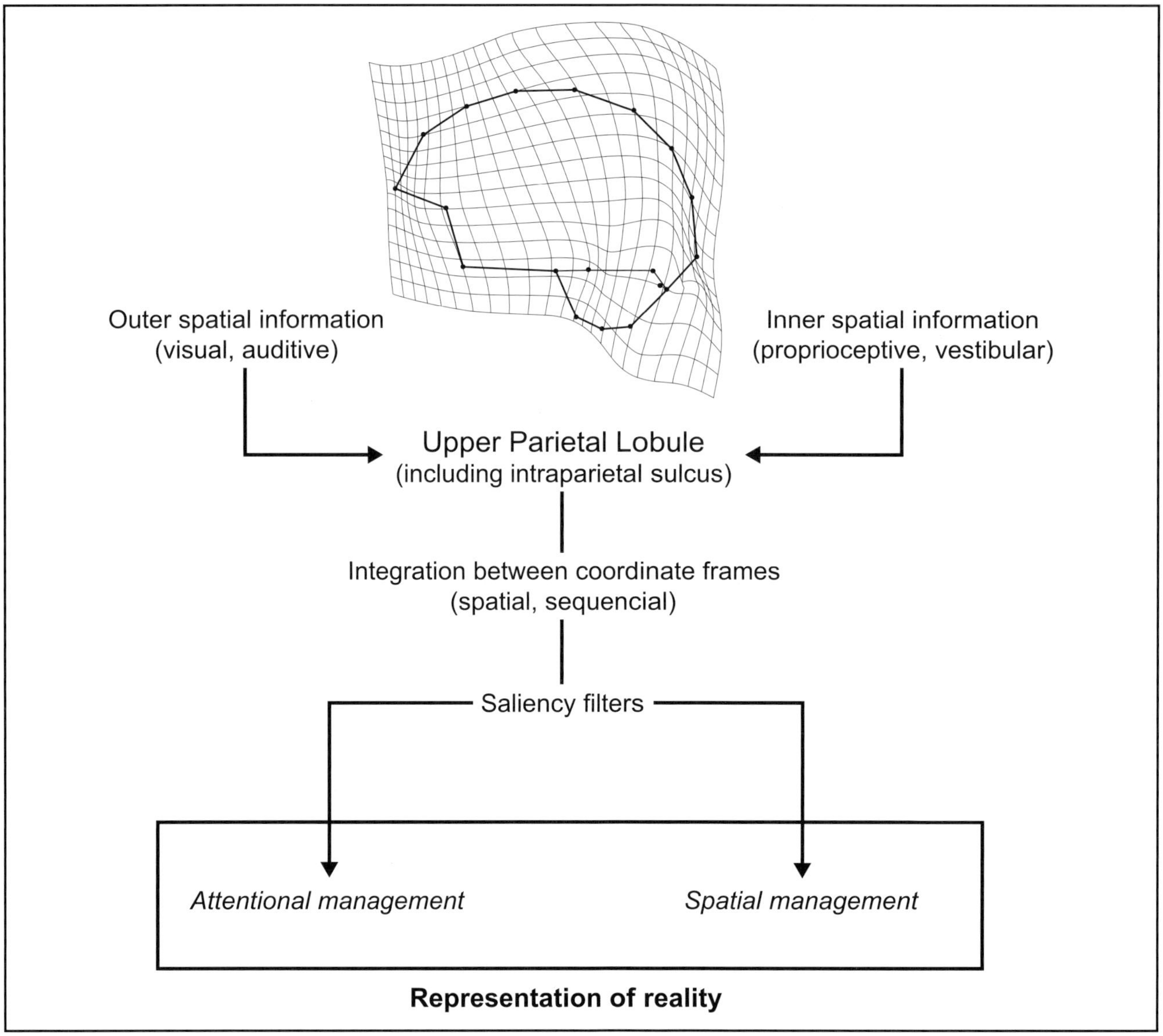

Figure 9. The upper parietal lobules (including the heterogeneous and specialised cortical surface deepened in the intraparietal sulcus) receive spatial information from the outer and inner environments, integrating the different coordinate frameworks in time and space, and producing a mental representation of both the self and the outer reality according to rules, priorities, and cues, associated with relevance, attention, and decision making processes. The upper parietal areas are the interface between mind and environment, reproducing and "handling" reality within a virtual and ordered frame.

sen, 1997; Hilgetag and Barbas, 2005; Toro and Burnod, 2005). In fact, neurons are not only part of the synaptical networks, but also physical anisotropic structures, with specific densities and strain distributions related to their biochemical composition. A change in the neural morphology or cellular organisation will influence the way size and shape changes can be directed throughout the anatomical components during the ontogenesis. Such structural frameworks linking geometry and morphogenesis are even more relevant in paleoneurology when considering that brain morphology also influences physiological variables like thermoregulation or connectivity. This last topic, being related to geometry (Sporns et al., 2002, 2004), should be further considered in paleoneurological studies. In fossils, soft tissues are gone, but the form of the endocranium still provides some information on their processes.

Naturally, the analysis of the parietal evolution has its neuropsychological and behavioural counterpart (Bruner, 2008b; Fig. 9). The studies on the parietal lobes have undergone a relevant development in the last decades (Mountcastle, 1995). The visuo-spatial integration processes associated with the upper parietal areas (including the deepened layers in the intraparietal sulcus) is aimed at receiving information from the inside (eyes, head, limbs) and outside (visual and acoustic stimuli) through different coordinate systems, generating one single coordinate frame able to represent the outer environment and the relationship between the environment and the self (Sakata et al., 1997; Wise et al., 1997). Such a representation is not "objective", being mediated by the personal experience which moulds saliency filters giving a different degree of relevance to different stimuli, and leading to important behavioural responses associated with decision-making and attention (Gootlieb et al., 1998; Rushworth et al., 2001; Andersen and Buneo, 2002; Wardak et al., 2005; Freedman and Assad, 2006). Finally, if the lower parietal areas are mostly linked with the temporal lobes and involved in speech functions, the upper parietal lobules are largely connected with the prefrontal dorsal districts interacting through re-entrant signalling (Battaglia-Mayer and Caminiti 2002; Battaglia-Mayer et al. 2006), opening to speculations on their reciprocal influence in functions associated with working memory and other high-order capabilities.

Most of the literature on the upper parietal areas focus on the intraparietal region as main centre of integration between the self and the outer environment, ranging from hand-eye coordination (that is, physical interaction) to "thought experiments" (that is, virtual interaction) (Andersen et al., 1997; Sakata et al., 1997; Rushworth et al., 2001; Andersen and Buneo, 2002; Bisley and Goldberg, 2003). The integration between self and non-self at the intraparietal sulcus directly leads to intention and goal organisation, including the interpetation of possible actions performed by other individuals (see Tunik et al., 2007 for a detailed review). The geometric comparisons preliminarily suggest that volumetric variation around the intraparietal area are compatible with the morphological differences observed between the modern and non-modern human endocasts. Although the intraparietal sulcus is hardly considered when dealing with the cortical surface, it represents a large volumetric percentage of the parietal cortex, being a rather deep structure, with a mean sulcal depth of 20 mm (Ebeling and Steinmetz, 1995). So, taking into account its functional role, its volumetric component, and the variation highlighted in the geometrical analyses, these areas should be carefully considered when dealing with the origin of the modern brain. It is worth also noting that the intraparietal sulcus is the main area of neural activation when a stone tool is produced (Stout and Chaminade, 2007), this process requiring a three dimensional virtual image of the raw object in mind, the future form visualised into it, shaping hands according to the outer reality, and a project.

In this regard, it must be once more stressed that also Neandertals showed a lateral widening of the upper parietal lobule (Bruner et al., 2003). This should be taken in mind when such a morphological change is associated with a technological one (Mousterian shared also with early modern humans), and a further difference is related to another cultural transition (Aurignacian associated with full modern humans). This leads to another very relevant question, of whether the "domed" appearance in rear view of the modern endocranium is derived from a "tent-like" morphology (maximum endocranial width at the upper temporal areas, like in *H. ergaster/erectus*, and maybe in *H. heidelbergensis* too) or from a "en-bombe" morphology (shared with Neandertals by means of a lateral widening of the upper parietal areas without any vertical and midsagittal enlargement).

Of course, one of the major cognitive proofs of some underlying neural evolution is art (Hodgson, 2006). Here, again we need to understand the spatial organisation of the outer reality, giving a differential importance to its components, to make a virtual projection of the outer scene, and to coordinate our hands and movements with our perceptions and with the relationship we have in mind to represent (i.e., a *simulation*). It is hence rather amazing that, since the early findings on the cave walls, such kinds of first evidence of the modern brain were called "parietal" art!

Acknowledgements

I am extremely grateful to Ralph Holloway, for having dedicated his career supporting the development of the studies of the fossil endocasts, and for his friendship and trust towards a young and improvised Italian paleoneurologist. Sometimes (often) personal support and motivation can help more than thousands of data. Michael Yuan, Doug Broadfield, Kathy Schick, and Nick Toth, thought, planned, and realised this volume on the brain evolution, and the incredible meeting from which it originated. The data on the endocasts were sampled at the Museum of Anthropology G. Sergi (Università La Sapienza, Roma) and at the Istituto Italiano di Paleontologia Umana (Roma), thanks to the collaboration with Giorgio Manzi. The endocasts of Jebel Irhoud and Skhul V were photographed at the Institut de Paleontologie Humaine, thanks to the collaboration with Dominique Grimaud-Hervé, and Henry and Marie-Antoniette de Lumley. Markus Bastir kindly provided comments and suggestions on an early version of this manuscript. I am also grateful to Alexandra Battaglia-Mayer and Roberto Caminiti (Università La Sapienza, Roma), Fred Coolidge (University of Colorado), and Sergio Vaccari (Stanford University) for their help. This paper was supported by the Istituto Italiano di Antropologia (www.isita-org.com).

References

Andersen, R.A., Buneo, C.A., 2002. Intentional maps in posterior parietal cortex. Annual Review of Neuroscience 25, 189-220.

Andersen, R.A., Snyder, L.H., Bradley, D.C., Xing, J., 1997. Multimodal representation of space in the posterior parietal cortex and its use in planning movements. Annual Review of Neuroscience 20, 303-330.

Bastir, M., Rosas, A., Kuroe, K., 2004a. Petrosal orientation and mandibular ramus breadth: evidence for an integrated petroso-mandibular developmental unit. American Journal of Physical Anthropology 123, 340-350.

Bastir, M., Rosas, A., O'Higgins, P., 2004b. Craniofacial levels and the morphological maturation of the human skull. Journal of Anatomy 209, 637-654.

Bastir, M., Rosas, A., Lieberman, D.E., O'Higgins, P., 2008. Middle cranial fossa anatomy and the origin of modern humans. Anatomical Record 291, 130-140.

Battaglia-Mayer, A., Caminiti, R., 2002. Optic ataxia as a result of the breakdown of the global tuning fields of parietal neurons. Brain 125, 225-237.

Battaglia-Mayer, A., Archambault, P.S., Caminiti, R., 2006. The cortical network for eye-hand coordination and its relevance to understaning motor disorders of parietal patients. Neuropsychologia 44, 2607-2620.

Bisley, J.W., Goldberg, M.E., 2003. Neuronal activity in the lateral intraparietal area and spatial attention. Science 299, 81-86.

Bookstein, F.L., 1991. Morphometric tools for landmark data. Cambridge: Cambridge University Press.

Bookstein, F., Schafer, K., Prossinger, H., Seidler, H., Fieder, M., Stringer, C., Weber, G.W., Arsuaga, J.L., Slice, D.E., Rohlf, F.J., Recheis, W., Mariam, A.J., Marcus, L.F., 1999. Comparing frontal cranial profiles in archaic and modern *Homo* by morphometric analysis. Anatomical Record 257, 217-224.

Bookstein, F.L., Gunz, P., Mitteroecker, P., Prossinger, H., Schaefer, K., Seidler, H., 2003. Cranial integration in *Homo*: singular warps analysis of the midsagittal plane in ontogeny and evolution. Journal of Human Evolution 44, 167-87.

Bruner, E., 2003a. Fossil traces of the human thought: paleoneurology and the evolution of the genus *Homo*. Riv Antropol 81, 29-56.

Bruner, E., 2003b. Computer tomography and paleoneurology: the Saccopastore Neandertals and brain evolution in the European Middle Pleistocene. Ph.D. Dissertation. University La Sapienza, Roma.

Bruner, E., 2004. Geometric morphometrics and paleoneurology: brain shape evolution in the genus *Homo*. Journal of Human Evolution 47, 279-303.

Bruner, E., 2007. Cranial shape and size variation in human evolution: structural and functional perspectives. Child's Nervous System 23, 1357-1365.

Bruner, E., 2008a. Comparing endocranial form and shape differences in modern humans and Neandertal: a geometric approach. Paleoanthropology (in press).

Bruner, E., 2008b. Morphological differences in the parietal lobes within the human genus: a neurofunctional perspective. In: Coolodge, F., Wynn, T., (Eds.), Working Memory and the Evolution of Modern Thinking. International Symposium series. Berg Publishers (in press).

Bruner, E., Manzi, G., 2008. Paleoneurology of an "early" Neanderthal: endocranial size, shape, and features of Saccopastore 1. Journal of Human Evolution doi:10.1016/j.jhevol.2007.08.014.

Bruner, E., Ripani, M., 2008. A quantitative and descriptive approach to morphological variation of the endocranial base in Modern Humans. American Journal of Physical Anthropology doi: 10.1002/ajpa.20837.

Bruner, E., Sherkat, S., 2008. The middle meningeal artery: from clinics to fossils. Manuscript.

Bruner, E., Manzi, G., Arsuaga, J.L., 2003. Encephalization and allometric trajectories in the genus *Homo*: evidence from the Neandertal and modern lineges. Proceedings of the National Academy of Sciences USA 100, 15335-15340.

Bruner, E., Saracino, B., Passarello, P., Ricci, F., Tafuri, M., Manzi, G., 2004. Midsagittal cranial shape variation in the genus *Homo* by geometric morphometrics. Collegium Antropologicum 28, 99-112.

Bruner, E., Mantini, S., Perna, A., Maffei, C., Manzi, G., 2005. Fractal dimension of the middle meningeal vessels: variation and evolution in *Homo erectus*, Neanderthals, and modern humans. European Journal of Morphology 42, 217-224.

Bruner, E., Manzi, G., Holloway, R., 2006. Krapina and Saccopastore: endocranial morphology in the pre-Würmian Europeans. Period Biolog 108, 433-441.

Cheverud, J.M., 1996. Developmental integration and the evolutino of pleiotropy. American Zoologist 36, 44-50.

Choi, H.J., Zilles, K., Mohlberg, H., Schleicher, A., Fink, G.R., Armstrong, E., Amunts, K., 2006. Cytoarchitectonic identification and probabilistic mapping of two distinct areas within the anterior ventral bank of the human intraparietal sulcus. Journal of Comparative Neurology 495, 53-69.

Coolidge, F., Wynn, T., 2005. Working memory, its executive functions, and the emergence of modern thinking. Cambridge Archaeological Journal 15, 5-26.

Ebeling, U., Steinmetz, H., 1995. Anatomy of the parietal lobe: mapping the individual pattern. Acta Neurochirurgica (Wien) 136, 8-11.

Eidelberg, D., Galaburda, A.M., 1984. Inferior parietal lobule. Archives of Neurology 41, 843-852.

Enlow, D.H., 1990. Facial Growth. Philadelphia: WB Saunders Company.

Falk, D., 1987. Hominid paleoneurology. Annual Review of Anthropology 16, 13-30.

Falk, D., 1993. Meningeal arterial pattern in great apes: implication for hominoid vascular evolution. American Journal of Physical Anthropology 92, 81-97.

Freedman, D.J., Assad, J.A., 2006. Experience-dependent representation of visual categories in parietal cortex. Nature 443, 85-88.

Gogtay, N., Giedd, J.N., Lusk, L., Hayashi, K.M., Greenstein, D., Vaituzis, A.C., Nugent III, T.F., Herman, D.H., Clasen, L.S., Toga, A.W., Rapoport, J.L., Thompson, P.M., 2004. Dynamic mapping of human cortical development during childhood through early adulthood. Proceedings of the Nationall Academy of Sciences USA 101, 8174-8179.

Gottlieb, J.P., Kusunoki, M., Goldberg, M.E., 1998. The representation of visual salience in monkey parietal cortex. Nature 391, 481-484.

Grimaud-Hervè, D., 1997. L'évolution de l'enchéphale chez *Homo erectus* et *Homo sapiens*. Paris : CNRS Editions.

Grimaud-Hervé, D., 2005. Apports de quelques caractères morphologiques endocraniens dans la determination taxinomique des hominidés de Jebel Ihroud (Maroc). Anthropologie XLIII: 101-112.

Gunz, P., Harvati, K., 2007. The Neanderthal "chignon": variation, integration, and homology. Journal of Human Evolution 52, 262-274.

Henderson, J.H., Longaker, M.T., Carter, D.R., 2004. Sutural bone deposition rate and strain magnitude during cranial development. Bone 34, 271-80.

Hilgetag, C.C., Barbas, H., 2005. Developmental mechanics of the primate cerebral cortex. Anatomy and Embryology 210, 411-417

Hodgson, D., 2006. Understanding the origin of the paleoart: the neurovisual resonance theory and brain functioning. Paleoanthropology 2006, 54-67.

Holloway, R.L., 1978. The relevance of endocasts for studying primate brain evolution. In: Noback, C.R., (Ed.), Sensory Systems of Primates. New York: Plenum Press. pp. 181-200.

Holloway, R.L., 1981. Exploring the dorsal surface of hominoid brain endocasts by stereoplotter and discriminant analysis. Philisophical Transactions of the Royal Society of London - Series B 393, 155 -166.

Holloway, R.L., 1995. Toward a synthetic theory of human brain evolution. In: Changeaux, J.P., Chavaillon, J., (Eds.), Origins of the Human Brain. Oxford: Clarendon Press. pp. 42-54.

Holloway, R.L., Broadfield, D.C., Yuan, M.S., 2003. Morphology and histology of chimpanzee primary visual striate cortex indicate that brain reorganization predated brain expansion in early hominid evolution. Anatomical Record A 273, 594-602.

Holloway, R.L., Broadfield, D.C., Yuan, M.S., 2004. Brain endocasts: The paleoneurological evidence. The Human fossil record (Vol III). Hoboken, New Jersey: Wiley-Liss.

Jiang, X., Iseki, S., Maxson, R.E., Sucov, H.M., Morriss-Kay, G.M., 2002. Tissue origins and interactions in the mammalian skull vault. Develop. Biol. 241, 106-116.

Lieberman, D.E., McBratney, B.M., Krovitz, G., 2002. The evolution and development of cranial form in Homo sapiens. Proceedings of the Nationall Academy of Sciences USA 99, 1134-9.

Manzi, G., Vienna, A., Hauser, G., 1996. Developmental stress and cranial hypostosis by epigenetic trait occurrence and distribution: an exploratory study on the Italian Neandertals. Journal of Human Evolution 30, 511-527.

Morriss-Kay, G.M., Wilkie, A.O., 2005. Growth of the normal skull vault and its alteration in craniosynostosis: insights from human genetics and experimental studies. Journal of Anatomy 207, 637-53.

Moss, M.L., 1959. The pathogenesis of premature cranial synostosis in man. Acta Anatomica 37, 351-370.

Moss, M.L., Young, R.W., 1960. A functional approach to craniology. American Journal of Physical Anthropology 18, 281-292.

Mountcastle, V.B., 1995. The parietal system and some higher brain functions. Cerebral Cortex 5, 377-390.

O'Laughlin, V., 1996. Comparative endocranial vascular changes due to craniosynostosis and artificial cranial deformation. American Journal of Physical Anthropology 101, 369-385.

Olson, E.C., Miller, R.L., 1958. Morphological integration. University of Chicago Press, Chicago.

Rakic, P., 2004. Genetic control of cortical convolutions. Science 303, 1983-1984.

Richtsmeier, J.T., Aldridge, K., de Leon, V.B., Panchal, J., Kane, A.A., Marsh, J.L., Yan, P., Cole, T.M., 2006. Phenotypic integration of neurocranium and brain. Journal of Experimental Zoology Part B: Molecular and Developmental Evolution) 306B, 360-378.

Rilling, J.K., 2006. Human and non-human primate brains: are they allometrically scaled versions of the same design? Evolutionary Anthropology 15, 65-67.

Rilling, J.K., Insel, T.R., 1999. The primate neocortex in comparative perspective using magnetic resonance imaging. Journal of Human Evolution 37, 191-223.

Rilling, J.K., Seligman, R.A., 2002. A quantitative morphometric comparative analysis of the primate temporal lobe. Journal of Human Evolution 42, 505-533.

Rohlf, F.J., 2004. tpsSplin 1.20. Ecology and Evolution, SUNY at Stony Brook, NY.

Rohlf, F.J., 2007. tpsRelw 1.45. Ecology and Evolution, SUNY at Stony Brook, NY.

Rushworth, M.F.S., Paus T., Sipila, P.K., 2001. Attentional system and organisation of the human parietal cortex. Journal of Neuroscience 21, 5262-5271.

Saban, R., 1979. Le systeme veineux méningé de l'homme fossile de Biache-Saint-Vaast Pas de Calais, d'apres le moulange endocranien. C R Acad Sci Paris D 289, 1129-1132.

Saban, R., 1982 Le systéme des veines méningées moyennes chez les hommes fossiles de Tchécoslovaquie, d'aprés le moulage endocranien. Anthropos 21, 281-295.

Sakata, H., Taira, M., Kusunoki, M., Murata, A., Tanaka, Y., 1997. The parietal association cortex in depth perception and visual control of hand action. Trends in Neurosciences 20, 350-357.

Semendeferi, K., Damasio, H., 2000. The brain and its main anatomical subdivisions in living hominoids using magnetic resonance imaging. Journal of Human Evolution 38, 317-332.

Semendeferi, K., Damasio, H., Frank, R., Van Hoesen, G.W., 1997. The evolution of the frontal lobes: a volumetric analysis based on three-dimensional reconstructions of magnetic resonance scans of human and ape brains. Journal of Human Evolution 32, 375-388.

Sergi, S., 1934. Ossicini fontanellari della regione del lambda nel cranio di Saccopastore e nei crani neandertaliani. Riv Antropol 30, 101-112.

Sergi, S., 1948. L'uomo di Saccopastore. Paleontographia Italica XLII, 25-164.

Slice, D.E., 2000. *Morpheus et al.* Ecology and Evolution, SUNY at Stony Brook, NY.

Sporns, O., Tononi, G., Edelman, G.M., 2002. Theoretical neuroanatomy and the connectivity of the cerebral cortex. Behavioural Brain Research 135, 69-74.

Sporns, O., Chialvo, D.R., Kaiser, M., Hilgetag, C.C., 2004. Organization, development and function of complex brain networks. Trends in Cognitive Sciences 8, 418-425.

Stout, D., Chaminade, T., 2007. The evolutionary neuroscience of tool making. Neuropsychologia 45, 1091-100.

Toro, R., Burnod, Y., 2005. A morphogenetic model for the development of cortical convolutions. Cerebral Cortex 15, 1900-1913.

Tunik, E., Rice, N.J., Hamilton, A., Grafton, S.T., 2007. Beyond grasping: representation of action in human anterior intraparietal sulcus. Neuroimage 36, T77-T86.

Van Essen, D.C., 1997. A tension-based theory of morphogenesis and compact wiring in the central nervous system. Nature 385, 313-318.

Wardak, C., Ben Hamed, S., Duhamel, J.R., 2005. Parietal mechanism of selective attention in monkeys and humans. In: Dehaene, S., Duhamel, J.R., Hauser, M.D., Rizzolatti, G., (Eds.), From monkey brain to human brain. Cambridge: MIT Press. pp. 273-291.

Wise, S., Boussaoud, D., Johnson, P.B., Caminiti, R., 1997. Premotor and parietal cortex: corticocortical connectivity and combinatorial computations. Annual Review of Neuroscience 20, 25-42.

Wynn, T., Coolidge, F., 2003. The role of working memory in the evolution of managed foraging. Before Farming 2, 1-16.

Wynn, T., Coolidge, F., 2006. The effect of enhanced working memory on language. Journal of Human Evolution 50, 230-231.

Wynn, T., Coolidge, F.L., 2004. The expert Neandertal mind. Journal of Human Evolution 46, 467-487.

Zelditch, M.L., Swidersky, D.L., Sheets, H.D., Fink, W.L., 2004. Geometric morphometrics for biologists. San Diego: Elsevier.

CHAPTER 7

Cerebellum and Brain Evolution in Holocene Humans

Anne H. Weaver

Abstract

Evolutionary development of the cerebellum and its implications for human cognitive evolution must be considered in the broader framework of hominin brain evolution–a gradual, complex process that involved heterochrony in response to genetic evolution, cultural innovation, population dynamics, and environmental challenges over an extended time. This chapter offers a chronology of relevant events and a proposed scenario that integrates the presently available data relating to cerebellar evolution in Late Pleistocene and Holocene humans (150,000 years BP to the present.)

Introduction

Overall brain volume in the genus *Homo* reached its maximum by the late Middle Pleistocene (>150,000 years BP), and then declined somewhat thereafter, probably in response to a decrease in body mass (Ruff, Trinkaus et al., 1997).

The relative proportions of the cerebellum and cerebral hemispheres continued to change as well, suggesting that cerebellar-neocortical interactions remained under selection. In recent humans, the cerebral hemispheres are relatively smaller and the cerebellum relatively larger than in terminal Pleistocene humans. The cerebellum did not reach modern proportions until the very late Pleistocene (some time after 28,000 years BP) postdating by several thousand years the appearance of modern humans in territories occupied by archaic human groups (Weaver, 2001; Weaver, 2005).

This pattern of continued cerebellar evolution is consistent with fossil, archeological and genetic evidence that human behavior and morphology, including brain morphology, have continued to evolve since the Late Pleistocene(Hawks, Wang et al., 2007).

Temporal correlations between divergent lines of evidence can stimulate causal hypotheses about the genetic and cultural dynamics that interacted to produce brain morphology and related cognitive patterns characteristic of living humans. The purpose of this paper is to outline chronological developments that may have contributed to cerebellar evolution, summarize the present state of the evidence, and suggest further directions for research.

Chronology for Archaic and Modern Human Populations

Dating of fossil and archeological remains during the crucial Late Pleistocene period has been subject to a flurry of recent revisions related to refined methodologies in analysis and calibration of Carbon-14 (^{14}C) dates. The chronology outlined below will rely on uncalibrated dates because they are the most accessible and consistent. However, the reader should note that uncalibrated dates tend to underestimate calendar years. For current discussions of the implications of reanalysis and redating of critical fossils and sites, refer to Trinkaus (Trinkaus, 2005), Mellars (Mellars, 2006), and Zihlão (Zilhão, 2006).

By the end of the Middle Pleistocene (250,000 years BP), Africa and Eurasia were populated by scattered groups of fairly large-brained, sturdy-bodied humans known as "early archaic *Homo sapiens*." They were descendants of *Homo erectus*, who had spread from Africa over the preceding 1.5 million years. Early archaic

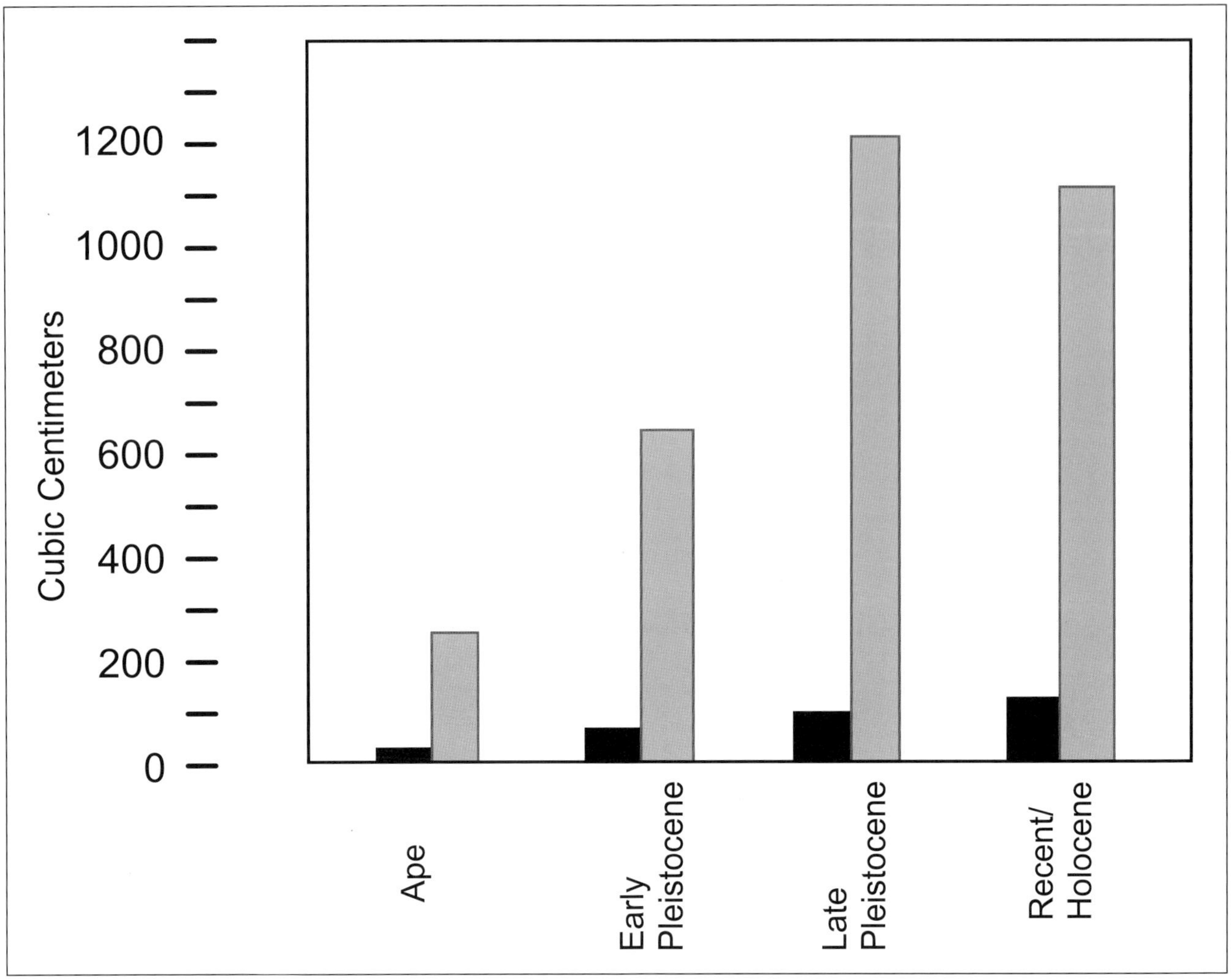

Figure. 1. Stages in cerebellar evolution. Light Gray = cerebral hemisphere volume; Dark Gray = cerebellum volume. Early–Middle Pleistocene includes Homo habilis, Homo erectus, *and early archaic* Homo sapiens. *Late Pleistocene includes Neandertals and Cro-Magnon 1.*

Homo sapiens gave rise to two populations of particular interest: Neandertals and anatomically modern humans.

Neandertals are a regional variant of late archaic *Homo sapiens* who inhabited Europe and Western Asia. Neandertals are distinguished from other late archaic *Homo sapiens* by their "cold adapted" body proportions, expanded nasal morphology, and other dental, skeletal and craniofacial features (Trinkaus, 2006). Individual distinctive Neandertal features are present in European fossils as old as 400-600,000 years BP (Bischoff, Shamp et al. 2003) but the "classic" Neandertal pattern did not coalesce until after 75,000 years BP. The most recent reliably dated diagnostic Neandertal remains have been radiocarbon dated to 32-33,000 years BP (Higham, Ramsey et al., 2006).

In Africa, archaic *Homo sapiens* gave rise to another distinct lineage, destined to become the ancestors of recent humans, approximately 200,000 years BP. Like Neandertals, they were large-brained. These hominins were characterized by their shorter faces, smaller incisors, narrower noses, and prominent chins, along with derived postcranial features especially notable in the shoulder and pelvis (Trinkaus, 2006) . Because Holocene humans share these characteristics, this group of fossils and later ones that resemble them are known as "early anatomically modern *Homo sapiens*." They are considered to be the direct ancestors of contemporary humans. Anatomically modern humans were apparently confined to Africa until 50,000-60,000 years BP, (with brief incursions into the Levant ca. 90,000 years BP) (Vandermeersch, 1981; Tillier, 1999).

Anatomically modern humans reached Australia by 50,000 years BP (Brown, 1992), Eastern Asia by 30,000 years BP (Trinkaus, 2005), and the Americas by at least 12,500 years BP (Dillehay, 1999).

The pattern of interaction between anatomically modern humans and the archaic populations they encountered as they spread out of Africa is ambiguous. A thorough analysis of the archeological, skeletal and genetic evidence precludes contact between anatomically modern humans and Neandertals prior to 43,000 years ago. Then, within a brief period of about 1,000 years, admixture, competition and conflict had resolved the distinctive modern and human population differences between the two groups. Neandertal skeletal features persisted in scattered hybrid descendants (Trinkaus 2006), but gradually decreased in frequency until few overt traces remained. To date, analyses of Neandertal DNA suggest that they may have made a minimal, as yet poorly defined, contribution to the modern human gene pool, perhaps through limited interbreeding between modern males and Neandertal females (Trinkaus, Moldovan et al., 2003; Green, Krause et al., 2006; Noonan, Coop et al., 2006; Zilhão, 2006). A recent study has identified *microcephalin* variant *MCPH1*, which appears to affect neocortical and cerebellar development (Trimborn, Bell et al., 2004), as an introgressive gene from late archaic humans–possibly Neandertals. (See discussion below.) (Evans, Mekel-Bobrov et al., 2006; Zilhão, 2006).

Endocranial Measurements in Fossils

Brain size in fossils is estimated from endocranial capacity. Relative brain size in fossil hominins is customarily expressed as an encephalization quotient, or "EQ," (Jerison, 1974) which represents logged values for actual/expected brain size for a given body mass. The best estimates of body mass and endocranial volumes for Late Pleistocene archaic and modern humans suggest that, despite their absolutely larger brains, mean EQ in Neandertals was slightly lower than that of anatomically modern humans, although there is considerable overlap at the lower end of the modern human range of variation (Ruff, Trinkaus et al., 1997; Rosenberg, Zune et al., 2006).

Neandertals, like other archaic humans, have elongated braincases, with low, sloping foreheads. However, their frontal lobe profiles, recorded on the inner table of the skull, do not differ significantly from modern humans (Bookstein, Schäfer et al., 1999; Holloway, Broadfield et al., 2004). Asymmetry of the cerebral hemispheres, well established in hominins since the early Pleistocene (Begun and Walker, 1993), is similar in Neandertals and anatomically modern humans, suggesting a similar pattern of functional lateralization. On the basis of differences in parietal proportions (at least with regard to chord measurements), have led Bruner (Bruner, 2004) to suggest that brain expansion in Neandertals and early modern humans may have followed independent evolutionary trajectories. However, as described below, the archeological record indicates that both groups engaged in similar subsistence behaviors and developed similar technical expertise over time (Mellars, 2006; Zilhão, 2006; Zilhao, d'Errico et al., 2006).

Relative Cerebellar Volume

Cerebellum size in fossil hominins can be estimated from the volume of the posterior cranial fossa, which is highly correlated with cerebellum volume (Weaver, 2005). To date, cerebellar volume estimates for fossils are available for a total of 18 fossil hominins. The Late Pleistocene sample includes one anatomically modern human (Cro-Magnon I; 28,000 years BP); and three classic Neandertals (La Chapelle I, La Ferrassie I, and Gilbraltar/Forbes Quarry (Weaver, 2005). The recent human comparative sample incorporates data for over 1,450 individuals, collected from multiple sources (Weaver, 2005). Such a large data set for modern humans permits a degree of statistical robusticity, despite the small size of the fossil sample. Nonetheless, it should be emphasized that the present fossil sample is very small and interpretations based on such a limited sample are tentative.

The data indicate that recent humans have cerebella that are both absolutely and relatively larger than *any* of the Late Pleistocene humans. As a consequence, *the cerebral hemispheres of contemporary humans are proportionately smaller than they are in Neandertals and Cro-Magnon 1* (Weaver, 2005).

The brain is a costly organ to maintain, despite its putative contribution to evolutionary fitness. It is likely that increased cerebellar volume represents a fitness advantage to Holocene and recent humans, reflected in a more useful or efficient cognitive strategy without an increase in overall brain mass.

Evolution in Genes That Contribute to Brain Development

Continued cerebellar evolution coincides with changes in relative frequency of a number of genes involved in regulating brain development. These genes appear to have been subject to positive Darwinian selection. Notable among them are *microcephalin*, *ASPM*, and *FOXP2* (Vargha-Khadem, Watkins et al., 1995; Vargha-Khadem, Watkins et al., 1998; Lai, Fisher et al., 2001; Enard, Przeworski et al., 2002; Evans, Anderson et al., 2004; Evans, Anderson et al., 2004; Shu, Cho et al., 2005; Evans, Mekel-Bobrov et al., 2006; Evans, Vallender et al., 2006). At present, the confidence intervals related to the timing of mutations in these genes are very broad. In addition, functional contributions of these genes to brain morphology and their possible contributions to cognitive behavior are poorly understood. However, temporal correlations between changes in these genes brain and behavioral changes are suggestive.

Microcephalin is of particular interest, because one variant (*MCPH1*) may have passed from Neandertals or other archaic humans into the anatomically modern gene pool (Evans, Anderson et al., 2004; Evans, Mekel-Bobrov et al., 2006). *Microcephalin* plays a critical role in brain size regulation. It appears to affect cerebellar size as well (Trimborn, Bell et al., 2004). Its influence on the cerebellum may be the result of a global developmental effect that manifests during the proliferation phase of neurogenesis. Analysis of the *MCPH1* allele suggests that it entered the gene pool of anatomically modern humans at or sometime before 37,000 (± 23,000) years BP from an archaic human population. As a result of positive selection, this allele has now reached a frequency of 70% in recent humans. The very high frequencies of this allele in eastern Siberia (98%) and in indigenous American populations (92-100%) seem to indicate that this allele was well established prior to the entry of humans into the Americas.

Unfortunately, both genetic and quantitative morphological data for Australian brain morphology are severely lacking. There is some evidence that cerebellar proportions in indigenous Australians are atypical. (Klekamp, Riedel et al., 1987; Klekamp, Riedel et al. 1989) Additional genetic and volumetric data for Australian populations could be highly informative in determining which alleles, if any, may have entered the modern human gene pool subsequent to human dispersal into Australia.

ASPM, another allele that is expressed in the cerebellum and also contributes to brain size may have affected brain evolution in hominins. *ASPM* appears to have been subject to positive selective pressure after the split of humans and chimpanzees (5-7 million years BP) but before modern humans left Africa (Zhang, 2003; Evans, Anderson et al., 2004). One variant of *ASPM*, haplotype 63, appeared ~5800 years BP (95% confidence interval 14,100 - 500 years), and may be subject to ongoing positive selection (Mekel-Bobrov, Gilbert et al., 2005).

Neither *microcephalin MCPH1* nor *ASPM* haplotype 63 has reached fixation in the human genome as a whole. Their present-day differential distribution across global populations may reflect alternate but equivalent cognitive strategies in living humans; mutations that were subject to positive selection in the past; or atypical historical population dynamics.

The *FOXP2* gene has received considerable attention due to its apparent role in regulating social communication in many organisms. Disruption of this gene in humans manifests in problems with grammar and speech production, non-verbal intelligence, and non-speech related movements of the mouth and face. *FOXP2*, a transcription factor, appears to affect development of several brain regions, most notably the caudate nucleus of the basal ganglia (Vargha-Khadem, Watkins et al., 1995; Vargha-Khadem, Watkins et al., 1998; Lai, Fisher et al., 2001; Shu, Cho et al., 2005). Experiments with mice suggest that *FOXP2* is involved in regulating cerebellar development and neural migration as well (Shu, Cho et al., 2005; Brumwell and Curran, 2006). This gene has apparently come under selection in recent human evolution, undergoing a selective sweep between 100,000 years and 10,000 years BP (Enard, Przeworski et al., 2002). *FOXP2* may indeed play a role in human cognitive evolution. However, it is important to emphasize that no single gene can possibly account for the emergence of the complex, highly distributed, redundant, and flexible set of cognitive behaviors involved in human language.

Cerebellar Evolution & Cognition

Many researchers have suggested that the specialized architecture of the cerebellum enables it to process neural data regardless of whether they arise in "sensory", "motor", or "association" areas of the neocortex (Leiner, Leiner et al., 1986; Ito, 1993; Houk and Wise, 1995; Arriada-Mendicoa, Otero-Silceo et al., 1999; Fox, Sitompul et al., 1999; Houk and Miller, 2001; Middleton and Strick, 2001; Houk, 2005). There is a growing consensus that the cerebellum makes a significant contribution to higher cognitive functions, including planning of future actions, working memory, visual perception, directed attention, and rule-based learning.

The archeological record supports the hypothesis that intensified demands related to these quintessentially human cognitive behaviors coincided with cerebellar evolution. Schoenemann (Schoenemann, 2006) summarizes social and ecological factors that may have intensified selective pressures for brain evolution during the course of hominid evolution. Following is a brief chronology focusing on behaviors that emerged as part of the archeological record of the Holocene, in comparison to those already in place by the Late Pleistocene. Again, however, it must be noted that temporal correlations are not causally definitive. While the data are thought-provoking, they are not conclusive.

By the Late Pleistocene, and prior to cerebellar expansion, both late archaic and anatomically modern humans had mastered a complex repertory of culturally-based subsistence techniques related to complex foraging, invented and maintained for thousands of years by large-brained hominins. For example, more than 80,000 years ago Neandertals had mastered a tightly-controlled and complex series of processes, including the advanced pyrotechnology, needed to manufacture adhesive pitch from birch bark. A fingerprint and wood fragments indicate that the pitch was used to fashion hafted tools (Koller, Baumer et al., 2001; Zilhão, 2006).

Like anatomically modern humans, Neandertals had the hunting skill to rank as top predators in diverse climates, adapting their hunting strategies to suit local circumstances (Richards, Pettitt et al., 2000; Bocherens, Drucker et al., 2005; Zilhão, 2006). Both anatomically modern humans and Neandertals left examples of non-figurative "concept-mediated markings" (Bednarick, 1995) on bone, rock, and gravestones (Peyrony, 1934). As early as 75,000 years ago, anatomically modern humans were stringing together beads made from drilled and pierced ostrich eggs and marine shells. By 35,000 years ago, Neandertals were making beads as well, favoring pierced and grooved animal teeth, bone, ivory, soft stone and even fossils (Peyrony, 1934; Bednarick, 1995; d'Errico, Zilhao et al., 1998; Zilhão, 2006). The earliest recognizable musical instruments–flutes made of bird bone, found at Geissenklösterle, have been dated between 33,500 and 37,000 years BP (Hahn and Münzel, 1995; Conard and Uerpmann, 2000).

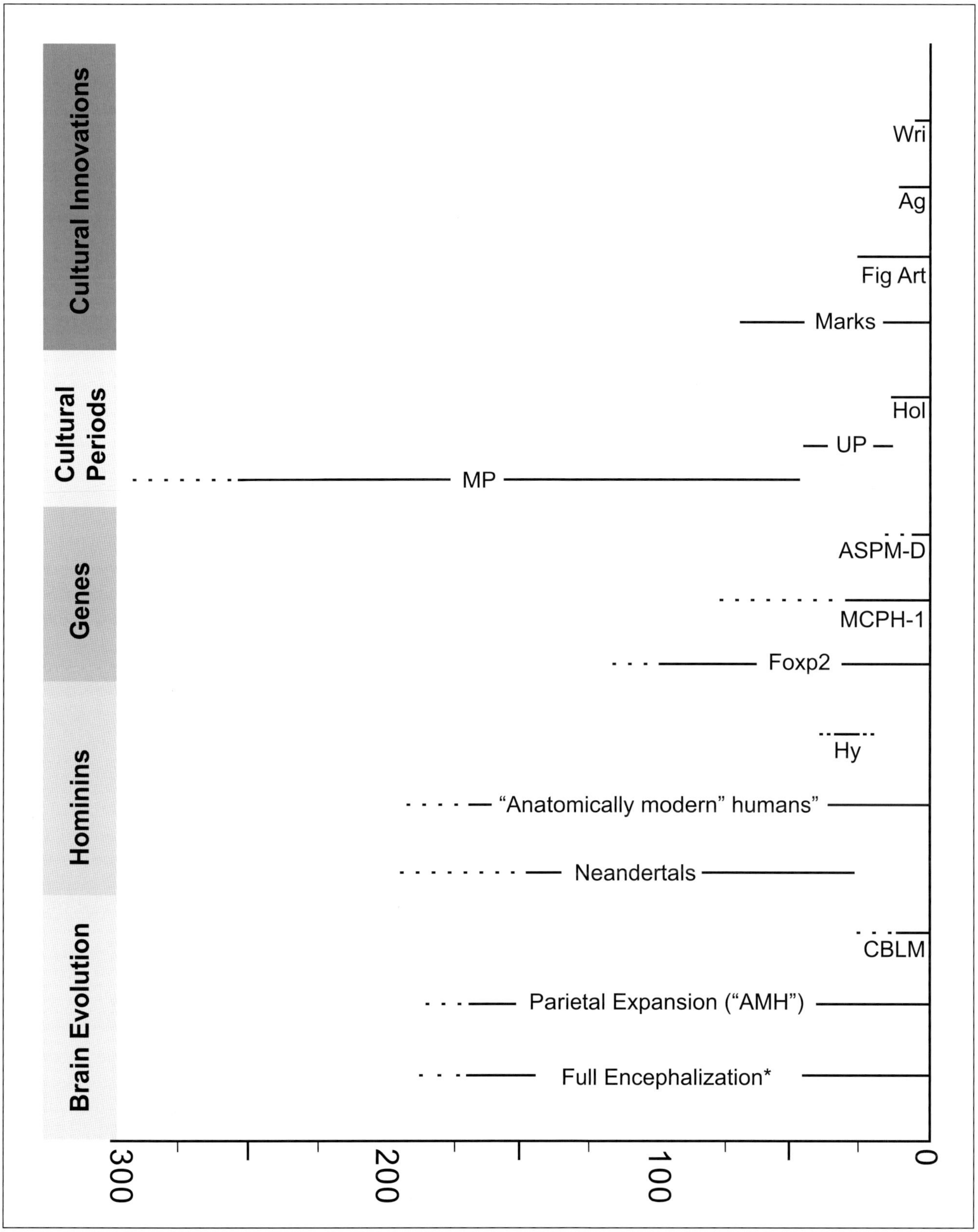

Figure. 2. Timetable of Cognitive Developments and Cultural Stages for the Late Pleistocene. Key to abbreviations: "AMH" = anatomically modern humans; "CBLM" = relative cerebellar expansion; "MCPH1" = *microcephalin variant* MCPH1*; "ASPM" = ASPM haplotype 63; "Hy" = period of probable hybridization between Neandertals and anatomically modern humans in Europe; "MP" = Middle Paleolithic cultural period; "UP" = Upper Paleolithic cultural period; "Hol" = Holocene epoch; "Marks" = appearance of "concept-mediated marking"(Bednarick 1995); "FigArt" = appearance of figurate/representational art; "Ag" = onset of agriculture in Western Asia; "Wri" = appearance of writing.*

That is to say, Neandertals and anatomically modern humans cannot be distinguished on the basis of their technological accomplishments or behavioral repertories. Despite marginal differences in encephalization and *possible* morphological differences related to differential parietal lobe development, there is no uncontroversial evidence that Late Pleistocene anatomically modern humans displayed "advanced" cognitive behavior relative to their Neandertal cousins. This is consistent with the similar cerebellar morphology characteristic of both groups. And it is in accord with the hypothesis that Neandertals and anatomically modern humans were members of the same species.

Introgression of the *microcephalin MCPH1* allele was roughly coincident with early encounters between Neandertals or other archaic humans with the emergence of anatomically modern humans from Africa 50,000–35,000 years BP. Shortly after that time (ca. 30,000 years BP), well-developed figurative art appears in the archeological record. Given the extensive use of ochre and occasional manifestations of "concept-mediated marking" dating back for tens of thousands of years, both in Africa and in Europe, it is reasonable to speculate that the sophisticated drawings, carvings, and small-scale sculptures produced after 30,000 years ago arose from a long-incipient human tendency for visual expression. Newly activated or continued selection acting on the *FOXP2* gene may have contributed to this flowering of symbolic activity. However, increased cerebellar capacity was apparently not a prerequisite for such expressions.

What, then, might account for continued brain evolution in humans after 30,000 years BP, including cerebellar expansion, continued selective pressure acting on *FOXP2*, *MCPH1,* and the appearance and selection associated with *ASPM* haplotype 63?

Climatic stress may have been an important selective agent. Beginning about 28,000 years ago, a period of severe global cooling ensued. Glaciers expanded, forests retreated, and by 18-20,000 years ago, northern Africa was an extreme tropical desert. Steppe-tundra covered those portions of southern Europe that remained unglaciated.

Proliferating technology kept pace with the intensifying cold and shift in ecology: the archeological record from the terminal Pleistocene in Europe includes cordage, woven goods; and eyed needles, fishhooks, weirs, traps, armatures, and projectile weapons such as the atl atl and, later, the bow (Straus, 1995; Straus, 1997; Soffer, 2004). Figurative statuary and cave art continued to flourish during this period as well. Innovations developed to meet extreme conditions were not lost when the glaciers retreated.

By the terminal Pleistocene, humans had achieved a degree of precise motor control, advanced concrete operational intelligence, and a sophisticated, complex, proto-modern material culture (Wynn, 1985; Wynn, 1991; Wynn, 1996). Even the most conservative researchers agree that terminal Pleistocene and early Holocene humans had already developed syntactical language(Conkey, 1987; Klein, 1992; Noble and Davidson, 1996).

However, they may not have fully developed the formal operational intelligence (characterized by abstract thinking)(Inhelder and Piaget 1958) that is achieved during adolescence and early adulthood in modern humans in contemporary societies. The capacity for formal operational intelligence (and concomitant enhancement of cortico-cerebellar circuitry) may have emerged as a cognitive tool in response to population pressures and/or subsistence strategies related to plant and animal domestication. Early evidence of pastoralism and horticulture are associated with the Younger Dryas, ca. 10,500 years BP (Bar-Yosef, 1998). The earliest towns began to coalesce about 8,000 years ago (Göktürk, Hillegonds et al., 2002). Complex, abstract "external memory devices," including calendars and notation systems, (Donald, 1993) appeared shortly thereafter.

The cultural context for human cognition changed slowly, but drastically. Formalized power structures, tribal networks, and kinship systems required greater impulse control, deferred gratification, and social cognition. The relentless, complex demands of growing crops and maintaining herds required *abstract* operational intelligence, impulse control and long-term planning that were qualitatively different from the *concrete* operational intelligence, patience, wiliness and observational acuity needed by foragers.

The balance between personal memory and culturally distributed memory shifted. Non-iconographic counting tokens from Sumeria date back 8,000 years, followed by fully developed writing systems by 5,300 years ago (Schmandt-Besserat, 1996). Calendars, accounting systems, formalized liturgies and rituals, legal systems, and epic literature blossomed thereafter.

The specialized cognitive demands placed on terminal Pleistocene and Holocene humans may have favored restructuring of neural computational networks, either ontologically, phylogenetically, or both.

Terminal Pleistocene and Holocene Cognitive Skills

If natural selection did in fact operate to change neural circuitry in terminal Pleistocene and Holocene humans, such selection would have been constrained by the need to preserve existing primary brain functions. A "language organ" or "abstract thinking organ" cannot be simply inserted into an unused gyrus or sulcus at an arbitrary point in development. New or enhanced functions emerge based on previously-established neural circuitry.

Phylogenetic changes in the rate or timing of organic development occur through an evolutionary dynamic known as heterochrony. Peramorphosis is a form of heterochrony that involves terminal extension of the latest phase of maturation. In the human brain, the period of late maturation involves vigorous synaptic pruning in the cerebral cortex, accompanied by an increase in subcortical volume due to myelination and expansion of subcortical circuitry. Peramorphosis is a credible mechanism to explain putative cognitive and neurological differences between recent humans and their Terminal Pleistocene and Holocene predecessors. For example, ontogenic shifts in the timing and duration of cortical growth and subsequent pruning can have significant effects on IQ (Shaw, Greenstein et al., 2006).

The protracted phase of brain maturation during adolescence in modern humans is correlated with a shift in cognitive and psychological strategies in which the cerebellum participates. This shift in cognitive strategies appears to reflect more efficient neural organization and sharing of information among cortical and subcortical brain regions (Katz and Steinmetz, 2002). Behaviorally, the late phase of brain maturation results in the capacity for formal operational thinking, greater voluntary control of behavior, cognitive flexibility, social identity formation, and an enhanced ability to fine-tune cortical activity (Giedd, Snell et al., 1996; Luna, Thulborn et al., 2001; Giedd, 2004; Blakemore and Choudhury, 2006; Shaw, Greenstein et al., 2006). In tasks that require voluntary control of reflexive/impulsive response tendencies, adults exhibit reduced reliance on the neocortex and greater recruitment of the cerebellum than adolescents do (Luna, Thulborn et al., 2001).

Summary

Given the present evidence, it is impossible to state exactly where or when cerebellar proportions changed during the last 28-30,000 years; whether the change was gradual or rapid; what specific cultural innovations may have been involved; or in what population(s) it first occurred.

However, the data we do have and the correlations that have been observed can be useful in generating hypotheses for future research. For example, cognitive shifts related to cultural behaviors in Terminal Pleistocene and Holocene humans can be explained by peramorphosis. Regulatory changes in genes such as *microcephalin MCPH1* and *ASPM* haplotype 63 that affect brain development might have led to an extended period of circuit-formation and neuronal pruning involving cortico-cerebellar circuitry in response to cognitive demands emerging from cultural intensification. In recent humans, the late adolescent period of brain development coincides with maturation of social cognition and conceptual thinking skills that appear to distinguish us from our terminal Pleistocene and early Holocene forebears.

Additional data are needed related to the timing of cerebellar development, circuit enhancement, neocortical pruning, and cerebellar plasticity. Further sampling will help to establish a range of variation in relative cerebellar volume in early modern humans and contemporary populations. Further genetic comparisons, based on the soon-to-be sequenced Neandertal genome, along with a greater understanding of how *microcephalin, ASPM, and FOXP2* affect cerebellar and neocortical development will be important in clarifying the genetic and cultural dynamics that nurtured modern human cognitive potentials.

Cerebellar evolution, regulated by genetic mutation under the influence of complex selective pressures, may well have enabled recent humans to achieve increased cognitive efficiency without a costly increase in overall brain size.

Acknowledgments

Thanks to A. Bergstrom, the Archeology Technologies Laboratory at NDSU, M. Braun, J. Clark, J. Csernansky, R.L. Holloway, J.J. Hublin, T. Insel, the L.S.B. Leakey Foundation, C. MacLeod, J. Rilling, K. Semendeferi, the University of New Mexico, L. Wang, M. Yuan, the Wenner-Gren Foundation, and, especially S. Thompson, E. Trinkaus, and R. Holloway for material support and invaluable input over the course of this research.

References

Arriada-Mendicoa, N., Otero-Silceo, E., Corona-Vazquez, T., 1999. Current concepts regarding the cerebellum and cognition. Revue Neurologique 29, 1075-1082.

Bar-Yosef, O., 1998. The Natufian culture in the Levant, threshold to the origins of agriculture. Evolutionary Anthropology 6, 159-177.

Bednarick, R.G., 1995. Concept-mediated markingi in the lower Palaeolithic. Current Anthropology 36, 605-634.

Begun, D.,Walker, A., 1993. The endocast. In: Walker, A., Leakey, R.E. (Eds.), The Nariokotome *Homo erectus* skeleton. Harvard University Press, Cambridge, pp. 326-358.

Bischoff, J.L., Shamp, D.D., Aramburu, A., Arsuaga, J.L., Carbonell, E., Bermudez de Castro, J.M., 2003. The Sima de los Huesos hominids date to beyond U/Th equilibrium (>350 kyr) and perhaps to 400-500 kyr: New radiometric dates. (Les dates des hominidés de la Sima de los Huesos au delà de l'équilibre U/Th (>350 000) et peut-être à 400-500 000 : des nouvelles dates radiométriques). Journal of Archaeological Science 30, 275-280.

Blakemore, S.-J., Choudhury, S., 2006. Brain development during puberty: state of the science. Developmental Science 9, 11-14.

Bocherens, H., Drucker, D.G., Billiou, D., Patou-Mathis, M., Vandermeersch, B., 2005. Isotopic evidence for diet and subsistence pattern of the Saint-Césaire Neandertal: review and use of a multi-source mixing model. Journal of Human Evolution 49, 71-87.

Bookstein, F., Schäfer, K., Prossinger, H., Seidler, H., Fieder, M., Stringer, C.B., Weber, G.W., Arsuaga, J.-L., Slice, D.E., Rohlf, F.j., Recheis, W., Mariam, A.J., Marchs, L.F., 1999. Comparing frontal cranial profiles in archaic and modern *Homo sapiens* by morphometric analysis. The Anatomical Record (New Anat.) 257, 217-224.

Brown, P., 1992. The origin of modern humans and the impact of chronometric dating. Philosophical Transactions: Biological Sciences 337, 235-242.

Brumwell, C.L., Curran, T., 2006. Developmental mouse brain gene expression maps. Journal of Physiology 575, 343-346.

Bruner, E., 2004. Geometric morphometrics and paleoneurology: brain shape evolution in the genus *Homo*. Journal of Human Evolution 47, 279-303.

Conard, N.J.,Uerpmann, H.-P., 2000. New Evidence for Paleolithic Rock Painting in Central Europe. Current Anthropology 41, 853-856.

Conkey, M.W., 1987. New approaches in the search for meaning? A review of research in "Paleolithic art". Journal of Field Archaeology 14, 413-430.

d'Errico, F., Zilhao, J., Baffier, D., Julien, M., Pelegrin, J., 1998. Neandertal acculturation in western Europe. Current Anthropology 39 (Supplement), S1-S44.

Dillehay, T.D., 1999. The late Pleistocene cultures of South America. Evolutionary Anthropology 7, 206-216.

Donald, M., 1993. Origins of the modern mind. Harvard University Press, Cambridge.

Enard, W., Przeworski, M., Fisher, S.E., Lai, C.S.L., Wiebe, V., Kitano, T., Monaco, A.P., Paabo, S., 2002. Molecular evolution of FOXP2, a gene involved in speech and language. Nature 418, 869-872.

Evans, P.D., Anderson, J.R., Vallender, E.J., Choi, S.S., Lahn, B.T., 2004. Reconstructing the evolutionary history of microcephalin, a gene controlling human brain size 10.1093/hmg/ddh126. Human Molecular Genetics 13, 1139-1145.

Evans, P.D., Anderson, J.R., Vallender, E.J., Gilbert, S.L., Malcom, C.M., Dorus, S., Lahn, B.T., 2004. Adaptive evolution of ASPM, a major determinant of cerebral cortical size in humans10.1093/hmg/ddh055. Human Molecular Genetics 13, 489-494.

Evans, P.D., Mekel-Bobrov, N., Vallender, E.J., Hudson, R.R., Lahn, B.T., 2006. Evidence that the adaptive allele of the brain size gene microcephalin introgressed into *Homo sapiens* from an archaic Homo lineage. Proceedings of the National Academy of Sciences 103, 18178-18183.

Evans, P.D., Vallender, E.J., Lahn, B.T., 2006. Molecular evolution of the brain regulator genes CDK5RAP2 and CENPJ. General Systems 375, 75-79.

Fox, E.A., Sitompul, A.F., van Schaik, C.P., 1999. Intelligent tool use in wild Sumatran orangutans. In: Parker, S.T., Miles, L., Mitchell, R. (Eds.), The mentality of gorillas and orang-utans. Cambridge University Press, Cambridge, pp. 99-116.

Giedd, J.N., 2004. Structural Magnetic Resonance Imaging of the Adolescent Brain 10.1196/annals.1308.009. Annals of the New York Academy of Sciences 1021, 77-85.

Giedd, J.N., Snell, J.W., Lange, N., Rajapakse, J.C., Casey, B.J., Kozuch, P.L., Vaituzis, A.C., Vauss, Y.C., Hamburger, S.D., Kaysen, D., Rapoport, J.L., 1996. Quantitative magnetic resonance imaging of human brain development: Ages 4-18. Cerebral Cortex 6, 551-560.

Göktürk, E.H., Hillegonds, D.J., Lipschutz, M.E., Hodder, I., 2002. Accelerator mass spectrometry dating at Catalhöyük. Radiochimica Acta 90, 407-410.

Green, R.E., Krause, J., Ptak, S.E., Briggs, A.W., Ronan, M.T., Simons, J.F., Du, L., Egholm, M., Rothberg, J.M., Paunovic, M., Pääbo, S., 2006. Analysis of one million base pairs of Neanderthal DNA. Nature 444, 330-336.

Hahn, J., Münzel, S., 1995. Knochenflöten aus dem Aurignacien des Geissenklösterle bei Blaubeuren, Alb-Donau-Kreis. Fundberichte aus Baden-Württemberg 20, 1-12.

Hawks, J., Wang, E.T., Cochran, G.M., Harpending, H.C., Moyzis, R.K., 2007. Recent acceleration of human adaptive evolution 10.1073/pnas.0707650104. Proceedings of the National Academy of Sciences, 0707650104.

Higham, T., Ramsey, C.B., Karavanic, I., Smith, F.H., Trinkaus, E., 2006. Revised direct radiocarbon dating of the Vindija G1 Upper Paleolithic Neandertals 10.1073/pnas.0510005103. Proceedings of the National Academy of Sciences 103, 553-557.

Holloway, R.L., Broadfield, D.C., Yuan, M.S., 2004. Brain endocasts: the paleoneurological evidence, Vol. 3. Wiley-Liss, New York.

Houk, J.C., 2005. Agents of the mind. Biological Cybernetics 92, 427-437.

Houk, J.C., Miller, L.E., 2001. Cerebellum: Movement Regulation and Cognitive Functions. Encyclopedia of Life Sciences, Nature Publishing Group. Levitan, I., Editor. http://www.els.net/ [doi:10.1038/npg.els.0006101].

Houk, J.C., Wise, S.P., 1995. Distributed modular architecture linking basal ganglia, cerebellum and cerebral cortex: Its role in planning and controlling action. Neocortex 5, 95-110.

Inhelder, B., Piaget, J., 1958. The growth of logical thinking from childhood to adolesence. Basic Books, New York.

Ito, M., 1993. Movement and thought: Identical control mechanisms by the cerebellum. Trends in Neuroscience 16, 448-454.

Jerison, H.J., 1974. Evolution of the Brain and Intelligence, Academic Press, New York.

Katz, D.B., Steinmetz, J.E., 2002. Psychological Functions of the Cerebellum. Behavior and Cognitive Neuroscience Reviews 1, 229-241.

Klein, R., 1992. The archeology of modern human origins. Evolutionary Anthropology 1, 5-14.

Klekamp, J., Riedel, A., Harper, C., Kretschmann, H.-J., 1987. A quantitative study of Australian aboriginal and Caucasian brains. Journal of Anatomy 150, 191-210.

Klekamp, J., Riedel, A., Harper, C., Kretschmann,H.-J., 1989. Morphometric study on the postnatal growth of non-cortical brain regions in Australian Aborigines and Caucasians. Brain Research 485, 79-88.

Koller, J., Baumer, U., Mania, D., 2001. High-tech in the middle Paleolithic: Neandertal-manufactured pitch identified. European Journal of Archaeology 4, 385-397.

Lai, C.S.L., Fisher, S.E., Hurst, J.A., Vargha-Khadem, F., Monaco, A.P., 2001. A forkhead-domain gene is mutated in a severe speech and language disorder. Nature 413, 465-66.

Leiner, H.C., Leiner, A.L., Dow, R.S., 1986. Does the cerebellum contribute to mental skills? Behavioral Neuroscience 100, 443-454.

Luna, B., Thulborn, K.R., Munoz, D.P., Merriam, E.P., Garver, K.E., Minshew, N.J., Keshavan, M.A., Genovese, C.R., Eddy, W.F., Sweeney, J.A., 2001. Maturation of Widely Distributed Brian Function Subserves Cognitive Development. Neuroimage 13, 786-793.

Mekel-Bobrov, N., Gilbert, S.L., Evans, P.D., Vallender, E.J., Anderson, J.R., Hudson, R.R., Tishkoff S.A., Lahn, B.T., 2005. Ongoing adaptive evolution of *ASPM,* a brain size determinant in *Homo sapiens*. Science 309, 1720-1722.

Mellars, P., 2006. Why did modern human populations disperse from Africa ca. 60,000 years ago? A new model10.1073/pnas.0510792103. Proceedings of the National Academy of Sciences 103, 9381-9386.

Middleton, F., Strick, P., 2001. Cerebellar projections to the prefrontal cortex of the primate. Journal of Neuroscience 21, 700-712.

Noble, W., Davidson, I., 1996. Human Evolution, Language and mind: a psychological and archaeological enquiry, Cambridge University Press.

Noonan, J.P., Coop, G., Kudaravalli, S., Smith, D., Kraus, J., Alessi, J., Chen, F., Platt, D., Pääbo, S., Pritchard, J.K., Rubin, E.M., 2006. "Sequencing and analysis of Neanderthal genomic DNA." Science 314, 1113-1118.

Peyrony, D., 1934. La Ferrassie. Préhistoire 3, 1-192.

Richards, M.P., Pettitt, P.B., Trinkaus, E., Smith, F.H., Paunovic, J., Karavanic, I., 2000. Neandertal diet at Vindija and Neandertal predation: The evidence from stable isotopes. Proceedings of the National Academy of Sciences 97, 7663-7666.

Rosenberg, K.R., Zune, L., Ruff, C.B., 2006. Body size, body proportions, and encephalization in a Middle Pleistocene archaic human from northern China 10.1073/pnas.0508681103. Proceedings of the National Academy of Sciences 103, 3552-3556.

Ruff, C.B., Trinkaus, E., Holliday, T.W., 1997. Body mass and encephalization in Pleistocene *Homo*. Nature 387, 173-176.

Schmandt-Besserat, D., 1996. How writing came about. University of Texas Press, Austin.

Schoenemann, P.T., 2006. Evolution of the size and functional areas of the human brain. Annual Review of Anthropology 35, 379-406.

Shaw, P., Greenstein, D., Lerch, J., Clasen, L., Lenroot, R., Gogtay, N., Evans, A., Rapoport, J., Giedd, J., 2006. Intellectual ability and cortical development in children and adolescents. Nature 440, 676-679.

Shu, W., Cho, J.Y., Jiang, Y., Zhang, M., Weisz, D., Elder, G.A., Schmeidler, J., De Gasperi, R., Sosa, M.A.G., Rabidou, D., Santucci, A.C., Perl, D., Morrisey, E., Buxbaum, J.D., 2005. Altered ultrasonic vocalization in mice with a disruption in the Foxp2 gene 10.1073/pnas.0503739102. Proceedings of the National Academy of Sciences 102, 9643-9648.

Soffer, O., 2004. Recovering perishable technologies through use wear on tools: preliminary evidence for Upper Paleolithic weaving and net making. Current Anthropology 45, 407-412.

Straus, L.G., 1995. The Upper Paleolithic: An overview. Evolutionary Anthropology 4, 4-16.

Straus, L.G., 1997. The Iberian situation between 40,000 and 30,000 b. p. in light of European models of migration and convergence. In: Clark, G.A., Willermet, C.M. (Eds.),Conceptual issues in modern human origins research. Aldine de Gruyter, Inc., New York.

Tillier, A.M., 1999. Les enfants Mousteriéns de Qafzeh. Interprétation phylogénique et paléauxologique. Paris.

Trimborn, M., Bell, S.M., Felix, C., Rashid, Y., Hussain, J., Griffiths, P.D., Luitgard, M.N., Krebs, A., Reis, A., Sperling, K., Neitzel, H., Jackson, A.P., 2004. Mutations in *microcephalin* cause abberrant regulation of chromosome condensation. American Journal of Human Genetics 75, 261-266.

Trinkaus, E., 2005. Early modern humans. Annual Review of Anthropology 34, 207-230.

Trinkaus, E., 2006. Modern Human versus Neandertal Evolutionary Distinctiveness. Current Anthropology 47, 597-620.

Trinkaus, E., Moldovan, O., Milota, S., Bilgar, A., Sarcina, L., Athreya, S., Bailey, S.E., Rodrigo, R., Mircea, G., Higham, T., Ramsey C.B., van der Plicht, J., 2003. An early modern human from the Pestera cu Oase, Romania 10.1073/pnas.2035108100." Proceedings of the National Academy of Sciences 100, 11231-11236.

Vandermeersch, B., 1981. Les hommes fossiles de Qafzeh (Israel)

Vandermeersch, B., 1981. Les Hommes Fossiles de Paris. Paris, C.N.R.S.

Vargha-Khadem, F., Watkins, K., Alcock, K., Fletcher, P, Passingham, R., 1995. Praxic and Nonverbal Cognitive Deficits in a Large Family with a Genetically Transmitted Speech and Language Disorder 10.1073/pnas.92.3.930. Proceedings of the National Academy of Sciences 92, 930-933.

Vargha-Khadem, F., Watkins, K.E., Price, C.J., Ashburner, J., Alcock, K.J., Connelly, A., Frackowiak, R.S.J., Friston, K.J., Pembrey, M.E., Mishkin, M., Gadian, D.G., Passingham, R.E., 1998. Neural basis of an inherited speech and language disorder. Proceedings of the National Academy of Sciences 95, 12695-12700.

Weaver, A.H., 2001. The cerebellum and cognitive evolution in Pliocene and Pleistocene hominids. Paleoanthropology Society Dissertations.<http://www.paleoanthro.org/dissertation_list.htm>.

Weaver, A.H., 2005. Reciprocal evolution of the cerebellum and neocortex in fossil humans. Procedings of the National Academy of Sciences 102, 3576-3580.

Wynn, T.G., 1985. Piaget, stone tools, and the evolution of human intelligence. World Archaeology 17, 32-53.

Wynn, T.G., 1991. Archeological evidence for modern intelligence. In: Foley, R.A., (Ed.), The origins of human behavior. Unwin Hyman, Ltd., Cambridge.

Wynn, T.G., 1996. The evolution of tools and symbolic behavior. In: Lock, A., Peters, C.R., (Eds.), Handbook of human symbolic evolution. Clarendon Press, Oxford.

Zhang, J., 2003. Evolution of the human ASPM gene, a major determinant of brain size. Genetics 165, 2063-2070.

Zilhão, J., 2006. Neandertals and moderns mixed, and it matters. Evolutionary Anthropology 15, 183-195.

Zilhao, J., d'Errico, F., Bordes, J.-G., Lenoble, A., Texier, J.-P., Rigaud, J.-P., 2006. Analysis of Aurignacian interstratification at the Chatelperronian-type site and implications for the behavioral modernity of Neandertals 10.1073/pnas.0605128103. Procedings of the National Academy of Sciences 103, 12643-12648.

CHAPTER 8

STUDY OF HUMAN BRAIN EVOLUTION AT THE GENETIC LEVEL

ERIC J. VALLENDER AND BRUCE T. LAHN

ABSTRACT

As a species, *Homo sapiens* is characterized by its uniquely large and complex brain. Comparative anatomists and paleoanthropologists have done much to elucidate the phenotypic changes of the human brain over evolutionary time. Here we review the emerging understanding of the genetic basis that underlies these phenotypic changes.

INTRODUCTION

People across virtually all disciplines – philosophers, sociologists, artists, preachers and scientists alike – have grappled with the question of what it means to be human. And while today mankind may be no closer to answering the metaphysical aspects of the question, the biological underpinnings of what it means to be human are gradually coming to light.

Perhaps no single feature is as salient or of greater importance in the evolution of *Homo sapiens* as the emergence of the modern human brain. The increase in brain size correlates with advances in cognitive capabilities and an increasingly complex behavioral repertoire including complex tool use, symbolic thought and language, and artistic expression. At the heart of it, it is this increased cognitive complexity that has allowed humans to develop society, culture, and indeed the ability to ask ourselves the philosophical question of what it means to be human.

From the birth of evolutionary theories, the relationship between humans and other primate species was apparent. And while differences among the primates are legion, it is the differences in brain size and complexity that are perhaps most difficult to comprehend. The human brain is roughly eight times the relative volume of New World monkeys and approximately three times that of chimpanzees (Falk, 1986). Further, the expansion of the human brain has not been proportional, rather certain regions, including the cerebral cortex, have seen size and complexity increases even relative to other human brain regions. In particular, the prefrontal cortex, which may play an important and unique role in social behavior, has seen significant enlargement (Semendeferi et al., 2002). Understanding the distinctiveness of humans means in part understanding these differences and the mechanisms that caused them to emerge.

As approaches to understanding the human condition have varied, so too have approaches to understanding the evolutionary development of the human brain. Primatologists have studied similarities and differences in the behaviors of human and non-human primates. Comparative anatomists have noted congruities and inconsistencies across the brain. Paleoanthropologists have identified a long history of hominids leading from humanity's last common ancestor with chimpanzees through to modern man with several dead-ends thrown in for good measure. The last century also saw the development of a new science which promises to add to the growing understanding of human origins, evolutionary genetics.

While the concept of evolutionary genetics has been around since the early 1900s, it was only in the last several decades that its use in understanding human species origins, and in particular the origins of the human brain, has blossomed (Vallender et al., 2008). This modern growth has been fueled by the ability of researchers to cheaply and quickly decipher sequence genetic information, from single genes to complete genomes. Using comparative genomics, it is possible to identify differences at the most fundamental, genetic, level between species and to probe the most basic mechanisms of evolutionary change and adaptation. Geneticists now have access to the actual and complete genomes of numerous species, human and non-human, primate and non-primate, mammal and non-mammal, and increasingly to the genomes, or at least select genotypes, of specific individuals within species. This influx in data has necessitated the development of associated tools, both for access and visualization of the information as well as techniques and methodologies for making sense of the immense quantities of data.

Coincident with the development of modern tools of genetic analysis has been the explosive growth of the neurosciences. Long studied, the complexity of the brain has refused to reveal its secrets easily. With the development of imaging, electrophysiology, genetic manipulation and whole hosts of other techniques, neuroscientists are gradually making headway into understanding the brain. In doing so, there has been increasing understanding of how specific genes contribute to specific aspects of brain development and function.

By coupling the functional information gained through the study of basic neurobiology with molecular evolution data gathered by comparative geneticists, slowly a new understanding of human origins is emerging where the random evolutionary mutations have lead to functional consequences, neurological and otherwise, that would eventually lead to the species-specific changes that characterize the emergence of the modern human brain. The subsequent sections offer a snapshot of these studies as they stand in the early part of the 21st century; an incomplete understanding to be sure, but the beginnings of the scientific, if not metaphysical, basis of what it means to be human.

Methodologies

Evolutionary change can occur by any number of mutational processes. Some of these changes are on a genetically large scale; chromosomes may come apart or fuse, as is the case for human chromosome 2 (Jauch et al., 1992), or there may be major inversions such as are seen on the Y chromosome (Lahn et al., 2001). There can be changes on a more moderate level, including the duplication or deletion of genes and genomic regions known as copy number variants. Most commonly, however, are the smallest mutations wherein only a handful of bases are added or deleted or the most common point mutations wherein only the identity of a base changes. Each of these mutations can have functional effects and has at some point in the history of human, primate, or mammalian evolution. However, they need not necessarily have functional effects. The translocation of a gene from one chromosome to another may result in a decoupling from a regulatory element or the positioning in a new and functionally relevant milieu or it may be nothing more than a change of scenery. Point mutations in particular are likely to be functionally silent (also referred to as "selectively neutral"). In order to differentiate between those mutations that are likely to be relevant to evolution and adaptation and those that are not, numerous methodologies have been developed.

Methodologies vary in many facets: what kinds of mutations they hope to identify, the timing of those mutations, and the functional nature of those mutations (Vallender, 2008; Zhai et al., 2009). Techniques devised for one type of mutation or selective event may not be relevant for others. Because of their ubiquity and a more complete understanding of their origins and downstream effects, many tests are designed to focus on point mutations. Within these tests two broad categories can be discerned, those that focus on inter-specific comparisons and those that focus on intra-specific comparisons. Inter-specific methodologies compare the fixed genetic differences between species while intra-specific methodologies utilize polymorphism data within a species to detect selective events.

Polymorphism-based approaches can come in many flavors (see for example Zeng et al., 2007). They may utilize the allele frequency spectrum (a measure of the frequency of SNPs in a population) (Fay and Wu, 2000; Fu and Li, 1993; Tajima, 1989), haplotype diversity and structure (Depaulis and Veuille, 1998; Fu, 1996; Hudson et al., 1994), linkage disequilibrium (Kelly, 1997; Sabeti et al., 2002; Slatkin and Bertorelle, 2001; Toomajian et al., 2003), population substructure (Lewontin and Krakauer, 1973), or any combination of these. In addition, the development of new tests aimed at detecting specific types of selective events in specific situations is ongoing. There are significant strengths and weaknesses to these tests even as they apply specifically to the changes associated with the emergence of the human brain. The major strength is that power is generally very strong and often specific functional mutations can be identified. Further, the tests are context independent and work equally well on coding regions and non-coding regions. The major weakness is that these tests half relative short half-lives, meaning that the selective event must have occurred fairly recently. These tests are very successful in identifying genetic changes that accompanied modern *Homo sapiens* dispersal into new environments and encounters with novel disease or the genetic and biological changes that occurred coincident with the emergence of civilization, but are less useful for identifying the genetic mutations that led to the emergence of modern humans (Vallender, 2008).

The reason for the inappropriateness of these tests in understanding the development of the human brain is fairly straightforward. These tests rely on differences within populations to identify selection, while by definition the genetic changes required for making a modern human brain are shared by all members of the species. Certainly there is some lag time wherein a signature of the selective event will still be present though the mutation itself has fixed, but these situations are often difficult to differentiate from demographic events. Also, in the case of the human brain, the change seems to have occurred even beyond what this lag time could hope to include.

Anatomically modern humans are believed to have emerged 100,000 to 200,000 years ago and paleoanthropologists tell us that the size of the human brain was largely fixed at that time. Indeed, even 500,000 years ago *Homo heidelbergensis*, one of *Homo sapiens* direct ancestors, appears to have a cranial capacity similar to those seen in extant humans (Neill, 2007). The most recent of the major growth spurts toward the human brain appears to have occurred during the transition from the Australopithecines to early *Homo* roughly two million years ago. The genetic changes associated with this anatomical step have thus been fixed for somewhere around 100,000 generations and their polymorphic signatures eroded.

With polymorphism-based tests unable to identify the genetic changes responsible for the development of the human brain, we turn to inter-specific divergence-based tests. Immediately divergence-based suffer a failing relative to their polymorphism-based brethren. Divergence-based tests require multiple functional categories of mutation. This commonly takes the form of functional versus neutral. Our current inability to *a priori* predict functional sites in non-coding regions of the genome has restricted the use of these tests to protein-coding regions where rates of change at amino acid changing sites (d_N or K_A) can be compared to those at synonymous or non-amino acid changing sites (d_S or K_S). This ratios (d_N/d_S, K_A/K_S, or ω) is an immediate and direct measure of selection (Miyata and Yasunaga, 1980). Equal to one indicates neutral evolution such as would be predicted in a pseudogene. Less than one is indicative of negative selection or functional constraint. Greater than one is evidence of positive selection, presumably (though not necessarily for reasons enumerated below) the primary source of adaptation in the human brain.

Difficulties abound even with these tests, however. Firstly, tests are usually conducted on genes as a whole and even when positive selection occurs at one position in a gene it is often balanced by negative selection at other locations. Indeed, for nearly all genes (the MHC is a specific counter-example) negative selection to maintain overall protein structure and function generates baseline ω values around 0.2. Selection needs to be exceptionally strong to have a significant detectable effect. Another issue is the fact that the time periods that can be studied are limited. Studies necessitate two extant species (or species that one can recover DNA from which today means the same thing) and lineages for studies cannot be shortened without adding a new species. In short, a selective event must be strong enough to overwhelm both negative selection at other positions in the gene as well as extended time periods that occurred without positive selection. To say this is a tall order would not be overstating it.

There remains another difficulty unique to human-specific traits, a short lineage problem. Looking for fixed differences between humans and chimpanzees is certainly possible and has been done several times in the past (2005; Clark et al., 2003). The difficulty with short lineages, however, is that stochastic variation in the mutational process can have too great an effect to be overcome. In essence stochastic variation in the rate of synonymous mutation can result in values that are low, resulting in false positives, or values that are too high, resulting in false negatives. The low signal to noise ratio in synonymous mutations can also make achieving statistical significance all but impossible.

Here we offer only a cursory discussion of these methodologies; more through undertakings can be found elsewhere (Vallender, 2008; Zhai et al., 2009). We present this to illustrate two points. The first is how our available methodologies have affected what has been found. Current methodologies are biased towards the identification of functional mutations in protein coding regions. King and Wilson famously hypothesized early on that many of the differences observed between humans and chimpanzees would be the result of non-coding, regulatory, changes (King and Wilson, 1975). This hypothesis has almost become dogma in the field and yet most studies still focus on protein-coding mutations. Methodological limitations are the explanation why. Secondly, it is important when reading the literature on the field to understand discrepancies in findings. It is all too easy to oversimplify the question, looking for recent human positive selection, when in fact the subjects of the study are actually a great deal more nuanced. As a result, while different studies of "recent human positive selection" may indeed produce different results, it is important to ensure that they were in fact designed to answer the same question.

Protein sequence change

For reasons described above, a large number of studies hoping to elucidate the changes leading to the human brain have focused on selection on proteins. For the most part, studies of this kind (and indeed most studies presented here) couple two findings to produce the hypothesis of functional relevance in the species-specific development of the human brain. The first is a brain related function for the gene and the second is evidence of positive selection. Taken together, these offer circumstantial evidence for selection acting on the brain phenotype of the gene. This is not necessarily the case, however, and it should be noted that functional evidence for a neurological effect of a specific mutation remains few and far between. Nevertheless, while the results produced by these studies still represent hypotheses, they are well-founded. The evidence for selection is not in doubt, nor is the evidence for neurological relevance.

Many studies have taken for their starting point genes that have been implicated in neurological diseases. This is particularly true of those diseases arising from developmental changes. One particularly well studied category is the genes that have been associated with microcephaly and, in particular, primary microcephaly. Microcephaly is a developmental affliction which is characterized by a severe reduction in brain size without any major abnormalities in brain structure or architecture (Dobyns, 2002; Mochida and Walsh, 2001; Woods et al., 2005). Primary microcephaly lacks any additional abnormalities as well. The microcephalic phenotype has been considered to be atavistic because in many ways it appears to recapitulate earlier hominid features. This similarity has led to significant exploration of the genes responsible for the disease as potential contributors to the evolutionary changes that lead to the modern human brain. Primary microcephaly is a genetically heterogeneous condition that has been mapped to six loci in the human genome with specific genes and mutations identified for four of these loci: *microcephalin* (*MCPH1*), *CDK5RAP2* (*MCPH3*), *ASPM* (*MCPH5*), and *CENPJ* (*MCPH6*) (Bond et al., 2002; Bond et al., 2005; Jackson et al., 2002).

ASPM and *microcephalin* were the first two genes to be mapped to primary microcephaly loci and several groups exploring each found evidence for positive selection. While *microcephalin* is characterized by a bout of positive selection in the lineage leading from the last common catarrhine ancestor to the great apes (Evans et al., 2004a; Wang and Su, 2004), *ASPM* bears signatures of positive selection along the entire lineage leading from early primates to extant humans (Evans et al., 2004b; Kouprina et al., 2004; Zhang, 2003). Both *ASPM* and *microcephalin*, as well as *CDK5RAP2* and *CENPJ*, show elevated ω values in primates relative to rodents, while *CDK5RAP2* additionally shows particularly high rates in the human and chimpanzee terminal lineages (Evans et al., 2006).

The function of these genes has only begun to emerge. *Microcephalin* appears to be involved in DNA damage control and condensation during mitosis (Evans et al., 2006; Trimborn et al., 2004; Wood et al., 2007; Xu et al., 2004). *ASPM*, *CDK5RAP2*, and *CENPJ* are also seemingly involved in mitotic spindle formation (Bond et al., 2005; Fish et al., 2006; Kouprina et al., 2005). Indeed, all four primary microcephaly-associated genes to date appear to be involved in cell cycle control and likely manifest developmental effects on the brain through the regulation of neural precursor cell proliferation. This is perhaps particularly relevant because of a widely held belief that the changes observed in the human brain may have resulted from increases in neural precursor division during neurogenesis (Kornack and Rakic, 1998).

Interestingly another gene involved in neural cell proliferation, *ADCYAP*1, has also been identified as harboring the signature of positive selection (Wang et al., 2005). This gene has one of the highest ratios of nonsynonymous to synonymous substitutions observed between humans and chimpanzees thus far. Although ADCYAP dysfunction results in many pathologies throughout the body, its role regulating the transition from proliferative to differentiated states offers the possibility of a role for this gene's evolution in the emergence of the human brain (Dicicco-Bloom et al., 1998; Suh et al., 2001). As before, however, it is important to note that the two lines of evidence, for selection and for neural function, remain to be formally conjoined.

While primary microcephaly may be an atavistic trait, other congenital brain malformations are not. One of these abnormalities, called holoprosencephaly, can be caused by mutations in the *sonic hedgehog* (*SHH*) gene. *SHH* encodes a highly studied signaling molecule that plays a role in the development and patterning of many tissues including the skeletal and nervous systems. The gene encodes a signaling molecule as well as an autocatalytic region. While the signaling molecule is extraordinarily conserved, the auto-catalytic domain shows a significantly increased rate of protein sequence change in primates compared to other animals (Dorus et al., 2006). In particular, the lineage leading to humans shows a rapid rate of evolution and a statistically non-random accumulation of serines and threonines, residues implicated in post-translational modifications. These findings raise again suggestions of ties to human-specific biology.

Joubert syndrome is another example of a neurological disorder where a causative dysfunctional gene has been shown to have an interesting evolutionary history. A syndrome with complex symptomologies, including cerebellar hypoplasia, on causative mutation in *AHI1* involves a protein involved in axon guidance from the brain to the spinal cord (Ferland et al., 2004). Like several of the other genes presented here, *AHI1* has been demonstrated to show accelerated rates of protein sequence change in humans since the last common ancestor with chimpanzees (Ferland et al., 2004).

Using behavioral variation as a substrate for identification of candidate genes has also proven fruitful. The X-linked *MAOA* gene encodes a protein that is responsible for the catabolism of many monoaminergic neurotransmitters including dopamine, serotonin, and norepinephrine. Variation in the gene has been associated with numerous behavioral consequences and neuropsychiatric disorders (Brunner et al., 1993; Cases et al., 1995; Kim et al., 1997; Shih et al., 1999; Sims et al., 1989). Intriguingly, in addition to variation currently segregating in humans, nonsynonymous mutations in the gene may have created a functional change in the enzyme as well (Andres et al., 2004). Of all behavioral changes, however, perhaps none is more obvious than the acquisition of language. It is unsurprising, therefore, that this significant step in the evolution of humans should also be a major emphasis for those seeking genetic correlates. While several of the microcephaly genes, *ASPM* and *microcephalin*, have been suggested to harbor roles in language (Dediu and Ladd, 2007), two others have demonstrated more prominent roles and offer intriguing possibilities for the evolution of speech. *SRPX2* has been associated with speech processing (Roll et al., 2006) and shows an accelerated rate of protein evolution along the human lineage (though as a result perhaps of short lineage effects it falls short of statistical significance) (Royer et al., 2007). The most interesting of genes in this category, however, was also one of the first to make headlines in the search for humanness genes, *FOXP2*.

FOXP2 is a gene that has been associated with developmental verbal dyspraxia, a disorder that is characterized by difficulties in the production of language and thought to be associated with defects in the brain translating intended speech into the complex muscle movements required (Lai et al., 2001). Since this early finding in humans, *FOXP2* has also been implicated in aural communication in mice and bird song (Haesler et al., 2007; Haesler et al., 2004; Shu et al., 2005; Teramitsu et al., 2004; Teramitsu and White, 2006). While *FOXP2* is nearly perfectly conserved in amino acid sequence across mammalian species, it has undergone two nonsynonymous mutations in the human lineage since the divergence from chimpanzees (Enard et al., 2002b). Because of the extraordinary conservation across mammals, these mutations contribute to statistically significant change in humans. The mere suggestion that these mutations may have played a role in the emergence of spoken language and all that accompanied it was enough to invigorate researchers and energize the field to its current flowering.

While the studies above have focused on specific candidate genes, there has also been genomic research taking a more broad approach (2005; Bustamante et al., 2005; Clark et al., 2003; Dorus et al., 2004; Nielsen et al., 2005). One early study focused on approximately 200 genes chosen for their neurological roles or association with neurological disease. Collectively, these genes showed an increase in their rate of protein change in primates as compared to rodents, an increase not seen in a companion set of more ubiquitously expressed genes (Dorus et al., 2004). Supporting a neurodevelopmental hypothesis of human adaptation, genes with roles during brain and nervous system development showed this acceleration more pronouncedly than genes involved in neurophysiological processes. While this finding was corroborated by a later study (Khaitovich et al., 2005), other studies failed to replicate the finding (Shi et al., 2006; Wang et al., 2007). While much has been made of the differences in results, it is important to note that rather than necessarily represent conflicting findings this may instead be a result of different methodologies answering different questions. Indeed there may be differences in the evolution of neurodevelopmental and neurophysiological genes that may be reflected in the findings even if not explicitly tested for. Similarly, studies may vary in the lineage and/or time scale that they test. Studies aimed at detecting positive selection that has occurred in the last hundred thousand years are unlikely to reveal the genes that contributed to human-specific characters prior to that point. While still not providing definitive answers, these studies nevertheless can offer insight and may be useful in revealing macro trends if carefully considered.

Gene gain and loss

Evolutionary changes in protein sequence are thought to tweak effects, somehow changing the existing functions of the protein. More drastic changes are possible, however. Losses of gene function can occur through point mutations that are difficult to detect, but usually genes without function undergo fairly rapid pseudogenization making their identification straightforward. While it can be counterintuitive to imagine the loss of a gene as adaptive, and indeed not all loses of genes are strictly beneficial, this can be the case. At the same time, while the addition of new genes can be more easily reconciled with adaptation, this occurrence is mechanistically more difficult than either point mutations or gene loss, making gene gain a fairly rare event. Despite these considerations, however, evidence has been found suggesting roles for each of these processes in the emergence of the human brain.

The most pronounced gene loss in humans, and indeed all primates, is in the olfactory system (Gilad et al., 2003a; Gilad et al., 2005a; Gilad et al., 2003b; Glusman et al., 2001; Young et al., 2002; Young and Trask, 2002). While rodents are estimated to have over twelve hundred functional olfactory receptor genes, the human genome appears to harbor between a third and a quarter that amount (Young et al., 2002; Young and Trask, 2002). Rampant pseudogenization has occurred in the olfactory gene family not only in humans, but across all primate species, though there are examples of specific gene losses in humans since the last common ancestor with chimpanzees. These losses should not be surprising given the shift to a primarily visual sensory focus in primates. While it is unclear if this shift merely rendered the ultra-complex olfactory system of ancestral mammals unnecessary or if there was an active benefit to the loss of these genes, suggestive evidence exists that positive selective pressures did, at least in part, shape the current human olfactory subgenome (Gilad et al., 2003a; Gilad et al., 2005a).

A more explicit and tantalizing example of gene loss playing a major role in human brain evolution comes from the gene encoding a myosin heavy chain protein, *MYH16*. This particular heavy chain is expressed uniquely in the muscles of the head including the masticatory apparatus. In humans this gene has undergone a pseudogenization event that has been attributed to positive selection (Stedman et al., 2004). Arguments in favor of this interpretation point to the loss of the sagittal crest in humans and expansions in cranial capacity coinciding with changes in diet and masticatory needs (Neill, 2007). This hypothesis has been challenged, however, because of a failure to reconcile the age of the mutation with the paleoanthropological data (Perry et al., 2005). While still unclear, it remains plausible that the loss of *MYH16* is related in some way to human evolution.

While gene loss occurs through genetically simple mechanisms, gene gain is more complex. Very rarely do genes emerge out of whole cloth, rather they are the result of duplication events (Ohno, 1970). With multiple copies of a gene an evolutionary relaxation of constraint occurs and, while usually resulting in a simple pseudogenization event, the duplicated gene may undergo neofunctionalization, wherein a new and unique function is imparted on the protein, or subfunctionalization, where multiple functions of the original protein are partitioned between its offspring (Force et al., 1999; Hughes, 1994; Lynch and Force, 2000; Lynch et al., 2001). Events such as these are not particularly common, but recent advances in genomic technologies have made their identification possible.

While not specific to humans, the emergence of trichromatic color vision in primates offers a striking example of duplication followed by neofunctionalization (Li et al., 1999). Most platyrrhines, and presumably ancestral primates, have dichromatic vision engendered by "blue" and "green" opsins. In the ancestor of catarrhines the X-linked "green" opsin was duplicated and neofunctionalization led to the emergence of a "red" opsin gene. This event led to the shift to trichromatic vision and has been argued to coincide with an increase in the importance of visual perception.

The *morpheus* gene family has undergone large scale duplication followed by positive selection and presumably neofunctionalization, in humans and great apes (Johnson et al., 2001). The functional changes in some members of this family are so strong as to obscure homology though studies of synteny confirm their origins. To this point, however, the function of the *morpheus* genes remain unknown. The function of the family of genes harboring the DUF1220 domain is likewise unknown, though their expression is limited to the brain and neurons (Popesco et al., 2006). Like the *morpheus* gene family, these genes have greatly expanded in number in primate species with evolutionary proximity to humans (Popesco et al., 2006). Although the functions of both of these gene families remain unclear, their dramatic and startling appearance in human lineages warrants further examination.

The most transparent example relating to changes in brain function was the duplication and subfunctionalization of the glutamate dehydrogenase genes (Burki and Kaessmann, 2004). Present in ancestral primates, *GLUD1* is responsible for the catabolism of glutamate, the chief excitatory neurotransmitter, and is broadly expressed through the body. Sometime during the lineage separating great apes from old world monkeys a retrotransposition event occurred creating a second glutamate dehydrogenase gene, *GLUD2* (Burki and Kaessmann, 2004). Unlike is parent gene, the expression of *GLUD2* is restricted to nerve tissues and the testis (Shashidharan et al., 1994). This subfunctionalization appears to have been followed by a period of positive selection optimizing enzymatic activity for its new milieu. While the phenotypic relevance of these changes remains shrouded, the evolutionary origins of this brain-specific great ape gene reveal an adaptive role.

Gene expression changes

The genetic differences between humans and chimpanzees pale in comparison with the phenotypic differences; the mutations that gave rise to these differences must have had hugely significant effects. This belief led to the hypothesis that the evolutionary action separating the species was due more to changes in gene expression and regulation rather than protein function (King and Wilson, 1975). More recently, the pleiotropic effects of mutations in brain genes have been invoked to support the same hypothesis (Carroll, 2005). While not excluding protein changes from the evolutionary process of human speciation from chimpanzees, it is clear that regulatory changes have played a significant and likely prominent role.

The primary difficulty with studying regulatory changes lies in our relative lack of understanding, borne out of the extreme lability of the cis-regulatory process and more generally a lack of *a priori* predictive power. Not only does this result in an inability to use traditional evolutionary tests of selection for fixed differences, but it often precludes even identifying potential changes. Rarely are researchers afforded the understanding of gene regulation necessary to make predictions of relevant evolutionary change. Indeed, when this is possible it is driven by an extraordinary interest in the gene itself for reasons almost never related to evolution. But while the genetics of regulatory differences are often difficult to tease apart, the readout of these effects, particularly in terms of mRNA expression is much more straightforward. Because of this an interesting dynamic has developed. While evolutionary studies of protein change focus on evidence for selection and often fall short of function, those analogous studies of regulatory change often demonstrate more clearly functional differences while struggling to prove evidence of selection.

Several disparate studies have approached this question by comparing gene expression in the brains of humans and other non-human primates (Khaitovich et al., 2006; Preuss et al., 2004). While some studies focus on specific regions of the brain, others are more broadly based (Caceres et al., 2003; Enard et al., 2002a; Khaitovich et al., 2004; Marvanova et al., 2003; Uddin et al., 2004). And though these differences in study design result in particular differences of result and may suffer from differing problems, two similar results continue to be found. Overall brain gene expression levels in humans are generally upregulated compared to chimpanzees and yet these expression patterns in the brain are more similar between the two species than gene expression profiles from other tissues. As with protein change, it seems that on the whole the brain is particularly conserved and yet specific changes necessarily have significant effects.

One difficulty with these studies comes not from the theoretical premises on which they rely, but rather on the difficulty in ensuring apples-to-apples comparisons; it can be very difficult to ensure homology between the samples being tested. This problem can take many forms. Firstly, pathological state of the samples must be determined. Because of ethical considerations and procedural difficulties, many samples used are from diseased animals or significantly aged individuals. Similarly, circumstances of death may result in confounding effects, for instance as related to circadian rhythms, seasonal differences, or menstrual cycles. More broadly, the environmental conditions in which the individual lived may profoundly affect gene expression and it goes without saying that humans and non-human primates in the best of situations live in very different environments (Myers et al., 2007). A second and related complication can be found in developmental timing. While comparing adult to adult seems straightforward, many of the most interesting and likely most important differences may be found in early development, possibly prenatal. Ensuring developmentally homologous time points is particularly difficult in non-human (and human) primates where the ages of the fetus for study cannot be controlled as in rodents. This, coupled with the general difficulties of generating cross-species timelines for development, especially when changes in this developmental timeline are precisely the variable under study, makes comparisons of developmental gene expression particularly daunting. In addition to developmental homology, anatomical homology must be considered. This is particularly relevant as regards the increasingly more refined anatomical substructures under study. As with differences in the developmental timeline between species, the issues surrounding complications that arise through changes in functional roles of specific brain regions must be addressed.

A separate, but equally important, issue that must be resolved is in the detection of mRNA levels themselves. While human array-based methodologies are largely well established and single gene studies using methods such as quantitative PCR can be developed across species, non-human primate array-based methodologies are less developed. Many large-scale studies of non-human primate gene expression rely upon xenohybridization, the hybridization of non-human primate mRNA to human probes. The relative effects of this cross-species hybridization can vary from platform to platform, gene to gene, and species to species, in all greatly complicating in unpredictable ways these studies (Gilad et al., 2005b). Luckily, these problems have a simple solution, the development of species-specific arrays, but one which still represents additional expenditures in time and money that may be difficult to overcome.

While studies which focus on the end phenotype, changes in mRNA expression, have flourished, there have also been a smaller number of studies that have proceeded from genotype to phenotype that have showed some success. The most notable among these is the evolution of an upstream cis-regulatory element in *PDYN*, a precursor of several endogenous opioidergic neuropeptides that have been implicated in many neural processes. This regulatory element shows an exceptionally rapid rate of evolutionary change in the human lineage since its divergence from chimpanzees, consistent with the effects of natural selection (Rockman et al., 2005). Further, in a cell culture system, the human regulatory element was demonstrated to significantly upregulate expression of a reporter gene compared to the orthologous chimpanzee sequence (Rockman et al., 2005). It remains to be seen whether the methodologies that were applied in the *PDYN* study will be successfully generalized, though it would appear unlikely as a perfect storm of prior knowledge, evolutionary timing, and functional assayability was necessary for its success.

It should be mentioned, however, that despite the difficulties involved, there are ongoing genomic efforts to identify regions of rapid evolution. Several genome-wide analyses have been preformed to identify regulatory regions that have undergone rapid change during human evolution (Bush and Lahn, 2008; Haygood et al., 2007; Pollard et al., 2006a; Prabhakar et al., 2006). While these studies have provided an excellent starting point and almost certainly will herald the beginning of a new focus for evolutionary genomics, at present their power for detecting positive selection, as opposed to relaxation of constraint or simply non-functional neutral evolution, is unclear. Similarly, like protein-coding changes, studies remain to be done showing the functional effects of regulatory changes. This is particularly important because, while changes in amino acid are relatively easy to visualize as having a functional effect, changes in conserved non-coding regions without clearly identified functions are not.

Before proceeding it is important to note one area of convergence between the studies of protein-coding change and regulatory evolution. Up until this point our discussion of the evolution of gene expression has focused on changes in the cis-regulatory elements themselves. Indeed, there are many reasons to believe that these changes should be most commonplace, not the least of which is their relative specificity in accomplishing a specific functional task without too many untoward side effects. And while it seems reasonable to believe that this will in fact be the substrate for major evolutionary change in gene expression, several genome-wide studies of protein change have identified a significant overrepresentation of transcription factors among genes likely to have undergone positive selection (Bustamante et al., 2005; Gibbs et al., 2007). Issues of pleiotropy raised more broadly against protein sequence evolution seem to be innumerably more relevant for transcription factors, however.

Evolution of gene expression will certainly prove to play an important role not only in the emergence of the human brain and other human-specific characters, but in adaptation broadly. While protein-coding changes remain a low-hanging fruit and an important in and of themselves, the efforts into understanding and identifying signatures of selection on gene expression and in cis-regulatory regions will only increase.

Other substrates for change

Changes in protein sequence have long been studied for their affect on phenotypic change during evolution. And while evolutionary studies of gene expression are relatively nascent, theories of their importance are fairly well-established. However, as our understanding of genetics develops so to do potential targets for natural selection and substrates for human-specific evolutionary changes. Among these, several are worthy of brief discussion: alternative splicing, epigenetics, post-translational modifications, and non-coding RNAs.

While whole-gene gain and loss has been considered here and has long been a topic of study in molecular evolutionary literature, the emergence and loss of alternative splice variants has received less attention. With total numbers of genes in mammalian genomes much lower than initially anticipated, the role of alternative splicing has taken on a renewed importance. The emergence of new alternative splice forms may offer a loophole for the lessening of pleiotropic effects. Unfortunately relatively little is known about the evolution of alternative splice forms though research is underway (Jin et al., 2008). Part of this has been the shift in focus to genomic DNA from mRNA. As comparable cDNA libraries from different species emerge it is likely that this research will develop rapidly. Of particular note in this regard are early studies comparing human and mouse cDNAs (Takeda et al., 2008). While still evolutionarily distant for identification of human specific changes, it is important to note that the human-mouse comparison was also the beginning point for many other studies of evolutionary change in humans and mammals.

Similar to single nucleotide point mutations in cis-regulatory regions, changes in epigenetic patterns may affect gene expression differences. In fact, it may be through these mechanisms that cis-regulatory evolution occurs (at least in part). Epigenetic gene silencing in particular is important during in utero development (Keverne and Curley, 2008), a period that has changed dramatically during human evolution and during which many of the brain developmental differences between humans and non-human primates are generated. As our understanding of epigenetics emerges, it seems likely that changes in epigenetic mechanisms will be discovered that have played an adaptive role in the human brain.

As epigenetic evolution may play a role in regulatory evolution more broadly, so too may post-translational modification evolution play a role in protein sequence evolution. The functional effects of protein sequence change are typically thought to be mediated through changes in protein structure, enzymatic activity, or ligand binding. Indeed, the science of understand why protein changes result in the functional affects they do is a major endeavor in it own right. Changes in protein sequence may also result in changes in post-translational modifications. Differences in dimerization can certainly be functional, as can differences in small molecule changes. Differences in sialic acid biology resulting in glycosylation differences were among the first changes to be noted between humans and chimpanzees (Chou et al., 1998; Muchmore et al., 1998). Also noted above is a significant evolution in humans of the autocatalytic region of SHH towards serines and threonines, common substrates for post-translational modifications (Dorus et al., 2006). While far from proven, the role of post-translational modifications must be considered when looking for the mechanisms underlying human-specific traits.

One last area that has only recently emerged yet shows great promise in developing importance is in non-coding RNA genes. These RNAs are relative newcomers to the scene and yet their importance has been immediately recognized. As a means of regulating gene expression they seem likely to play a role in the processes considered here. Evolutionary changes in these genes suffer from the same pros and cons as cis-regulatory changes, and methodologies designed for one often apply to the other. Indeed, the first putatively positively selected non-coding RNA was discovered in the course of a genome-wide study of non-coding DNA. *HAR1* is an RNA gene of unknown function, yet it is expressed in the neurons of the developing neocortex (Pollard et al., 2006b). Although only 118 base pairs in length, there are 18 changes between the human and chimpanzee orthologs, roughly ten times the neutral rate (Pollard et al., 2006b). This difference is even more striking when viewed in light of the chicken-chimpanzee comparison, only two changes (Pollard et al., 2006b). It seems inconceivable that changes of this magnitude do not have some effect, and yet what that effect is remains elusive. Just as we await functional verification of protein changes, so too do we now wait functional verification of non-coding RNA changes.

Conclusion

Understanding the evolution of the human brain will not be easy. The function of the brain is so complex and such a scientifically daunting task by itself, and yet we hope to overlay on top of this another layer of complexity, evolutionary change. It is certainly not a trivial task. Yet we continue to strive to achieve this seemingly insurmountable goal because in doing so we strive to better understand ourselves. We approach the question from many angles, multiple scientific disciplines, using diverse methodologies and techniques. Evolutionary genetics is but one of many of these.

The progress that has been made in identifying and understanding the genetic differences between humans and our closes primate relatives over the last decade has been astounding. The substrates of evolutionary change have expanded, from proteins to regulatory regions and beyond. Techniques have improved and the scale upon which these questions are considered has broadened. Yet much remains to be done.

Some of those questions that are still outstanding are resource-driven: more species, more individuals, more spatial and temporal time points, greater throughput. The question that dominates all others, however, is functional relevance. How do we demonstrate the functional relevance of the putatively human-important changes? In vitro protein functional assays may be useful for some changes, for some proteins. Cell culture based assays may give additional insight though caution must be taken in interpretation. Transgenic animals, particularly rodents, are likely to provide some clues but again contextual differences may be relevant.

There may be no simple answer to demonstrating unequivocally functional importance, but as a field this must be the goal to which we aspire. A systems biology approach to evolutionary change is difficult to envision, yet it has begun already as we consider the implications of pleiotropy on our hypotheses and theories. Dobzhansky famously said, "Nothing in biology makes sense except in the light of evolution." As the field of molecular evolution matures, we must not forget the biology underlying it.

References

(2005) Initial sequence of the chimpanzee genome and comparison with the human genome. Nature *437:*69-87.

Andres AM, Soldevila M, Navarro A, Kidd KK, Oliva B, and Bertranpetit J (2004) Positive selection in MAOA gene is human exclusive: determination of the putative amino acid change selected in the human lineage. Hum Genet *115:*377-86.

Bond J, Roberts E, Mochida GH, Hampshire DJ, Scott S, Askham JM, Springell K, Mahadevan M, Crow YJ, Markham AF, Walsh CA, and Woods CG (2002) ASPM is a major determinant of cerebral cortical size. Nat Genet *32:*316-20.

Bond J, Roberts E, Springell K, Lizarraga S, Scott S, Higgins J, Hampshire DJ, Morrison EE, Leal GF, Silva EO, Costa SM, Baralle D, Raponi M, Karbani G, Rashid Y, Jafri H, Bennett C, Corry P, Walsh CA, and Woods CG (2005) A centrosomal mechanism involving CDK5RAP2 and CENPJ controls brain size. Nat Genet *37:*353-5.

Brunner HG, Nelen M, Breakefield XO, Ropers HH, and van Oost BA (1993) Abnormal behavior associated with a point mutation in the structural gene for monoamine oxidase A. Science *262:*578-80.

Burki F, and Kaessmann H (2004) Birth and adaptive evolution of a hominoid gene that supports high neurotransmitter flux. Nat Genet *36:*1061-3.

Bush EC, and Lahn BT (2008) A genome-wide screen for noncoding elements important in primate evolution. BMC Evol Biol *8:*17.

Bustamante CD, Fledel-Alon A, Williamson S, Nielsen R, Hubisz MT, Glanowski S, Tanenbaum DM, White TJ, Sninsky JJ, Hernandez RD, Civello D, Adams MD, Cargill M, and Clark AG (2005) Natural selection on protein-coding genes in the human genome. Nature *437:*1153-7.

Caceres M, Lachuer J, Zapala MA, Redmond JC, Kudo L, Geschwind DH, Lockhart DJ, Preuss TM, and Barlow C (2003) Elevated gene expression levels distinguish human from non-human primate brains. Proc Natl Acad Sci U S A *100:*13030-5.

Carroll SB (2005) Evolution at two levels: on genes and form. PLoS Biol *3:*e245.

Cases O, Seif I, Grimsby J, Gaspar P, Chen K, Pournin S, Muller U, Aguet M, Babinet C, Shih JC, and et al. (1995) Aggressive behavior and altered amounts of brain serotonin and norepinephrine in mice lacking MAOA. Science *268:*1763-6.

Chou HH, Takematsu H, Diaz S, Iber J, Nickerson E, Wright KL, Muchmore EA, Nelson DL, Warren ST, and Varki A (1998) A mutation in human CMP-sialic acid hydroxylase occurred after the Homo-Pan divergence. Proc Natl Acad Sci U S A *95:*11751-6.

Clark AG, Glanowski S, Nielsen R, Thomas PD, Kejariwal A, Todd MA, Tanenbaum DM, Civello D, Lu F, Murphy B, Ferriera S, Wang G, Zheng X, White TJ, Sninsky JJ, Adams MD, and Cargill M (2003) Inferring nonneutral evolution from human-chimp-mouse orthologous gene trios. Science *302:*1960-3.

Dediu D, and Ladd DR (2007) Linguistic tone is related to the population frequency of the adaptive haplogroups of two brain size genes, ASPM and Microcephalin. Proc Natl Acad Sci U S A *104:*10944-9.

Depaulis F, and Veuille M (1998) Neutrality tests based on the distribution of haplotypes under an infinite-site model. Mol Biol Evol *15:*1788-90.

Dicicco-Bloom E, Lu N, Pintar JE, and Zhang J (1998) The PACAP ligand/receptor system regulates cerebral cortical neurogenesis. Ann N Y Acad Sci *865:*274-89.

Dobyns WB (2002) Primary microcephaly: new approaches for an old disorder. Am J Hum Genet *112:*315-7.

Dorus S, Anderson JR, Vallender EJ, Gilbert SL, Zhang L, Chemnick LG, Ryder OA, Li W, and Lahn BT (2006) Sonic Hedgehog, a key development gene, experienced intensified molecular evolution in primates. Hum Mol Genet *15:*2031-7.

Dorus S, Vallender EJ, Evans PD, Anderson JR, Gilbert SL, Mahowald M, Wyckoff GJ, Malcom CM, and Lahn BT (2004) Accelerated evolution of nervous system genes in the origin of Homo sapiens. Cell *119:*1027-40.

Enard W, Khaitovich P, Klose J, Zollner S, Heissig F, Giavalisco P, Nieselt-Struwe K, Muchmore E, Varki A, Ravid R, Doxiadis GM, Bontrop RE, and Paabo S (2002a) Intra- and interspecific variation in primate gene expression patterns. Science *296:*340-3.

Enard W, Przeworski M, Fisher SE, Lai CS, Wiebe V, Kitano T, Monaco AP, and Paabo S (2002b) Molecular evolution of FOXP2, a gene involved in speech and language. Nature *418:*869-72.

Evans PD, Anderson JR, Vallender EJ, Choi SS, and Lahn BT (2004a) Reconstructing the evolutionary history of microcephalin, a gene controlling human brain size. Hum Mol Genet *13:*1139-45.

Evans PD, Anderson JR, Vallender EJ, Gilbert SL, Malcom CM, Dorus S, and Lahn BT (2004b) Adaptive evolution of ASPM, a major determinant of cerebral cortical size in humans. Hum Mol Genet *13:*489-94.

Evans PD, Vallender EJ, and Lahn BT (2006) Molecular evolution of the brain size regulator genes CDK5RAP2 and CENPJ. Gene *375:*75-9.

Falk D (1986) Hominid Evolution. Science *234:*11.

Fay JC, and Wu CI (2000) Hitchhiking under positive Darwinian selection. Genetics *155:*1405-13.

Ferland RJ, Eyaid W, Collura RV, Tully LD, Hill RS, Al-Nouri D, Al-Rumayyan A, Topcu M, Gascon G, Bodell A, Shugart YY, Ruvolo M, and Walsh CA (2004) Abnormal cerebellar development and axonal decussation due to mutations in AHI1 in Joubert syndrome. Nat Genet *36:*1008-13.

Fish JL, Kosodo Y, Enard W, Paabo S, and Huttner WB (2006) Aspm specifically maintains symmetric proliferative divisions of neuroepithelial cells. Proc Natl Acad Sci U S A *103:*10438-43.

Force A, Lynch M, Pickett FB, Amores A, Yan YL, and Postlethwait J (1999) Preservation of duplicate genes by complementary, degenerative mutations. Genetics *151:*1531-45.

Fu YX (1996) New statistical tests of neutrality for DNA samples from a population. Genetics *143:*557-70.

Fu YX, and Li WH (1993) Statistical tests of neutrality of mutations. Genetics *133:*693-709.

Gibbs RA, Rogers J, Katze MG, Bumgarner R, Weinstock GM, Mardis ER, Remington KA, Strausberg RL, Venter JC, Wilson RK, Batzer MA, Bustamante CD, Eichler EE, Hahn MW, Hardison RC, Makova KD, Miller W, Milosavljevic A, Palermo RE, Siepel A, Sikela JM, Attaway T, Bell S, Bernard KE, Buhay CJ, Chandrabose MN, Dao M, Davis C, Delehaunty KD, Ding Y, Dinh HH, Dugan-Rocha S, Fulton LA, Gabisi RA, Garner TT, Godfrey J, Hawes AC, Hernandez J, Hines S, Holder M, Hume J, Jhangiani SN, Joshi V, Khan ZM, Kirkness EF, Cree A, Fowler RG, Lee S, Lewis LR, Li Z, Liu YS, Moore SM, Muzny D, Nazareth LV, Ngo DN, Okwuonu GO, Pai G, Parker D, Paul HA, Pfannkoch C, Pohl CS, Rogers YH, Ruiz SJ, Sabo A, Santibanez J, Schneider BW, Smith SM, Sodergren E, Svatek AF, Utterback TR, Vattathil S, Warren W, White CS, Chinwalla AT, Feng Y, Halpern AL, Hillier LW, Huang X, Minx P, Nelson JO, Pepin KH, Qin X, Sutton GG, Venter E, Walenz BP, Wallis JW, Worley KC, Yang SP, Jones SM, Marra MA, Rocchi M, Schein JE, Baertsch R, Clarke L, Csuros M, Glasscock J, Harris RA, Havlak P, Jackson AR, Jiang H, et al. (2007) Evolutionary and biomedical insights from the rhesus macaque genome. Science *316:*222-34.

Gilad Y, Bustamante CD, Lancet D, and Paabo S (2003a) Natural selection on the olfactory receptor gene family in humans and chimpanzees. Am J Hum Genet *73:*489-501.

Gilad Y, Man O, and Glusman G (2005a) A comparison of the human and chimpanzee olfactory receptor gene repertoires. Genome Res *15:*224-30.

Gilad Y, Man O, Paabo S, and Lancet D (2003b) Human specific loss of olfactory receptor genes. Proc Natl Acad Sci U S A *100:*3324-7.

Gilad Y, Rifkin SA, Bertone P, Gerstein M, and White KP (2005b) Multi-species microarrays reveal the effect of sequence divergence on gene expression profiles. Genome Res *15:*674-80.

Glusman G, Yanai I, Rubin I, and Lancet D (2001) The complete human olfactory subgenome. Genome Res *11:*685-702.

Haesler S, Rochefort C, Georgi B, Licznerski P, Osten P, and Scharff C (2007) Incomplete and inaccurate vocal imitation after knockdown of FoxP2 in songbird basal ganglia nucleus Area X. PLoS Biol *5:*e321.

Haesler S, Wada K, Nshdejan A, Morrisey EE, Lints T, Jarvis ED, and Scharff C (2004) FoxP2 expression in avian vocal learners and non-learners. J Neurosci *24:*3164-75.

Haygood R, Fedrigo O, Hanson B, Yokoyama KD, and Wray GA (2007) Promoter regions of many neural- and nutrition-related genes have experienced positive selection during human evolution. Nat Genet *39:*1140-4.

Hudson RR, Bailey K, Skarecky D, Kwiatowski J, and Ayala FJ (1994) Evidence for positive selection in the superoxide dismutase (Sod) region of Drosophila melanogaster. Genetics *136:*1329-40.

Hughes AL (1994) The evolution of functionally novel proteins after gene duplication. Proc Biol Sci *256:*119-24.

Jackson AP, Eastwood H, Bell SM, Adu J, Toomes C, Carr IM, Roberts E, Hampshire DJ, Crow YJ, Mighell AJ, Karbani G, Jafri H, Rashid Y, Mueller RF, Markham AF, and Woods CG (2002) Identification of microcephalin, a protein implicated in determining the size of the human brain. Am J Hum Genet *71:*136-42.

Jauch A, Wienberg J, Stanyon R, Arnold N, Tofanelli S, Ishida T, and Cremer T (1992) Reconstruction of genomic rearrangements in great apes and gibbons by chromosome painting. Proc Natl Acad Sci U S A *89:*8611-5.

Jin L, Kryukov K, Clemente JC, Komiyama T, Suzuki Y, Imanishi T, Ikeo K, and Gojobori T (2008) The evolutionary relationship between gene duplication and alternative splicing. Gene *427:*19-31.

Johnson ME, Viggiano L, Bailey JA, Abdul-Rauf M, Goodwin G, Rocchi M, and Eichler EE (2001) Positive selection of a gene family during the emergence of humans and African apes. Nature *413:*514-9.

Kelly JK (1997) A test of neutrality based on interlocus associations. Genetics *146:*1197-206.

Keverne EB, and Curley JP (2008) Epigenetics, brain evolution and behaviour. Front Neuroendocrinol *29:*398-412.

Khaitovich P, Enard W, Lachmann M, and Paabo S (2006) Evolution of primate gene expression. Nat Rev Genet *7:*693-702.

Khaitovich P, Hellmann I, Enard W, Nowick K, Leinweber M, Franz H, Weiss G, Lachmann M, and Paabo S (2005) Parallel patterns of evolution in the genomes and transcriptomes of humans and chimpanzees. Science *309:*1850-4.

Khaitovich P, Muetzel B, She X, Lachmann M, Hellmann I, Dietzsch J, Steigele S, Do HH, Weiss G, Enard W, Heissig F, Arendt T, Nieselt-Struwe K, Eichler EE, and Paabo S (2004) Regional patterns of gene expression in human and chimpanzee brains. Genome Res *14:*1462-73.

Kim JJ, Shih JC, Chen K, Chen L, Bao S, Maren S, Anagnostaras SG, Fanselow MS, De Maeyer E, Seif I, and Thompson RF (1997) Selective enhancement of emotional, but not motor, learning in monoamine oxidase A-deficient mice. Proc Natl Acad Sci U S A *94:*5929-33.

King MC, and Wilson AC (1975) Evolution at two levels in humans and chimpanzees. Science *188:*107-16.

Kornack DR, and Rakic P (1998) Changes in cell-cycle kinetics during the development and evolution of primate neocortex. Proc Natl Acad Sci U S A *95:*1242-6.

Kouprina N, Pavlicek A, Collins NK, Nakano M, Noskov VN, Ohzeki J, Mochida GH, Risinger JI, Goldsmith P, Gunsior M, Solomon G, Gersch W, Kim JH, Barrett JC, Walsh CA, Jurka J, Masumoto H, and Larionov V (2005) The microcephaly ASPM gene is expressed in proliferating tissues and encodes for a mitotic spindle protein. Hum Mol Genet *14:*2155-65.

Kouprina N, Pavlicek A, Mochida GH, Solomon G, Gersch W, Yoon YH, Collura R, Ruvolo M, Barrett JC, Woods CG, Walsh CA, Jurka J, and Larionov V (2004) Accelerated evolution of the ASPM gene controlling brain size begins prior to human brain expansion. PLoS Biol *2:*E126.

Lahn BT, Pearson NM, and Jegalian K (2001) The human Y chromosome, in the light of evolution. Nat Rev Genet *2:*207-16.

Lai CS, Fisher SE, Hurst JA, Vargha-Khadem F, and Monaco AP (2001) A forkhead-domain gene is mutated in a severe speech and language disorder. Nature *413:*519-23.

Lewontin RC, and Krakauer J (1973) Distribution of gene frequency as a test of the theory of the selective neutrality of polymorphisms. Genetics *74:*175-95.

Li WH, Boissinot S, Tan Y, Shyue SK, and Hewett-Emmett D (1999) Evolutionary genetics of primate color vision: recent progress and potential limits to knowledge. Evol Biol *32*.

Lynch M, and Force A (2000) The probability of duplicate gene preservation by subfunctionalization. Genetics *154:*459-73.

Lynch M, O'Hely M, Walsh B, and Force A (2001) The probability of preservation of a newly arisen gene duplicate. Genetics *159:*1789-804.

Marvanova M, Menager J, Bezard E, Bontrop RE, Pradier L, and Wong G (2003) Microarray analysis of nonhuman primates: validation of experimental models in neurological disorders. Faseb J *17:*929-31.

Miyata T, and Yasunaga T (1980) Molecular evolution of mRNA: a method for estimating evolutionary rates of synonymous and amino acid substitutions from homologous nucleotide sequences and its application. J Mol Evol *16:*23-36.

Mochida GH, and Walsh CA (2001) Molecular genetics of human microcephaly. Curr Opin Neurol *14:*151-6.

Muchmore EA, Diaz S, and Varki A (1998) A structural difference between the cell surfaces of humans and the great apes. Am J Phys Anthropol *107:*187-98.

Myers AJ, Gibbs JR, Webster JA, Rohrer K, Zhao A, Marlowe L, Kaleem M, Leung D, Bryden L, Nath P, Zismann VL, Joshipura K, Huentelman MJ, Hu-Lince D, Coon KD, Craig DW, Pearson JV, Holmans P, Heward CB, Reiman EM, Stephan D, and Hardy J (2007) A survey of genetic human cortical gene expression. Nat Genet *39:*1494-9.

Neill D (2007) Cortical evolution and human behaviour. Brain Res Bull *74:*191-205.

Nielsen R, Bustamante C, Clark AG, Glanowski S, Sackton TB, Hubisz MJ, Fledel-Alon A, Tanenbaum DM, Civello D, White TJ, J JS, Adams MD, and Cargill M (2005) A scan for positively selected genes in the genomes of humans and chimpanzees. PLoS Biol *3:*e170.

Ohno S (1970) Evolution by Gene Duplication. New York: Springer-Verlag.

Perry GH, Verrelli BC, and Stone AC (2005) Comparative analyses reveal a complex history of molecular evolution for human MYH16. Mol Biol Evol *22:*379-82.

Pollard KS, Salama SR, King B, Kern AD, Dreszer T, Katzman S, Siepel A, Pedersen JS, Bejerano G, Baertsch R, Rosenbloom KR, Kent J, and Haussler D (2006a) Forces shaping the fastest evolving regions in the human genome. PLoS Genet *2:*e168.

Pollard KS, Salama SR, Lambert N, Lambot MA, Coppens S, Pedersen JS, Katzman S, King B, Onodera C, Siepel A, Kern AD, Dehay C, Igel H, Ares M, Jr., Vanderhaeghen P, and Haussler D (2006b) An RNA gene expressed during cortical development evolved rapidly in humans. Nature *443:*167-72.

Popesco MC, Maclaren EJ, Hopkins J, Dumas L, Cox M, Meltesen L, McGavran L, Wyckoff GJ, and Sikela JM (2006) Human lineage-specific amplification, selection, and neuronal expression of DUF1220 domains. Science *313:*1304-7.

Prabhakar S, Noonan JP, Paabo S, and Rubin EM (2006) Accelerated evolution of conserved noncoding sequences in humans. Science *314:*786.

Preuss TM, Caceres M, Oldham MC, and Geschwind DH (2004) Human brain evolution: insights from microarrays. Nat Rev Genet *5:*850-60.

Rockman MV, Hahn MW, Soranzo N, Zimprich F, Goldstein DB, and Wray GA (2005) Ancient and recent positive selection transformed opioid cis-regulation in humans. PLoS Biol *3:*e387.

Roll P, Rudolf G, Pereira S, Royer B, Scheffer IE, Massacrier A, Valenti MP, Roeckel-Trevisiol N, Jamali S, Beclin C, Seegmuller C, Metz-Lutz MN, Lemainque A, Delepine M, Caloustian C, de Saint Martin A, Bruneau N, Depetris D, Mattei MG, Flori E, Robaglia-Schlupp A, Levy N, Neubauer BA, Ravid R, Marescaux C, Berkovic SF, Hirsch E, Lathrop M, Cau P, and Szepetowski P (2006) SRPX2 mutations in disorders of language cortex and cognition. Hum Mol Genet *15:*1195-207.

Royer B, Soares DC, Barlow PN, Bontrop RE, Roll P, Robaglia-Schlupp A, Blancher A, Levasseur A, Cau P, Pontarotti P, and Szepetowski P (2007) Molecular evolution of the human SRPX2 gene that causes brain disorders of the Rolandic and Sylvian speech areas. BMC Genet *8:*72.

Sabeti PC, Reich DE, Higgins JM, Levine HZ, Richter DJ, Schaffner SF, Gabriel SB, Platko JV, Patterson NJ, McDonald GJ, Ackerman HC, Campbell SJ, Altshuler D, Cooper R, Kwiatkowski D, Ward R, and Lander ES (2002) Detecting recent positive selection in the human genome from haplotype structure. Nature *419:*832-7.

Semendeferi K, Lu A, Schenker N, and Damasio H (2002) Humans and great apes share a large frontal cortex. Nat Neurosci *5:*272-6.

Shashidharan P, Michaelidis TM, Robakis NK, Kresovali A, Papamatheakis J, and Plaitakis A (1994) Novel human glutamate dehydrogenase expressed in neural and testicular tissues and encoded by an X-linked intronless gene. J Biol Chem *269:*16971-6.

Shi P, Bakewell MA, and Zhang J (2006) Did brain-specific genes evolve faster in humans than in chimpanzees? Trends Genet *22:*608-13.

Shih JC, Chen K, and Ridd MJ (1999) Monoamine oxidase: from genes to behavior. Annu Rev Neurosci *22:*197-217.

Shu W, Cho JY, Jiang Y, Zhang M, Weisz D, Elder GA, Schmeidler J, De Gasperi R, Sosa MA, Rabidou D, Santucci AC, Perl D, Morrisey E, and Buxbaum JD (2005) Altered ultrasonic vocalization in mice with a disruption in the Foxp2 gene. Proc Natl Acad Sci U S A *102:*9643-8.

Sims KB, de la Chapelle A, Norio R, Sankila EM, Hsu YP, Rinehart WB, Corey TJ, Ozelius L, Powell JF, Bruns G, and et al. (1989) Monoamine oxidase deficiency in males with an X chromosome deletion. Neuron *2:*1069-76.

Slatkin M, and Bertorelle G (2001) The use of intraallelic variability for testing neutrality and estimating population growth rate. Genetics *158:*865-74.

Stedman HH, Kozyak BW, Nelson A, Thesier DM, Su LT, Low DW, Bridges CR, Shrager JB, Minugh-Purvis N, and Mitchell MA (2004) Myosin gene mutation correlates with anatomical changes in the human lineage. Nature *428:*415-8.

Suh J, Lu N, Nicot A, Tatsuno I, and DiCicco-Bloom E (2001) PACAP is an anti-mitogenic signal in developing cerebral cortex. Nat Neurosci *4:*123-4.

Tajima F (1989) Statistical method for testing the neutral mutation hypothesis by DNA polymorphism. Genetics *123:*585-95.

Takeda J, Suzuki Y, Sakate R, Sato Y, Seki M, Irie T, Takeuchi N, Ueda T, Nakao M, Sugano S, Gojobori T, and Imanishi T (2008) Low conservation and species-specific evolution of alternative splicing in humans and mice: comparative genomics analysis using well-annotated full-length cDNAs. Nucleic Acids Res *36:*6386-95.

Teramitsu I, Kudo LC, London SE, Geschwind DH, and White SA (2004) Parallel FoxP1 and FoxP2 expression in songbird and human brain predicts functional interaction. J Neurosci *24:*3152-63.

Teramitsu I, and White SA (2006) FoxP2 regulation during undirected singing in adult songbirds. J Neurosci *26:*7390-4.

Toomajian C, Ajioka RS, Jorde LB, Kushner JP, and Kreitman M (2003) A method for detecting recent selection in the human genome from allele age estimates. Genetics *165:*287-97.

Trimborn M, Bell SM, Felix C, Rashid Y, Jafri H, Griffiths PD, Neumann LM, Krebs A, Reis A, Sperling K, Neitzel H, and Jackson AP (2004) Mutations in microcephalin cause aberrant regulation of chromosome condensation. Am J Hum Genet *75:*261-6.

Uddin M, Wildman DE, Liu G, Xu W, Johnson RM, Hof PR, Kapatos G, Grossman LI, and Goodman M (2004) Sister grouping of chimpanzees and humans as revealed by genome-wide phylogenetic analysis of brain gene expression profiles. Proc Natl Acad Sci U S A *101:*2957-62.

Vallender EJ (2008) Exploring the origins of the human brain through molecular evolution. Brain Behav Evol *72:*168-77.

Vallender EJ, Mekel-Bobrov N, and Lahn BT (2008) Genetic basis of human brain evolution. Trends Neurosci *31:*637-44.

Wang HY, Chien HC, Osada N, Hashimoto K, Sugano S, Gojobori T, Chou CK, Tsai SF, Wu CI, and Shen CK (2007) Rate of evolution in brain-expressed genes in humans and other primates. PLoS Biol *5:*e13.

Wang YQ, Qian YP, Yang S, Shi H, Liao CH, Zheng HK, Wang J, Lin AA, Cavalli-Sforza LL, Underhill PA, Chakraborty R, Jin L, and Su B (2005) Accelerated evolution of the pituitary adenylate cyclase-activating polypeptide precursor gene during human origin. Genetics *170:*801-6.

Wang YQ, and Su B (2004) Molecular evolution of microcephalin, a gene determining human brain size. Hum Mol Genet *13:*1131-7.

Wood JL, Singh N, Mer G, and Chen J (2007) MCPH1 functions in an H2AX-dependent but MDC1-independent pathway in response to DNA damage. J Biol Chem *282:*35416-23.

Woods CG, Bond J, and Enard W (2005) Autosomal recessive primary microcephaly (MCPH): a review of clinical, molecular, and evolutionary findings. Am J Hum Genet *76:*717-28.

Xu X, Lee J, and Stern DF (2004) Microcephalin is a DNA damage response protein involved in regulation of CHK1 and BRCA1. J Biol Chem *279:*34091-4.

Young JM, Friedman C, Williams EM, Ross JA, Tonnes-Priddy L, and Trask BJ (2002) Different evolutionary processes shaped the mouse and human olfactory receptor gene families. Hum Mol Genet *11:*535-46.

Young JM, and Trask BJ (2002) The sense of smell: genomics of vertebrate odorant receptors. Hum Mol Genet *11:*1153-60.

Zeng K, Mano S, Shi S, and Wu CI (2007) Comparisons of site- and haplotype-frequency methods for detecting positive selection. Mol Biol Evol *24:*1562-74.

Zhai W, Nielsen R, and Slatkin M (2009) An investigation of the statistical power of neutrality tests based on comparative and population genetic data. Mol Biol Evol *26:*273-83.

Zhang J (2003) Evolution of the human ASPM gene, a major determinant of brain size. Genetics *165:*2063-70.

CHAPTER 9

Brain Reorganization in Humans and Apes

Katerina Semendeferi, Nicole Barger and Natalie Schenker

Abstract

This paper examines evidence from comparative neuroanatomical studies ofhumans and apes to address Holloway's ideas (Holloway, 1968, 1979, 2001) about reorganization in hominoid and hominid evolution. Specifically, work accomplished mostly in our laboratory supports the theory of reorganization for some brain areas but not others. Quantifiable parameters of selected gross anatomical regions, individual cortical areas, and subcortical nuclei point to selective reorganization in both human and great ape brains. Much of this accumulated evidence is based on species differences between gross measures of the frontal lobe (e.g., overall volume or the amount of white matter or cortex) and functionally relevant subregions within it (such as the dorsal, mesial, and orbital sectors). Further support is derived from analyses of regions neighboring the frontal lobe, such as the frontal insular cortex, using histological measurements of size of cortical areas and neuronal densities. Evidence for reorganization in the temporal lobe includes distribution of white matter, and the organization of a major subcortical structure, the amygdala. These findings are a starting point for studying and understanding reorganization in these and other parts of the brain.

Introduction

Most regions of the primate brain conform to regular scaling relationships (Jerison, 1973; Finlay et al., 2001), but there are some exceptions to this regularity across species. For example, Holloway hypothesized (1968) that the human brain is not an enlarged ape brain, and that ape and monkey brains are not enlarged or reduced versions of each other. It is likely that different regions of the brain are differentially increased or decreased depending on thc adaptive forces present in the evolutionary environment. For example, animals that rely heavily on a particular sensory modality (e.g. audition, vision, somatosensory) have differentially enlarged cortical territories (Figure 1) that are involved in such functions (Krubitzer and Kahn, 2003). Although information supporting Holloway's hypothesis is now available for smaller mammals, comparisons of humans and apes remain scarce. Neuroanatomical comparisons of extant ape and human brains can identify features that are either shared among hominoids, i.e., humans and apes, or are unique to humans and thus may have arisen specifically during hominid evolution. Our laboratory has focused on these issues and we have modified and expanded upon previous hypotheses, many of which were originally based on limited empirical information. While advanced human cognitive abilities might be attributable in part to an increase in total brain size (Gibson, 2002; Passingham, 2002), they might also arise from discrete modifications in the relative size of specific neural systems that accompany absolute brain size increases. Species-specific cognitive abilities might also exist independently of changes in absolute brain size (Figure 2) and thus may be differentially present in primates that have similarly sized brains (e.g. great apes). To address this latter possibility, our studies have been aimed at understanding the organization and size of individual cortical areas and subcortical regions across hominoids.

In human and ape species few detailed evolutionary studies of the brain exist and even fewer studies include analyses of isolated regions of the brain. Unlike

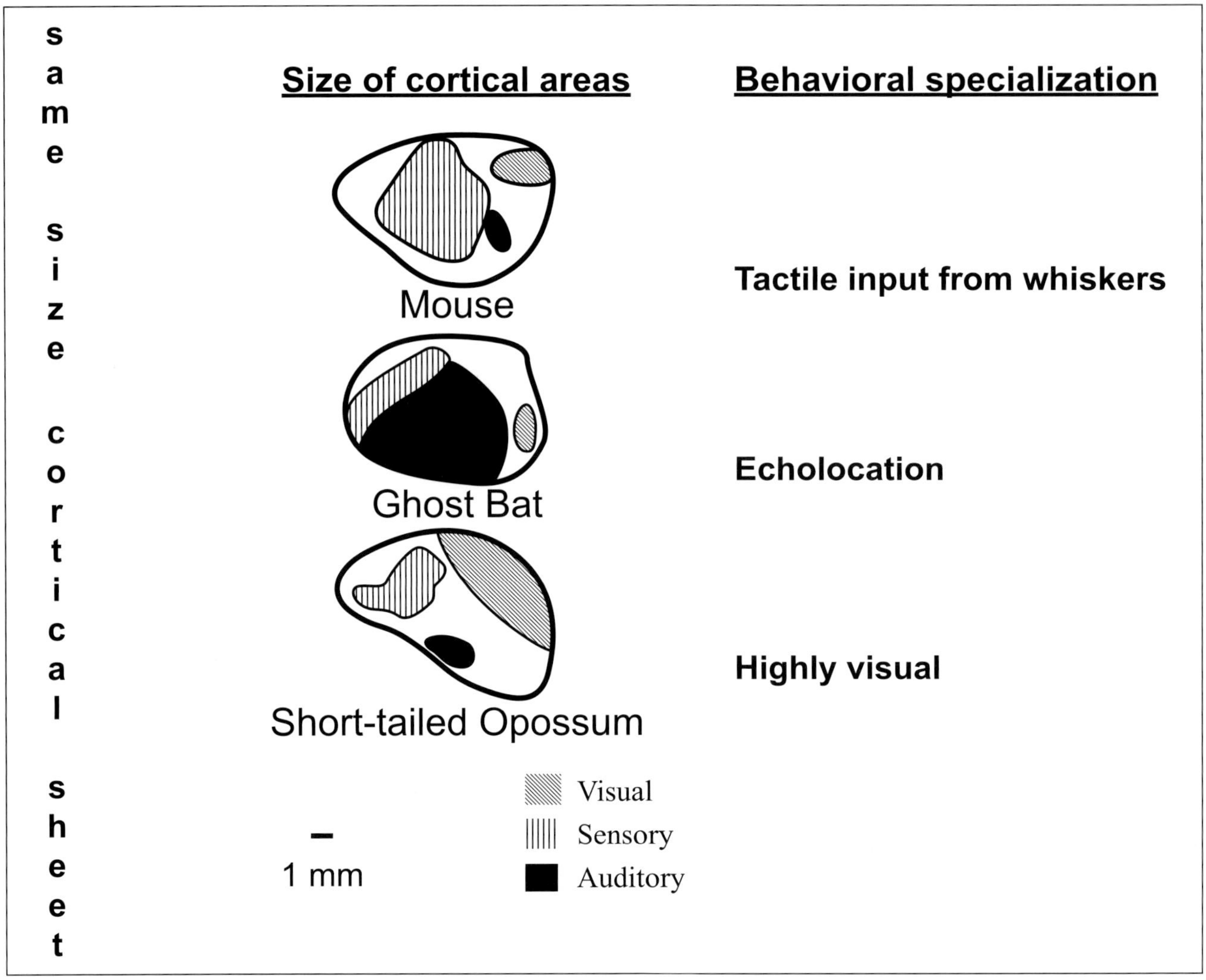

Figure 1. Three mammalian species with cortical sheets of similar size representing different sensory adaptations (modified from Krubitzer & Kahn, 2003)

the majority of previous analyses, our data are drawn from a sample that includes multiple individuals of each hominoid species. We use both in vivo structural magnetic resonance images (MRI) and histological sections of postmortem specimens (donated by zoological and research facilities) in our analyses. With these samples we isolated different cortical areas to determine the extent to which they have expanded or become diminished in each hominoid species, resulting in distinct neural organizations. Differences in the size and organization of the whole brain and specific subregions might reflect species-specific adaptations, functional specializations, and/or major evolutionary events relating to changes in the organization of the hominoid brain (Armstrong, 1990). From a neuroscientific perspective, comparisons of specific neural circuits or cortical areas in closely related species are necessary for understanding the species-specific adaptations and neural circuitry underlying behavior. Studies of relationships such as total brain size to body size are insufficient for the understanding of species-specific adaptations in behavior and underlying neural circuitry.

With a diverse sample of hominoids, we tested the assumptions of longstanding hypotheses about human brain evolution using larger samples and more rigorous experimental techniques. The frontal lobe has long been considered the most likely candidate for evolutionary expansion in the human line, because it plays a central role in higher order cognitive functions like planning for the future, abstract symbolic processing, and categorization (Damasio, 1985). Until recently, many evolutionary reconstructions assumed an enlarged frontal cortex in humans (Deacon, 1997). Early data from Brodmann (1909) and Blinkov and Glezer (1968) supported the notion that human frontal lobes were disproportionately enlarged compared to other primates. Unfortunately, it is difficult to accept these data as reliable, because only a few hominoid species were represented, and the sample size often included only one or two hemispheres per species. Furthermore, the organization of the primate frontal cortex is complex (Goldman-Rakic, 1984; Barbas and Pandya, 1989), comprising many diverse territories with distinct functional properties. Addressing the frontal

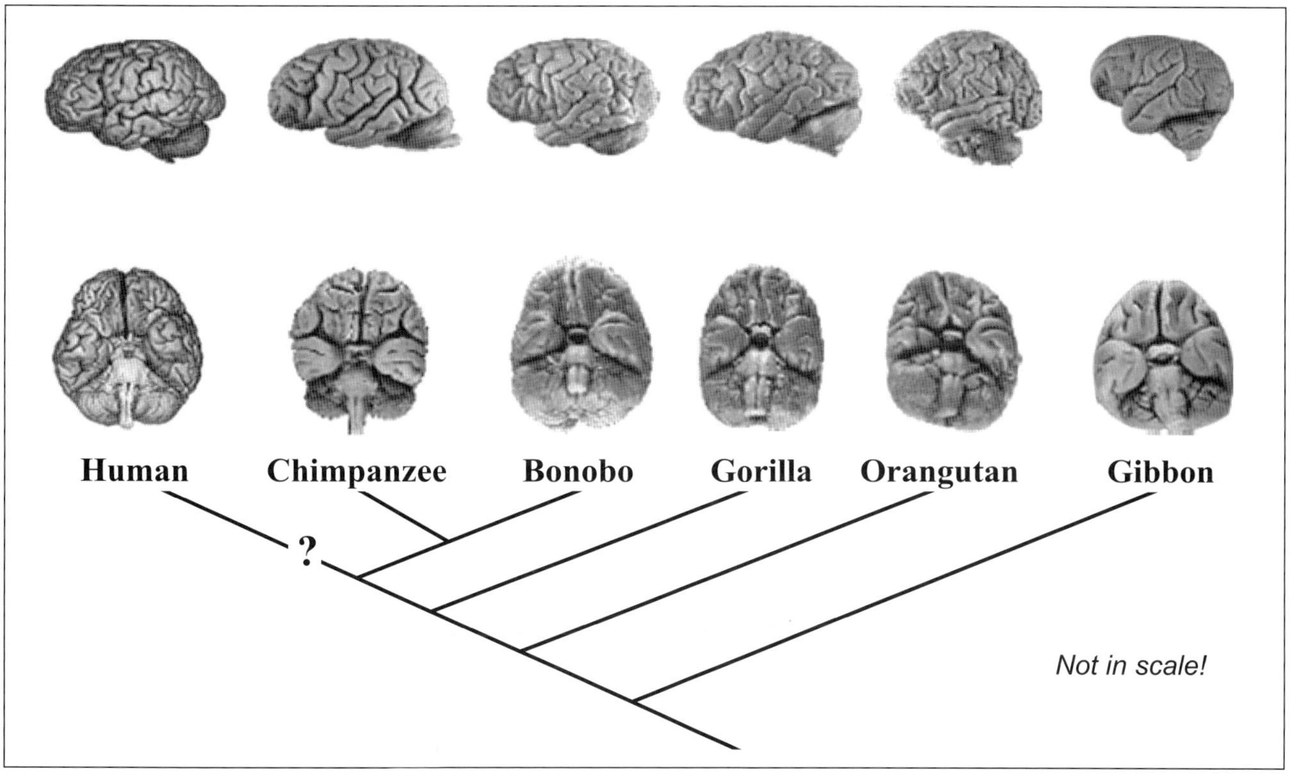

Figure 2. Lateral and orbital views of ape and human brains in a phylogenetic tree. Brains are not in scale, but are either enlarged (gibbon) or reduced (human) to match in size.

lobe only as a whole elides this functional heterogeneity. Few studies have addressed comparative morphometric differences in subregions of the frontal lobe or the organization of this sector from an evolutionary perspective (Fuster, 1997; Holloway, 1968; Jerison, 1997). We have used this novel approach in the laboratory and we discuss our findings below.

Another theoretical tenet in the study of brain evolution is that the limbic system is a conserved region and therefore less likely to exhibit evolutionary change. This notion has been questioned by recent comparative research. The limbic system, also referred to as the "paleomammalian" brain (MacLean, 1990), has long been considered to mediate primarily basic functions such as drives and emotions. Because of this association, it is sometimes viewed as primitive and thus unlikely to be evolutionarily effected in the human brain, in contrast to the proposed enlargement of association areas like the prefrontal cortex that are known to be involved in abstract thinking and language (Foundas, 2001). Contrary to this prediction, several limbic structures exhibit evolutionary reorganization in humans, specifically, the thalamus, the orbitofrontal cortex, and the amygdala. These latter two structures are also considered central components in the neural system subserving social affiliation. In humans and several species of primates, lesions of the amygdala and certain frontal cortical regions alter normal social behaviors (Kling and Brothers, 1992). Humans with frontal trauma are impaired in their ability to process socio-emotional information and to follow social rules (Adolphs, 1999). Nonhuman primates with amygdala lesions show atypical behaviors in social interactions such as the tendency to be either socially disinhibited (Emery, et al., 2001) or socially fearful (Kling, 1986). The fact that the portions of the neural system subserving social behavior show evolutionary reorganization in humans and apes is striking given the increasing popularity of arguments which posit that the pressure of living in complex social environments has had considerable influence on primate brain evolution. Despite an increased interest in social cognition across scientific fields, though, our understanding of the evolution of associated neural structures in humans and the apes is in its infancy.

Thus, there is some evidence to support the idea that, beyond the presumed influence of gross increases in overall isocortical or frontal lobe volume, systems and regions of the brain that are explicitly associated with certain behaviors, especially social behaviors, have been the targets of evolutionary reorganization. Still, it is unclear whether and to what extent neural systems involved in social cognition vary among primate species. We are just beginning to address how the neural systems that determine how individuals relate to conspecifics and make decisions about social interactions are organized and reorganized through evolutionary processes. Here we will examine evidence from our laboratory that supports Holloway's ideas on mosaic evolution and brain reorganization in limbic and isocortical structures associated with social behavior. We will also present findings suggesting that regions in the frontal lobe show a more conservative pattern of evolution across hominoids.

In Vivo Investigations of the Frontal Lobe

The central role assigned to the frontal lobe in human evolutionary reconstructions is based largely on its involvement in complex cognitive functions such as symbolic thought, cognitive planning, decision making, and language production (Owen et al., 1996; Bechara et al., 2000; Foundas, 2001; Pochon et al., 2001). The frontal lobe is also associated with perception, response selection, working memory, problem solving (Owen, 1997; Bechara et al., 1998; Petrides, 2000; Pochon et al., 2001), processing emotional stimuli, the production of affective responses (Cummings, 1993; Rezai et al., 1993), planning and initiation of voluntary motor sequences (Tanji and Mushiake, 1996), theory of mind (Fletcher et al., 1995; Gallagher et al., 2000; Stuss et al., 2001), attention management (Carter et al., 1999; Dagher et al., 1999), and the evaluation of actions based on emotional reinforcers (Damasio, 1994; Stone et al., 1998; Rolls, 2000). Large lesions of the frontal lobe produce the delayed response deficit, which is characterized by a lack of initiative or, in other words, the impairment of "interest and hence sustained attention and initiative" (Sanides, 1964; Harlow et al., 1964; Rosvold et al., 1964).

Whole frontal lobe

The frontal lobe is located anterior to the parietal and temporal lobes (Figure 3) and is bounded in all primates by the central and lateral sulci. In a series of comparative morphometric investigations we used three dimensional reconstructions of MRI scans (Semendeferi et al., 1997, Semendeferi and Damasio, 2000) to investigate the size of the frontal lobe in postmortem and living humans (n=11), great apes (n=19), and other primates (*Hylobates* sp. n=5; Macaca mulatta n=1). Although in absolute volume the human frontal lobe is larger than the frontal lobe of other primates, allometric analyses suggest that the human frontal lobe is as large as expected for an anthropoid brain of human size (Semendeferi et al., 1997; Semendeferi and Damasio, 2000). Because both humans and great apes have a large frontal lobe relative to the rest of the brain, human brain evolution and the evolution of complex cognitive capabilities cannot simply be attributed to differential enlargement of the frontal lobe as a whole.

We hypothesized that differences in the neural substrates underlying complex cognitive functions in humans may instead be present in subregions of the frontal lobe at gross anatomical and/or microscopic levels. The frontal lobe includes parts of subcortical structures such as the basal ganglia as well as gray matter (cortex) and white matter (connective fibers) (Figure 3). A number of morphometric studies have addressed the size of individual gross anatomical regions of the frontal lobe in normal (Caviness et al., 1996) and pathological human brains (Andreasen et al., 1994; Mitelman et al., 2003; Carper and Courchesne, 2005). Using landmarks that are consistently present across hominoids, we identified three such regions: the dorsal, mesial, and orbital sectors of the frontal lobe (Semendeferi et al., 1997; Schenker et al., 2005). We made volumetric assessments of the frontal lobe and these three sectors. We also segmented them into both gray matter and white matter for separate quantifications.

Frontal cortex

Like the frontal lobe as a whole, the frontal cortex in humans is as large as expected for an ape of human brain size (Semendeferi et al., 2002) (Figure 4). It occupies a similar proportion of total cerebral cortex in humans and in great apes, but a smaller proportion in smaller-brained primates (Figure 5). This is also true for

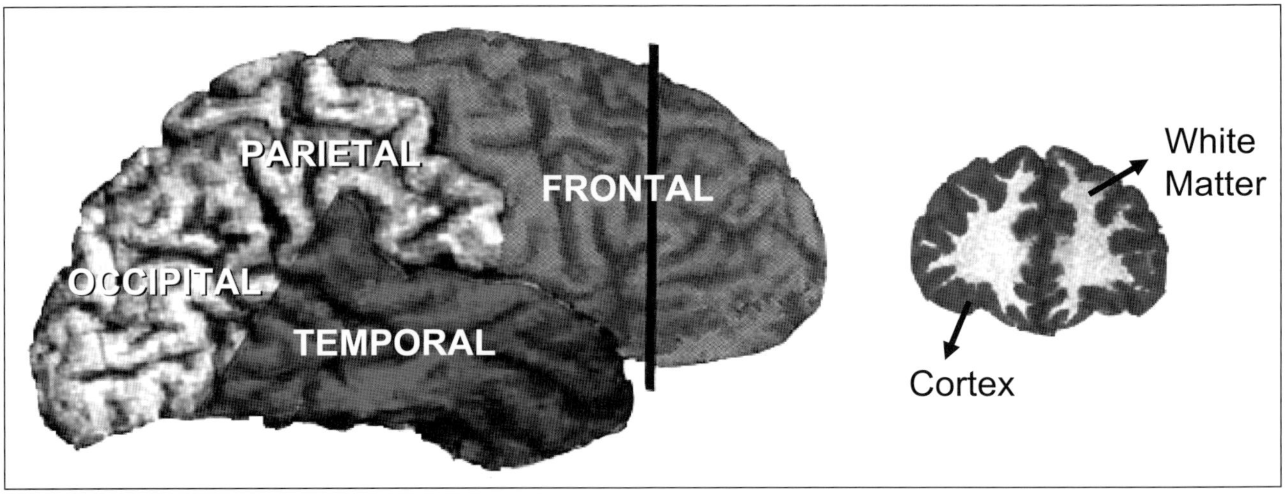

Figure 3. Lateral view of a human brain highlighting the frontal and temporal lobes that were targeted in our comparative studies. Vertical bar represents the location of a cross section (shown on the right) which reveals the presence of the cortex (gray) and white matter (white).

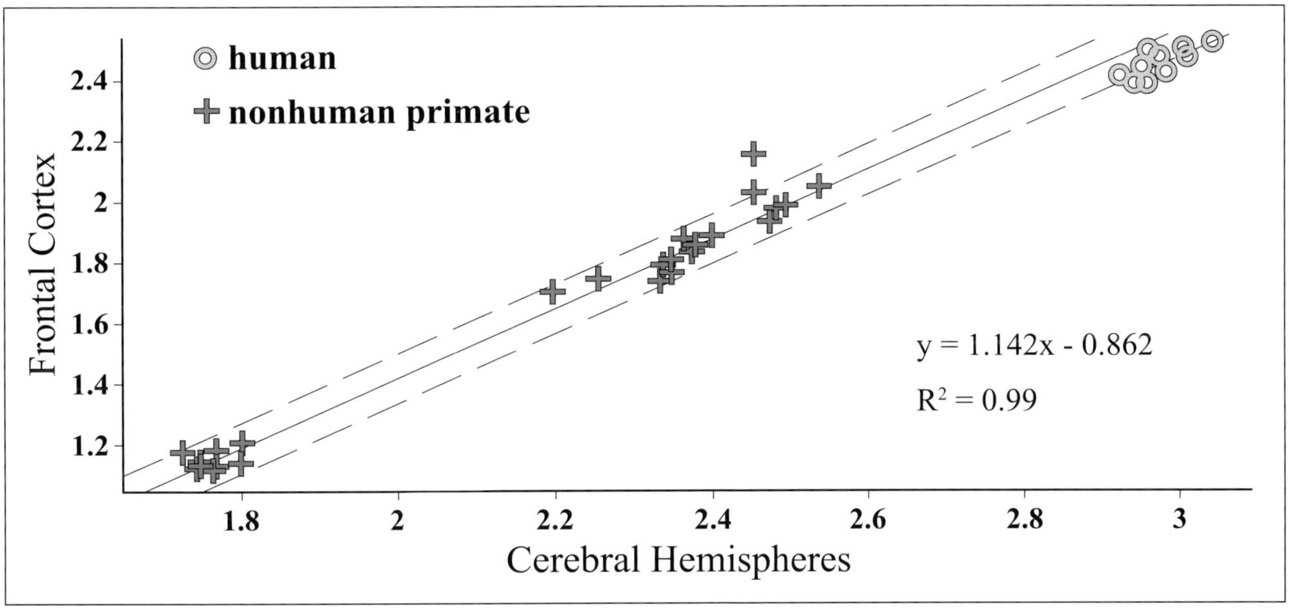

Figure 4. Volume of frontal cortex against volume of cerebral hemispheres in human and nonhuman primates (modified from Semendeferi et al., 2002).

the orbital, mesial, and dorsal sectors of the human frontal lobe (Figure 6). These three sectors maintain a lawful relationship with increasing brain size in humans and in most hominoids.

White matter

Classical investigations of the frontal lobe focused mostly on measurements of the cortex, but more recent studies have recognized that species differences in white matter volumes may be of functional and morphological importance. We identified and measured the volume of the white matter in the frontal lobe (Figure 3) and found that, as expected, white matter has a positive allometric relationship with increasing brain size (Schenker et al., 2005). Like the frontal cortex, total white matter volume in the frontal lobes are as large as expected in hominoids (Schenker et al., 2005). Our findings support the long established idea that larger brains tend to have a larger percentage of white matter than smaller brains and are consistent with expected values for humans and apes in gross frontal lobe measures (Frahm et al., 1982; Prothero and Sundsten, 1984; Hofman, 1989; Rilling and Insel, 1999; Zhang and Sejnowski, 2000; Bush and Allman, 2004).

Nonetheless, because white matter is a complex territory comprising numerous, distinct fiber systems, humans might show variation in the distribution of white matter amongst these subsectors. Commissural systems such as the corpus callosum provide communicative pathways between the hemispheres. Projection fiber systems such as the corona radiata link isocortex with non-isocortical neural structures and the periphery. Long association fibers mediate intrahemispheric communication between distant cortical areas. There are also short association fibers that function as connections between topographically adjacent regions. The overall spatial arrangement of the fiber systems reflects the topographic relationship between the origins and targets of the connections (Dejerine, 1895; Heimer, 1995). While it is not possible to distinguish all specific fiber systems in gross brain scans, a number of morphometric studies in humans (using structural MRI scans) have parcellated the white matter to measure the volume of its subdivisions (Makris et al., 1997; Herbert et al., 2004). One subdivision is the gyral white matter which includes the white matter immediately underlying cortical territories; another subdivision is the core, which includes all remaining white matter in the frontal lobe (Figure 7). The gyral white matter immediately underlying cortical territories generally includes, in addition to long projection fibers, also short corticocortical association fibers that link neighboring cortical regions (Heimer, 1995). Gyral white matter volume approximates the size of connections between regions that lie in close proximity, whereas core white matter reflects more the degree of connectivity between distant regions.

The relationship between gyral white matter and core white matter is different in humans compared to other primates (Schenker, et al., 2005). Most human values for gyral white matter are larger than expected in the frontal lobe (Figure 8), while the average percent residuals of human values for the frontal gyral white matter are more than 26%. Enlarged gyral white matter in humans suggests increased interconnectivity within and between neighboring cortical regions, many of which support complex cognitive behaviors. Thus an increased ratio of gyral white matter to core white matter is likely to reflect an increased emphasis on short, intrahemispheric, cortico-cortical associations. Nevertheless, the volume of the white matter, whether core or gyral, can increase as

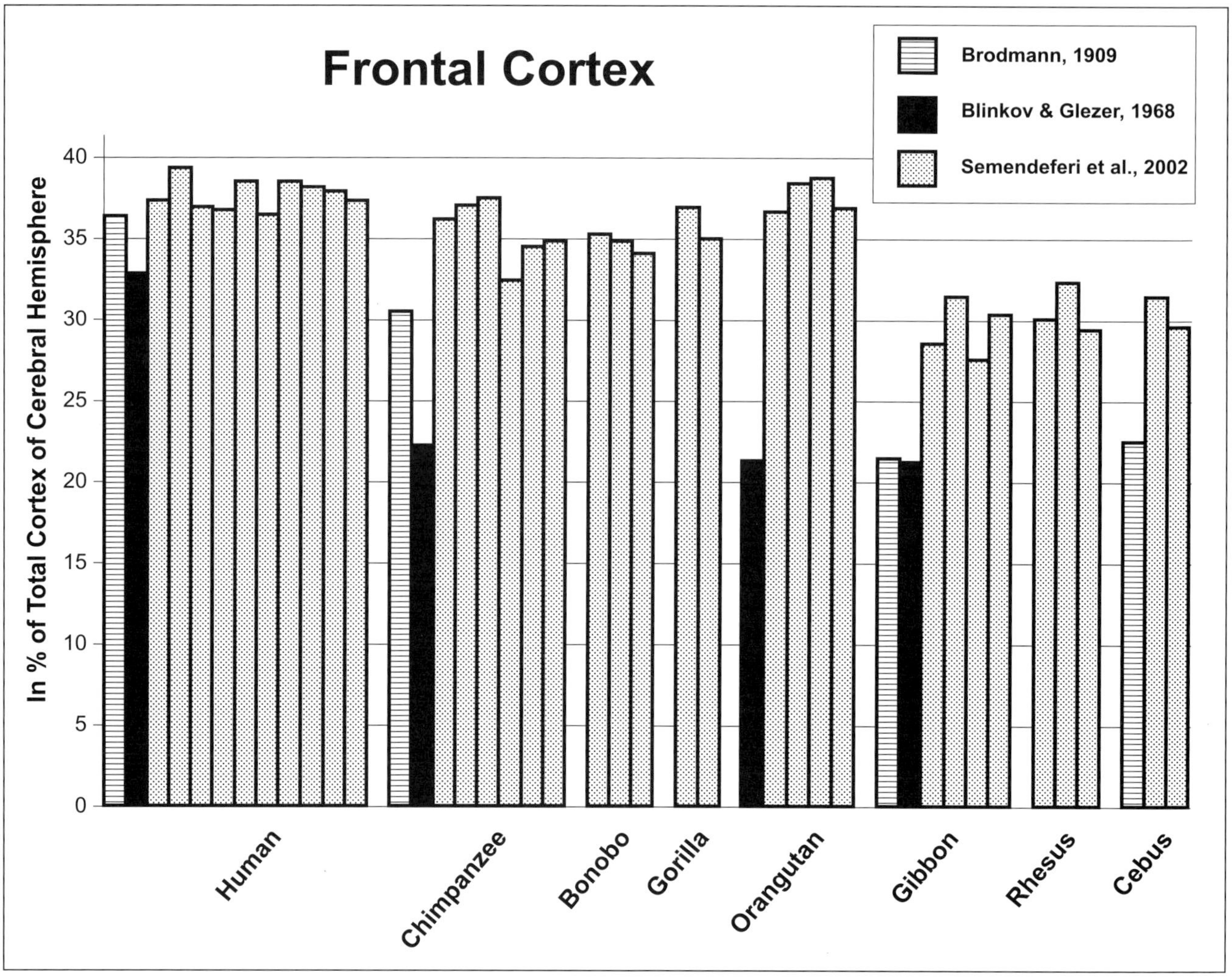

Figure 5. The size of the frontal cortex as a percent of total cortex. Each bar represents individual human and nonhuman primate specimens in the studies listed in legend.

a function of several parameters, including numbers of fibers, size of fibers and/or degree of myelination (Harrison et al., 2002).

Developmentally there are differences in the myelination of gyral versus core white matter. White matter in closer proximity to the cortex myelinates after core white matter, and occurs late in the first year of life or later, especially in the frontal cortex (Yakovlev and LeCours, 1967). Similar differences might exist across species, but no cross species comparisons exist regarding the degree of myelination. Thus, speciesspecific differences might include increased numbers of axons, increased crosssectional area of axons, or increased myelination between cortical areas resulting in faster information processing, facilitating complex cognitive function (Harrison et al., 2002). Other studies of fiber systems based on structural MRI and diffusion tensor imaging (Rilling and Insel, 1999; Glasser and Rilling, 2008) also suggest that some fiber systems are represented in different proportions in selected primates, supporting the idea of the presence of selective reorganization in the white matter.

Reorganization and the human frontal lobe

There is no evidence for brain reorganization based on the size of the frontal lobe and the frontal cortex as a whole. Homogeneity in the scaling relations of some of the major sectors in the brain was suggested previously (Jerison, 1973; Clark, et al., 2001; Finlay and Darlington, 1995), and our results show that homogeneity also exists with respect to the size of the hominoid frontal cortex. Our findings support Bonin's (1948), Holloway's (1968), and Jerison's (1973) long standing ideas regarding the evolution of the frontal cortex and frontal lobe as a whole, which received additional support more recently (Bush and Allman, 2004), namely that "man has precisely the frontal cortex which he deserves" (Bonin, 1948). Nevertheless our morphometric results challenge the two data sets collected by Brodmann (1912) and Blinkov and Glezer (1968) and the dominant dogma derived from their data that views human cognitive evolution as largely driven by a differential enlargement of the frontal lobe and frontal cortex (Deacon, 1997; Fuster,

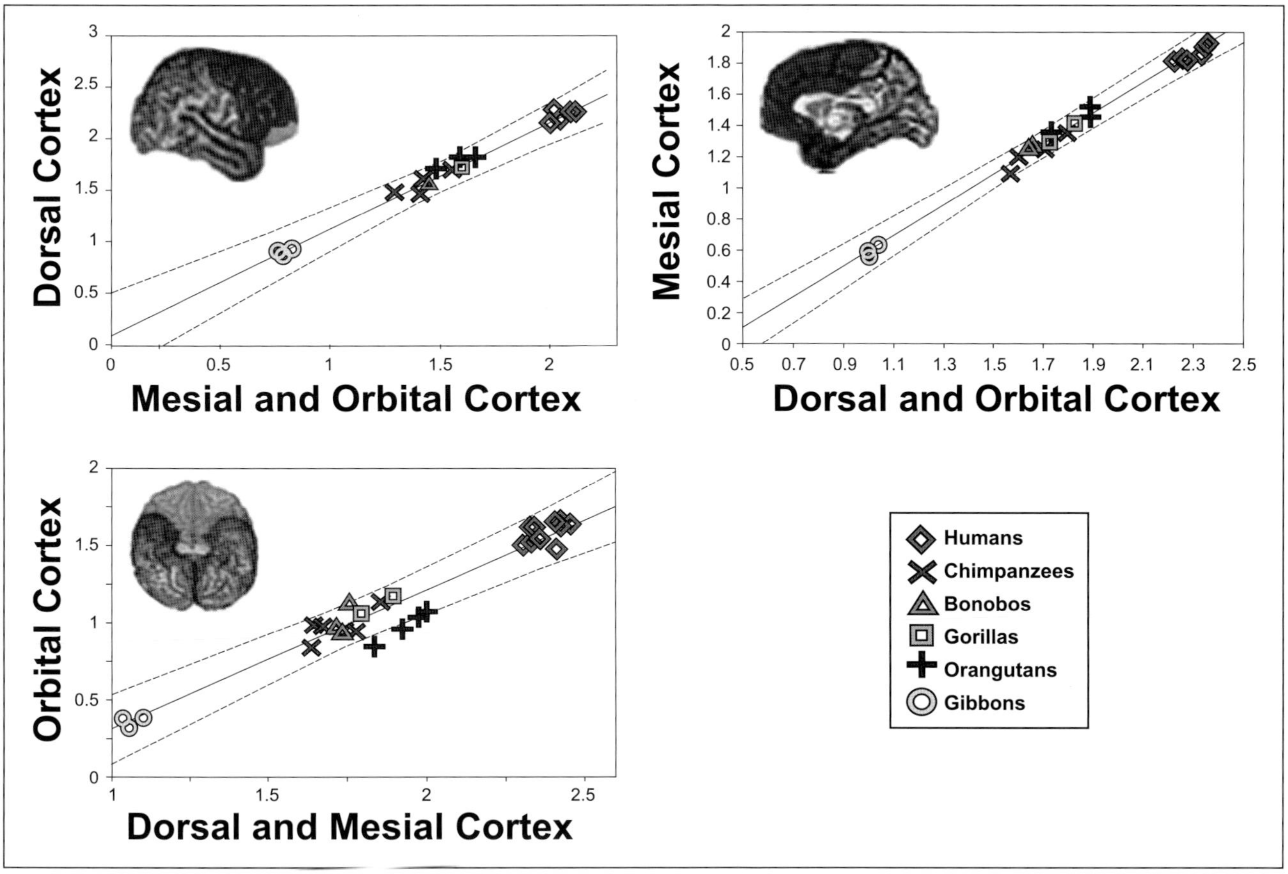

Figure 6. Log-log regressions through humans and apes of dorsal cortex, mesial cortex, and orbital cortex in the frontal lobe (modified from Schenker et al., 2005).

1997). How can we account for the difference between our morphometric results and those of Brodmann (1912) and Blinkov and Glezer (1968)?

The frontal cortex includes prefrontal cortex, motor cortex, and premotor cortex. Brodmann (1912) measured the frontal cortex as a whole and also its subcomponent, the prefrontal cortex. He reported that both the frontal cortex and the prefrontal cortex are larger in the human brain than in the chimpanzee. Our findings are directly comparable to what Brodmann (1912) and Blinkov and Glezer (1968) reported for frontal cortex. The frontal cortex can be measured and directly compared across individuals and species in a consistent manner using reliable and reproducible landmarks. Additionally, use of MRI scans of living individuals provides a solid basis for morphometric studies widely used today, free of concerns regarding shrinkage or manipulation of post-mortem tissue (which was the case with all quantitative studies in the early and mid-20th century). According to Brodmann, the frontal cortex occupies 6% more of the surface of the brain in the human than in the chimpanzee. His data set includes a single human hemisphere and a single chimpanzee hemisphere with no representation of the other great apes (Figure 5). Our data set includes several individuals per species (see Figure 5) and the range of values for each species is considerable. An examination of our sample reveals that if only one human from the upper end of the species range were compared to only one chimpanzee from the lower end of the species range then the presence of species differences could be supported. This could easily have been the case with the individual subjects used in previous studies, and the previous findings may therefore reflect a sampling bias.

The limited number of primate species and an under-representation of the great apes might also account for the discrepant results between our findings and previous findings. Previous studies focused on human versus nonhuman primate comparisons, while our studies focus on how humans compare to our closest relatives, the great apes. If our studies had excluded most of the great apes, the results would have been more consonant with the previous findings, because gibbons and monkeys do, in fact, have relatively smaller frontal cortices than humans (Figure 5).

Additionally, we calculated volumes for the cortical regions of interest, while earlier studies (Brodmann, 1912; Blinkov and Glezer, 1968) used surface estimates. This is unlikely to account for much of the variance, however, because cortical volume is highly correlated with the surface of the cortical sheet. We conclude that sample size, sample composition, and the presence of intraspecific variation across species are the most likely factors underlying the differences between previous and recent studies of the frontal cortex.

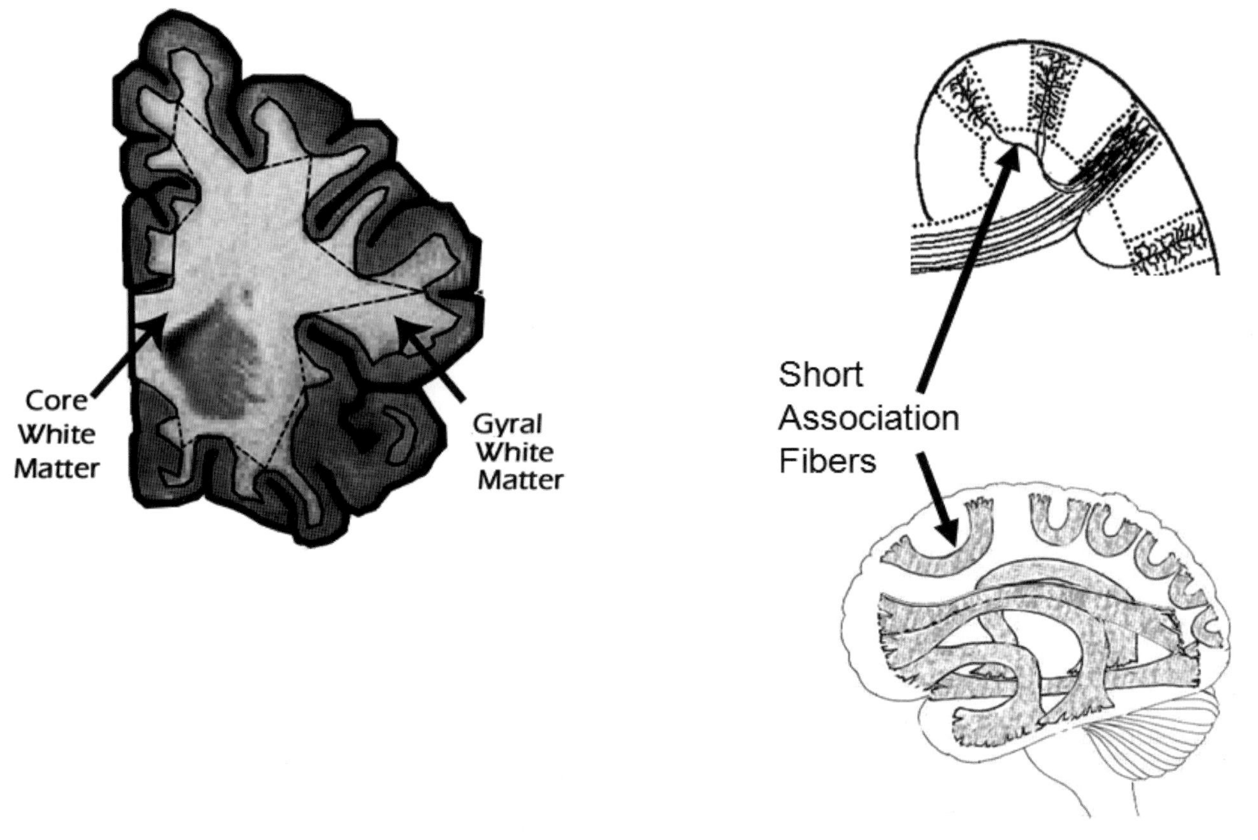

Figure 7. Left: Coronal section through the frontal lobe. The dashed line indicates the boundary between gyral white matter and core white matter; the thin solid line indicates the boundary between cortex and white matter Right: Diagram showing short association fibers in a cross section of a gyrus (upper) and in the entire cerebral hemisphere (lower).

The fact that the relative size of the frontal cortex is similar in humans and great apes does not mean that the frontal cortex is less critical to hominid cognitive specialization than has been suggested (Goldman-Rakic, 1984; Fuster, 1997; Damasio, 1994). The frontal cortex could support the outstanding cognitive capabilities of humans without undergoing a disproportionate overall increase in size because: 1) mere differences in absolute brain size could provide an explanation (Passingham, 2002); 2) subsectors of the frontal lobe may be reorganized and differentially enlarged. In comparative and evolutionary studies of the brain, large anatomical territories have been commonly treated as uniform entities, despite their heterogeneity. The frontal cortex is not functionally homogenous and comprises anatomical subdivisions with distinct functional attributes. These anatomical subdivisions are identifiable at various levels of analysis. We have no direct findings from *in vivo* studies that address what Brodmann (1912) and Blinkov and Glezer (1968) reported for prefrontal cortex because the boundaries of the prefrontal cortex are difficult to define on MR images. It is likely, though, that their calculations for a subregion of the frontal cortex (the prefrontal cortex) may be affected by the same sampling biases as their calculations for whole frontal cortex. In the following section we provide empirical support for the idea that selected parts of the human prefrontal cortex are enlarged.

Reorganization in the frontal cortex of other hominoids

The relative enlargement of the frontal cortex may be one of several reorganizational features present in Plio-Pleistocene hominoid precursors, which distinguish the brains of extant hominoids from those of the smaller primates, such as gibbons and monkeys including baboons (McBride, et al., 1999). The smaller primates have a smaller percentage of their total cortex devoted to frontal cortices (Figure 5), and the range of their values does not overlap with the values of the larger hominoids. These findings point to a possible great ape/human specialization for an enlarged frontal cortex that may set larger hominoids apart from other anthropoid primates (Radinsky, 1974).

Notable anatomical variations are present in selected hominoid species. For example, all orangutans in our studies have significantly smaller orbital sectors than predicted for an ape brain of their size. In spite of having some of the largest brains in our sample of great apes,

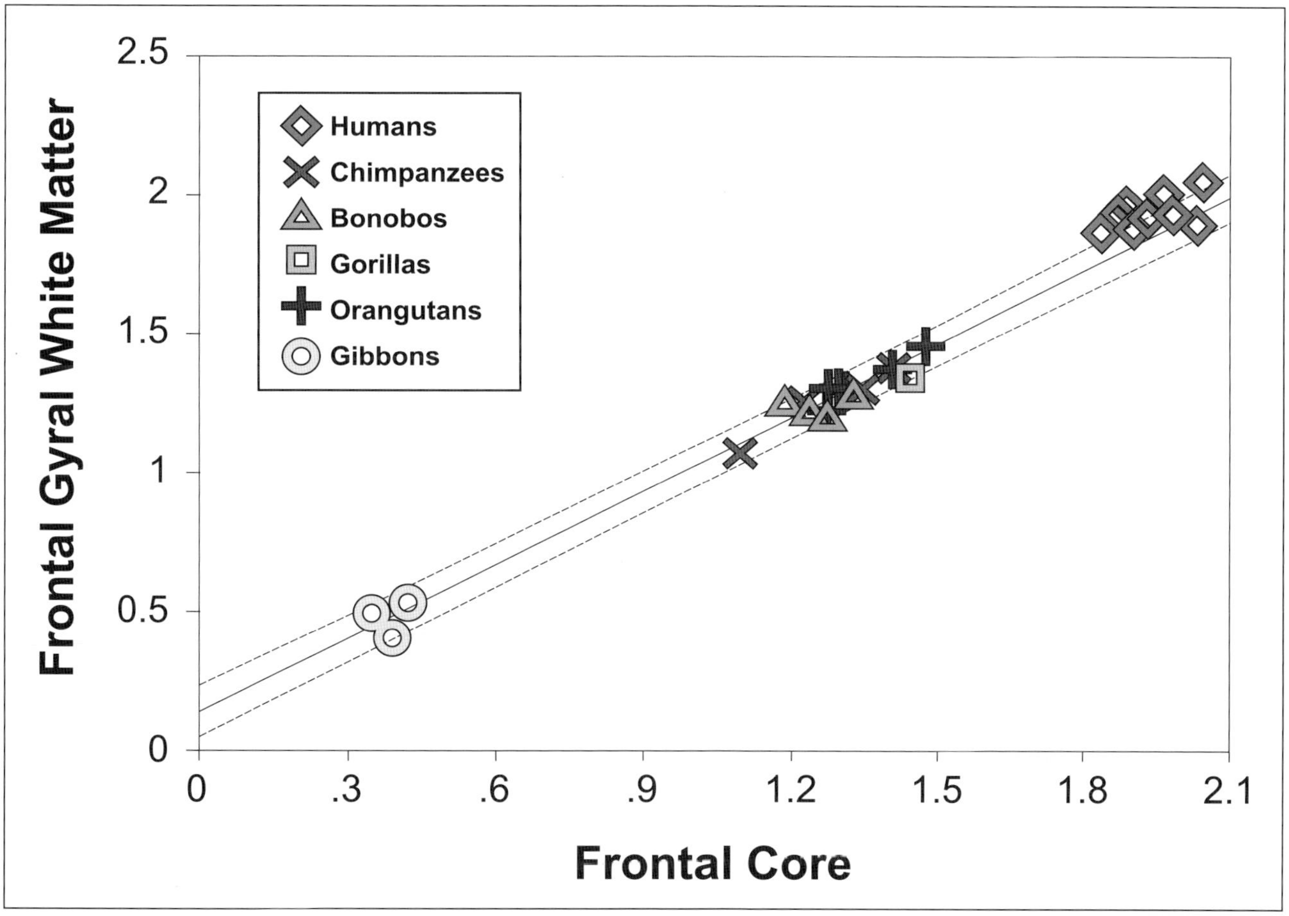

Figure 8. Gyral white matter versus core for the frontal lobe; solid and dashed lines represent the regression lines and their confidence intervals, respectively.

orangutans have some of the smallest absolute values for the orbital sector and the smallest relative size, with no overlap in individual values with any of the other ape species (Figure 9). Additional observations come from previous studies involving postmortem material and parameters such as total length of orbitofrontal cortex (Semendeferi, 1994; Semendeferi et al., 1997). Overall we examined this sector in a total of 12 hemispheres of orangutan brains, and compared them to a total of 32 hemispheres representing the other great apes in our sample (Semendeferi et al., 1997, 1998, 2001; Schenker et al., 2005). Orangutans stand out consistently across studies for their smaller orbital sector.

The ratio of gyral white matter to cortex within each sector does not distinguish humans from other apes, but the ratio in the dorsal sector distinguishes apes from one another. Within the dorsal sector, the ratio of gyral white matter to cortex is larger in Pan (Figure 10) than in either *Gorilla* or *Pongo*. While it is expected for primates with larger brains to have an increased ratio of white to gray matter, chimpanzee and bonobo brains are smaller than gorilla and orangutan brains, not larger. Furthermore, this relationship is only present in the dorsal sector of the frontal lobe. Additionally, within *Pan*, chimpanzees have a relatively larger dorsal sector than bonobos. This difference is largely driven by an increased amount of gyral white matter (Figure 11), not cortex (Schenker et al., 2005). Orangutans also have a relatively large dorsal cortex and gyral white matter, but individual values overlap with those of other apes.

Our findings further suggest that although most gross anatomical sectors in the frontal lobe are as large as expected, morphometric differences exist in selected sectors and selected species. Our data suggest that the inclusion of closely related species and larger numbers of individuals per species may reveal that certain areas of the brain present modifications expressed even at the gross level (de Winter and Oxnard, 2001). The traditional notion that disproportionately large frontal lobes and frontal cortices are the hallmark of hominid brain evolution is not supported by our findings, but there is evidence that the frontal lobe has undergone reorganization. Apes are distinguishable from each other in the orbital sector and the dorsal sector. Humans exhibit unique patterns of white matter distribution in the frontal lobe, suggesting that specializations relating to connectivity occurred during hominid evolution. The rest of the paper will review evidence supporting the hypothesis that specific subsectors have been reorganized.

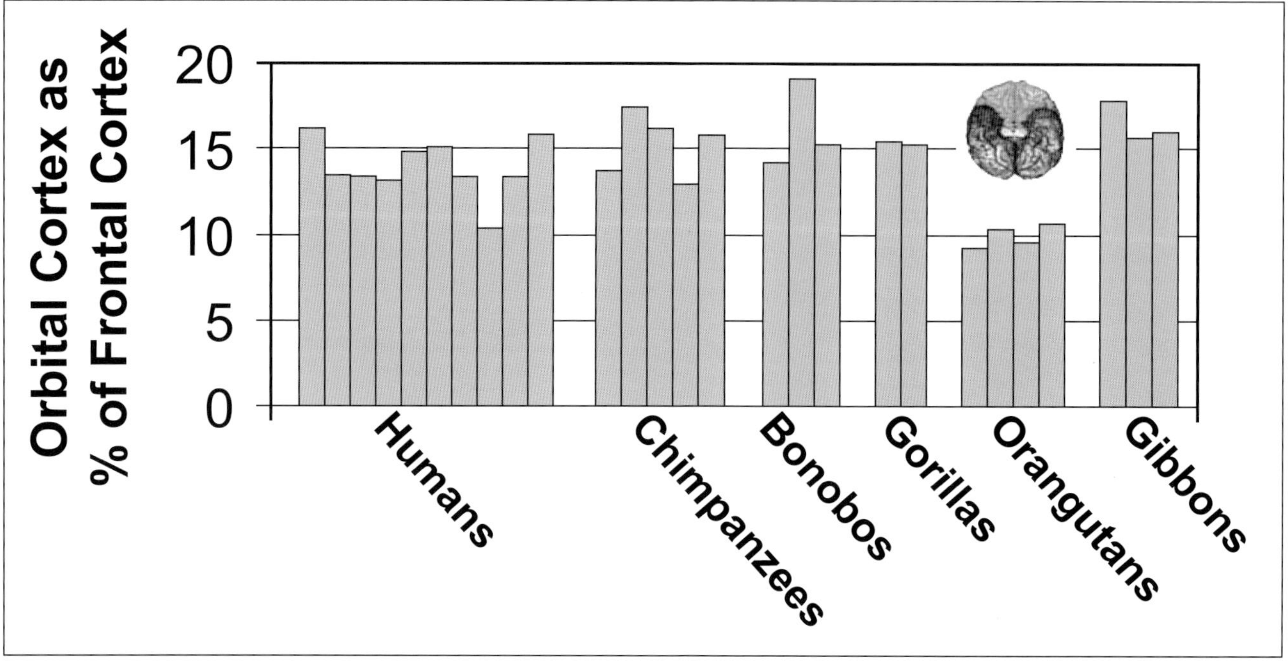

Figure 9. Relative volumes of the orbital cortex as a percentage of frontal cortex in humans and apes.

Prefrontal Cortex

At the cytoarchitectonic level, the frontal cortex (Figure 12) can be subdivided into motor related cortex (primary and supplementary) and prefrontal cortex. The prefrontal cortex is located anterior to the motor and premotor cortices. It is also called the frontal association cortex or the frontal granular cortex, referring to its functional and structural attributes, respectively. The primate prefrontal cortex has been the focus of a host of studies for the past century. Like the rest of the cortex, it has been further subdivided qualitatively into smaller architectonic regions on the basis of their distinct neuronal organization, such as the number and size of the cortical layers, the size, shape, and density of the neurons, and the degree of axon myelination. In addition, support for this more refined cortical parcellation comes from the distinct connections of cortical areas with the various subdivisions of the mediodorsal nucleus of the thalamus and other cortical and subcortical structures (e.g., the temporal and parietal lobes, the hypothalamus, the amygdala, and the hippocampal formation) (Rempel-Clower and Barbas, 2000). The prefrontal cortex includes Brodmann's areas 8, 9, 10, 11, 44, 45, 47, and 46 ("frontal region", Brodmann 1909). It is bordered on the dorsolateral surface by Brodmann's areas 4 and 6 ("precentral region"), on the mesial surface by Brodmann's areas 24, 32, 33, 25, 23, and 31 ("cingulate region") as well as the mesial extension of areas 4 and 6, and on the orbital and lateral surface by the "anterior insular region".

Since Brodmann's map of the human cortex was originally published, efforts have been made to remap and characterize selected cortical areas in the human brain (Amunts et al., 1995; Amunts et al., 1996; Amunts et al., 1999; Amunts et al., 2003; Bailey and von Bonin, 1951; Blinkov and Glezer, 1968; Braak, 1980; Donoghue and Sanes, 1994; Hof et al., 1995; Öngür et al., 2003; Petrides and Pandya, 1999; Petrides and Pandya, 2002; Rademacher et al., 2001; Rajkowska and Goldman-Rakic, 1995a,b; Semendeferi et al., 1998; Semendeferi et al., 2001; Vogt et al., 1995; von Economo, 1929; Zilles et al., 1995). Multiple efforts have been made to map the frontal cortex in the commonly used experimental primates, mostly rhesus monkeys and other species of macaques (Barbas and Pandya, 1987; Burman et al., 2006; Carmichael and Price, 1994; Donoghue and Sanes, 1994; Dusser de Barenne et al., 1941; Fogassi et al., 1994; Gebhard et al., 1995; Matelli et al., 1985, 1986; Petrides and Pandya, 1999; Petrides and Pandya, 2002; Preuss and Goldman-Rakic, 1991; Rosabal, 1967; Stepniewska et al., 1993; von Bonin and Bailey, 1947; Walker, 1940; Watanabe- Sawaguchi et al., 1991; Zilles et al., 1982, 1986). A few studies have also demonstrated the organization of the frontal cortex in apes at the histological level (Bailey et al., 1950; Hakeem et al., 2004; Semendeferi et al., 1998, 2001; Raghanti et al., 2008).

The size of the prefrontal cortex (after excluding motor, premotor and limbic cortices from the rest of the frontal lobe) is a subject of long debate. Brodmann (1912) reports as much as a 12% enlargement of the human prefrontal cortex compared to the chimpanzee (the only ape included in his data set). Despite the longstanding debate, no new data have been collected on the size of the prefrontal cortex as a whole in more than five decades. Deacon (1997), based on the quantitative measures obtained earlier in the 20th century (Brodmann, 1912; Blinkov and Glezer, 1968) argues that the relative size of the prefrontal cortex in humans is 202% more

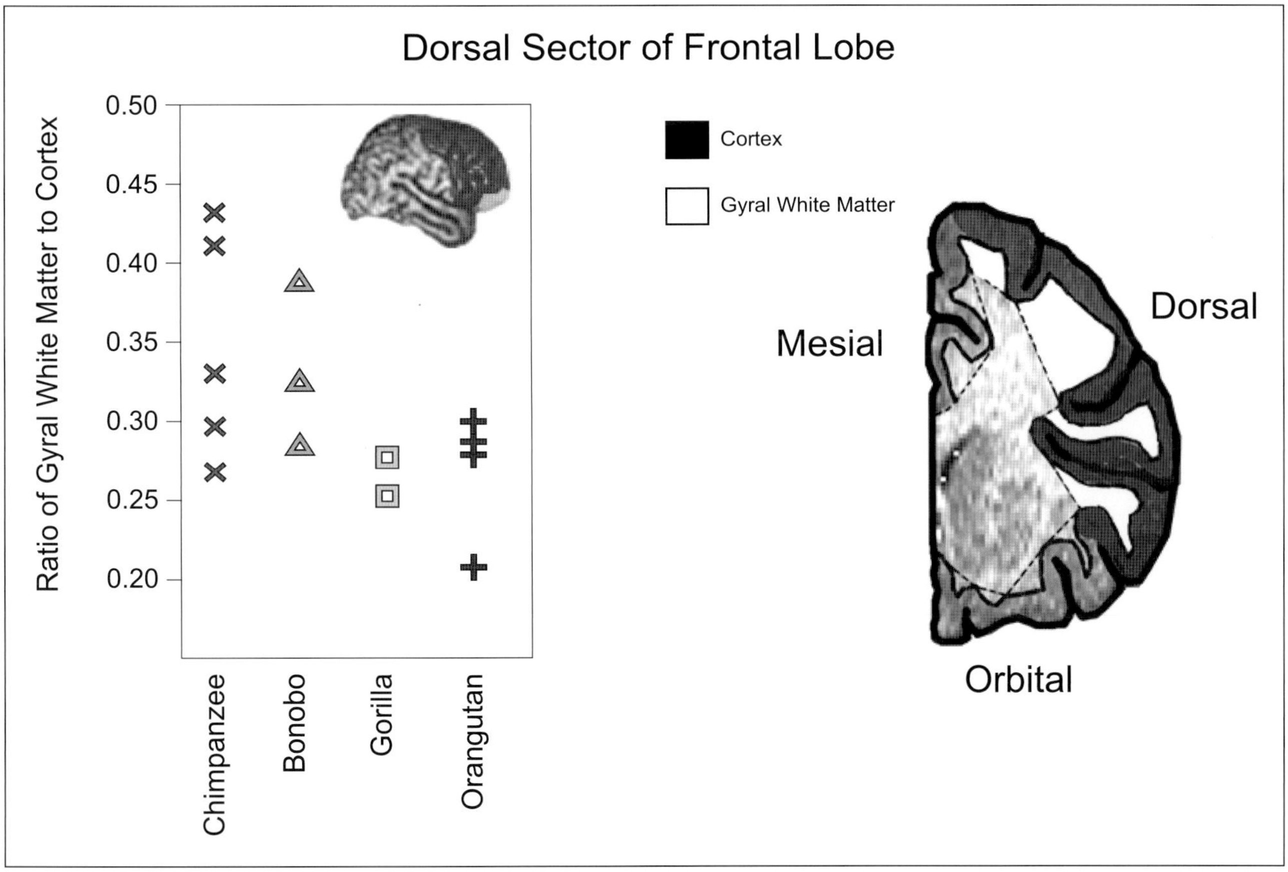

Figure 10. Ratio of gyral white matter to cortex in the dorsal sector of the frontal lobe.

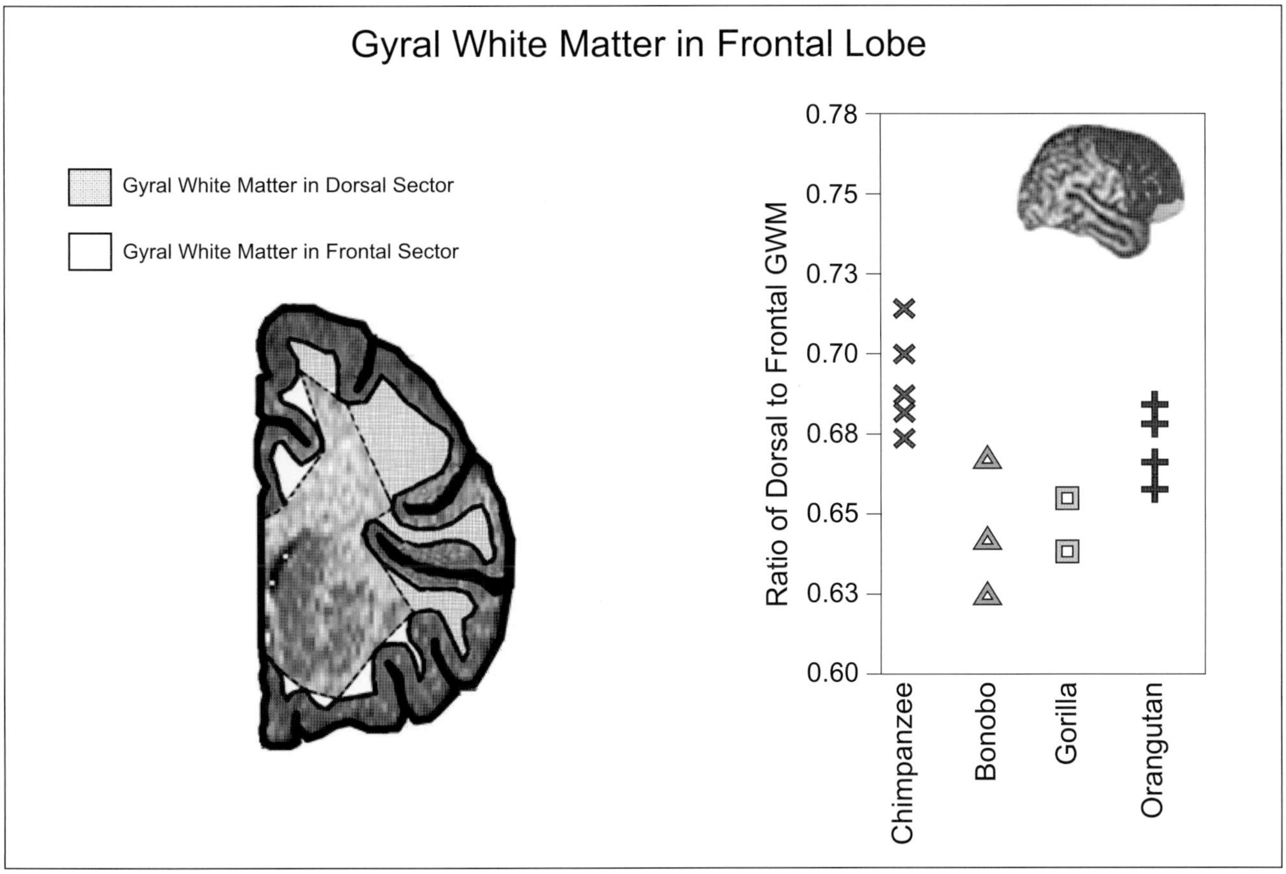

Figure 11. Ratio of gyral white matter in the dorsal sector in relation to the rest of frontal lobe.

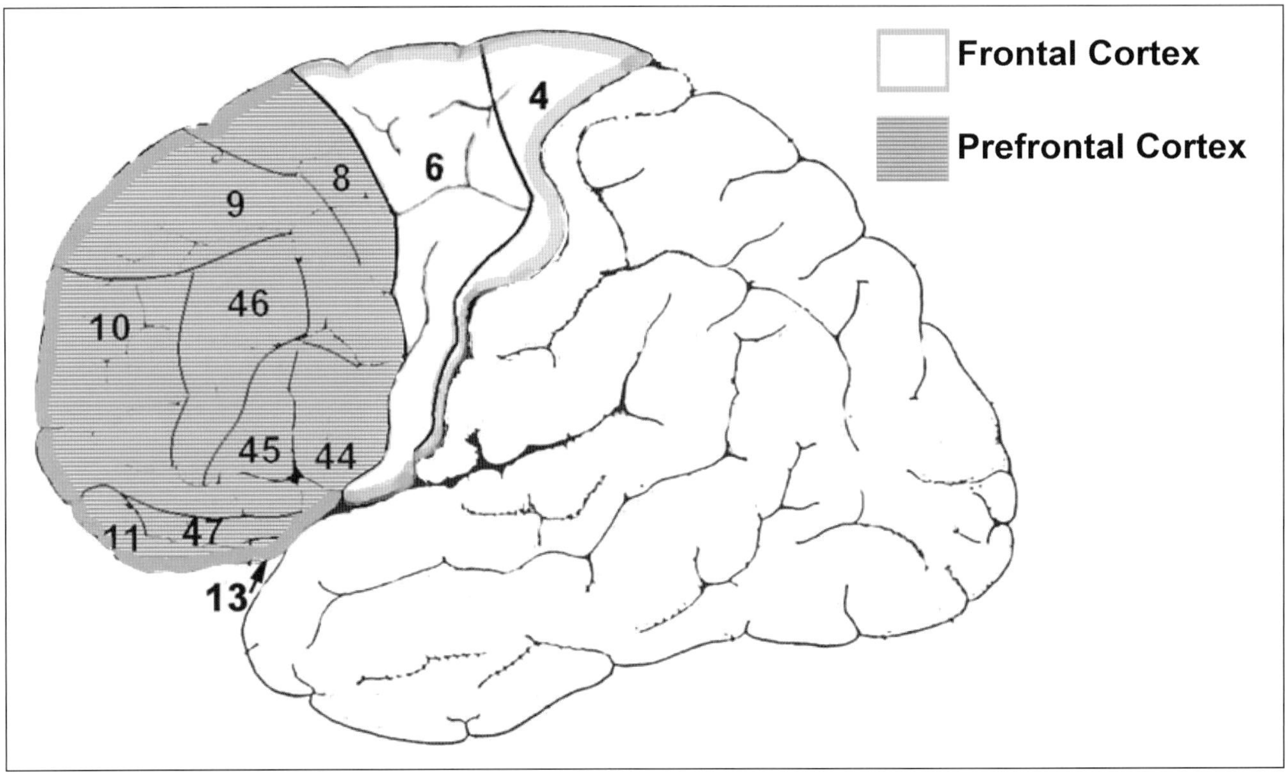

Figure 12. Lateral view of the human brain showing extent of the frontal and prefrontal cortex including cortical areas as assigned by Brodmann.

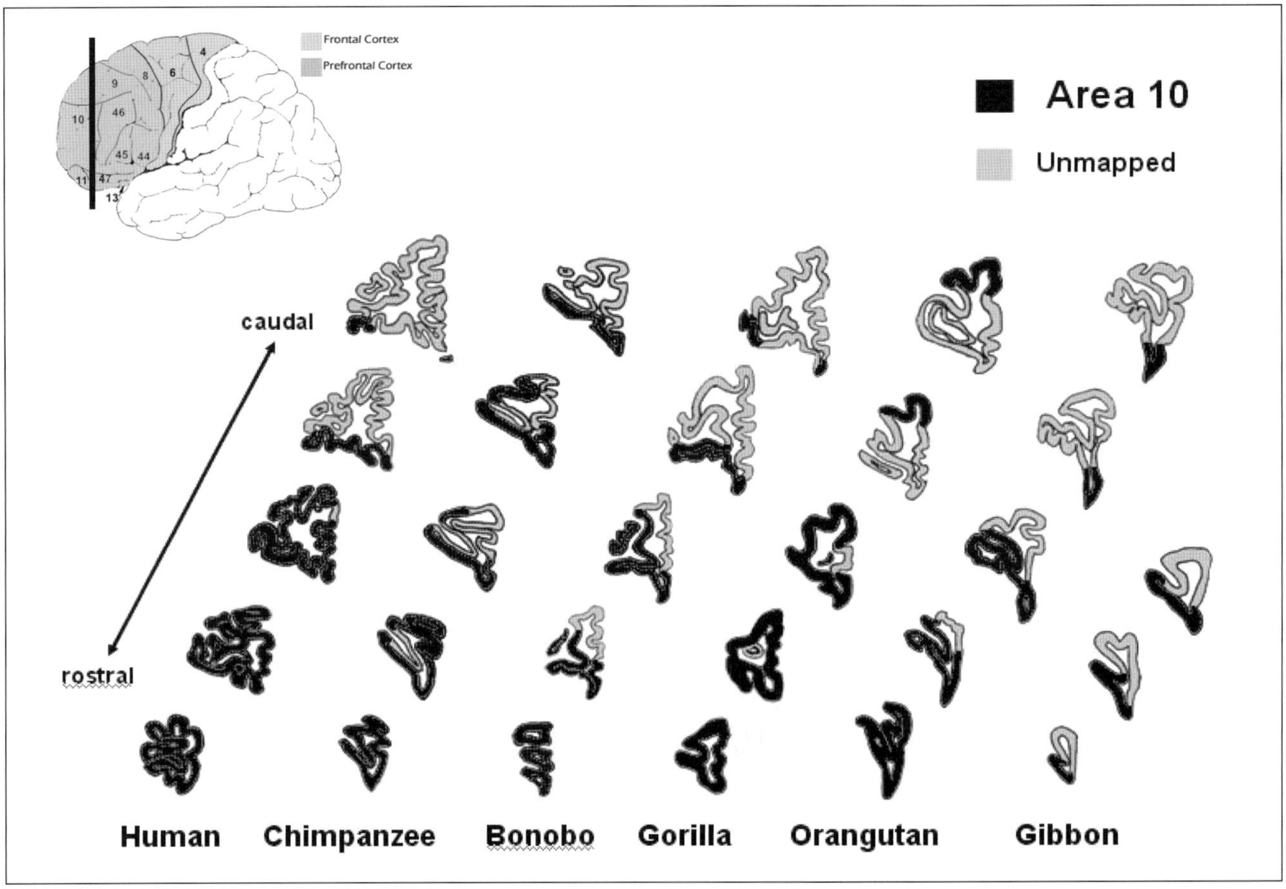

Figure 13. Upper left: Lateral view of the human brain with vertical bar cutting through Brodmann's area 10. Lower part: Cross sections through the prefrontal cortex in the human and ape brains showing location and extent of Brodmann's area 10.

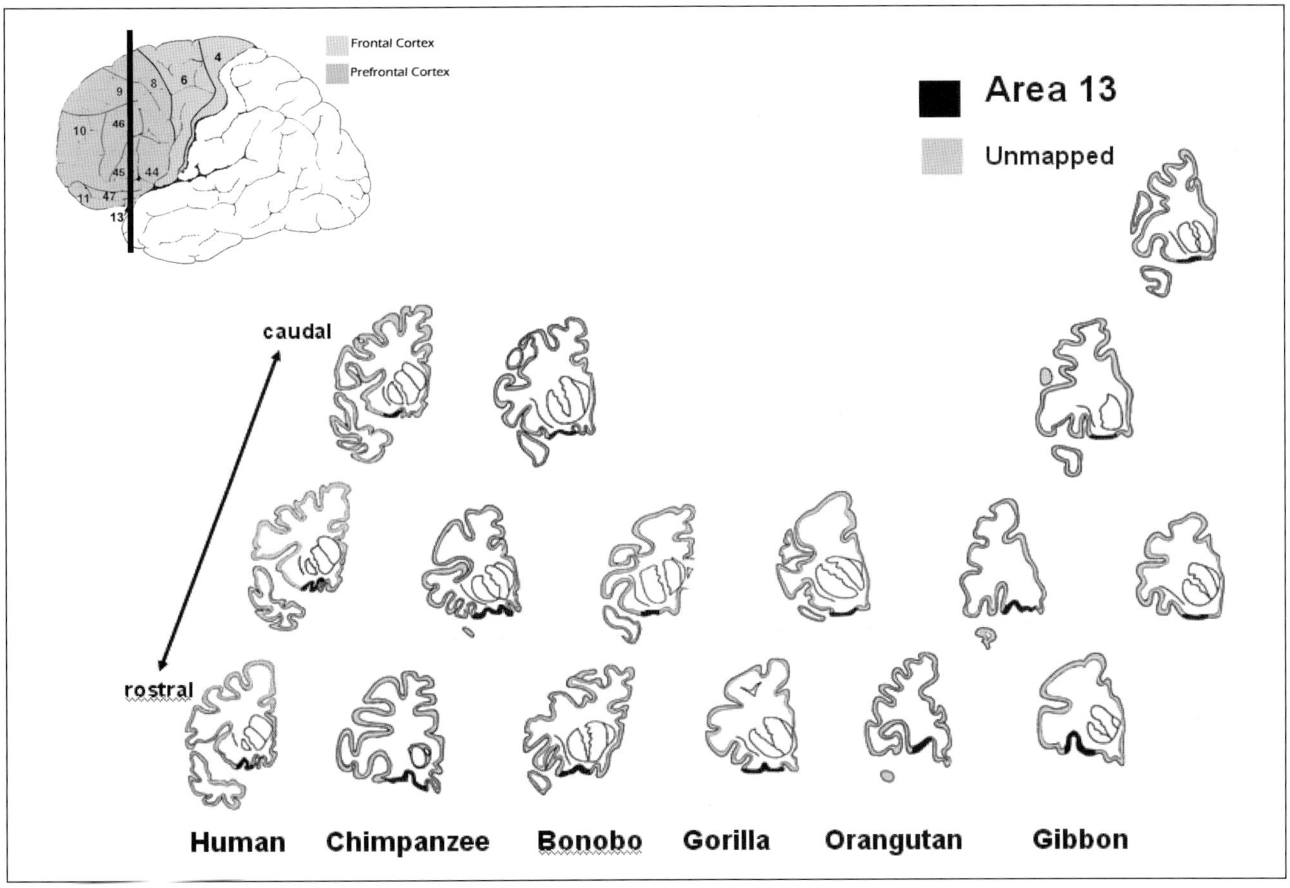

Figure 14. Upper left: Lateral view of the human brain with vertical bar cutting through Brodmann's area 13. Lower part: Cross sections through the prefrontal cortex in the human and ape brains showing location and extent of Brodmann's area 13.

than expected for a nonhuman primate brain of the human size. Bonin (1948) and Holloway (1968) analyzed the same data sets and instead conclude that humans have a frontal lobe as large as expected for a primate brain of human size.

Proper identification of the prefrontal cortex requires analysis of the microscopic features that define its boundaries with neighboring areas. For that, histological investigations are necessary. No gross morphological criteria are sufficiently accurate to replicate Brodmann's definition, because no sulcal landmarks can reliably establish the transition of prefrontal cortex to premotor cortex or the borders of individual cortical areas. Consequently, this issue can only be resolved using cytoarchitectonic criteria in combination with quantitative studies based on histological sections. To date, no such studies exist.

The only segment of the frontal cortex that can be identified reliably based on gross morphology, and that if removed, can bring the size of the remaining frontal cortex closer to prefrontal cortex, is the precentral gyrus, which is largely occupied by Brodmann's area 4 (Figure 12). We found that even after removing the precentral gyrus from our analysis, the remaining human frontal cortex is as large as expected for an ape of their brain size (Semendeferi et al., 2002). The relative size of the frontal cortex after excluding the precentral gyrus ranges from 28.8% to 33% in humans, 25.5% to 29.7% in great apes, and 22.0% to 23.8% in gibbons. All African ape species overlap with humans, and there is extensive overlap among the great apes themselves. While definitive statements about the size of the prefrontal cortex can come only from comparative cytoarchitectonic studies of the brains of extant ape species and humans, small sample size and under-representation of great apes in the studies of Brodmann (1912) and Blinkov and Glezer (1968) place any definitive conclusions in favor of an enlarged prefrontal cortex in humans into question. The frontal lobe as a whole, unlike what the above studies suggest, is not enlarged in humans. Finally, if the quantitative data for the whole frontal cortex in the above studies cannot be replicated, then the data provided by the same sources for a fraction of the whole frontal, i.e. for the prefrontal cortex are likely not reliable either. A large-scale cytoarchitectonic study is necessary to resolve this issue.

While maps and comparisons of the primate cortex are generally qualitative in their approach, several recent studies have also quantified different areas (Dombrowski et al., 2001; Hof et al., 1995; Semendeferi et al., 1998). These quantification efforts target selected neurobiological parameters, including numbers and densities of neurons (Rajkowska and Goldman-Rakic, 1995a,b;

Semendeferi et al., 1998; Semendeferi & Damasio, 2000; Uylings et al., 2006), density of subpopulations of immunoreactive neurons (Raghanti et al., 2008), density of glia cells (Sherwood et al., 2006), size of neurons (Nimchinsky et al. 1999), as well as size of individual cortical areas (Semendeferi et al., 1998, 2001). Of particular interest are studies that document the strong quantitative signatures of individual cortical regions. Amunts and colleagues (Amunts et al., 1999; Scheperjans et al., 2008; Schleicher and Zilles, 1999) used automated quantitative techniques based on statistical algorithms to identify microstructural boundaries between cortical areas. Because the boundaries are defined using a computer algorithm that analyzes structural density patterns, the boundaries are observer-independent and reproducible. Furthermore, their consistency across adjacent sections provides evidence that these boundaries are not the result of random events. Dombrowski et al. (2001) demonstrated that quantitative architecture can be used to distinguish prefrontal cortical areas in rhesus monkeys. The prefrontal cortical areas they analyzed are well characterized by their connectivity patterns within and outside the frontal lobe (Rempel-Clower and Barbas, 2000). This suggests that quantitative cytoarchitecture provides reproducible criteria for the identification of cortical systems relevant to function. This approach does not require the invasive procedures involved in acquiring tissue from experimental animals or the use of multiple immunohistochemical markers to characterize a cortical territory and thus is a useful approach in the study of the human and ape brains.

Brodmann's areas 10 and 13

Even though the frontal lobe and frontal cortex as a whole are not differentially enlarged in humans, some of the lobe's constituent areas seem to vary differentially in size across the hominoids. Morphometric studies of histological sections using cytoarchitectonic criteria suggest that the size and organization of individual cortical areas in the hominoid prefrontal cortex, and not the prefrontal cortex as a whole, may set humans apart from the great apes. Within the prefrontal cortex, selected areas are differentially enlarged or diminished (Semendeferi et al., 1998, 2001). We examined two areas, areas 10 and 13, in terms of size and structural organization at the histological level. Our findings provide evidence for mosaic evolution within the prefrontal cortex.

Brodmann's area 10 is involved with the planning of future actions, the undertaking of initiatives, and working memory and attention (Okuda et al., 1998; Lepage et al., 2000; Daffner et al., 2000). In humans, lesions of the anterior portion of the prefrontal cortex that include area 10 are associated with impairment in highercognitive abilities that facilitate extraction of meaning from ongoing experiences, the organization of mental contents that control creative thinking and language, and the artistic expression and planning of future actions (Damasio, 1985). In contrast, area 13, part of the posterior orbital region, supports behaviors related to responses to social stimuli and complex aspects of social cognition, (Damasio and Van Hoesen, 1983). In macaque monkeys, changes in emotional states and disinhibition of emotional reactions are associated with lesions including area 13. Removal of this cortex enhances aversive reactions and reduces aggressive reactions in threatening situations. These emotional alterations have been interpreted on the basis of the close relationships between the posterior orbital cortex and limbic structures, especially the mediodorsal nucleus of the thalamus and the amygdala (Butter and Snyder, 1972).

Brodmann's areas 10 and 13 form the frontal pole and the core posterior orbitofrontal region respectively. Their cytoarchitecture and connectivity have been well described in the macaque monkey (Barbas and Pandya, 1989; Preuss and Goldman-Rakic, 1991; Morecraft et al., 1992; Carmichael and Price, 1994). Less was known about them in the human brain (area 13 was not identified as a separate cortical area by Brodmann) or the brain of the apes prior to our comparative studies (Semendeferi et al., 1998, 2001). In Asian and African large-bodied hominoids, area 10 is present in orangutans, chimpanzees, bonobos, and humans and occupies the entire frontal pole (Semendeferi et al., 2001). In smaller primates such as gibbons, area 10 occupies only a restricted location in the orbital part of the frontal pole (Figure 13). In contrast, area 13 shares a similar topographic and topological location across all of the species examined (Figure 14).

Area 10 in the human brain presents some specialized features (Semendeferi et al., 2001); one such feature involves the considerable increase in its overall size in the human brain (Figure 15). Area 10 is larger in the human brain than in the other hominoids, even in relative terms. It is twice as large in the human brain (1.2% of brain volume) than in the brains of great apes (0.46–0.74% of brain volume). Although the increase is considerable in terms of the percentage the area occupies in the brain, more data are required to test whether the area is larger than expected for a brain of human size (Holloway, 2002). Area 13 is present in all hominoids, but is *reduced* in humans and bonobos (Semendeferi et al., 1998). This reduction might be the result of a proliferation of other cortical areas or subdivisions within the orbitofrontal cortex of these species. In orangutans, however, the orbitofrontal cortex is relatively homogeneous, with area 13 occupying a larger portion of this region than in the other great apes. The total volume of area 13 in the right hemisphere of the great apes and humans is very similar, ranging from 269.9 mm3 in the chimpanzee to 366.2 mm3 in the human. An exception is the bonobo, which has a volume of 110.5 mm3. However, relative to brain size, the human and bonobo have a small area 13 (0.03%), while in the gibbon and the other great apes area 13 occupies a greater percentage of the brain (0.06-0.09%). Across hominoid species, areas 10 and 13 do not vary in size in coordination with one another.

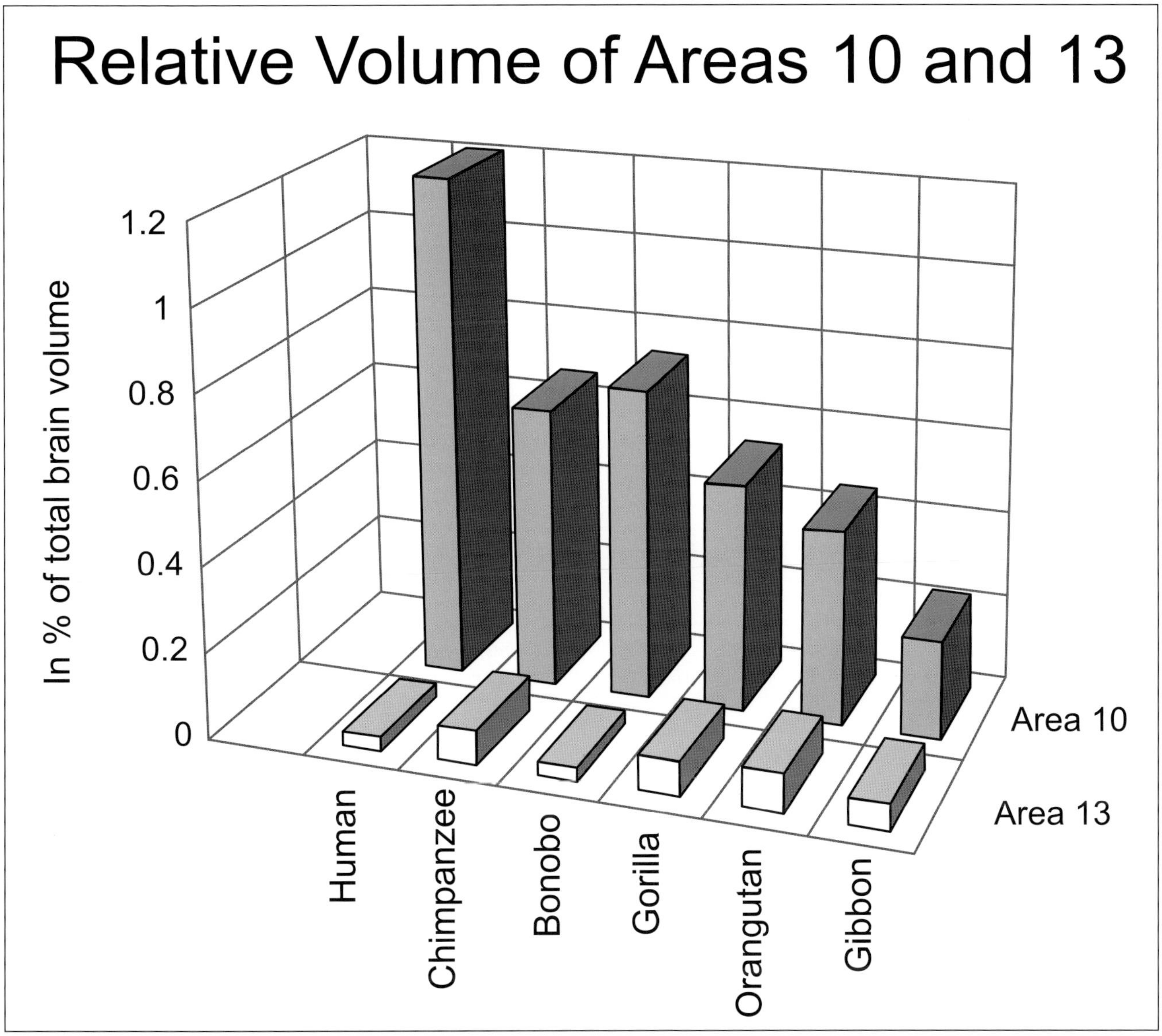

Figure 15. The volume of Brodmann's areas 10 and 13 as a percent of total brain volume in the human and ape brains.

Another specialized feature of area 10 involves the neuropil, or the space between cell bodies, which is largely devoted to axons and dendritic processes within the cortex (Semendeferi et al., 1998, 2001). Unlike area 13, neuropil is increased in area 10 of the human brain in layers that are primarily connected to other higher order cortical areas of the same and opposite hemispheres (Figure 16). This is of interest given that the volume of gyral white matter is larger than expected in humans and primarily represents connectivity between closely located higher order cortical areas. Other lines of evidence also support the idea that parts of the prefrontal cortex have increased dendritic arborization in humans compared to macaques (Elston et al., 2006).

Studies in macaques have demonstrated an anterior-posterior gradient in neuronal density across frontal cortical areas (Dombrowski et al., 2001). Our data also demonstrate this trend (Figure 17). Across species, with the exception of the gorilla, the density of neurons in area 10, located in the frontal pole, is greater than the density in area 13, located in posterior orbital cortex. In both areas, the density of neurons is considerably lower in humans than in the apes. However, within the apes, neuron densities in area 10 and area 13 do not vary in coordination with each other. For instance, while the orangutan has the lowest neuron density in area 13, it has the second highest density in area 10. Variation among species in the neuron densities in these areas provides some evidence for the mosaic evolution of cortical areas in the prefrontal cortex.

Area 13 is relatively conserved, particularly in terms of its absolute size. A hypothetical reconstruction of the Plio-Pleistocene hominoid brain would place area 13 in a restricted area, occupying the most posterior parts of the medial orbital gyrus and the posterior orbital gyrus, with structural features similar to those in the extant species. In general, area 10 in the human brain appears to be specialized in size and organization, which suggests that

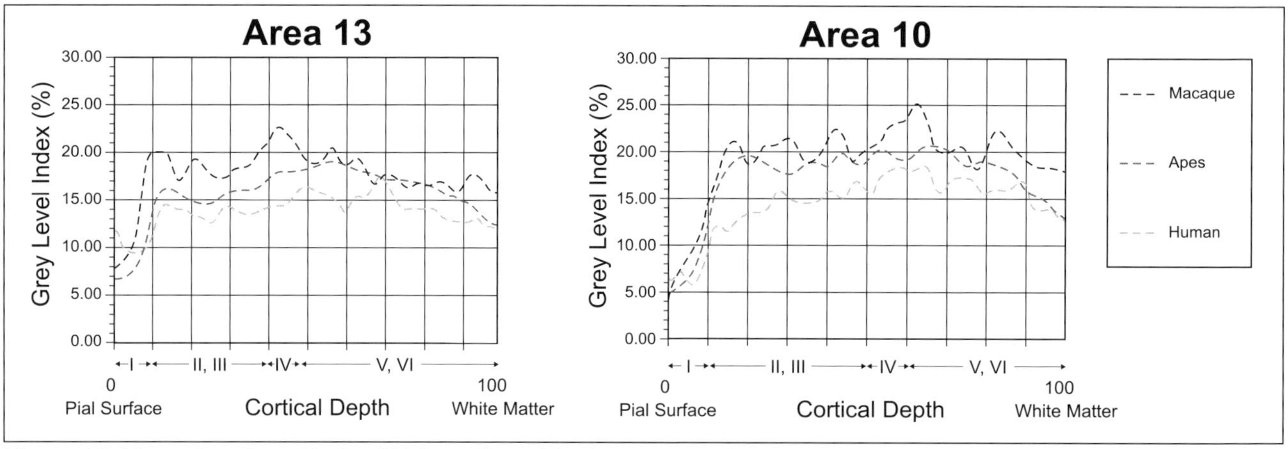

Figure 16. Mean Grey-Level Index (GLI) profiles of the human, ape (chimpanzee, bonobo, gorilla, orangutan, and gibbon), and macaque cortex in Brodmann's areas 10 and 13. Valleys in the profiles represent areas occupied to a larger extent by space available for connections and to a lesser extent by cell bodies.

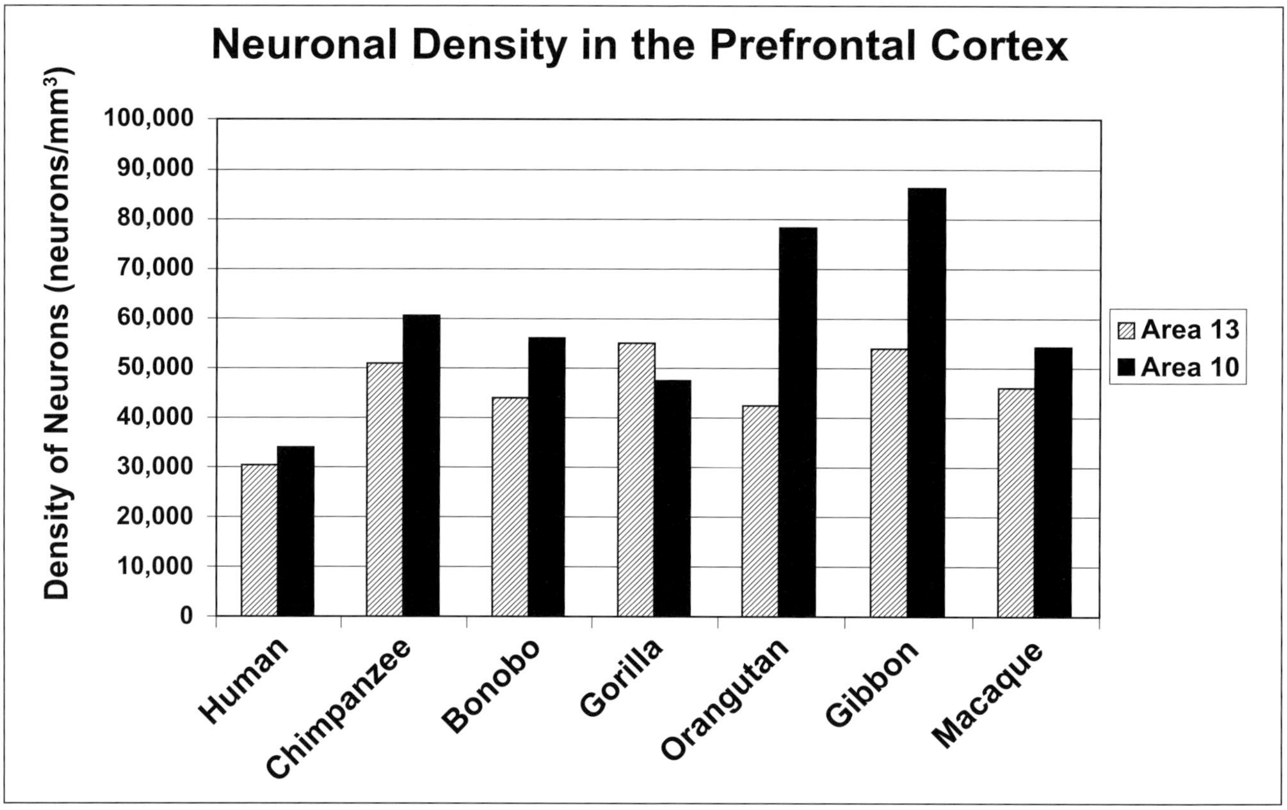

Figure 17. Neuronal denisity in areas 10 and 13 (human and ape data from Semendeferi et al., 1998 and 2001; macaque data from Dombrowski et al. 2001)

functions associated with this part of the cortex have become particularly important during hominid evolution. Planning and initiating actions are hallmarks of human behavior, and although these features are present to some extent in other hominoids and possibly other primates, they became fully expressed in the Plio-Pleistocene hominids.

There is variation in the size and in aspects of the organization of the frontal lobes among the hominoids (Semendeferi et al., 1997). These differences might reflect species-specific adaptations, functional specializations, and/or major evolutionary events relating to changes in the organization of the hominoid brain. Relationships such as total brain size to body size is not sufficient for understanding species-specific adaptations in behavior and underlying neural circuitry. An analysis of two regions of prefrontal cortex reveals differences in the organization of parts of the limbic frontal cortex, involved in social cognition, between species that have very similar absolute brain sizes (orangutan versus chimpanzee) (Semendeferi and Damasio, 2000). It is clear that specific neural circuits or cortical areas have to be compared among closely related species.

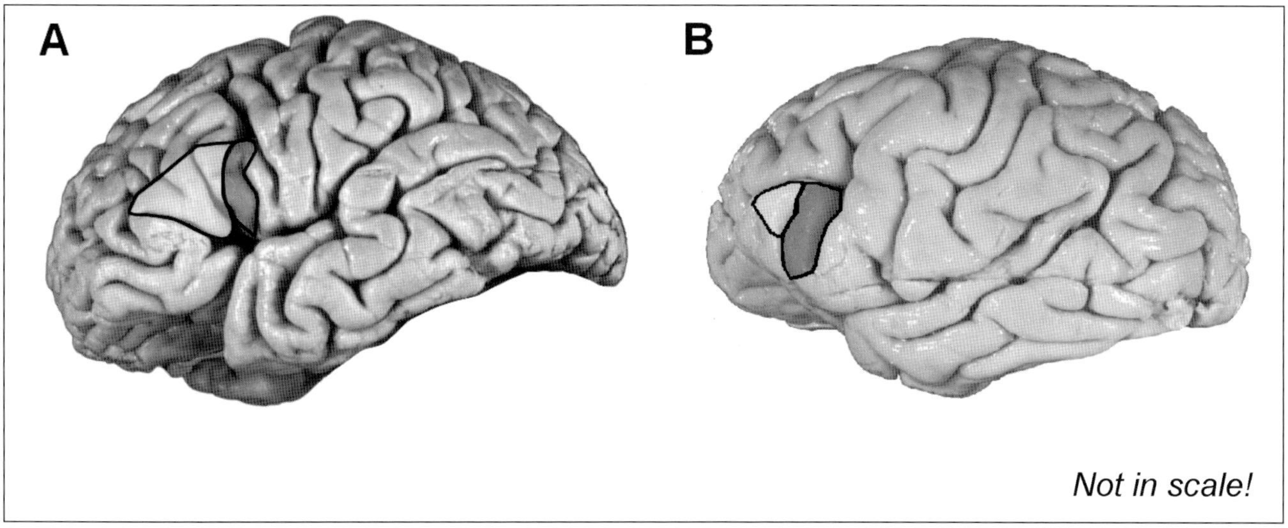

Figure 18. Photographs of (A) a human brain and (B) a chimpanzee brain showing the location of Broca's area, including Brodmann's areas 45 (light gray) and 44 (dark gray).

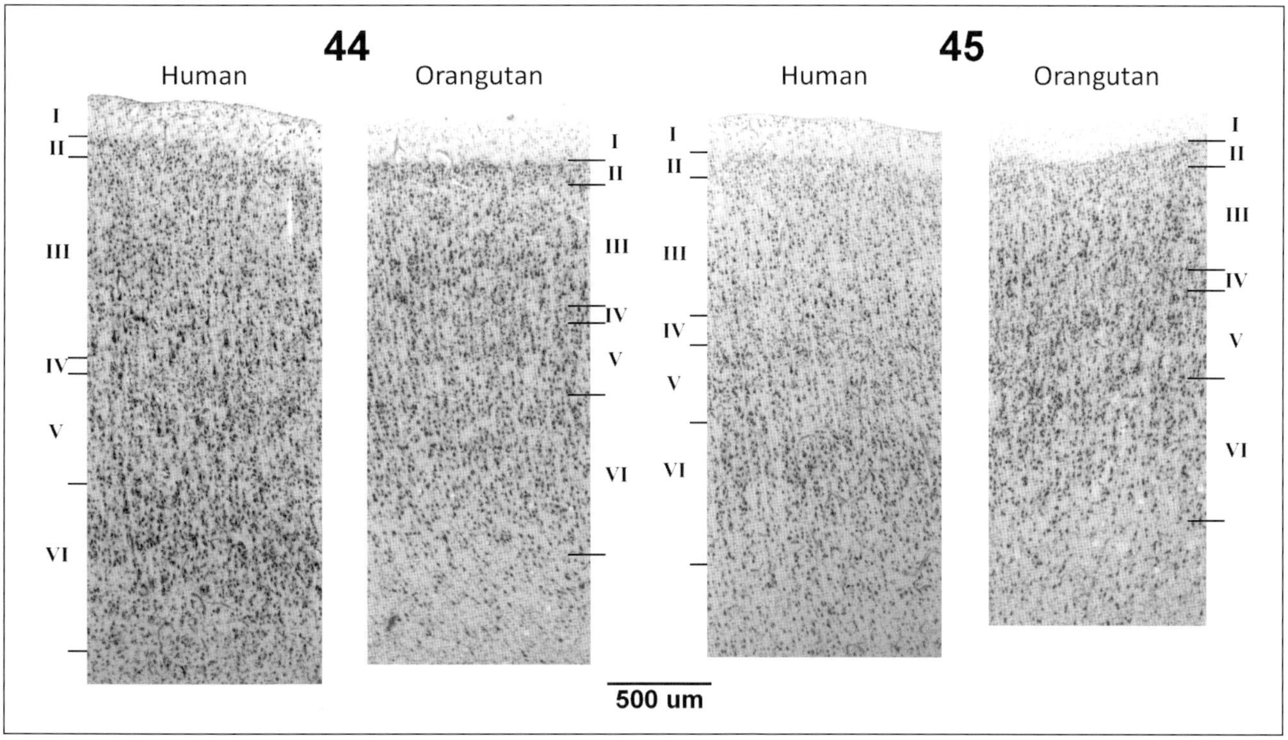

Figure 19. The cytoarchitecture of Brodmann's areas 44 and 45 in a human and orangutan brain. Layers I through VI are marked on the microphotographs.

Brodmann's areas 44 and 45 (Broca's area)

Two other cytoarchitectonic areas in the prefrontal cortex, Brodmann's areas 44 and 45, comprise what is known as Broca's area, and typically occupy part of the inferior frontal gyrus in the human brain (Figure 18). In all hominoids, Brodmann's areas 44 and 45 are located within the inferior frontal gyrus, anterior to the inferior precentral sulcus (Schenker et al., 2008). The two areas can be distinguished by differences in the prominence of layer IV and the total thickness of cortex (Figure 19). They also exhibit certain differences in overall cortical thickness and in relative laminar width among species, with gorillas and orangutans displaying less difference in thickness between the two areas than chimpanzees and bonobos.

Within areas 44 and 45, we investigated the minicolumnar organization of the cortex (Schenker et al., 2008). Minicolumns are vertically-oriented aggregates of cells with strong vertical interconnections among layers, forming fundamental structural and functional units within cortex (Douglas and Martin, 1992; Mountcastle,

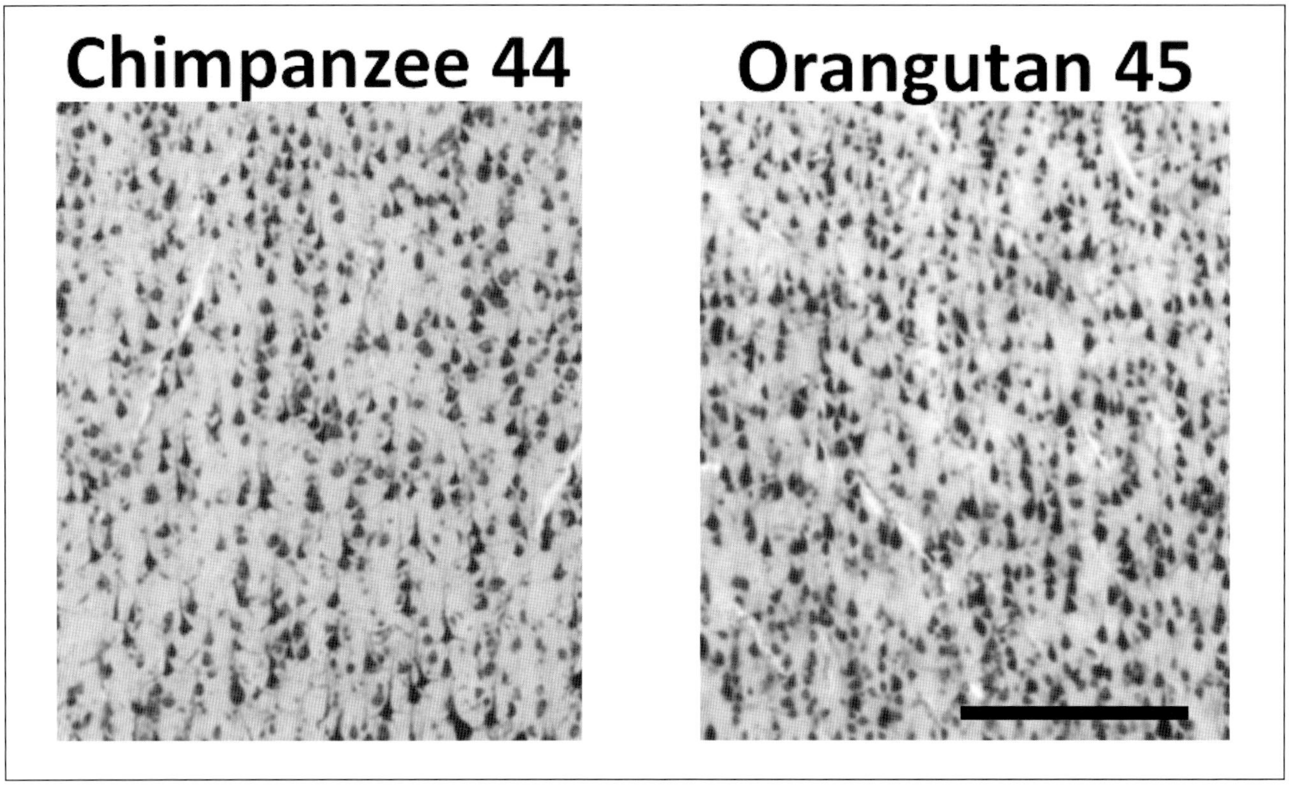

Figure 20. Minicolumnar organization in the cortex of Brodmann's areas 44 and 45 in a chimpanzee and orangutan brain. Scale bar equals 100 μm.

1997). Their organization in the adult brain is thought to be derived from ontogenetic columns and the migration of cells into radial columns during development (Rakic, 1995). Minicolumns comprise rows of neurons traversing layers II-VI (Fig. 20) (Buxhoeveden and Casanova, 2002; Mountcastle, 1997) and are assumed to be one cell wide in layers III, V, and VI (Seldon, 1981).

Minicolumns in Broca's area are larger in humans than in great apes (Figure 21). This pattern is similar to the pattern reported previously for the planum temporale (area Tpt, Buxhoeveden et al., 2001). Nevertheless, even though spacing between minicolumns in humans is larger in absolute terms, it is smaller relative to total brain volume (Figure 22). This indicates that despite the increased width of minicolumns, a human's cortex contains more minicolumns than the cortex of a great ape.

It is not yet clear whether larger minicolumns in humans is a specific characteristic of cortical areas involved in language function (Broca's area and area Tpt), or if larger minicolumns exist throughout the human cortex regardless of functional attributes. To date, comparative studies of minicolumns in humans and great apes have focused on cortical regions that are active in linguistic functions. Therefore, these findings may be evidence either of differential changes in the inferior frontal gyrus and the superior temporal gyrus or a cortex-wide difference in humans. Further conclusions await data from minicolumns in additional regions of cortex.

Frontoinsular cortex

Neuroanatomists in the late 19th and early 20th centuries described the presence of an unusual cell in the human cortex (von Economo and Koskinas, 1925). These cells were named spindle neurons based on their characteristic bipolar shape and large size, which is approximately four times larger than pyramidal neurons. Spindle neurons were identified in a specific layer (layer Vb) of two areas of the frontal part of the brain, the anterior cingulate and the frontoinsular cortex (von Economo & Koskinas, 1925 or "anterior insular" according to Brodmann, 1909).

Contemporary studies replicated the early reports in the human brain and also identified the presence of this neuronal phenotype in the anterior cingulate cortex of some mammals. Spindle cells were found in the human, gorilla, bonobo, chimpanzee and orangutan anterior cingulate cortex, but were not found in 22 other primate species or 30 other mammals examined (Nimchinsky et al., 1995; Nimchinsky et al., 1999). Spindle neurons are larger in humans and chimpanzees and smaller in gorillas and orangutans. Spindle neuron volume correlates with encephalization, while the volume of other neurons in layers V (pyramidal) and VI (fusiform) does not (Nimchinsky et al., 1999).

We have used stereological sampling to determine the number of spindle neurons in the "anterior insular region" (Brodmann, 1909) also known as the frontoinsular cortex (area FI) (von Economo and Koskinas, 1925),

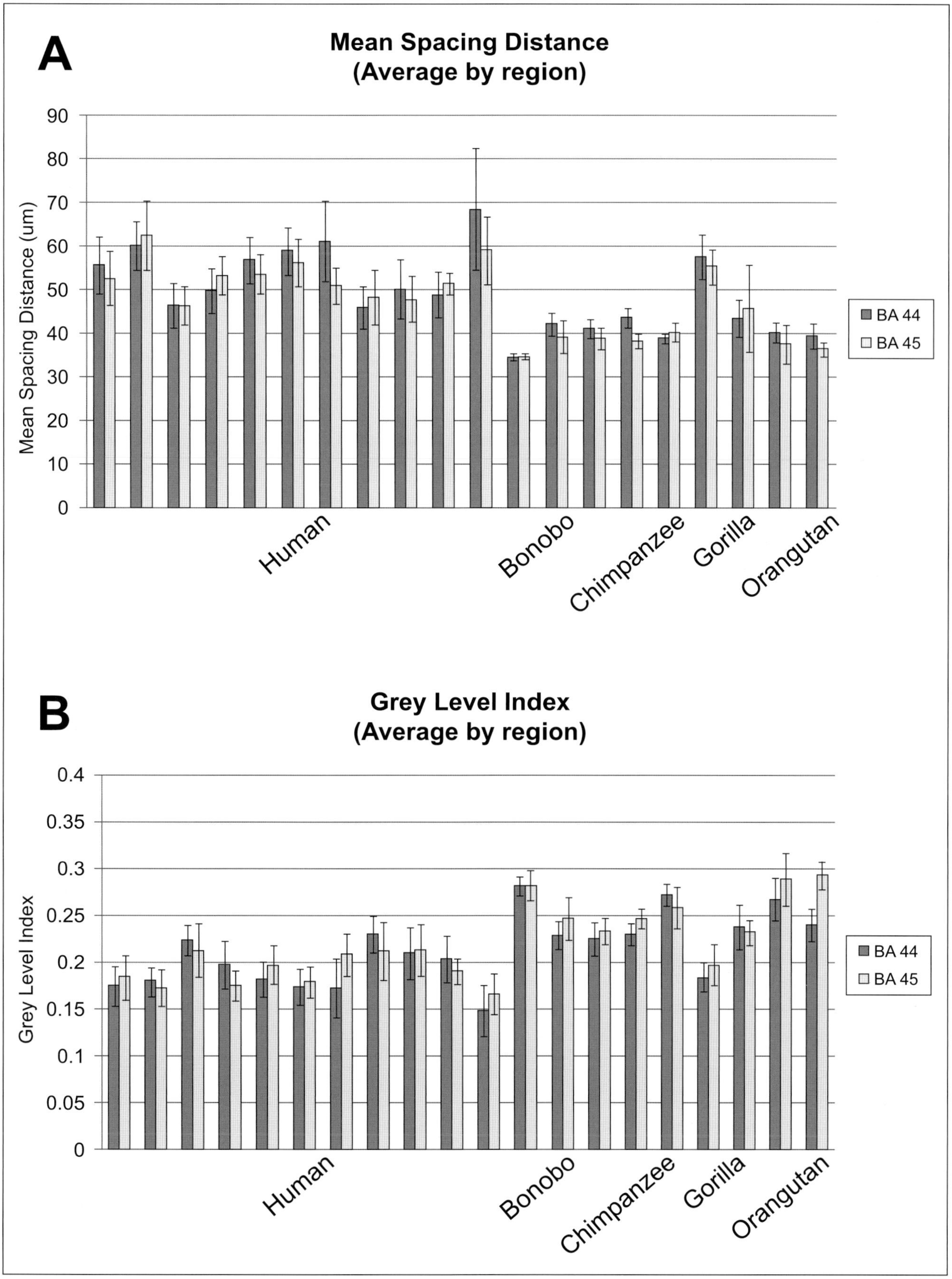

Figure 21. Mean horizontal spacing distance (A) and gray level index (B) by individual and area for the human and ape brains in Brodmann's areas 44 and 45 (modified from Schenker et al., 2008)

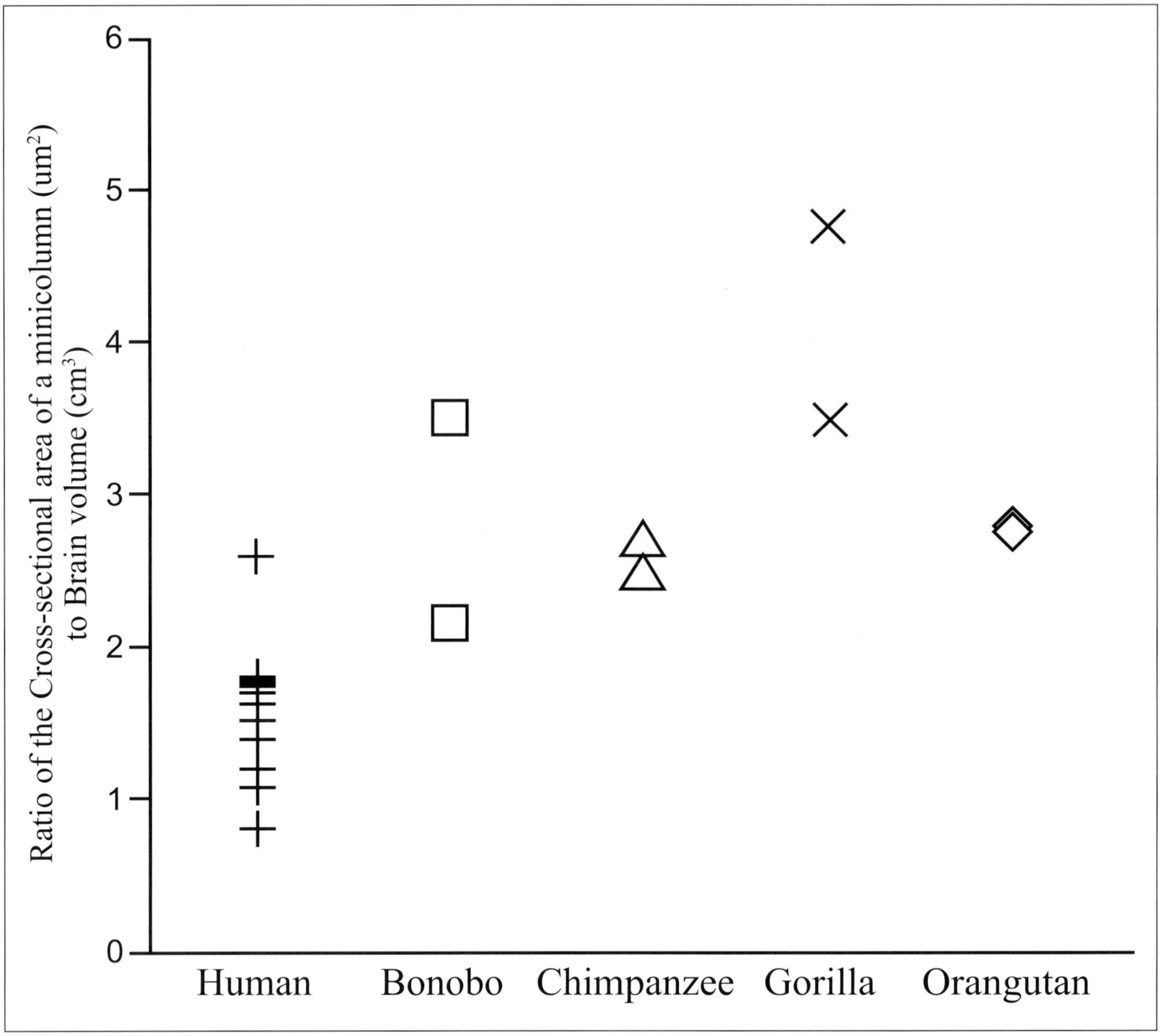

Figure 22. Estimated cross-sectional area of a minicolumn relative to brain volume (modified from Schenker et al., 2008)

in humans and African apes (Hakeem et al., 2004; Kennedy et al., 2007). The frontoinsular cortex is agranular cortex that makes up the anterior portion of the insula (von Economo & Koskinas, 1925; Mesulam & Mufson, 1982). Two cytoarchitectural features unmistakably define area FI (von Economo and Koskinas, 1925). First, FI is agranular cortex, with very few cells in layer II creating a discontinuous appearance, and a near complete or complete absence of layer IV. Second, layer Vb contains large bipolar spindle neurons (Fig. 23), which make identification of the boundaries of FI unambiguous.

After close examination of the orbitofrontal and insular cortex in 25 primate species, only humans and African apes exhibit spindle neurons. We did not identify spindle neurons in the Asian apes (orangutans and gibbons), in Old and New World monkeys, or in prosimians (Hakeem et al., 2004). The FI spindle neurons are approximately 30% more numerous in the right hemisphere of both humans and apes. It is very likely that spindle neurons are a specialization found in humans and great apes, and their presence in area FI is a phylogenetic specialization of the clade comprised more specifically of humans and African apes. Since they are present in all members of this clade, they are likely to have been present in the last common ancestors of the clade, which lived less than 10 million years ago.

Given this evolutionary scenario and the functional properties of the areas containing spindle neurons, this hemispheric specialization might have arisen before the evolution of language and might be relevant to the domains of emotion and social cognition. Based on the functions normally attributed to the rostral anterior cingulate and FI and the evolutionary uniqueness of spindle neurons, researchers have suggested that these neurons might play a key role in socioemotional and higher-order cognitive processing (Watson et al., 2007; Allman et al., 2005; Nimchinsky et al., 1999), leading many to speculate that they may be dysfunctional in autism (Allman et

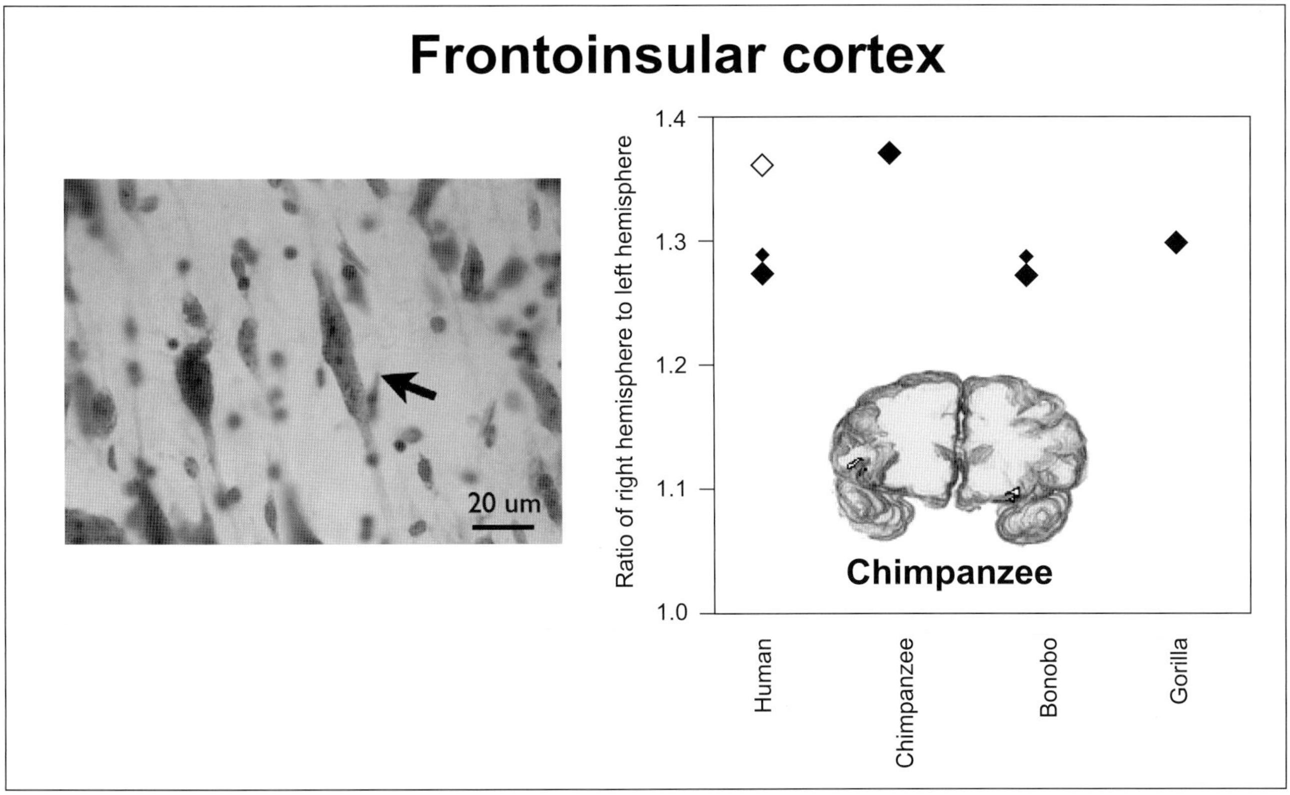

Figure 23. Left: Arrow points to presence of spindle neurons in area FI of the human brain. Right: Graph shows the ratio of right to left hemisphere numbers of spindle neurons and a reconstruction of a chimpanzee brain showing location of area FI (dark gray) (modified from Kennedy et al., 2007; Hakeem et al., 2004).

al., 2005; Courchesne and Pierce, 2005; Mundy, 2003). Allman and colleagues (2005) proposed that the spindle neurons relay to other brain structures signals concerning value judgments in situations involving risk or uncertainty, especially in social bonding and economic decision-making. Furthermore, several species of whales have been shown to possess spindle neurons, and this might be an example of convergent evolution (Hof and Van Der Gucht, 2006).

Although the human frontal cortex as a whole is not differentially enlarged compared to apes, there is variation in aspects of the organization of the frontal cortex among the hominoids. There is some support for the idea that brain enlargement has been accompanied by a reorganization of specific connectivity patterns and that individual species may process information differently. Instead of an overall enlargement of this part of the brain relative to the rest of the brain, specific cortical areas have changed in size. Some areas are enlarged in humans, while others are smaller; similar differences are present across the apes. Additionally

Temporal Lobe

While the frontal lobe is featured in the study of human evolution because of its functional properties, increasing comparative evidence suggests that temporal cortical and subcortical structures are undergoing considerable evolutionary change, perhaps even more so than the frontal lobe. The temporal lobe (Figure 3) is recruited in many essential cognitive processes such as the formation and processing of declarative memory (Squire et al., 2004), auditory processing (Poremba et al., 2003), selfrecognition (Kircher et al., 2001), visual processing (Mishkin et al., 1983), and the detection of biological motion that underlies theory of mind (Frith and Frith, 1999). Similarly, while species-specific vocalizations activate only cells in the superior temporal sulcus in nonhuman primates, language processing occurs throughout the temporal lobe (Damasio et al., 1996; Price, 2000; Gorno-Tempini and Price, 2001; Grabowski et al., 2001; MacSweeney et al., 2002; Rilling and Seligman, 2002).

MRI data from two independent studies, Semendeferi and Damasio (2000) and Rilling and Seligman (2002), indicate that the human temporal lobe is, on average, larger than would be predicted for an ape of human brain size. Using at least two individuals per species, both studies concluded that human residuals were predominantly positive and in many cases significantly so. In Rilling and Seligman's (2002) study, the average temporal lobe volume, as well as the temporal cortical surface area of their six human brains was larger than predicted by the regression line drawn through the apes (Figure 24). Similarly, all 10 of the human specimens measured by Semendeferi and Damasio (2000) fall above the ape regression line, and the mean human value

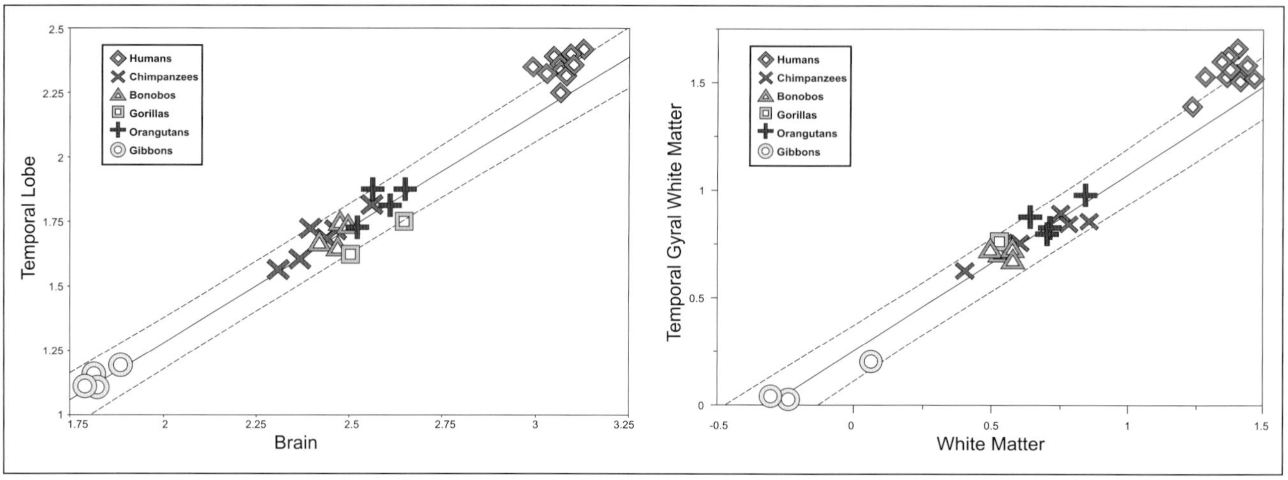

Figure 24. Log-log graphs of temporal lobe versus brain (left) and temporal lobe gyral white matter versus core (right). Solid and dashed lines represent the regression lines and their confidence intervals, respectively (based on data from Schenker et al., 2005).

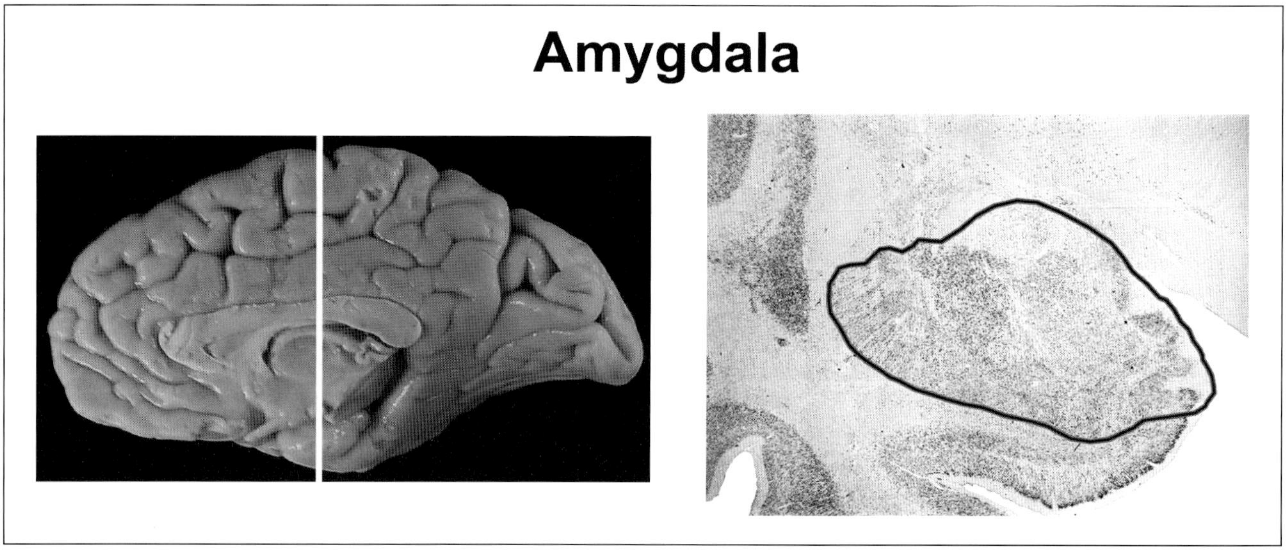

Figure 25. Left: Mesial view of a gorilla hemisphere. Vertical white bar represents location of the cross section shown on the right. Right: Cross section through the amygdala of a gorilla hemisphere.

falls outside of the prediction interval (Figure 24). Regressing the ape temporal lobe volumes from Semendeferi and Damasio (2000) against brain volumes produces a slope of less than one, suggesting temporal lobe volume and whole brain volume are negatively allometric in the apes. When humans are added to the ape sample, the slope of the line closely approaches isometry (slope = 0.979). In the apes, then, the rate of temporal lobe increase lags behind expansion of the entire brain, but temporal lobe size increases at approximately the same rate as the whole brain when human data are included. Thus, these studies suggest that, in contrast to the frontal lobe, the temporal lobe as a whole shows evolutionary expansion in the human lineage.

In contrast to findings in human temporal cortex, our data suggest that temporal lobe white matter has not undergone evolutionary expansion (Schenker et al., 2005). Larger brains contain a greater ratio of white to gray matter than smaller brains (Frahm et al., 1982; Prothero and Sundsten, 1984; Hofman, 1989; Rilling and Insel, 1999; Zhang and Sejnowski, 2000; Bush and Allman, 2004), and total temporal lobe white matter volume shows a positive allometric relationship with temporal lobe volume across hominoids. The ratio of temporal white matter to cortex is not greater in humans than would be expected based on temporal lobe volume; however, analyses of subdivisions of the white matter suggest reorganization similar to the frontal lobes. As with the frontal lobes, the ratio of gyral to core white matter in the human temporal lobe is larger than would be predicted from ape values (Figure 24). The human values for this measure fall above the confidence interval in half of the cases and show average percent residuals greater than 45% (Schenker, et al., 2005). A differentially enlarged volume of gyral white matter, as opposed to core white matter, in humans allows for increased interconnectivity via short association fibers which might contribute to increased human cognitive capabilities.

Amygdala

The temporal lobe comprises subcortical structures essential to both simple and complex behaviors, including central components of the limbic system. The idea that the production and mediation of complex behavior falls exclusively under the purview of the isocortex has been challenged. The challenge comes from the perspective that interactions between multiple cortical territories, including both "basic" emotional processing mediated by limbic structures and "higher order" isocortical cognitive processing, are important for the production of complex behaviors, especially in the social domain (e.g., Damasio, 1994). One such structure, the amygdala or amygdaloid complex (Figure 25), has traditionally been associated with emotional regulation but has more recently received scientific attention for its central role in mediating social cognition and affiliation (Kling, 1986; Brothers, 1990; Adolphs, 1999; Brothers and Ring, 1992; Kling and Brothers, 1992; Adolphs, 2003). The amygdala modulates emotional, neural, and bodily responses to external stimuli and directs an individual's attention based on the emotional significance of the stimulus to produce a context appropriate response (Adolphs, 1999). Although implicit associative learning, attending to salient stimuli, memory consolidation (Phelps, 2005), and environmental appraisal (Emery, 2000) are undoubtedly central to many social cognitive skills, the amygdala is also associated with mediating and evaluating explicitly social stimuli. Some examples from neuroimaging studies include the processing of emotional vocal, facial, and full body expressions (Yang et al., 2002; Hadjikhani and de Gelder, 2003; Glascher et al., 2004; Sander et al., 2005), evaluating trustworthiness in others (Grezes et al., 2004; Singer et al., 2004), and deciding whether to conform to peers' suggestions (Berns et al., 2005). At the cellular level, neurons in the macaque amygdala are activated by both dynamic social behaviors such as social interaction (Brothers and Ring, 1992, 1993) and social approach (Kling et al., 1979) and static representations such as images of faces (Brothers, 1990). In the rapidly changing, complex social environments inhabited by primates (Humphrey, 1988; Whiten and Byrne, 1988; Dunbar, 2003), the sorts of processes subserved by the amygdala might provide an individual with essential tools for evaluating conspecifics and navigating the social milieu. Given its importance in social cognition, a possible "prime mover" in primate cognitive evolution, the amygdala could be a target for evolutionary change or reorganization. Moreover, mounting evidence suggests that not only isocortical but also subcortical or allocortical regions of the human brain undergo evolutionary adaptation. For example, the nuclei of the thalamus, another subcortical limbic structure, show differential volumetric increase across primate species which suggests evolutionary reorganization (Armstrong, 1986).

The amygdala is a heterogenous structure comprising numerous highly interconnected nuclei. While a significant number of these nuclei share connections with non-isocortical structures, the lateral, basal, and accessory basal nuclei have strong reciprocal connections with the isocortex. These three nuclei are collectively referred to as the basolateral division of the amygdaloid complex (Figure 26). The basolateral division, together with its extensive interconnections with the isocortex, is important for pairing affective values with incoming stimuli, associative learning (Sah et al., 2003), and memory consolidation (McIntyre et al., 2003). Although little is known about the evolution of the amygdala in primates, previous research suggests that a portion of the amygdala that includes the basolateral division shows evolutionary increase in primates when compared with other regions of the amygdala (Stephan and Andy, 1977; Stephan, et al., 1987). This differential change has been attributed to the influence of isocortical expansion on the connected basolateral nuclei. The volume of the expanded region correlates with social group size, and thus it is likely that this mosaic pattern in the evolution of the amygdaloid subcomponents is driven by social evolutionary pressures (Barton, et al., 2003; Barton & Aggleton, 2000). Unfortunately, these studies of the amygdala included few hominoid species, making it difficult to assess the role the amygdala has played in human evolution. To better understand the importance of the amygdala and its constituent nuclei in human and ape evolution in particular, we performed a morphometric analysis of the amygdala as a whole and also of the basolateral division (the accessory basal, basal, and lateral nuclei) using specimens from all hominoid species.

In the human brain, the amygdala is as large as expected in overall volume, although absolutely it is more than three times the size of the chimpanzee amygdala (Figure 27). In contrast, the basolateral division shows a unique pattern of organization. The lateral nucleus is clearly the largest of the basolateral nuclei in humans (Figure 28), while the basal nucleus is the largest in apes. Stereological analyses of the amygdala and the basolateral division performed on a large sample of human brains confirm our findings in humans (Schumann and Amaral, 2005). The basolateral nuclei of macaques (Amaral et al., 1992; control data in Emery et al., 2001) are organized more like apes, suggesting that humans may be derived compared with other Old World primates. The lateral nucleus in the human brain is also larger than would be predicted for an ape of human brain size (Figure 29). Thus, our data suggest that the organization of the human basolateral division is distinguished by a volumetric increase in the lateral nucleus, and that the lateral nucleus may be evolutionarily emphasized in the human amygdala.

It is very likely that the differential expansion of amygdaloid regions is influenced by the expansion of neuroanatomical regions that are strongly connected with the nuclei, as previously suggested for larger sub-

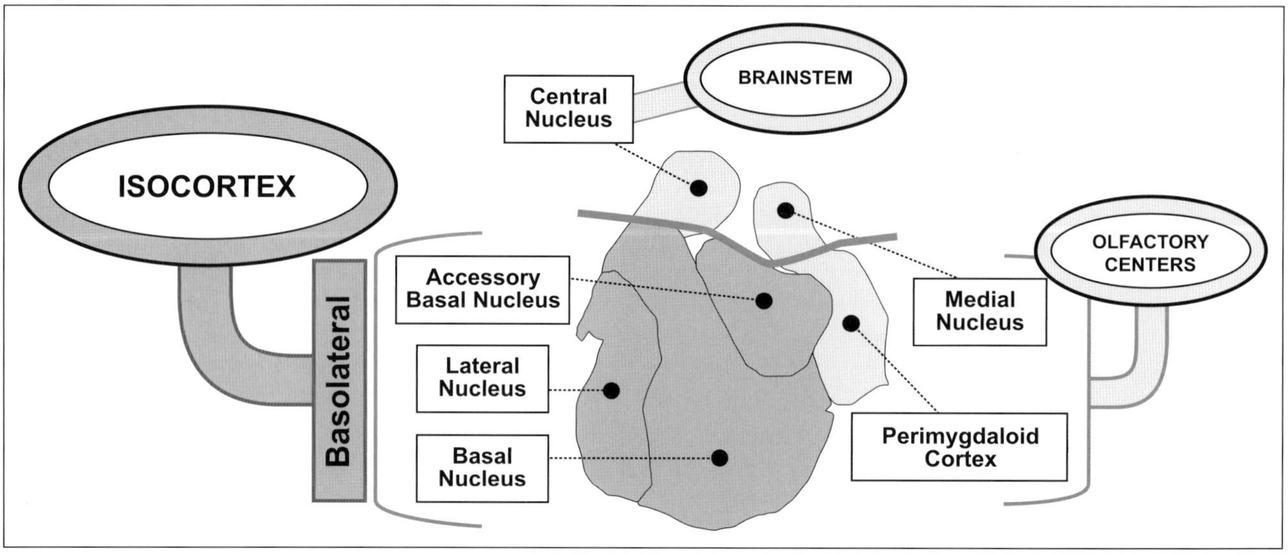

Figure 26. Diagram shows major components of the amygdala and related connections with rest of the brain.

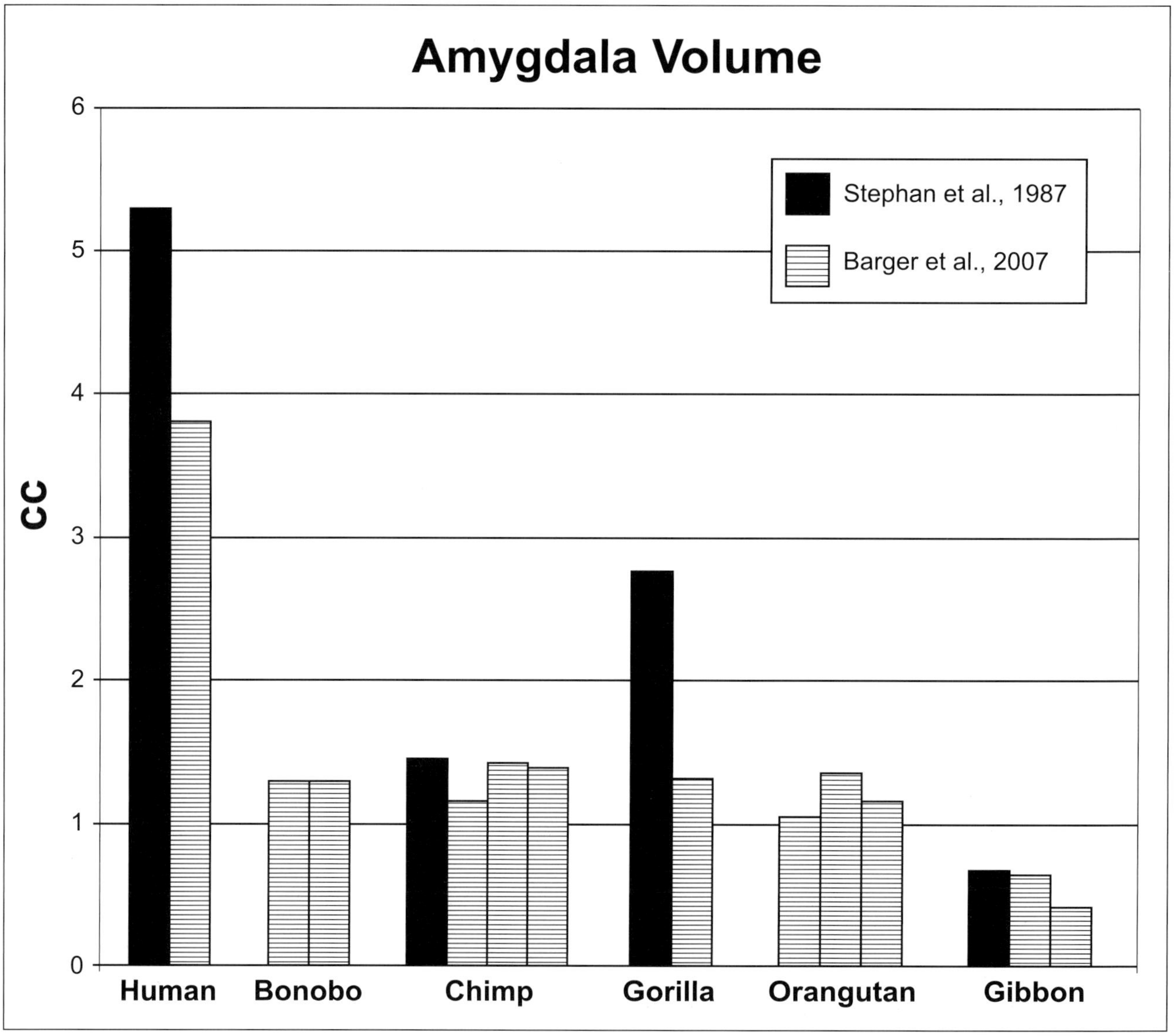

Figure 27. Volume of the amygdala across species. Bars represent individual specimens used in the two studies presented.

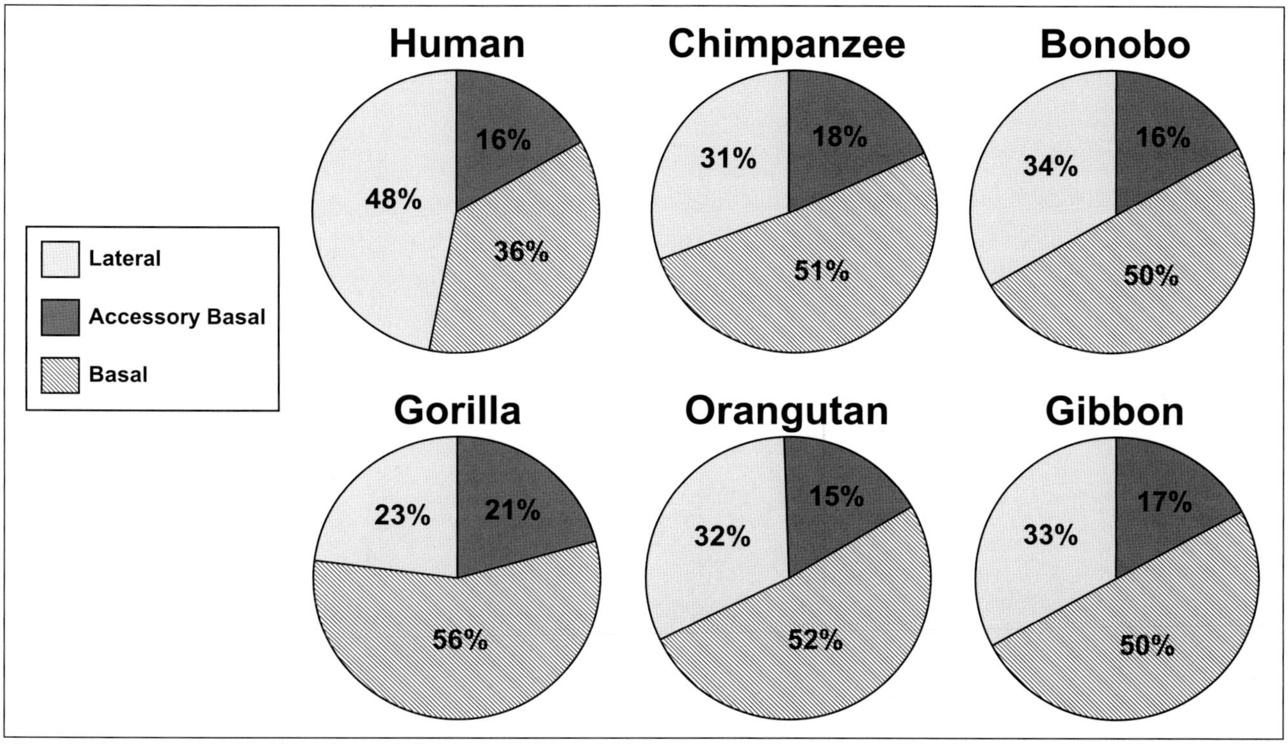

Figure 28. Percentage of basolateral division of the amygdala occupied by the lateral, basal, and accessory basal nuclei in each species. (Data from Barger et al., 2007).

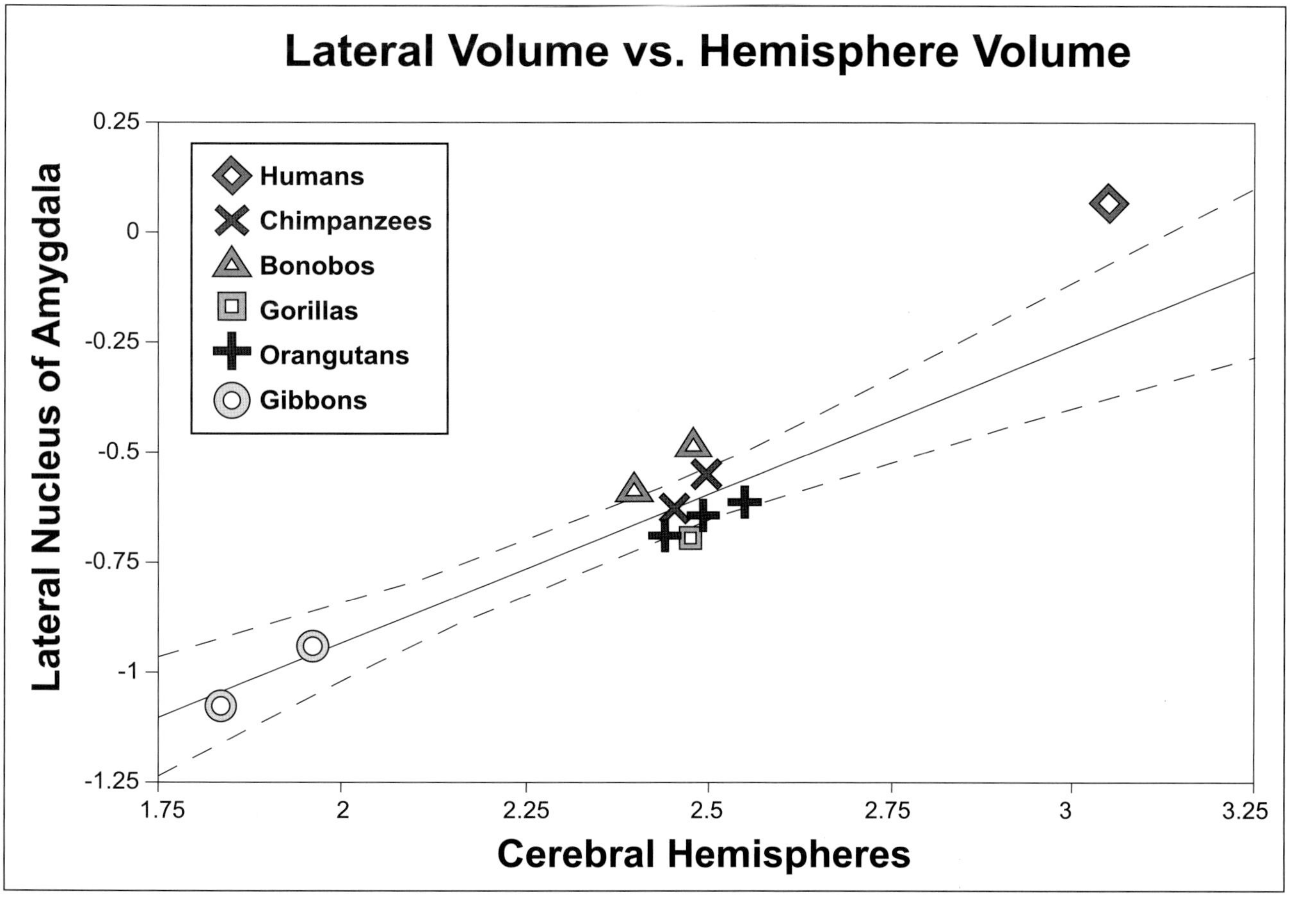

Figure 29. Regression of hemisphere volume against the volume of the lateral nucleus of the amygdala (modified from Barger et al., 2007)

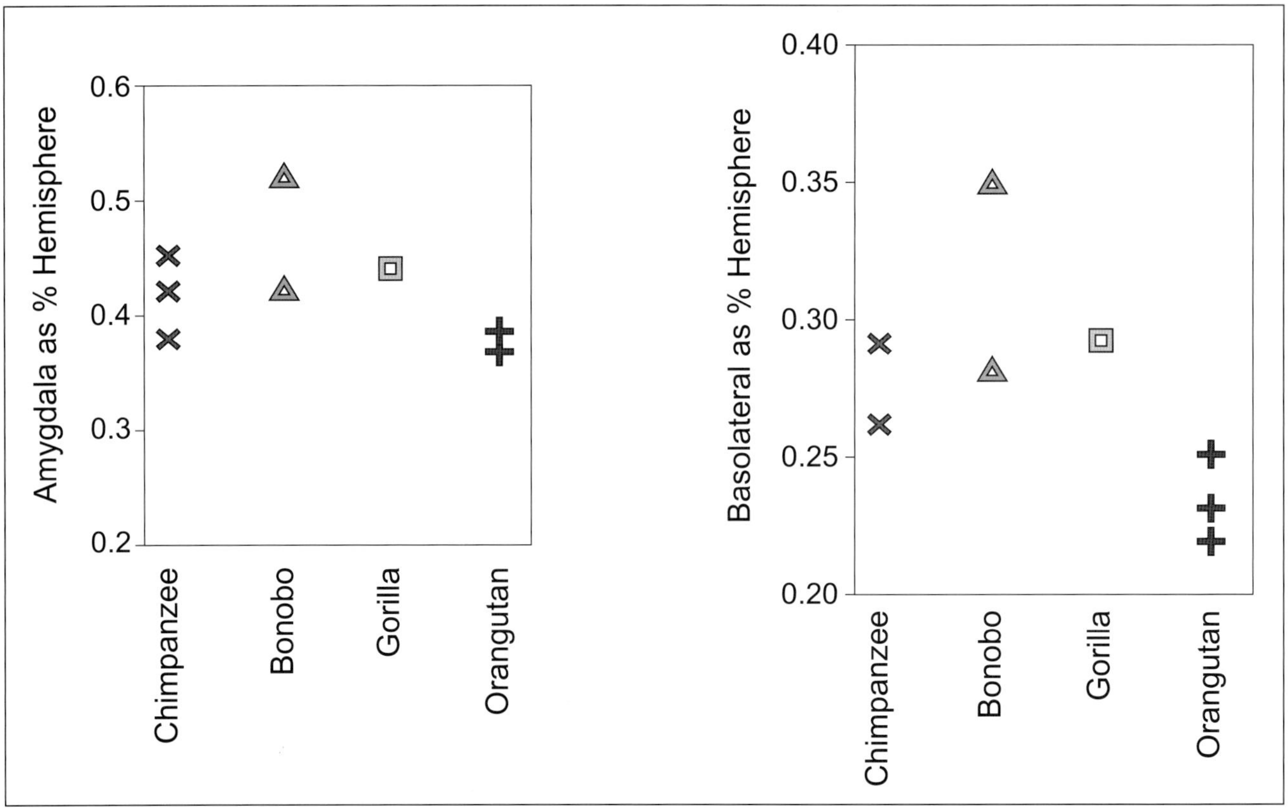

Figure 30. Left: Relative volume of the amygdala (amygdaloid complex) across great apes. Right: Relative volume of the basolateral division across great apes (modified from Barger et al., 2007)

components of this structure (Stephan and Andy, 1977). While information about the function of individual amygdaloid nuclei in the primate brain is limited, it has been hypothesized (Emery and Amaral, 2000; Stefanacci and Amaral, 2002) that polymodal and unimodal sensory information from the temporal cortex first enters the lateral nucleus, where it is received and categorized. This information passes to the basal nucleus where it is paired with information about the social context of the signal through its extensive connections with the orbitofrontal cortex. This highly processed information is then transferred to the striatum and the central nucleus (which subsequently targets hypothalamic and brainstem nuclei) to initiate the production of a context appropriate response. In humans, a preponderance of information from the elaborated temporal lobe would be flowing into the human amygdala via the lateral nucleus, increasing processing demands within the basolateral division. Given the functional and connective relationships between the temporal lobe and the lateral nucleus, it is likely that the unique organization of the human basolateral division is driven by information flowing from the enlarged human temporal lobe into the lateral nucleus (Stefanacci and Amaral, 2002).

Among the apes, the basolateral division of the gorilla and orangutan are most specialized (Barger, et al. 2007). The basal and accessory basal nucleus are exceptionally large in the gorilla, while the lateral nucleus is diminished, a more extreme manifestation of the pattern found among the other apes (Figure 28). If connected isocortical regions influence the elaboration of amygdaloid subcomponents, then this might account for the smaller than expected size of the gorilla temporal lobe (Semendeferi and Damasio, 2000; Rilling and Seligmann, 2002), which is an inverse of the human pattern. Orangutans show more extensive differences in amygdala volume compared to other apes. They have uniquely smaller total amygdala and basolateral division volumes compared to the other great apes (Figure 30). Within the basolateral division, the accessory basal nucleus is smaller in orangutans than would be predicted based on volumes in other hominoids (Figure 31). The basolateral division and especially the accessory basal nucleus receive considerable projections from the orbitofrontal cortex, including area 10 (Ghashghaei and Barbas, 2002; Stefanacci and Amaral, 2002), which is also smaller in orangutans, as discussed above (Figure 32). These data from the amygdala together with existing data on other neural structures provide new perspectives on the evolution of the amygdala and related neural systems (Barton and Harvey, 2000). While amygdaloid nuclei volumes in humans and gorillas might be influenced by the respective increase and decrease of predicted temporal lobe volumes, diminution in orangutan amygdala and amygdaloid nuclei volumes parallel decreases in the size of interconnected, limbic orbitofrontal regions.

Overall our findings largely support hypotheses of amygdala evolution that highlight the importance of

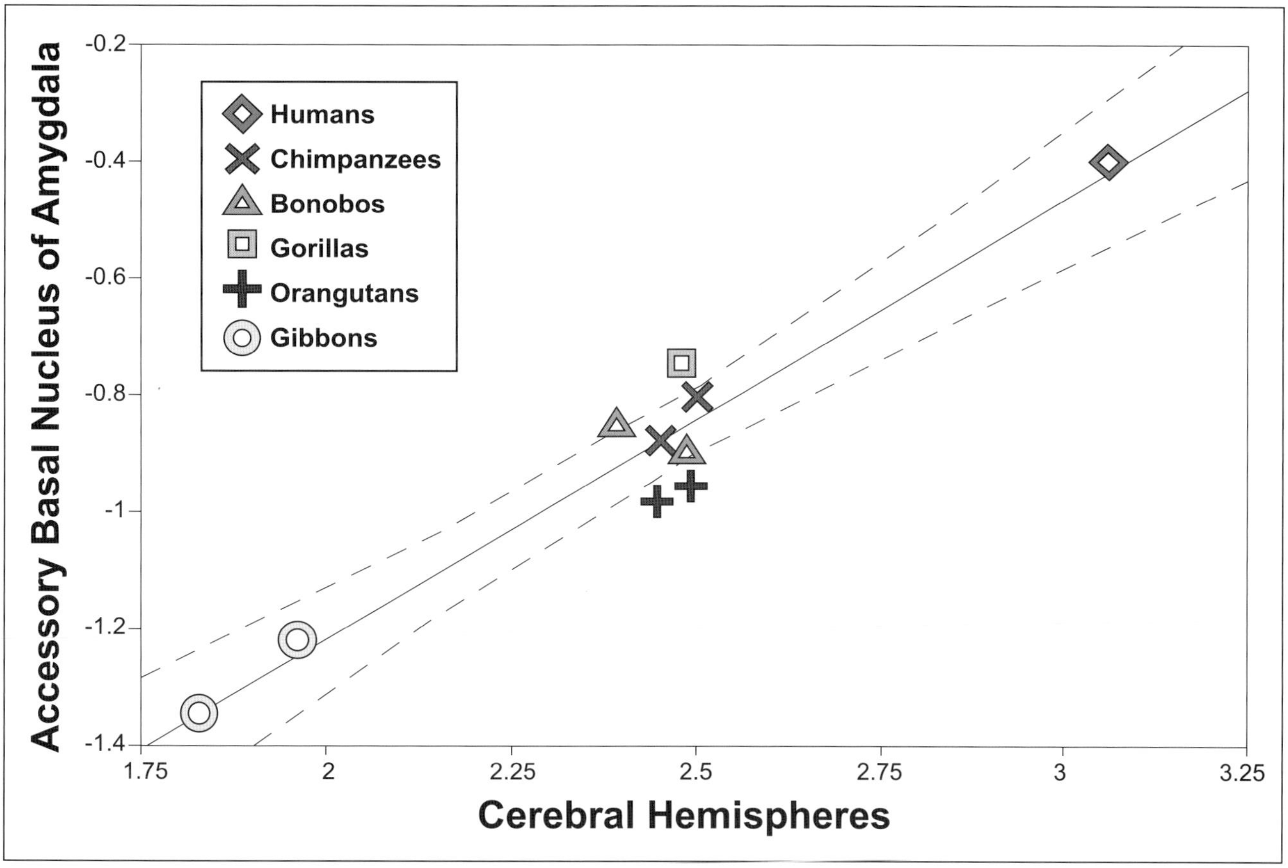

Figure 31. Regression of hemisphere volume against the volume of the accessory nucleus of the amygdala (modified from Barger et al., 2007)

functional networks within the brain and point to the importance of interconnected networks that might be influenced in a mosaic fashion characteristic of evolutionary reorganization. We found variation in limbic structures among hominoids, suggesting that parts of the human limbic system might be highly specialized. Along with the reorganization of the limbic orbitofrontal sectors of the isocortex, humans also show reorganization of the amygdala. These factors reinforce the idea that human emotional processes are not primitive relics of our evolutionary past but instead are highly evolved systems that complement higher order cognitive processes. Moreover, the associations between temporal lobe and amygdala expansion point to importance of social information processing in human brain evolution.

Discussion

While there are clear cognitive differences between humans and apes, the neural underpinnings of these differences are considerably less obvious. How do differences in the size and organization of the human brain produce the cognitive specializations found in our species? Have specific functional circuits been acted upon by evolutionary processes, or is overall size increase the only hallmark of human brain evolution? Anatomical studies comparing the human brain with the ape brain may gives us a sense of the associated cognitive changes that might have occurred in ancestral hominids during the Plio-Pleistocene.

The frontal lobes: Reorganization over enlargement

Early students of hominid brain evolution identified the frontal lobe as a candidate region for evolutionary expansion given its involvement in higher order executive functions such as abstract thinking and planning and also its involvement in language production. Thus, unique human cognitive capabilities were thought to result from an overall increase in human brain volume accompanied by a disproportionate increase in frontal cortex volume, and this assumption received early empirical support from Brodmann's comparative studies. We addressed this question more recently using modern morphometric techniques and larger samples of hominoids, and found that complex human cognition could not be attributed to a relative increase in human frontal lobe volume. Although the size of the human frontal lobe is larger in absolute terms, frontal lobe volume is remarkably similar across hominoids when whole brain size is factored out. It is likely, then, that the relative size of the frontal lobes has not changed significantly during hominid evolution. These results contradict deep-rooted assumptions about the evolution of the human brain.

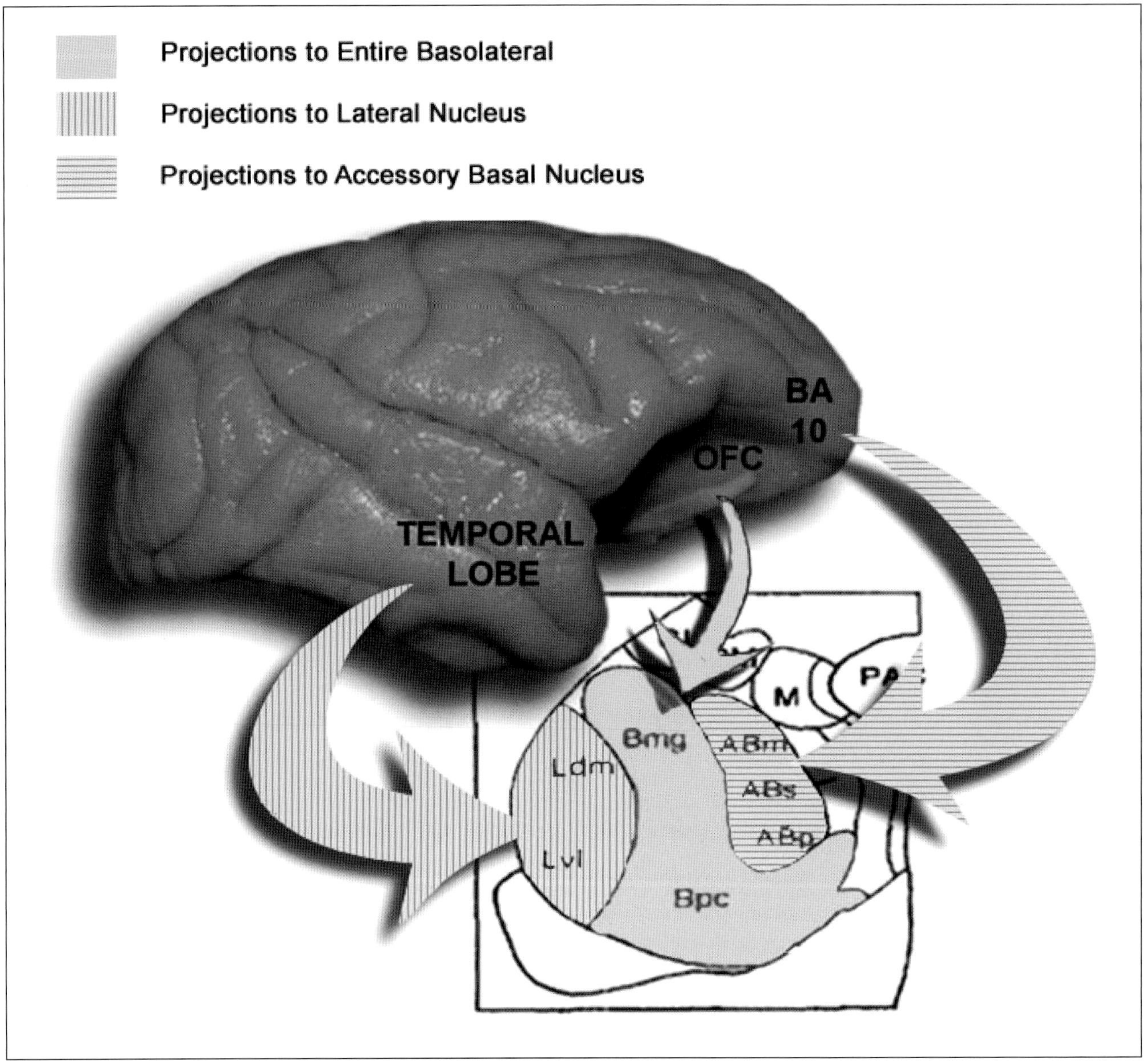

Figure 32 Diagram shows major connections of the amygdala and selected nuclei discussed here.

Moreover, the degree of inter- and intra-specific variation we found in our own large sample merits a cautionary statement when interpreting the results of studies that rely exclusively on one or two hemispheres of only a few species. Similarly, frontal lobe measurements from individual hominid endocasts might represent extreme values from a highly variable territory. While expanded frontal lobes are not a unique human characteristic, other anthropoid primates have relatively smaller frontal lobes than both apes and humans, suggesting that this could be a hominoid specialization.

In regard to human evolution, specifically, our results demonstrate the need for new directions of investigation. While the frontal lobe as a whole is not differentially enlarged in humans, it might instead exhibit internal reorganization. This might be manifest at either the gross or the histological level. At the gross level, reorganization can be identified in the distribution of white matter and gray matter across species. At the histological level, reorganization of the frontal lobe may be indicated by the presence or absence of particular cortical areas or by species-specific variation in their size or structural features. Portions of the frontal lobe also exhibit modifications in local circuitry reflected in specific cytoarchitectonic patterns, as seen in areas 10 and 13. Like the entire frontal lobe, the relative size of the dorsal, mesial, and orbital sectors of the human frontal lobe do not stand apart from the apes. In contrast, discrete cytoarchitectonic regions evidence frontal lobe reorganization. Frontal polar area 10, which has dorsal and orbital components, is increased in humans. In contrast, area 13 in the posterior orbitofrontal cortex is reduced. It is possible that anterior components of the orbitofrontal cortex have become more emphasized in human evolution over more posterior portions. Likewise, because the relative size of the orbitofrontal cortex is similar in humans and apes but area 13 is smaller in humans, the human orbitofrontal cortex might be more heterogeneous than the orbitofrontal cortex of most great apes.

The distribution of white matter in human brains is consistently different from the white matter of the apes. This suggests that differences in connectivity might contribute to human cognitive specializations. More white matter underlies gyral convolutions in humans than in other hominoids, possibly to allow for closely associated areas to communicate more efficiently with one another. Humans have more neuropil in the frontal lobe than other primates, indicating that more space for connectivity is available at the microstructural level. There is increased neuropil in the superficial layers (II/III) of frontal polar area 10, and in Broca's areas, areas 44 and 45, as indicated by the size of minicolumns and gray level index measurements. Such differences in neuropil might not be uniformly distributed across cortical layers or regions as the data from apes and humans cluster together elsewhere in the cortex (Semendeferi et al., 2001).

Frontal lobe reorganization does not solely occur in humans; it also characterizes other hominoids. Like humans, bonobos have a relatively small area 13 and might have increased complexity in the orbitofrontal cortex due to an increase in the numbers of cytoarchitectonically distinct areas. In contrast, orangutans exhibit a particularly small orbital sector, which is more homogenous and predominantly comprises area 13. The cytoarchitecture of area 13 in orangutans appears more "prefrontal" than limbic, though the most "prefrontal" region of the orbital surface, area 10, is also much smaller in orangutans than in other apes. It is possible that portions of the orangutan orbital sector have shifted toward the lateral surface of the frontal lobe, because the dorsal cortex is relatively enlarged in orangutans. Because the functions of the orbitofrontal cortex are vital to social cognitive processing, the orangutan's smaller orbitofrontal cortex might reflect the species's less gregarious social structure and might be related to a reduced emphasis on limbic cortices in orangutan neuroanatomy. In the gorilla, area 10 is somewhat distinct from the other apes and appears to be either selectively reorganized or shifted to a different position in the frontal lobe. Thus, humans as well as several other hominoid species show differential increase in particular portions of the frontal lobe, despite the fact that measures of the whole frontal lobe are not relatively increased in any of the species. Contrary to long-standing dogma, evidence suggests that internal reorganization and not gross volumetric increase characterizes hominid frontal lobe evolution. It appears, in fact, that the most extensive evolutionary expansion has not occurred in the human frontal lobe but in the temporal lobe.

Beyond the frontal cortex

Despite these new findings for the frontal cortex, reorganization of the whole brain appears to have occurred over the course of human evolution. The human brain is not simply a scaled up version of an ape brain. At the gross level, the temporal lobe and temporal cortex are larger in relation to whole brain size than would be predicted for a primate of human brain size. From an evolutionary perspective, this is striking given the lack of a parallel increase in the frontal lobe or the primary visual cortex (Holloway, 1968). Unfortunately, it is difficult to conclude whether the temporal lobe is uniquely expanded, because there is little comparative information about the hominoid parietal lobe. Nevertheless, our preliminary unpublished observations on the size and morphology of the parietal lobes in apes and humans support Holloway's (1968) conclusions favoring a differential expansion and reorganization of the inferior parietal lobule.

Within the temporal lobe, the amygdala exhibits intrinsic reorganization which appears to reflect the reorganization occurring in the whole brain. This pattern of coordinated change may indicate the influence of evolutionary pressures on specific networks, suggesting mosaic evolution (Barton, et al., 2003; Barton and Harvey, 2000). The basolateral division of the amygdala has expanded over the course of primate evolution, and this is probably due to its strong connective relationship with the particularly enlarged primate isocortex (Stephan and Andy, 1977; Stephen, et al., 1987). In contrast, the more conserved amygdaloid nuclei, which lie outside of the basolateral division, are interconnected with more conserved brainstem and olfactory regions. Strikingly, we found that humans are distinguished by an intrinsic reorganization of the basolateral division. This finding is consistent with our initial impressions that isocortical increases drive volumetric increases in the basolateral division of the amygdala. The ape basolateral division is characterized by a large basal nucleus followed in size by the lateral nucleus and the accessory basal nucleus, respectively. This organization is similar to the organization of the basolateral division of macaque monkeys. In humans the basolateral division is comparatively reorganized. The lateral nucleus is largest, followed in size by the basal and accessory basal nuclei, respectively. The three nuclei that constitute the basolateral division have distinct connections with specific portions of the isocortex. While the basal and accessory basal nuclei have a stronger connective relationship with the orbitofrontal cortex, the lateral nucleus shares the majority of its connections with the enlarged human temporal lobe (Stefanacci and Amaral, 2002). Both the temporal lobe and the lateral nucleus are enlarged in humans, while nuclei connected to regions that are not disproportionately enlarged in humans are not enlarged. As such, evolutionary mosaics present at gross levels of the human brain are reflected in the organization of the basolateral nuclei. This finding provides intriguing evidence that evolution might be acting upon neural systems rather than discrete structures yielding neuroanatomical mosaics.

While none of the ape specimens exhibited basolateral reorganization, there were some exceptional volumetric differences among the apes. Orangutans have a smaller total amygdala and a smaller basolateral division than the other great apes (Barger, et al. 2007). The basal and accessory basal nuclei are smaller than those of the other great apes, and the accessory basal nucleus is smaller than predicted for a hominoid of orangutan brain size. The basal and accessory basal nuclei are more connected to the orbitofrontal cortex than the lateral nucleus (Carmichael and Price, 1995; Stefanacci and Amaral, 2002), which is also smaller in orangutans. Given that these regions are limbic in nature, it is likely that limbic structures have been deemphasized in orangutan evolution.

In the inverse of the human pattern, the gorilla has a particularly small lateral nucleus relative to the other great apes and also has the smallest temporal lobe of the apes (Semendeferi and Damasio, 2000; Rilling and Seligman, 2002). The gorilla's smaller lateral nucleus is associated with a larger accessory basal nucleus. Because the accessory basal nucleus shares strong connections with both the orbitofrontal cortex and the superior temporal gyrus (Stefanacci and Amaral, 2002), it is more difficult to resolve the issue of whether connectivity influences volume. While the temporal lobe is smaller in gorillas, the superior temporal gyrus is not similarly reduced (Rilling and Seligman, 2002). Thus, the superior temporal gyrus might be larger than expected based on overall temporal volume in gorillas. Similarly, the orbitofrontal cortex is only slightly larger in gorillas compared to other apes (Semendeferi et al., 1997), but portions of the orbitofrontal cortex contain a greater number of neurons than other apes (Semendeferi et al., 1998), indicating that there may be more information transfer between the orbitofrontal cortex and other portions of the brain such as the accessory basal nucleus. The amygdala has not only been reorganized over the course of hominid and hominoid evolution, but it also reflects other reorganizational events in functionally related portions of the brain.

Functional implications and neuroecological considerations

The results of the new body of comparative research presented here challenge long held theories about which cognitive properties drive human brain evolution. Limbic structures, traditionally held to be conserved, show evolutionary change in the human brain, while regions that subserve many higher-order cognitive functions, such as the frontal lobe, have not enlarged as a whole. Likewise, neither the frontal cortex nor the amygdala is disproportionately represented in the hominoid brain, although they might be expected to be larger or smaller in humans, respectively, due to their functional properties. Instead, there appears to be a species-specific reorganization of their circuitry, and our data suggest that these regions have evolved in a mosaic fashion.

Within the human frontal lobe, the increased size of area 10 (the frontal pole) and decreased size of area 13 (orbitofrontal limbic cortex) suggest an increased emphasis on executive functions. This reorganization supports previous hypotheses about human brain evolution, which predict human differences in portions of the brain that control executive functions. At the same time, however, a considerable portion of area 10 comprises the orbitofrontal cortex, which is complex, heterogeneous, and is characterized as limbic cortex. In humans, limbic cortices are not reduced appreciably in size in relation to the apes. This finding suggests that regions long associated with emotional processing are not diminished in humans. Furthermore, available data suggest that the size of the prefrontal sector, involved in higher order cognitive functions, may not be as large in humans as once thought. It is difficult to assess the importance of prefrontal cortex-mediated functions in human brain evolution because little histological data is available for the areas contained within the prefrontal cortex, although the results suggest that the story is more complex than previously envisioned. The available evidence indicates that the frontal lobe may have played a key role in hominoid brain evolution, while playing a supporting one in hominid evolution.

Our data suggest that volumetric increase or reorganization of specific neural areas (Semendeferi et al., 1998, 2001) are associated with concomitant changes in heavily interconnected areas (Barger, et al., 2007), and that these mosaic changes occur in specific neural systems. Specifically, portions of the brain that are critical for complex social behavior, such as the limbic frontal cortex and amygdaloid nuclei, show volumetric increase, reorganization, or both. Studies of social cognition among primates (DeWaal and Aureli, 1996; Ingmanson, 1996; Byrne, 1996; Van Schaik and Van Hooff, 1996) emphasize species-specific patterns in how individuals deal with conspecifics and the importance of this behavior for survival and reproduction. In humans, damage to the orbital and mesial frontal sector, i.e., the limbic frontal cortex, is associated with a variety of deficits in social behavior (Damasio, 1994). While orbitofrontal lesions do not necessarily impair basic cognitive abilities such as memory, attention, and language (Stuss and Benson, 1984; Damasio, 1994), the ability to learn social rules and engage in socially affiliative behaviors can be diminished. Numerous analyses suggest a connection between the orbitofrontal cortex and the evaluation of social context based on emotional reinforcers (Bechara et al., 2000; Northoff et al., 2000; Rolls, 2000; Schoenbaum and Setlow, 2001). An extensive body of neuroimaging research has linked both orbitofrontal cortical activation and amygdala activation with social appraisal processes (Emery, 2000; Emery and Amaral, 2000). Normal frontal lobe and amygdala development and function are compromised in autism, which is predominantly characterized by impaired socioemotional behavior. In autism, the frontal lobe and amygdala show abnormal overgrowth during early development (Schumann et al., 2004; Carper and Courchesne, 2005). It is striking that these regions are all targeted in human evolution, suggesting that systems that influence social behavior are under selective pressure. Area FI has also been associated with socioemotional processing and it is tempting to view the unique cell types (spindle cells) found in this region as adaptations related to social processing. While the entire temporal lobe is not a dedicated social brain "module", it is considerably entrenched in the processing of species-specific visual and auditory information and in encoding the identity of conspecifics (Kircher et al., 2001; Mishkin et al., 1983; Frith and Frith, 1999). The temporal polar region has also long been hypothesized to be an essential component of the neural system subserving primate social behavior in association with the orbitofrontal cortex and amygdala (Brothers, 1990). Neurons in the superior temporal gyrus respond to species-specific vocalizations, and portions of the superior temporal sulcus are involved in processing facial gestures. Visual information about conspecifics flows through the inferior temporal sulcus through the "what/who" pathway, conveying information about the identity of a stimulus, and complex multimodal social information is processed in the temporal polar region (Brothers, 1990). Further, Rilling and Seligman (2002) point to the temporal lobe's extensive role in language processing and suggest that the human temporal lobe increase reflects this particular social communicative adaptation.

From this perspective, our orangutan findings are particularly relevant because orangutans are the only semi-solitary ape (Delgado and van Schaik, 2000). Both of the structures that are most reduced in orangutans, the orbitofrontal cortex and the amygdala, are limbic structures that have long been considered central components of the neural system subserving primate social behavior (Adolphs, 1999; Brothers, 1990). Our neuroanatomical findings give rise to behavioral predictions that can be tested by comparative studies of socioemotional differences across ape species. Although to date such studies are rare, one relevant case found that orangutans behave less impulsively than chimpanzees in a numerical ordination task that required them to evaluate edible stimuli (Shumaker et al., 2001). The authors suggested that this difference is related to reduced feeding competition in the orangutan's social environment compared to the chimpanzee. While orangutans are gregarious and share close bonds with group mates in rehabilitation centers (Russon, 2000, 2002), the high costs of feeding competition in the wild keeps party sizes small, averaging around two individuals per party (Delgado and van Schaik, 2000). Increased emphasis on limbic components such as the orbitofrontal cortex and amygdala might be necessary in situations where competition is high, and less so for situations involving reduced competition. In support of this idea, amygdala volume is actually increased in early autism. An enlarged amygdala might account for the increased social anxiety experienced by autistic children (Schumann et al., 2004). It is possible that less impulsivity or reduced anxiety about the behavior of conspecifics might be adaptive for a solitary great ape species, given that social learning plays a critical role in hominoid development and the opportunities to do so would be reduced in a solitary species.

Our accumulated neuroanatomical findings provide support for the idea that even though most brain components enlarge in a lawful manner across species of various brain sizes, selected neural systems have been reorganized in a mosaic fashion in ways that might reflect the evolutionary socioecology of each species. While addressing hypotheses about human brain evolution such as the theory of neural reorganization put forth by Holloway, this research also has implications for emerging fields such as neuroecology, which attempts to connect the neural substrates of behavior with species-specific socioecological adaptations. In some ways, Holloway's ideas on reorganization were a prelude to neuroecology, and our data suggest that both perspectives may serve as useful frameworks for the study of human brain evolution. Although further empirical investigations are necessary to fully test his early predictions, a considerable amount of information is fast accruing in favor of Holloway's pioneering ideas about primate brain evolution made almost a half a century ago.

Acknowledgements

We thank J. Moore and S. Strum for helpful discussions and Lisa Stefanacci for editorial assistance. Support from the Chancellor's Interdisciplinary Collaboratory Scholarship and the Kavli Institute for Brain and Mind at the University of California at San Diego helped make this work possible.

References

Adolphs, R., 1999. Social cognition and the human brain. Trends in Cognitive Sciences 3, 469-479.

Adolphs, R., 2003. Investigating the cognitive neuroscience of social behavior. Neuropsychologia 41, 119-126.

Allman, J.M., Watson, K.K., Tetreault, N.A., Hakeem, A.Y., 2005. Intuition and autism: A possible role for Von Economo neurons. Trends in Cognitive Sciences 9, 367-373.

Amaral, D.G., Price, J.L., Pitkänen, A., Carmichael, S.T., 1992. Anatomical organization of the primate amygdaloid complex. In: Aggleton, J.P., (Ed.), The Amygdala. New York: Wiley-Liss. pp. 1-66.

Amunts, K., Istomin, V., Schleicher, A., Zilles, K., 1995. Postnatal development of the human primary motor cortex: a quantitative cytoarchitectonic analysis. Anatomy and Embryology (Berl) 192, 557-571.

Amunts, K., Schlaug, G., Schleicher, A., Steinmetz, H., Dabringhaus, A., Roland, P.E., Zilles, K., 1996. Asymmetry in the human motor cortex and handedness. Neuroimage 4, 216-222.

Amunts, K., Schleicher, A., Burgel, U., Mohlberg, H., Uylings, H.B., Zilles, K., 1999. Broca's region revisited: cytoarchitecture and intersubject variabililty. Journal of Comparative Neurology 412, 319-341.

Amunts, K., Schleicher, A., Ditterich, A., Zilles, K., 2003. Broca's region: cytoarchitectonic asymmetry and developmental changes. Journal of Comparative Neurology 465, 72-89.

Andreasen, N.C., Flashman, L., Flaum, M., Arndt, S., Swayze, V., O'leary, D.S., Ehrhardt, J.C., Yuh, W.T., 1994. Regional brain abnormalities in schizophrenia measured with magnetic resonance imaging. JAMA 272, 1723-1769.

Armstrong, E., 1986. Enlarged limbic structures in the human brain: the anterior thalamus and medial mamillary body. Brain Research 362, 394-397.

Armstrong, E., 1990. Evolution of the brain. In: Paxinos, G., (Ed.), The human nervous system. New York: Academic Press. pp. 1-16.

Bailey, P., Bonin, G., McCulloch, W.S., 1950. The Isocortex of the Chimpanzee. Urbana: University of Illinois Press. p. 440.

Bailey, P., von Bonin, G., 1951. The Isocortex of Man. Urbana: University of Illinois Press.

Barbas, H., Pandya, D.N., 1987. Architecture and frontal cortical connections of the premotor cortex (area 6) in the rhesus monkey. Journal of Comparative Neurology 256, 211-228.

Barbas, H., Pandya, D.N., 1989. Architecture and intrinsic connections of the prefrontal cortex in the rhesus monkey. Journal of Comparative Neurology 286, 353-375.

Barger, N., Stefanacci, L., Semendeferi, K., 2007. A comparative volumetric analysis of the amygdaloid complex and basolateral division in the human and ape brain. American Journal of Physical Anthropology 134, 392-403.

Bartels, A., Zeki, S., 2000. The neural basis of romantic love. NeuroReport 11, 3829-3834.

Barton, R.A., Harvey, P.H., 2000. Mosaic evolution of brain structure in mammals. Nature 405, 1055-1058.

Barton, R.A., Aggleton, J.P., 2000. Primate evolution and the amygdala. In: Aggleton, J.P., (Ed.), The amygdala: a functional analysis. Oxford: Oxford University Press. pp. 479-508.

Barton, R.A., Aggleton, J.P., Grenyer, R., 2003. Evolutionary coherence of the mammalian amygdala. Proc Biol Sci 270, 539-543.

Bechara, A., Damasio, H., Damasio, A.R., 2000. Emotion, decision making and the orbitofrontal cortex. Cerebral Cortex 10, 295-307.

Bechara, A., Damasio, H., Tranel, D., Anderson, S.W., 1998. Dissociation of working memory from decision making within the human prefrontal cortex. Journal of Neuroscience 18, 428-437.

Berns, G.S., Chappelow, J., Zink, C.F., Pagnoni, G., Martin-Skurski, M.E., Richards, J., 2005. Neurobiological correlates of social conformity and independence during mental rotation. Biological Psychiatry 58, 245-253.

Blinkov, S.M., Glezer, II., 1968. The Human Brain in Figures and Tables: A Quantitative Handbook. New York: Plenum Press.

Braak, H., 1980. Architectonics of the Human Telecephalic Cortex. Berlin: Springer-Verlag.

Brodmann, K., 1909. Vergleichende Lokalisationslehre der Grosshirnrinde. Leipzig: Barth.

Brodmann, K., 1912. Neue Ergebnisse über die vergleichende histologische Lokalisation der Grosshirnrinde mit besonderer Berücksichtigung des Stirnhirns. Anatomischer Anzeiger 41, 157-216.

Brothers, L., 1990. The social brain: a project for integrating primate behavior and neurophysiology in a new domain. Concepts in Neuroscience 1, 27-51.

Brothers, L., Ring, B., 1992. A neuroethological framework for the representation of minds. Journal of Cognitive Neuroscience 4, 107-118.

Brothers, L., Ring, B., 1993. Mesial temporal neurons in the macaque monkey with responses selective for aspects of social stimuli. Behavioural Brain Research 57, 53-61.

Burman, K.J., Palmer, S.M., Gamberini, M., Rosa, M.G., 2006. Cytoarchitectonic subdivisions of the dorsolateral frontal cortex of the marmoset monkey (Callithrix jacchus), and their projections to dorsal visual areas. Journal of Comparative Neurology 495, 149-172.

Bush, E.C., Allman, J.M., 2004. The scaling of frontal cortex in primates and carnivores. PNAS 101, 3962-3966.

Butter, C., Snyder, D.R., 1972. Alterations in aversive and aggressive behaviors following orbital frontal lesions in rhesus monkeys. Acta Neurobiologiae Experimentalis 32, 525-565.

Buxhoeveden, D., Casanova, M.F., 2002. The minicolumn hypothesis in neuroscience. Brain 125, 935-951.

Buxhoeveden, D.P., Switala, A.E., Roy, E., Litaker, M., Casanova, M.F., 2001. Morphological differences between minicolumns in human and nonhuman primate cortex. American Journal of Physical Anthropology 115, 361-371.

Byrne, R.W., 1996. The misunderstood ape: cognitive skills of the gorilla. In: Russon, A.E., Bard, K.A., Parker, S.T., (Eds.), Reaching Into Thought: The Minds of the Great Apes. Cambridge: Cambridge University Press, pp. 111-130.

Carmichael, S.T., Price, J.L., 1994. Architectonic subdivision of the orbital and medial prefrontal cortex in the macaque monkey. Journal of Comparative Neurology 346, 36-402.

Carmichael, S.T., Price, J.L., 1995. Limbic connections of the orbital and medial prefrontal cortex in macaque monkeys. Journal of Comparative Neurology 363, 615-641.

Carper, R.A., Courchesne, E., 2005. Localized enlargement of the frontal cortex in early autism. Biological Psychiatry 57, 126-133.

Carter, C.S., Botvinick, M.M., Cohen, J.D., 1999. The contribution of the anterior cingulate cortex to executive processes in cognition. Reviews in the Neurosciences 10, 49-57.

Caviness, V.S., Meyer, J., Makris, N., Kennedy, D.N., 1996. MRI-based topographic parcellation of human neocortex: an anatomically specified method with estimate of reliability. Journal of Cognitive Neuroscience 8, 566-587.

Clark, D.A., Mitra, P.P., Wang, S.S.H., 2001. Scalable architecture in mammalian brains. Nature 411,189-193.

Courchesne, E., Pierce, K., 2005. Why the frontal cortex in autism might be talking only to itself: local over-connectivity but long-distance disconnection. Current Opinion in Neurobiology 15, 225-230.

Cummings, J.L., 1993. Frontal-subcortical circuits and human behavior. Archives of Neurology 50, 873-880.

Daffner, K.R., Mesulam, M.M., Scinto, L.F., Acar ,D., Calvo, V., Faust, R., Chabrerie, A., Kennedy, B., Holcomb, P., 2000. The central role of the prefrontal cortex in directing attention to novel events. Brain 123, 927-939.

Dagher, A., Owen, A.M., Boecker, H., Brooks, D.J., 1999. Mapping the network for planning: a correlational PET activation study with the tower of London task. Brain 122, 1973-1987.

Damasio, A.R., 1985. The frontal lobes. In: Heilman, K., Valenstein, E., (Eds.), Clinical neuropsychology. Oxford: Oxford University Press. pp. 339-375.

Damasio, A.R., 1994. Decartes' error. New York: Avon Publishers.

Damasio, H., Grabowski, T.J., Tranel, D., Hichwa, R.D., Damasio, A.R., 1996. A neural basis for lexical retrieval. Nature 380, 499-505.

Damasio, A.R., Van Hoesen, G.W., 1983. Emotional disturbances associated with focal lesions of the frontal lesions of the frontal lobe. In: Heilman, K., Satz, P., (Eds.), Neuropsychology of Human Emotion: Recent Advances. New York: Guilford Press, pp. 85-110.

Damasio, H., Grabowski, T.J., Tranel, D., Hichwa, R.D., Damasio, A.R., 1996. A neural basis for lexical retrieval. Nature 380, 499-505.

Deacon, T.W., 1997. The Symbolic Species. New York: Norton.

Dejerine, J., 1895. Anatomie Des Centres Nerveux. Rueff et Cie, Paris.

Delgado, R.A., Van Schaik, C.P., 2000. The behavioral ecology and conservation of the orangutan (Pongo pygmaeus): a tale of two islands. Evolutionary Anthropology 9, 201-218.

De Waal, F.B.M., Aurelli, F., 1996. Consolation, reconciliation, and a possible cognitive difference between macaques and chimpanzees. In: Russon, A.E., Bard, K.A., Parker, S.T., (Eds.), Reaching Into Thought: The Minds of the Great Apes. Cambridge: Cambridge University Press. pp. 80-110.

de Winter, W., Oxnard, C.E., 2001. Evolutionary radiations and convergences in the structural organization of mammalian brains. Nature 409, 710-714.

Dombrowski, S., Hilgetag, C.C., Barbas, H., 2001. Quantitative architecture distinguishes prefrontal cortical systems in the rhesus monkey. Cerebral Cortex 11, 975-988.

Donoghue, J.P., Sanes, J.N., 1994. Motor areas of the cerebral cortex. Journal of Clinical Neurophysiology 11, 382-396.

Douglas, R., Martin, K., 1992. Exploring cortical microcircuits: a combined anatomical, physiological, and computational approach. In: McKenna, T., Davies, J., Zornetzer, S., (Eds.), Single neuron computation. San Diego: Academic Press, pp. 381-412.

Dunbar, R., 2003. Evolution of the social brain. Science 302, 1160-1161.

Dusser de Barenne, J.G., Garol, H.W., McCulloch, W.S., 1941. The "motor" cortex of the chimpanzee. Journal of Neurophysiology 4, 287-304.

Elston, G.N., Benavides-Piccione, R., Elston, A., Zietsch, B., Defelipe, J., Manger, P., Casagrande, V., Kaas, J.H., 2006. Specializations of the granular prefrontal cortex of primates: implications for cognitive processing. The Anatomical Record Part A: Discoveries in Molecular, Cellular, and Evolutionary Biology 288, 26-35.

Emery, N., 2000. The eyes have it: the neuroethology, function and evolution of social gaze. Neuroscience and Biobehavioral Reviews 24, 581-604.

Emery, N., Amaral, D., 2000. The role of the amygdala in primate social cognition. In: Lane, R.D., Nadel, L., (Eds.), Cognitive neuroscience of emotion. Oxford: Oxford Univ Press. pp. 156-191.

Emery, N.J., Capitanio, J.P., Mason, W.A., Machado, C.J., Mendoza, S.P., Amaral, D.G., 2001. The effects of bilateral lesions of the amygdale on dyadic social interactions in rhesus monkeys (Macaca mulatta). Behavioral Neuroscience 115, 515-544.

Finlay, B.L., Darlington, R.B., 1995. Linked regularities in the development and evolution of mammalian brains. Science 268, 1578-1583.

Finlay, B.L., Darlington, R.B., Nicastro, N., 2001. Developmental structure in brain evolution. Behavioral and Brain Sciences 24, 263-278.

Fletcher, P.C., Happe, F., Frith, U., Baker, S.C., Dolan, R.J., Frackowiak, R.S.J., Frith, C.D., 1995. Other minds in the brain a functional imaging study of theory of mind in story comprehension. Cognition 57, 109-128.

Fogassi, L., Gallese, V., Gentilucci, M., Luppino, G., Matelli, M., Rizzolatti, G., 1994. The frontoparietal cortex of the prosimian Galago: patterns of cytochrome oxidase activity and motor maps. Behavioural Brain Research 60, 91-113.

Foundas, A.L., 2001. The anatomical basis of language. Topics in Language Disorders 21, 1-19.

Frahm, H.D., Stephan, H., Stephan, M., 1982. Comparison of brain structure volumes in Insectivora and Primates.1. Neocortex. Journal für Hirnforschung 23, 375-389.

Frith, C.D., Frith, U., 1999. Interacting minds a biological basis. Science 286, 1692-1695.

Fuster, J.M., 1997. The Prefrontal Cortex. Anatomy, Physiology, and Neuropsychology of the Frontal Lobe. Philadelphia: Lippincott-Raven.

Gallagher, H.L., Happe, F., Brunswick, N., Fletcher, P.C., Frith, U., Frith, C.D., 2000. Reading the mind in cartoons and stories: an fMRI study of 'theory of mind' in verbal and nonverbal tasks. Neuropsychologia 38, 11-21.

Gebhard, R., Zilles, K., Schleicher, A., Everitt, B.J., Robbins, T.W., Divac, I., 1995. Parcellation of the frontal cortex of the New World monkey Callithrix jacchus by eight neurotransmitter-binding sites. Anatomy and Embryology (Berl) 191, 509-517.

Ghashghaei, H.T., Barbas, H., 2002. Pathways for emotion: interactions of prefrontal and anterior temporal pathways in the amygdala of the rhesus monkey. Neuroscience 115, 1261-1279.

Gibson, K.R., 2002. Evolution of human intelligence: the roles of brain size and mental construction. Brain, Behavior and Evolution 59, 10-20.

Glascher, J., Tuscher, O., Weiller, C., Buchel, C., 2004. Elevated responses to constant facial emotions in different faces in the human amygdala: an fMRI study of facial identity and expression. BMC Neuroscience 5, 45. Also available at http://www.biomedcentral.com/1471-2202/5/45.

Glasser, M.F., Rilling, J.K., 2008. DTI tractography of the human brain's language pathways. Cerebral Cortex 18, 2471-2482.

Goldman-Rakic, P.S., 1984. The frontal lobes: uncharted provinces of the brain. Trends in Neurosciences. 7, 425-429.

Gorno-Tempini, M.L., Price, C.J., 2001. Identification of famous faces and buildingsda functional neuroimaging study of semantically unique items. Brain 124, 2087-2097.

Grabowski, T.J., Damasio, H., Tranel, D., Ponto, L.L.B., Hichwa, R.D., Damasio, A.R., 2001. A role for left temporal pole in the retrieval of words for unique entities. Human Brain Mapping 13, 199-212.

Grezes, J., Frith, C., Passingham, R.E., 2004. Brain mechanisms for inferring deceit in the actions of others. Journal of Neuroscience 24, 5500-5505.

Hadjikhani, N., de Gelder, B., 2003. Seeing fearful body expression activates the fusiform cortex and amygdala. Current Biology 13, 2201-2205.

Hakeem, A., Allman, J., Tetreault, N., Semendeferi, K., 2004. The spindle neurons of frontoinsular cortex (area FI) are unique to humans and African apes. American Journal of Physical Anthropology 36 (Suppl.), 106.

Harlow, H.F., Akert, K., Schiltz, K..A., 1964. The effects of bilateral prefrontal lesions on learned behavior of neonatal, infant and preadolescent monkeys. In: Warren, J.M., Akert, K., (Eds.), The frontal granular cortex and behavior. New York: McGraw-Hill Book Co. pp. 126-148.

Harrison, K., Hof, P.R., Wang, S.H., 2002. Scaling laws in the mammalian neocortex: does form provide clues to function? Journal of Neurocytology 31, 289-298.

Heimer, L., 1995. The Human Brain and Spinal Cord: Functional Neuroanatomy and Dissection Guide. New York: Springer-Verlag.

Herbert, M.R., Ziegler, D.A., Makris, N., Filipek, P.A., Kemper, T.L., Normandin, J.J., Sanders, H.A., Kennedy, D.N., Caviness, V.S., Jr., 2004. Localization of white matter volume increase in autism and developmental language disorder. Annals of Neurology 55, 530-540.

Hof, P.R., Mufson, E.J., Morrison, J.H., 1995. Human orbitofrontal cortex: cytoarchitecture and quantitative immunohistochemical parcellation. Journal of Comparative Neurology 359, 48-68.

Hof, P., Van Der Gucht, E., 2007. Structure of the cerebral cortex of the humpback whale, Megaptera novaeangliae (Cetacea, Mysticeti, Balaenopteridae). The Anatomical Record 290, 1-31.

Hofman, M.A., 1989. On the evolution and geometry of the brain in mammals. Progress in Neurobiology 32, 137-158.

Holloway, R.L., 1964. Some quantitative relations of the primate brain. Ph.D. thesis. University of Michigan. Available on microfilm.

Holloway, R.L., 1968. The evolution of the primate brain: some aspects of quantitative relations. Brain Research 7, 121-172.

Holloway, R.L., 1979. Brain size, allometry, and reorganization: toward a synthesis. In: Hahn, M.E., Jensen, C., Dudek, B.C., (Eds.), Development and Evolution of Brain Size: Behavioral Implications. New York: Academic Press pp. 59-88.

Holloway, R.L., 2001. Does allometry mask important brain structure residuals relevant to species-specific behavioral evolution? Behavioral and Brain Sciences 24, 286-287.

Holloway, R.L., 2002. Brief Communication: How Much Larger Is the Relative Volume of Area 10 of the Prefrontal Cortex in Humans? American Journal of Physical Anthropology 118, 399-401.

Humphrey N. 1988. The social function of intellect. In: Byrne, R., Whiten, A., (Eds.), Machiavellian intelligence. Oxford: Oxford Univ Press. pp. 13-26.

Ingmanson, E.J., 1996. Tool-using behavior in wild Pan paniscus: Social and ecological considerations. In: Russon, A.E., Bard, K.A., Parker, S.T., (Eds.), Reaching Into Thought: The Minds of the Great Apes. Cambridge: Cambridge University Press. pp. 190-210.

Jerison, H., 1973. Evolution of the Brain and Intelligence. New York: Academic Press.

Jerison, H.J., 1997. Evolution of the Prefrontal Cortex. In: Krasnegor, N.A., Lyon, R., Goldman-Rakic, P.S., (Eds.), The Development of Prefrontal Cortex: Evolution, Neurobiology, and Behavior. Baltimore: Brooks. pp. 9-26.

Karama, S., Lecours, A.R., Leroux, J., Bourgouin, P., Beaudoin, G., Joubert, S., Beuregard, M., 2002. Areas of brain activation in males and females during viewing of erotic film excerpts. Human Brain Mapping 16, 1-13.

Kennedy D, Semendeferi, K, Courchesne E. 2007. No reduction of spindle neuron number in frontoinsular cortex in autism. Brain and Cognition 64, 124-129.

Kircher, T.T., Senior, C., Phillips, M.L., Rabe-Hesketh, S., Benson, P.J., Bullmore, E.T., Brammer, M., Simmons, A., Bartels, M., David, A.S., 2001. Recognizing one's own face. Cognition 78, B1-B15.

Kling, A.S., 1986. Neurological correlates of social behavior. Ethol Sociobiol 7, 175-186.

Kling, A.S., Brothers, L.A., 1992. The amygdala and social behavior. In: Aggleton, J.P., (Ed.), The amygdala. New York: Wiley-Liss. pp. 353-377.

Kling, A., Steklis, H.D., Deutsch, S., 1979. Radiotelemetered activity from the amygdala during social interactions in the monkey. Experimental Neurology 66, 88-96.

Krubitzer, L., Kahn, D.M., 2003. Nature versus nurture revisited: an old idea with a new twist. Progress in Neurobiology 70, 33-52.

Lepage, M., Ghaffar, O., Nyberg, L., Tulving, E., 2000. Prefrontal cortex and episodic memory retrieval mode. PNAS 97, 506-511.

MacLean, P.D., 1990. The triune brain in evolution. New York: Plenum.

MacSweeney, M., Woll, B., Campbell, R., McGuire, P.K., David, A.S., Williams, S.C.R., Suckling, J., Calvert, G.A., Brammer, M.J., 2002. Neural systems underlying British sign language and audio-visual english processing in native users. Brain 125, 1583-1593.

Makris, N., Worth, A.J., Sorensen, A.G., Papadimitriou, G.M., Wu, O., Reese, T.G., Wedeen, V.J., Davis, T.L., Stakes, J.W., Caviness, V.S., Kaplan, E., Rosen, B.R., Pandya, D.N., Kennedy, D.N., 1997. Morphometry of in vivo human white matter association pathways with diffusionweighted magnetic resonance imaging. Annals of Neurology 42, 951-962.

Matelli, M., Camarda, R., Glickstein, M., Rizzolatti, G., 1986. Afferent and efferent projections of the inferior area 6 in the macaque monkey. Journal of Comparative Neurology 251, 281-298.

Matelli, M., Luppino, G., Rizzolatti, G., 1985. Patterns of cytochrome oxidase activity in the frontal agranular cortex of the macaque monkey. Behavioural Brain Research 18, 125-136.

McBride, T., Arnold, S.E., Gur, R.C., 1999. A comparative volumetric analysis of the prefrontal cortex in human and baboon MRI. Brain, Behavior and Evolution 54, 159-166.

McIntyre, C.K., Power, A.E., Roozendaal, B., McGaugh, J.L., 2003. Role of the basolateral amygdala in memory consolidation. Annals of the New York Academy of Sciences 985, 273-293.

Mesulam, M., Mufson, E.J., 1982. Insula of the Old-World Monkey. 1. Architectonics in the Insulo-Orbito-Temporal Component of the Paralimbic Brain. Journal of Comparative Neurology 212, 1-22.

Mishkin, M., Ungerleider, L.G., Macko, K.A., 1983. Object vision and spatial vision: 2 cortical pathways. Trends in Neuroscience 6, 414-417.

Mitelman, S.A., Shihabuddin, L., Brickman, A.M., Hazlett, E.A., Buchsbaum, M.S., 2003. MRI assessment of gray and white matter distribution in Brodmann's areas of the cortex in patients with schizophrenia with good and poor outcomes. American Journal of Psychiatry 160, 2154-2168.

Morecraft, R.J., Geula, C., Mesulam, M.M., 1992. Cytoarchitecture and neural afferents of orbitofrontal cortex in the brain of the monkey. Journal of Comparative Neurology 323, 341-358.

Mountcastle, V., 1997. The columnar organization of the neocortex. Brain 120, 701-722.

Mundy, P., 2003. Annotation: The neural basis of social impairments in autism: the role of the dorsal medial-frontal cortex and anterior cingulate system. Journal of Child Psychology and Psychiatry 44, 793-809.

Nimchinsky, E.A., Gilissen, E., Allman, J.M., Perl, D.P., Erwin, J.M., Hof, P.R., 1999. A neuronal morphologic type unique to humans and great apes. Proceedings of the National Academy of Sciences USA 96, 5268-5273.

Nimchinsky, E.A., Vogt, B.A., Morrison, J.H., Hof, P.R., 1995. Spindle neurons of the human anterior cingulate cortex. Journal of Comparative Neurology 355, 27-37.

Northoff, G., Richter, A., Gessner, M., Schlagenhauf, F., Fell, J., Baumgart, F., Kaulisch, T., Kotter, R., Stephan, K.E., Leschinger, A., Hagner, T., Bargel, B., Witzel, T., Hinrichs, H., Bogerts, B., Scheich, H., Heinze, H.J., 2000. Functional dissociation between medial and lateral prefrontal cortical spatiotemporal activation in negative and positive emotions: a combined fMRI/meg study. Cerebral Cortex 10, 93-107.

Okuda, J., Fujii, T., Yamadori, A., Kawashima, R., Tsukiura, T., Fukatsu, R., Suzuki, K., Ito, M., Fukuda, H., 1998. Participation of the prefrontal cortices in prospective memory: evidence from a PET study in humans. Neuroscience Letters 253, 127-130.

Öngür, D., Ferry, A.T., Price, J.L., 2003. Architectonic subdivision of the human orbital and medial prefrontal cortex. Journal of Comparative Neurology 460, 425-449.

Owen, A.M., 1997. The functional organization of working memory processes within human lateral frontal cortex: the contribution of functional neuroimaging. European Journal of Neuroscience 9, 1329-1339.

Owen, A.M., Doyon, J., Petrides, M., Evans, A., 1996. Planning and spatial working memory examined with positron emission tomography. European Journal of Neuroscience 9, 353-364.

Passingham, R.E., 2002. The frontal cortex: does size matter? Nature Neuroscience 5, 190-192.

Petrides, M. 2000. Dissociable roles of mid-dorsolateral prefrontal and anterior inferotemporal cortex in visual working memory. Journal of Neuroscience 20, 7496-7503.

Petrides, M., Pandya, D.N., 1999. Dorsolateral prefrontal cortex: comparative cytoarchitectonic analysis in the human and the macaque brain and corticocortical connection patterns. European Journal of Neuroscience 11, 1011-1036.

Petrides, M., Pandya, D.N., 2002. Comparative cytoarchitectonic analysis of the human and the macaque ventrolateral prefrontal cortex and corticocortical connection patterns in the monkey. European Journal of Neuroscience 16, 291-310.

Phelps, E., 2006. Emotion and cognition: Insights from studies of the human amygdala. Annual Review of Psychology 57, 27-53.

Pochon, J.B., Levy, R., Poline, J.B., Crozier, S., Lehericy, S., Pillon, B., Deweer, B., Le Bihan, D., Dubois, B., 2001. The role of dorsolateral prefrontal cortex in the preparation of forthcoming actions: an fMRI study. Cerebral Cortex 11, 260-266.

Poremba, A., Saunders, R.C., Crane, A.M., Cook, M., Sokoloff, L., Mishkin, M., 2003. Functional mapping of the primate auditory system. Science 299, 568-572.

Preuss, T.M., Goldman-Rakic, P.S., 1991. Myelo- and cytoarchitecture of the granular frontal cortex and surrounding regions in the strepsirhine primate Galago and the anthropoid primate Macaca. Journal of Comparative Neurology 310, 429-474.

Price, C.J., 2000. The anatomy of language: contributions from functional neuroimaging. Journal of Anatomy 197, 335-359.

Prothero, J.W., Sundsten, J.W., 1984. Folding of the cerebral cortex in mammals a scaling model. Brain, Behavior and Evolution 24, 152-167.

Radinsky, L., 1974. The fossil evidence of anthropoid brain evolution. American Journal of Physical Anthropology 41, 15-28.

Raghanti, M.A., Sherwood, C.C., Hof, P.R., Erwin, J.M., 2008. Evolutionary changes in cortical neuromodulator innervation within the frontal cortex of humans, chimpanzees, and macaque monkeys. American Journal of Primatology 70 (Suppl. 1), 32.

Rajkowska, G., Goldman-Rakic, P.S., 1995a. Cytoarchitectonic definition of prefrontal areas in the normal human cortex: I. Remapping of areas 9 and 46 using quantitative criteria. Cerebral Cortex 5, 307-322.

Rajkowska, G., Goldman-Rakic, P.S., 1995b. Cytoarchitectonic definition of prefrontal areas in the normal human cortex: II. Variability in locations of areas 9 and 46 and relationship to the Talairach coordinate system. Cerebral Cortex 5, 323-337.

Rakic, P., 1995. Radial versus tangential migration of neuronal clones in the developing cerebral-cortex. PNAS 92, 11323-11327.

Rempel-Clower, N.L., Barbas, H., 2000. The laminar pattern of connections between prefrontal and anterior temporal cortices in the Rhesus monkey is related to cortical structure and function. Cerebral Cortex 10, 851-861.

Rezai, K., Andreasen, N.C., Alliger, R., Cohen, G., Swayze, V., Oleary, D.S., 1993. The neuropsychology of the prefrontal cortex. Archives of Neurology 50, 636-642.

Rilling, J., Insel, T., 1999. The primate neocortex in comparative perspective using magnetic resonance imaging. Journal of Human Evolution 37, 191-223.

Rilling, J.K., Seligman, R.A., 2002. A quantitative morphometric comparative analysis of the primate temporal lobe. Journal of Human Evolution 42, 505-533.

Rolls, E.T., 2000. The orbitofrontal cortex and reward. Cerebral Cortex 10, 84-294.

Rosabal, F., 1967. Cytoarchitecture of the frontal lobe of the squirrel monkey. Journal of Comparative Neurolology 130, 87-108.

Rosvold, H., Enger, A., Szwarcbart, I.G., 1964. Neural structures involved in delayed response performance. In: Warren, J.M., Akert, K., (Eds.), The frontal granular cortex and behavior. New York: McGraw-Hill Book Co. pp. 1-15.

Russon, A.E., 2000. Orangutans: Wizards of the Rain Forest. Buffalo: Firefly Books Ltd.

Russon, A.E., 2002. Return of the native: cognition and site specific expertise in orangutan rehabilitation. International Journal of Primatology 23, 461-478.

Sah, P., Faber, E.S.L., De Armentia, M.L., Power, J., 2003. The amygdaloid complex: anatomy and physiology. Psychological Review 83, 803-834.

Sanides, F., 1964. Structure and function of the human frontal lobe. Neuropsychology 2, 209-219.

Sander, K., Scheich, H., 2005. Left auditory cortex and amygdala, but right insula dominance for human laughing and crying. Journal of Cognitive Neuroscience 17, 1519-1531.

Schenker, N., Desgouttes, A.M., Semendeferi, K., 2005. Neural connectivity and cortical substrates of cognition in hominoids. Journal of Human Evolution 49, 547-569.

Schenker, N.M., Buxhoeveden, D.P., Amunts, K., Zilles, K., Semendeferi, K., 2008. A comparative quantitative analysis of cytoarchitectonics and minicolumnar organization in Broca's area in humans and great apes. Journal of Comparative Neurolology 510, 117-128.

Scheperjans, F., Hermann, K., Eickhoff, S.B., Amunts, K., Schleicher, A., Zilles, K., 2008. Observer-independent cytoarchitectonic mapping of the human superior parietal cortex. Cerebral Cortex 18, 846-867.

Schleicher, A., Zilles, K., 1989. A quantitative approach to cytoarchitectonics: analysis of structural inhomogeneities in nervous tissue using an image analyzer. Journal of Microscopy 157, 367-381.

Schoenbaum, G., Setlow, B., 2001. Integrating orbitofrontal cortex into prefrontal theory: common processing themes across species and subdivisions. Learning and Memory 8, 134-147.

Schumann, C., Amaral, D.G., 2005. Stereological estimation of the number of neurons in the human amygdaloid complex. Journal of Comparative Neurology 491, 320-329.

Schumann, C., Hamstra, J., Goodlin-Jones, B.L., Lotspeich, L., Kwon, H., Buonocore, M.H., Lammers, C.R., Reiss, A.L., Amaral, D.G., 2004. The amygdala is enlarged in children but not adolescents with autism; the hippocampus is enlarged at all ages. Journal of Neuroscience 24, 6392-6401

Semendeferi, K., 1994. Evolution of the hominoid prefrontal cortex: a quantitative and image analysis of area 13 and 10. Ph.D. Dissertation, University of Iowa.

Semendeferi, K., Armstrong, E., Schleicher, A., Zilles, K., Van Hoesen, G.W., 1998. Limbic frontal cortex in hominoids: a comparative study of area 13. American Journal of Physical Anthropology 106, 129-155.

Semendeferi, K., Armstrong, E., Schleicher, A., Zilles, K., Van Hoesen, G.W., 2001. Prefrontal cortex in humans and apes: a comparative study of area 10. American Journal of Physical Anthropology 114, 224-241.

Semendeferi, K., Damasio, H., 2000. The brain and its main anatomical subdivisions in living hominoids using magnetic resonance imaging. Journal of Human Evolution 38, 317-332.

Semendeferi, K., Damasio, H., Frank, R., Van Hoesen, G.W., 1997. The evolution of the frontal lobes: a volumetric analysis based on three-dimensional reconstructions of magnetic resonance scans of human and ape brains. Journal of Human Evolution 32, 375-388.

Semendeferi, K., Lu, A., Schenker, N., Damasio, H., 2002. Humans and great apes share a large frontal cortex. Nature Neuroscience 5, 272-276.

Sherwood, C., Stimpson, C.D., Raghanti, M.A., Wildeman, D.E., Uddin, M., Grossman, L.I., Goodman, M., Redmond, J.C., Bonar, C.J., Erwin, J.M., Hof, P.R., 2006. Evolution of increased glia-neuron ratios in the human frontal cortex. PNAS 103, 13606-13611.

Singer, T., Kiebel,, S.J., Winston, J.S., Dolan, R., Frith, C., 2004. Brain responses to the acquired moral status of faces. Neuron 41, 653-662.

Squire, L.R., Stark, C.E.L., Clark, R.E. 2004. The medial temporal lobe. Annual Review of Neuroscience 27, 279-306.

Stefanacci, L., Amaral, D.G., 2002. Some observations on cortical inputs to the macaque monkey amygdala: An anterograde tracing study. Journal of Comparative Neurology 451, 301-323.

Stephan, H., Andy, O.J., 1977. Quantitative comparisons of the amygdala in insectivores and primates. Acta Anatomica (Basel) 98, 130-153.

Stephan, H., Frahm, H.D., Baron, G., 1987. Comparison of brain structure volumes in Insectivora and primates: VII. Amygdaloid components. Journal für Hirnforschung 28, 571-584.

Stepniewska, I., Preuss, T.M., Kaas, J.H., 1993. Architectonics, somatotopic organization, and ipsilateral cortical connections of the primary motor area (M1) of owl monkeys. Journal of Comparative Neurology 330, 238-271.

Stone, V.E., Baron-Cohen, S., Knight, R.T., 1998. Frontal lobe contributions to theory of mind. Journal of Cognitive Neuroscience 10, 640-656.

Stuss, D.T., Benson, D.F., 1984. Neuropsychological studies of the frontal lobes. Psychological Bulletin 95, 3-28.

Stuss, D.T., Gallup, G.G., Alexander, M.P., 2001. The frontal lobes are necessary for 'theory of mind.' Brain 124, 279-286.

Tanji, J., Mushiake, H., 1996. Comparison of neuronal activity in the supplementary motor area and primary motor cortex. Cognitive Brain Research 3, 143-150.

Uylings, H.B.M., van Eden, C.G., 1990. Qualitative and quantitative comparison of the prefrontal cortex in rat and in primates, including humans. Progress in Brain Research 85, 31-62.

Van Schaik, C.P., Van Hooff, J.A.R.A.M., 1996. Toward an understanding of the orangutan's social system. In: McGrew, W.C., Marchant, L.F., Nishida, T., (Eds.), Great Ape Societies. Cambridge: Cambridge University Press, pp. 3-15.

von Bonin, G., 1948. The frontal lobe of primates: cytoarchitectural studies. Res Publ Ass nerv ment Dis 27, 67-83.

von Bonin, G., Bailey, P., 1947. The neocortex of Macaca mulatta. Urbana: University of Illinois Press.

von Economo, C., 1929. The Cytoarchitectonics of the Human Cerebral Cortex. London: Oxford University Press.

von Economo, C., Koskinas, G.N., 1925. Die Cytoarchitectonik der Hirnrinde des Erwachsenen Menschen. Berlin: Verlag von Julius Springer.

Walker, A.E., 1940. A cytoarchitectural study of the prefrontal area of the macaque monkey. Journal of Comparative Neurology 73, 59-86.

Watanabe-Sawaguchi, K., Kubota, K., Arikuni, T., 1991. Cytoarchitecture and intrafrontal connections of the frontal cortex of the brain of the hamadryas baboon (Papio hamadryas). Journal of Comparative Neurology 311, 108-133.

Watson, K.K., Matthews, B.J., Allman, J.M., 2007. Brain activation during sight gags and language-dependent humor. Cerebral Cortex 17, 314-324.

Whiten, A., Byrne, R., 1988. The Machiavellian intelligence hypothesis. In: Byrne, R., Whiten, A., (Eds.), Machiavellian intelligence. New York: Oxford Univ Press. pp. 50-65.

Yakovlev, P., LeCours, A., 1967. The myelogenetic cycles of regional maturation of the brain. In: Minkowski, A., (Ed.), Regional Development of the Brain in Early Life. Oxford: Blackwell. pp. 3-65.

Yang, T.T., Menon, V., Eliez, S., Blasey, C., White, C.D., Reid, A.J., Gotlib, I.H., Reiss, A.L., 2002. Amygdalar activation associate with positive and negative facial expressions. NeuroReport 13, 1737-1741.

Zhang, K., Sejnowski, T.J., 2000. A universal scaling law between gray matter and white matter of cerebral cortex. Proceedings of the National Academy of Sciences USA 97, 5621-5626.

Zilles, K., Armstrong, E., Gottfried, S., Schleicher, A., 1986. Quantitative cytoarchitectonics of the posterior cingulate cortex in primates. Journal of Comparative Neurology 253, 514-524.

Zilles, K., Schlaug, G., Matelli, M., Luppino, G., Schleicher, A., Qu, M., Dabringhaus, A., Seitz, R., Roland, P.E., 1995. Mapping of human and macaque sensorimotor areas by integrating architectonic, transmitter receptor, MRI and PET data. Journal of Anatomy 187, 515-537.

Zilles, K., Stephan, H., Schleicher, A., 1982. Quantitative cytoarchitectonics of the cerebral cortices of several prosimian species. In: Armstrong, E., Falk, D., (Eds.), Primate Brain Evolution: Methods and Concepts. New York: Plenum Press. pp. 177-202.

CHAPTER 10

Searching for Human Brain Specializations with Structural and Functional Neuroimaging

James K. Rilling

Abstract

The comparative study of living primate brains is one method for elucidating the neurobiological changes that evolved to support human cognitive specializations. We have been using non-invasive neuroimaging to compare brain structure and function in humans, chimpanzees and rhesus macaques. Specifically, we have used diffusion tensor imaging (DTI) to compare the size and trajectory of the arcuate fasciculus fiber tract to provide insights into the evolution of the neural substrates supporting human language. Results suggest that the human arcuate fasiculus is considerably larger and has more widespread projections to both temporal and frontal cortex than the arcuate fasiculus of either chimpanzees or macaques. We have also used [^{18}F]-fluorodeoxyglucose Positron Emission Tomography (PET) imaging to compare resting brain glucose metabolism in humans and chimpanzees in order to shed light on resting state cognition in the two species. Results show that like humans, chimpanzees show high levels of activity in a network of areas implicated in mental self-projection. Humans, but not chimpanzees, also show strongly left-lateralized activity in cortical areas involved in language and conceptual processing. These results imply both similarity and difference in resting state cognition between the two species. Comparative primate neuroimaging is one of many available tools that will help us to flesh out the specifics of Professor Holloway's early recognition that brain reorganization was a critical component of the evolution of the human brain and mind.

Introduction

Ralph Holloway was instrumental in demonstrating that hominin cranial capacity approximately tripled in size over the past 3 million years (Holloway 1970; Holloway 1973; Holloway 2000; Holloway et al. 2008). This fact is undoubtedly of importance in explaining the evolution of human intelligence. However, from very early in his career, Holloway emphasized that human evolution was also characterized by fundamental reorganization of the brain (Holloway 1968). That is, the human brain is not just a scaled up version of an ape brain, rather it is qualitatively different. To support this claim, he pointed to evidence that human microcephalics with chimpanzee sized brains were capable of behavior patterns that were often more human-like than pongid-like. He also famously observed that the anterior border of primary visual cortex (as estimated by the lunate sulcus) was in a human-like as opposed to a chimpanzee-like position in early australopithecine endocasts that had cranial capacities comparable to living chimpanzees (Holloway 1983; Holloway 1985; Holloway et al. 2003; Holloway and Kimbel 1986). Thus, reorganization appeared to have preceeded large scale encephalization in the hominin lineage. Finally, he turned to the comparative study of human and living non-human primate brains for indirect evidence of brain reorganization. For example, he showed that human primary visual cortex is significantly smaller than one would expect for a typical primate of human brain size (Holloway 1992). This raised the possibility that the adjacent posterior parietal association cortex enlarged disproportionately in human evolution, a prediction that has received support in subsequent research (Orban et al. 2006; Orban et al.

2004). The comparative study of living brains enables researchers to investigate the evolution of a wide variety of neurobiological traits that are not preserved in the fossil record, and has the potential to dramatically expand our knowledge of human brain evolution.

In our own work, we are using non-invasive neuroimaging techniques to compare brain structure and function in humans and non-human primates. We particularly emphasize the human-chimpanzee comparison, given that chimpanzees are our closest living primate relative and that, without chimpanzee data, it is not possible to make inferences about human brain specializations or human brain evolution. In this paper, we will discuss two of our most recent comparative neuroimaging studies. The first uses diffusion tensor imaging (DTI) to investigate the evolution of the neural substrates supporting language by comparing white matter fiber tracts involved in human language with their homologues in chimpanzees and rhesus macaques. The second uses Positron Emission Tomography (PET) imaging to investigate the evolution of resting-state cognition by comparing resting state brain activity in humans and chimpanzees.

Evolution of the neural substrates supporting language

Among the most distinguishing features of the human species is our capacity for language. How and why language evolved in humans is one of the greatest mysteries in anthropology. The human brain must have been significantly modified to support this highly specialized and complex skill.

How does the human brain process and produce language? The classic model, as summarized by Geschwind (Geschwind 1970), postulates that there is a region of cerebral cortex in the left posterior superior temporal gyrus, Wernicke's area, that is responsible for speech comprehension, and a region in the left inferior frontal cortex, Broca's area, that is responsible for speech production (figure 1). Broca's area encompasses two gyri, pars opercularis (BA 44) posteriorly and pars triangularis (BA 45) anteriorly. The model further postulates that Wernicke's and Broca's areas are linked by a white matter fiber tract known as the arcuate fasciculus that originates in Wernicke's area and curves around the sylvian fissure to project to Broca's area. The putative function of this tract is to convey information from Wernicke's to Broca's area, for example during the repetition of spoken language.

However, recent evidence from functional neuroimaging studies as well as from brain damaged patients suggests that cortical areas involved in language extend well beyond Wernicke's and Broca's areas. For example, virtually the entire surface of the left temporal lobe is involved in either phonetic or lexical-semantic processing (Damasio et al. 1996; Hickok and Poeppel 2004; Price 2000; Sakai 2005; Vigneau et al. 2006). Still, Wernicke's

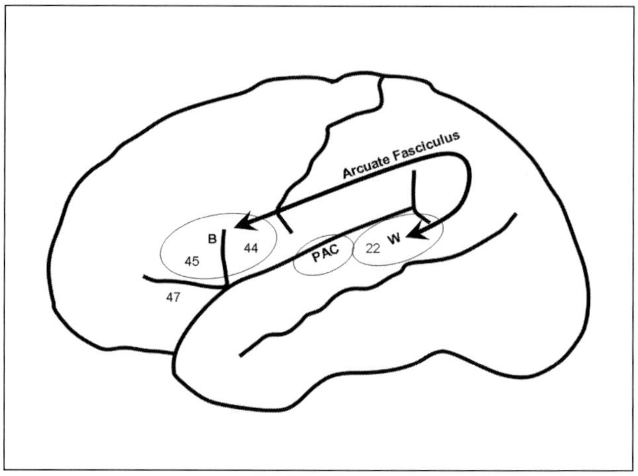

Figure 1: Classic model of human brain language processing as proposed by Geschwind (1970). PAC = primary auditory cortex, W = Wernicke's area, B = Broca's area.

and Broca's areas remain critical nodes in the network supporting language, so we might ask whether there is evidence that homologues of these brain regions exist in non-human primates. Indeed, based on location, cytoarchitecture and shared non-linguistic functional properties, putative homologues to Wernicke's and Broca's areas have been identified in macaques (Preuss 2004) (figure 2). But are Wernicke's and Broca's homologues connected in non-human primates as they are in humans? Studies using neuronal tracer injections suggest that the arcuate fasciculus of macaque monkeys links posterior STG (Wernicke's area homologue) with posterior dorsolateral prefrontal cortex, rather than Broca's area homologue in the inferior frontal cortex (Petrides and Pandya 2002) (figure 3). These findings suggest that there may be differences in the trajectory of the arcuate fasciculus between humans and macaques. However, the arcuate fasciculus has not yet been compared in humans and nonhuman primates using the same method. Moreover, the it has not been explored in our closest living primate relative, the chimpanzee.

The recent advent of diffusion tensor imaging (DTI), which can track white-matter pathways noninvasively, makes it possible to compare patterns of connectivity in humans and non-human primates. Although standard MRI protocols can image white matter, they do not permit identification of specific fiber tracts within white matter. DTI, however, enables tracking and identification of fiber pathways (Basser and Jones 2002; Mori and Van Zijl 2002). DTI measures the direction and magnitude of water diffusion in brain tissue. Within white matter, water will preferentially diffuse parallel to axons that compose fiber tracts because the myelin that coats the axons is hydrophobic and restricts diffusion perpendicular to the direction of the axon. Thus, in white matter, water diffusion is highly directional. On the other hand, in gray matter, diffusion is less restricted. For each

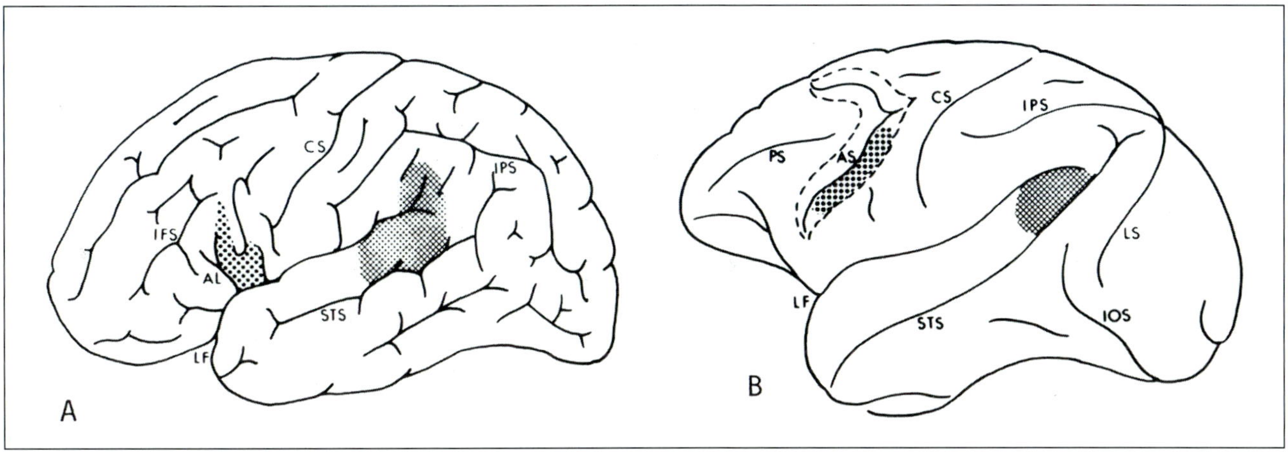

Figure 2: Wernicke's and Broca's areas in humans (left) and their putative homologues in macaques (right). From figure 5 (p.212) in: Galabdura AM, and Pandya DN. 1982. Role of architectonics and connections in the study of primate brain evolution. In: Falk EAaD, editor. Primate Brain Evolution: Methods and Concepts. New York: Plenum Press. p 203-216.

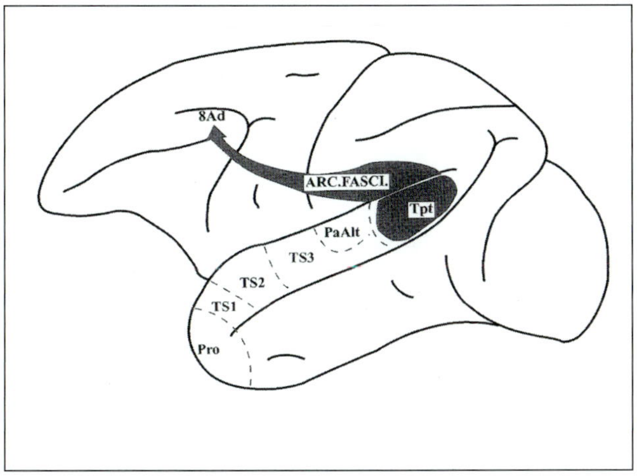

Figure 3: Trajectory of macaque arcuate fasciculus based on neuronal tracer study. From figure 3-8 (p.38) in: Petrides M, and Pandya DN. 2002. Association pathways of the prefrontal cortex and functional observations. In: Stuss DT, and Knight RT, editors. Principles of Frontal Lobe Function. New York: Oxford University Press. p 31-50.

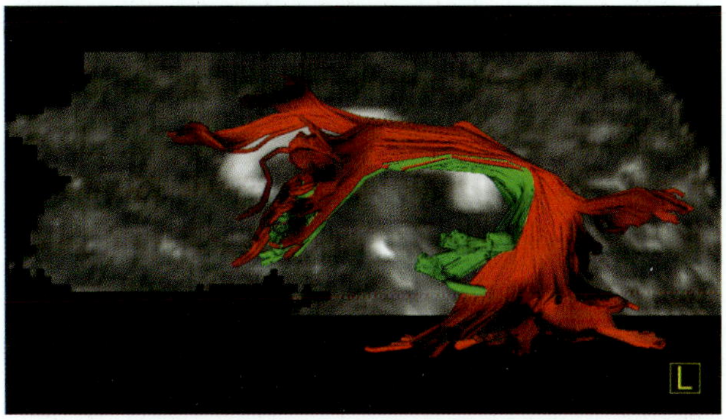

Figure 4: Reconstruction of arcuate fasciculus in human brain using diffusion tensor imaging (DTI) and a deterministic tractography algorithm. Green portion of the tract terminates in the superior temporal gyrus. The red portion terminates in the middle and inferior temporal gyri, as well as the angular gyrus. A mid-sagittal non diffusion-weighted image is shown in the background.

brain voxel of an image, a diffusion tensor can be calculated that describes the direction and magnitude of water diffusion. Tractography algorithms can then use the information from these diffusion tensors to reconstruct fiber tracts in the brain (figure 4) (Basser and Jones 2002; Behrens et al. 2003; Mori and Van Zijl 2002).

We acquired DTI brain scans from human, chimpanzee and rhesus macaque brains in order to compare the size and trajectory of the arcuate fasciculus across these three species using the same method (Rilling et al. 2008). Specifically, scans were acquired from ten live human subjects, three postmortem chimpanzee brains, one live chimpanzee subject, two postmortem macaque brains and one live macaque subject. Protocol parameters for each scan are listed in table 1.

The principal direction of water diffusion in each voxel of a diffusion tensor image can be represented with colors (figure 5). Typically, red, green and blue are used to represent diffusion in the x (medial-lateral), y (anterior-posterior) and z (superior-inferior) directions, respectively. Therefore, for example, voxels in the corpus callosum that carry fibers passing from one cerebral hemisphere to the other through the midline of the brain, are colored red. On the other hand, voxels in the posterior limb of the internal capsule, that carry fibers projecting from motor cortex to the spinal cord, are colored blue. The arcuate fasiculus is one of the largest fiber tracts in the human brain and can be easily visualized in parasagittal sections of a principle diffusion direction color map. The dorsal portion of the arcuate, which travels in an anterior-posterior direction, as indicated by its green color, transitions into blue where the pathway descends into the temporal lobe, and turns green again as it moves anteriorly in the temporal lobe. This is the situation in the human brain. However, in chimpanzees, a small region of red (medio-laterally directed fibers) interrupts the transition from green to blue in the hook of the arcuate. In macaques, the red area is considerably expanded, and the color map in the region of the arcuate bears little resemblance to human or chimpanzee color maps. Thus, only in the human brain is a continuous uninterrupted arcuate pathway evident in the color map of the principal diffusion direction. It is possible, however, that in chimpanzees, at least, the arcuate actually does pass into the temporal lobe, but that this pathway is not the dominant pathway in the region of the hook of the arcuate. Standard tractography algorithms, which consider only the principal diffusion direction, cannot follow it through a region where it intermingles with a larger, medio-laterally oriented pathway. For this reason, we utilized a newly developed algorithm designed to track through crossing fibers by also considering the secondary diffusion direction (Behrens et al. 2007).

We used this technique to track the arcuate fasciculus along with two additional pathways that convey fibers between frontal and parietal-temporal cortex, the superior longitudinal fasciculus and the extreme capsule. These pathways can be clearly identified in a coronal section through the color map at the level of the precentral sulcus (figure 6a). In all three species, we tracked between a coronal region of interest (ROI) that encompassed these three pathways and an ROI in the white matter underlying the superior, middle and inferior temporal gyri, as well as the inferior parietal lobule (figure 6b).

Below, we first describe the tractography results, and then interpret them and discuss their significance.

Tractography Results (see figure 7 and 8)

Macaque tractography revealed posterior terminations in posterior superior temporal gyrus (STG, 22) and anterior inferior parietal cortex (area 7a). Anteriorly, terminations were found in the frontal operculum, insular cortex and the inferolateral margin of the frontal lobe (area 6), including the extreme ventral aspects of areas 44 and 45 in the arcuate sulcus (figure 7). The pathway of highest probability ran deep to the insula in the vicinity of the extreme capsule and projects most strongly to area 45. Weaker pathways ran both dorsal and lateral to the insula (figure 8). The dorsal pathway was in the location of SLFII and the arcuate fasciculus, and the lateral pathway was in the location of SLFIII (Petrides and Pandya 2006). Thus, these DTI results are compatible with tracer studies that found extreme capsule projections from posterior superior temporal gyrus to area 45 and SLFIII projections from area PF to area 44 (Petrides and Pandya 2002), as well as a DTI study showing that the extreme capsule and SLF pathways projected with highest probability to areas 45 and 44, respectively (Croxson et al. 2005). Tractography also revealed projections to dorsolateral prefrontal cortex (DLPFC), but terminations are only observed with lower thresholds (see methods).

Chimpanzee tractography revealed posterior terminations in the posterior superior temporal gyrus (STG, 22), the supramarginal gyrus (SMG, 40) and the angular gyrus (AG, 39), with minimal connectivity to the superior temporal sulcus and the middle temporal gyrus (MTG). Anteriorly, the pathway projected with high probability to the inferolateral margin of the frontal lobe (ventral 6, 44), extending into the ventral most aspect of pars opercularis (possible 44) and the cortex just rostral to the fronto-orbital sulcus (possible 44 or 45) (Sherwood et al. 2003). Connections also reached dorsolateral prefrontal and dorsal premotor cortex, specifically the superior (6) and middle frontal gyri (8, 46) (figure 7). Terminations were also found in insular cortex as well as the frontal operculum. In contrast to macaques, the pathway that runs dorsal to the insula is stronger than the extreme capsule pathway running deep to the insula (figure 8). In chimpanzees, this dorsal pathway was dominated by connections with the inferior parietal lobe, including both SMG and AG.

In humans, tractography results revealed posterior terminations in posterior superior temporal gyrus (STG, BA 22), middle temporal gyrus (MTG, BA 21 and 37), inferior temporal gyrus (ITG, BA 20), as well as the angular (BA 39) and supramarginal gyri (BA 40) of the parietal lobe. Anteriorly, the pathways reached the insular cortex, frontal operculum, pars opercularis (BA 44), pars triangularis (BA 45), pars orbitalis (BA 47) and the inferior frontal gyrus (BA 46 and 10) rostral to pars triangularis (figure 7). There was also a small projection to dorsolateral prefrontal cortex, dorsal to the inferior frontal sulcus. As with chimps, the dorsal pathway is dominant to the extreme capsule pathway, but in humans the temporal projections from the arcuate fasciculus make a much greater contribution to the dorsal pathway (figure 8).

Two noteworthy asymmetries were observed in humans. Angular gyrus terminations were stronger in the right hemisphere, whereas temporal lobe terminations were stronger and more widespread in the left hemisphere, particularly within the middle temporal gyrus (figure 7). Limited sample sizes in chimpanzees and macaques preclude conclusions about the presence or absence of asymmetries in these species.

Our results show that in macaques, the strongest link between auditory cortex in the STG and frontal cortex is via the more ventral extreme capsule pathway. This pathway has been implicated in auditory object recognition, analogous to the role of the ventral visual stream in visual object recognition (Petrides and Pandya 2002; Romanski et al. 1999). Thus, the pathway may be involved in processing the identity of an object based on its sound, and it is particularly relevant that cells in area 45, where the pathway terminates, respond to monkey vocalizations (Romanski et al. 2005). This pathway may

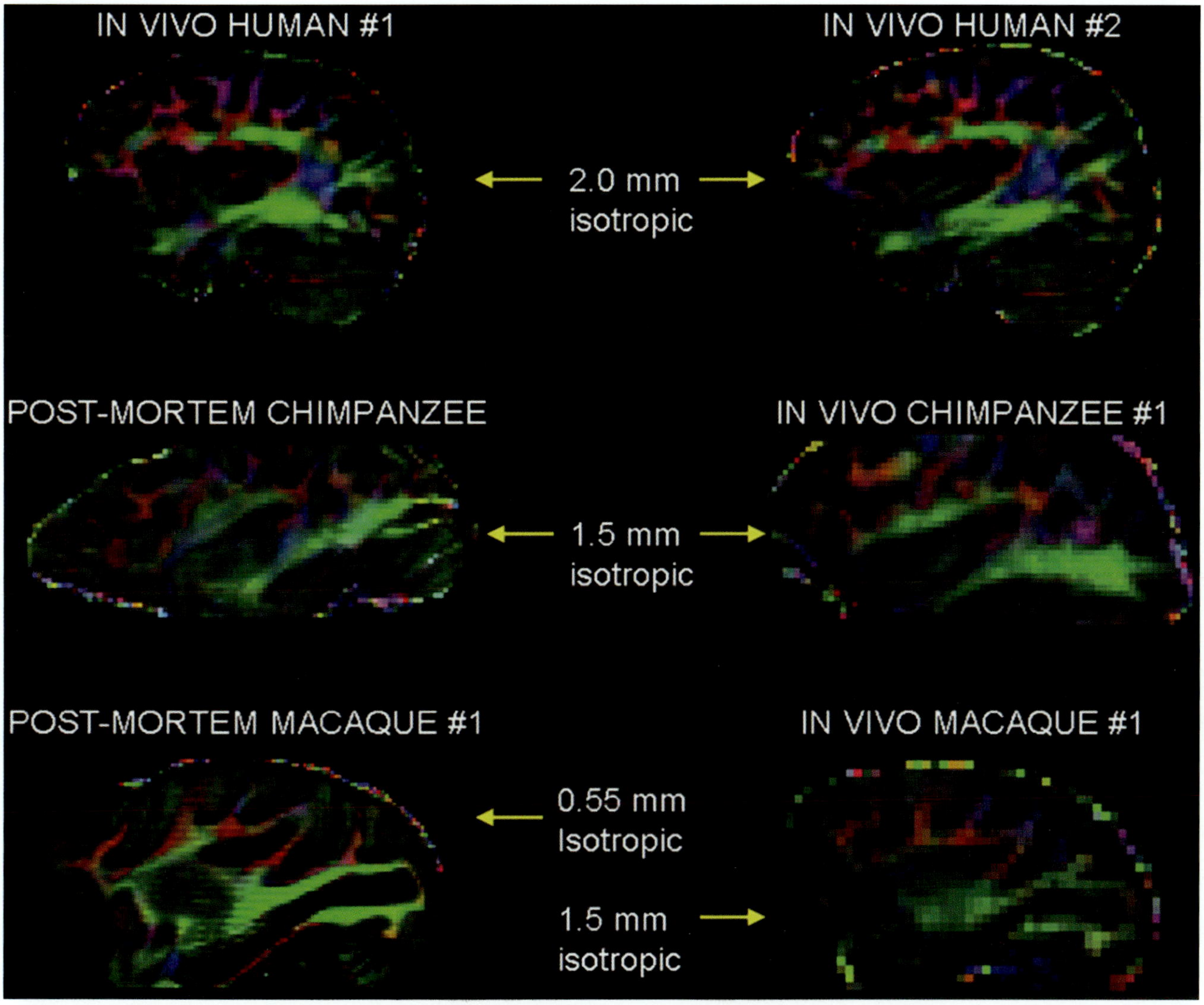

Figure 5: DTI color maps of the principle diffusion direction in humans, chimpanzees and rhesus macaques. Parasagittal sections through the arcuate fasciculus are shown for all three species.

be involved in identifying a caller.

Relative to macaques, chimpanzees have a much stronger dorsal pathway that projects with high probability to the inferior parietal lobe. This pathway likely includes SLFII and SLFIII. Although the function of this pathway has not been investigated in chimpanzees, it is of interest that in humans, a network consisting of inferior parietal and inferior frontal cortices is involved in self-recognition and self-awareness, as well as action understanding through simulation (Uddin et al. 2005). The strong dorsal pathway of chimpanzees could therefore provide part of the substrate of mirror self-recognition, a capacity they share with humans but not macaques(Gallup 1970; Povinelli et al. 1997). Given the pathway's role in understanding the actions of others, perhaps via simulation, it might also help to explain the greater sophistication of chimpanzee social cognition compared with macaques.

Humans differ from chimpanzees and macaques in having much stronger terminations posteriorly in the middle temporal gyrus, as well as stronger terminations anteriorly in pars opercularis and pars triangularis, particularly in their more dorsal aspects. Also, in humans terminations extend further anteriorly into BA 46 and even area 10. Humans also differ in having terminations in pars orbitalis (BA 47). What are the specific functions of these regions of expanded connectivity in humans? Substantial evidence indicates that the middle temporal and angular gyri are involved in lexical-semantic processing (Price 2000), and that pars triangularis (BA 45) and pars orbitalis (BA 47) are involved in syntactic processes of sentence comprehension (Sakai 2005). To explore whether these two regions involved in higher aspects of linguistic processing were specifically connected with one another, we quantified the probability of connectivity between MTG/AG and pars opercularis (BA 44) on the one hand, and between MTG/AG and pars triangularis and orbitalis combined (BA 45 and 47)

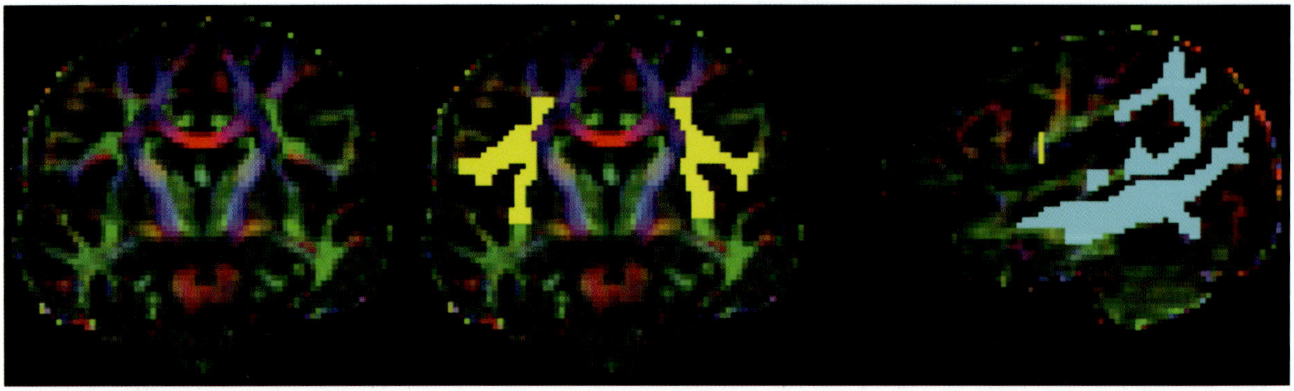

Figure 6: Tractography method for a human brain, illustrating a) anterior coronal ROI, with (right) and without (left) mask, b) parasagittal section showing posterior ROI in white matter of temporal and parietal lobes (blue), along with anterior ROI (yellow). Probabilistic tractography was used to track between these two ROIs in each scan for each species.

Species	Status	sequence	b value	diff. directions	# averages	duration	voxel size
Human	in vivo	EPI	1000	60	3	45 minutes	2.0 isotropic
Chimpanzee	in vivo	segmented EPI	1000	60	5	115 minutes	1.5 isotropic
Chimpanzee	post-mortem	spin echo	4500	60	2	24 hours	1.5 isotropic
Macaque	in vivo	segmented EPI	1000	30	4	32 minutes	1.5 isotropic
Macaque	post-mortem	spin echo	2000	60	3	72 hours	0.55 isotropic

Table 1

on the other hand. In both hemispheres, MTG/AG had a higher probability of connectivity with pars triangularis and pars orbitalis, suggesting that the expanded pathway in humans supports the transmission of lexical-semantic information stored in MTG/AG to pars triangularis for sentence comprehension. Thus, in contrast to macaques in which the predominant pathway from auditory responsive cortex in the temporal lobes to the frontal lobe travels ventrally and conveys information about object identity, in humans the predominant pathway travels dorsally via the arcuate fasciculus and conveys information about the conceptual and semantic meaning of what is heard.

This observation of an expanded projection from MTG/AG to lateral inferior frontal cortex in humans is consistent with other comparative evidence. Human temporal lobes are significantly larger than predicted for a primate of human brain size, and the difference is most pronounced within the white matter of the temporal lobes (Rilling and Seligman 2002), particularly the gyral white matter as opposed to the core white matter (Schenker et al. 2005). Furthermore, visual cortical areas in humans are in a more posterior and ventral location compared with visual cortical areas in macaques, perhaps to accommodate expansion of language cortex on the lateral surface of the left temporal lobe (Orban et al. 2004; Preuss 2004; Ungerleider et al. 1998). Although there is some debate, considerable evidence also suggests that prefrontal cortex is disproportionately large in humans (Avants et al. 2006; Brodmann 1912; Deacon 1997; Passingham 1973; Preuss 2004; Rilling and Insel 1999; Schoenemann et al. 2005; Semendeferi et al. 2002; Sherwood et al. 2005), again particularly in the gyral white matter (Schenker et al. 2005).

In humans, angular gyrus terminations were found to be stronger in the right hemisphere, an asymmetry that could relate to right hemispheric specialization for self-recognition (Uddin et al. 2005), theory of mind (Saxe and Wexler 2005), or visuospatial attention (Mort et al. 2003). Humans also exhibited a leftward asymmetry in the connection probability and spatial extent of terminations in the middle temporal gyrus. This result is consistent with functional imaging evidence suggesting that lexical-semantic processing is lateralized to the left middle temporal and angular gyri (Price 2000), and with previous studies reporting leftward asymmetries in the human arcuate fasciculus as a whole (Glasser and Rilling 2008; Nucifora et al. 2005; Powell et al. 2006).

Thus, we observe human-specific differences within brain regions involved in the two domains of language believed to distinguish humans from non-human primates: symbolic(Deacon 1997) and syntactic (Hauser et al. 2002; Pinker 2000) processing. These significant modifications within language-related cortex challenge earlier suggestions that human language evolved as an incidental by-product of selection for general brain size enlargement (Gould 1991), instead suggesting that lexical-semantic and syntactic processing were specific

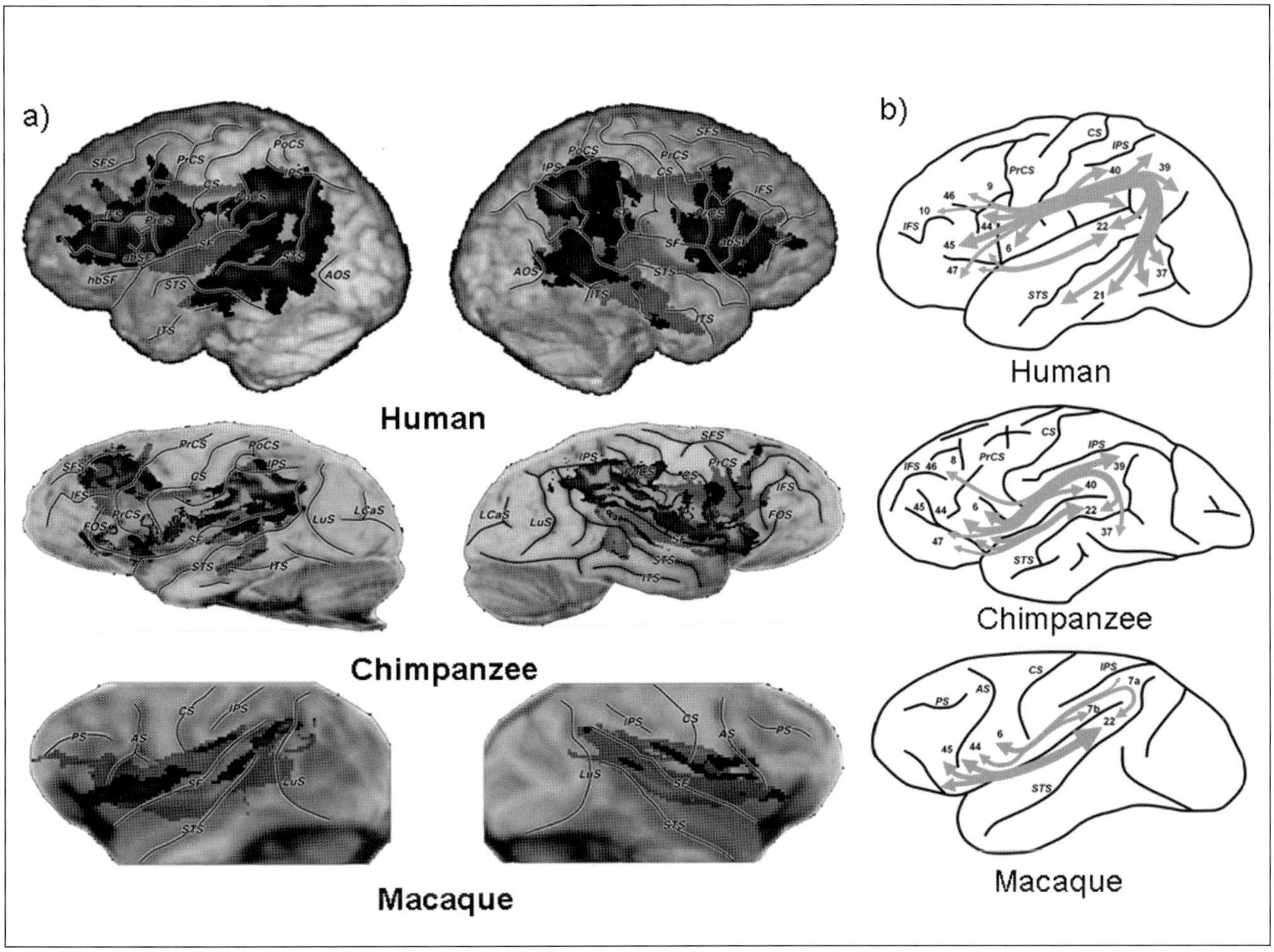

Figure 7: Tractography results. a) Average results for humans, chimpanzees and rhesus macaques, b) schematic summary of results. abSF, ascending branch of the Sylvian fissure; AOS, anterior occipital sulcus; AS, arcuate sulcus; CS, central sulcus; FOS, fronto-orbital sulcus; hbSF, horizontal branch of the Sylvian fissure; IFS, inferior frontal sulcus; IPS, intraparietal sulcus; ITS, inferior temporal sulcus; LCaS, lateral calcarine sulcus; LuS, lunate sulcus; PoCS, postcentral sulcus; PrCS, precentral sulcus; PS, principal sulcus; SF, Sylvian fissure; SFS, superior frontal sulcus; STS, superior temporal sulcus. From figure 2 in Rilling JK, Glasser MF, Preuss TM, Ma X, Zhao T, Hu X, and Behrens TE. 2008. The evolution of the arcuate fasciculus revealed with comparative DTI. Nat Neurosci 11(4):426-428.

targets of natural selection.

Comparison of resting state brain activity in humans and chimpanzees

One of the remarkable aspects of human cognition is our ability to mentally project ourselves into other times and places so that we are not limited to thinking about the immediate here and now (Buckner and Carroll 2007; Tulving 2005). In other words, we can simulate alternative worlds that are separate from the one being directly experienced. We can project ourselves into the past to remember things that have happened to us, into the future to formulate and rehearse plans, and even into the mind of others to understand their mental states (Buckner and Carroll 2007). How do they feel? What do they know?

Experimental evidence suggests that chimpanzees may also be capable of some degree of mental self-projection. For example, a capacity to project into the future is suggested by the fact that they will transport tools for future use (Mulcahy and Call 2006). However, others have argued that the ability to mentally travel into the past and future is unique to humans (Suddendorf and Corballis 1997; Tulving 2005). There has also been considerable debate as to whether chimpanzees can understand the mental states of others. Anecdotal evidence of deception in field studies raises the possibility that they can (Byrne and Whiten 1992); however, this has been difficult to definitively demonstrate in experimentally controlled laboratory studies (compare, e.g., (Hare et al. 2006; Povinelli et al. 2000)).

In humans, each of these forms of self-projection, remembering, prospection and theory of mind, seems to rely on a common neural network, consisting of medial prefrontal cortex as well as medial and lateral parietal cortex, and in many cases the hippocampus (figure 8). Interestingly, a very similar network, known as the default mode network, is tonically active at rest, that is,

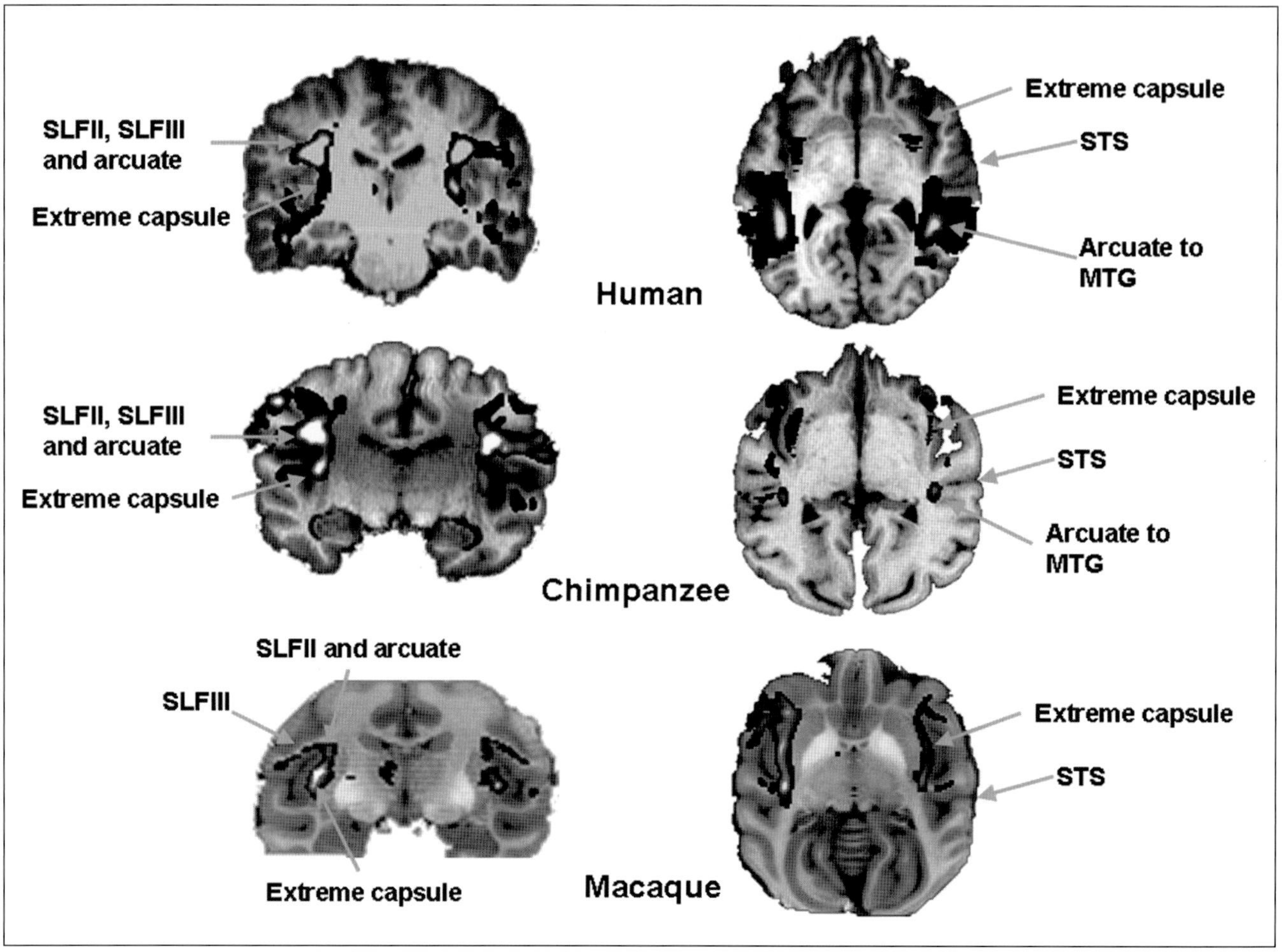

Figure 8: Two-dimensional tractography results. Coronal (a) and axial (b) sections from an individual human, chimpanzee and macaque, illustrating the relative strength of the dorsal and ventral pathways. SLFII and SLFIII, superior longitudinal fasciculus II and III. From figure 3 in Rilling JK, Glasser MF, Preuss TM, Ma X, Zhao T, Hu X, and Behrens TE. 2008. The evolution of the arcuate fasciculus revealed with comparative DTI. Nat Neurosci 11(4):426-428.

when subjects are lying awake in the scanner, but are not engaged in an attention demanding tasks. This observation suggests that people may engage in mental self-projection when resting (Buckner and Carroll 2007). Consistent with this hypothesis, subject self-reports suggest that much of this time is spent reflecting on past social interactions and planning or rehearsing future social interactions (Andreasen et al. 1995; Christoff et al. 2004; Ingvar 1979). These mental exercises may prove useful in clarifying the meaning of past interactions and practicing pending future interactions so they can be exercised more skillfully. Planning in the non-social domain would be similarly adaptive, for example, planning to save or store currently available food and water so that it can be used to survive a future drought. These abilities may be fundamental to the current and past success of our species.

To shed light on the question of whether chimpanzees are capable of mental self-projection, we used functional neuroimaging to define resting state brain activity in chimpanzees, and we compare these results with those of a human sample.

For this study, we used [^{18}F]-fluorodeoxyglucose PET ([^{18}F]-FDG PET) imaging, which makes it possible to image resting state brain activity in awake subjects outside the scanner. Adult humans (n=8) and adult chimpanzees (n=5) received a dose of [^{18}F]-FDG, a radioactively-labeled, chemically modified glucose molecule. After entering the bloodstream, [^{18}F]-FDG accumulates and becomes trapped in neurons at a rate proportional to their glucose metabolic rate (Phelps and Mazziotta 1985). During this extended period of cellular [^{18}F]-FDG uptake (~45 minutes in humans and 75 minutes in chimpanzees), human subjects rested quietly by themselves in a private room adjacent to the PET scanner, and chimpanzee subjects rested quietly in their home cages. Chimpanzee subjects were dosed in the late morning hours when they typically interact minimally with either their cagemates or the animal care staff. After the uptake period, subjects received a PET scan to image the distribution of [^{18}F]-FDG in the brain. Variation across the brain in the resulting image results from regional differences in glucose metabolism during the period of [^{18}F]-FDG uptake. Human subjects were scanned awake, whereas

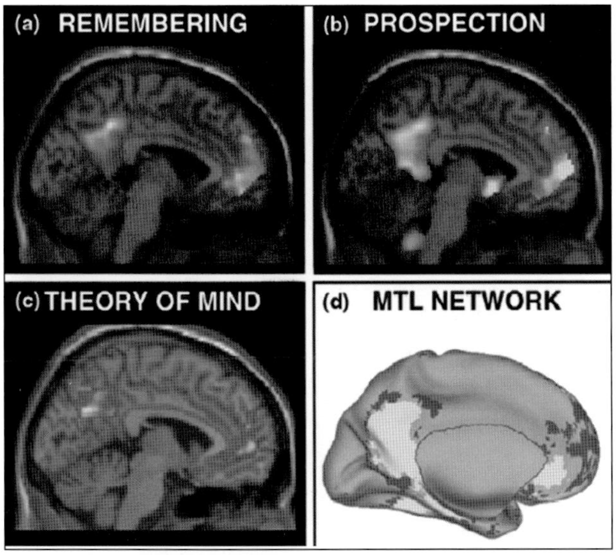

Figure 9: Common neural network activated during different types of mental self-projection, including remembering, prospection and theory of mind. Reproduced from: Figure 2 in Buckner RL, Carroll DC (2007): Self-projection and the brain. Trends Cogn Sci 11:49-57.

chimpanzee subjects were sedated and scanned in the anesthetized state. It is important to recognize, however, that since [^{18}F]-FDG uptake is largely complete prior to sedation, and since it leaves cells at a very slow rate, the resulting images reflect brain metabolism during the uptake period when the animal was awake, and not real-time activity in the anesthetized state.

Although the homologies of chimpanzee and human cortical areas have in some cases not been definitively established, we reasoned that if the chimpanzee pattern of resting brain activity differs substantially from that found in humans, it is unlikely that they are engaged in the same mental processes as humans are at rest. On the other hand, if chimpanzee and human patterns of activation are similar, one possible explanation is that there are similarities in their resting-state cognition.

Behavioral Results

Chimpanzee subjects were videotaped during the [^{18}F]-FDG uptake period to verify that we had attained a reasonable "resting state". An ethogram was used to quantify each subject's behavior during the uptake period. None of the five chimpanzees spent any time in physical contact with their cagemate during the uptake period. Subjects spent the overwhelming majority of their time lying down or sitting in what we characterized as a "neutral" state of attention, as opposed to "alert", "watching" or "moving" (see (Rilling et al. 2007) for further details).

Imaging Results

In both humans and chimpanzees, we identified the 5% most metabolically active voxels (figure 9). In humans, this included the classic default-mode regions, including dorsomedial prefrontal cortex, and medial and lateral parietal cortex. Humans also showed strongly left-laterlalized activity in lateral frontal, temporal and parietal cortices, as well as in several subcortical structures, including the striatum and thalamus. Finally, there was activity in visual cortex, as expected given that our subjects rested with their eyes open during the uptake period.

The lateralized activity we observed in left posterior temporal and inferior parietal areas is consistent with previous studies reporting this area to be more active at rest compared with various active task conditions (Binder et al. 1999; Shulman et al. 1997). In combination with left frontal lobe activity, these areas may form a conceptual processing network that is involved in semantic memory retrieval and its manipulation in working memory for the purposes of planning, organization and problem solving (Binder et al. 1999; Christoff et al. 2004; Shulman et al. 1997). The left-lateralized cortical activity overlaps extensively with the human brain language network (as discussed above), raising the prospect that, even in the resting state, humans can't help but think with words. Language is essential to human thought.

In sum, the pattern of brain activity observed in our human subjects is similar to that reported in previous resting state studies, and this pattern of activity is consistent with a resting state involving mental self-projection, conceptual and semantic processing, and inner speech.

Like humans, chimpanzees exhibited high levels of activity in default mode areas, including medial prefrontal cortex, as well as medial and lateral parietal cortex. If these regions have a similar function in humans and chimpanzees, then our results are consistent with the possibility that chimpanzees engage in mental self-projection in the resting state.

There were also some subtle differences between humans and chimpanzees in activity within the default mode network. Within medial prefrontal cortex, humans showed the highest level of activity in more dorsal areas, whereas chimpanzees showed more widespread activity, including activity in more ventral areas. Recently, it has been suggested that different subdivisions of medial prefrontal cortex are related to different aspects of mentalizing (Amodio and Frith 2006; Frith and Frith 2006; Mitchell et al. 2006), with more dorsal regions being involved with thinking about others' thoughts as well as person knowledge, and more ventral regions being involved with monitoring emotion in self and others or emotional processing more generally. Thus, it is possible that the chimpanzee resting state is imbued with a stronger emotional tone than the human resting state, perhaps including greater reflection on emotional states as opposed to thoughts. However, given that other studies have found high levels of activity within ventromedial PFC in human subjects (Raichle et al. 2001), it is possible that the lack of high levels of activity in this area in our human sample relates to differences in the exact

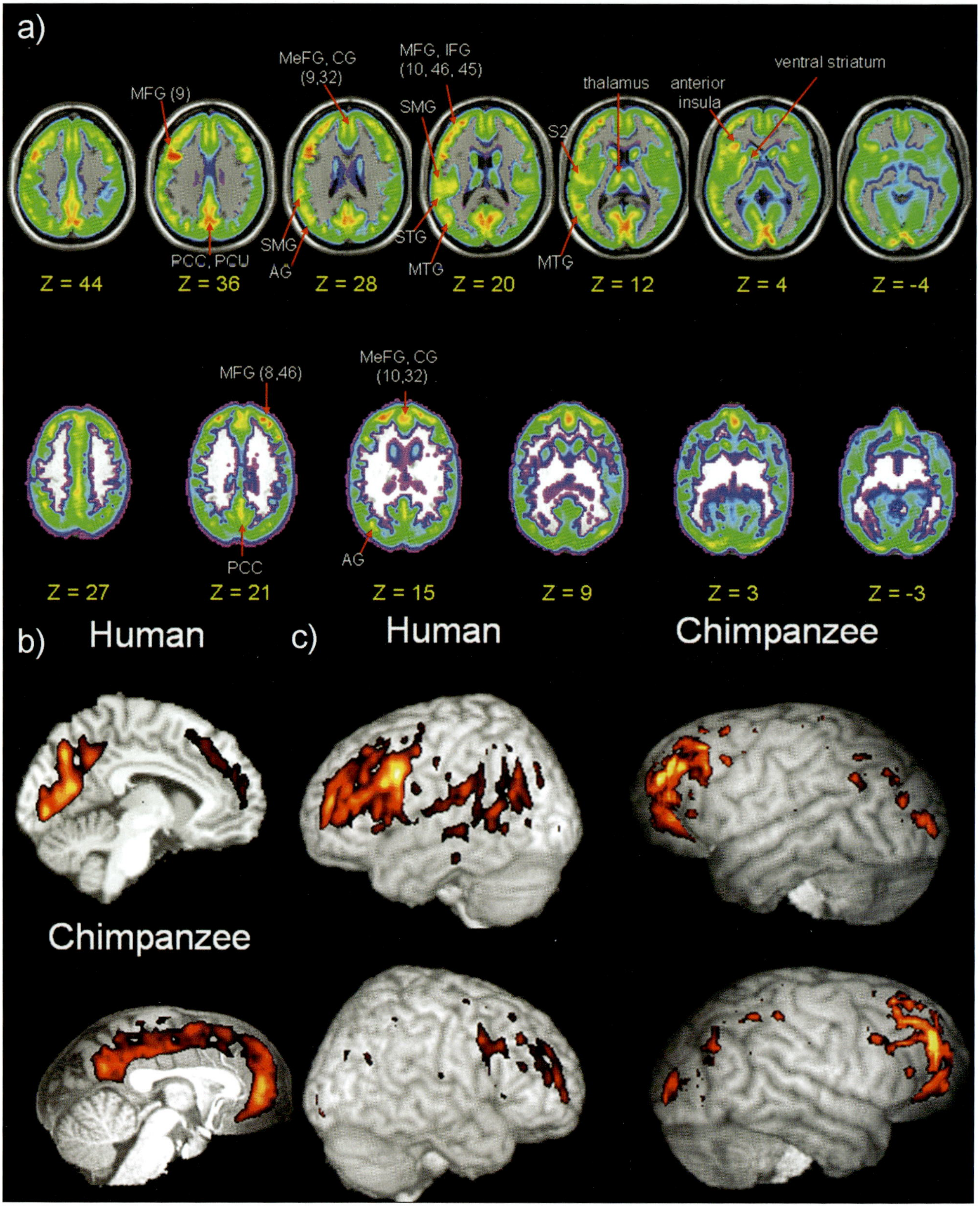

Figure 10: Resting-state brain activity in humans and chimpanzees. a) Average human (n=8) and chimpanzee (n=5) PET images in horizontal section. Regions of highest metabolic activity are colored yellow to red. The five percent most active voxels in each species are shown in b) midsagittal and c) lateral views. Modified from figures 1 and 2 in Rilling JK, Barks SK, Parr LA, Preuss TM, Faber TL, Pagnoni G, Bremner JD, and Votaw JR. 2007. A comparison of resting-state brain activity in humans and chimpanzees. Proc Natl Acad Sci U S A 104(43):17146-17151.

nature of the resting state condition rather than genuine species differences. For example, thinking about familiar and unfamiliar others has been localized to ventral and dorsal aspects of MPFC, respectively (Mitchell et al. 2006). The fact that chimpanzees, unlike the humans, were surrounded by familiar others during the [^{18}F]-FDG PET uptake period could explain the higher levels of activity ventrally in chimpanzee images.

Unlike humans, chimpanzees did not show left lateralized activity in frontal, temporal and parietal regions involved in language and conceptual processing. These results suggest that one major difference between humans and chimpanzees is that human resting state cognition is linked with language. The left lateralized areas that are active in humans but not chimpanzees have also been implicated more generally in conceptual processing involving semantic knowledge retrieval, representation in awareness, and directed manipulation of represented knowledge for organization, problem-solving and planning (Binder et al. 1999). Thus, organization, planning and problem-solving may be other aspects of resting state cognition that differentiate humans from chimpanzees.

In conclusion, our results imply some degree of commonality in resting state cognition between humans and chimpanzees, possibly including a tendency to mentally project oneself into other times, places or mental perspectives. However, left lateralized activity in humans that is absent in chimpanzees, may mean that humans are engaged in a greater degree of conceptual processing than chimpanzees at rest, and that humans think with words when in a resting state.

Overall Conclusion

With the new methods of neuroimaging, we can begin to non-invasively compare both the structure and function of human and non-human primate brains, in the quest to identify the unique features of the human brain that evolved since we shared a last common ancestor with chimpanzees. These techniques will help us to flesh out the specifics of Professor Holloway's early recognition that reorganization was a critical component of the evolution of human brain and mind.

Acknowledgements

I thank the many co-authors on our published research described in this paper, and I thank Todd Preuss and Matthew Glasser for assistance with various aspects of this paper. I also thank Nick Toth and Kathy Schick for hosting the symposium, as well as Doug Broadfield and Michael Yuan for their invitation to participate in the symposium.

References

Amodio DM, and Frith CD. 2006. Meeting of minds: the medial frontal cortex and social cognition. Nat Rev Neurosci 7(4):268-277.

Andreasen NC, O'Leary DS, Cizadlo T, Arndt S, Rezai K, Watkins GL, Ponto LL, and Hichwa RD. 1995. Remembering the past: two facets of episodic memory explored with positron emission tomography. Am J Psychiatry 152(11):1576-1585.

Avants BB, Schoenemann PT, and Gee JC. 2006. Lagrangian frame diffeomorphic image registration: Morphometric comparison of human and chimpanzee cortex. Med Image Anal 10(3):397-412.

Basser PJ, and Jones DK. 2002. Diffusion-tensor MRI: theory, experimental design and data analysis - a technical review. NMR Biomed 15(7-8):456-467.

Behrens TE, Berg HJ, Jbabdi S, Rushworth MF, and Woolrich MW. 2007. Probabilistic diffusion tractography with multiple fibre orientations: What can we gain? Neuroimage 34(1):144-155.

Behrens TE, Woolrich MW, Jenkinson M, Johansen-Berg H, Nunes RG, Clare S, Matthews PM, Brady JM, and Smith SM. 2003. Characterization and propagation of uncertainty in diffusion-weighted MR imaging. Magn Reson Med 50(5):1077-1088.

Binder JR, Frost JA, Hammeke TA, Bellgowan PS, Rao SM, and Cox RW. 1999. Conceptual processing during the conscious resting state. A functional MRI study. Journal of Cognitive Neuroscience 11(1):80-95.

Brodmann K. 1912. Neue Ergebnisse uber die vergleichende histologische Localisation der Grosshirnrinde mit besonderer Berucksichtigung des Stirnhirns. Anat Anz Supplement 41:157-216.

Buckner RL, and Carroll DC. 2007. Self-projection and the brain. Trends Cogn Sci 11(2):49-57.

Byrne RW, and Whiten A. 1992. Cognitive evolution in primates: evidence from tactical deception. Man 27:609-627.

Christoff K, Ream JM, and Gabrieli JD. 2004. Neural basis of spontaneous thought processes. Cortex 40(4-5):623-630.

Croxson PL, Johansen-Berg H, Behrens TE, Robson MD, Pinsk MA, Gross CG, Richter W, Richter MC, Kastner S, and Rushworth MF. 2005. Quantitative investigation of connections of the prefrontal cortex in the human and macaque using probabilistic diffusion tractography. J Neurosci 25(39):8854-8866.

Damasio H, Grabowski TJ, Tranel D, Hichwa RD, and Damasio AR. 1996. A neural basis for lexical retrieval. Nature 380:499-505.

Deacon T. 1997. The Symbolic Species. New York: W.W. Norton.

Frith CD, and Frith U. 2006. The neural basis of mentalizing. Neuron 50(4):531-534.

Gallup G. 1970. Chimpanzees: self-recognition. Science 167:86-87.

Geschwind N. 1970. The organization of language and the brain. Science 170(961):940-944.

Glasser MF, and Rilling JK. 2008. DTI Tractography of the Human Brain's Language Pathways. Cereb Cortex.

Gould SJ. 1991. Exaptation: a crucial tool for an evolutionary psychology. Journal of Social Issues 47(3):43-65.

Hare B, Call J, and Tomasello M. 2006. Chimpanzees

deceive a human competitor by hiding. Cognition 101(3):495-514.

Hauser MD, Chomsky N, and Fitch WT. 2002. The faculty of language: what is it, who has it, and how did it evolve? Science 298(5598):1569-1579.

Hickok G, and Poeppel D. 2004. Dorsal and ventral streams: a framework for understanding aspects of the functional anatomy of language. Cognition 92(1-2):67-99.

Holloway RL. 1968. The evolution of the primate brain: some aspects of quantitative relations. Brain Research 7:121-172.

Holloway RL. 1970. New endocranial values for the australopithecines. Nature 227(5254):199-200.

Holloway RL. 1973. New endocranial values for the East African early hominids. Nature 243(5402):97-99.

Holloway RL. 1983. Cerebral brain endocast pattern of Australopithecus afarensis hominid. Nature 303(5916):420-422.

Holloway RL. 1985. The past, present and future significance of the lunate sulcus in early hominid evolution. In: Tobias PV, editor. Hominid Evolution: Past, Present and Future. New York: Alan R. Liss. p 47-62.

Holloway RL. 1992. The failure of the gyrification index (GI) to account for volumetric reorganization in the evolution of the human brain. Journal of Human Evolution 22:163-170.

Holloway RL. 2000. Brain. In: Delson E, editor. Encyclopedia of Human Evolution and Prehistory. New York: Garland Publishing. p 141-149.

Holloway RL, Broadfield DC, and Yuan MS. 2003. Morphology and histology of chimpanzee primary visual striate cortex indicate that brain reorganization predated brain expansion in early hominid evolution. Anatomical Record 273A(1):594-602.

Holloway RL, and Kimbel WH. 1986. Endocast morphology of Hadar hominid AL 162-28. Nature 321(6069):536-537.

Holloway RL, Sherwood CC, Rilling JK, and Hof PR. 2008. Evolution, of the brain: in humans – paleoneurology. In: M.D. Binder NH, U. Windhorst, M.C. Hirsch, editor. Encyclopedia of Neuroscience: Springer-Verlag.

Ingvar DH. 1979. "Hyperfrontal" distribution of the cerebral grey matter flow in resting wakefulness; on the functional anatomy of the conscious state. Acta Neurol Scand 60(1):12-25.

Mitchell JP, Macrae CN, and Banaji MR. 2006. Dissociable medial prefrontal contributions to judgments of similar and dissimilar others. Neuron 50(4):655-663.

Mori S, and Van Zijl PC. 2002. Fiber tracking: principles and strategies - a technical review. NMR Biomed 15(7-8):468-480.

Mort DJ, Malhotra P, Mannan SK, Rorden C, Pambakian A, Kennard C, and Husain M. 2003. The anatomy of visual neglect. Brain 126(Pt 9):1986-1997.

Mulcahy NJ, and Call J. 2006. Apes save tools for future use. Science 312(5776):1038-1040.

Nucifora PG, Verma R, Melhem ER, Gur RE, and Gur RC. 2005. Leftward asymmetry in relative fiber density of the arcuate fasciculus. Neuroreport 16(8):791-794.

Orban GA, Claeys K, Nelissen K, Smans R, Sunaert S, Todd JT, Wardak C, Durand JB, and Vanduffel W. 2006. Mapping the parietal cortex of human and non-human primates. Neuropsychologia 44(13):2647-2667.

Orban GA, Van Essen D, and Vanduffel W. 2004. Comparative mapping of higher visual areas in monkeys and humans. Trends Cogn Sci 8(7):315-324.

Passingham RE. 1973. Anatomical differences between the neocortex of man and other primates. Brain, Behavior and Evolution 7:337-359.

Petrides M, and Pandya DN. 2002. Association pathways of the prefrontal cortex and functional observations. In: Stuss DT, and Knight RT, editors. Principles of Frontal Lobe Function. New York: Oxford University Press. p 31-50.

Petrides M, and Pandya DN. 2006. Efferent association pathways originating in the caudal prefrontal cortex in the macaque monkey. J Comp Neurol 498(2):227-251.

Phelps ME, and Mazziotta JC. 1985. Positron emission tomography: human brain function and biochemistry. Science 228(4701):799-809.

Pinker S. 2000. The language instinct: how the mind creates language. New York: Perennial Classics. 525 p.

Povinelli DJ, Bering JM, and Giambrone S. 2000. Toward a science of other minds: escaping the argument by analogy. Cognitive Science 24(3):509-541.

Povinelli DJ, Gallup GGJ, Eddy TJ, Bierschwale DT, Engstrom MC, and Perilloux HK. 1997. Chimpanzees recognize themselves in mirrors. Animal Behavior 53:1083-1088.

Powell HW, Parker GJ, Alexander DC, Symms MR, Boulby PA, Wheeler-Kingshott CA, Barker GJ, Noppeney U, Koepp MJ, and Duncan JS. 2006. Hemispheric asymmetries in language-related pathways: a combined functional MRI and tractography study. Neuroimage 32(1):388-399.

Preuss TM. 2004. What is it like to be a human? In: Gazzaniga MS, editor. The Cognitive Neurosciences. Third ed. Cambridge, MA: MIT Press. p 5-22.

Price CJ. 2000. The anatomy of language: contributions from functional neuroimaging. J Anat 197 Pt 3:335-359.

Raichle ME, MacLeod AM, Snyder AZ, Powers WJ, Gusnard DA, and Shulman GL. 2001. A default mode of brain function. Proc Natl Acad Sci U S A 98(2):676-682.

Rilling JK, Barks SK, Parr LA, Preuss TM, Faber TL, Pagnoni G, Bremner JD, and Votaw JR. 2007. A comparison of resting-state brain activity in humans and chimpanzees. Proc Natl Acad Sci U S A 104(43):17146-17151.

Rilling JK, Glasser MF, Preuss TM, Ma X, Zhao T, Hu X, and Behrens TE. 2008. The evolution of the arcuate fasciculus revealed with comparative DTI. Nat Neurosci 11(4):426-428.

Rilling JK, and Insel TR. 1999. The primate neocortex in comparative perspective using magnetic resonance imaging. J Hum Evol 37(2):191-223.

Rilling JK, and Seligman RA. 2002. A quantitative morphometric comparative analysis of the primate temporal lobe. J Hum Evol 42(5):505-533.

Romanski LM, Averbeck BB, and Diltz M. 2005. Neural representation of vocalizations in the primate ventrolateral prefrontal cortex. J Neurophysiol 93(2):734-747.

Romanski LM, Tian B, Fritz J, Mishkin M, Goldman-Rakic PS, and Rauschecker JP. 1999. Dual streams of auditory afferents target multiple domains in the primate prefrontal cortex. Nat Neurosci 2(12):1131-1136.

Sakai KL. 2005. Language acquisition and brain development. Science 310(5749):815-819.

Saxe R, and Wexler A. 2005. Making sense of another mind: the role of the right temporo-parietal junction. Neuropsychologia 43(10):1391-1399.

Schenker NM, Desgouttes AM, and Semendeferi K. 2005. Neural connectivity and cortical substrates of cognition in hominoids. J Hum Evol 49(5):547-569.

Schoenemann PT, Sheehan MJ, and Glotzer LD. 2005. Prefrontal white matter volume is disproportionately larger in humans than in other primates. Nature Neuroscience 8(2):242-252.

Semendeferi K, Lu A, Schenker N, and Damasio H. 2002. Humans and great apes share a large frontal cortex. Nature Neuroscience 5(3):272-276.

Sherwood CC, Broadfield DC, Holloway RL, Gannon PJ, and Hof PR. 2003. Variability of Broca's area homologue in African great apes: implications for language evolution. Anat Rec A Discov Mol Cell Evol Biol 271(2):276-285.

Sherwood CC, Holloway RL, Semendeferi K, and Hof PR. 2005. Is prefrontal white matter enlargement a human evolutionary specialization? Nat Neurosci 8(5):537-538; author reply 538.

Shulman GL, Fiez JA, Corbetta M, Buckner RL, Miezin FM, Raichle ME, and Petersen SE. 1997. Common blood flow changes across visual tasks: II. Decreases in cerebreal cortex Joutrnal of Cognitive Neuroscience 9(5):648-663.

Suddendorf T, and Corballis MC. 1997. Mental time travel and the evolution of the human mind. Genet Soc Gen Psychol Monogr 123(2):133-167.

Tulving E. 2005. Episodic memory and autonoesis: uniquely human? In: Metcalfe HSTaJ, editor. The Missing Link in Cognition: Origins of Self-Reflective Consciousness. Oxford: Oxford University Press. p 3-56.

Uddin LQ, Kaplan JT, Molnar-Szakacs I, Zaidel E, and Iacoboni M. 2005. Self-face recognition activates a frontoparietal "mirror" network in the right hemisphere: an event-related fMRI study. Neuroimage 25(3):926-935.

Ungerleider LG, Courtney SM, and Haxby JV. 1998. A neural system for human visual working memory. Proc Natl Acad Sci U S A 95(3):883-890.

Vigneau M, Beaucousin V, Hervé PY, Duffau H, Crivello F, Houde O, Mazoyer B, and Tzourio-Mazoyer N. 2006. Meta-analyzing left hemisphere language areas: phonology, semantics, and sentence processing. Neuroimage 30(4):1414-1432.

CHAPTER 11

STRUCTURAL AND DIFFUSION MRI OF A GORILLA BRAIN PERFORMED *EX VIVO* AT 9.4 TESLA

JASON A. KAUFMAN, J. MICHAEL TYSZKA, FRANCINE "PENNY" PATTERSON, JOSEPH M. ERWIN, PATRICK R. HOF, JOHN M. ALLMAN

ABSTRACT

Data on brain connectivity in great apes are difficult to obtain because of the lack of reliable *ex vivo* chemical tracers and the preclusion of terminal experimentation using *in vivo* tracers. A new method for obtaining connectivity data, called diffusion-weighted MRI, is a variant of conventional MRI that allows researchers to measure the coherence and orientation of fiber tracts within an entire brain. From these imaging data, tract-tracing algorithms have been developed to conduct non-invasive tractography. Here we apply high-field diffusion-weighted MRI and probabilistic tractography on an isolated, fixed gorilla brain. To test the reliability of this method, we attempt to reconstruct two well-known fiber pathways: the visual (retino-geniculo-striate) pathway, and the corticospinal pathway. The results produced excellent reproductions of these pathways, but also included "false-positive" pathways caused by "tract-jumping" among adjacent fiber pathways. We conclude that diffusion-weighted MRI constitutes an important new tool for studying brain connectivity in humans and great apes, but researchers must be vigilant for false-positive pathways.

INTRODUCTION

> "To say that the white matter is but a uniform substance like wax in which there is no hidden contrivance, would be too low an opinion of nature's finest masterpiece. We are assured that wherever in the body there are fibers, they everywhere adopt a certain arrangement among themselves, created more or less according to the functions for which they are intended…all the diversity of our sensation and our movements depends upon this."
>
> -Nicolaus Steno (1671)

The heterogeneity of the brain's white matter was apparent even to 17th century anatomists whose dissections of large-scale fiber bundles were painstakingly described and illustrated [for a comprehensive review see (Schmahmann and Pandya, 2006)]. But the functional significance of these pathways remained largely mysterious until the development of methods for localizing particular functions to particular regions of the brain. White matter connectivity between functional regions was inferred first from lesion/degeneration studies (e.g., Damasio and Damasio, 1989; Goldman et al., 1971; Pribram and Mishkin, 1955) and later from chemical neuronal tracers (e.g., Barbas and Pandya, 1987; Van Essen et al., 1986). However, the terminal nature of these experiments restricts their application to laboratory animals and precludes their use in humans or apes. For this reason there has been an absence of comparative data on structural brain connectivity for humans and our ape relatives–data that could contain important phylogenetic signals on the evolution of the brain's wiring scheme.

Diffusion-weighted magnetic resonance imaging (DW-MRI) provides a new means for obtaining quantitative data on white matter connectivity. DW-MRI is sensitive to the magnitude and spatial orientation of Brownian water diffusion in tissue, which occurs more readily along axon tracts than across them (Basser et al., 1994; LeBihan et al., 2001). The method is non-invasive, and has been used successfully to trace known fiber pathways in laboratory animals and in humans (Basser et al., 2000; Conturo et al,. 1999; Mori et al., 1999; Xue et al., 1999), and to quantify the coherence of white matter fiber tracts in healthy and diseased individuals (Barnea-Goraly et al., 2004; Michael, 2002; Moseley et al., 2002; Neil et al., 2002; Nguyen et al., 2005; Ramnani et al., 2004; Sundgren et al., 2004). Technical advances in DW-MRI have significantly improved the angular resolution of diffusion scanning (Jones et al., 1999), and an analytical method has been developed to characterize the uncertainty associated with DW-MRI tract-tracing (Behrens et al., 2003b). This process, called probabilistic tractography, has also recently been expanded to model crossing-fibers within voxels (Behrens et al., 2007).

Our purpose here is to demonstrate the feasibility of conducting DW-MRI tractography on an isolated, fixed brain using a high-field experimental imaging system. Fixed tissue is entirely compatible with DW-MRI, and there are distinct advantages for imaging *ex vivo* as opposed to *in vivo*: fixed tissue allows for extended scanning periods which substantially boosts the signal-to-noise ratio; and fixed tissue can be processed histologically following scanning. (Obviously, functional imaging (fMRI) is not possible *ex vivo*.) In this chapter, we present results from an *ex vivo* structural–(i.e., grey-white contrast) and diffusion–MRI experiment on a gorilla brain conducted on a 9.4 Tesla MRI system. To test the reliability of the probabilistic tractography method, we attempt to reconstruct two well-understood white matter pathways: the visual (retino-geniculo-striate) pathway and the motor (corticospinal) pathway.

Methods

The brain is that of an adult male gorilla, named Michael, who at the age of 27 succumbed to a myocardial disease characterized by deterioration of the heart's electrical conduction pathway. It is particularly important to note that careful prior preparations were made for quick and precise extraction of the brain with a minimal post-mortem interval. This type of preparation yields especially good tissue preservation, and should serve as a model for the compassionate use of tissue from great apes who have died from natural causes.

Within four hours post-mortem the brain was immersion fixed in 4% paraformaldehyde (freshly depolymerized) and subsequently stored in phosphate-buffered-saline with 0.01% sodium azide added as a preservative.

An acrylic canister was constructed so that the brain could be submerged in a high-viscosity perfluoropolyether (Galden, Solvay Solexis) that has a zero MR signal. The zero MR signal yields images of the brain in "black" or "empty" space, thereby aiding in subsequent tissue segmentation. The brain was wrapped in thin Teflon® (DuPont) which, as a fluorinated polymer, also has a zero MR signal, and positioned in the canister using sponges. The canister fits snugly inside a 180mm birdcage RF coil in our 9.4 Tesla Bruker imaging system (Bruker Biospin Ltd.).

Two series of scans were performed. The first was a high-resolution 3D FLASH sequence, lasting approximately 16 hours, with an anatomical resolution of 250 microns isotropic. The second series were high-angular-resolution diffusion scans (PGHE) weighted isotropically along 72 directions. A series of 6 diffusion scans, lasting approximately 36 hours in total, were averaged together in order to boost the signal-to-noise ratio, with a final resolution of 1mm isotropic.

All post-processing of the images were performed using Amira (Mercury, San Diego) and the FSL suite of MRI applications (Smith et al., 2004). For the diffusion imaging, post-processing steps included corrections for eddy-current distortions, followed by a computation termed Bayesian Estimation of Diffusion Parameters Obtained using Sampling Techniques (known as a BEDPOST operation). This computation involves a Markov Chain Monte Carlo resampling technique to construct probability distributions for diffusion parameters within each voxel (Behrens et al., 2003b). The BEDPOST operation requires complex computations for every voxel within the brain, and the computing time can be substantial: in this case it took approximately two weeks of computing time on a high-end Linux workstation to complete the BEDPOST operation. For each tractography trial, 10,000 samples were drawn from the global probability density function for each voxel in the seed mask.

Probabilistic tractography of the retino-geniculo-striate pathway was performed using the method of Behrens et al. (Behrens et al., 2003a). Specifically, seeds were placed at the grey-white border surrounding the lateral geniculate nucleus (LGN). This single-mask approach should identify projections both in the anterior direction (afferent fibers from the retina) and posterior direction (efferent fibers to the striate cortex).

Tractography of the corticospinal tract was performed using the method of Ciccarelli et al. (Ciccarelli et al., 2006). First, the neocortex is segmented and the boundaries of primary motor cortex are defined both morphologically (the posterior bank of the precentral gyrus) and histologically by the presence of Betz cells in Nissl-stained coronal sections. The M1 region-of-interest is then mapped into MRI space using an affine registration algorithm (Jenkinson and Smith, 2001). Next, a connectivity-based seed classification is performed on an axial mask through the internal capsule. This step identifies voxels within the posterior limb of the internal capsule mask that are most-likely to connect with primary motor cortex. Probabilistic tractography is then performed from the motor cortex seed mask, with the condition that only projections passing through the internal capsule mask are retained. In other words, the region of the internal capsule most likely to connect with primary motor cortex is defined as a waypoint mask for reconstructing the corticospinal tract. We also performed trials in which the medullary pyramids were masked as waypoints to further constrain the fibers to the corticospinal tract.

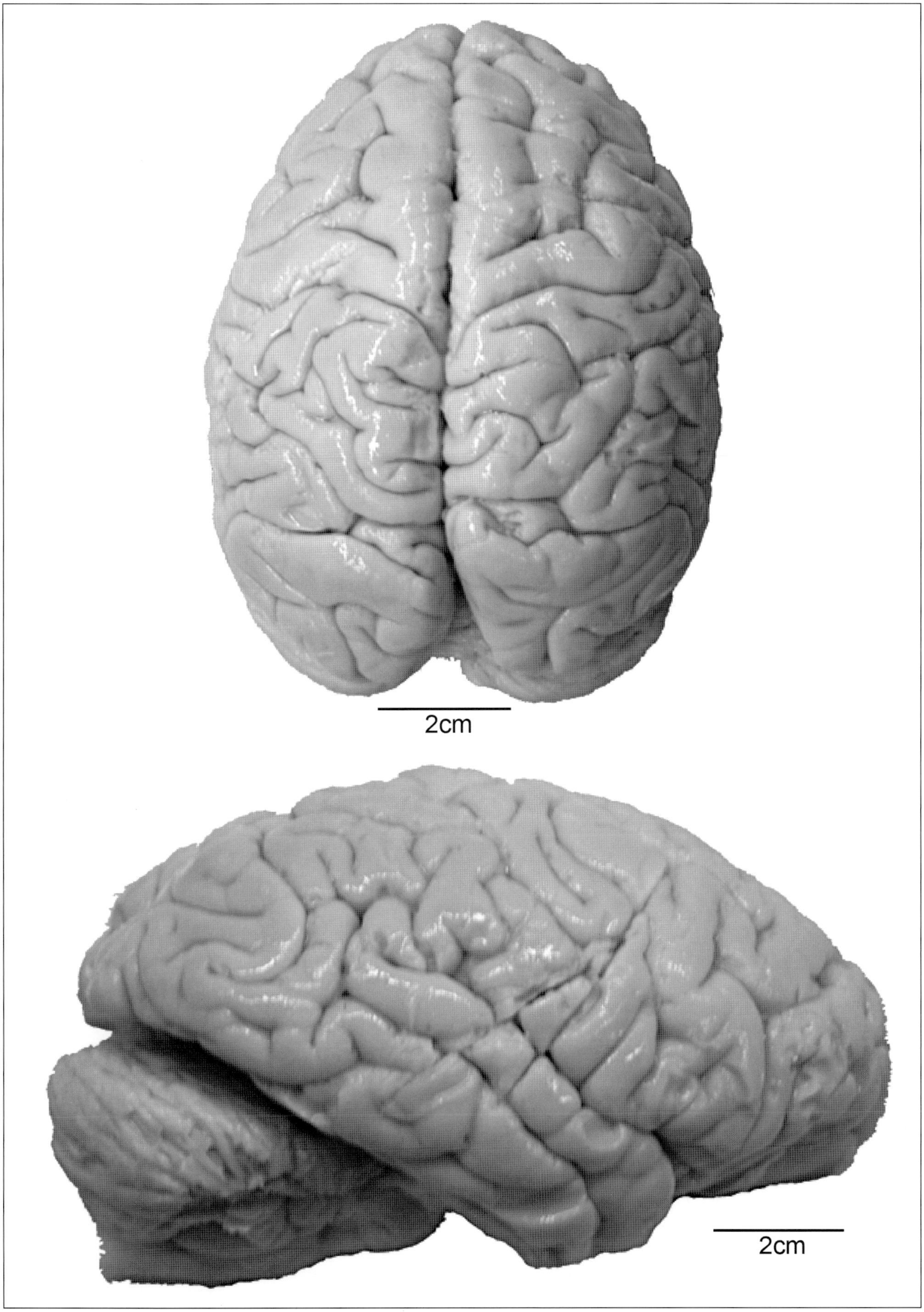

Figure 1. Photographs of the gorilla brain used in this experiment. Top: dorsal view. Bottom: right lateral view.

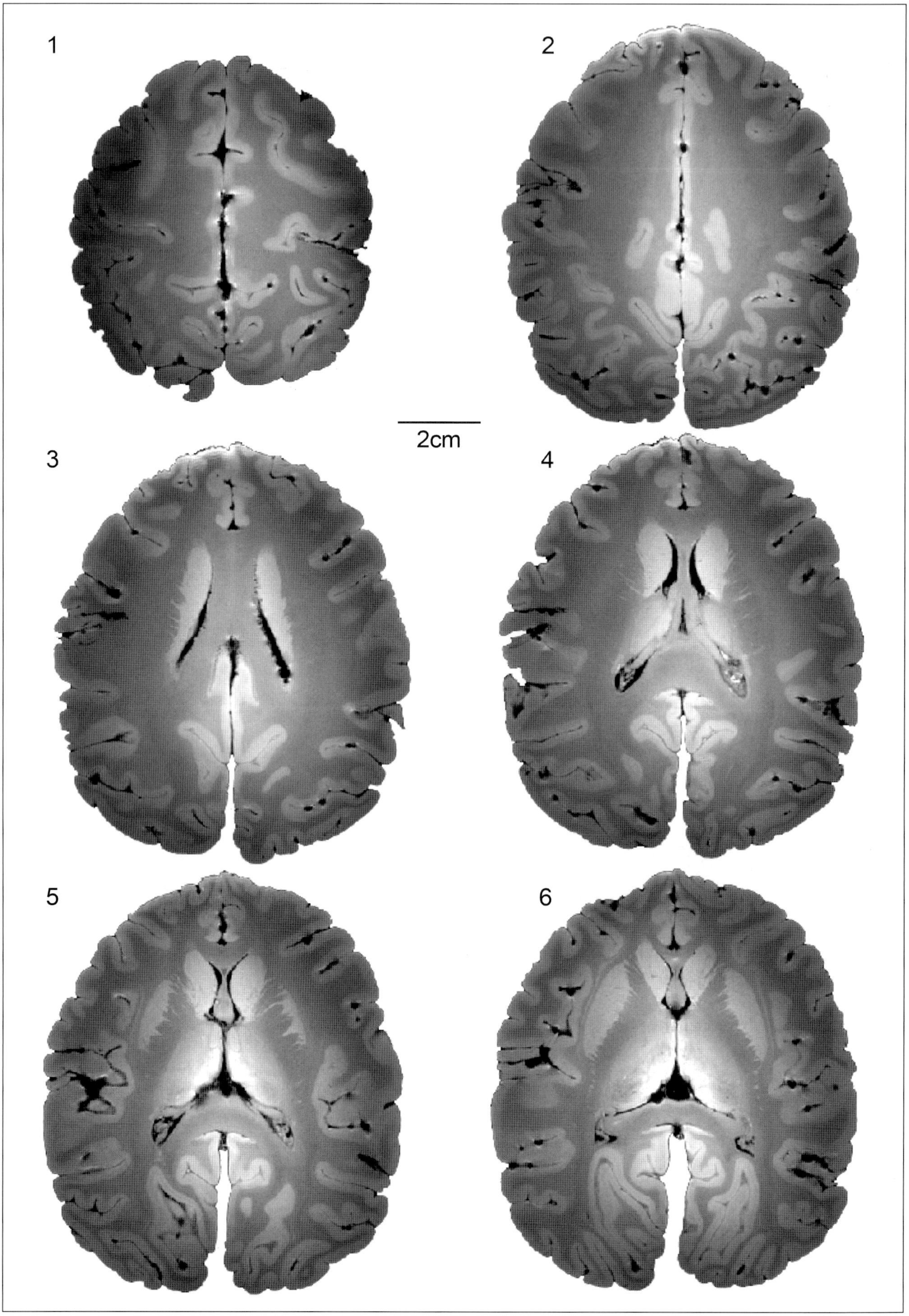
1
2
2cm
3
4
5
6

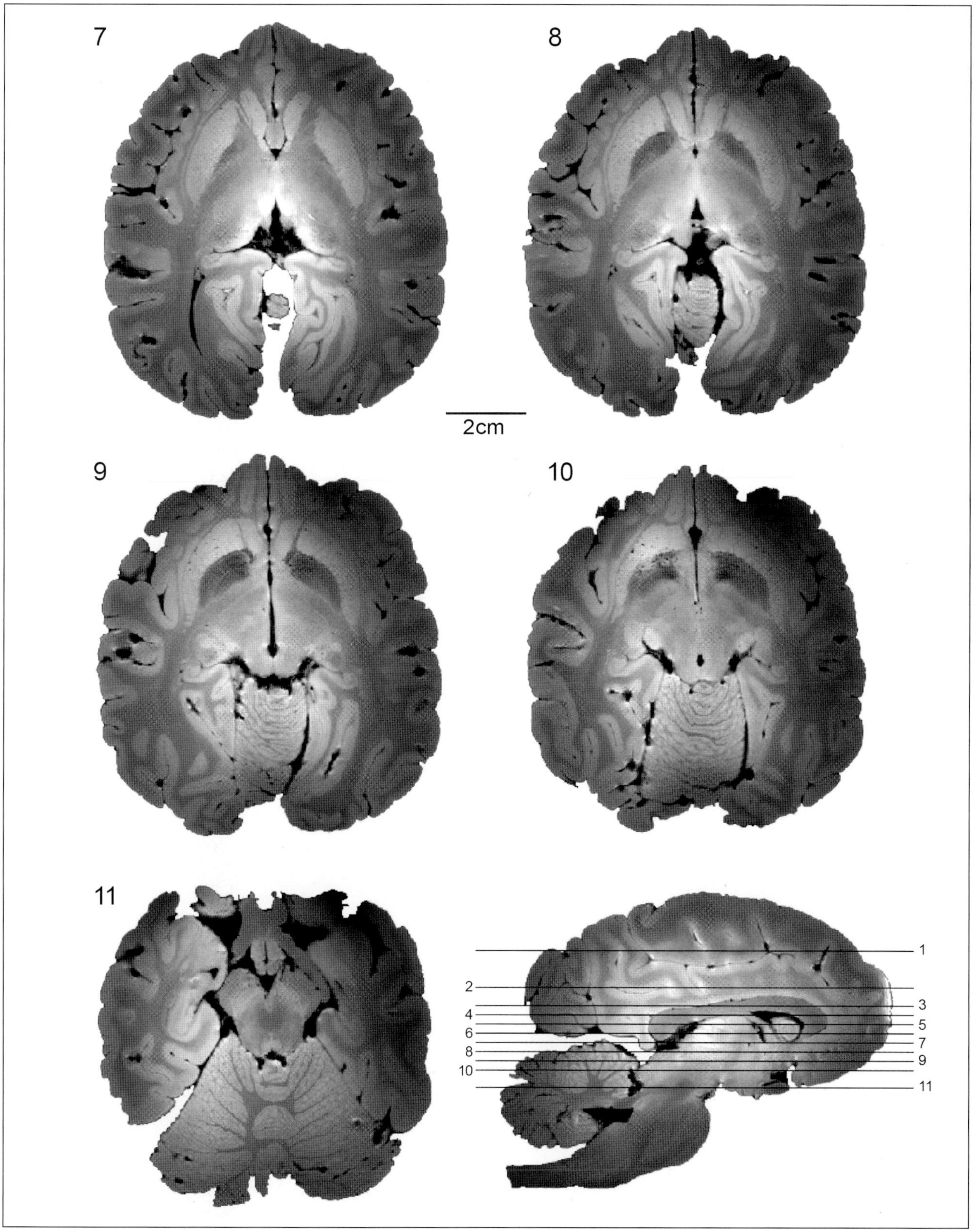

Figure 2. Axial slices through the structural MR image (resolution = 250μm isotropic). Legend in lower right depicts the position of individual slices.

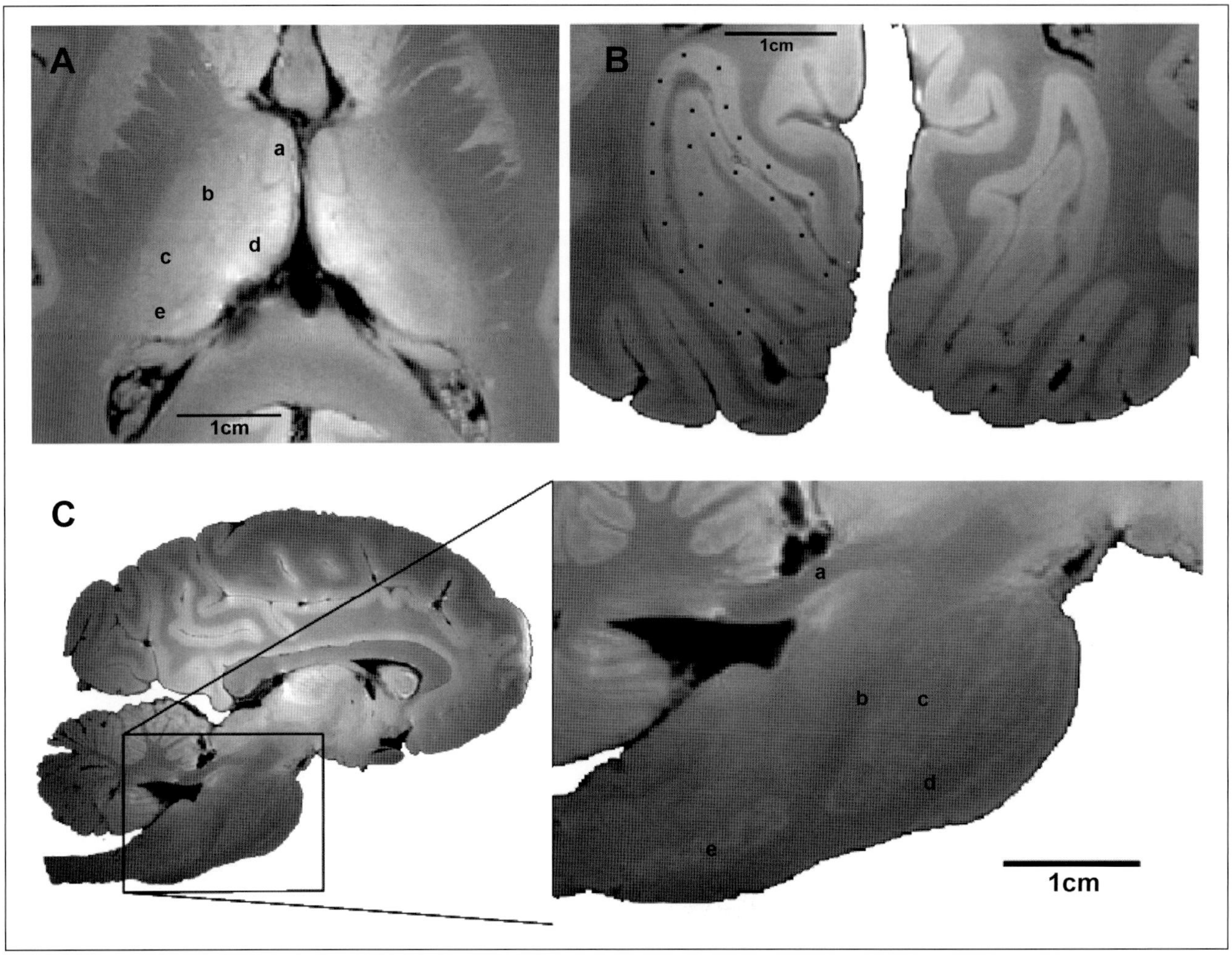

Figure 3. Close-up views of anatomical details evident in the structural MR image. A: Individual nuclei are visible in the thalamus (a: anterior nucleus; b: lateral anterior nucleus; c: lateral posterior nucleus; d: dorsomedial nucleus; e: pulvinar). B: In primary visual cortex, the Stria of Gennari is visible (indicated on the left side by black squares). C: In the brainstem, several anatomical features are evident, including a: the superior cerebellar peduncle; b: the medial lemniscus, c: crossing fibers of the pons, d: the descending pyramidal tract, and e: the inferior olivary nucleus.

Results

The 3D FLASH structural scan yielded excellent anatomical imaging with a resolution of (Figure 2). Minor artifacts such as "ringing" occurred at the extreme edges of the field-of-view, or coincided with small bubbles trapped within cortical convolutions. Notwithstanding these small artifacts, the 9.4 Tesla imaging system produced anatomical scans with 4-6 times the resolution of most large-brain structural MRI experiments (1.5 Tesla clinical imaging systems usually yield a resolution of approximately 1mm).

Several anatomical structures are noteworthy as they are not normally visible on large-brain MR images (Figure 3). Subdivisions of the thalamus are visible, often with clear delineations of thalamic nuclei, such as the internal medullary laminae separating the anterior thalamic nucleus from the lateral-anterior and dorsomedial nuclei (Figure 3a). In the occipital cortex, many image slices display Stria of Gennari, a cytoarchitectural feature delineating the boundaries of primary visual cortex (Figure 3b). And many features are visible in the brainstem, including the clearly undulating surface of the inferior olivary nucleus (Figure 3c).

The high-field system produced equally impressive results for the diffusion imaging sequences. The final voxel size for the DW-MRI volume was 1mm isotropic with an angular sampling of 72 spatial directions. Figure 4 illustrates a subset of the data on the principle-diffusion-direction, which is represented both as vector fields as well as red-green-blue color-coding. FSL is able to model multiple-fiber orientations; we chose a two-fiber model as illustrated in Figure 4c.

Tractography experiments using the two-fiber model produced generally good reproductions of known fiber pathways, but not without some errors (usually caused by the tendency for tract-tracing to "jump" onto adjacent tracts). The first experiment, in which a seed mask was placed around the right lateral geniculate nucleus, yielded a very good representation of the retino-

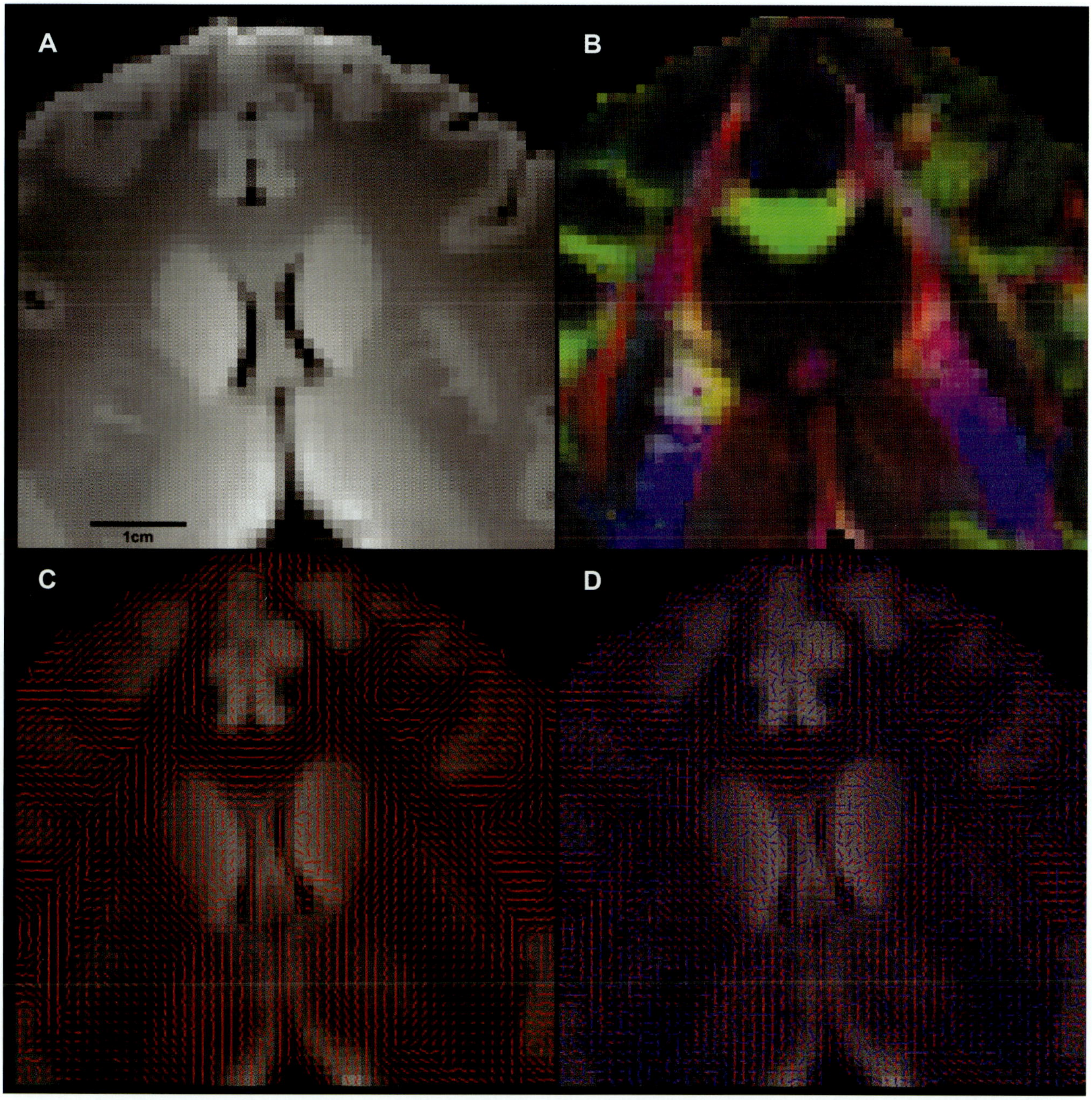

Figure 4. Representation of raw diffusion data (resolution = 1mm isotropic). A: Structural image resampled to 1mm voxel size. B: Red-Green-Blue pseudo-color image representing fiber orientation with each voxel. Voxels labeled green contain fibers oriented axially; voxels labeled red contain fibers oriented longitudinally; and voxels labeled blue contain fibers oriented dorso-ventrally. C. Representation of fiber orientation by vector mapping. Red lines depict the principle diffusion direction. D. The two-fiber model allows identification of regions containing crossing fibers. Red lines depict the principle diffusion direction, while blue lines indicate the minor diffusion direction (i.e., red lines correspond to the first eigenvector of the diffusion tensor and blue lines correspond to the second eigenvector of the diffusion tensor).

geniculo-striate pathway (Figure 5a), but not without errors. The pathway was traced anteriorly through the optic tract and arrived at the optic chiasm. Within the chiasm itself, a tract-jumping error is evident in which the algorithm traces a U-turn to join the optic tract on the opposite side. While this is a reasonable connection through the vector dataset, it is not anatomically correct.

Caudally from the LGN, the algorithm traces the optic radiation and terminates within cortical areas corresponding to the occipital pole and lateral-occipital areas (Figure 5a). However, the algorithm fails to find a pathway to the medial striate cortex. Additionally, the inclusion of connections with lateral occipital cortex anterior to V1 likely represents a "tract-jumping" error involving adjacent V2 projections that also travel through the optic radiation.

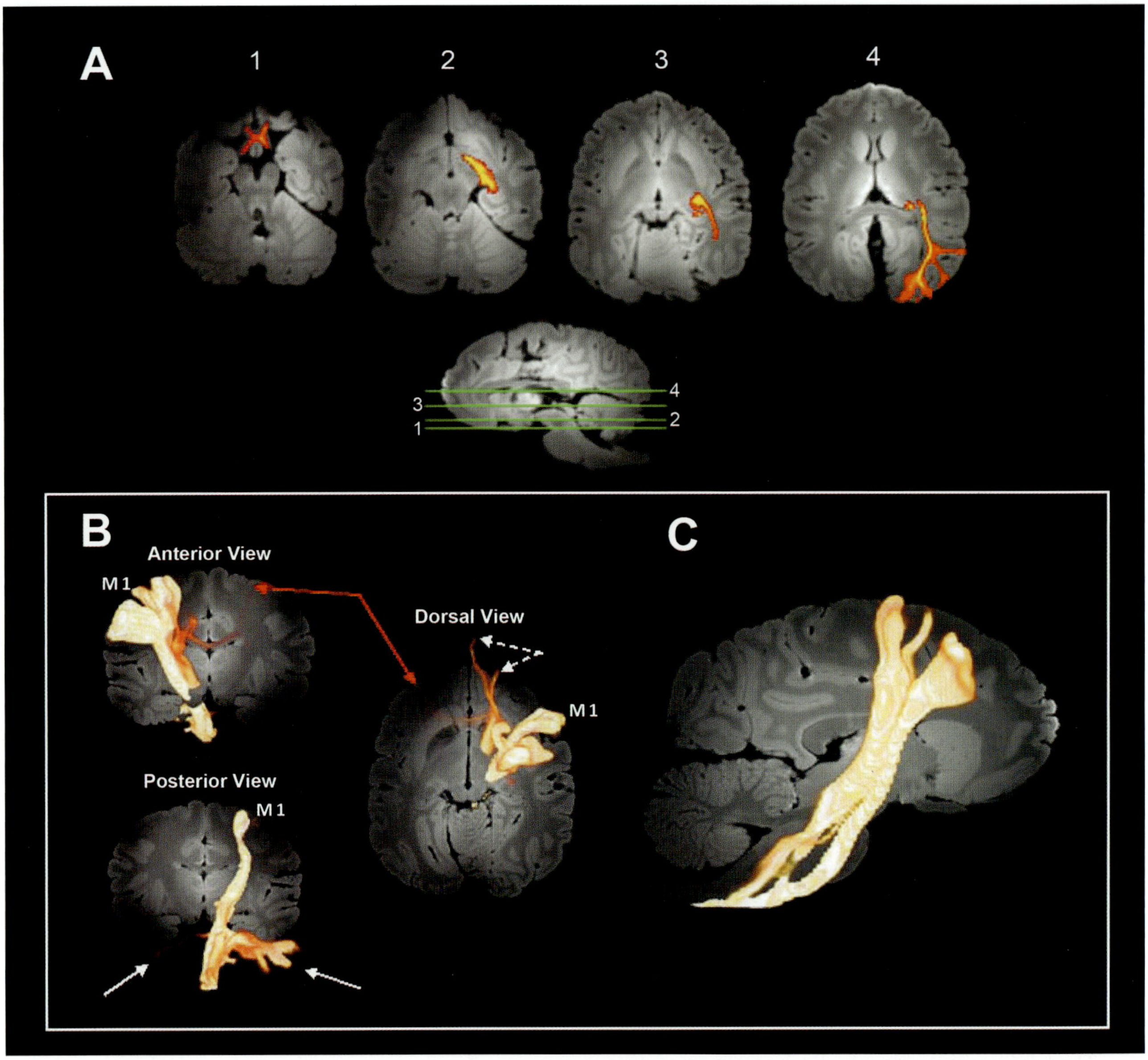

Figure 5. Results of pathway reconstructions by means of probabilistic tractography. A: Tractography of the retino-geniculo-striate pathway reconstructed by seeding in the lateral geniculate nucleus. Bright orange represents pathways of higher-probability. The algorithm successfully reconstructs the visual pathway in both the anterior and posterior direction, but does not reach the medial striate cortex. B: Three-dimensional rendering of the motor pathway reconstructed by seeding from M1 and passing through the posterior limb of the internal capsule. Red arrows point to fibers projecting to the contralateral forebrain. Dashed white arrows point to ipsilateral projections to the forebrain. Solid white arrows point to cerebellar projections. D: Three-dimensional rendering of the corticospinal tract reconstructed by seeding from M1 and including waypoint masks at the posterior limb of the internal capsule and the medullary pyramids. False-positive tracts of the medial and lateral lemniscus and central tegmental tract are visible posterior to the true fibers of the corticospinal tract descending through the pons.

The second tractography experiment on the corticospinal tract also yielded generally good results (Figure 5b). Tractography from the M1 seed mask (with the internal capsule waypoint mask) traced fibers gathering into the posterior limb of the internal capsule and descending through the cerebral peduncles. The tractography algorithm generates terminations in the cerebellum (including a small projection to the contralateral cerebellum) and the medullary pyramids. However, at the level of the ventral thalamus, the algorithm "jumps tracts" and follows thalamic projections to the dorsal prefrontal cortex, including a projection which crosses the corpus callosum and terminates in the contralateral dorsolateral prefrontal cortex. And again, at approximately the level of the red nucleus the tractography bifurcates to include not only the corticospinal tract but also the medial- and lateral-lemniscus, and the central tegmental tract. These latter tracts are false positives (the medial lemniscus carries somatosensory fibers, while the lateral lemniscus and central tegmental tract connect brainstem nuclei with the mesencephalon). Again, it is likely that these errors arise from "tract jumping".

When the medullary pyramids are used as waypoint masks, the cerebellar and frontal connections are discarded, leaving the descending corticospinal tract but still including the false positives: medial- and lateral-lemniscus and central tegmental tract (Figure 5c).

Discussion

Our application of high-field MR imaging to an isolated *ex vivo* gorilla brain produced excellent structural imaging with remarkably high anatomic clarity. The raw diffusion data also appeared to be of excellent quality, with more than twice the resolution normally seen in DTI studies. We performed two sets of trial tractography experiments on known cerebral pathways to test the reliability of probabilistic tractography: the first experiment was intended to trace the retino-geniculo-striate pathway while the second experiment was intended to trace the corticospinal tract. The application of probabilistic tractography to these pathways produced mixed results: the intended pathway was always reconstructed, but additional false-positive pathways were also produced.

The corticospinal tract, in particular, is difficult to reconstruct because it is long and there are many opportunities for errors involving crossing fibers or "kissing fibers" to produce "tract-jumping" false-positives (Holodny et al., 2005). Studies in which the corticospinal tract is successfully isolated from the corticobulbar or other adjacent tracts must use multiple levels of waypoint masks and exclusion masks (Aoki et al., 2005; Reich et al., 2006), or concentrate only on certain levels of the corticospinal tract (e.g. brainstem (Chen et al., 2007)).

The false positives identified during our reconstructions of the retino-geniculo-striate pathway and the corticospinal pathway all occur at areas where the "true" fiber tract (the one we intend to trace) lies adjacent to an unrelated pathway with much the same orientation (a situation termed "kissing fibers"). The result is "tract jumping", and clearly it is important that investigators are vigilant for these errors (Dauguet et al., 2007). The tractography model must be progressively refined (using exclusion masks) to weed out unwanted pathways. When available, the combined use of DW-MRI and fMRI can improve reconstruction results (Guye et al., 2003; Kamada et al., 2005; Staempfli et al., 2008).

Methods for analyzing connectivity from fMRI data (Bartels and Zeki, 2005; Logothetis et al., 1999) are currently impractical for large primates, and there are no reliable long-distance post-mortem neuronal tracers. Robust, quantitative data on global brain connectivity are therefore sparse for humans, and do not exist at all for the apes. DW-MRI fills an important technical gap in our ability to measure human and ape brain connectivity, but it is not without error and must be applied with progressive refinement. As with any method, diffusion tractography has both advantages and disadvantages. Although DW-MRI does not have the high degree of spatial resolution that chemical neuronal tracers can provide it does have the advantage of providing connectivity data for the brain as a whole, which would be impossible to achieve using chemical tracers. At the same time, the application of probabilistic tractography shown here demonstrates that prior knowledge about the position of fiber bundles is essential for weeding out tracts that are false positives.

Conclusions

High-field MR imaging yielded excellent structural and diffusion data even for a fixed, isolated brain. The application of probabilistic tractography to the diffusion data produced mixed results: in both trial experiments the intended pathways were reconstructed, along with false-positive pathways. Prior knowledge of the fiber tracts was necessary to eliminate "tract jumping" errors. DW-MRI fills an important methodological gap for measuring brain connectivity, but future refinement of tractography algorithms will be important for accurate analysis of diffusion data.

Acknowledgements

The authors wish to thank Dr. Ralph L. Holloway for inviting us to participate in this important symposium. We are honored to be a part of this legacy. This research was funded by a Discovery Grant from the Gordon and Betty Moore Foundation.

References

Aoki, S., Iwata, N.K., Masutani, Y., Yoshida, M., Abe, O., Ugawa, Y., Masumoto, T., Mori, H., Hayashi, N., Kabasawa, H., et al., 2005. Quantitative evaluation of the pyramidal tract segmented by diffusion tensor tractography: feasibility study in patients with amyotrophic lateral sclerosis. Radiation Medicine 23, 195-199.

Barbas, H., Pandya, D.N., 1987. Architecture and frontal cortical connections of the premotor cortex (area 6) in the rhesus monkey. Journal of Comparative Neurology 256, 211-228.

Barnea-Goraly, N., Kwon, H., Menon, V., Eliez, S., Lotspeich, L., Reiss, A.L., 2004. White matter structure in autism: preliminary evidence from diffusion tensor imaging. Biological Psychiatry 55, 323-326.

Bartels, A., Zeki, S., 2005. The chronoarchitecture of the cerebral cortex. Philosophical Transactions of the Royal Society London B Biological Sciiences 360, 733-750.

Basser, P.J., Mattiello, J., LeBihan, D., 1994. Estimation of the effective self-diffusion tensor from the NMR spin echo. Journal of Magnetic Resonance, Series B 103, 247-254.

Basser, P.J., Pajevic, S., Pierpaoli, C., Duda, J., Aldroubi, A., 2000. In vivo fiber tractography using DT-MRI data. Magnetic Resonance in Medicine 44, 625-632.

Behrens, T.E., Berg, H.J., Jbabdi, S., Rushworth, M.F., Woolrich, M.W., 2007. Probabilistic diffusion tractography with multiple fibre orientations: What can we gain? Neuroimage 34, 144-155.

Behrens, T.E.J., Johansen-Berg, H., Woolrich, M.W., Smith, S.M., Wheeler-Kingshott, C.A.M., Boulby, P.A., Barker, G.J., Sillery, E.L., Sheehan, K., Ciccarelli, O., et al., 2003a. Non-invasive mapping of connections between human thalamus and cortex using diffusion imaging. Nature Neuroscience 6, 750-757.

Behrens TEJ, Woolrich MW, Jenkinson M, Johansen-Berg H, Nunes RG, Clare S, Matthews PM, Brady JM, and Smith SM. 2003b. Characterization and propagation of uncertainty in diffusion-weighted MR imaging. Magnetic Resonance in Medicine 50(5):1077-1088.

Chen, X., Weigel, D., Ganslandt, O., Buchfelder, M., Nimsky, C., 2007. Diffusion tensor imaging and white matter tractography in patients with brainstem lesions. Acta Neurochirurgica 149, 1117-1131.

Ciccarelli, O., Behrens, T.E., Altmann, D.R., Orrell, R.W., Howard, R.S., Johansen-Berg, H., Miller, D.H., Matthews, P.M., Thompson, A.J., 2006. Probabilistic diffusion tractography: a potential tool to assess the rate of disease progression in amyotrophic lateral sclerosis. Brain 129, 1859-1871.

Conturo, T.E., Lori, N.F., Cull, T.S., Akbudak, E., Snyder, A.Z., Shimony, J.S., McKinstry, R.C., Burton, H., Raichle, M.E., 1999. Tracking neuronal fiber pathways in the living human brain. Proceedings of the National Academy of Sciences 96, 10422-10427.

Damasio, H., Damasio, A.R., 1989. Lesion Analysis in Neuropsychology. Oxford, UK: Oxford University Press.

Dauguet, J., Peled, S., Berezovskii, V., Delzescaux, T., Warfield, S.K., Born, R., Westin, C-.F., 2007. Comparison of fiber tracts derived from in-vivo DTI tractography with 3D histological neural tract tracer reconstruction on a macaque brain. NeuroImage 37, 530-538.

Goldman, P.S., Rosvold, H.E., Vest, B., Galkin, T.W., 1971. Analysis of the delayed-alternation deficit produced by dorsolateral prefrontal lesions in the rhesus monkey. Journal of Comparative and Physiological Psychology 77, 212.

Guye, M., Parker, G.J.M., Symms, M., Boulby, P., Wheeler-Kingshott, C.A.M., Salek-Haddadi, A., Barker, G.J., Duncan, J.S., 2003. Combined functional MRI and tractography to demonstrate the connectivity of the human primary motor cortex in vivo. NeuroImage 19, 1349-1360.

Holodny, A.I., Gor, D.M., Watts, R., Gutin, P.H., Ulug, A.M., 2005. Diffusion-tensor MR tractography of somatotopic organization of corticospinal tracts in the internal capsule: initial anatomic results in contradistinction to prior reports. Radiology 234, 649-653.

Jenkinson, M., Smith, S., 2001. A global optimisation method for robust affine registration of brain images. Medical image analysis 5, 143-156.

Jones, D.K., Horsfield, M.A., Simmons, A., 1999. Optimal strategies for measuring diffusion in anisotropic systems by magnetic resonance imaging. Magnetic Resonance in Medicine 42, 515.

Kamada, K., Sawamura, Y., Takeuchi, F., Kawaguchi, H., Kuriki, S., Todo, T., Morita, A., Masutani, Y., Aoki, S., Kirino, T., 2005. Functional identification of the primary motor area by corticospinal tractography. Neurosurgery 56, 98-109; discussion 198-109.

LeBihan, D., Mangin, J.F., Poupon, C., Clark, C.A., Pappata, S., Molko, N., Chabriat, H., 2001. Diffusion tensor imaging: concepts and applications. Journal of Magnetic Resonance Imaging 13, 534-546.

Logothetis, N.K., Guggenberger, H., Peled, S., Pauls, J., 1999. Functional imaging of the monkey brain. Nature Neuroscience 2, 555

Michael, M., 2002. Diffusion tensor imaging and aging–a review. NMR in Biomedicine 15, 553-560.

Mori, S., Crain, B.J., Chacko, V.P., van Zijl, P.C., 1999. Three-dimensional tracking of axonal projections in the brain by magnetic resonance imaging. Annals of Neurology 45, 265-269.

Moseley, M., Bammer, R., Illes, J., 2002. Diffusion-tensor imaging of cognitive performance. Brain and Cognition 50, 396.

Neil, J., Miller, J., Mukherjee, P., Hüppi, P.S., 2002. Diffusion tensor imaging of normal and injured developing human brain - a technical review. NMR in Biomedicine 15, 543-552.

Nguyen, T.H., Yoshida, M., Stievenart, J.L., Iba-Zizen, M.T., Bellinger, L., Abanou, A., Kitahara, K., Cabanis, E.A., 2005. MR tractography with diffusion tensor imaging in clinical routine. Neuroradiology 47, 334.

Pribram, K.H., Mishkin, M., 1955. Simultaneous and successive visual discrimination by monkeys with inferotemporal lesions. Journal of Comparative and Physiological Psychology 48, 198.

Ramnani, N., Behrens, T.E.J., Penny, W., Matthews, P.M., 2004. New approaches for exploring anatomical and functional connectivity in the human brain. Biological Psychiatry 56, 613.

Reich, D.S., Smith, S.A., Jones, C.K., Zackowski, K.M., van Zijl, P.C., Calabresi, P.A., Mori, S., 2006. Quantitative Characterization of the Corticospinal Tract at 3T. American Journal of Neuroradiology 27, 2168-2178.

Schmahmann, J.D., Pandya, D.N., 2006. White Matter Pathways in Early Neuroscience. Fiber Pathways of the Brain. Oxford: Oxford University Press. p 7-37.

Smith, S.M., Jenkinson, M., Woolrich, M.W., Beckman, C.F., Behrens, T.E.J., Johansen-Berg, H., Bannister, P.R., De Luca, M., Drobnjak, I., Flitney, D.E., et al., 2004. Advances in functional and structural MR image analysis and implementation as FSL. NeuroImage 23, 208-219.

Staempfli, P., Reischauer, C., Jaermann, T., Valavanis, A., Kollias, S., Boesiger, P. 2008. Combining fMRI and DTI: A framework for exploring the limits of fMRI-guided DTI fiber tracking and for verifying DTI-based fiber tractography results. NeuroImage 39, 119-126.

Steno, N., 1671. Dissertatio de cerebri anatome, spectatissimis viris dd Societatis apud dominum Thevenot collectae, dictata, atque è gallico exemplari. Parisiis edito 1669: Latinate donatat, opera & studio Guidonis Fanosii.

Sundgren, P.C., Dong, Q., Gómez-Hassan, D., Mukherji, S.K., Maly, P., Welsh, R., 2004. Diffusion tensor imaging of the brain: review of clinical applications. Neuroradiology 46, 339-350.

Van Essen, D.C., Newsome, W.T., Maunsell, J.H., Bixby, J.L., 1986. The projections from striate cortex (V1) to areas V2 and V3 in the macaque monkey: asymmetries, areal boundaries, and patchy connections. Journal of Comparative Neurology 244, 451-480.

Xue, R., van Zijl, P.C., Crain, B.J., Solaiyappan, M., Mori, S., 1999. In vivo three-dimensional reconstruction of rat brain axonal projections by diffusion tensor imaging. Magnetic Resonance in Medicine 42, 1123-1127.

CHAPTER 12

The Role of Vertical Organization in the Encephalization and Reorganization of the Primate Cortex

Daniel P. Buxhoeveden

Introduction

While cortical enlargement dominated the thinking of hominid evolution and paleoneurology, it was not until the last few decades that the mechanisms responsible for this were made known. The answer was provided by the radial unit hypothesis as revealed by the seminal work of Pasko Rakic (1972, 1978). The significance of this work to the field of paleoneurology cannot be overstated and it is now gaining the attention it deserves. The model provides insights into the relationship between cortical size and re-organization, and it sheds light on the proliferation of cortical regions and the relationship between surface area and cortical depth. I have chosen to address two main topic areas based on the radial unit hypothesis. The first considers the relationship between cortical enlargement, reorganization, and minicolumn size. The second section briefly considers what is known about the size of minicolumns in the primate order and suggests possible implications.

Defining the Micro-Vertical Organization of the Cortex

The minicolumn is a particular feature of cortical organization; one based on vertical components of cortical function at a spatially small scale. It does not disregard horizontal organization and recognizes that the complexity of the brain allows for multiple ways of processing information. The use of the vertical organization of the cortex is an attempt to find unifying principles in cortical organization which integrate horizontal lamina and intrinsic circuits into a testable model. There is arguably substantial evidence of functionality at this level of organization, and as a computer model minicolumns demonstrate self-organizing and other functional properties that are sometimes surprising (Amirikan and Georgopoulos, 2003; Favorov and Kelly, 1996; Hasselmo, 2005; Johannsson and Lansner, 2007; Kohn et al, 1997; Lucke, 2004; Lucke and Malburg, 2004; Mountcastle, 1997, 2003; Rao et al, 1999; Sugimoto et al., 1997). However, it is also important to recognize there is considerable debate and conflicting evidence regarding the ubiquity and functionality of the adult anatomical elements, where various approaches sometimes yield different conclusions (Catania, 2002; Jones, 2000; Kreiger et al, 2007; Rockland, 2004; Swindale, 1990), and species specific differences complicate the picture, though if anything, the primate cortex may display a heightened columnar organization.

Variation in neuronal types and connectivity at the microcircuit level may rule out a rigid over aching definition of the minicolumn (and the larger cortical column). The minicolumn appears to be a common template rather than a stereotypical component in all brains and regions (Buxhoeveden and Casanova, 2002; Mountcastle, 2003; Silberberg et al., 2002). Nonetheless there are local and species-specific examples of repeating configurations of minicolumns and Mountcastle (2003) noted that "The important point is that columnar organization depends upon a certain set of properties common to all neurons in the elementary unit, but that other properties may vary between different neurons in the same minicolumn." Mountcastle provides a conceptual basis to variability upon the basic template by stating that "differences in afferent input are convolved with different intrinsic operations in different cortical areas to produce what we call different functions." Silberberg et al. (2002) also concludes that despite the great range in microcircuitry, stereotypical features exist nonetheless at multiple levels indicating a deterministic basis for them and suggests that all neocortical microcircuits may be subtle variations of a common template (see also Jin et al., 2001; Kisvarday et al., 2002; Kosloski, et al., 2001). Thus, a broader conception of the minicolumn is to see it as a 'template' for a shared set of properties of a given set of neurons across several or more lamina. It seems to be a general principle that cortical neurons with similar stimulus selection properties are found in close proximity to each other (Reich et al., 2001) and the minicolumn is the vertical component of that association. Further, in addition to being a dynamic component (both anatomically and physiologically), it is important to think of the minicolumn as part of larger organizing units in the cortex and not an end in itself.

The minicolumn in the adult traces its foundations to the development of the cortex itself, a cortex which contains highly visible ontogenetic cell arrays and from which the adult cortex will emerge. There is evidence that the ontogenetic units become the adult components of vertical organization (below). The anatomical components are often conspicuous features of cortex across taxa, and metabolic and physiological evidence have helped to provide evidence of functionality at this level (Mountcastle, 1997). Because horizontal lamina within the vertical organization maintain functional specialization, it is not surprising that the activity of columns may be seen at levels that encompass several layers only, and not the entire depth. This speaks to the flexibility of the system and not against the concept of narrow vertical organization. Some of the major questions surrounding minicolumns are the extent to which they are present throughout the cortex, the different forms they may acquire in diverse cortical regions or species, and whether functionality is always present at the narrowest level of vertical organization. The last question addresses the possibility that the minicolumn (and cortical column) may represent one type of functional unit among others; a system that may be activated for selected purposes but is not a general processor of information. I suspect that to the extent this is the case, it is the rare and probably not descriptive of the primate cortex, but the jury is still not in.

The anatomical minicolumn has at least three basic characteristics that when combined, set it apart from other elements in the cortex. These are *vertical organization, periodicity*, and *interconnected multiple components*. *Vertical organization* describes the interconnections of neurons within the vertical plane that crosses several lamina. This may not always refer to all six layers. In fact, as Rockland and Ichinohe (2004) have noted, there is no single anatomical element that we know of which actually encompasses all six layers. The closest to this are the long apical dendrite bundles that extend from Layer V to layer I, but even here, layer VI is excluded. However, the interrelated sharing between the different components does result in a vertical physiology that can cross all of the layers. Intrinsic optical studies for example, display a narrow vertical interconnectivity across the depth of the cortex (Kohn et al, 1997)

Periodicity refers to anatomical components that are located next to each other in a repeating fashion within a region or on a larger scale up to the entire cortex. This does not infer clone-like identical units, nor does it mean the spatial distances or physiological properties or anatomical elements are exactly the same. This repetition occurs within a very narrow size range, with the majority of spacing distances falling within 30-60 microns. These two characteristics comprise the most fundamental aspects of cortical vertical units. The reality of these features in neocortex is generally not controversial; especially if there is recognition of variability (Mountcastle, 2003).

The third characteristic is a combination of the first two; repeating *multiple vertical components* that share an anatomical relationship. One of the problems associated with the minicolumn is that it is composed of many parts that are not readily visible at the same time. The six-layer minicolumn is the product of interconnected sub-systems. The specificity of lamina and intracolumnar inhibition, means that the entire unit would rarely, if ever, be active at precisely the same moment, though delayed metabolic activation of these units across layers may be observed by intrinsic optical signaling as noted above (Kohn et al., 2002, 1997; Tommerdahl et al., 1993). The individual cells within a column are integrated by the interaction of multiple overlapping subsystems, and it is this which makes them a unit, and not a single anatomical entity.

The anatomical elements that typically comprise vertical organization include three fiber systems and two anatomical cell types. They are the (long) apical dendrites, myelinated axons, double bouquet cell axons, pyramidal cells in layers III, V, VI, and double bouquet cells. The double bouquet cell axon bundles may be a component of function for minicolumn inhibition, but do not appear to be as ubiquitous as the others. The apical dendrite bundles contain at least two main 'systems' that can vary within cortex and species (Rockland and Ichinohe, 2004). The 'long' system begins in layer V and terminates in Layer I, containing apical dendrites from pyramidal cells of layers V, III, and II, and is visible throughout the cortex. A shorter one extends from layer VI pyramidal cells and terminates in Layer IV. There is evidence of regional specificity regarding the beginning and termination of these bundles (Rockland and Ichinohe, 2004), and there are interesting specializations within the bundles themselves (Vercelli et al., 2004). However, apical dendrites of pyramidal cells seem to always bundle together and are present in a repetitive fashion. Myelinated axons bundle together as well, becoming prominent in the infragranula layers. In these instances, vertically oriented periodicity is the constant feature whereas the specifics are not. This pattern can also be said about other basic features of the cortex, such as pyramidal cells, which vary in size, distribution, neurotransmitters, and connectivity. The brain utilizes the 'template' of the pyramidal cell in numerous ways and therefore it is a building block of cortex. The fact that narrow vertical units consists of somewhere around one hundred cells make it a more powerful functional entity than the single cell, in the same way that the larger cortical column has more measurable physiological effects than the subunits within it.

Based on the discussion above, I prefer to use the term 'reiterative micro-vertical organization' because it is a descriptive term restricted to defining observed phenomena. Those are the characteristics of periodicity, vertical orientation, and at the micro-anatomical scale, which distinguish it from the larger metabolic or cortical columns. The term is applicable to cell arrays, various forms of apical dendrite bundles and their pyramidal cells, myelinated axon bundles, double bouquet cells and their axons, and output minicolumns (Vercelli et al, 2003). Vercelli et al (2004) coined the term 'output minicolumn' based on a detailed examination of apical dendrite bundles in rat V1. These bundles are present early in development and the cells from which they derive are probably clonally related (Rakic, 1988). The 'output minicolumns' describes segregated bundles within the minicolumn based on their projections. Certain projections bundle together as a subset within the main bundle. The only separate bundles based on output are those going to the dorsal lateral geniculate nucleus and are found in layer VI, a system that was already described by Escobar et al, (1986). This fascinating discovery demonstrates the basic vertical template on the one hand, with intra-columnar specificity on the other, and is an example of a term that describes a specific organization and anatomical relationship.

While we have been traditionally focusing on vertical organization as a processor of information, LaBerge (2001, 2006, 2007) argues that two types of mental activity take place within the cortical column; information processing and subjective experience. He posits that sustained attention is expressed in a cortical column by repeated surges of current that are found in the long layer V apical dendrite bundles (i.e., the micro-vertical unit or minicolumns). Information processing requires input and initiates a response in the form of output. On the other hand, with subjective experience, the activation of the long apical dendrites is the goal itself, and not a particular output. The input impulses are said to be converted into waves, which act as repeated surges of current within the apical dendrite shafts which forms the wave activity measured at the scalp as EEG oscillations. If true, this reveals another dimension of function at the level of the apical dendrite bundle, how single bundles contribute to the overall capacity of the larger cortical column. It also means that the conventional analysis of connectivity does not necessarily describe the functionality of the vertical system in total. A striking aspect of the morphology of layer V apical dendrites is that they bundle together and have a very long length versus diameter ratio. The apical dendrites are so long compared to their typical diameter that it is the equivalent of a 100 meter-long tube that is 16.66cm in diameter which yields a length-diameter ratio of 600:1. The result is that most inputs (except for those close to the soma) would decay before arriving to the soma. Those that do arrive lose their temporal and rate information. Supragranula apical dendrites, while not as long as those in layer V, still have a lengthy ratio. By comparison, basal dendrites typically have about a 5:1 ratio and are considered ideal for the processing of input information. Basal dendrites are in a much better position to relay direct information, or to do information processing. Furthermore, basal dendrites have many side branches while apical dendrites, whose orientation is vertical, have only a few. Other potential changes that may be occurring in these bundles have not been tested. These include a narrowing of the spacing between them, changing the diameter-length ratio of individual dendrites and bundles, and changes in the number of dendrites per bundle.

The relationship between the apical dendrite anatomy and the mental states alluded to above, can only be speculated. However, it provides a theoretical basis as to how alterations in the morphology of the apical dendrites can have effects on attention and other mental states. Properties of the wave form would potentially have a relationship with the number of long apical dendrite bundles per unit area as well as the intensity of their individual activity, which is based on length and number of cells within the circuit and the distance between them.

CorticoNeurogenesis

The genesis of the cortex occurs in the ventricles by a series of symmetrical and asymmetrical divisions (Rakic and Korack, 2001). In the first phase, cells located in the ventricular zone produce two additional progenitor cells with each mitotic cell division (Rakic, 1988). This symmetrical division is responsible for the number of founder cells which controls the total number of ontogenetic columns that will be produced in the cortex. According to the radial unit hypothesis, it is the number of these ontogenetic columns that determines the cortical surface area (Rakic and Kornac, 2001). At some point, progenitor cells begin to divide asymmetrically, producing one daughter cell that becomes a neuron and will move out into the cortical plate, and which will not undergo further division. The second phase is responsible for the number of cells within a column and the thickness of the cortex. Several clones of neurons that share a common site of origin in the ventricular zone use a common migratory pathway along the fascicles of the radial glial cells to settle within the same column in the cortical plate (Rakic, 2003). Radial glial cells create long fascicles that extend from the ventricular zone to the top of the cortical plate so that they span the entire width of the cerebral wall during corticoneurogenesis. New born nerve cells use these to traverse the cortical plate. Though there are small differences between radial glial cells among mammals, overall they are very similar in morphology and chemistry.

On the other hand, some cortical interneurons do not originate from the ventricular zone and migrate in a radial fashion. In rodents, this is most notable as the majority of cortical interneurons originate from the ganglionic eminence of the ventral telecephalon and migrate tangentially to the cortical plate (Marin and Rubenstein, 2001). In mice, up to 25% of all cortical neurons migrate non-radially, whereas in humans this percentage is less than 10% of the total (Letinic et al., 2002). Thus there are taxonomic specializations associated with this process.

The total amount of radial units that will be present in the cortex are controlled during embryogenesis by a few regulatory genes, while the final pattern and size of cytoarchitectonic regions is thought to be the work of a different set of genes (Rakic and Kornac, 2001).The final configuration of columns within a cytoarchitectonic area, is therefore the result of the genetic influences described above and epigenetic factors such as interactions of cells, inhibitory neurons, and afferent systems. It is clear to see that alterations in these genes or their influences can have profound effects on the cortex. The increase in founder cell number is exponential and not linear, so that a small prolongation of cell division or changes in length of the cell cycle would result in significant increases in the number of ontogenetic units produced.

The importance of adult vertical organization is based on its connection to the ontogenetic cell column. This relationship may either be a direct one, that is, the ontogenetic units and adult minicolumn are the same (see below), or the ontogenetic unit is the template upon which the adult cortex might possibly overlay new circuits according to regional and species requirements. Direct confirmation that a given ontogenetic column becomes an adult one in the same animal, is not possible using post-mortem studies since that requires different sets of animals for each age group. However, studies examining the size of fetal columns and fiber bundles in post-mortem tissue and early interconnectivity between pyramidal cells support the hypothesis that they are, at the very least, the basic pyramidal cell core described above, remains intact in the adult cortex (Buxhoeveden et al., 1996; Curtetti et al., 2002; Krmpotic-Nemanic et al., 1984; Lohmann and Koppen, 1995; LoTurco and Kriegstein, 1991; Ong and Carey, 1990; Peinado et al., 1993; Vercelli et al., 2004). In the early cortex, prospective pyramidal neurons are clustered into vertical columns which are also coupled by gap junctions (LoTurco and Kriegstein, 1991; Peinado et al.,1993).

Summary

Despite recent advances, fundamental questions about the cortex such as the number of cell types in the cortex, or the convergence of inputs to cells in the cortex, remain elusive (DeFelipe et al., 2002a). Perhaps the most cautious approach to micro-vertical organization is one that avoids oversimplification. Evidence supports the physiological basis for sub-cortical column organization in areas as diverse as motor, barrel cortex, and prefrontal cortex (Amirikian and Georgpoulos, 2004; Bruno et al., 2003; Georgpoulos et al., 2007; Ohki et al., 2005; Vercelli et al., 2004; Rao et al., 1999). Precisely defining how the minicolumn is anatomically and physiologically organized for different regions of the cortex remains a complex question (Ohki et al., 2005). Vertical organization appears capable of functioning at many different levels and the suggestion of 'structures at multiple spatial scales' is certainly plausible (Rockland and Ichinohe, 2004). The proposition that narrow vertical organization performs two distinct generalized functions (LaBerge, 2001, 2006) opens up new perspectives on the role of the narrow vertical unit that have yet to be explored. The ontogenetic column unit, as a template on which the adult cortex is built, may undergo more transformation in some regions of cortex than others, but the unifying feature seems to be in the outline and not the details (DeFelipe et al., 2002b). The fundamental structure would be defined as consisting of anatomical (and physiological) elements that are spatially narrow in size, demonstrate a vertical component to organization, and that can be found repeatedly within a cortical area. To the extent that this can be found in a given brain, the term 'reiterative micro-vertical organization' is one way of describing this template.

Models for Evolutionary Change in the Cortex

Mutational events occurring on regulatory genes that control the number of founder cells could easily result in a substantial increase in the number of ontogenetic columns above the amount normally produced for a given region. These in turn would create more initial ontogenetic units and potentially more adult minicolumns. Provided that there has not been an increase in total afferents to the region, the presence of additional ontogenetic columns means there will be more units to compete for the same input, thus altering the ratio of column units to afferent. It is reasonable under this condition to envision a decrease in the amount of neuropil space per column which would result in the phenomena of smaller than normal minicolumns (See Figure 1).

If the ratio between new ontogenetic columns far exceeds that of existing afferents, it might be expected that pronounced cell death would result, causing severe disruption of ontogenetic units. From this perspective it would be very difficult to add new ontogenetic units to the cortex during evolution because it would seem to require a match between additional columns and the afferent input. However, it appears that this is not required. The majority of synaptic input to cells in the cortex derives from intracortical circuits and the thalamic afferents contributes only a small portion of the mean number of synapses. A vertical unit of cells comprised of all the layers would thus have very only a small percentage of its synapses from the thalamus. The predominance of ispi and contralateral synaptic inputs found in the cortex can only help sustain new column units. However, this does not mean that thalamic input does not exert a strong influence on the response properties of cells and columns, which it does.

This means that rather than causing a strain on existing synaptic terminals, additional ontogenetic units immediately contribute to local and long distant circuits. The highest density of synaptic connections for a given neuron may be found within a relatively short distance of a parent neuron (Budd and Kisvardy, 2001; Elston 2000; Elston and Rosa, 2000), so that additional ontogenetic columns reciprocally connect to each other and become a major source for synaptogenesis. It is the subcortical and long distance afferent input that would have to be redistributed among the additional ontogenetic units and it is here that a drop in overall synapses per column might occur. In instances where there has been a significant increase in new ontogenetic units, without an increase in either subcortical or long distance input, the number of contacts per column would have to decrease as the ontogenetic units compete for these limited contacts during development. The afferent inputs would be distributed to a more units than before, resulting in form of signal divergence. On the other hand, the total number of synapses from local connections might be expected to undergo less of a drop, if any. The result would be a change in the ratio of intrinsic local synapses versus those from subcortical and other regions of cortex.

If there is a narrowing of cell columns that result from the assimilation of newer ontogenetic units, this would offset to some degree the expected increase in cortical surface area. Viewed in this manner, additional columns immediately become part of the cortical system, contributing synapses and receiving input in return. If there is significant cell death due to the sudden addition of too many new columns, this could lead to a rearrangement of connections between the affected region and its targets. This is one way that corticoneurogenesis could result in a re-reorganization that does not require an increase in brain size. An interesting result of adding significant numbers of columns units in one area would be on the efferent side, where the additional column units would give rise to an increase in axonal connections to their target regions. In these target areas this would result in more inputs. Hence, a change in one region would effect other areas even if they did not undergo alterations in the number of ontogenetic columns

Scenarios for Re-Organization and Encephalization Based on Ontogenetic Columns

It is important to note that the following scenarios are highly simplistic models of corticoneurogenesis and do not take account of numerous other factors. The emphasis is solely on the impact of new ontogenetic columns on circuits and connections. I will examine four possible relationships (figures 1-4). In the first there is a substantial increase in the number of additional ontogenetic columns—without an increase in afferent input. In the second, the number of columns is stable but there is an increase in afferent input. In the third one there are more column units created but there is a corresponding increase in afferent input. In the last example there is an increase in the number of columns produced and an even larger increase in afferent input coming into that region.

Additional Ontogenetic Columns without an Increase in Afferent Input (Figure 1).

This is a situation in which more columns are produced in one part of the cortex only. Thus, the amount of afferent input from subcortical and cortical areas is presumed to be unchanged. This means the additional columns must compete for the same number of afferents as the 'normal' contingent of columns units did before. In order for the columns to survive as whole units, there would have to be a reduction in the total number of connections per column unit (but not necessarily in the intrinsic connections). The resultant fewer synapses per column would lead to a reduction in the neuropil space. Depending on the actual relationships that develop, it is possible in this instance for there to be no change in overall surface area in this particular region of the cortex because though there has been an increase in column units. The decrease in neuropil space compensates for this and the result is stasis in regards to cortex size. This is one way in which additional units can be added to cortex without there necessarily being a concomitant change in surface area. Variations in column size have been found across primate species, regions, hemispheres, and disease states.

No Change in Number of Ontogenetic Units with an Increase in Afferent Input (Figure 2).

In this example, there is no change in the number of column units produced but there is an increase in afferent input. This would presumably result in rich synaptic areas that would increase the neuropil space and thus the distance between columns. This is an example of where a region may increase in size without an increase in ontogenetic units. Both of these examples demonstrate the need to measure column size as well as cortical region. The larger columns would become more generalized processors of information (Gufstassen 1997, 2004) than they were before, signaling a change in function.

More Column Units and a Matching Increase in Afferent Input (Figure 3).

In this model there is an increase in the number

of ontogenetic units and incoming afferents. This represents an instance where others regions of cortex may have supernumerary columns that are sending out more axons and/or increases could be coming from subcortical regions as well, or a combination thereof. This model is one where the added columns and inputs balance out so the size of the columns in that region remain the same as before, but now the size of the cortical region has undergone an increase because of the additional ontogenetic units. Whether this is the more typical scenario in evolution remains to be seen. The reported differences in column size, like cortical depth, are small compared to surface area but significant nonetheless.

In this instance the increase from incoming fibers is disproportionately greater than the increase in columns. This could be due to a significant increase in cell columns in other regions (scenario #1) resulting in especially large amounts of ipsi and contralateral connectivity, or events in subcortical regions that give rise to new cells and more connections, or both. It would lead to both an increase in column size and surface area. This may reflect the human condition (except for V1) where humans display larger columns and larger cortical regions. The behavioral success and selection pressures created by tool making could feed regions pertinent to those activities (i.e., somatosensory, motor, higher order) whereas other selection pressures derived from socialization, deception, theory of mind, etc., could have been fueling this kind of thing in higher order association cortex. The small columns found in visual cortex may reflect a relative homeostasis as regards initial visual processing, where differences between human and nonhuman primates is emphasized farther down the processing chain.

In all of these it must be considered that a change in the number of cells produced during the second phase will affect cortical depth and hence the size of the columns along the y axis. Columns can add more cells to each unit when there has been an increase in the depth of the cortex. This allows for changes in intrinsic complexity without increasing the diameter. This also creates the potential for more cells per column without increasing density. Because changes in cortical depth have been small compared to surface area, this aspect tends to be overlooked. However, a mere 10% increase in depth, spread throughout the cortex, can signify considerable increases in processing capacity per column and total number of new cells.

Summary

The addition of significant numbers of ontogenetic units in one region of the cortex with no increase in cortical or subcortical projections, would place all the units (in the affected region) at higher risk of increased cell death. Neurons must compete to attain enough synaptic connections to survive. If each new column contains about 80-100 neurons, then a 10% increase in ontogenetic units in a region containing 5000 minicolumns, means 500 new columns or about 4000-5000 new neurons would be added that have to find a home. The added ontogenetic columns have to compete with the 'existing' inputs for the limited amount of connections. The size of adult minicolumns would have to be based in some part on the interaction between the number of ontogenetic units created during neurogenesis, the amount of input to a region, and consequential intrinsic circuitry.

Based on the descriptions given above, it may be possible to make the following predictions regarding changes in the surface area. The first scenario would result in little or no change in surface area. The second would result in a modest enlargement of surface area. The third might also show a modest enlargement of surface area, and the last would result in the greatest increase in surface area. Further, all scenarios would probably tend towards some degree of change in circuitry and function. When coupled with other neurological changes (cell types, membrane properties, inhibition-excitation, up and down regulation, neurotransmitter quantities and subtypes, cell numbers, etc.), corticogenesis and the developmental period that follows can be envisioned as a time that is favorable to modification. However, most of it can be expected to account for individual variability rather than evolutionary events.

Did Minicolumns Get Smaller in Primate Evolution and What is the Functional Significance?

Traditionally, the number of column units produced has received the most attention because of the vast differences in surface area of the cortex. In the scientific literature, the size of minicolumns typically refers to their horizontal width or diameter. This is because the scale of variation for column size among species pales in comparison to that of the surface area. Nonetheless, the three-dimensional size of minicolumns does vary across species and area and may play a role in organization.

In primates, columns in visual cortex (V1) are notably small, both in absolute and relative terms when compared with data for other small mammals (Table 1). Even humans have smaller minicolumns than reported for animals like the cat or rat. The functional significance may be related to the species-specific complexity of primate vision and suggests that smaller columns may represent enhanced processing complexity (Peters and Sethares, 1996, 1997). Differences in the size of columns can represent functional differences and circuits, and Seldon's (1981) study of lateralization of minicolumns in human auditory cortex demonstrated some of the ways in which this might occur. Basically, functional connectivity is the result of the relationship between the size of the columns vis-à-vis the amount of extrinsic and intrinsic fiber terminals. If there is no change in an afferent terminal system but columns are much smaller in one brain compared to another, the distribution of the inputs will be different so that the incoming signal will be broken down into more units than in the former brain. This would theoretically

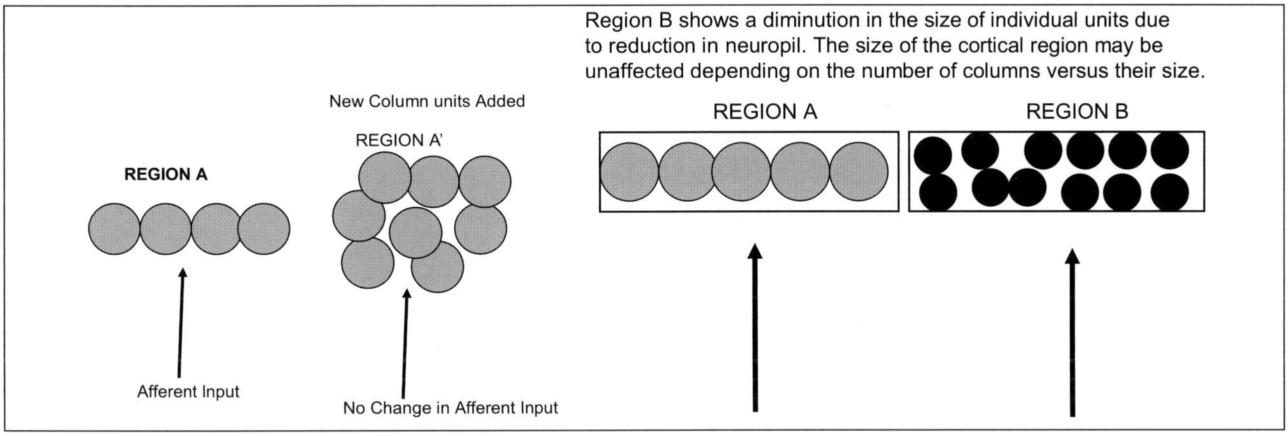

Figure 1. An increase number of ontogenetic units in a region that has no increase in afferent input. Region A (far left) represents the region in its normal configuration of ontogenetic columns and A within a box represents the normal region with developing minicolumns. Region A' is the same region with additional units added as a result of a prolongation of symmetrical cell division. The columns that survive in part by local connections (see text) would have fewer long distance connections and there would be a decrease in neuropil space. This would result in an increase in column number but no substantive change in surface areas.

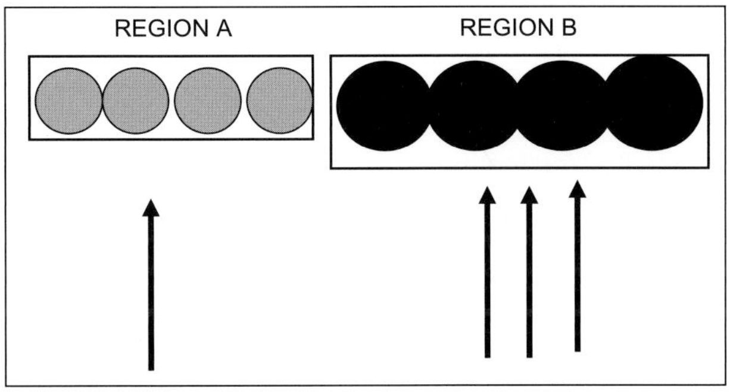

Figure 2. Same Contingent of Ontogenetic Units with an Increase in Afferent. This is the reverse of A. Region B has the the same number of units but is now exposed to more incoming fibers. This should increase the neuropil space resulting in an increase in column size. In this scenario, the surface would increase without an increase in column numbers. Conversely, in Figure 1 there would be an increase in column number without an increase in cortical surface area. Changes in cortical depth have not been figured into these scenarios but could play an role as well by allowing for more cells per unit without requiring a change in their diameter.

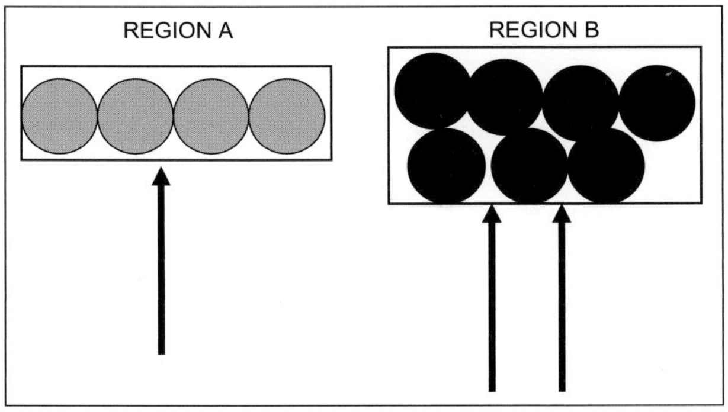

Figure 3. More Column Units and a Matching Increase in Afferent Input. This should result in an increase in surface area by virtue of additional ontogenetic units, but not in the size of the individual columns. Changes of this nature favor stability in regards to the amount of neuropil space per column.

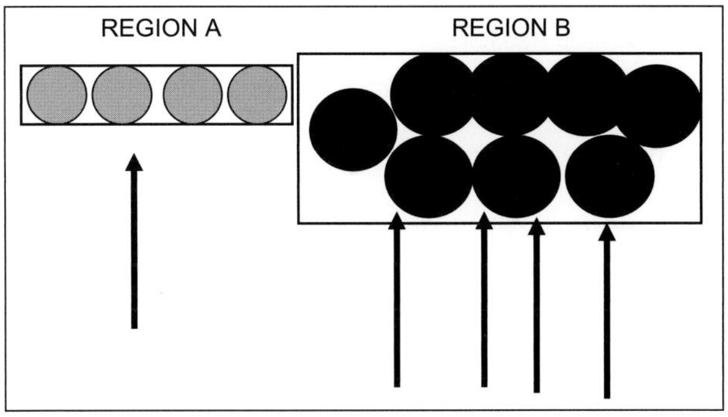

Figure 4. Additional Columns with Disproportionate Increase in Afferent Input. This figure demonstrates a condition where there is an increase in the number of columns and a proportionately greater increase in afferent input. The expected results would be to see an increase in neuropil space per column and thus an increase in their size. The combination of more columns and larger ones would cause the most significant increases in surface area of any of these proposed scenarios.

result in greater resolution or specificity of information processing (Gufstassen, 1997).

The 'size' of a minicolumn or ontogenetic cell column is usually defined according to the horizontal spacing distance between them, which can be measured on the basis of their pyramidal cells or fiber bundles. The major determinant of minicolumn size is the neuropil space that separates them in the horizontal plane (Seldon, 1981a). In the fetal cortex cells are packed tightly together and during development, the 'non-cell' space between them increases both in the vertical and horizontal axis. Thus, once the ontogenetic cell columns are in place, the emphasis on the expansion of this space causes the cells and their interconnected fiber systems to grow farther apart. A study of cell column development in humans showed that the neuropil space increases disproportionately to the column size during development (Buxhoeveden et al, 1996; unpublished data). Therefore, it is the increase in neuropil space that accounts for the majority of the enlargement of the individual columns. While other factors such as cell size, thickness of individual axonal or dendritic fibers, and bundle thickness contribute as well, this is more of a factor across species and brains of vastly different size.

Even though the horizontal spacing of minicolumns is a rather simplistic measure, differences found at this level represent profound changes in cortical development and organization. Changing the size of minicolumns affects the relationship between afferents, the intrinsic anatomy, and the physiology (Gufstasson, 2004; Seldon, 1981a). While the individual size of any given minicolumn varies according to extrinsic and intrinsic factors, the total number of cell columns is determined during corticoneurogenesis, so the *overall* number, and therefore the *mean* size of the columns, cannot change, provided their integrity remains intact (below).

A caveat must precede any discussion of data compiled for minicolumns size across species. The lack of uniformity in method and tissue preparation makes it difficult to make accurate comparisons across studies since this requires a stringent control of method, shrinkage, and preparation. Nonetheless, there is a degree of consistency in the results provided by the scientific literature that permits the making of certain generalizations. First, it can be seen that the size of minicolumns is not uniform but differs between species, within species, and within regions of the same brain (Buxhoeveden and Casanova, 2002, 2005). Secondly, there is no linear correlation between columns size and brain size for animals with diverse evolutionary history (Buxhoeveden and Casanova, 2002). However, it is possible that there may be some degree of correlation between column size and brain size within closely related taxonomic groups (Table 1).

Table 1, though limited in scope, represents data obtained by the use of very similar or identical methods and material preparation, which makes it more reliable than using results from disparate tissue, methods, and morphological elements. Even though the selection of species is small, a great number of primates are represented including all the greater and lesser apes. It is already apparent from this table that there is no correspondence between brain size and column size across diverse taxonomic categories. With the exception of humans, primates as a whole standout as having small columns in absolute size, and all primates examined so far including humans, display small columns in primary visual cortex.

The results are tantalizing because they suggest that columns are absolutely smaller in primates compared with other mammals studied thus far. The exception would be humans and possibly some overlap with the gorilla, but more samples will be needed. Even here, the column size in human matches that seen in small brain mammals, but does not exceed it. The data suggests that in the course of hominid evolution columns were getting larger along with the cortex. Of course when considered for brain size, all columns in primates are relatively small.

A dramatic example of the relative and absolute small size of minicolumns in primates is the Siamiri, which has a brain weight many times larger than that of the other small mammals examined, and yet their minicolumns are the smallest measured to date. Compared to the mouse, the brain is about 60x greater and yet it has smaller minicolumns. The complete answer to the question of column size variation, and whether they got smaller in the primate order, is a doable task but will have to await future research that includes large brained land mammals of similar or greater size than that of humans, as well as systematic analysis of many more mammals and primate species including prosimians. The

Table 1. Comparison of cell columns based on same or similar method of analysis. These areas do not contain data on area V1. In primates V1 is always smaller than found in other mammals so far tested and typically have mean values of ~30um. For a general comparison between other mammals using diverse methods and vertical anatomy, see Buxhoeveden and Casanova, 2002b.

Animal	Typical Brain Weights	Minicolumn Size
Primates		
Siamiri	25gms	20um
OWM	70-100gms	30+um
Great Apes	250-500gms	30-40+um
Humans	1350gms	40-50+um
Other Mammals		
Mouse	0.4gms	~26um
Rat	2gms	~40um
Rabbit	10gms	~40um
Cetaceans	350-3000gms	25-34um

especially small minicolumns in area V1 in primates is interesting because vision is a keystone of the initial primate radiation. Could the re-organization that occurred in primate visual cortex have affected the organization and hence size of columns in other regions as well? Finally, the cetaceans are interesting and they demonstrate very small columns in line with those of the primates, but their very thin cortex and unique aquatic evolution make direct comparisons to land mammals difficult.

Column size and brain size: Functional implications?

If larger brains contain more ontogenetic units and more cortical regions, this creates a target rich environment for additional columns to establish reciprocal connections with. Thus, large cortex may be better able to assimilate additional columns compared to smaller brains because they are being placed in an environment that has an abundance of cells, columns, and cortical regions. Presumably there are more target areas for the new columns to connect with and to receive input from. The implication is that the process of encephalization would proceed more slowly in a small brain and become easier as the brain enlarged.

There may be some relationship between processing complexity and column size. This is a model that could be tested but it would have to be done in the context of the total number of columns in a given brain. Small columns may be an indication of enhanced processing complexity based on increased interconnectivity between them. Or it could at least be representative of functional specialization. Some of the rationale for this is derived from the comparison of minicolumn size in primary visual cortex above (Peters and Yilmaz, 1993). The increased complexity is attained by having more columnar interconnections, more cells, and greater density of cells per column. This is assisted by the increased length (cortical depth) of the columns in primate brains that permits more cells to be placed in a narrower unit. Added to this is the fact that the total cortical volume devoted to V1 is much larger in primates, resulting in a huge increase in the total number of processing units. The approach taken by evolution of the primate brain is to have more column units, which increases the number of interconnection and the ability of each column to process more specific information. The alternative is to have fewer columns with more intrinsic connectivity and less interconnectivity. The combination of having smaller cell columns and more cortex devoted to a particular function, results in an enhancement of the resolution (based on narrower columns that 'break down' the input into more discrete properties), and it also allows more interconnectedness between these more specialized units.

Gustafsson (2004) proposes several scenarios that could lead to narrow columns. However, it must be noted that these arguments are based on a set number of already existing ontogenetic columns. One stems from neural network theory where self-organizing networks, columns in this instance, are formed when lateral feedback synaptic strength is a function of lateral distance as shaped by the Mexican hat model. If the inhibitory synaptic strengths increase the columns become narrower while the reverse is also true (Favorov and Kelly, 1994a,b; Gustafsson, 1997). This can be expected to occur during development. It is also possible for columnar organization to emerge without the usual lateral excitatory-inhibitory feedback mechanism. A basic organization can be laid down before the lateral feedback connections are developed so that when they do arise, they fine-tune or maintain the columnar organization. Others have reported that neural columns would be narrower if levels of nitric oxide (NO) were reduced so that given the same stimulus drive the column size varied according to the level of NO (Gally et al., 1990; Krekelberg and Taylor, 1996). It is also found to be involved in the metasynaptic organization of the frontal cortex in primate, but had no effect in visual cortex.

Finally, the ability to add more columns and connect to more regions enhances the opportunity for variability in larger brains. The variation in column size and brain size seems especially noticeable in human brain. How much of this is relative needs to be clarified and it remains to be seen whether animals with small columns and a small cortex have relatively less variation in the size of the columns and cortex than do large brained primates.

Summary

The process of normal encephalization cannot be the cause for a narrowing of cell columns. If this were the case, then minicolumns would have become progressively smaller in the millions of years of evolution which is counter to the evidence and which would hit a biological wall at some point since there must be a limit to how small a minicolumn can be. Cell density and column size across mammals is similar enough (though not identical) to demonstrate that as cortex enlarged by adding ontogenetic units, the 'new' units assumed the general size configuration of the host brain. A Darwinian model of cortical evolution would reflect incremental changes with a balance between the selective pressures for more columns on the one hand, and more afferents on the other. The result would be additional columns of similar size so that the presence of more columns results in a larger cortical area. At any one time the addition of 'new' columns can be expected to be limited with little or no change in mean column or cortical size. It can even be predicted that the process of adding new columns is so gradual that it would be difficult to measure significant differences from one generation to another.

Even though the horizontal spacing of minicolumns is a rather simplistic measure, differences found at this level represent profound changes in cortical development and organization. Changing the size of narrow vertical units (minicolumns) affects the relationship between columns and afferents, and alters the intrinsic anatomy and physiology (Gufstasson, 2004; Seldon, 1981a). While the individual size of any given minicolumn varies according to extrinsic and intrinsic factors, the total number of cell columns is determined during corticoneurogenesis, so the *overall* number, and therefore the *mean* size of the columns, cannot change, provided their integrity remains intact.

The process of encephalization that occurred in mammalian evolution is thought to arise from the addition of more ontogenetic units which is the basis for increased cortical surface area (Rakic and Kornac, 2001). Ontogenetic column number determines cortical surface area, whereas cortical cell numbers within them account for cortical depth (above). Since surface area has increased a thousand-fold (comparing mouse to human), while cortical depth has only increased around 3-4 times, the major impetus for cortical enlargement has been the addition of new ontogenetic units. Therefore, it can be expected that the addition of more ontogenetic cell columns should normally result in an increase in cortical surface area and white matter. However, this is based on the increase in number of columns of *similar size*. If the additional columns were to become smaller or larger, then this would alter the expected outcome in proportion to that change.

In summary, the size of minicolumns in adult cortex is at least partly the outcome of the number of ontogenetic units formed during development. If a significant number of additional columns are produced in one region the effects will be different than if additional columns occur simultaneously in several *interconnected* regions, where the cortico-cortical connections from each will help sustain the presence of the additional neurons. One can see how the distribution of synaptic connections can change as well. For example, if interconnected regions both incur a significant increase in ontogenetic columns, but not in thalamic input, then the ratio of thalamic to cortico-cortical input will presumably undergo change. The thalamic input, which is constant in number, will have to be distributed to more column units thereby lowering the number of inputs per column, whereas the number of cortico-cortical inputs will not decrease, and may even increase in one area if there is a disproportionate growth between the two regions. Furthermore, the number of intrinsic connections may also maintain their numbers as described above, which would result in a relative decrease of thalamic input compared to intrinsic and long distance connections. This is a theoretical concept that assumes all other factors are constant, but it demonstrates potential re-configuration of cortex due to changes in the numbers of ontogenetic units

Conclusions and Hypotheses

The elegance of the ontogenetic column lies in its explanatory power across a wide range of topics in brain evolution, comparative neuroanatomy, and anomalies of the brain (Buxhoeveden et al, 2006a,b, 2004; Casanova et al., 2003). The mechanisms described by the radial unit hypothesis are powerful tools in general neurobiology and especially so in the field of paleoneurology, and it is hoped that future work will further consider the potential applications associated with the radial unit hypothesis.

1. The ontogenetic unit is the main genetic determinant for the size of the cortex and is the template upon which later neurological events act. Thus it is a pertinent morphological and physiological object for the study of brain evolution.
2. The mutational events that initiate new columns and cells link developmental processes to re-organization and encephalization.
3. From the perspective of micro-vertical columns, it would seem that reorganization can occur without a demonstrated increase in brain size. This means that in hominid evolution it would not have been necessary for the hominid cortex to demonstrate significant enlargement from that of apes to prove it had undergone reorganization.
4. The result of these processes is to enhance heterogeneity in the configuration of the cortex, both across and within species.
5. It may be easier to induce increases in cortical size in a larger brain than a smaller one. Cortical enlargement proceeds faster in larger brains until constrained by other factors (i.e., pelvis, white/grey matter ratio, metabolics, etc).
6. There may be more variability among minicolumns in larger brains due to the increase in number of regions and regional specialization.
7. Cell columns may have become absolutely smaller in the evolution of the primate order. On the other hand, columns in humans are the largest among primates and may reflect both significant increases in additional minicolumns and in afferent input coming into those columns.
8. Smaller minicolumns may represent a reorganization that favors increase complexity based on maximizing specificity and enhanced resolution.
9. Rather than making the argument for a clone-like homogeneity of the cortex, the micro-vertical organization of cortex is a template upon which cortical heterogeneity is played out, one that can result in diverse modular configurations in the adult animal.

References

Amirikian, B., Georgopoulos, A.P., 2003. Modular organization of directionally tuned cells in the motor cortex: is there a short-range order? Proceedings of the National

Academy of Sciences of the U.S.A. 100, 12474-12479.

Bruno, R.M., Kharti, V., Land, P.W., Simons, D.J., 2003. Thalamocortical angular tuning domains within individual barrels of rat somatosensory cortex. The Journal of Neuroscience. 23, 9565-9574.

Budd, J., Kisvarday, Z.F., 2001. Local lateral connectivity of inhibitory clutch cells in layer 4 of cat visual cortex (area 17). Experimental Brain Research 140, 245-250.

Buxhoeveden, D., Semendeferi, K., Schenker, N., Switzer, R., Ph.D., Courchesne, E., 2006. Reduced minicolumns in the frontal cortex of patients with autism. Neuropathology and Applied Neurobiology 32, 483-91.

Buxhoeveden, D., Hasselrot, U., Buxhoeveden, N.E., Booze, R.M., Mactutus, C.F., 2006. Microanatomy in 21 day rat brains exposed prenatally to cocaine. International Journal of Developmental Neuroscience 24, 335-341.

Buxhoeveden, D., Casanova, M.F., 2005. The cell column in comparative anatomy, Chap 5. Encephalization, minicolumns, and hominid evolution, Chap 6. In: Casanova, M.F., (Ed.) Neocortical Modularity and the Cell Minicolumn. Hauppauge, New York: Nova Science Publishers.

Buxhoeveden, D., Casanova, M.F., 2004. Accelerated maturation in brains of patients with Downs syndrome. Journal of Intellectual Disability 48,705-706.

Buxhoeveden, D., Casanova, M.F., 2002. The minicolumn and evolution of the brain. Brain, Behavior, and Evolution 60, 125-151

Buxhoeveden, D.P., Lefkowitz, W., Loats, P., Armstrong, E., 1996. The linear organization of cell columns in human and nonhuman anthropoid Tpt cortex. Anatomy and Embryology 194, 23-36.

Casanova, M., Buxhoeveden, D., Gomez, J., 2003. Disruption in the inhibitory architecture of the cell minicolumn: implications for autism. The Neuroscientist 9, 496-507.

Catania, K.C., 2002. Barrels, stripes and fingerprints in the brain-implications for theories of cortical organization. Journal of Neurocytology 31, 347-358.

Curtetti, R., Garbossa, D., Vercelli, A., 2002. Development of dendritic bundles of pyramidal neurons in the rat visual cortex. Mechanism of Ageing and Development 123, 473-479.

DeFelipe, J., Alonso-Nanclares, L., Arellano, J.I., 2002. Microstructure of the neocortex: comparative aspects. Journal of Neurocytology 31, 299-316.

Elston, G.N., 2000a. Pyramidal cells of the frontal lobe: all the more spinous to think with. Journal of Neuroscience 20(RC95), 1-4.

Elston, G.N., Rosa, M.G.P., 1997. The occipitoparietal pathway of the macaque monkey: comparison of pyramidal cell morphology in layer III of functionally related cortical visual areas. Cerebral Cortex 7, 432-452.

Elston, G.N., Rosa, M.G.P., 2000b. Pyramidal cells, patches, and cortical columns: a comparative study of infragranular neurons in TEO, TE, and the superior temporal polysensory areas of he macaque monkey. The Journal of Neuroscience 20(RC117), 1-5.

Escobar, M.I., Pimenta, H., Caviness Jr., V.S., Jacobson, M., Crandall, J.E., Kosik, K.S., 1986. Architecture of apical dendrites in the murine neocortex: dual apical dendritic systems. Neuroscience 17, 975-989.

Favorov, O.V., Kelly, D.G., 1996. Stimulus-response diversity in local neuronal populations of the cerebral cortex. NeuroReport 7, 2293-2302.

Favorov, O.V., Kelly, D.G., 1994a. Minicolumnar organization within somatosensory cortical segregates I: development of afferent connections. Cerebral Cortex 4, 408-427.

Favorov, O.V., Kelly, D.G., 1994b. Minicolumnar organization within somatosensory cortical segregates II: emergent functional properties. Cerebral Cortex 4, 428-442.

Gustafsson, L., 2004. Comment on "Disruption in the inhibitory architecture of the cell minicolumn: implications for autism". Neuroscientist 10, 189-191.

Gustafsson, L., 1997. Inadequate cortical feature maps: a neural circuit theory of autism. Biological Psychiatry 42, 1138-1147

Hasselmo, M.E., 2005. A model of prefrontal cortical mechanisms for goal-directed behavior. Journal of Cognitive Neuroscience 17, 1115-1129

Jin, X., Mathers, P.H., Szabo, G., Katarova, Z., Agmon, A., 2001. Vertical bias in dendritic trees of non-pyramidal neocortical neurons expressing GAD67-GFP *In Vitro*. Cerebral Cortex 11, 666-678.

Johansson, C., Lansner, A., 2007. Towards cortex sized artificial neural systems. Neural Networks 20, 48-61.

Jones, E.G., 2000. Microcolumns in the cerebral cortex. Proceedings of the National Academy of Sciences of the U.S.A. 97, 5019-5021.

Kisvarday, Z.F., Ferecsko, A.S., Kovacs, K., Buzas, P., Budd, J.M., Eysel, U.T., 2002. One axon-multiple functions: Specificity of lateral inhibitory connections between large basket cells. Journal of Neurocytology 31, 255-264.

Krieger, P., Funer, T., Sakmann, B., 2007. Synaptic connections between Layer 5B pyramidal neurons in mouse somatosensory cortex are independent of apical dendrite bundling. Journal of Neuroscience 27, 11473-11482.

Kozloski, J., Hamzei-Sichani, F., Yuste, R., 2001. Stereotyped position of local synaptic targets in neocortex. Science 293, 868-872.

Kohn, A., Pinheiro, A., Tommerdahl, M.A., Whitsel, B.L., 1997. Optical imaging *in vitro* provides evidence for the minicolumnar nature of cortical response. NeuroReport 8, 3513-3518.

Kohn, A., Metz, C., Tommerdahl, M.A., Whitsel, B.L., 2002. Stimulus-evoked modulation of sensorimotor pyramidal neuron EPSPs. Journal of Neurophysiology 88, 3331-3347.

Krmpotic-Nemanic, J., Kostovic, I., Nemanic, D., 1984. Prenatal and perinatal development of radial cell columns in the human auditory cortex. Acta Oto-laryngologica 97, 489-495.

LaBerge, D., 2001. Attention, consciousness, and electrical wave activity within the cortical column. International Journal of Psychophysiology. 43, 5-24.

LaBerge, D., 2006. Apical dendrite activity in cognition and consciousness. Consciousness and Cognition 15, 235-257. Epub 2005 Nov 8. Review.

Laberge, D., Kasevich, R., 2007. The apical dendrite theory of consciousness. Neural Networks 20, 1004-1020. Epub 2007 Sep 11.

Letinic, K., Zoncu, N., Rakic, P., 2002. Origin of GABAergic neurons in the human neocortex. Nature 417, 645-649.

LoTurco, J.J., Blanton, M.G., Kriegstein, A.R., 1991. Initial expression and endogenous activation of NMDA channels in early neocortical development. The Journal of Neuroscience 11, 792-799.

Lucke, J., von der Malsburg, C., 2004. Rapid processing and unsupervised learning in a model of the cortical macrocolumn. Neural Computation 16, 501-533.

Lucke, J., 2004. Hierarchical self-organization of minicolumnar receptive fields. Neural Networks 17, 1377-1389.

Krmpotic-Nemanic, J., Kostovic, I., Nemanic, D., 1984. Prenatal and perinatal development of radial cell columns in the human auditory cortex. Acta Oto-laryngologica 97, 489-495.

Lohmann, H., Koppen, H.J., 1995. Postnatal development of pyramidal dendritic and axonal bundles in the visual cortex of the rat. J Hirnforsch 36, 101-111.

Mountcastle, V.B., 1997. The columnar organization of the neocortex. Brain 120, 701-722.

Mountcastle, V.B., 2003. Introduction. Computation in cortical columns. Cerebral Cortex 13, 2-4.

Ohki, K., Chung, S., Ch'ng, Y.H., Kara, P., Reid, R.C., 2005. Functional imaging with cellular resolution reveals precise micro-architecture in visual cortex. Nature 433, 597-603. Epub 2005 Jan 19

Ong, W.Y., Carey, L.J., 1990. Neuronal architecture of the human temporal cortex. Anatomy and Embryology 181, 351-369.

Peinado, A., Yuste, R., Katz, L.C., 1993. Gap junctional communication and the development of local circuits in neocortex. Cerebral Cortex 3, 488-498. [Review]

Peters, A., Sethares, C., 1997. The organization of double bouquet cells in monkey striate cortex. Journal of Neurocytology 26, 779-797.

Peters, A., Sethares, C., 1996. Myelinated axons and the pyramidal cell modules in monkey primary visual cortex. Journal of Comparative Neurology 365, 232-55.

Peters, A., Yilmaz, E., 1993. Neuronal organization in area 17 of cat visual cortex. Cerebral Cortex 3, 49-68.

Rakic, P., 2003. Developmental and evolutionary adaptations of cortical radial glia. Cerebral Cortex 13, 541-549. Review.

Rakic, P., Kornack, D.R., 2001. Neocortical expansion and elaboration during primate evolution: a view from neuroembryology. In: Falk, D., Gibson, K.R., (Eds.). Evolutionary Anatomy of the Primate Cerebral Cortex. Cambridge (England): Cambridge University Press, pp. 30-56.

Rakic, P., 1988. The specification of cerebral cortical areas: the radial unit hypothesis. Science 241, 928-931.

Rakic, P., 1978. Neuronal migration and conduct guidance in primate telencephalon. Postgraduate Medical Journal 54, 25-40.

Rakic, P., 1972. Mode of Cell Migration to the superficial layers of fetal monkey cortex. The Journal of Comparative Neurology 145, 61-84.

Rao, S.G., Williams, G.V., Goldman-Rakic, P.S., 1999. Isodirectional tuning of adjacent interneurons and pyramidal cells during working memory: evidence for microcolumnar organization in PFC. Journal of Neurophysiology 81, 1903-1916.

Reich, D.S., Mechler, F., Victor, J.D., 2001. Independent and redundant information in nearby cortical neurons. Science 294, 2566-2568.

Rockland, K.S., Ichinohe, N., 2004. Some thoughts on cortical minicolumns. Experimental Brain Research 158, 265-277.

Silberberg, G., Gupta, A., Markram, H., 2002. Stereotypy in neocortical microcircuits. Trends in Neurosciences 25, 227-230. [Review]

Sugimoto, S., Sakurada, M., Horikawa, J., Taniguchi, I., 1997. The columnar and layer-specific response properties of neurons in the primary auditory cortex of Mongolian gerbils. Hearing Research 112, 175-185.

Swindale, N.V., 1990. Is the cerebral cortex modular? Trends in Neurosciences 13, 487-492.

Tommerdahl, M., Favorov, O., Whitsel, B.L., Nakhle, B., Gonchar, Y.A., 1993. Minicolumnar activation patterns in cat and monkey S1 cortex. Cerebral Cortex 3, 399-411.

Vercelli, A.E., Garbossa, D., Curtetti, R., Innocenti, G.M., 2004. Somatodendritic minicolumns of output neurons in the rat visual cortex. European Journal of Neuroscience 20, 495-502.

CHAPTER 13

The Evolution of Cortical Neurotransmitter Systems Among Primates and Their Relevance to Cognition

Mary Ann Raghanti, Patrick R. Hof and Chet C. Sherwood

Abstract

The neurotransmitters dopamine, serotonin, and acetylcholine are known to exert modulatory effects on prefrontal cognitive functions, such as learning and memory processes, by regulating neuronal activity. In addition, all three neurotransmitters are implicated in neurodegenerative processes to which humans appear to be uniquely susceptible. Taken together, these facts suggest that neuromodulatory supply to the prefrontal cortex may have been modified during human evolution to support human-specific cognitive and behavioral specializations. This chapter reviews recent molecular, genetic, anatomical, and behavioral evidence concerning the contributions of neuromodulatory transmitter systems to the evolution of human and nonhuman primate brains.

Introduction

One of the most compelling and complicated questions yet to be answered in the study of human evolution involves identifying the neuroanatomical bases of human behavior and cognition. One hallmark of the modern human condition is an enlarged brain. However, while expansion of the brain in the course of human evolution has certainly made an important contribution to our species' intelligence, total brain size alone may not fully explain the origin of human-specific cognitive abilities. Indeed, despite its expanse, the human brain matches up on a macroscopic level, nearly part for part, to that of a macaque monkey, with similar nuclei, neocortical areas, and axon pathways (e.g., Petrides and Pandya, 1994; Preuss and Goldman-Rakic, 1989). What, then, underlies our species' elaborated capacity for reason and symbolic thinking?

As Ralph Holloway put it in 1968, "This preoccupation [with overall brain size] led to the use of these mass aspects as explanations *in themselves* of behavioral differences of quite specific natures, such as 'memory', 'insight', 'forethought', 'symbolization' etc. It should be obvious that such correlations are not causal analyses, and that a parameter such as brain weight in grams, or volume in ml, or area in sq. mm. cannot explain the differences in behavior which are observed" (Holloway, 1968). Throughout his career, Ralph Holloway championed the notion that "reorganization" is equally as important as encephalization in understanding the evolution of the human brain. The evidence that he marshaled in favor of this proposal concerned both macro- and microanatomical changes (Holloway, 1996). Some aspects of gross structural reorganization can be seen in the fossil record, including redistribution of neocortical area volumes as indicated by sulcal positions and the asymmetry of cerebral hemispheres (e.g., Holloway, 1985; Holloway and De La Costelareymondie, 1982). Other structural changes, however, are not recorded in paleontological remains because they are at the microscopic or molecular level. Such evolutionary modifications to neuroanatomy, nonetheless, can comprise a very significant mechanism for encoding behavioral variation within and between species.

Recent comparative research into the neuroanatomical microstructure of primates has begun to yield tantalizing clues regarding uniquely human specializations (Buxhoeveden et al., 2001; Dorus et al., 2004; Hof et al., 2001; Nimchinsky et al., 1999; Preuss and Coleman, 2002; Sherwood et al., 2007). It is becoming clear that a real understanding of human evolution requires that we supplement traditional volumetric neuroanatomical studies with comparative data on the infinitesimal aspects of our species' brain histology, connectivity, and gene expression patters in order to see how they differ from those of our closest living relatives (i.e., apes and monkeys; see, e.g., Preuss, 2000; Preuss, 2006). Several researchers have noted that the human brain is not merely an enlarged chimpanzee brain (Penn et al., in press; Premack, 2007) and Ralph Holloway made the prescient statement more than 40 years ago that "One c.c. of chimpanzee cortex is not equivalent to one c.c. of human cortex" (Holloway, 1966, p. 108).

This conclusion is supported by numerous findings indicating that the histology and molecular composition of the modern human neocortex differs from that of other primates, including chimpanzees. For example, humans demonstrate histological differences from other apes in having a distinctive patterned arrangement of dendrites and interneurons in layer IVA of primary visual cortex, possibly translating into functional differences in how the visual pathway processes motion-related cues (Preuss and Coleman, 2002). There is also a unique neuronal subtype, the Von Economo neurons (VENs), found in great apes and humans with a restricted cortical distribution within anterior cingulate and frontoinsular cortex (Nimchinsky et al., 1999; Nimchinsky et al., 1995). Human VENs differ from those of great apes in being more numerous and having larger somata. These spindle-shaped neurons are projection neurons that are enriched with dopaminergic D3 and serotonergic 2b receptors (Allman et al., 2005). The unique localization and biochemical phenotype of VENs indicate that they may have played a critical role in mediating intuitive processes that synthesize a homeostatic representation of the self with social information, a key component of the capacity to attribute mental states to others and to envision oneself projected into alternative scenarios (Allman et al., 2005). Humans may also be distinguished from nonhuman primate species in having increased population-level asymmetry of neuropil across several cortical areas, including area Tpt and the primary motor cortex representation of the hand (Buxhoeveden et al., 2001; Sherwood et al., 2007). Other studies have identified potential species differences in the functional biochemistry of the cortex using genomic and molecular approaches. For example, adaptations for increased neuronal activity and energy production appear to have occurred in human evolution through the evolution and upregulation of genes involved in the aerobic metabolic pathway (Cáceres et al., 2003; Uddin et al., 2004) and proliferation of glial cells (Sherwood et al., 2006).

Additional compelling candidates for microscopic modifications during human evolution include the serotonergic, cholinergic, and dopaminergic systems of the brain. These neurotransmitters are key components of higher cognitive functions, including learning and memory processes, language comprehension, and overall intelligence (Azmitia, 1999; Goldman-Rakic, 1998; Hasselmo, 1995; Herremans et al., 1995; Previc, 1999; Sarter and Bruno, 1997). Moreover, these systems are selectively compromised in human-specific neuropathological conditions that result in devastating cognitive deficits, including Alzheimer's disease, Parkinson's disease, and schizophrenia (Akil et al., 1999; Mega, 2000; Naughton et al., 2000; Roth et al., 2004; Venator et al., 1999; Whitehouse, 1992). Finally, the innervation patterns of these neurotransmitters are regionally distinct (i.e., different cortical areas receive variant levels of input), with the primate neocortex receiving denser and more complex patterns of innervation relative to other mammals (e.g., Berger et al., 1991; Berger et al., 1988). The evidence of neurotransmitter regional heterogeneity, their roles in higher cognitive functions, and their deficits in human neurodegenerative diseases all collectively raise the question of how human neuromodulator systems differ from those of other primates. Is it possible that an increased reliance on neuromodulators has contributed to human-specific intellectual advances, and that this in turn has rendered humans uniquely susceptible to neurodegenerative diseases? This chapter will explore this question by reviewing the properties and effects of dopaminergic, serotonergic, and cholinergic systems with a concentration on their roles in cognitive functions mediated by the prefrontal cortex (PFC). In addition, we summarize the results of our recent comparative studies of the innervation pattern of these neuromodulators across the frontal cortex of macaque monkeys, chimpanzees and humans.

The PFC lies rostral to the premotor and primary motor regions of the frontal lobes (Uylings and van Eden, 1990) and is involved in many of our higher order functions, including personality, working memory, attentional processing, mental state attribution (also known as "theory of mind"), behavioral inhibition, and planning and executing actions (Fuster, 1997; Goldberg, 2001). Humans and great apes together share an enlarged frontal cortex relative to other primate species (Semendeferi et al., 2002). Further, human PFC appears to be disproportionately larger compared to great apes, as primary motor (area 4) and premotor cortex (area 6) occupy a smaller proportion of the frontal lobe when compared with other species (Deacon, 1997; Preuss, 2004; Rilling, 2006). The PFC is comprised of a network of many cytoarchitecturally and functionally distinct regions that send and receive projections from virtually all other cortical regions as well as subcortical structures (Brodmann, 1909; Fuster, 1997). Through these extensive connections, the PFC synthesizes information from motor, sensory, and limbic areas of the brain, thus integrating the functions of all other brain regions, leading Goldberg (2001) to liken this brain region to the conductor of an orchestra.

What is a Neuromodulator?

Neurotransmitters regulate neuronal communication through actions mediated by various receptor subtypes; depending on the effect at the postsynaptic target, neurotransmitters may also act as neuromodulators. Classically defined, neurotransmitters have effects that are immediate and short-term, usually engaging an ion channel to allow for the influx or efflux of charged ions. In these instances, the neurotransmitter receptors form ion channels, with no associated downstream metabolic consequences (Gu, 2002; von Bohlen und Halbach and Dermietzel, 2006). In contrast, neuromodulatory actions are slower, of longer duration, and are more spatially diffuse (Hasselmo, 1995). Receptors mediating the neuromodulatory actions of neurotransmitters belong to the family of G-protein linked receptors that, once activated, may signal multiple signal transduction pathways. In this way, neurotransmitters have long-term effects on the processing characteristics of cortical networks by influencing synaptic transmission and pyramidal cell adaptation (Dreher and Burnod, 2002; Goldman-Rakic, 1998; Hasselmo, 1995).

The separate neuromodulatory transmitter systems share several defining characteristics (e.g., Gu, 2002). First, the cortical neuromodulator systems that target the cerebral cortex are derived from discrete subcortical neuron populations that have long projection axons. The effects of neuromodulators may be either excitatory or inhibitory in nature, depending upon the postsynaptic receptor complex, and neuromodulators further have the capacity to mediate their own release via autoreceptors located at the presynaptic site. In addition to the ascending projection axons that innervate neurons of the PFC, there are also reciprocal connections between the PFC and the subcortical neuronal populations that modulate neurotransmitter systems through descending projections. Further, the separate systems also interact with one another to fine-tune cellular excitability (for a review, see Briand et al., 2007). It is likely that multiple systems act in concert with one another, rather than being recruited separately, to support cognition. However, as will be discussed further below, while acting synergistically, each of the transmitter system functions to support discrete components of cognition. For example, Robbins and Roberts (2007) recently reviewed the differential contributions of DA, ACh, 5HT and norepinephrine within the PFC in the performance of attentional set-shifting tasks that assess cognitive flexibility and perseverative deficits, highlighting the distinct and separate components of the task supported by each transmitter system. In this task, it was shown that DA supports set formation, 5HT mediates reversal learning, and ACh is necessary for serial reversal learning (Robbins and Roberts, 2007). This evidence illustrates the highly orchestrated organization of neuromodulatory transmitter systems in regulating cognitive processing, and highlights the importance of understanding each system's contribution to the evolution of human intellectual abilities.

Dopamine (DA)

Dopaminergic systems originate in the midbrain and include the mesostriatal system that sends projections to many subcortical areas (e.g., striatum, nucleus accumbens), and the mesocortical system that innervates the frontal, piriform, and entorhinal cortices (Fuster, 1997; Squire et al., 2003). It is the mesocortical DAergic system that innervates the PFC, sending out long projection axons from cell bodies located in the nucleus parabrachialis pigmentosus of the ventral tegmental area (VTA) (Goldman-Rakic et al., 1989; Smiley et al., 1999). DA does not act as an excitatory or inhibitory neurotransmitter in the PFC (e.g., González-Burgos et al., 2002). Rather, DA acts as a neuromodulator, targeting its G-protein linked receptors on apical and basal dendritic shafts and spines of pyramidal glutamatergic cells (Benavides-Piccione et al., 2005; Goldman-Rakic et al., 1989; Seamans and Yang, 2004) and on the dendrites of parvalbumin-containing GABAergic interneurons involved in inhibitory processes (Goldman-Rakic et al., 1989; Sesack et al., 1995; Sesack et al., 1998). Thus, DA is capable of moderating the signal-to-noise ratio within the PFC by preventing interruptions in the active maintenance of information (Dreher and Burnod, 2002; Kulisevsky, 2000; Winterer and Weinberger, 2004).

At least five known receptor subtypes interact with DA (D1 - D5), each being functionally distinct (Seamans and Yang, 2004). The five receptor subtypes are classified into two groups, D1-like (D1 and D5) and D2-like (D2, D3, and D4) (von Bohlen und Halbach and Dermietzel, 2006), with D1-like receptors activating (G_s) and the D2 group inhibiting (G_i) adenylate cyclase. D1 and D2 receptors mediate DAergic actions within the PFC and exhibit lamina- and region-specific distributions, with D1 receptors concentrated in supragranular layers and D2 receptors preferentially located in layer V in macaques and humans (Goldman-Rakic et al., 1990; Goldman-Rakic et al., 1992; Lidow et al., 1989). Further, D1 and D2 receptors have differential binding affinities for DA, with D2 receptors demonstrating an increased sensitivity to low concentrations of DA compared to D1 receptor binding affinity (Grace, 2000).

Through its neuromodulatory actions in the PFC, DA is well known for regulating working memory, the capacity to hold a finite amount of information "on-line" in order to comprehend and plan actions (Abi-Dargham, 2004; Goldman-Rakic, 1998; Kulisevsky, 2000). Blocking the actions of DA impairs performance on working memory tasks (Brozoski et al., 1979; Cools et al., 2002; González-Burgos et al., 2002; Robbins, 2000; Sawaguchi and Goldman-Rakic, 1991), and agonists are associated with improved performance (Akil et al., 1999; Brozoski et al., 1979). In addition, the extent of cortical DAergic innervation correlates with behaviors involved in planning voluntary actions that invoke working memory functions, indicating that an increase in DAergic afferents allows for these functional capacities (Nieoullon, 2002). Additional executive functions that rely on DAergic input to the PFC include language comprehension, reasoning, and overall intelligence (Arnsten et al., 1995; Boshes and Arbit, 1970; Dreher and Burnod, 2002; Goldman-Rakic, 1998; Sawaguchi and Goldman-Rakic, 1991).

Available evidence indicates that there are phylogenetic differences in cortical DA innervation. Berger et al. (1991) qualitatively compared cortical DA innervation, as measured by tyrosine hydroxyase-immunoreactive (TH-ir) axon density, in rodents (rats) and primates (rhesus macaques, long-tailed macaques, and humans) and found that the primates exhibited denser and more extensive DAergic innervation within the cerebral cortex. In fact, humans and other primates receive DAergic input to all cortical areas. This is in contrast to rodents who have little or no DAergic innervation in the motor, premotor, and supplementary motor areas, or to the parietal, temporal, and posterior cingulate cortex. There are also differences in laminar distribution, with widespread dense innervation of layer I in primates, whereas in rodents, dense innervation of the superficial layers occurs only in a few select areas, such as the anterior cingulate cortex and entorhinal cortex (Berger et al., 1991). Humans and other primates demonstrate different regional patterns, with TH-containing fibers in all cortical layers of agranular cortices, and a bilaminar pattern, with the densest innervation in layers I and V-VI, in granular somatosensory and association cortex (Berger et al., 1991; Gaspar et al., 1989; Lewis et al., 2001). The differences between primates and rodents in both the organization of the frontal cortex and in DAergic innervation of this area are striking and suggest evolutionary changes that paralleled increases in both the size and functional differentiation of the cerebral cortex (Berger et al., 1991; Gaspar et al., 1989; Lewis et al., 2001; Preuss, 1995; Sesack et al., 1995; van Eden et al., 1987; Williams and Goldman-Rakic, 1998).

A broader view of phylogenetic differences was provided by Hof and colleagues in an analysis of cortical TH-ir axon distribution in the harbor porpoise and pilot whale (Hof et al., 1995). Their findings revealed a different pattern of innervation of cetacean auditory and visual cortex compared to that of other mammals. Most other mammals share in common a sparser DAergic innervation in primary sensory cortex relative to all other cortical areas. This is particularly true of the primary visual cortex, where DAergic axons are rarely evident in rodents, and present only in layer I of human and nonhuman primates. In contrast, the cetacean primary visual cortex is innervated throughout all layers and is more densely innervated than the auditory cortex, whereas the reverse is true for other mammals. Such phylogenetic differences strongly suggest a potent role for this neurotransmitter in brain evolution.

Additional lines of evidence suggest that DAergic systems may have been further altered during human evolution. DAergic system dysfunction plays an important role in a number of neuropsychiatric disorders presenting with cognitive deficits that appear to be exclusive to humans (Akil et al., 1999; Ciliax et al., 1999; Sutoo et al., 2001; Venator et al., 1999; Winterer and Weinberger, 2004). In fact, Previc (1999) has argued that an expansion of DAergic innervation in the human cerebral cortex is singularly responsible for the origin of human intelligence.

Raghanti et al. (submitted) recently tested the hypothesis that humans have an increased DAergic input to prefrontal cortical areas relative to chimpanzees and macaque monkeys. In this study, dorsolateral prefrontal area 9 (involved in inductive reasoning and specific components of working memory) and medial prefrontal area 32 (involved in "theory of mind") were examined, with primary motor cortex (area 4) serving as a control region, as it is not associated with higher cognitive functions (Figure 1). TH-ir axon length density to neuron density (i.e., innervation per neuron) was quantitatively assessed in layers I, II, III, and V/VI of each cortical area using computer-assisted stereology (MBF Biosciences, Williston, VT). Species differences were not detected in the primary motor cortex, but several differences were detected in the prefrontal areas. Humans exhibited a sublaminar pattern of innervation in layer I of areas 9 and 32, meaning that DA-containing axons were restricted to the bottom of layer I rather than being distributed evenly through the layer, as observed in other species. This pattern of sublaminar innervation of the molecular layer has been reported in agranular cortices of long-tailed macaques (Berger et al., 1988), but may have been extended in humans to include both agranular and granular cortices, a pattern that has been suggested to have evolutionary implications (Gaspar et al., 1989). Humans and chimpanzees together deviated from macaques in having an increased density of DAergic afferents in layers III and V/VI of the prefrontal cortical areas. This is particularly interesting in light of the lamina-specific decrease in TH-ir axons in layer VI of area 9 reported for schizophrenic subjects (Akil et al., 1999), a deficit that may potentially underlie the working memory deficits associated with this disease (Abi-Dargham, 2004; Akil et al., 1999). It is conceivable that an increase in DAergic innervation to the PFC infragranular layers is critical to the integrity of executive functions governed by these brain regions, and this may be a point of vulnerability for the progression of neuropathological processes.

In addition, morphological specializations in the form of TH-ir axon coils (Figure 2), were found in human and chimpanzee cortex, to the exclusion of macaques. Although TH-ir coils were previously described in humans, the functional significance of these structures is unknown (Benavides-Piccione and DeFelipe, 2003; Gaspar et al., 1989). However, analogous axonal configurations have been reported in human cortex immunostained for cholinergic axons, with the suggestion that they may represent local events of cortical plasticity or local circuit alterations (Mesulam et al., 1992), as will be discussed further below.

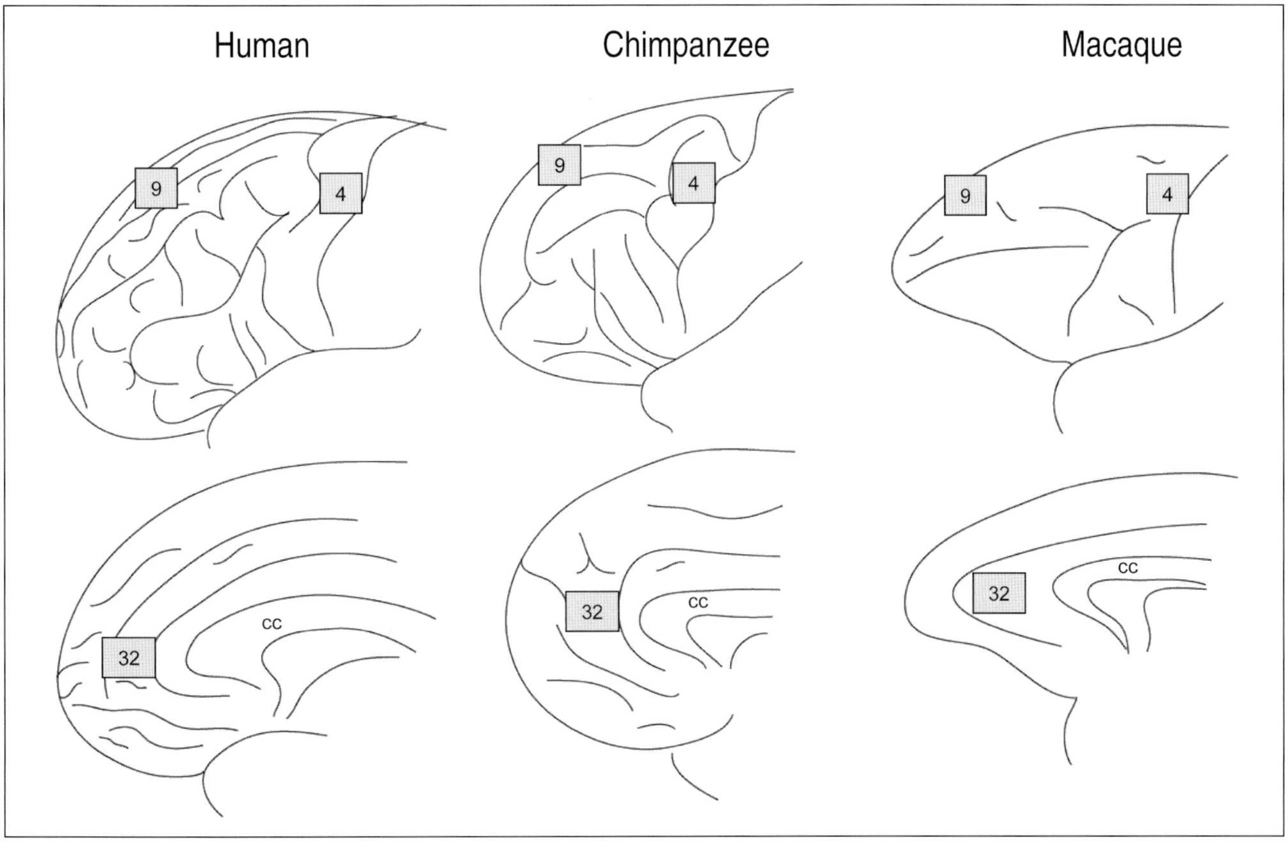

Figure 1. Lateral (upper) and medial (lower) views of human, chimpanzee, and macaque brains. The cortical regions sampled are labeled with their respective numerical designations.

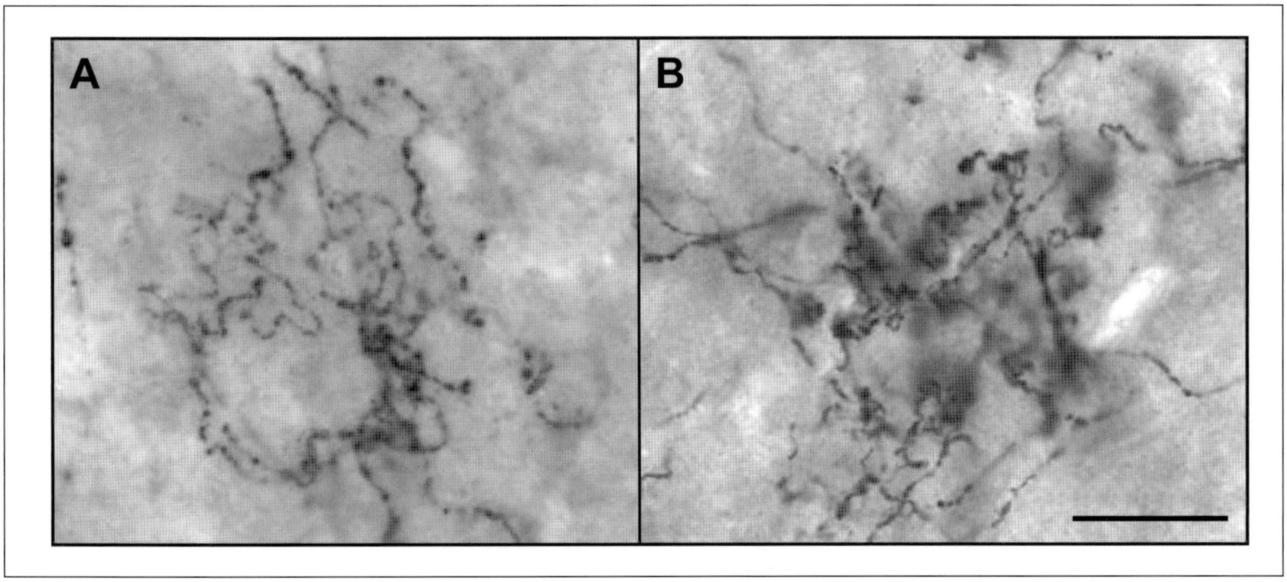

Figure 2. TH-ir axon coils in human (A) and chimpanzee (B). Scale bar = 25 µm.

The presence of TH-ir neurons in the neocortex has been noted in several vertebrate species (for review see Smeets and González, 2000). These cortical neurons have been classified as aspiny non-pyramidal cells (Benavides-Piccione and DeFelipe, 2003) and demonstrate considerable species-specific variation in their location and distribution within the cortical mantle. For example, TH-ir neurons were noted in the lower portion of layer I in two cetacean species (Hof et al., 1995), whereas rats express TH-ir cells in all layers of the cortex, with the highest density occurring in layers II and III (Kosaka et al., 1987). We noted the presence of TH-ir cells in lower layer VI and in the white matter immediately adjacent to layer VI in all frontal cortical fields of the Moor macaque (*Macaca maura;* unpublished data). In the human neocortex, TH-ir interneurons are found almost exclusively

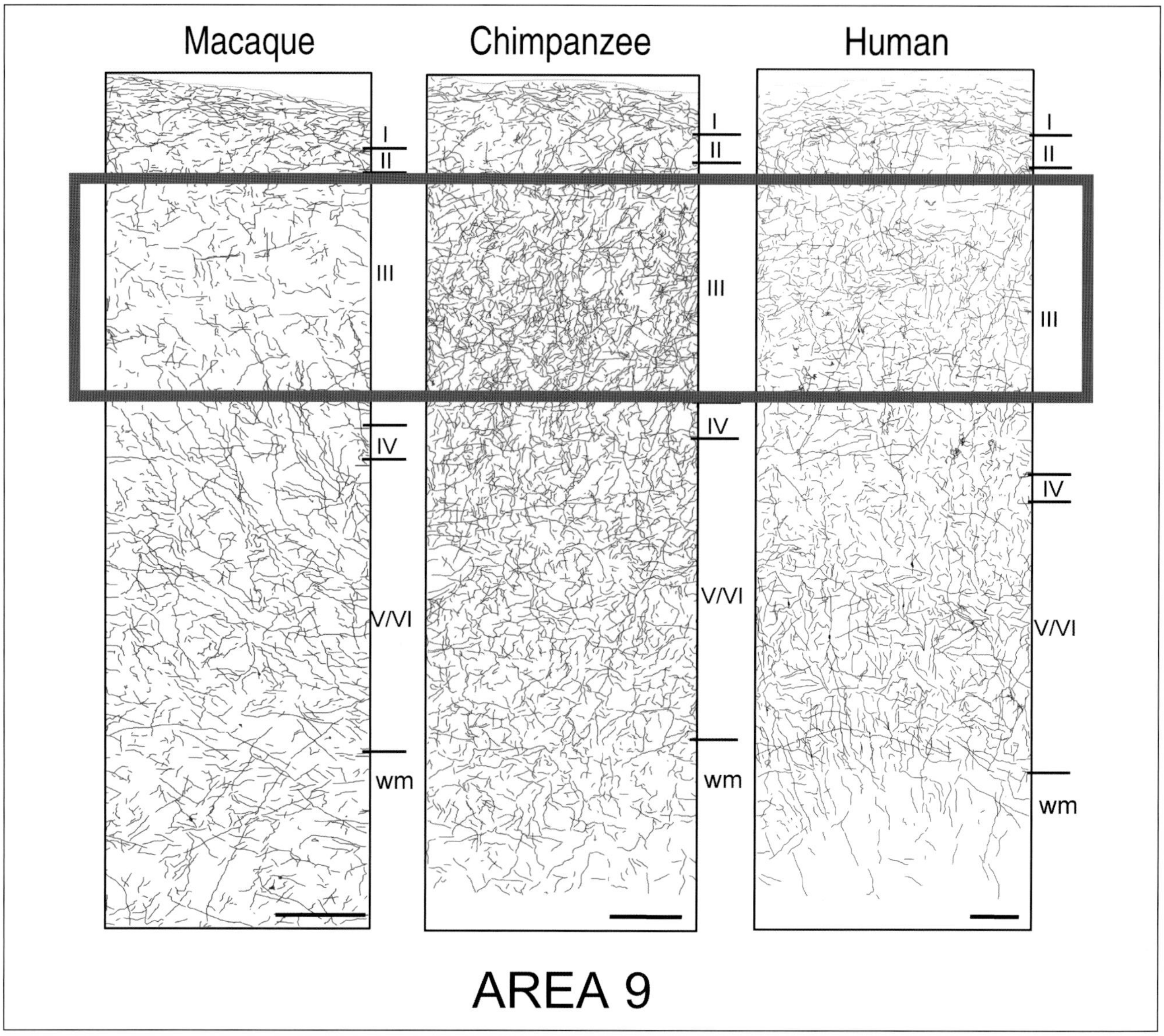

Figure 3. TH-ir axon tracings in PFC area 9 for macaque, chimpanzee, and human. Scale bar = 250 μm, 'wm' = white matter.

in the infragranular layers and subjacent white matter (Benavides-Piccione and DeFelipe, 2003; Gaspar et al., 1987; Hornung et al., 1989), and humans are the only species to express these cells in all cortical areas examined (Benavides-Piccione and DeFelipe, 2007).

Although these neurons have been noted in several species, little is known regarding their function. However, recent studies have illustrated a significant decrease in cortical TH-ir neuron numbers in individuals afflicted by dementia with Lewy bodies (Marui et al., 2003) and in individuals with Parkinson's disease (Fukuda et al., 1999). Benavides-Piccione and DeFelipe (2007) recently assessed the density and distribution of TH-ir neurons across eleven cortical areas in humans. They reported significant regional differences in TH-ir neuron density, and have proposed an event in human evolution that involved the utilization of this intrinsic source of cortical DA to support function-specific cortical circuits.

In humans, we have also observed TH-ir neurons mostly in layers V/VI, and rarely in layer III of cortical areas 9, 32, and 4 (unpublished data). Interestingly, we have not observed TH-ir neurons within any neocortical region of great apes, including chimpanzee, bonobo, gorilla, and orangutan. This is intriguing given that the TH-ir axon length density to neuron density ratio in layer III of chimpanzee areas 9 and 32 was twice that of humans or macaques (Raghanti et al., submitted) (Figure 3). Perhaps the DAergic afferent input to layer III increased in chimpanzees and other great apes to compensate for the loss of TH-ir cells within the cortical mantle.

Serotonin

Serotonergic neurons are located in the dorsal and median raphe nuclei of the brainstem and their projections are widely distributed throughout the brain (Azmi-

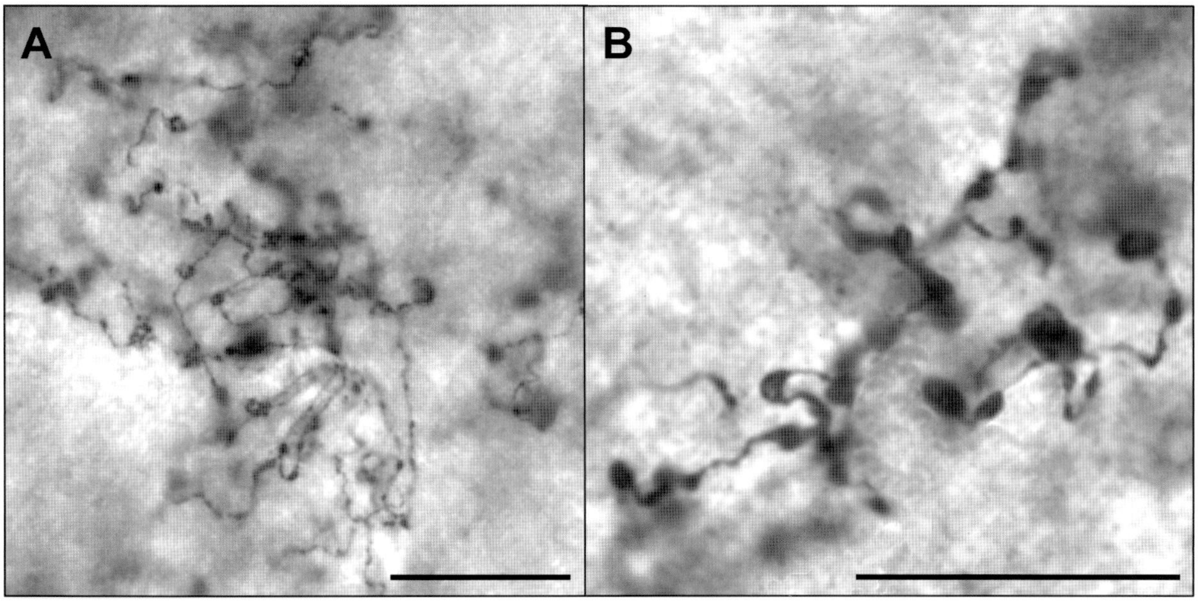

Figure 4. SERT-ir axon coils in human (A) and chimpanzee (B). Scale bar = 25 μm

tia, 1999; Bradshaw, 2003; Kandel et al., 1995). There exists a duality of serotonergic innervation within the mammalian cerebral cortex, with type 1 axons (thin with small ovoid varicosities) originating from the dorsal raphe nuclei and type 2 axons (thin with large spherical varicosities) arising from the median raphe nuclei (Hornung et al., 1990; Kosofsky and Molliver, 1987; Miner et al., 2000; Mulligan and Törk, 1988; Trottier et al., 1996; Wilson and Molliver, 1991). This duality of innervation and the variations observed in the local patterns of cortical serotonergic afferents suggests that the separate classes of axons target different populations of cortical neurons, thereby selectively affecting specific elements of cortical networks (Hornung et al., 1990; Wilson and Molliver, 1991).

Serotonin is known to interact with many different receptor types and subtypes, many of which are still incompletely understood (Bradshaw, 2003; Buhot, 1997). Currently, there are at least fourteen different 5HTergic G-protein linked receptors recognized within the central nervous system, classified according to their second-messenger system, location, and binding affinity (Buhot, 1997; von Bohlen und Halbach and Dermietzel, 2006). There is also one ligand-gated ion channel receptor ($5HT_3$) (Jakab and Goldman-Rakic, 2000). The diverse array of 5HT receptors may support cell- and circuit-specific postsynaptic effects, allowing 5HT to modulate many different functions at once (Briand et al., 2007; von Bohlen und Halbach and Dermietzel, 2006). The effect of 5HT on cortical neurons may yield either enhanced or inhibited activity, depending on the post-synaptic receptor type (Buhot, 1997).

The localization of 5HT receptors in the PFC facilitates 5HT's contributions to memory and cognition (Azmitia, 1999; Buhot, 1997; Marek and Aghajanian, 1998). 5HT is also involved in behavioral inhibition, and its action may be to moderate the influences of synapses from intrinsic local circuits in comparison to extrinsic sources, allowing for learning (Buhot, 1997; Harrison et al., 1999). 5HT acts as a neuromodulator, targeting its receptors on dendritic shafts and the perisomatic region of pyramidal cells as well as several subclasses of inhibitory interneurons (Buhot, 1997; Jakab and Goldman-Rakic, 2000). Thus, 5HT is able to alter directly the excitability of pyramidal cells by targeting their dendrites or indirectly through inhibitory interneurons (Jakab and Goldman-Rakic, 2000).

Several lines of evidence illustrate the role of 5HT in cognitive functions. For example, 5HT levels are positively correlated with accuracy of performance on an attention task in rats (Puumala and Sirviö, 1998). Moreover, drugs that increase central 5HT concentrations, such as 5HT uptake blockers, improve attention, visual and verbal memory, working memory, and processing speed in intact, healthy rodents as well as in macaques and patients with schizophrenia (Buchanan et al., 2003; Meneses and Hong, 1995; Williams et al., 2002). Also, prefrontal depletion of 5HT in marmosets results in impaired performance on serial discrimination reversal tasks and decreased cognitive flexibility (Clarke et al., 2004; Clarke et al., 2005). Dysfunction of 5HTergic systems contribute to the cognitive disturbances associated with depression and suicide, obsessive-compulsive disorder, anxiety disorders, and impulse-control disorders (Austin et al., 2002; Bradshaw, 2003; Noguchi et al., 2001). In addition, cortical depletion of 5HT has been noted in human neurodegenerative diseases including schizophrenia, Parkinson's disease, and Alzheimer's disease (e.g., Naughton et al., 2000; Vergé and Calas, 2000). For example, Thomas et al. (2006) reported a 47% decrease in serotonin transporter (SERT) density in the PFC of Alzheimer's disease patients regardless of depressive symptoms.

Reports on the distribution of serotonergic afferents within the primate frontal cortex have been published for marmosets (*Callithrix jacchus*) (Hornung et al., 1990), long-tailed macaques (*Macaca fascicularis*) (Azmitia and Gannon, 1986; Berger et al., 1988; Wilson and Molliver, 1991), rhesus macaques (*Macaca mulatta*) (Smiley and Goldman-Rakic, 1996; Wilson and Molliver, 1991), vervets (*Chlorocebus aethiops*), and humans (Trottier et al., 1996; Varnäs et al., 2004). Recent comparative research indicates that the cortical 5HTergic system has been substantially reorganized in humans and chimpanzees relative to macaques (Raghanti et al., 2007). As with DA, coils of 5HTergic axons were found in humans and chimpanzees but were absent in macaques (Raghanti et al., 2007) (Figure 4). Additionally, humans and chimpanzees together deviated from macaques in having denser 5HTergic input in layers V/VI of cognitive prefrontal areas 9 and 32, with no species differences detected in the primary motor cortex (Raghanti et al., 2007). Of interest in this context, Hornung et al. (1990) noted that the only consistent difference between earlier macaque studies and their analysis of the marmoset cerebral cortex was a weaker 5HT innervation of the infragranular layers in marmosets. Although one cannot assume that a single species represents a larger phylogenetic clade, it is tempting to speculate that within the order Primates there might be a shift across New World monkeys, Old World monkeys, and hominoids towards increasing innervation of cortical output layers within the PFC. Additionally, the infragranular layers of rats are sparsely innervated relative to the layers I-IV of cats (DeFelipe et al., 1991). Although carnivores are only distantly related to primates, these findings raise the possibility that primates diverged from other mammalian species in having an increased reliance on 5HTergic afferents in cortical output functions. The fact that humans and chimpanzees have emphasized this difference is particularly notable when considering that Austin et al. (2002) found a specific reduction in 5HT transporter density in layer VI (but not in layers II or IV) of dorsolateral prefrontal cortical area 46 in depressed subjects who committed suicide. This layer-specific deficit in 5HT transmission may contribute to the cognitive deficits, such as the disruption of working memory processes, that are characteristic of major depression (Pelosi et al., 2000).

An additional species differences in cortical serotonergic input includes the presence of pericellular arrays (or 'baskets') formed by type 2 axons and surrounding nonpyramidal neurons in the supragranular layers. Pericellular arrays have been reported in cats (DeFelipe et al., 1991; Mulligan and Törk, 1987), marmosets (Hornung et al., 1990), rhesus, long-tailed, and Moor macaques (Foote and Morrison, 1984; Raghanti et al., 2007; Smiley and Goldman-Rakic, 1996; Wilson et al., 1989), and chimpanzees (Raghanti et al., 2007). However, no such morphologies were observed in the human neocortex (Raghanti et al., 2007). Interestingly, both morphological specializations, pericellular arrays in cats and nonhuman primates and axon coils in humans and chimpanzees, are comprised of type 2 serotonergic axons that originate from the median raphe nuclei. There is evidence of a greater number of 5HTergic neurons in the median raphe of cats and primates relative to that of rodents (Azmitia and Gannon, 1986; Jacobs et al., 1984). This putative increase in cell number may be correlated with the incidence of pericellular baskets, as this feature has not been detected in rodents (Audet et al., 1989). It has been suggested that axons arising from the dorsal raphe nuclei (type 1 axons) play a specific role in prefrontal cognitive control because this is the most abundant axon type found in the PFC (Briand et al., 2007). However, albeit type 2 axons are not as abundant within the PFC as type 1 axons, this axon type does form morphological specializations and therefore may play a critical role in supporting cognitive flexibility and learning capabilities.

As noted, 5HT has a role in regulating behavioral inhibition (Soubrié, 1986), and differences in the capacity for behavioral inhibition have been reported among primate species. Chimpanzees and humans share the capacity to demonstrate self-control in delay-maintenance tasks in order to maximize rewards (Beran and Evans, 2006; Evans and Beran, 2007a). In contrast, rhesus macaques fail to demonstrate a consistent ability for similar self-control (Evans and Beran, 2007b). Cools et al. (2007) recently posited that 5HT actions within the orbital PFC mediate emotional processing and behavioral output by facilitating descending projections, presumably originating within the infragranular layers (i.e., layers V/VI). While quantitative comparative data have not been reported for regions of the orbital PFC, it is likely that, similar to areas 9 and 32, humans and chimpanzees share an increased serotonergic innervation in the infragranular layers of orbital PFC regions. If this is the case, then the increased density of serotonergic afferents within infragranular layers of human and chimpanzee orbital PFC may facilitate their enhanced behavioral inhibition capacities in delay of gratification tasks.

Acetylcholine

Cholinergic axons originate from the magnocellular neurons of the nucleus basalis of Meynert of the basal forebrain and project to all regions of the cerebral cortex, with a substantial degree of regional heterogeneity (Ichikawa and Hirata, 1986; Lehmann et al., 1984; Lewis, 1991; Lysakowski et al., 1986; Mesulam and Geula, 1991; Mesulam, 2004; Mesulam et al., 1992; Mesulam et al., 1986). Additionally, subpopulations of cholinergic cells within the nucleus basalis preferentially project to specific areas within the PFC (Ghashghaei and Barbas, 2001). Although cortical cholinergic input is ubiquitous, the existence of regional differences in cholinergic innervation of cortical areas and distinct laminar preferences support the concept that cholinergic systems have specific local circuit processing properties. Cholinergic axons (as measured by ChAT-immunoreactive axons) synapse on glutamatergic pyramidal neurons as well as on layer-specific populations of GABAergic interneurons (Chadhuri et al., 2005; Mesulam, 2004; Mrzljak et al., 1995).

The actions of ACh are mediated either by nicotinic or muscarinic receptors. Nicotinic receptors are ligand-gated ion channels that have immediate excitatory effects with no associated second messenger systems (Gu, 2002; von Bohlen und Halbach and Dermietzel, 2006). There are five genes for muscarinic receptors (M1-M5), each belonging to the class of G-protein linked receptors (Bonner et al., 1987). As with the other G-protein linked receptors discussed, muscarinic receptors may have excitatory or inhibitory effects depending on the postsynaptic complex and each receptor subtype appears to have region- and layer-specific concentrations within the cortical mantle (Bozkurt et al., 2005; Mrzljak et al., 1998; Mrzljak et al., 1993; Rasmusson, 2000).

The role of ACh in cognition was initially demonstrated in studies using ACh receptor antagonists in humans and rats (Deutsch, 1971; Drachman, 1977). ACh projections to the PFC act to enhance input processing in attentional contexts and to facilitate memory encoding (Blokland, 1996; Harder et al., 1998; Levin and Simon, 1998; Sarter and Parikh, 2005). ACh accomplishes this by amplifying the influence of synapses from outside the cortex relative to those from other cortical pyramidal cells (Hasselmo, 1995). The input provided by ACh to the PFC is critical to the learning process (Levin and Simon, 1998; Sarter and Parikh, 2005; Steckler and Sahgal, 1995) and is also important in cognitive flexibility and working memory (Levin and Simon, 1998; Sarter and Parikh, 2005; Steckler and Sahgal, 1995). The administration of scopolamine, a cholinergic antagonist, eliminates the capacity to form episodic memories and diminishes the ability to analyze information or acquire semantic knowledge in macaque monkeys (Harder et al., 1998). Furthermore, lesions or drugs that deplete ACh cortical innervation in primates and rodents impair learning and memory in the acquisition and performance on discrimination tasks that challenge attentional processes (Fine et al., 1997; Harder et al., 1998; Irle and Markowitsch, 1987; Levin and Simon, 1998; McGaughy et al., 2000; Sarter and Parikh, 2005), and learning and memory deficits are ameliorated with ACh agonists (Levin and Simon, 1998; Wu et al., 2000).

Cortical ACh has been suggested as a potential marker for intelligence due to its fundamental role in attentional processes, learning, and memory (Gray and Thompson, 2004). Interindividual variation in cortical cholinergic innervation in the hippocampus, caudate nucleus, and frontoparietal cortex was positively correlated with performance on learning tasks in mice (Durkin et al., 1977), and it has been suggested that variations in cholinergic innervation of human and nonhuman primates would reflect individual differences in learning abilities (Mesulam et al., 1986). Several researchers have found that neurodegenerative diseases such as Alzheimer's disease, Parkinson's disease, and schizophrenia demonstrate reduced ACh activity in prefrontal cortical areas implicated in learning and memory (Mega, 2000; Sarter and Parikh, 2005; Whitehouse, 1992), as measured by ChAT-immunoreactive (ChAT-ir) fibers (Katzman et al., 1988; Mechawar et al., 2000; Mega, 2000; Sigle et al., 2003). For example, Beach et al. (1997) reported significant decreases in cholinergic fiber densities in both the entorhinal cortex and inferior temporal gyrus associated with the preclinical stage of Alzheimer's disease. Furthermore, Ikonomovic et al. (2007) demonstrated a loss of both cholinergic fiber and varicosity densities in prefrontal cortical area 9 in mild to moderate Alzheimer's disease that correlated with impaired cognitive function. Recent evidence also indicates an accelerated rate of protein evolution in primates relative to rodents for three cholinergic receptor subtypes (Dorus et al., 2004), suggesting that natural selection has modified the postsynaptic function of this system along the lineage leading to humans.

A comparative study of cortical ACh input involved measuring the amount of potassium-induced ChAT activity in mouse versus human neocortical slices (Sigle et al., 2003). Relative to mice, only a very low concentration of potassium was required to induce ChAT activity in humans. Several studies have analyzed cortical cholinergic afferents in different mammalian species, and comparisons across species can be made using these data. ChAT-ir axons are present in all cortical areas and layers of rat neocortex (Ichikawa and Hirata, 1986; Lysakowski et al., 1986; Mechawar et al., 2000) with the frontal cortex receiving the densest complement of fibers and layer I having the highest laminar density (Mechawar et al., 2000). In contrast, Old World primates exhibit a rostrocaudal gradient of innervation, with the most rostral areas of the frontal cortex demonstrating fewer fibers than the caudal motor and premotor areas (Lewis, 1991; Mesulam et al., 1992; Mesulam et al., 1986).

Our recent quantitative comparative analysis of ChAT-ir axons in the frontal cortex found that humans and chimpanzees together demonstrated a pattern of innervation that emphasized cholinergic input to layers III and V/VI of prefrontal cortical areas 9 and 32 relative to macaques, whereas no species differences were detected in the primary motor cortex (Raghanti et al., 2008). Further, clusters (or coils) of cholinergic fibers were present in humans and chimpanzees, but not in macaques (Figure 5). Differences among primates have also been reported for the localization of galanin relative to cholinergic neurons in the basal forebrain (Benzing et al., 1993; Kordower et al., 1992). Galanin is an inhibitory modulator of ACh in rats (Elvander and Ögren, 2005; Laplante et al., 2004) and galanin-ir fibers are hyper-

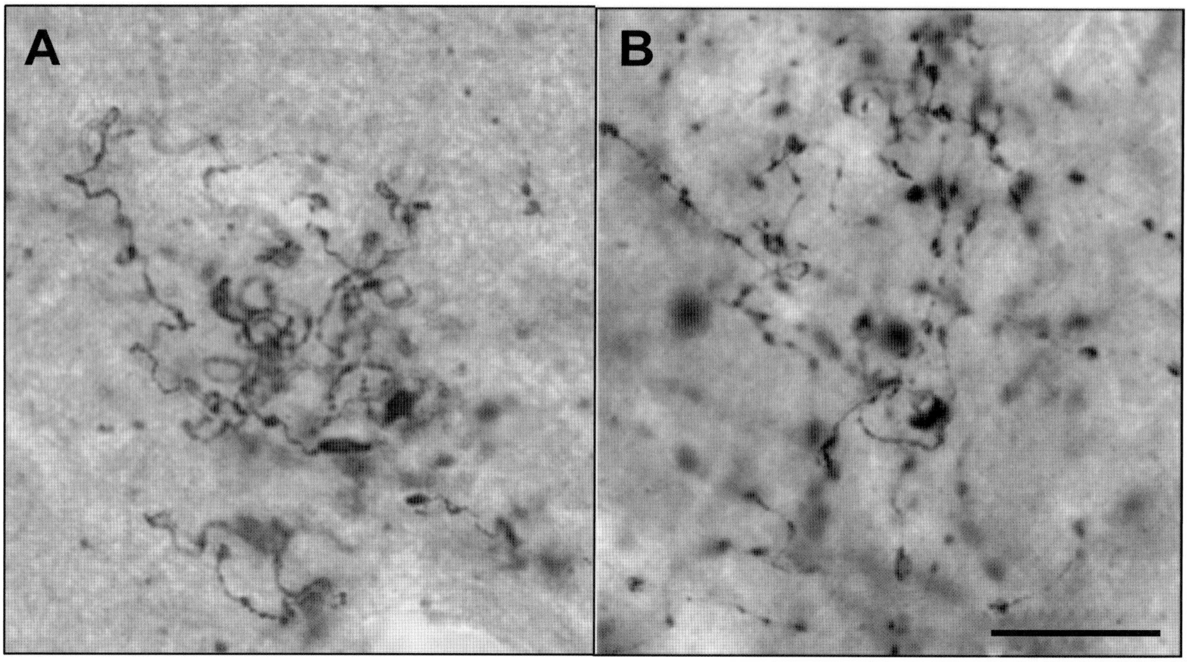

Figure 5. ChAT-ir axon coils in human (A) and chimpanzee (B). Scale bar = 25 μm

trophied in Alzheimer's disease (Mufson et al., 2000). Galanin hyperfunction is associated with cholinergic hypofunction and likely contributes to the associated learning and memory deficits characteristic of Alzheimer's patients (Chan-Palay, 1988). Among primates, hominoids (gibbons, chimpanzees, and gorillas) displayed a distinctively different localization of galanin immunoreactivity relative to monkeys (brown capuchins, rhesus macaques, and baboons) (Benzing et al., 1993). Taken together, it appears that both cholinergic innervation and the modulation thereof were altered in the evolution of apes and humans.

Neuromodulatory Transmitter Summary

As discussed, the DAergic, 5HTergic, and cholinergic neuromodulatory systems play important roles in the regulation of specific cognitive functions. It is likely that these neuromodulator systems do not act exclusively, but several neuromodulator systems may have evolved in concert with one another to support higher cognitive specializations in humans, including language. Previous research revealed that phylogenetic differences exist among humans and other primate species in neuromodulator axon length density relative to neuron density (DA, 5HT, and ACh) in prefrontal cortical areas involved in cognition (areas 9 and 32), but not in primary motor cortex (Raghanti et al., 2007; Raghanti et al., 2008; Raghanti et al., submitted). Additionally, the unique morphological appearance of coils, highly varicose axons surrounding specific cells, is suggestive of a distinct functional specialization. This morphological specialization of neuromodulatory axons is seen only in humans and great apes. Neuromodulatory transmitters have well described functions in cortical plasticity, modifying cortical neuron response properties as mediated by numerous receptor subtypes for each neuromodulator (Gu, 2002; von Bohlen und Halbach and Dermietzel, 2006). Specific effects include long-term potentiation and long-term inhibition, depending on the properties of the post-synaptic element involved. Both long-term potentiation and inhibition alter the response properties of neurons, characteristic of cortical plasticity. The distinctive morphology of coils indicates that they may be involved in cortical plasticity events or local circuit rearrangement (Mesulam et al., 1992). If this is indeed true, coils of neuromodulatory system axons in prefrontal cortical areas may contribute to the increased cognitive and behavioral flexibility shared by humans and great apes and may represent a special neuroanatomical substrate for supporting advanced learning capacities. These neural adaptations might relate to some of the shared cognitive abilities displayed by humans and great apes, such as the diffusion of social learning through regional traditions, a capacity for self-awareness, enhanced attention to the gaze of others, increased social tolerance, and the ability to manufacture tools (Boesch, 1993; Heyes, 1994; Keller, 2004; Povinelli and Bering, 2002; Povinelli and Preuss, 1995; Suddendorf and Whiten, 2001).

Human Brain Evolution: Advantages and Afflictions

The human capacity for reasoning, behavioral and cognitive flexibility, and language, just to name a few, is unparalleled among other species (Neill, 2007; Premack, 2007). However, the evolution of these abilities and the neuroanatomical substrates that support them is not without cost. Humans are also uniquely susceptible to neuropathological and neurodegenerative disorders, including Alzheimer's disease, Parkinson's disease, and schizophrenia. Each is characterized by some common cognitive manifestations that are indicative of diminished cortical function. In addition, their incidence among human populations is universal and at a relatively high percentage (e.g., approximately 1% of human populations are schizophrenic and 5% of individuals over 65 years of age are diagnosed with Alzheimer's disease) (Molnar, 2006). Most hypotheses regarding the etiology of these diseases involve the disruption of PFC functioning via dysregulation of one or more of the neuromodulatory systems described in this review (Briand et al., 2007). It is conceivable that these neuropathological processes may represent a byproduct resulting from the evolution of human intellect. A further possibility is that human cognitive advances rely on specialized neuromodulatory moderation of neuronal communication and that this increase in functional responsibility has made a significant contribution to our vulnerability to neurodegenerative processes.

Early researchers of schizophrenia, including Emil Kraepelin, believed that this pathology was intimately linked to the acquisition of higher intellectual functions of humans (Goldberg, 2001). More recently, researchers have described schizophrenia as a byproduct of human brain evolution, particularly as a consequence of language acquisition and/or complex social relations (Burns, 2006a; Crow, 2000; Kuttner et al., 1967; Randall, 1998). Genetic studies are now providing some support for these hypotheses. Although schizophrenia is likely a complex, polygenic disorder, many genes have been identified that appear to moderate its incidence and progression. It has been shown that some of the genes or gene combinations that contribute to the development of schizophrenia have undergone positive selection during human evolution, hence accounting for the high incidence of this disease in human populations. Crespi et al. (2007) reported that 28 of 76 genes underlying schizophrenia were subjected to positive selection during human evolution. DA receptor (D4), ACh receptor (muscarinic) and SERT genes are included among these candidate genes (Bustamante et al., 2005; Crespi et al., 2007; Dorus et al., 2004). The negative symptoms and cognitive deficits associated with schizophrenia are linked to DAergic hypofunction within the cortex while a subcortical excess of DA has been implicated in the positive symptoms (Abi-Dargham, 2004). Additionally, disruptions of the cortical 5HTergic and cholinergic systems make further contributions to the cognitive disturbances associated with schizophrenia, with therapeutic recommendations to include appropriate agonists/antagonists for effective treatment (Roth et al., 2004; Stip et al., 2005).

Here, we suggest that not only is schizophrenia a possible side-effect of human encephalization, but other neurodegenerative diseases, including Alzheimer's and Parkinson's diseases, may be as well. Whereas this claim in and of itself is not novel (e.g., Burns, 2006b; Seeley et al., 2007), we suggest one possible neuroanatomical basis for this postulate. Neuromodulatory systems regulate communication within cortical circuits, and their roles in higher cognitive functions and cognitive pathologies are well documented. Our recent research has revealed significant alterations in neuromodulatory transmitter systems within the cortex of the lineage leading to humans, including chimpanzees (Raghanti et al., 2007; Raghanti et al., 2008; Raghanti et al., submitted). While an increased dependence upon these systems may have contributed to cognitive specializations in humans, a trade-off with neurodegenerative pathologies may have taken place.

Acknowledgements

We thank C. D. Stimpson and Drs. J. L. Marcinkiewicz and J. M. Erwin for expert technical assistance, help, and discussion, and Dr. Randy Blakely for providing the serotonin transporter antibody. This research was supported by the National Science Foundation (BCS-0515484 and BCS-0549117), the Wenner-Gren Foundation for Anthropological Research, and the James S. McDonnell Foundation (22002078). The brain materials used in the studies referenced here were loaned by the Great Ape Aging Project (USPHS/NIH grant AG14308, "A Comparative Neurobiology of Aging Resource," J. Erwin, PI), the Foundation for Comparative and Conservation Biology, and the Northwestern University Alzheimer's Disease Center Brain Bank (NADC grant P30 AG13854).

References

Abi-Dargham, A., 2004. Do we still believe in the dopamine hypothesis? New data bring new evidence. International Journal of Neuropsychopharmacology 7, S1-S5.

Akil, M., Pierri, J.N., Whitehead, R.E., Edgar, C.L., Mohila, C., Sampson, A.R., Lewis, D.A., 1999. Lamina-specific alterations in the dopamine innervation of the prefrontal cortex in schizophrenic subjects. The American Journal of Psychiatry 156, 1580-1589.

Allman, J., Watson, K., Tetreault, N., Hakeem, A., 2005. Intuition and autism: a possible role for Von Economo neurons. Trends in Cognitive Sciences 9, 367-373.

Arnsten, A., Cai, J., Steere, J., Goldman-Rakic, P.S., 1995. Dopamine D2 receptor mechanisms contribute to age-related sognitive decline: the effects of quinpirole on memory and motor performance in monkeys. The Journal of Neuroscience 15, 3429-3439.

Audet, M.A., Descarries, L., Doucet, G., 1989. Quantified regional and laminar distribution of the serotonin distribution in the anterior half of the adult rat cerebral cortex. Journal of Chemical Neuroanatomy 2, 29-44.

Austin, M.C., Whitehead, R.E., Edgar, C.L., Janosky, J.E., Lewis, D.A., 2002. Localized decrease in serotonin transporter-immunoreative axons in the prefrontal cortex of depressed subjects committing suicide. Neuroscience 114, 807-815.

Azmitia, E.C., 1999. Serotonin neurons, neuroplasticity, and homeostasis of neural tissue. Neuropsychopharmacology 21, 33S-45S.

Azmitia, E.C., Gannon, P.J., 1986. The primate serotonergic system: a review of human and animal studies and a report on *Macaca fascicularis*. Advances in Neurology 43, 407-468.

Beach, T.G., Honer, W.G., Hughes, L.H., 1997. Cholinergic fibre loss associated with diffuse plaques in the non-demented elderly: the preclinical stage of Alzheimer's disease? Acta Neuropathologica 93, 146-153.

Benavides-Piccione, R., Arellano, J.I., DeFelipe, J., 2005. Catecholaminergic innervation of pyramidal neurons in the human temporal cortex. Cerebral Cortex 15, 1584-1591.

Benavides-Piccione, R., DeFelipe, J., 2003. Different populations of tyrosine-hydroxylase-immunoreactive neurons defined by differential expression of nitric oxide synthase in the human temporal cortex. Cerebral Cortex 13, 297-307.

Benavides-Piccione, R., DeFelipe, J., 2007. Distribution of neurons expressing tyrosine hydroxylase in the human cerebral cortex. Journal of Anatomy 211, 212-222.

Benzing, W.C., Kordower, J.H., Mufson, E.J., 1993. Galanin immunoreactivity within the primate basal forebrain: evolutionary change between monkeys and apes. The Journal of Comparative Neurology 336, 31-39.

Beran, M.J., Evans, T.A., 2006. Maintenance of delay of gratification by four chimpanzees (*Pan troglodytes*): the effects of delayed reward visibility, experimenter presence, and extended delay intervals. Behavioural Processes 73, 315-324.

Berger, B., Gaspar, P., Verney, C., 1991. Dopaminergic innervation of the cerebral cortex: unexpected differences between rodents and primates. Trends in Neurosciences 14, 21-27.

Berger, B., Trottier, S., Verney, C., Gaspar, P., Alvarez, C. 1988. Regional and laminar distribution of the dopamine and serotonin innervation in the macaque cerebral cortex: A radioautographic study. The Journal of Comparative Neurology 273, 99-119.

Blokland, A., 1996. Acetylcholine: a neurotransmitter for learning and memory? Brain Research Reviews 21, 285-300.

Boesch, C., 1993. Aspects of transmission of tool-use in wild chimpanzees. In KR Gibson and T Ingold (Eds.), Tools, language and cognition in human evolution. Cambridge University Press, Cambridge.

Bonner, T.I., Buckley, N.J., Young, A.C., Brann, M.R., 1987. Identification of a family of muscarinic acetylcholine receptor genes. Science 237, 527-532.

Boshes, B., Arbit, J., 1970. A controlled study of the effect of L-DOPA upon selected cognition and behavioral functions. Transactions of the American Neurological Association 95, 59-63.

Bozkurt, A., Zilles, K., Schleicher, A., Kamper, L., Sanz Arigita, E., Uylings, H.B.M., Kötter, R., 2005. Distributions of transmitter receptors in the macaque cingulate cortex. NeuroImage 25, 219-229.

Bradshaw, C.M., 2003. Neuropsychopharmacology. In: Halligan, P.W., Kischka, U., Marshall, J.C., (Eds.), Handbook of Clinical Neuropsychology. Oxford University Press, New York, pp. 445-469.

Briand, L.A., Gritton, H., Howe, W.M., Young, D.A., Sarter, M., 2007. Modulators in concert for cognition: Modulator interactions in the prefrontal cortex. Progress in Neurobiology 83, 69-91.

Brodmann, K., 1909. Vergleichende Lokalisationslehre der Großhirnrinde. Leipzig: Barth (reprinted as Brodmann's *Localisation in the Cerebral Cortex*, translated by L.J. Garey, London: Smith-Gordon, 1994).

Brozoski, T.J., Brown, R.M., Rosvold, H.E., Goldman, P.S., 1979. Cognitive deficit caused by regional depletion of dopamine in prefrontal cortex of rhesus monkey. Science 205, 929-32.

Buchanan, R., Summerfelt, A., Tek, C., Gold, J., 2003. An open-labeled trial of adjunctive donepezil for cognitive impairment in patients with schizophrenia. Schizophrenia Research 59, 29-33.

Buhot, M., 1997. Serotonin receptors in cognitive behaviors. Current Opinion in Neurobiology 7, 243-254.

Burns, J., 2006a. The social brain hypothesis of schizophrenia. World Psychiatry 5, 77-81.

Burns, J.K., 2006b. Psychosis: a costly by-product of social brain evolution in *Homo sapiens*. Progress in Neuro-psychopharmacology and Biological Psychiatry 30, 797-814.

Bustamante, C.D., Fledel-Alon, A., Williamson, S., Nielson, R., Hubisz, M.T., Glanowski, S., Tanenbaum, D.M., White, T.J., Sninsky, J.J., Hernandez, R.D., Civello, D., Adams, M.D., Cargill, M., Clark, A.G., 2005. Natural selection on protein-coding genes in the human genome. Nature 437, 1153-1157.

Buxhoeveden, D.P., Switala, A.E., Roy, E., Litaker, M., Casanova, M.F., 2001. Morphological differences between minicolumns in human and nonhuman primate cortex. American Journal of Physical Anthropology 115, 361-71.

Cáceres, M., Lachuer, J., Zapala, M.A., Redmond, J.C., Kudo, L., Geschwind, D.H., Lockhart, D.J., Preuss, T.M., Barlow, C., 2003. Elevated gene expression levels distinguish human from non-human primate brains. Proceedings of the National Academy of Sciences USA 100, 13030-13035.

Chadhuri, J.D., Hiltunen, M., Nykänen, M., Ylä-Herttuala, S., Soininen, H., Miettinen, R., 2005. Localiztion of m2 muscarinic receptor protein in parvalbumin and calretinin containing cells of the adult rat entorhinal cortex using two complementary methods. Neuroscience 131, 557-566.

Chan-Palay, V., 1988. Galanin hyperinnervates surviving neurons of the human basal nucleus of Meynert in dementias of Alzheimer's and Parkinson's disease. A hypothesis for the role of galanin in accentuating cholinergic dysfunction in dementia. The Journal of Comparative Neurology 273, 543-557.

Ciliax, B.J., Drash, G.W., Staley, J.K., Haber, S., Mobley, C.J., Miller, G.W., Mufson, E.J., Mash, D.C., Levey, A.I., 1999. Immunocytochemical localization of the dopamine transporter in human brain. The Journal of Comparative Neurology 409, 38-56.

Clarke, H.F., Dalley, J.W., Crofts, H.S., Robbins, T.W., Roberts, A.C., 2004. Cognitive inflexibility after prefrontal serotonin depletion. Science 304.

Clarke, H.F., Walker, S.C., Crofts, H.S., Dalley, J.W., Robbins, T.W., Roberts, A.C., 2005. Prefrontal serotonin depletion affects reversal learning but not attentional set shifting. The Journal of Neuroscience 25, 532-538.

Cools, R., Roberts, A.C., Robbins, T.W., 2007. Serotoninergic regulation of emotional and behavioral control processes. Cell doi: 10.1016/j.tics.2007.10.011:31-40.

Cools, R., Stefanova, E., Barker, R.A., Robbins, T.W., Owen, A.M., 2002. Dopaminergic modulation of high-level cognition in Parkinson's disease: the role of the prefrontal cortex revealed by PET. Brain Research Reviews 125, 584-594.

Crespi, B., Summers, K., Dorus, S., 2007. Adaptive evolution of genes underlying schizophrenia. Proceedings of the Royal Society B 2007, 2801-2810.

Crow, T.J., 2000. Schizophrenia as the price that *Homo sapiens* pays for language: a resolution of the central paradox in the origin of the species. Brain Research Reviews 31, 118-29.

Deacon, T.W., 1997. The Symbolic Species: The Co-evolution of Language and the Brain. W.W. Norton & Company, Inc., New York.

DeFelipe, J., Hendry, S.H., Hashikawa, T., Jones, E.G., 1991. Synaptic relationships of serotonin-immunoreactive terminal baskets on GABA neurons in the cat auditory cortex. Cerebral Cortex 1, 117-133.

Deutsch, J.A., 1971. The cholinergic synapse and the site of memory. Science 174, 788-794.

Dorus, S., Vallender, E.J., Evans, P.D., Anderson, J.R., Gilbert, S.L., Mahowald, M., Wyckoff, G.J., Malcom, C.M., Lahn, B.T., 2004. Accelerated evolution of nervous system genes in the origin of *Homo sapiens*. Cell 119, 1027-40.

Drachman, D.A., 1977. Memory and cognitive function in man: does the cholinergic system have a specific role? Neurology 27:783-790.

Dreher, J., Burnod, Y., 2002. An integrative theory of the phasic and tonic modes of dopamine modulation in the prefrontal cortex. Neural Networks 15, 583-602.

Durkin, T., Ayad, G., Ebel, A., Mandel, P., 1977. Regional acetylcholine turnover rates in the brains of three inbred strains of mice: correlation with some interstrain behavioral differences. Brain Research 136, 475-486.

Elvander, E., Ögren, S.O., 2005. Medial septal galanin and acetylcholine: influence on hippocampal acetylcholine and spatial learning. Neuropeptides 39, 245-248.

Evans, T.A., Beran, M.J., 2007a. Chimpanzees use self-distraction to cope with impulsivity. Biology Letters 3, 599-602.

Evans, T.A., Beran, M.J., 2007b. Delay of gratification and delay maintenance by rhesus macaques (*Macaca mulatta*). The Journal of General Psychology 134, 199-216.

Fine, A., Hoyle, C., Maclean, C.J., Levatte, T.L., Baker, H.F., Ridley, R.M., 1997. Learning impairments following injection of a selective cholinergic immunotoxin, ME20.4 IgG-Saporin, into the basal nucleus of Meynert in monkeys. Neuroscience 81, 331-343.

Foote, S.L., Morrison, J.H., 1984. Postnatal development of innervation patterns by monoamine fibers in monkey (*Macaca fascicularis*) primary visual cortex. The Journal of Neuroscience 4, 2667-2680.

Fukuda, T., Takahashi, J., Tanaka, J., 1999. Tyrosine hydroxylase-immunoreactive neurons are decreased in number in the cerebral cortex of Parkinson's disease. Neuropathology 19, 10-13.

Fuster, J.M., 1997. The Prefrontal Cortex: Anatomy, Physiology, and Neuropsychology of the Frontal Lobe. Lippincottt-Raven Publishers, Philadelphia.

Gaspar, P., Berger, B., Febvret, A., Vigny, A., Henry, J.P., 1989. Catecholamine innervation of the human cerebral cortex as revealed by comparative immunohistochemistry of tyrosine hydroxylase and dopamine-beta-hydroxylase. The Journal of Comparative Neurology 279, 249-271.

Gaspar, P., Berger, B., Febvret, A., Vigny, A., Krieger-Poulet, M., Borri-Voltattorni, C., 1987. Tyrosine hydroxylase-immunoreactive neurons in the human cerebral cortex: a novel catecholaminergic group? Neuroscience Letters 80, 257-262.

Ghashghaei, H.T., Barbas, H., 2001. Neural interaction between the basal forebrain and functionally distinct prefrontal cortices in the rhesus monkey. Neuroscience 3, 593-6614.

Goldberg, E., 2001. The Executive Brain; Frontal Lobes and the Civilized Mind. Oxford University Press, New York.

Goldman-Rakic, P.S., 1998. The cortical dopamine system: Role in memory and cognition. Advances in Pharmacol ogy 42, 707-711.

Goldman-Rakic PS, Leranth C, Williams SM, Mons N, and Geffard M (1989) Dopamine synaptic complex with pyramidal neurons in primate cerebral cortex. Proceedings of the National Acadamy of Sciences USA 86, 9015-9019.

Goldman-Rakic, P.S., Lidow, M.S., Gallager, D.W., 1990. Overlap of dopaminergic, adrenergic, and serotonergic receptors and complementarity of their subtypes in primate prefrontal cortex. The Journal of Neuroscience 10, 2125-2138.

Goldman-Rakic, P.S., Lidow, M.S., Smiley, J.F., Williams, M.S., 1992. The anatomy of dopamine in monkey and human prefrontal cortex. Journal of Neural Transmission 36, 163-177.

González-Burgos, G., Kröner, S., Krimer, L.S., Seamans, J.K., Urban, N.N., Henze, D.A., Lewis, D.A., Barrionuevo, G. 2002. Dopamine modulation of neuronal function in the monkey prefrontal cortex. Physiology & Behavior 77, 537-543.

Grace, A.A., 2000. The tonic/phasic model of dopamine system regulation and its implications for understanding alcohol and psychostimulant craving. Addiction 95 (8S2), S119-128.

Gray, J.R., Thompson, P.M., 2004. Neurobiology of intelligence: science and ethics. Nature Reviews Neuroscience 5, 471-482.

Gu, Q., 2002. Neuromodulatory transmitter systems in the cortex and their role in cortical plasticity. Neuroscience 111, 815-835.

Harder, J.A., Baker, H.F., Ridley, R.M., 1998. The role of central cholinergic projections in cognition: Implications of the effects of scopolamine on discrimination learning by monkeys. Brain Research Bulletin 45, 319-326.

Harrison, A.A., Everitt, B.J., Robbins, T.W., 1999. Central serotonin depletion impairs both the acquisition and performance of a symmetrically reinforced go/no-go conditional visual discrimination. Behavioural Brain Research 100, 99-112.

Hasselmo, M.E., 1995. Neuromodulation and cortical function: modeling the physiological basis of behavior. Behavioural Brain Research 67, 1-27.

Herremans, A.H.J., Hijzen, T.H., Oliver, B., Slangen, J.L., 1995. Serotonergic drug effects on a delayed conditional discrimination task in the rat: involvement of the 5-HT_{1A} receptor in working memory. Journal of Psychopharmacology 9, 242-250.

Heyes, C.M., 1994. Social congition in primates: Animal learning and cognition. Academic Press, New York, pp. 281-305.

Hof, P.R., Glezer, II., Revishchin, A.V., Bouras, C., Charnay, Y., Morgane, P.J., 1995. Distribution of dopaminergic fibers and neurons in visual and auditory cortices of the harbor porpoise and pilot whale. Brain Research Bulletin 36, 275-284.

Hof, P.R., Nimchinsky, E.M., Perl, D.P., Erwin, J.M., 2001. An unusual population of pyramidal neurons in the anterior cingulate cortex of hominids contains calcium-binding protein calretinin. Neuroscience Letters 307,139-142.

Holloway, R.L., 1966. Cranial capacity, neural reorganization, and hominid evolution: a search for more suitable parameters. American Anthropologist 68, 103-121.

Holloway, R.L., 1968. The evolution of the primate brain: some aspects of quantitative relations. Brain Research 7, 121-172.

Holloway, R.L., 1985. The past, present, and future significane of the lunate sulcus in early hominid evolution. In: Tobias, P.V., (Ed.), Hominid evolution: past, present, and future. Alan R. Liss, New York, pp. 47-62.

Holloway. R.L., 1996. Evolution of the human brain. In: Lock, A., Peters, C.R., (Eds.): Handbook of human symbolic evolution. Oxford University Press, Oxford, pp. 74-114.

Holloway, R.L., De La Costelareymondie, M.C., 1982. Brain endocast asymmetry in pongids and hominids: some preliminary findings on the paleontology of cerebral dominance. American Journal of Physical Anthropology 58, 101-10.

Hornung, J.P., Fritschy, J-.M., Törk, I., 1990. Distribution of two morphologically distinct subsets of serotonergic axons in the cerebral cortex of the marmoset. The Journal of Comparative Neurology 297, 165-181.

Hornung, J.P., Tork, I., De Tribolett, N., 1989. Morphology of tyrosine hydroxylase-immunoreactive neurons in the human cerebral cortex. Experimental Brain Research 76, 12-20.

Ichikawa, T., Hirata, Y., 1986. Organization of choline acetyltransferase-containing structures in the forebrain of the rat. The Journal of Neuroscience 6, 281-292.

Ikonomovic, M.D., Abrahamson, E.E., Isanski, B.A., Wuu, J., Mufson, E.J., DeKosky, S.T., 2007. Superior frontal cortex cholinergic axon density in mild cognitive impairment and early Alzheimer disease. Archives of Neurology 64, 1312-1317.

Irle, E., Markowitsch, H.H., 1987. Basal forebrain-lesioned monkeys are severely impaired in tasks of association and recognition memory. Annals of Neurology 22, 735-743.

Jacobs, B.L., Gannon, P.J., Azmitia, E.C., 1984. Atlas of serotonergic cell bodies in the cat brainstem: an immunocy to chemical analysis. Brain Research Bulletin 13, 1-31.

Jakab, R.L., Goldman-Rakic, P.S., 2000. Segregation of serotonin 5-HT2A and 5-HT3 receptors in inhibitory circuits of the primate cerebral cortex. The Journal of Comparative Neurology 417, 337-348.

Kandel, E.R., Schwartz, J.H., Jessell, T.M., 1995. Essentials of Neural Science and Behavior. Appleton & Lange, Norwalk.

Katzman, R., Terry, R., DeTeresa, R., Brown, T., Davies, P., Fuld, P., Renbing, X., Peck, A., 1988. Clinical, pathological, and neurochemical changes in dementia: a subgroup with preserved mental status and numerous neocortical plaques. Annals of Neurology 23, 138-144.

Keller, J.D., 2004. Human cognitive ecology: an instructive framework for comparative primatology. American Journal of Primatology 62, 229-241.

Kordower, J.H., Piecinski, P., Rakic, P., 1992. Neurogenesis of the amygdaloid nuclear complex in the rhesus monkey. Brain Research. Developmental Brain Research 68, 9-15.

Kosaka, T., Hama, K., Nagatsu, I., 1987. Tyrosine hydroxylase-immunoreactive intrinsic neurons in the rat cerebral cortex. Experimental Brain Research 68, 393-405.

Kosofsky, B.E., Molliver, M.E., 1987. The serotonergic innervation of cerebral cortex: different classes of axon terminals arise from dorsal and median raphe nuclei. Synapse 1, 153-168.

Kulisevsky, J., 2000. Role of dopamine in learning and memory: Implications for the treatment of cognitive dysfunction in patients with Parkinson's disease. Drugs & Aging 16, 365-379.

Kuttner, R.E., Lorincz, A.B., Swan, D.A., 1967. The schizophrenia gene and social evolution. Psychological Reports 20, 407-412.

Laplante, F., Crawley, J.N., Quirion, R., 2004. Selective reduction in ventral hippocampal acetylcholine release in awake galanin-treated rats and galanin-overexpressing trangenic mice. Regulatory Peptides 122, 91-98.

Lehmann, J., Struble, R.G., Antuono, P.G., Coyle, J.T., Cork, L.C., Price, D.L., 1984. Regional heterogeneity of choline acetyltransferase activity in primate neocortex. Brain Research 322, 361-364.

Levin, E.D., Simon, B.B., 1998. Nicotinic acetylcholine involvement in cognitive function in animals. Psychopharmacology 138, 217-230.

Lewis, D.A., 1991. Distribution of choline acetyltransferase-immunoreactive axons in monkey frontal cortex. Neuroscience 40, 363-374.

Lewis, D.A., Melchitzky, D.S., Sesack, S.R., Whitehead, R.E., Auh, S., Sampson, A., 2001. Dopamine transporter immunoreactivity in monkey cerebral cortex: regional, laminar, and ultrastructural localization. The Journal of Comparative Neurology 432, 119-36.

Lidow, M.S., Goldman-Rakic, P.S., Rakic, P., Innis, R.B., 1989. Dopamine D2 receptors in the cerebral cortex: distribution and pharmacological characterization with [^{3}H] raclopride. Proceedings of the National Academy of Sciences USA 86, 6412-6416.

Lysakowski, A., Wainer, B.H., Rye, D.B., Bruce, G., Hersh, L.B., 1986. Cholinergic innervation displays strikingly different laminar preferences in several cortical areas. Neuroscience Letters 64, 102-108.

Marek, G.J., Aghajanian, G.K., 1998. The electrophysiology of prefrontal serotonin systems: therapeutic implications for mood and psychosis. Biological Psychiatry 44, 1118-1127.

Marui, W., Iseki, E., Kato, M., Kosaka, K., 2003. Degeneration of tyrosine hydroxyalse-immunoreactuve neurons in the cerebral cortex and hippocampus of patients with dementia with Lewy bodies. Neuroscience Letters 340, 185-188.

McGaughy, J., Everitt, B.J., Robbins, T.W., Sarter, M., 2000. The role of cortical cholinergic afferent projections in cognition: impact of new selective immunotoxins. Behavioural Brain Research 115, 251-263.

Mechawar, N., Cozzari, C., Descarries, L., 2000. Cholinergic innervation in adult rat cerebral cortex: A quantitative immunocytochemical description. The Journal of Comparative Neurology 428, 305-318.

Mega, M.S., 2000. The cholinergic deficit in Alzheimer's disease: impact on cognition, behavior and function. International Journal of Neuropsychpharmacology 3, S3-S12.

Meneses, A., Hong, E., 1995. Effect of fluoxetine on learning and memory involves multiple 5-HT systems. Pharmacology Biochemistry and Behavior 52, 341-346.

Mesulam, M., Geula, C., 1991. Differential distribution of a neurofilament protein epitope in acetylcholinesterase-rich neurons of human cerebral neocortex. Brain Research 544, 169-73.

Mesulam, M.M., 2004. The cholinergic innervation of the human cerebral cortex. Progress in Brain Research 145, 67-78.

Mesulam, M.M., Hersh, L.B., Mash, D.C., Geula, C., 1992. Differential cholinergic innervation within functional subdivisions of the human cerebral cortex: a choline acetyltransferase study. The Journal of Comparative Neurology 318, 316-328.

Mesulam, M.M., Volicer, L., Marquis, J.K., Mufson, E.J., Green, R.C., 1986. Systematic regional differences in the cholinergic innervation of the primate cerebral cortex: distribution of enzyme activities and some behavioral implications. Annals of Neurology 19, 144-151.

Miner, L.H., Schroeter, S., Blakely, R.D., Sesack, S.R., 2000. Ultrastructural localization of the serotonin transporter in superficial and deep layers of the rat prelimbic prefrontal cortex and its spatial relationship to dopamine terminals. The Journal of Comparative Neurology 427, 220-234.

Molnar, S., 2006. Human Variation: Races, Types, and Ethnic Groups, sixth edition. Pearson Prentice Hall, Upper Saddle River, NJ.

Mrzljak, L., Levey, A.I., Belcher, S., Goldman-Rakic, P.S. 1998. Localization of the m2 muscarinic acetylcholine receptor protein and mRNA in cortical neurons of the normal and cholinergically deafferented rhesus monkey. The Journal of Comparative Neurology 390, 112-132.

Mrzljak, L., Levey, A.I., Goldman-Rakic, P.S., 1993. Association of m1 and m2 muscarinic receptor proteins with asymmetric synapses in the primate cerebral cortex: morphological evidence for cholinergic modulation of excitatory neurotransmission. Proceedings of the Nationall Academy of Sciences USA 90, 5194-5198.

Mrzljak, L., Pappy, M., Leranth, C., Goldman-Rakic, P.S. 1995. Cholinergic synaptic circuitry in the macaque prefrontal cortex. The Journal of Comparative Neurology 357, 603-617.

Mufson, E.J., Deecher, D.C., Basile, M., Izenwasse, S., Mash, D.C., 2000. Galanin receptor plasticity within the nucleus basalis in early and late Alzheimer's disease: an in vitro and autoradioraphic analysis. Neuropharmacology 39, 1404-1412.

Mulligan, K.A., Törk, I., 1987. Serotonergic axons form basket-like terminals in cerebral cortex. Neuroscience Letters 81, 7-12.

Mulligan, K.A., Törk, I., 1988. Serotonergic innervation of the cat cerebral cortex. The Journal of Comparative Neurology 270, 86-110.

Naughton, M., Mulrooney, J.B., Leonard, B.E., 2000. A review of the role of serotonin receptors in psychiatric disorders. Human Psychopharmacology 15, 397-415.

Neill, D., 2007. Cortical evolution and human behavior. Brain Research Bulletin 74, 191-205.

Nieoullon, A., 2002. Dopamine and the regulation of cognition and attention. Progress in Neurobiology 67, 53-83.

Nimchinsky, E.A., Gilissen, E., Allman, J.M., Perl, D.P., Erwin, J.M., Hof, P.R., 1999. A neuronal morphologic type unique to humans and great apes. Proceedings of the Nationall Academy of Sciences USA 96, 5268-73.

Nimchinsky, E.A., Vogt, B.A., Morrison, J.H., Hof, P.R., 1995. Spindle neurons of the human anterior cingulate cortex. The Journal of Comparative Neurology 355, 27-37.

Noguchi, T., Yoshida, Y., Chiba, S., 2001. Effects of psychological stress on monoamine systems in subregions of the frontal cortex and nucleus accumbens of the rat. Brain Research 916, 91-100.

Pelosi, L., Slade, T., Blumhardt, L.D., Sharma, V.K., 2000. Working memory dysfunction in major depression: an event-related potential study. Clinical Neurophysiology 111, 1531-1543.

Penn, D.C., Holyoak, K.J., Povinelli, D.J., (in press). Darwin's mistake: Explaining the discontinuity between human and nonhuman minds. Behavioral and Brain Sciences.

Petrides, M., Pandya, D.N., 1994. Comparative architectonic analysis of the human and macaque frontal cortex. In: Boller, F., Grafman, J., (Eds.), Handbook of Neuropsychology. Amsterdam: Elsevier, pp. 17-58.

Povinelli, D.J., Bering, J.M., 2002. The mentality of apes revisited. Current Directions in Psychological Science 11, 115-119.

Povinelli, D.J., Preuss, T.M., 1995. Theory of mind: evolutionary history of a cognitive specialization. Trends in Neurosciences 18, 418-424.

Premack, D., 2007. Human and animal cognition: Continuity and discontinuity. Proceedings of the National Academy of Sciences USA 104, 13861-13867.

Preuss, T.M., 1995. Do rats have prefrontal cortex? Journal of Cognitive Neuroscience 7, 1-24.

Preuss, T.M., 2000. What's human about the human brain? In: Gazzaniga, M.S., (Ed.), The New Cognitive Neurosciences. MIT Press, Cambridge, pp. 1219-1234.

Preuss, T.M., 2004. What is it like to be a human? In: Gazzaniga, M.S., (Ed.), The Cognitive Neurosciences III. MIT Press, Cambridge, MA, pp. 5-22.

Preuss, T.M., 2006. Who's afraid of *Homo sapiens*? Journal of Biomedical Discovery and Collaboration 1, 17 doi:10.1186/1747-5333-1-17.

Preuss, T.M., Coleman, G.Q., 2002. Human-specific organization of primary visual cortex: alternating compartments of dense Cat-301 and calbindin immunoreactivity in layer 4A. Cerebral Cortex 12, 671-691.

Preuss, T.M., Goldman-Rakic, P.S., 1989. Connections of the ventral granular frontal cortex of macaques with perisylvian premotor and somatosensory areas: anatomical evidence for somatic representation in primate frontal association cortex. The Journal of Comparative Neurology 282, 293-316.

Previc, F.H., 1999. Dopamine and the origins of human intelligence. Brain and Cognition 41, 299-350.

Puumala, T., Sirviö, J., 1998. Changes in activities of dopamine and serotonin systems in the frontal cortex underlie poor choice accuracy and impulsivity of rats in an attention task. Neuroscience 83, 489-499.

Raghanti, M.A., Stimpson, C.D., Marcinkiewicz, J.L., Erwin, J.M., Hof, P.R., Sherwood, C.C., 2008. Differences in cortical serotonergic innervation among humans, chimpanzees, and macaque monkeys: a comparative study. Cerebral Cortex 18, 584-597, doi*: 10.1093/cercor/bhm089*.

Raghanti, M.A., Stimpson, C.D., Marcinkiewicz, J.L., Erwin, J.M., Hof, P.R., Sherwood, C.C., 2008. Cholinergic innervation of the frontal cortex: Differences among humans, chimpanzees, and macaque monkeys. The Journal of Comparative Neurology 506, 409-424.

Raghanti, M.A., Stimpson, C.D., Marcinkiewicz, J.L., Erwin, J.M., Hof, P.R., Sherwood, C.C., (submitted) Cortical dopaminergic innervation among humans, chimpanzees, and macaque monkeys: A comparative study. Neuroscience.

Randall, P.L., 1998. Schizophrenia as a consequence of brain evolution. Schizophrenia Research 30, 143-8.

Rasmusson, D.D., 2000. The role of acetylcholine in cortical synaptic plasticity. Behavioural Brain Research 115, 205-218.

Rilling, J.K., 2006. Human and nonhuman primate brains: Are they allometrically scaled versions of the same design? Evolutionary Anthropology 15, 65-77.

Robbins, T.W., 2000. Chemical neuromodulation of frontal-executive functions in humans and other animals. Experimental Brain Research 133, 130-138.

Robbins, T.W., Roberts, A.C., 2007. Differential regulation of fronto-executive function by the monoamines and acetylcholine. Cerebral Cortex 17 Suppl 1, i151-i160.

Roth, B.L., Hanizavareh, S.M., Blum, A.E., 2004. Serotonin receptors represent highly favorable molecular targets for cognitive enhancement in schizophrenia and other disorders. Psychopharmacology 174, 17-24.

Sarter, M., Bruno, J.P., 1997. Cognitive functions of cortical acetylcholine: toward a unifying hypothesis. Brain Research Reviews 23, 28-46.

Sarter, M., Parikh, V., 2005. Choline transporters, cholinergic transmission and cognition. Nature Reviews Neuroscience 6, 48-56.

Sawaguchi, T., Goldman-Rakic, P.S., 1991. D1 dopamine receptors in prefrontal cortex; involvement in working memory. Science 251, 947-950.

Seamans, J.K., Yang, C.R., 2004. The principal features and mechanisms of dopamine modulation in the prefrontal cortex. Progress in Neurobiology 74, 1-57.

Seeley, W.W., Allman, J., Carlin, D.A., Crawford, R.K., Macedo, M.N., Greicius, M.D., DeArmond, S.J., Miller, B.L., 2007. Divergent social functioning in behavioral variant frontotemporal dementia and Alzheimer disease: reciprocal connections and neuronal evolution. Alzheimer's Disease and Associated Disorders 21, S50-S57.

Semendeferi, K., Lu, A., Schenker, N., Damasio, H., 2002. Humans and great apes share a large frontal cortex. Nature Neuroscience 5, 272-6.

Sesack, S.R., Bressler, C.N., Lewis, D.A., 1995. Ultrastructural associations between dopamine terminals and local circuit neurons in the monkey prefrontal cortex: a study of calretinin-immunoreactive cells. Neuroscience Letters 200, 9-12.

Sesack, S.R., Hawrylak, V.A., Melchitzky, D.S., Lewis, D.A., 1998. Dopamine innervation of a subclass of local circuit neurons in monkey prefrontal cortex: ultrastructural analysis of tyrosine hydroxylase and parvalbumin immunoreactive structures. Cerebral Cortex 8, 614-22.

Sherwood, C.C., Stimpson, C.D., Raghanti, M.A., Wildman, D.E., Uddin, M., Grossman, L.I., Goodman, M., Redmond, J.C., Bonar, C.J., Erwin, J.M., Hof, P.R., 2006. Evolution of increased glia-neuron ratios in the human frontal cortex. Proceedings of the National Academy of Sciences USA 103, 13606-13611.

Sherwood, C.C., Wahl, E., Erwin, J.M., Hof, P.R., Hopkins, W.D., 2007. Histological asymmetries of primary motor cortex predict handedness in chimpanzees (*Pan troglodytes*). The Journal of Comparative Neurology 503, 525-537.

Sigle, J.P., Zander, J., Ehret, A., Honegger, J., Jackisch, R., Feuerstein, T.J., 2003. High potassium-induced activation of choline-acetyltransferase in human neocortex: implications and species differences. Brain Research Bulletin 60, 255-262.

Smeets, W.J.A.J., González, A., 2000. Catecholamine systems in the brain of vertebrates: new perspectives through a comparative approach. Brain Research Reviews 33, 308-379.

Smiley, J.F., Goldman-Rakic, P.S., 1996. Serotonergic axons in monkey prefrontal cortex synapse predominantly on interneurons as demonstrated by serial section electron microscopy. The Journal of Comparative Neurology 367, 431-443.

Smiley, J.F., Subramanian, M., Mesulam, M.M., 1999. Monoaminergic-cholinergic interactions in the primate basal forebrain. Neuroscience 93, 817-829.

Soubrié, P., 1986. Reconciling the role of central serotonin neurons in human and animal behavior. Behavioral and Brain Sciences 9, 319-364.

Squire, L.R., Bloom, F.E., McConnell, S.K., Roberts, J.L., Spitzer, N.C., Zigmond, M.J., 2003. Fundamental Neuroscience, Second Edition. Academic Press, San Diego.

Steckler, T., Sahgal, A., 1995. The role of serotonergic-cholinergic interactions in the mediation of cognitive behavior. Behavioural Brain Research 67, 165-199.

Stip, E., Chouinard, S., Boulay, L.J., 2005. On the trail of a cognitive enhancer for the treatment of schizophrenia. Progress in Neuropsychopharmacology and Biological Psychiatry 29, 219-232.

Suddendorf, T., Whiten, A., 2001. Mental evolution and development: evidence for secondary representation in children, great apes, and other animals. Psychological Bulletin 127, 629-650.

Sutoo, D., Akiyama, K., Yabe, K., 2001. Quantitative imaging of tyrosine hydroxylase and calmodulin in the human brain. Journal of Neuroscience Research 63, 369-376.

Thomas, A.J., Hendriksen, M., Piggott, M., Ferrier, I.N., Perry, E., Ince, P., O'Brien, J.T., 2006. A study of the serotonin transporter in the prefrontal cortex in late-life depression and Alzheimer's disease with and without depression. Neuropathology and Applied Neurobiology 32, 296-303.

Trottier, S., Evrand, B., Vignal, J., Scarabin, J., Chauvel, P. 1996. The serotonergic innervation of the cerebral cortex in man and its changes in focal cortical dysplasia. Epilepsy Research 25, 79-106.

Uddin, M., Wildman, D.E., Liu, G., Xu, W., Johnson, R.M., Hof, P.R., Kapatos, G., Grossman, L.I., Goodman, M., 2004. Sister grouping of chimpanzees and humans as revealed by genome-wide phylogenetic analysis of brain gene expression profiles. Proceedings of the National Academy of Sciences USA 101, 2957-62.

Uylings, H.B.M., van Eden, C.G., 1990. Qualitative and quantitative comparison of the prefrontal cortex in rats and primates, including humans. Progress in Brain Research 85, 31-62.

van Eden, C., Hoorneman, E., Buijs, R., Matthijssen, M., Geffard, M., Uylings, H.B.M., 1987. Immunocytochemical localization of dopamine in the prefrontal cortex of the rat at the light and electron microscopical level. Neuroscience 22, 849-862.

Varnäs, K., Halldin, C., Hall, H., 2004. Autoradiographic distribution of serotonin transporters and receptor subtypes in human brain. Human Brain Mapping 22, 246-260.

Venator, D.K., Lewis, D.A., Finlay, J.M., 1999. Effects of parietal dopamine loss in the medial prefrontal cortex on local baseline and stress-evoked extracellular dopamine concentrations. Neuroscience 93, 497-505.

Vergé, D., Calas, A., 2000. Serotonergic neurons and serotonin receptors: gains from cytochemical approaches. Journal of Chemical Neuroanatomy 18, 41-56.

von Bohlen und Halbach, O., Dermietzel, R., 2006. Neurotransmitters and Neuromodulators. Handbook of Receptors and Biological Effects. Second Edition. Wiley-VCH Verlag GmbH & Co, Weinheim, Germany.

Whitehouse, P.J., 1992. Alzheimer's disease: relationship of cognition and behavior to neurochemistry. International Psychogeriatrics 4, 71-78.

Williams, G.V., Rao, S.G., Goldman-Rakic, P.S., 2002. The physiological role of 5-HT2A receptors in working memory. The Journal of Neuroscience 22, 2843-2854.

Williams, S., Goldman-Rakic, P.S., 1998. Widespread origin of the primate mesofrontal dopamine system. Cerebral Cortex 8, 321-345.

Wilson, M.A., Molliver, M.E., 1991. The organization of serotonergic projections to cerebral cortex in primates: regional distribution of axon terminals. Neuroscience 44, 537-553.

Wilson, M.A., Ricaurte, G.A., Molliver, M.E., 1989. Distinct morphologic classes of serotonergic axons in primates exhibit differential vulnerability to the psychotropic drug 3,4-methylenedioxymethamphetamine. Neuroscience 28, 121-137.

Winterer, G., Weinberger, D.R., 2004. Genes, dopamine and cortical signal-to-noise ratio in schizophrenia. Trends in Neuroscience 27, 683-690.

Wu, C., Hersh, L.B., Geula, C., 2000. Cyto- and chemoarchitecture of basal forebrain cholinergic neurons in the common marmoset (*Callithrix jacchus*). Experimental Neurology 165, 306-326.

CHAPTER 14

SEX DIFFERENCES IN THE CORPUS CALLOSUM OF *MACACA FASCICULARIS* AND *PAN TROGLODYTES*

DOUGLAS C. BROADFIELD

ABSTRACT

In 1982 Ralph Holloway along with his student Kitty deLacoste-Utamsing published a paper asserting that there are sex differences in the brains of humans. While this was not the first paper on sex differences in the brain, it was one of the most prominent and controversial, setting off an area of neuroscience research that continues to today. While the extent and meaning of sexual dimorphism in the human corpus callosum has been investigated countless times over the past 30 years, what this structure is like in our closest relatives, the living apes, has not been approached. This paper investigates whether sex differences are present within two primate species, *Pan troglodytes* and *Macaca fascicularis*, addressing several issues important to neurology, paleoneurology, and human evolution. Looking at the morphological and histological aspects of these species demonstrates that there is not a statistically significant difference between males and females of *P. troglodytes* and *M. fascicularis* with regard to total and regional midsagittal area of the corpus callosum or with regard to axon density/100μm^2, overall axon numbers, or within any of the axonal diameter classes in the splenium of the corpus callosum in either species. These results strongly suggest that dimorphism of the brain and corpus callosum arose later in hominin evolution, possibly not until the arrival of *Homo sapiens*.

KEY WORDS

Brain evolution, corpus callosum, sex differences, chimpanzee, macaque

INTRODUCTION

The cerebral cortex has undergone a dramatic evolution during hominin history. Progressing from a small, chimpanzee-like brain in *Australopithecus*, and presumably *Ardipithecus*, the human brain has come to be capable of linguistic, mathematical, abstract, and behavioral elements apparently unobtainable by other primate groups. An additional aspect of this evolution has been the emergence of sex differences in cognitive behaviors. The existence of sex differences is not unheard of in primates (e.g., Philips et al., 2007), but it has been difficult to document in primate cognition (e.g, Hellner-Burris et al., 2010). Anatomical distinctions between nonhuman primates and modern humans have become more difficult as we have come to appreciate our evolutionary history. It is possible that during the course of primate evolution sex differences in the brain developed in early sexually dimorphic clades such as the cercopithecoids. This scenario is plausible due to the presence of sexually dimorphic skeletal morphology and group behaviors. Females behave differently from males, possibly due to different reproductive strategies. If sex differences occur in such phylogenetically distant taxa such as *Macaca* and *Papio*, it is possible that sex differences became even more distinct in a more recent common ancestor to humans such as *Pan*. The presence of sex differences in the brain of modern human's closest living relative would indicate that sex differences were already present in the earliest hominins. This would suggest that sex differences exhibited in modern humans are not unique, but merely an extension of *Homo*'s evolutionary past.

An alternative hypothesis suggests that sex differences in the modern human brain are unique to modern

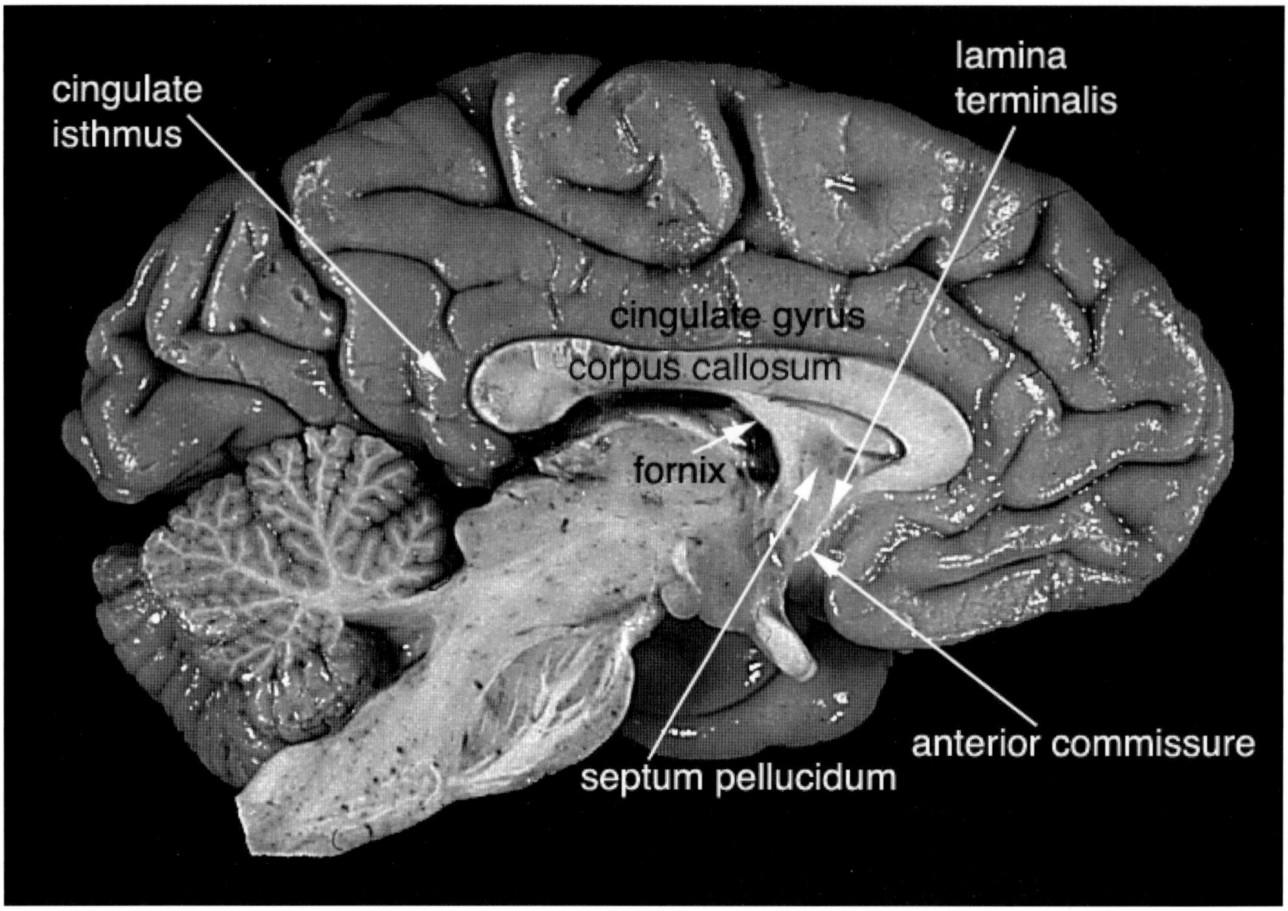

Fig. 1. Midsagittal view of brain of Pan troglodytes. A: anterior. P: posterior.

humans and did not occur until late in hominin evolution, possibly not until the advent of our own species, *Homo sapiens*. Although some similarity exists in the brains of *Pan* and modern humans, these similarities have not exposed any common sex differences between these two groups. Studies on modern human brains, however, have exposed a number of sex differences, albeit these discoveries occurred within non-neocortical structures. The presence of sex differences in cognition has also been observed. These results suggest that sex differences in the brain and in cognition did not occur until late in human evolution.

This study focuses on the second hypothesis that sex differences in the telencephalon occurred late in hominin evolution. The cerebral cortex represents one of the most complex and costly structures humans possess. The complexity of this structure has evolved over 3-5 million years of hominin history to allow modern humans to perform complex cognitive tasks not seen in other animal groups. In addition, humans have evolved the cerebral areas responsible for these tasks such that males excel at certain tasks while female excel at others. Males for example perform better at tasks of mental rotation while females do better on tests of verbal richness. There is little information about how these differences develop or within which specific cerebral structures they reside. One cerebral structure, however, has shed light on the presence of sex differences in the brain, the corpus callosum (Fig. 1). As the major interhemispheric pathway of the brain, the corpus callosum provides a point at which to begin to examine cerebral sex differences. Since morphological sex differences have been noted in this structure, the question of when these sex differences developed in human evolution can be asked. If the presence of sex differences in the corpus callosum represents an epiphenomenon of primate brain evolution associated with the advent of the Catarrhini, then sex differences in this structure should manifest in *Macaca*. If these differences don't occur until the evolution of the Hominoidea, then *Pan* would exhibit this trait. If sex differences in this structure did not occur until after the ape-human split, then it would represent an autapomorphic character of the hominin clade.

Considerable current controversy surrounds the existence of identifiable sexual differences in the nonhuman primate and human brain. An area that has come under increasingly greater focus is the corpus callosum, the principal neocortical commissure. For example, many recent studies on humans have demonstrated morphological differences between the sexes in callosal measures (de Lacoste-Utamsing and Holloway, 1982; Wium, 1984; de Lacoste et al., 1986; Holloway and de Lacoste, 1986; Holloway, 1990; Holloway et al., 1993; Davatzikos and Resnick, 1998; Oka et al., 1999; Sullivan et al., 2001; Allen et al., 2003; Lee et al., 2009) as well as fiber composition of the corpus callosum (Tomasch, 1954; Aboitiz et al., 1992a,b,c; Liu et al., 2010). Comparable data from nonhuman primates has, however, been generally lacking (e.g., Le May, 1976; de Lacoste and Woodward, 1988; LaMantia and Rakic, 1990a,b; Holloway and Heilbroner, 1991; Dunham and Hopkins, 2006; Phillips et al., 2007). The paucity of information on the primate corpus callosum has prevented further exploration of the origin, evolution, and functional significance of sex differences in the primate brain. Nevertheless, the limited information that is available for human and nonhuman primates provides provocative data concerning the above issues. In addition, the present study adds to the current knowledge of and provides new information on sex differences of this structure in *M. fascicularis* and *P. troglodytes*.

Sex differences in the corpus callosum of humans

The human corpus callosum has been the subject of extensive study relating to its involvement in a number of diseases such as: Down's syndrome (Wang et al., 1992; Kivitie-Kallio et al., 1998), epilepsy (e.g. Khanna et al., 1994; Hermann et al., 2003), amyotrophic lateral sclerosis (Yamauchi et al., 1995), Alzheimer's (Vermersch, 1996; Thompson et al., 1998), attention-deficit hyperactivity disorder (Baumgardner et al., 1996; Lyoo et al., 1996), autism (Piven et al., 1997; Manes et al., 1999), schizophrenia (e.g., Coger and Serafetinides, 1990; Raine et al., 1990; Hoff et al., 1994; Cowell et al., 1996; McCarley et al., 1999; Meisenzahl et al., 1999; Narr et al., 2000; Panizzon et al., 2003), Williams syndrome (Schmitt et al., 2001), Marchiafava-Bignami disease (Shiota et al., 1996), Tourette syndrome (Baumgardner et al., 1996; Mostofsky et al., 1999), dyslexia (Rumsey et al., 1996; Robichon and Habib, 1998) and other speech associated deficiencies seen when the corpus callosum is sectioned (Kaga et al., 1990; Davidson and Hugdahl, 1995). These studies, however, have done little to discern the sex differences associated with this structure.

Early studies on the corpus callosum found no differences in sex based on size and shape (Bean, 1906; Mall, 1909). However, subsequent research of this kind on the corpus callosum remained dormant until de Lacoste-Utamsing and Holloway (1982) re-addressed the issue. De Lacoste-Utamsing and Holloway took into account what Mall (1909) had stressed earlier, namely that brain size must be considered when suggesting dimorphism in brain morphology. They concluded that while the area differences between males and females may be small, they are nevertheless significant. Subsequent studies in this area have produced varying results. Some studies have suggested that there is sexual dimorphism in the corpus callosum (de Lacoste, 1981; de Lacoste-Utamsing and Holloway, 1982; Witelson, 1985; Holloway and de Lacoste, 1986; Yoshii et al., 1986; Reinarz et al., 1988; Clarke et al, 1989; Hayakawa et al., 1989; Witelson, 1989; Holloway, 1990; Elster et al., 1990; Allen et al., 1991; Clarke and Zaidel, 1994; Johnson et al., 1994, 1996; Driesen and Raz, 1995; Steinmetz et al., 1992, 1995, 1996; Salat et al., 1997; Davatzikos and Resnick, 1998; Oka et al., 1999; Achiron et al., 2001; Sullivan et al., 2001; Mitchell et al., 2003; Dubb et al., 2003; Westerhausen et al., 2004; Shin et al., 2005; Yokota et al., 2005; Lee et al., 2009). Others report that it lacks dimorphism (Bell and Variend, 1985; Weber and Weis, 1986; Kertesz et al., 1987; Oppenheim et al., 1987; Byne et al., 1988; Demeter et al., 1988; O'Kusky et al., 1988; Weis et al., 1989; Going and Dixson, 1990; Prokop et al., 1990; Denenberg et al., 1991a,b; Emory et al., 1991; Habib et al., 1991; Aboitiz et al., 1992c; Steinmetz et al., 1992; Zaidel et al., 1995; Constant and Ruther, 1996; Koshi et al., 1997; Matano and Nakano, 1998; Luders et al., 2003; Morton and Rafto, 2006). However, several of these latter studies did not consider sexual dimorphism in brain size, and thus did not analyze relative callosal measurements (i.e., taking brain size into account), but with the exception of Luders et al. (2003). Holloway et al. (1993) reexamined some of these results, and concluded that when brain size is taken into account sexual dimorphism in the corpus callosum is indeed indicated by such studies as: Witelson (1985), Weber and Weis (1986), Yoshii et al. (1986), Kertesz et al. (1987), Oppenheim et al. (1987), Byne et al. (1988), Demeter et al. (1988), Elster et al. (1990), Going and Dixson (1990), Habib et al. (1991), and Steinmetz et al. (1992).

In one meta-analysis, Bishop and Wahlsten (1997; see also Fitch and Denenberg, 1998) suggested that there are no sexual differences in callosal shape or size. It should be noted, though, that Bishop and Wahlsten downplay the effect of allometric scaling in the brain, proposing that it is not an appropriate way to analyze cortical data. This is contradicted by a more recent meta-analysis by Smith (2005), which demonstrated the importance of allometric considerations in comparative data, concluding that the corpus callosum of human females *is* relatively larger than that of males. Despite Smith's (2005) analysis there is still disagreement in the literature as to the validity of relative comparisons within species.

Reviews by McGlone (1980), Kimura (1980, 1983, 1987, 2000), Witelson (1983), Davidson and Hugdahl (1995), and Smith (2005) among others, confirm that there are sex differences in the brains of humans. Through cognitive studies on visuospatial tasks (see McGlone, 1980) and speech tasks such as speed of articulation, fluency within a language, and grammar (Hutt, 1972; LeDoux, 1982; Ross et al., 1997), it has been suggested that the adult male brain is more asymmetrical than the adult female brain with regard to verbal functions (Hutt, 1972; McGlone, 1977; LeDoux, 1982; Zaidel et al., 1995; Grimshaw, 1998), spatial functions (Witelson, 1977, 1983; Corsi-Cabrera et al., 1997), or both (Hutt, 1972; Springer and Deutsch, 1989). This information has led to the suggestion that the structure of the corpus callosum is responsible for certain sex differences in cerebral lateralization (de Lacoste-Utamsing and Holloway, 1982; Witelson and Kigar, 1987; Witelson, 1989; Holloway, 1990; Pulvermuller and Mohr, 1996; Funnell et al., 2000b).

Few studies have addressed the nature of the human corpus callosum on a histological level (Tomasch, 1954; Aboitiz et al., 1989, 1992b; Highley et al., 1999). Tomasch (1954) conducted the first study focused on the fiber composition of the corpus callosum. While he did not include any females in his study, Tomasch established the corpus callosum as the primary interhemispheric pathway. Later, Aboitiz et al. (1989, 1992a,b,c) reexamined the topic of fiber composition of the human corpus callosum. Unlike Tomasch (1954), Aboitiz et al. (1989, 1992a,b) included females in their sample. This allowed for a comparison of fiber numbers and types between sexes. From their examination of ten males and ten females they concluded that any differences in either total fiber number or fiber type were not statistically significant. While it was found that females posses more large myelinated fibers (> 3μm) than males, this difference was statistically insignificant. In addition, males were found to have more small myelinated fibers (< 3μm), yet this difference was also statistically insignificant. These results suggest that sex differences in the corpus callosum are not evident in the overall fiber composition of this structure. Although they do not specifically propose that sex differences in fiber composition may occur within certain callosal subsections, their data suggest that such differences may occur within certain regions such as the isthmus and midbody (Aboitiz et al., 1992a, 1996). While the above studies by Tomasch (1954) and Aboitiz et al., (1992a,b,c, 1996) have led to a greater understanding of the neuronal contribution to the corpus callosum, there is still a gap in studies on sexual differences that explain the cognitive differences seen between human males and females.

With regard to sex differences and pathology, Highly et al. (1999) found that there is a significant sex difference in the density of callosal fibers in normal and schizophrenic subjects. In the normal sample midsagittal area of the corpus callosum was not significantly different between males and females. However, normal females had a statistically significant greater density of callosal axons than males, especially in the splenium. The converse was found in schizophrenics. Male schizophrenics had a greater axon density in all callosal regions, especially in the splenium. A sex specific trend that occurs in schizophrenia is that along with a general reduction in brain size females exhibit a concordant reduction in fiber density in the corpus callosum, while males do not show a significant change. Why females show a dramatic reduction in the density of fibers passing through the corpus callosum, although the overall size of the corpus callosum, save the splenium, is not reduced from normal subjects, is difficult to discern. Highly et al. (1999) and Crow et al. (1998) conjecture that one variable that may account for the differences discussed above (i.e., the significant reduction in the size of the corpus callosum and fiber density in schizophrenic females) is the presence of increased lateralization or impairment of hemispheric communication in schizophrenia. Moreover, the significant difference in fiber density between normal males and females may explain certain cognitive differences between the sexes.

More recently, Westerhausen et al. (2003) did a study to see if gender is associated with microstructural differences in the human corpus callosum. They did find sex differences in the microarchitecture of the callosal pathways. This study was the first to find sex differences in the anisotropy of the corpus callosum (Westerhausen et al., 2003). Westerhausen et al. (2003) found a higher anisotropy value in the corpus callosum of the male subjects, which could result from fewer myelinated fibers and a lower density of fibers in the males. In a newer study by Westerhausen et al. (2004), the results were basically the same with the male subjects showing higher anisotropy than females and with higher anisotropy values in the posterior third as compared to the genu region. The males had a larger midsagittal area of the corpus callosum and a larger callosal area consisting of myelin than females (Westerhausen et al., 2004). In a recent study by Shin et al. (2005), they found decreased fractional anisotropy in the female corpus callosum as compared with that of the male, and conclude that the corpus callosum is a region of sex differences.

Sex differences in the corpus callosum of nonhuman primates and rodents

While studies such as Aboitiz et al. (1992a,b,c) and Highly et al. (1999) on humans have begun to address the question of the reality of gender-related differences in the corpus callosum, they have not completed the journey. Beginning with deLacoste-Utamsing and Holloway (1982), there have been many studies coming down on either side of the question. Obtaining an unambiguous answer is important, since the corpus callosum plays such an important role in lateralization of function in the brain, most importantly vision (Demeter et al., 1990; Payne, 1990; Krubitzer and Kaas, 1993; Vercelli and Innocenti, 1993; Intriligator et al., 2000) and speech (Kertesz et al., 1987; O'Kusky et al., 1988; Kaga et al., 1990; Galaburda, 1995; Preis et al., 2000). While all aspects of the human corpus callosum cannot be gleaned from studies on other mammals, examinations of this structure, however, in two particular mammalian groups, rodents and nonhuman primates, have provided clues about the function of the corpus callosum and its regions.

In response to the supposition of sex differences in the splenium of the corpus callosum in humans by de Lacoste and Holloway (1982), Juraska and Kopcik (1988) began to examine the development of sex differences in the corpus callosum of rats to determine the stimuli required to produce sex differences in this structure. In the first of a series of studies on the rat corpus callosum, Juraska and Kopcik (1988) found no sex differences in the size of the corpus callosum in rats that had either been raised in a complex environment or isolation, albeit they used only gross measurements. They did, however, find that females possessed more unmyelinated axons than males regardless of environment. In addition, females that were raised in a complex environment had more myelinated axons than similarly raised males, although males tended to have larger myelinated axons passing through the corpus callosum. The relevance of this study was to show that although morphological sex differences may not exist in the midsagittal area of the corpus callosum in humans, it is possible that axonal differences do exist. It also demonstrated that environmental conditions may influence the composition of this structure.

Subsequent studies on the corpus callosum have revealed sex differences in the fiber composition of the splenium. While there are no significant sex differences in the total number of axons passing through the splenium, there are sex differences in the types of axons in it. Females tend to possess more unmyelinated axons than males. In contrast, males possess larger myelinated axons than females (Kopcik et al., 1992; Mack et al., 1995; Kim et al., 1996). The production of the differences is currently a subject of debate. Are the differences merely environmental and thus developmental (Juraska and Kopcik, 1988; Kopcik et al., 1992; Kim and Juraska, 1997; Nuñez et al., 2000), or are they based purely on hormonal influences (Fitch et al., 1991; Mack et al., 1996; Bishop and Wahlsten, 1999; Bimonte et al., 2000)?

Many studies have focused on the sexual dimorphism of the human brain (Mall, 1909; Kimura, 1992; see McGlone, 1980; Falk, 1997 for reviews), but few have examined the issue in nonhuman primates (Le May, 1976; de Lacoste and Woodward, 1988; Falk et al., 1999; Franklin et al., 2000; Dunham and Hopkins, 2006; Phillips et al., 2007). At the same time most of the research that has been performed on sexual dimorphism in the brain of primates has had more to do with morphology than with the actual composition of this organ. The distribution of callosal fibers in nonhuman primates has been demonstrated several times (Seltzer and Pandya, 1983; Gould et al., 1986; O'Kusky et al., 1988; LaMantia and Rakic, 1990a,b; Beck and Kaas, 1994). Although LaMantia and Rakic (1990) approached gender differences in the course of their study, differences between the sexes with regard to fiber composition have yet to be sufficiently and specifically addressed.

De Lacoste and Woodward (1988) examined the midsagittal area of the corpus callosum in pongids, cercopithecoids, cebids, and strepsirhines. They found sex differences in the size of the corpus callosum and the width of the splenium relative to brain size in pongids. They also found sex differences in the size of the corpus callosum relative to brain size in strepsirhines. While these results would suggest that sex differences in the corpus callosum exist in certain primate groups, it should be noted that the four primate groups used in the above study (pongids, cercopithecoids, ceboids, and strepsirhines) are comprised of thirty-four species. Thus, their results are merely suggestive of sex differences in primate groups and not specific species. Other primate studies in which species were not combined show less sexual dimorphism in the corpus callosum than was previously suspected. Holloway and Heilbroner (1992) report that there are no sex differences in the corpus callosum or its subsections relative to brain size in *M. mulatta*, *M. fascicularis*, *Callithrix jacchus*, or *Saguinus oedipus*. Only *M. mulatta* demonstrated a slight sexual difference in the width of the splenium, with males being larger than females. Separately, Franklin et al. (2000) suggested that the total area of the corpus callosum is larger in *M. mulatta* males than females. They also showed that females possess a larger splenium. While these results are contrary to those of Holloway and Heilbroner (1992), it should be noted that the results of Franklin et al., (2000) are based on raw data and not relative measurements. Thus, these results merely serve to complicate the issue of sex differences in the corpus callosum. More recently, Phillips et al. (2007) demonstrated that female capuchin monkeys possess a larger corpus callosum compared to males with regard to overall size and posterior subregional measurements.

While the above studies have sought to determine sex differences in the corpus callosum of nonhuman primates based on total callosal area or subsectional areas, few studies have attempted to address the question of fiber differences in this important structure. Seltzer and Pandya (1983), Gould et al. (1986), O'Kusky et al. (1988), and Beck and Kaas (1994) have examined the topography of the nonhuman primate corpus callosum; however, these studies did not address the issue of sex differences. LaMantia and Rakic (1990a) also examined the development and topography of the nonhuman primate corpus callosum. In addition to their primary data, they also include anecdotal data on sex differences in the fiber composition of the corpus callosum in *M. mulatta*. In a comparison of two age- and brain weight-matched individuals, the male possessed 10 million more axons than the female, although the female's corpus callosum was larger. While this difference appears large, they suggest that the disparity could quickly disappear with a larger sample, since the corpus callosum normally contains fifty to sixty million axons in *M. mulatta*.

In general there is a paucity of data on sex differences in the corpus callosum of nonhuman primates. While the above studies have provided intriguing clues to the lack of definitive sex differences in this structure, the disparity of their results mandates the need for additional data, especially in species such as *Pan*. This includes information on the relative size of the corpus callosum in individual species as well as supplementary data on the fiber composition of this structure. Such data are important to understanding the function, development and evolution of human and nonhuman primate brains.

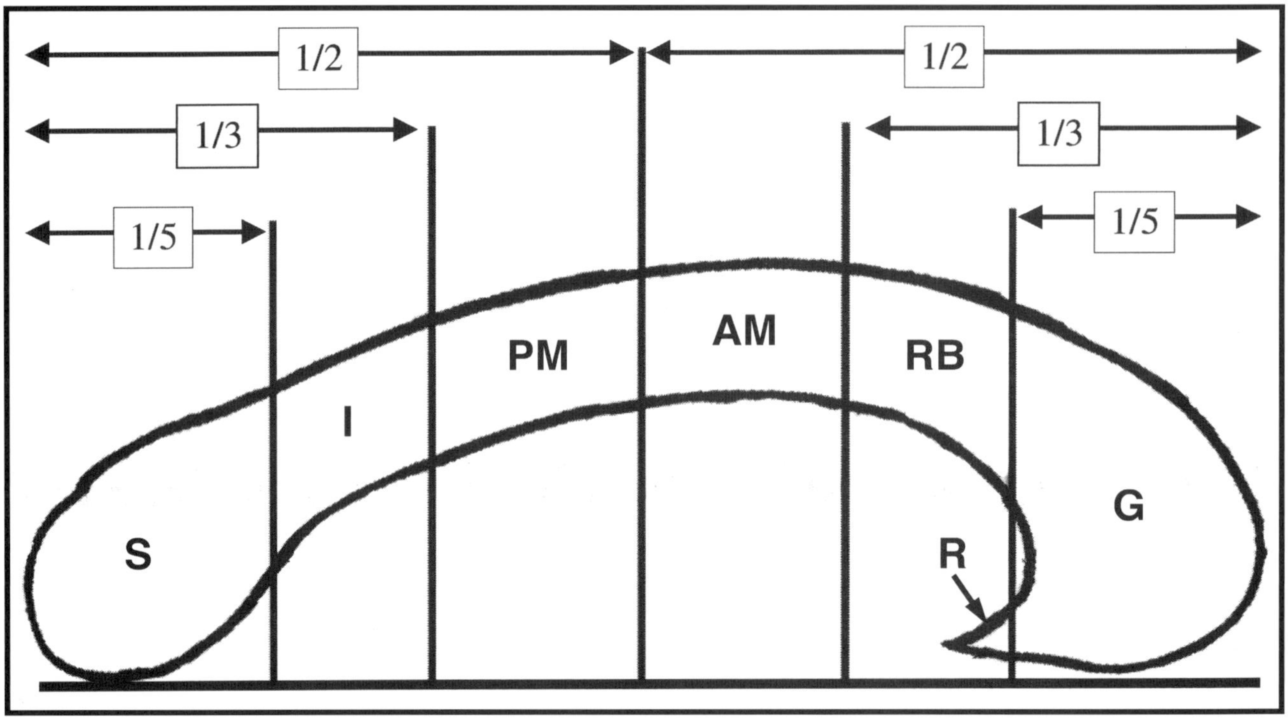

Fig. 2. Diagram of the midsagittal view of the corpus callosum of an adult human, showing the regional subdivisions. S: splenium, I: isthmus, PM: midbody, posterior midbody, AM: midbody, anterior midbody, RB: genu, rostral body, G: genu, R: rostrum. (after Witelson, 1989)

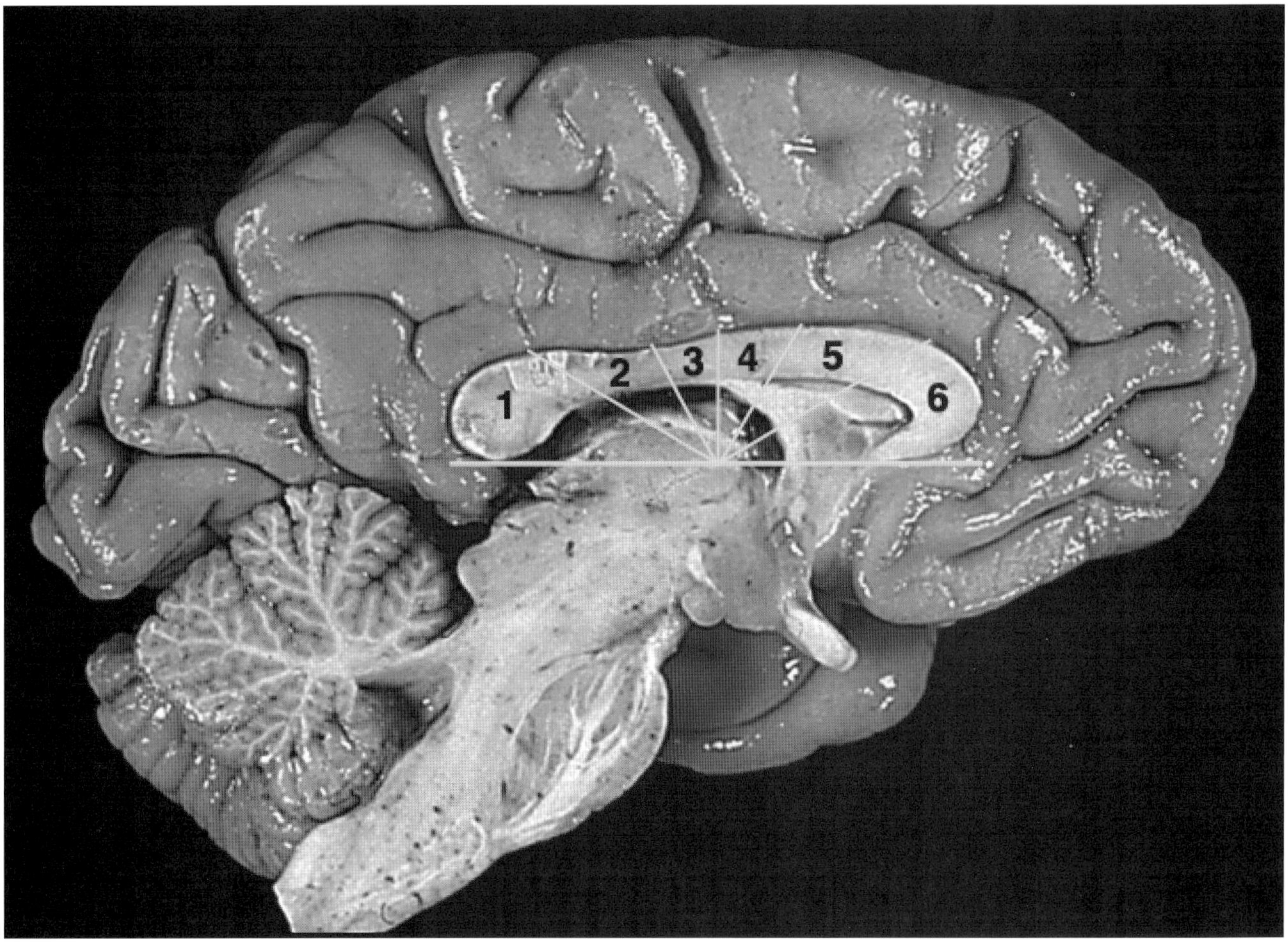

Fig. 3. Midsagittal view of the brain of Pan troglodytes, showing the radial-line method of callosal division. 1: splenium, 2: isthmus, 3: posterior midbody, 4: anterior midbody, 5: rostral body, 6: genu/rostrum.

Methods

Corpus callosal measurements were performed using postmortem specimens of *M. fascicularis* (male n = 20, female n = 20) derived from the collections of Dr. Patrick Gannon (then housed in the Department of Otolaryngology, The Mount Sinai School of Medicine) and the laboratory of Dr. Ralph Holloway, Department of Anthropology, Columbia University. Brains of *P. troglodytes* (male n = 11, female n = 12) for this study were obtained from Yerkes Regional Primate Center, Emory University (n = 7 brain tissue, n = 6 MRI), the Department of Mammals at the National Museum of Natural History, Smithsonian Institution (n = 6), and the collection of Dr. Ralph Holloway (n = 4).

Morphological analysis

The corpus callosum has been traditionally parcelled into five regions. Although there are no anatomical or histological landmarks defining each region, they can be defined according to a straight rostrocaudal length, dividing the corpus callosum into thirds and fifths to delineate each region (Mall, 1909; de Lacoste and Holloway, 1982; Witelson, 1989; Aboitiz et al., 1992a,b). The different callosal regions defined by this method are (i) rostrum (anterior one-third); (ii) genu (area between the anterior one-fifth and anterior one-third); (iii) midbody (middlc onc-third); (iv) isthmus (area between the posterior one-third and posterior one-fifth); (v) splenium (posterior one-fifth) (Fig. 2). Further, some researchers (Aboitiz et al., 1992a) divide the midbody into an anterior midbody (area between the anterior one-third and one-half) and posterior midbody (area between the posterior one-half and one-third) (Fig. 3). As a result of these differences in allocating the callosal subregions both methods of the dividing the corpus callosum into either five (straight-line method) or six (radial line) parts were used.

Absolute measurements on the total callosal area and the areas of its regions were recorded using SigmaScan Pro (SPSS Science). In addition, statistical calculations were performed using absolute measurements. The final analyses of the areas calculated on the corpus callosum and the conclusions drawn from these analyses were, however, conducted using only standardized measurements, utilizing Jerison's (1973) slope.

$$(\text{CC measure})/(\text{Brain Weight})^{2/3} \quad (1)$$

Histological analysis

Millions of axons of different diameters traverse the corpus callosum. Given the large number of axons occupying any given area in the corpus callosum, sex differences may be manifested in subtle aspects unavailable through purely morphological measurements. While it is prudent to assume that there are no sex differences in the fiber composition of the corpus callosum in the subjects examined here given the considerable lack of sex differences in the morphological dimensions of this structure, it was determined that a histological study of a small number of select individuals would be performed in order to demonstrate this assumption in a timely manner. The region chosen for this portion of the study is the splenium, since it is this callosal region that has arguably undergone the most significant evolutionary change and is most sexually dimorphic in modern humans (de Lacoste and Holloway, 1982; Holloway, 1990).

The splenium in four *M. fascicularis*, two individuals of each sex (male n = 2, female n = 2), and five *P. troglodytes* (male n = 2, female n = 3) were embedded in Epon and sliced into ultrathin sections (< 0.5μm). The sections were then stained with toluidine bluse and examine using a Zeiss Axioskop light microscope. Using bright field emission and a 40x objective lens, histological samples were examined for myelin and cellular integrity. Samples which did not meet specific criteria for myelin integrity were rejected. Each of the *M. fascicularis* individuals chosen for study was deemed appropriate for study, since none of these individuals exhibited any myelin or cellular degradation. Three of the five *P. troglodytes* specimens (YN88-256, YN92-115, YN95-60) were rejected, since all exhibited significant and severe myelin and cellular degradation. Thus, a single *P. troglodytes* female (YN94-67) and single *P. troglodytes* male (YN97-139) were selected. Once the integrity of the specimens was determined, splenial fiber counts were achieved using a 100x oil immersion objective and digital capture. Analysis of the total number of fibers in the splenium was determined using IPLab 3.1 (Scanalytics, Inc.).

Results

Macaca

For *M. fascicularis*, males on average possess an absolutely larger corpus callosum than females. However, when these results are standardized according to brain weight the differences between males and females for total callosal area disappears and are statistically insignificant. Such a result indicates that any cognitive differences between males and females of this species are probably caused by the overall structure of the corpus callosum. In addition, they would also suggest that sex differences in this structure as seen in modern humans (see de Lacoste and Holloway, 1982) likely do not have their origins within the cercopithecoid clade. However, this conclusion may be somewhat presumptuous at this stage when discussing the results of this research, because it assumes that evolution of the corpus callosum results in an overall change in this structure rather than a mosaic alteration.

M. fascicularis does not exhibit any statistically significant difference between males and females either in the absolute or relative size of the genu. The lack of sex differences in this area is unsurprising, since the genu is thought to connect portions of the motor cortex as well as areas within the prefrontal cortex (Pandya et al., 1971; de Lacoste, 1981; Seltzer and Pandya, 1983; Barbas and Pandya, 1984). Differences in this region should not be suspected, simply because there is little information to suggest that male and female members of *M. fascicularis* differ from each other with regard to motor skills. In addition, while fibers traversing this region of the corpus callosum connect portions of the orbitofrontal cortex, which is important for both memory and behavior (Parker et al., 1997), there is little information in *Macaca* suggesting that males and females differ significantly in these tasks (Lacreuse et al., 1999).

The anterior and posterior midbodies represent the two divisions of the midbody defined using the straight and radial line methods. The anterior portion of this region contains interhemispheric fibers connecting the primary, secondary, and supplementary motor cortices, while the posterior portion connects primary and secondary sensory areas (Pandya et al., 1969; de Lacoste, 1981; Pandya and Seltzer, 1986). In addition, the posterior portion of the midbody possesses fibers connecting the postcentral and posterior parietal lobe as well as portions of the superior and inferior temporal lobes (de Lacoste, 1981; Seltzer and Pandya, 1983; LaMantia and Rakic, 1990a). Despite the complexity of the connections passing through this region, the midbody areas for males and females of *M. fascicularis* do not differ significantly from each other. Albeit these results are not unusual when compared to the human data (Oppenheim et al., 1987; Allen et al., 1991; Witelson, 1989; Matano and Nakano, 1998), they are somewhat unexpected given the role of callosal axons passing through the midbody in sexually dimorphic tasks.

The splenium represents the region of the corpus callosum that has often been found to exhibit sexual dimorphism in humans (Holloway, 1990; Holloway et al., 1993; Davatzikos and Resnick, 1998). Moreover, area differences between males and females do not appear to be the result of isometric expansion of the splenium in one sex versus the other. Instead the relatively larger splenium of human females is also more bulbous than that of males (de Lacoste and Holloway, 1982; Holloway et al., 1993; Davatzikos and Resnick, 1998). Since the splenium is responsible for connecting occipital, temporal, and posterior parietal areas of the brain (Pandya and Seltzer, 1986; Gazzaniga, 2000), it is possible that these area and form differences may be related to sex differences in visuospatial, language, and somatosensory cognitive functions.

The splenium of the *M. fascicularis* sample used in this study did not exhibit any sex differences. While males possessed absolutely larger splenia using the straight and radial-line methods, this difference was eliminated when brain size was taken into account. In fact, the samples overlap entirely. Since the composition and form of the corpus callosum appears to be the result of cortical size and function, it is unlikely that the hypothesis that there are sex differences in the corpus callosum of cercopithecines would be true.

Pan

Female *P. troglodytes* possess an absolutely and relatively larger corpus callosum. However, these differences are statistically insignificant. Despite the lack of a significant difference between males and females, it is worth noting that unlike *M. fascicularis* the corpus callosum in *P. troglodytes* trends toward being larger in females. Measurements of total callosal area provide some information regarding the presence of sexual differences of the structure, but they do not provide specific information that may be useful for the assessment of possible lobular or cognitive differences in the brain. To gain insight into such differences when examining the midline profile of the corpus callosum it is necessary to examine callosal regions. Below the results for each region are discussed.

The genu as defined using the straight-line method is roughly equivalent to the genu and rostral body as defined through the radial-line method. As such these areas occupying the anterior one-fifth of the corpus callosum will be referred to as the genu here. For both the straight and radial-line methods females possess an absolutely larger genu than males. However, the relative values of this structure do not indicate any difference between males and females. In addition, there is no statistical difference between males and females in the genu. The lack of a significant difference between males and females in the genu means that sex differences in this region as displayed in humans (Witelson, 1989) must have evolved after the ape-human split.

Both the absolute and relative values for midbody area differences between males and females are statistically insignificant for *P. troglodytes*. The averages for the anterior and posterior midbody using the straight-line and radial-line methods display significant overlap, such that there is no apparent trend towards one sex possessing a slightly larger midbody than the other. For example, the greatest difference between males and females occurs when the averages of the relative size of the anterior midbody as defined using the straight-line method are compared. The average relative size of the anterior midbody is 0.0095cm^2 for females and 0.0086cm^2 for males. However, the standard deviation of the sample is large, and thus there is a significant degree of overlap. The lack of sexual differences in this area, though, is expected, since the areas of the brain connected by fibers passing through this region have not become highly specialized over the course of primate evolution. Moreover, humans do not display any sexual differences in this area (Witelson, 1989), and as such it is not expected that *Pan* would.

The isthmus of the human corpus callosum displays sexual dimorphism with females possessing a relatively larger isthmus than males (Witelson, 1989; Steinmetz et al., 1992; see also Davatzikos and Resnick, 1998). Due to this relationship it is hypothesized that female *P. troglodytes* may also possess a relatively larger isthmus. Indeed females possess an absolutely and relatively larger isthmus on average as defined using the straight-line method. However, there is a large degree of overlap between the two samples, and thus there is not a statistically significant relationship between sex and isthmus size. This means that any statistically significant sex differences in this region are unique to humans, and must have evolved after the ape-human split. Alternately, chimps may have retained the earliest trends toward such a dimorphism.

The splenium of the corpus callosum has been an area of intense interest in human studies (de Lacoste and Holloway, 1982; Oppenheim et al., 1986; Holloway, 1990, 1993). Bean (1906) had first described the splenium of females as being different from males. Later studies found similar differences and described the female splenium as more bulbous. This general description has become useful in identifying the corpus callosum of human females, although the functional significance of this morphology has not been deciphered. Some researchers, though, have suggested that despite this general morphological dissimilarity between males and females sex differences in the human splenium do not exist (e.g., Witelson, 1989). Due to the disparity of splenial data from humans it is not possible to predict the presence of sex differences in this region in *P. troglodytes*. Indeed, there is not a significant difference between male and female *P. troglodytes* for area measurements of the splenium using the straight or radial line method.

Histology

Regional and total area measurements of the corpus callosum provide useful information of the overall structure of this interhemispheric highway. In their study of the composition of the corpus callosum, LaMantia and Rakic (1990a) found no appreciable difference between a single male and female *M. mulatta* for the total number of fibers comprising this structure. Concurring with LaMantia and Rakic (1990a), this current study found no difference in the number of fibers comprising the splenium as well as the total number of fibers inferred to compose the entire corpus callosum in both *M. fascicularis* and *P. troglodytes*. This is consistent with results on morphological measurements of total and regional callosal area (LaMantia and Rakic, 1990a; Aboitiz et al., 1992a; see also Highley, 1999). However, contra to LaMantia and Rakic, (1990a) who suggest that males have slightly more axons in the corpus callosum than females, this study found that females possess slightly more axons than males. From this it can be assumed that there is a large degree of variability expressed in *Macaca* with regard to the total number of axons in the corpus callosum. In addition, this study concurs with the conclusion of Aboitiz et al. (1992a), which states that if the overall area of a callosal region does not demonstrate sexual dimorphism then one would not expect to find a difference in the number of fibers comprising that area.

In humans, Aboitiz et al. (1992a) found no appreciable difference between males and females in the number of fibers comprising the corpus callosum. This would indicate that even apparent sexual differences in the size of the corpus callosum do not impart any correlation to its composition. Aboitiz et al. (1992a) predict that the area of the corpus callosum is a good indicator of the number of fibers contained in it. However, they go on to acknowledge that this predictive hypothesis may not be accurate for estimating the number of gigantic fibers (> 3μm in diameter). For this reason it is not possible to propose the presence or lack of sex difference in the corpus callosum by merely estimating the total numbers of fibers it contains. Instead it is necessary to additionally account for the types of fibers comprising the corpus callosum. Thus, counting the total number of fibers in the corpus callosum is only one step to the conclusion of assessing sexual dimorphism in this structure.

There is no significant difference between *M. fascicularis* males and females with regard to fiber type (Fig. 4; Table 1). While these results combined with those from other aspects of this study conclusively show that there are no sex differences in the corpus callosum of this species, they can be discussed descriptively to provide information that may be useful for drawing a hypothesis on the evolution of sex differences in the brain. Males of this species tend to possess more medium, large, and very large axons than females. Females conversely tend to possess more small axons than males. While the differences between males and females are not statistically significant, this descriptive information does offer some insight into relative differences between males and females.

The differences between the male and female *P. troglodytes* sampled do not appear to be significant (Table 2). Proportionally, the female possesses a greater number of fibers than the male, but based on data from macaques (LaMantia and Rakic, 1990a) and humans (Aboitiz et al., 1992a; Highley et al., 1999) this type of variation between individuals for total callosal axon number is not unusual. While the difference in the total number of axons in the corpus callosum between the male and female sampled demonstrate that sex differences in this structure most likely do not exist, it is, nevertheless, possible to discuss the general differences in the types of fibers found in the splenium of these individuals.

The male *P. troglodytes* sampled possesses more small and large diameter axons than the female, while the female possesses more medium, very large, and giant axons than the male. While this cannot be tested statistically due to the small sample size, it can be assumed by examining the number of axons in each category and their percentage to the total number of axons that there is not sexual dimorphism with regard to types of fibers in the splenium of the corpus callosum. This data can provide basic descriptive information regarding possible sex differences in this species and the evo-

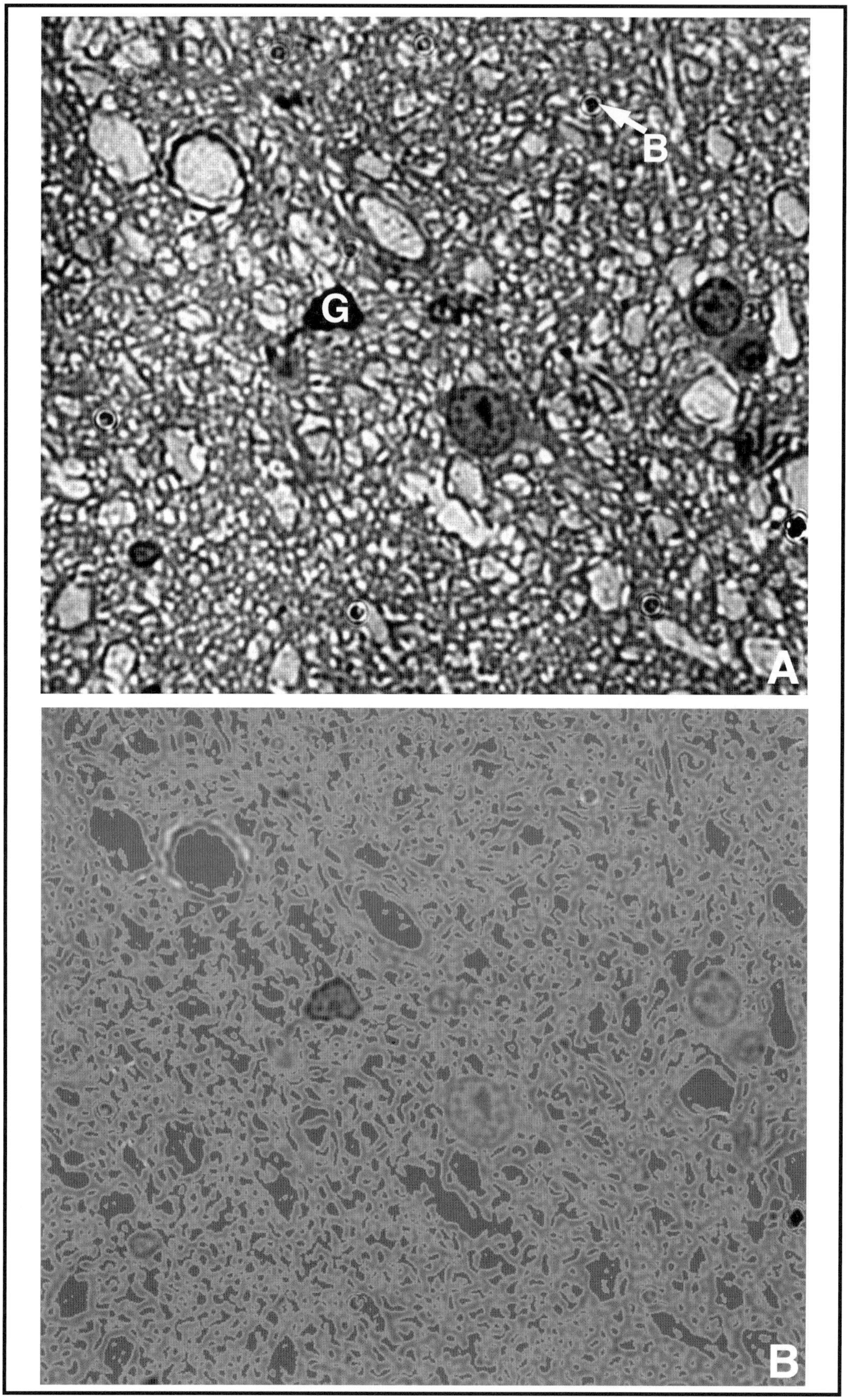

Fig. 4. Histological section (1000x) from the splenium of Macaca fascicularis, showing the counting regime of the IPLab software. unsampled section. Figure B is the same section indicating the cells counted by IPLab. G: glial cell, B: air bubble. The latter features are manually removed before the end count is made.

Table 1. Splenial axon number based on axonal size for Macaca fascicularis[1]

Specimen	Sex	Very large axons (≥ 2.5μm)	Large axons (1 - 2.5μm)	Medium axons (0.4 – 0.99μm)	Small axons (< 0.4μm)
PGM 40	F	7073 (.09)	20768 (.28)	27250 (.36)	19956 (.26)
PGM 54	F	4875 (.13)	10831 (.30)	13001 (.36)	7447 (.21)
PGM 43	M	6017 (.12)	14050 (.28)	16136 (.32)	13915 (.28)
PGM 45	M	6220 (.13)	14999 (.31)	16606 (.35)	9906 (.21)

1. Total number of axons for each axon category. Percentage to the total number of axons in the sampled area is listed in parentheses. Percentages are rounded up. M = male, F = female.

Table 2. Splenial axon number based on axonal size for Pan troglodytes1

Specimen	Sex	Giant axons (≥5μm)	Very large axons (2.5 - 5μm)	Large axons (1 - 2.5μm)	Medium axons (0.4 – 0.99μm)	Small axons (< 0.4μm)
YN94-67	F	62 (.06)	197 (.18)	277 (.25)	370 (.34)	181 (.24)
YN97-139	M	50 (.03)	276 (.18)	438 (.28)	495 (.32)	284 (.22)

1. Total number of axons for each axon category. Percentage to the total number of axons in the sampled area is listed in parentheses. Percentages are rounded up.

lution of sex differences in general. For example, the female *P. troglodytes* possessed more medium and very large fibers than the male, while the female *M. fascicularis* were found to possess only more small axons than the males. This difference as expressed in *Pan* is similar to the result obtained by Aboitiz et al. (1992a) for humans, speculatively implying that the structure of the corpus callosum in *Pan* is more similar to humans than to cercopithecoids.

Discussion

The conclusion of this study is that based on measurements of the total midsagittal area of the corpus callosum, midsagittal regional areas of the corpus callosum, and the number and type of axons in the splenium of the corpus callosum, there are no sex differences in this structure in *M. fascicularis* or *P. troglodytes*. Indeed, neither species exhibits a statistical trend, indicating that one sex may possess a larger callosum, more axons, or more of a particular type of axon. From these results it is also possible to conclude that modern humans are the only extant primate group that exhibits any sexual dimorphism in the corpus callosum or its regions. In some ways these results are consistent with the literature suggesting specialized lateralization of the human brain and sex differences exhibited in lateralized cortical processes (e.g., Witelson, 1977; Kimura, 1980, 1983; Hugdahl et al., 1993; Eviatar et al., 1997; Crucian and Berenbaum, 1998; Halpern et al., 1998; Hausmann and Gunturkun, 1999; Vallortigara et al., 1999; Amunts et al., 2000; also see review by McGlone 1980). This is because many lateralized processes often are related to functions of speech and language, which have never been isolated in nonhuman primates.

Despite the apparent lack of lateralization in the nonhuman primate brain with regard to language, there have been other studies that indicate the brain of nonhuman primates may be lateralized (e.g., Gannon et al., 1998). However, many of these studies depend on correlations between handedness and a given task (Note: the author disagrees with the usage of the term handedness as it has been applied in many of the following psychological studies and prefers the term hand preference). For example, Bard et al. (1990) found that *P. troglodytes* displays a general right hand preference during feeding behaviors. At the same time Hopkins (1990) found that *P. troglodytes* and *Pongo* display a general right hand preference in an experimental model requiring subjects to manipulate a joystick (see also review by Hopkins and Morris, 1993). Later, Hopkins and his colleagues have correlated hand preference to birth order (Hopkins and Dahl, 2000), gestural communication (Hopkins and Leavens, 1998), and other manipulation tasks (Hopkins and Pearson, 2000). Although these particular studies do not provide definitive data on the lateralization of the nonhuman primate brain, they do provide a means to understand the origins of laterality.

Recent anatomical asymmetries have been noted in the brains of great apes but not Old World or New World monkeys (Hopkins and Rilling, 2000; Hopkins and Marino, 2000). In their study on petalial patterns in primates using left and right anterior frontal, posterior frontal, parietal, and occipital cerebral width measurements on axial magnetic resonance images, Hopkins and Marino (2000) found that the great apes (*Pan*, *Gorilla*, *Pongo*) display a right-frontal, left-occipital directional asymmetry or petalia pattern. While there was an individual from each taxon that displayed the converse asymmetry, the results for these genera were more consistent than for other groups. That is, Old and New World genera did not display directional asymmetry, albeit certain individuals within the *M. mulatta* sample did. Working from the same dataset Hopkins and Rilling (2000) report that measured asymmetries in neocortical surface area and brain volume indicate that the brains of the great apes are more asymmetrical than those of Old and New World monkeys. Moreover, this particular study suggests that individuals that possess a more leftward asymmetric brain had a smaller corpus callosum than those individuals that displayed rightward or no asymmetry. Handedness (hand preference) data collected by Hopkins (1995) and Westergaard et al. (1998) suggest that there is a general shift in primates from population-level left-hand preference to population right-handedness for quadrupedal and bipedal reaching such that *Pan* more often displays a preference for right handed reaching and manipulation than Old and New World primate groups. Moreover, individuals that display right-handedness or right hand preference possess a smaller corpus callosum as a function of neocortical surface area and brain volume (Hopkins and Rilling, 2000). While this finding cannot confirm the presence of lateralized brain function in any of these species studied, especially *Pan*, it does suggest an early evolution for the development of lateralization.

Experiments designed to test cognitive skills in nonhuman primates, such as handedness, provide important data that can be used to formulate hypotheses concerning the origins of brain lateralization as well as the development of sex differences in the brain. In addition to handedness or hand preference studies, other behavioral experiments have been report that may enhance these evolutionary and cognitive hypotheses. Data collected from memory and cognitive performance studies on nonhuman primates indicate that certain male-female differences occur. In particular, several studies have found that male and female *M. mulatta* differ from each other with regard to facial discrimination tasks (Buccafusco et al., 1999; Lacreuse et al., 1999; Parr et al., 2000). For example, Buccafusco et al. (1999) reports that male *M. mulatta* performed better on memory-related tasks compared to females, although these tasks required simple memory recall, and not recall of complex subjects.

Complex subject recall requires the individual to not only recall specific subject matter, but also associated features of the item in question. In humans such complex tasks are usually associated with language tasks (Hugdahl et al., 1993; Hadar et al., 1998; Hausmann and Gunturkun, 1999). For example, when an individual is required to recognize familiar faces prefrontal and lateral temporal regions are bilaterally activated. However, when an individual is exposed to newly learned or unfamiliar faces hippocampal, parahippocampal, parietal and anterior temporal activation is observed (Clark et al., 1998; Leveroni et al., 2000). Observations such as these are significant, since these tasks, except for the hippocampal responses, require the participation of callosal axons. Moreover, males and females are dissimilar from each other for these and many cognitive tasks involving language areas (Shaywitz et al., 1995; Levin et al., 1996; Gur et al., 1999; see also Kimura, 1983, 1987).

Although macaques do not possess cognitive abilities approaching those of humans, studies on these nonhuman primates indicate that they possess some ability to perform tasks such as facial recognition and recognition of facial cues (Vermeire et al., 1998; Parr et al., 2000). While it is not currently possible to test nonhuman primates with PET or fMRI to determine the specific functional areas of their brain, it is possible to use topographic studies to draw some correlations between cortical anatomy and possible cognitive functions. Work by de Lacoste (1981), Pandya and his colleagues (Pandya et al., 1969; Seltzer and Pandya, 1983; Barbas and Pandya, 1984), and LaMantia and Rakic (1990a,b) indicate that humans and macaques share many functional areas within the cerebral cortex. From such correlations it is possible to hypothesize that if sex differences exist with regard to certain cognitive functions that males and females may demonstrate differences in the callosal fibers associated with those tasks. For facial recognition tasks these fibers likely, in part, pass through the midbody of the corpus callosum. Thus, it is probable that the midbody would be different between males and females. The data presented here, though, concur with measurements on humans indicating there is no difference between males and females in the area of the midbody of the corpus callosum (Oppenheim et al., 1987; Allen et al., 1991; Witelson, 1989; Matano and Nakano, 1998).

The above behavioral studies are restricted to *Macaca*, but other data also provide important information suggesting the presence of sex differences in the brains, and possibly the corpus callosum, of nonhuman primates. Two recent studies involving *P. troglodytes* suggest that this species possesses memory and recall abilities that exceed those displayed by *Macaca mulatta*. In the first study, Menzel (1999) reports the ability of a single female *P. troglodytes* that retained the ability to recall the locations of randomly hidden objects for up to sixteen hours. In the second study, Parr et al. (2000) report that *P. troglodytes* displays a greater recall of conspecifics facial features than *Macaca mulatta*. In this last report chimpanzee individuals were required to match similar pictures of conspecifics. While both the *Macaca mulatta* individuals and chimpanzees displayed an equal ability to discriminate conspecifics, the *Macaca mulatta* individuals required significantly more trials to be able to perform the task successfully. Although these reports could be described as rudimentary behavioral studies, they do still suggest the possible presence of specialization (and possibly lateralization) in the nonhuman primate brain.

Three final studies that are more relevant to the current study than many of those discussed above include spatial experiments performed on *Macaca mulatta*. This is because both of the following studies not only discuss the likely presence of lateralized function in parts of the nonhuman primate brain, but also the presence of sex differences on spatial tasks. In an experiment on twenty-six split-brain *Macaca mulatta*, Vermeire et al. (1998) found that faces were better remembered by the right hemisphere than the left. In addition, they also found that females monkeys were more lateralized for learning to discriminate faces than were males. A later study by Kavcic et al. (2000) agrees with the above findings that left hemisphere dominance for certain visual-memory tasks occurs in *Macaca*. Finally, work by Lacreuse et al. (1999) shows that *Macaca mulatta* displays sex differences with regard to spatial ability. However, it should be noted that Lacreuse et al. (2000) found a decline in spatial ability among males as they age, such that old males perform no better than old females. Yet for any given age class, except this late one, males outperform females in spatial cognitive tasks.

The studies discussed above report provocative results that suggest the presence of lateralization for certain tasks in nonhuman primates. While chimpanzees seem to possess greater asymmetry and cognitive abilities than macaques, macaques do appear to exhibit some lateralization in cognitive function. Moreover, males and females differ in some of these functional tasks. This later point, though, is contradicted by the results of this study and those of reports such as Hopkins and Rilling (2000). Hopkins and Rilling (2000) suggest that the brains of macaques are not as lateralized as those of chimpanzees. This would imply that spatial, memory, or other cognitive tasks are not lateralized in Old World monkeys. In addition, the information provided here suggests that males and females should not perform differently for these tasks. However, these hypotheses assume that the corpus callosum must be integral to all cognitive tasks. This, though, is not the case.

First, the various reports that suggest lateralization of the nonhuman primate brain rely upon what has been described as handedness (more properly hand preference) and visual capabilities. While tasks related to these features may be useful in understanding cognitive tasks and callosal function, there is no known study that adequately demonstrates the existence of higher cognitive processes in nonhuman primates. Because of this disparity between human and nonhuman primate studies, many of the results that suggest laterality in function may be explained as proving not the existence of complex pathways traversing the corpus callosum or specific lateralization of the neocortex, but as lateralization in basic mammalian cognitive tasks involving more primitive pathways such as the superior colliculus, anterior commissure, and hippocampal commissure, all of which are capable of carrying the type of information investigated in the afore mentioned reports.

Secondly, the studies that report sex differences in cognitive performance utilize visual information. While the splenium of the corpus callosum is important for relaying visual information, the type of visual discrimination described by Kavcic et al. (2000) and Vermeire et al. (1998) can occur via the superior colliculus (Wright and Craggs, 1976; Sommer and Wurtz, 1998). In addition, results showing sex differences in throwing among capuchin monkeys (Watson, 2001) may occur via sex differences in the anterior commissure (see Noonan et al., 1998). Although this does not eliminate the likelihood of lateralization of visual and motor components of the cerebral cortex in nonhuman primates, the possibility that these sex differences occur as the result of other hemispheric pathways explains why it is possible to suggest lateralization of and sex differences in the brain of nonhuman primates, yet to not find sex differences in the corpus callosum.

In general, there is a wealth of information that implies the presence of lateralized function within the brains of macaques and chimpanzees (see above discussion). These studies, though, lack the sophistication to ally simple visual and motor functions of the nonhuman primate brain with higher cognitive processes involving the integration of data as seen in humans. It is probable that some lateralization exists within the nonhuman primate brain, albeit not at the level present in modern humans. Indeed, the results of Hopkins and Rilling (2000) study would say that the degree of lateralization is different between macaques, chimpanzees and humans with humans displaying the most asymmetric brains in this group and macaques the least. However, the question still remains, is the level of asymmetry seen in great ape brains sufficient to produce human-like cognitive functions?

Based on behavioral data the answer remains unresolved. A lack of cerebral laterality in nonhuman primates, though, does not preclude one from suggesting that the corpus callosum would not be expected to display sexual dimorphism in either midsagittal area or axonal composition until the brain is sufficiently lateralized in function. This can be assumed because none of the above studies examines cognitive functioning at a level sufficient to assume the corpus callosum has been co-opted for the task of interhemispheric integration of cognitive information. Such information could only be approached through invasive retrograde histology or PET and fMRI studies. To conclude, the above studies are useful in understanding the evolution of the brain and sex differences within it, but they do not contradict the results of this study, which concludes that sex differences do not exist in the midsagittal area or axonal composition of the corpus callosum of nonhuman primates.

The uniqueness of the human brain has been discussed for thousands of years since the times of the Egyptians, Aristotle, and Descartes with little resolution (see Finger, 1994). Moreover, its has been a contentious topic in anthropology since the days of eminent neuroscientists/anatomists/anthropologists such as Broca, Smith, Dart, and Anthony (see Holloway, 1997). There is, however, still disagreement concerning the advent of human-like features in the brain, which eventually led to human cognitive abilities. Recently, Ambrose (2001) has revived an idea first proposed by Holloway (1970) and later revisited by Calvin (1983, 1993) and Wilson (1998) hypothesizing that the need for accurate throwing and tool making skills created selective pressures for advancement of the hominin brain, and in turn the development of sex differences in the cerebral cortex. These selective pressures also aided the development of sex differences in the modern human brain. While there are other hypotheses for the evolution of the human brain (e.g., Tobias, 1971; Jerison, 1973; Gould, 1977; Gould and Lewontin; 1979; Falk, 1990), few have been visited as frequently as Holloway's "throwing theory". This, though, has not quelled the debate of general human brain evolution or the development of sex differences in the brain, since the data that may be used for such studies is merely corroborative. The paucity of endocasts in the fossil record and the limitations of endocasts restrict their ability to provide conclusive answers of primate brain evolution. In addition, behavioral data on human and nonhuman primate subjects can provide information on cortical and cognitive functions of extant brains. However, an examination of both types of data, fossil and living, can be used to develop robust theories of brain evolution. In the case of this study it is possible to propose a hypothesis about the advent of sex differences in the corpus callosum of the primate brain.

The results of this study indicate that sex differences in the corpus callosum did not develop until after the ape-human split some 7 – 5 million years ago. Indeed, sex differences in this interhemispheric pathway may not have developed until the advent of our own species some 200,000 years ago. Neither *M. fascicularis* nor *P. troglodytes* display sex differences in total callosal area or the area of individual callosal regions. Moreover, neither species shows a difference between males and females for the number or types of fibers comprising the splenium of the corpus callosum. One would be inclined to conclude that these statements are possible, since the results do not exhibit statistical significance or a statistical trend.

From the results obtained here it seems apparent that sex differences in certain cognitive features represent an evolutionarily recent phenomenon. However, the finality of these results should be questioned, since it is difficult to assume that sex differences in the corpus callosum and cognition must be statistically significant. While the results reported here are not significant, lending confidence to the conclusions discussed above, the general patterning of sex differences in *M. fascicularis* versus *P. troglodytes* may provide important clues as to when sex differences resulting in differences in cognitive performance came about. The results for *M. fascicularis* show that there is complete overlap in the relative size of the corpus callosum and its regions between males and females. From this it is possible to conclude that the corpus callosum is not wholly responsible for the differences between males and females in the performance of certain tasks. The results for *P. troglodytes*, though, do show a tendency for females to possess a slightly larger corpus callosum, genu and isthmus than males, albeit these distinctions are not statistically significant. In addition, distribution of the types of axons passing through the splenium in *P. troglodytes* is similar to the distribution seen in modern humans in that the female possesses more medium, very large, and giant axons than males (Aboitiz et al., 1992a). While this does not suggest that the corpus callosum of humans and chimpanzees are similar in their composition and fiber distribution, it does pose an interesting question. What level of uniqueness in the human corpus callosum is required to separate its features of form, function, and sexual dimorphism from that of chimpanzees?

The corpora callosa of great apes and humans are smaller relative to neocortical surface area and brain volume. From this it is assumed that the brains of great apes and humans are more lateralized than either Old or New World monkeys (Rilling and Insel, 1999; Hopkins and Rilling, 2000; see also Gannon et al., 1998). In addition, the findings of Hopkins and Marino (2000) suggest that the great apes possess a torque pattern similar to modern humans. Despite these general comparisons, though, these results do not imply that the brains of great apes and humans are alike. More importantly they indicate that the evolution of the human brain has been largely the result of a long, continuous evolution throughout primate history, and not rapid punctuated change, albeit this is conjecture. These studies as well as those testing for lateralization of the brain for certain cognitive and motor functions do suggest that *Pan* possesses a more lateralized brain than its cercopithecoid relatives. However, data on *Pan* behavioral, motor, and visual tasks do not suggest that *Pan* possesses a degree of lateralization in the cerebral cortex that would permit cognitive functioning beyond the level of a modern human two year old child (Deacon, 1997; Savage-Rumbaugh et al., 1998). The fact that *Pan* may possess a degree of lateralization approaching but not mimicking the human condition helps to explain why *Pan* would display a callosal morphology and composition similar to humans yet not possess similar cognitive characteristics. This observation that the brain and corpus callosum of *Pan* are similar but not the same as those of modern humans also explains why one does not find sex differences in the corpus callosum. That is, the brain of *Pan* has not become sufficiently specialized at the species level to permit the development of measurable sex differences in neocortical components and the corpus callosum.

There are several cognitive differences between males and females. These include differences with regard to visuospatial, motor, and language skills. While it is likely that visuospatial and motor skills contributed to the expansion and reorganization of the hominin brain (Holloway, 1970), one can argue that the most significant consequence of human evolution in general and human brain evolution specifically has been the development of complex language abilities.

The similarities between the brain and corpus callosum of *Pan* and humans can be used to express the uniqueness of each species. As discussed above, *Pan* appears to approach the neocortical condition of humans but does not mimic it. This explains why sex differences in the brain and corpus callosum of *Pan* do not approach statistical significance. It also explains why certain brain structures such as the planum temporale and petalial pattern may display asymmetry in *Pan* but do not confer human-like cognitive functioning (Gannon et al., 1998; Hopkins and Leavans, 1998; Rilling and Insel, 1999; Hopkins and Marino, 2000; Hopkins and Rilling, 2000). This difference between human nonhuman primates is best understood by examining the issue of language.

Several studies have attempted to assign some level of language to *Pan* (Savage-Rumbaugh et al., 1998). However, regardless of the displayed "intelligence" of study subjects, none have ever been able to express communicative abilities beyond those capable in a normal two and a half year old child. This is not to imply that *Pan* does not express some level of intelligence but instead indicates the mere differences between the brain of *Pan* and the brains of modern humans. For example, Gannon et al. (1998) found that human-like asymmetry can be found in the planum temporale of *P. troglodytes*. While this level of asymmetry in humans is thought to result in or represent a product of the laterality of language, the authors do not express any intent to align the language skills of *Pan* and humans. This is because it is difficult to assign advanced cognitive functions such as language to asymmetry in one single structure. In this case asymmetry in the planum temporale may confer laterality in certain cognitive processes in both *Pan* and humans, but it does not presume language in both species.

The role and relationship of the corpus callosum in speech and language has been well established (O'Kusky et al., 1988; Zaidel et al., 1995; Rumsey et al., 1996; Moffat et al., 1998; Gazzaniga, 2000; Habib, 2000; Preis et al., 2000; Shevtsova and Reggia, 2000). The size of the corpus callosum has been shown to be related to the lateralization of language function (Witelson, 1995; Zaidel et al., 1995). In addition, women, who are thought to be less lateralized than men for language, possess a larger corpus callosum and more bulbous splenium (de Lacoste, 1981; de Lacoste and Holloway, 1982; Kimura and Harshman, 1984; Witelson, 1991, 1995; Holloway et al., 1993; Moffat et al., 1998). The presence of continued argument as to the existence of sex differences in the corpus callosum of humans attest to the degree of difference between males and females, which in some cases is small. However, it is still uncertain how much of a difference must occur between the brains of two individuals or the sexes to obtain significant differences in cognitive features. For example, it is generally accepted that males and females differ from each other in certain cognitive skills (Kimura, 1987; Hugdahl et al., 1993; Halpern et al., 1998; de Courten-Myers, 1999; Hausmann and Gunturkun, 1999; Amunts et al., 2000). Yet, each sexually dimorphic skill does not correlate to an equally sexually dimorphic neuroanatomical area, albeit certain areas such as the motor cortex do exhibit direct correlations (de Courten-Myers, 1999; Amunts et al., 2000). Nevertheless, these gaps in human research leave the question of how sexually dimorphic the splenium must be to permit one sex to possess greater integrative capabilities with regard to language and visuospatial skills remains unresolved. Without the resolution of these particular issues the specific role of sex differences in the corpus callosum will remain uncertain.

The when, where, why and how of the evolution of language are questions that are not easily answered. This is because data relevant to these questions must be derived from at least three mutually exclusive categories: living nonhuman primates, living humans, and endocasts of fossils. As mentioned above communicative information in nonhuman primates like that being produced by Sue Savage-Rumbaugh and others attest to the level of skill in species such as *P. troglodytes* and *P. paniscus*. However, these studies do not specifically prove the existence of language or language areas in nonhuman primates or *Pan* in particular. They do, though, shed some light on the development of language. Based on these behavioral studies and the anatomical studies mentioned above, it is possible that *Pan* possesses certain brain structures and a degree of cerebral lateralization that permit *Pan* to communicate at a level beyond other nonhuman primates. Though this level of cerebral and cognitive development is not the same as displayed by humans, it does provide provocative evidence for the existence of a cerebral archetype early in human evolution rather than the arise of areas such as Broca's and Wernicke's *de novo* in *Homo sapiens*.

The fossil record appears to support this claim. Although endocasts of australopiths do not appear to be significantly different from *Pan*, later species such as *Homo habilis* and *Homo erectus* do begin to display human-like proportions and features (Tobias, 1975; Holloway, 1981a,b; Broadfield et al., 2001). The presence, though, of human-like features does not necessarily confer the capacity for modern human speech and language on any species other than modern humans. However, they do indicate that the development of neuroanatomical features related to speech, language, and visuospatial skills may have existed long before the arrival of *Homo sapiens*. As to the role of these features for the development of sex differences in the corpus callosum, in particular the splenium, the development of certain higher cognitive features of the brain should precede the development of sex differences in those functions (speech, language, and visuospatial skills) as well as sex differences in the neuroanatomical structures related to those functions. Sex differences in the corpus callosum would thus not be expected in taxa such as *Pan* and *Macaca*, since neither species possesses the neuroanatomical substrate for modern human speech, language, and visuospatial skills or the degree of lateralization of the cerebral cortex required to produce the specialized features of language. Due to the role of the splenium in connecting modern human language areas, it is suspected that if a particular species is to possess communicative features comparable to humans then this area may display sex differences as it does in humans. However, since *Pan*, as mentioned above, does not possess human communicative abilities, visuospatial skills, or the neuroanatomical substrate that would lead one to propose the ability for human-like communication or visuospatial skills, one would not expect to find sex differences in this particular callosal region. Humans, therefore, are unique among living primate taxa in possessing a highly lateralized, sexually dimorphic brain and corpus callosum.

References

Aboitiz F, Scheibel AB, Zaidel E. 1989. Variability in fiber composition in different regions of the corpus callosum in humans. Anat Rec 223:6A.

Aboitiz F, Scheibel AB, Fisher RS, Zaidel E. 1992a. Fiber composition of the human corpus callosum. Brain Res 598:143-153.

Aboitiz F, Scheibel AB, Zaidel E. 1992b. Morphometry of the sylvian fissure and the corpus callosum, with emphasis on sex differences. Brain 115:1521-1541.

Aboitiz F, Scheibel AB, Fisher RS, Zaidel E. 1992c. Individual differences in brain asymmetries and fiber composition in the human corpus callosum. Brain Res 598:154-161.

Aboitiz F, Rodríguez E, Olivares R, Zaidel E. 1996. NeuroReport. 7:1761-1764.

Achiron R, Lipitz S, Achiron A. 2001. Sex-related differences in the development of the human fetal corpus callosum: in utero ultrasonographic study. Prenat Diagn 21: 116-120.

Allen LS, Richey MF, Chai YM, Gorski RA. 1991. Sex differences in the corpus callosum of the living human being. J Neurosci 11:933-942.

Allen JS, Damasio H, Grabowski TJ, Bruss J, Zhang W. 2003. Sexual dimorphism and asymmetries in the gray-white composition of the human cerebrum. NeuroImage 18: 880-894.

Ambrose SH. 2001. Paleolithic technology and human evolution. Science 291:1748-1753.

Amunts K, Jäncke L, Mohlberg H, Steinmetz H, Zilles K. 2000. Interhemispheric asymmetry of the human motor cortex related to handedness and gender. Neuropsychologia 38:304-312.

Barbas H, Pandya DN. 1984. Topography of commissural fibers of the prefrontal cortex in the rhesus monkey. Exp Brain Res 55:187-191.

Bard KA, Hopkins WD, Fort C. 1990. Lateral bias in infant chimpanzees (*Pan troglodytes*). J Comp Psychol 104:309-321.

Baumgardner TL, Singer HS, Denckla MB, Rubin MA, Abrams MT, Colli MJ, Reiss AL. 1996. Corpus callosum morphology in children with Tourette syndrome and attention deficit hyperactivity disorder. Neurology 47:477-482.

Bean RB. 1906. Some racial peculiarities of the Negro brain. Am J Anat 5: 353-432.

Beck PD, Kaas JH. 1994. Interhemispheric connections in neonatal owl monkeys (*Aotus trivirgatus*) and galagos (*Galago crassicaudatus*). Brain Res 651:57-75.

Bell AD, Variend S. 1985. Failure to demonstrate sexual dimorphism of the corpus callosum in childhood. J Anat 143:143-147.

Bishop KM, Wahlsten D. 1997. Sex differences in the human corpus callosum: Myth or reality? Neurosci Biobehav Rev 21:581-601.

Bishop KM, Wahlsten D. 1999. Sex and species differences in mouse and rat forebrain commissures depend on the method of adjusting for brain size. Brain Res 815:358-366.

Broadfield DC, Holloway RL, Mowbray K, Silvers A, Yuan MS, Marquez S. 2001. The endocast of Samungmacan 3 (Sm 3): A new *Homo erectus* from Indonesia. Anat Rec 262:369-379.

Buccafusco JJ, Jackson WJ, Jonnala RR, Terry AV. 1999. Differential improvement in memory-related task performance with nicotine by aged male and female rhesus monkeys. Behav Pharmacol 10:681-690.

Byne W, Bleier R, Houston L. 1988. Variations in human corpus callosum do not predict gender: A study using magnetic resonance imaging. Behav Neurosci 102:222-227.

Calvin W. 1983. A stone's throw and its launch window: timing precision and its implications for language and hominid brains. J Theor Biol 104:121-135.

Calvin W. 1993. The unitary hypothesis: a common neural circuitry for novel manipulations, language, plan-ahead, and throwing? In: Gibson KR, Ingold T, editors. Tools, Language, and Cognition in Human Evolution. New York: Cambridge University Press. p. 230-250.

Clark VO, Maisog JM, Haxby JV. 1998. fMRI study of face perception and memory using random stimulus sequences. J Neurophysiol 79:3257-3265.

Clarke JM, Zaidel E. 1994. Anatomical-behavioral relationships: corpus callosum morphometry and hemispheric specialization. Behav Brain Res 64:185-202.

Clarke S, Kraftsik R, VanDer Loos H, Innocenti GM. 1989. Forms and measures of adult and developing human corpus callosum: Is there sexual dimorphism? J Comp Neurol 280:213-230.

Coger RW, Sarafetinides EA. 1990. Schizophrenia, corpus callosum, and interhemispheric communication: a review. Psychiatry Res 34:163-184.

Constant D, Ruther H. 1996. Sexual dimorphism in the human corpus callosum? a comparison of methodologies. Brain Res 727:99-106.

Cooke BM, Tabibnia G, Breedlove SM. 1999. A brain sexual dimorphism controlled by adult circulating androgens. Proc Natl Acad Sci USA 96:7538-7540.

Corsi-Cabrera M, Arce C, Ramos J, Guevara MA. 1997. Effect of spatial ability and sex on inter- and intrahemispheric correlation of EEG activity. Electroencephalogr Clin Neurophysiol 102:5-11.

Cowell PE, Kostianovsky DJ, Gur RC, Turetsky BI, Gur RE. 1996. Sex differences in neuroanatomical and clinical correlations in schizophrenia. Am J Psychiatry 153:799-805.

Crow TJ, Crow LR, Done DJ, Leask SJ. 1998. Relative hand skill predicts academic ability: global deficits of the point of hemispheric indecision. Neuropsychologia 36:1275-1282.

Crucian GP, Berenbaum SA. 1998. Sex differences in right hemisphere tasks. Brain Cogn 36:377-389.

Davatzikos C, Resnick SM. 1998. Sex differences in anatomic measures of interhemispheric connectivity: correlations with cognition in women but not men. Cereb Cortex 8:635-40.

Davidson RJ, Hugdahl K, editors. 1995. Brain Asymmetry. Cambridge: The MIT Press.

de Courten-Myers GM. 1999. The human cerebral cortex: gender differences in structure and function. J Neuropathol Exp Neurol 58:217-226.

de Lacoste-Lareymondie. 1981. Anatomical and morphological aspects of the human corpus callosum. PhD Dissertation, Columbia University.

de Lacoste-Utamsing MC, Holloway RL. 1982. Sexual dimorphism in the human corpus callosum. Science 216:1431-1432.

de Lacoste MC, Holloway RL, Woodward DJ. 1986. Sex differences in the fetal human corpus callosum. Hum Neurobiol 5:93-96.

de Lacoste MC, Woodward D. 1988. The corpus callosum in nonhuman primates: determinates of size. Brain Behav Evol 31:318-323.

Deacon TW. 1997. The Symbolic Species: the Co-evolution of Language and the Brain. New York: WW Norton.

Demeter S, Ringo JL, Doty RW. 1988. Morphometric analysis of the human corpus callosum and anterior commisure. Hum Neurobiol 6: 219-226.

Demeter S, Rosene DL, Van Hoesen GW. 1990. Fields of origin and pathways of the interhemispheric commissures in the temporal lobe of macaques. J Comp Neurol 302:29-53.

Denenberg VH, Kertesz A, Cowell PE. 1991a. A factor analysis of the human's corpus callosum. Brain Res 548:126-132.

Denenberg, V.H., Cowell, P.E., Fitch, R.H., Kertesz, A., and Kenner, G.H. (1991b) Corpus callosum: multiple parameter measurements in rodent and humans. Physiol Behav 49:433-437.

Descartes R. 1664. Les Traites de L'homme et de la Formation du Foetus. Paris: Chez Nicholas Le Gras.

Driesen NR, Raz N. 1995. Sex-, age-, and handedness-related differences in human corpus callosum observed in vivo. Psychobiology 23:240-247.

Dubb A, Gur R, Avants B, Gee J. 2003. Characterization of sexual dimorphism in the human corpus callosum. NeuroImage 20: 512-519.

Dunham LA, Hopkins WD. 2006. Sex and handedness effects on corpus callosum morphology in chimpanzees (*Pan troglodytes*). Behav Neurosci 120: 1025-1032.

Elster AD, DePersio DA, Moody DM. 1990. Sexual dimorphism of the human corpus callosum studies by magnetic resonance imaging: Fact, fallacy and statistical confidence. Brain Dev 12:321-325.

Emory LE, Williams DH, Cole CM, Amparo EG, Meyer WJ. 1991. Anatomic variation of the corpus callosum in persons with gender dysphoria. Arch Sex Behav 20:409-417.

Eviatar Z, Hellige JB, Zaidel E. 1997. Individual differences in lateralization: effects of gender and handedness. Neuropsychol 11:562-576.

Falk D. 1990. Brain evolution in *Homo*: the "radiator" theory. Behav Brain Sci 13:333-381.

Falk D. 1997. Brain evolution in females: an answer to Mr Lovejoy. In: Hager LD, editor. Women in Human Evolution. New York: Routledge, p 114-136.

Falk D, Froese N, Sade DS, Dudek BC. 1999. Sex differences in brain/body relationships of rhesus monkeys and humans. J Hum Evol 36:233-238.

Finger S. 1994. Origins of Neuroscience: a History of Explorations into Brain Function. New York: Oxford University Press.

Fitch RH, Denenberg VH. 1998. A role for ovarian hormones in sexual differentiation of the brain. Behav Brain Sci 21:311-327; discussion 327-352.

Fitch RH, Cowell PE, Schrott LM, Denenberg VH. 1991. Corpus callosum: ovarian hormones and feminization. Brain Res 542:313-317.

Franklin MS, Kraemer GW, Shelton SE, Baker E, Kalin NH, Uno H. 2000. Gender differences in brain volume and size of the corpus callosum and amygdala of rhesus monkey measured from MRI images. Brain Res 852:263-267.

Funnell MG, Corballis PM, Gazzaniga MS. 2000b. Insights into the functional specificity of the human corpus callosum. Brain 123:920-926.

Galaburda AM. 1995. Anatomic basis of cerebral dominance. In: Davidson RJ, Hugdahl K, editors. Brain Asymmetry. Cambridge: The MIT Press, p 51-73.

Gannon PJ, Holloway RL, Broadfield DC, Braun AR. 1998. Asymmetry of chimpanzee planum temporale: humanlike pattern of Wernicke's brain language area homolog. Science 279:220-222.

Gazzaniga MS. 2000. Cerebral specialization and interhemispheric communication: does the corpus callosum enable the human condition? Brain 123:1293-1326.

Going JJ, Dixson A. 1990. Morphometry of the adult human corpus callosum: lack of sexual dimorphism. J Anat 171:163-167.

Gould III HJ, Cusik CG, Pons TP, Kaas JK. 1986. The relationship of corpus callosum connections to electrical stimulation maps of motor, supplementary motor, and the frontal eye fields in owl monkeys. J Comp Neurol 247:297-325.

Gould SJ. 1977. Ontogeny and Phylogeny. Cambridge MA: Bellknap Press.

Gould SJ, Lewontin RC. 1979. The spandrels of San Marco and the Panglossian program: a critique of the adaptationist program. Proc R Soc Lond (Biol) 205:281-288.

Grimshaw GM. 1998. Integration and interference in the cerebral hemispheres: relations with hemispheric specialization. Brain Cog 36:108-127.

Gur RC, Turetsky BI, Matsui M, Yan M, Bilker W, Hughett P, Gur RE. 1999. Sex differences in brain gray and white matter in healthy young adults: correlations with cognitive performance. J Neurosci 19:4065-4072.

Habib M, Gayraud D, Oliva A, Regis J, Salamon G, Khalil R. 1991. Effects of handedness and sex on the morphology of the corpus callosum: A study with brain magnetic resonance imaging. Brain Cogn 16:41-61.

Hadar U, Wenkert-Olenik D, Krauss R, Soroker N. 1998. Gesture and the processing of speech: neuropsychological evidence. Brain Lang 62:107-126.

Halpern DF, Haviland MG, Killian CD. Handedness and sex differences in intelligence: evidence form the medical college admission test. Brain Cogn 38:87-101.

Hausmann M, Gunturkun O. 1999. Sex differences in functional cerebral asymmetries in a repeated measures design. Brain Cogn 41:263-275.

Hayakawa K, Konishi Y, Matsuda T, Kuriyama M, Konishi K, Yamashita K, Okumura R, Hamanaka D. 1989. Development and aging of brain midline structures: assessment with MR imaging. Radiology 172:171-177.

Hellner-Burris K, Sobieski CA, Gilbert VR, Phillips KA. 2010. Prey capture efficiency in brown capuchin monkeys (*Cebus apella*) is influenced by sex and corpus callosum morphology. Am J Primatol 72: 502-8.

Hermann B, Hansen R, Seidenberg M, Magnotta V, O'Leary D. 2003. Neurodevelopmental vulnerability of the corpus callosum to childhood onset localization-related epilepsy. NeuroImage 18: 284-292.

Highley JR, Esiri MM, McDonald B, Cortina-Borja M, Herron BM, Crow TJ. 1999. The size and fibre composition of the corpus callosum with respect to gender and schizophrenia: a post-mortem study. Brain 122:99-110.

Hoff AL, Neal C, Kushner M, DeLisis LE. 1994. Gender differences in the corpus callosum size in first-episode schizophrenics. Biol Psychiatry 35:913-919.

Holloway RL. 1970. Neural parameters, hunting, and the evolution of the human brain. In: Noback CR, Montagna W, editors. Advances in Primatology, Volume 1: The Primate Brain. New York: Appleton-Century-Crofts, p 299-310.

Holloway RL. 1981a. The Indonesian *Homo erectus* brain endocasts revisited. Am J Phys Anthropol 55:503-521.

Holloway RL. 1981b. The evidence from endocasts: Preliminary studies from stereoplotting the dorsal surface. Philos Trans R Soc London (Biol) 292:155-166.

Holloway RL. 1990. Sexual dimorphism in the human corpus callosum: Its evolutionary and clinical implications. In: Sperber G, editor. From Apes to Angels: Essays in Anthropology in Honor of Phillip V. Tobias. New York: Wiley-Liss, p 221-228.

Holloway RL. 1997. Neuroanatomy, comparative. In: Spencer, F, editor. History of Physical Anthropology: An Encyclopedia, Volume 2 (M-Z). New York: Garland Publishing, p 732-43.

Holloway RL, de Lacoste MC. 1986. Sexual dimorphism in the human corpus callosum: an extension and replication study. Hum Neurobiol 5:87-91.

Holloway RL, Heilbroner P. 1992. Corpus callosum in sexually dimorphic and nondimorphic primates. Am J Phys Anthropol 87:349-357.

Holloway RL, Anderson PJ, Defendini R, Harper C. 1993 Sexual dimorphism of the human corpus callosum from three independent samples: relative size of the corpus callosum. Am J Phys Anthropol 92:481-498.

Hopkins WD. 1990. Handedness and laterality in monkeys and apes. In: Ehara A, Kimura T, Takenaka O, Iwamoto M, editors. Primatology Today. Amsterdam: Elsevier, p 271-274.

Hopkins WD. 1995. Hand preferences for a coordinated bimanual task in 110 chimpanzees: cross-sectional analysis. J Comp Psychol 109:291-297.

Hopkins WD, Morris RD. 1993. Handedness in great apes: a review of findings. Int J Primatol 14:1-26.

Hopkins WD, Leavens DA. 1998. Hand use and gestural communication in chimpanzees (*Pan troglodytes*). J Comp. Psychol 112:95-99.

Hopkins WD, Dahl JF. 2000. Birth order and hand preference in chimpanzees (*Pan troglodytes*): implications for pathological models of handedness in humans. J Comp Psychol 114:302-306.

Hopkins WD, Marino L. 2000. Asymmetries in cerebral width in nonhuman primate brains as revealed by magnetic resonance imaging (MRI). Neuropsychologia 38:493-499.

Hopkins WD, Pearson K. 2000. Chimpanzee (*Pan troglodytes*) handedness: variability across multiple measures of hand use. J Comp Psychol 114:126-135.

Hopkins WD, Rilling JK. 2000. A comparative MRI study of the relationship between neuroanatomical asymmetry and interhemispheric connectivity in primates: implication for the evolution of functional asymmetries. Behav Neurosci 114:739-748.

Hugdahl K, Iversen PM, Johnsen BH. 1993. Laterality for facial expressions: does the sex of the subject interact with the sex of the stimulus face? Cortex 29:325-331.

Hutt C. 1972. Males and Females. Middlesex, England: Penguin Books.

Intriligator J, Hénaff MA, Michel F. 2000. Able to name, unable to compare: the visual abilities of a posterior split-brain patient. Cog Neurosci Neuropsychol 11:2639-2642.

Jerison HJ. 1973. Evolution of the Brain and Intelligence. New York: Academic Press.

Johnson SC, Farnworth T, Pinkston JB, Bigler ED, Blatter DD. 1994. Corpus callosum surface area across the human adult life span: effect of age and gender. Brain Res Bull 35:373-377.

Johnson SC, Pinkston JB, Bigler ED, Blatter DD. 1996. Corpus callosum morphology in normal controls and traumatic brain injury: sex differences, mechanisms of injury, and neuropsychological correlates. Neuropsychology 10:408-415.

Juraska JM, Kopcik JR. 1988. Sex and environmental influencees on the size and ultrastructure of the rat corpus callosum. Brain Res 450:1-8.

Kaga K, Shindo M, Gotoh O, Tamura A. 1990. Speech perception and auditory P300 potentials after section of the posterior half of the truncus of the corpus callosum. Brain Topogr 3:175-181.

Kavcic V, Fei R, Hu S, Doty RW. 2000. Hemispheric interaction, metacontrol, and mnemonic processing in split-brain macaques. Behav Brain Res 111:71-82.

Kertesz A, Polk M, Howell J. 1987. Cerebral dominance, sex, and callosal size on MRI. Neurology 37:1385-1388

Khanna S, Chugani HT, Messa C, Curran JG. 1994. Corpus callosum agenesis and epilepsy: PET findings. Pediatr Neurol. 10:221-227.

Kim JH, Juraska JM. 1997. Sex differences in the development of axon number in the splenium of the rat corpus callosum from postnatal day 15 through 60. Brain Res Dev Brain Res 102:77-85.

Kimura D. 1980. Sex differences in intra-hemispheric organization of speech. Behav Brain Sci 3:240-241.

Kimura D. 1983. Sex differences in cerebral organization of speech and praxic functions. Canad J Psychol 37:19-35.

Kimura D. 1987. Are men's and women's brains really different? Can Psychol 28:133-147.

Kimura D. 1992. Sex differences in the brain. Sci Am 267:118-125.

Kimura D. 2000. Sex and Cognition. Cambridge, MA: MIT Press.

Kimura D, Harshman RA. 1984. Sex differences in brain organization for verbal and nonverbal functions. Prog Brain Res 61:423-441.

Kivitie-Kallio S, Autti T, Salonen O, Norio R. 1998. MRI of the brain in the Cohen syndrome: a relatively large corpus callosum in patients with mental retardation and microcephaly. Neuropediatrics 29:298-301.

Koshi R, Koshi T, Jeyaseelan L, Vettivel S. 1997. Morphology of the corpus callosum in human fetuses. Clin Anat 10:22-26.

Kopcik JR, Seymoure P, Schneider SK, Kim-Hong J, Juraska JM. 1992. Do callosal projection neurons reflect sex differences in axon number? Brain Res Bull 29:493-497.

Krubitzer LA, Kaas JH. 1993. The dorsomedial visual area of owl monkeys: connections, myeloarchitecture, and homologies in other primates. J Comp Neurol 334:497-528.

Lacreuse A, Herndon JG, Killiany RJ, Rosene DL, Moss MB. 1999. Spatial cognition in rhesus monkeys: male superiority declines with age. Horm Behav 36:70-76.

LaMantia AS, Rakic P. 1990a. Cytological and quantitative characteristics of four cerebral commissures in the rhesus monkey. J Comp Neurol 291:520-537.

LaMantia AS, Rakic P. 1990b Axon overproduction and elimination in the corpus callosum of the developing rhesus monkey. J Neurosci 10:2156-2175.

LeDoux, J.E. (1982) Neuroevolutionary mechanisms of cerebral asymmetry in man. Brain Behav Evol 20:196-212.

Lee BY, Sohn JH, Choi MH, Lee SJ, Kim HS, Yang JW, Choi JS, Kim HS, Yi JH, Tack GR, Chung SC. 2009. A volumetric stduy of the corpus callosum in 20s and 40s Korean people. Brain Struct Funct 213: 463-467.

LeMay, M. (1976) Morphological cerebral asymmetries of modern man, fossil man, and nonhuman primates. Ann N Y Acad Sci 280:349-366.

Leveroni CL, Seidenberg M, Mayer AR, Mead LA, Binder JR, Rao SM. 2000. Neural systems underlying the recognition of familiar and newly learned faces. J Neurosci 20:878-886.

Levin JM, Ross MH, Yurgelun-Todd DA. 1996. Age and gender differences in fMRI with photic stimulation. (Abstract) NeuroImage 3:S578.

Liu F, Vidarsson L, Winter JD, Tran H, Kassner A. 2010. Sex differences in the human corpus callosum microstructure: A combined T(2) myelin-water and diffusion tensor magnetic resonance imaging study. Brain Res 1343C: 37-45.

Luders E, Rex DE, Narr KL, Woods RP, Jancke L, Thompson PM, Mazziotta JC, Toga AW. 2003. Relationships between sulcal asymmetries and corpus callosum size: gender and handedness effects. Cereb Cortex 13: 1084-1093.

Lyoo IK, Noam GG, Lee CK, Lee HK, Kennedy BP, Renshaw PF. 1996. The corpus callosum and lateral ventricles in children with attention- deficit hyperactivity disorder: a brain magnetic resonance imaging study. Biol Psychiatry 40:1060-1063.

Mack CM, Fitch RH, Hyde LA, Seaman AJ, Bimonte HA, Wei W, Denenberg VH. 1996. Lack of activational influence of ovarian hormones on the size of the female rat's corpus callosum. Physiol Behav 60:431-434.

Mall FP. 1909. On several anatomical characters of the human brain, said to vary according to race and sex. Am J Anat 9: 1-32.

Manes F, Piven J, Vrancic D, Nanclares V, Plebst C, Starkstein SE. 1999. An MRI study of the corpus callosum and cerebellum in mentally retarded autistic individuals. J Neuropsychiatry Clin Neurosci 11:470-474.

Matano S, Nakano Y. 1998. Size comparison of the male and female human corpus callosum from autopsy samples. Z Morphol Anthropol 82:67-73.

McCarley RW, Wible CG, Frumin M, Hirayasu Y, Levitt JJ, Fischer IA, Shenton ME. 1999. MRI anatomy of schizophrenia. Biol Psychiatry 45:1099-1119.

McGlone J. 1977. Sex differences in the cerebral organization of verbal functions in patients with unilateral brain lesions. Brain 100:775-793.

McGlone J. 1980. Sex differences on human brain asymmetry: a critical survey. Behav Brain Sci 3:215-263.

Meisenzahl EM, Frodl T, Greiner J, Leinsinger G, Maag KP, Heiss D, Hahn K, Hegerl U, Moller HJ. 1999. Corpus callosum size in schizophrenia--a magnetic resonance imaging analysis. Eur Arch Psychiatry Clin Neurosci 249:305-312.

Menzel CR. 1999. Unprompted recall and reporting of hidden objects by a chimpanzee (*Pan troglodytes*) after extended delays. J Comp Psychol 113:426-434.

Mitchell TN, Free SL, Merschhemke M, Lemieux L, Sisodiya SM, Shorvon SD. 2003. Reliable callosal measurement: population normative data confirm sex-related differences. AJNR Am J Neuroradiol 24: 410-418.

Moffat SD, Hampson E, Lee DH. 1998. Morphology of the planum temporale and corpus callosum in left handers with evidence of left and right hemisphere speech representation. Brain 121:2369-2379.

Mostofsky SH, Wendlandt J, Cutting L, Denckla MB, Singer HS. 1999. Corpus callosum measurements in girls with Tourette syndrome. Neurology 53:1345-1347.

Morton BE, Rafto SE. 2006. Corpus callosum is linked to dichotic deafness and hemisphericity, not sex or handedness. Brain Cogn 62: 1-8.

Narr KL, Thompson PM, Sharma T, Moussai J, Cannestra AF, Toga AW. 2000. Mapping morphology of the corpus callosum in schizophrenia. Cereb Cortex 10:40-49.

Noonan M, Smith MA, Kelleher K, Sanfilippo MA. 1998. Sex differences in anterior commissure of the rat. Brain Res Bull 45:101-104.

Nuñez JL, Nelson J, Pych JC, Kim JHY, Juraska JM. 2000. Myelination in the splenium of the corpus callosum in adult male and female rats. Brain Res Dev Brain Res 120:87-90.

Panizzon MS, Hoff AL, Nordahl TE, Kremen WS, Reisman B, Wieneke M, Harris D, Goodman C, Espinoza S, Liu W, Lim K. 2003. Sex differences in the corpus callosum of patients with schizophrenia. Schizophr Res 62: 115-122.

Oka S, Miyamoto O, Janjua NA, Honjo-Fujiwara N, Ohkawa M, Nagao S, Kondo H, Minami T, Toyoshima T, Itano T. 1999. Re-evaluation of sexual dimorphism in human corpus callosum. NeuroReport 10:937-940.

O'Kusky J, Strauss E, Kosaka B, Wada J, Li D, Druhan M, Petrie J. 1988. The corpus callosum is larger with right-hemisphere cerebral speech dominance. Ann Neurol 24: 379-383.

Oppenheim JS, Lee BC, Nass R, Gazzaniga M. 1987. No sex-related difference in human corpus callosum based on magnetic resonance images. Ann Neurol 21: 604-606.

Pandya DN, Gold D, Berger T. 1969. Interhemispheric connections of the precentral motor cortex in the rhesus monkey. Brain Res 15:594-596.

Parr LA, Winslow JT, Hopkins WD, de Waal FB. 2000. Recognizing facial cues: individual discrimination by chimpanzees (*Pan troglodytes*) and rhesus monkeys (*Macaca mulatta*). J Comp Psychol 114:47-60.

Payne BR. 1990. Function of the corpus callosum in the representation of the visual field in cat visual cortex. Vis Neurosci 5:205-211.

Phillips KA, Sherwood CC, Lilak AL. 2007. Corpus callosum morphology in capuchin monkeys is influenced by sex and handedness. PLoS One 2: e792.

Piven J, Bailey J, Ranson BJ, Arndt S. 1997. An MRI study of the corpus callosum in autism. Am J Psychiatry 154: 1051-1056.

Preis S, Steinmetz H, Knorr U, Jäncke L. 2000. Corpus callosum size in children with developmental language disorder. Cogn Brain Res 10:37-44.

Prokop A, Oehmichen M, Zilles K. 1990. Geschlechtsdimorphismus des corpus callosum? [Sexual dimorphism of the corpus callosum]. Beitr Gerichtlich Med 48: 263-270.

Pujol J, Vendrell P, Junqué C, Martí-Vilalta JL, Capdevila A. 1993. When does human brain development end? Evidence of corpus callosum growth up to adulthood. Ann Neurol 34:71-75.

Pulvermüller F, Mohr B. 1996. The concept of transcortical cell assemblies: a key to the understanding of cortical lateralization and interhemispheric interaction. Neurosci Biobehav Rev 20:557-566.

Raine A, Harrison GN, Reynolds GP, Sheard C, Cooper JE, Medley I. 1990. Structural and functional characteristics of the corpus callosum in schizophrenics, psychiatric controls, and normal controls: A magnetic resonance imaging and neuropsychological evaluation. Arch Gen Psychiatry 47: 1060-1064.

Reinarz SJ, Coffman CE, Smoker WR, Godersky JC. 1988. MR imaging of the corpus callosum: normal and pathologic findings and correlation with CT. Am J Roentgenol. 151:791-798.

Rilling JK, Insel TR. 1999. Differential expansion of neural projection systems in primate brain evolution. NeuroReport 10:1453-1459.

Robichon F, Habib M. 1998. Abnormal callosal morphology in male adult dyslexics: relationships to handedness and phonological abilities. Brain Lang 62:127-46.

Rumsey JM, Casanova M, . 1996. Corpus callosum morphology, as measured with MRI, in dyslexic men. Biol Psychiatry 39:769-775.

Salat D, Ward A, Kaye JA, Janowsky JS. 1997. Sex differences in the corpus callosum with aging. Neurobiol Aging 18:191-197.

Savage-Rumbaugh S, Shanker SG, Taylor TJ. 1998. Apes, Language, and the Human Mind. New York: Oxford University Press.

Schmitt JE, Eliez S, Warsofsky IS, Bellugi U, Reiss AL. 2001. Corpus callosum morphology of Williams syndrome: relation to genetics and behavior. Dev Med Child Neurol 43:155-159.

Seltzer B, Pandya DN. 1983. The distribution of posterior parietal fibres in the corpus callosum of the rhesus monkey. Exp Brain Res 49: 147-150.

Shaywitz BA, Shaywitz SE, Pugh KR, Constable RT, Skudlarski P, Fulbright RK. 1995. Sex differences in the functional organization of the brain for language. Nature 373:607-609.

Shevtsova N, Reggia JA. 2000. Interhemispheric effects of simulated lesions in a neural model of letter identification. Brain Cogn 44:577-603.

Shin YW, Kim DJ, Hyon T, Park HJ, Moon WJ, Chung EC, Lee JM, Kim IY, Kim SI, Kwon JS. 2005. Sex differences in the human corpus callosum: diffusion tensor imaging study. NeuroReport 16: 795-798.

Shiota JY, Nakano I, Kawamura M, Hirayama K. 1996. An autopsy case of Marchiafava-Bignami disease with peculiar chronological CT changes in the corpus callosum: neuroradiopathological correlations. J Neurol Sci 136:90-93.

Sommer MA, Wurtz RE. 1998. Frontal eye field neurons orthodromically activated from the superior colliculus. J Neurophysiol 80:3331-3333.

Springer SS, Deutsch G. 1989. Left Brain, Right Brain. New York: Freeman.

Steinmetz H, Janke L, Kleinschmidt A, Schlaug MD, Volkmann J, Huang Y. 1992. Sex but no hand difference in the isthmus of the corpus callosum. Neurology 42:749-752.

Steinmetz H, Staiger JF, Schlaug G, Huang Y, Jancke L. 1995. Corpus callosum and brain volume in women and men. NeuroReport 6:1002-1004.

Steinmetz H, Staiger JF, Schlaug SG, Huang Y, Jancke L. 1996. Inverse relationship between brain size and callosal connectivity. Naturwissenschaften 83:221.

Sullivan EV, Rosenbloom MJ, Desmond JE, Pfefferbaum A. 2001. Sex differences in corpus callosum size: relationship to age and intracranial size. Neurobiol Aging 22: 603-611.

Thompson PM, Moussai J, Zohoori S, Goldkorn A, Khan AA, Mega MS, Small GW, Cummings JL, Toga AW. 1998. Cortical variability and asymmetry in normal aging and Alzheimer's disease. Cereb Cortex 8:492-509.

Tobias PV. 1971. The Brain in Hominid Evolution. New York: Columbia University Press.

Tobias PV. 1975. Brain evolution in the Hominoidea. In: Tuttle RH, editor. Primate Functional Morphology and Evolution. The Hague: Mouton, p 353-392.

Tomasch J. 1954. Size, distribution, and numbers of fibers in the human corpus callosum. Anat Rec 119:119-135.

Vallortigara G, Rogers LJ, Bisazza A. 1999. Possible evolutionary origins of cognitive brain lateralization. Brain Res Rev 30:164-175.

Vercelli A, Innocenti GM. 1993. Morphology of visual callosal neurons with different locations, contralateral targets or patterns of development. Exp Brain Res 94:393-404.

Vermeire BA, Hamilton CR, Erdmann AL. 1998. Right-hemispheric superiority in split-brain monkeys for learning and remembering facial discriminations. Behav Neurosci 112:1048-1061.

Vermersch P, Roche J, Hamon M, Daems-Monpeurt C, Pruvo JP, Dewailly P, Petit H. 1996. White matter magnetic resonance imaging hyperintensity in Alzheimer's disease: correlations with corpus callosum atrophy. J Neurol 243:231-234.

Wang PP, Doherty S, Hesselink JR, Bellugi U. 1992. Callosal morphology concurs with neurobehavioral and neuropathological findings in two neurodevelopmental disorders. Arch Neurol 49: 407-411.

Watson NV. 2001. Sex differences in throwing: monkeys having a fling. Trends Cogn Sci 5:98-99.

Weber G, Weis S. 1986. Morphometric analysis of the human corpus callosum fails to reveal sex-related differences. J Hirnforsch 27: 237-240.

Weis S, Weber G, Wenger E, Kimbacher M. 1989. The controversy about a sexual dimorphism of the human corpus callosum. Int J Neurosci 47:169-173.

Westergaard GC, Kuhn HE, Suomi SJ. 1998. Bipedal posture and hand preference in humans and other primates. J Comp Psychol 112:56-63.

Westerhausen R, Kreuder F, Dos Santos Sequeira S, Walter C, Woerner W, Wittling RA, Schweiger E, Wittling W. 2004. Effects of handedness and gender on macro- and microstructure of the corpus callosum and its subregions: a combined high-resolution and diffusion-tensor MRI study. Cog Brain Res 21: 418- 426.

Wilson FR. 1998. The Hand. How Its Use Shapes the Brain, Language, and Human Culture. New York: Pantheon Books.

Witelson SF. 1977. Anatomic asymmetry in the temporal lobes: its documentation, phylogenesis, and relationship to functional asymmetry. Ann N Y Acad Sci 299:328-354.

Witelson SF. 1983. Bumps on the brain: left-right asymmetry in brain anatomy and function. In: Segalowitz S, editor. Language Functions and Brain Organization. New York: Academic Press, p 117-143.

Witelson SF. 1985. The brain connection: the corpus callosum is larger in left-handers. Science 229:665-668.

Witelson SF. 1989. Hand and sex differences in the isthmus and genu of the human corpus callosum. Brain 112:799-835.

Witelson SF. 1991. Sex differences in neuroanatomical changes with aging. N Engl J Med 325:211-212.

Witelson SF. 1995. Neuroanatomical bases of hemispheric functional specialization in the human brain: possible developmental factors. In: Kitterle FL, editor. Hemispheric Communication: Mechanisms and Models. Hillsdale, NJ: Lawrence Erlbaum Associates, p 61-84.

Witelson SF, Kigar DL. 1987. Individual differences in the anatomy of the corpus callosum: sex, hand preference, schizophrenia and hemisphere specialization. In: Glass A, editor. Individual Differences in Hemispheric Specialization. NATO ASI Series A: 130. New York: Plenum Press, p 55-92.

Wium BM. 1984. Sexual dimorphism in the splenium of the corpus callosum: preliminary observations. S Afr J Sci 80: 434.

Wright JJ, Craggs MD. 1976. Visual attention in split-brain monkeys. Nature 261:580-581.

Yamauchi H, Fukuyama H, Ouchi Y, Nagahama Y, Kimura J, Asato R, Konishi J. 1995. Corpus callosum atrophy in amyotrophic lateral sclerosis. J Neurol Sci 134:189-196.

Yokota Y, Kawamura Y, Kameya Y. 2005. Callosal shapes at the midsagittal plane: MRI differences of normal males, normal females, and GID. Conf Proc IEEE Eng Med Biol Soc 3: 3055-3058.

Yoshii F, Barber W, Apicella J. 1986. Measurements of the corpus callosum (CC) on magnetic resonance (MR) scans: Effects of age, sex, handedness, and disease. Neurology Supp 1:133.

Zaidel E, Aboitiz F, Clarke J, Kaiser D, Matteson R. 1995. Sexual differences in interhemispheric relations for language. In: Kitterle FL. Hemispheric Communication: Mechanisms and Models. Hillsdale, NJ: Lawrence Erlbaum Associates, p 85-175.

CHAPTER 15

Dental Maturation, Middle Childhood and the Pattern of Growth and Development in Earlier Hominins

Janet Monge and Alan Mann

The direction of research outlined in this paper owes a great deal to the life-long research and publications of Ralph Holloway. His amassing and interpretation of large comparative data sets of hominin and hominid brain endocasts have provided science with a normative basis for the collection and analysis of many other human biological complexes, including the dentition. In this research, he moved into uncharted areas: not only to identify those unique aspects of the brain that developed in our lineage, but also to employ knowledge gleaned from studies in brain research to integrate these morphological changes with behavior. As with dental development studies, diverse data sets need to be synthesized in order to fully understand the nature of "humanness". For all his contributions to our understanding of the evolution of the human neurological system and the emergence of human cognition, our discipline is deeply in his debt; we are very pleased to have this paper included in a volume honoring Professor Holloway.

Abstract

Recent research indicates that human dental development and eruption are much more variable than had been previously thought. Data collected on wild chimpanzees shows their eruption patterns are significantly retarded in comparison to that of captive animals. These data imply that considerable caution must be exercised in using modern dental standards to reconstruct growth and development in extinct hominins. There is, however, one aspect of human development that may have significant implications for our understanding of the emergence of human cognition. Between the eruption of the initial permanent teeth, the two incisors and first molar, there is a time of about three years, from about seven to ten years of age, when no teeth erupt. This time, termed the 'Quiescent period' is followed by the eruption of the two premolars, canine and second molar. The Quiescent period in dental maturation appears to be coincident with the developmental age known as middle childhood, a time when a youngster's ability to utilize the cultural norms of its society emerges. Examination of the dentition of immature fossil hominin specimens, including australopithecines and members of *Homo*, reveals the presence of the Quiescent period, whereas dental development in chimpanzees lacks this time. Using the models of neurological reorganization, especially of the inferior parietal cortex, described by Holloway in a series of publications, it is suggested that middle childhood evolved very early in hominin evolution, perhaps prompted by the need for enhanced foraging abilities in seasonally variable mosaic environments.

Variation in Human Dental Development

Over the last twenty years, a number of original research projects dealing with aspects of human dental growth and development have been published (see, for example Thompson et al., 2003; Bogin, 1999; Minugh-Purvis and McNamara, 2002; Hawkes and Paine, 2006; Robson and Wood, 2008). These have vastly increased our knowledge of many parts of dental development that were not known before this. For example, by the beginning of 2009, over 200 genes had been identified that are expressed during the complex processes involved in tooth development (De Coster et al., 2009). Considering the potential complexity and interactions of these gene pathways, including possible cascading effects of each, it is no wonder that variation occurs at both the histologic and developmental levels. For example, on the micro-structural level, Smith and Tafforeau (2008) summarized recent research and concluded that human variation in dental histologic development is substantial. Finally, in this context, Liversidge (2003 and 2008) reviewed developmental dental studies primarily employing x-rays, emphasizing the degree of developmental variation that exists in living *Homo sapiens*.

One of the major problems with these studies is that it is difficult or impossible to resolve or integrate the results of these analyses to each other. Further, it is troublesome to assign recognized variations to the level of the individual, population, sub-species, or species. While it does appear that a certain amount of variation is patterned, given the limitations inherent in each data set, it is not currently possible to determine at which level these patterns are significant.

In order to move beyond the purely descriptive or comparative nature of studies it is necessary to more precisely focus on those specific features that identify the dental development correlates of the period of prolonged growth and maturation that has often been described as a unique characteristic of our species (see, for example, Bogin 1999). There are several sources of data for this analysis including dental development in wild chimpanzees (Zihlman et al., 2004) and the much larger data sets produced in the last decade on human dental development (summarized in Liversidge, 2003). After delineating the possible species specific pattern of maturation as displayed in the dentition, we apply the very same indentifiers to a sample of extinct fossil hominin forms.

Finally, we attempt to understand the human dental development pattern in terms of unique aspects of human behavior and biology. Assuming that dental development is tied to other aspects of growth and development, not only can differences in dental development be tied to issues of population or taxonomic distinctiveness, but to fundamental growth trajectory changes that perhaps are associated with lineages.

This work is informed by the life-long research of Ralph Holloway. He set the bar in the collection and interpretation of large comparative data sets on human brain evolution. In this research, he moved into uncharted areas: to identify unique aspects of the brain in our lineage, but more importantly, to the translation of these morphological changes to behavior as associated with state of the knowledge in brain research. As with dental development studies, very diverse types of data need to be synthesized in order to fully understand the nature of "humanness".

Dental Development - the State of it All

An overview of studies on dental development must include a discussion of both histological dental development as well as measures of developmental chronologic events usually performed with imaging techniques (x-ray or CT analyses) or, on a less refined level, using dental eruption timing. In all but a few cases (see, for example, Kuykendall, 2003 and Skinner and Wood, 2006), most studies ultimately direct discussion of the overarching issues concerning growth and development to comparison within the confines of each data set. It has become increasingly difficult to bring these sometimes conflicting data sets into a kind of synchrony in the evaluation of both living *Homo sapiens* and species of the common chimpanzee, for which we have the most complete information. Adding more complexity, are the resolution of issues surrounding dental growth and development of fossil hominins. A full review of the literature is not necessary to highlight some of the emerging difficulties in the application of these methods to living and extinct forms.

It has become increasingly clear that histological studies of enamel formation in modern humans indicate that there is substantial variation in all detailed parameters (cuspal, cervical enamel, enamel extension rates and periodicity) associated with enamel formation. The recent expansion of histological findings on the composition and structure of Neandertal molar enamel highlights some of the interpretive difficulties. In a discussion of the enamel thickness and histology on the fossil from Lakonis, Greece, Smith and colleagues (2009) summarize the information accumulated from several studies undertaken on Neandertal enamel. In some aspects of molar and incisor enamel histology including projections to enamel formation timing, Neandertals appear to overlap the known modern human range. In other aspects, this fossil form appears unique (summarized in Guatelli-Steinberg, 2009). The question becomes: what, if any of these differences, are significant in projections to growth and development patterning? For example, does the conclusion based on microstructure that Neandertals formed molar enamel in something like 100 days shorter (approximately 3 months) than a limited sample of modern humans, have any meaning in the extrapolation to life history variables? Are these representations of population or taxon differences? Certainly some of these differences are a reflection of enamel thickness, cusp morphology, and even crown height (Dean, 2000).

As critically and importantly, certain enigmas emerge as these histological studies move away from description to extrapolations of time frames of dental development. Some of these problems could certainly be resolved with more information on root formation timing since a much larger proportion of overall dental development depends on this portion of the tooth. For example, Beynon et al. (1998) completed a study of incisor histology in the chimpanzee in relation to the timing of development. These authors concluded that in the genus *Pan* incisor enamel is formed in 4.5 to 5.6 years. Since chimpanzees, based on radiographic studies, erupt the incisors at just under 6 years, and with what appears to be approximately 3 years of root formation (root 3/4th complete), it is hard to reconcile these two pieces of data. While it is true that radiography is notoriously variable in its ability to resolve fine details of tooth mineralization, it is difficult to imagine how a radiograph could both underestimate crown formation times by half while at the same time overestimating root formation by a magnitude of well over double. This same inconsistency, resulting from histologic versus radiographic data for the molar teeth (Reid et al., 1998) which led to a re-evaluation of histologic methods to bring crown formation times more into line with radiographic studies (Smith et al., 2007).

In the end, the power of growth and development studies depends increasingly on an understanding of the developmental timing of individual teeth in conjunction with the order and time frame of initial enamel differentiation. The relative order of dental development still relies entirely on radiographs with comparative standards based on a broad range of populations and species. While the time frame of development of individual teeth can be obtained from radiographic data on dental development of children, it can be used in conjunction with histologically derived time frames. Based on multiple studies of histologically derived time frames, it seems clear that there is a remarkable amount of overlap in the time frame of molar development among hominoids including many extinct hominin species (summarized in Kuykendall, 2003). In addition, this synchrony of results appears to well match the data derived from radiographic analyses. It is probably fair to say that hominoids appear to develop molar tooth crowns in approximately 2 to 3 years (not including the M3 with longer crown formation times). The differences in incisor formation appears broader (Smith and Tafforeau, 2008).

As interpretive and methodological problems exist in histological studies, so too do difficulties arise in the extraction of data from radiographs (cross sectional and mixed longitudinal) of the developing dentition of both modern humans and other primates. Many of these problems are outlined in (Liversidge, 2003 and 2006) and include:

- methods used to score the individual teeth
- variation in the x-ray equipment and ability to judge relative opacity or translucency of regions of the teeth
- statistical methods used to analyze the data.

Based on the population sample including sample size and methodological variation in assessment presented in the literature, it is almost impossible to compare studies to each other.

In her critical assessment of the literature, Liversidge (2008) does report on population-based differences as a well documented phenomenon in the crown and root formation timing of the M3, the only tooth in the radiographic sequence that can be fully evaluated from crown initiation to apex closure. She concludes that the population-based differences may result differences in the architecture of the face including the mandible between populations, an aspect of dental development explored earlier by Simpson et al. (1990) on fossil hominin forms. De Coster et al. (2009), in a similar way, speculate that some of the differences observed in dental development in modern human samples especially in reference to the permanent premolar sequence may be a consequence of more effective preventive dentistry and the longer retention of healthy deciduous molars. Finally, Liversidge (2008) argues that dental development is minimally influenced by environmental factors and under strong genetic control in comparison to other growth and development systems.

These complex data sets, from histologic and radiographic analyses of the developing dentition, have together forced a newer synthesis of dental development and to ask the question: What do we know about dental development in living Homo sapiens and in the common chimpanzee? Can this be applied to fossil forms and what are the limits and limitations?

Synthesis of Dental Development Data

It is likely that lengthened molar crown formation times, in the range of 2.5 to 3 years, is the primitive condition for all hominoids. These assumptions appear to be supported by both histologic and radiographic studies and appears to be the case for Miocene fossil hominoids (Keeley, 2002). This appears to be confirmed from the limited histologic studies on australopithecine molar crown formation time (summarized in Kuykendall, 2003).

In addition, it is reasonable to infer that increase in crown formation time occurs in the sequence from M1 to M3 (Dean, 2000) except perhaps in the genus *Pan* where crown formation time in the M3 appears to decrease from CFT in the M2 (Smith et al., 2007).

What remains are the discussions of the time frames of full dental development since by all measures, virtually all of these analyses attempt to resolve issues of unique features of either individual tooth growth (which may be applied to issues of taxonomy or taxonomic affinity) or unique features in total patterns of growth and development. Newer data sets on the developing dentition in both humans (for example, Monge et al., 2007; Nadler, 1998; and Rousset et al., 2003) and wild chimpanzees (Zihlman et al., 2004) and have blurred the chronologic age differences representing distinctive growth and development patterns between these two living forms. Thus, the question may be raised: are there growth and development differences in the dentition that can uniquely identify unique patterns in humans and chimpanzees? And can these be applied to the fossil record of human evolution?

We previously published information on a large data set derived from US populations living within the city confines of Philadelphia (preliminary details of this

	WILD CHIMPANZEES	*CAPTIVE CHIMPANZEES* *10–90% range*
dC	</– 1.5	0.8–1.4
M1	4.1 (2.6 < x </– 4.9)	2.7–4.1
I1	6.3–8.4 (5.7 < x </– 10.2)	4.7–6.5
I2	7.4–8.6 (6.5 < x </– 10.2)	5.3–6.9
M2	8.2–8.4 (8.2 x </– 10.2)	5.3–7.3
C	10.1–10.8 (8.5 < x </– 14.2)	7.9
M3	12.4 (10.8 < x </– 14.2)	10.5

Table 1. All information on wild chimpanzees is from Zihlman et. al. (2004). For a further summary of data on captive chimpanzees, see Kuykendall et al. (1992). For the most part, wild chimpanzees appear to show a delayed pattern of dental maturation over their captive counterparts. In some cases, wild animal emergence timing is outside the maximum limit of the recorded range of emergence in captive animals. Clearly environmental factors have an influence on dental development both in humans and non-human primates.

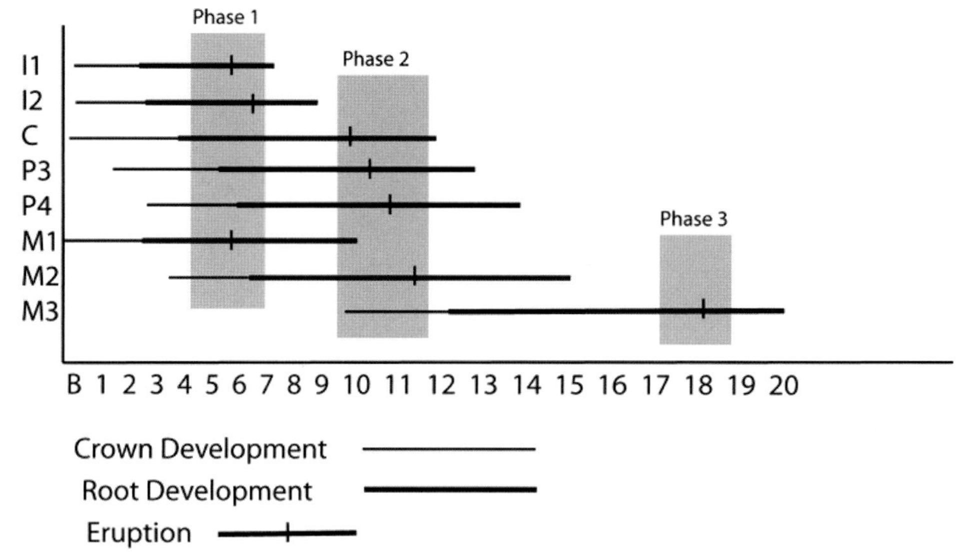

Figure 1. The three major Phases of human dental emergence. Tanner and Eveleth outlined three Phases based on the eruptive cycles of the human dentition. Phase 1 includes the Is and M1s; Phase 2, the Cs, Ps, and M2s. In Phase 3, the most varied in eruptive times, includes only the M3s, standing alone and thus marking the end of dental maturation. On the other hand, chimpanzees (both captive and wild born and raised) have no distinct phases of dental emergence and presumed calcification. In fact, based on the work in sequence polymorphisms by Conroy and Mahoney (1991), what clearly are Phase 1 teeth in humans, mix in with eruption of Phase 2 teeth in the genus Pan. (See TABLE 2 for sequence polymorphisms from Conroy and Mahoney 1991.)

study are published in Monge et al., 2007). This data set, along with internal comparisons of the European-American and African-American subsamples, appear to suggest that there is a trend towards reduction in the chronological time frames associated with M1 and M2 development (Blankenstein et al., 1990; Harris and McKee, 1990; Liversidge et al., 1999; Liversidge and Speechly, 2001; Olze et al., 2004). Thus, this data set appears to reflect 2 distinct patterns:

1. there are significant population differences in the chronology of dental developmental events, and 2. there appears to be a significant reduction overall in the time frame of dental development since the original dental standards were established (Liversidge, 2008 and Nadler, 1998; for a comprehensive listing of dental calcification and emergence studies, see Liversidge, 2003). Others (for example Rousset et al., 2003; many studies summarized in Liversidge, 2003) have also noted this developmental timing shift but the bulk of this data centers on eruption rather than calcification staging of the dentition.

Similarly, dental development schedules derived from radiographs, show variation between 2 captive common chimpanzee groups (Kuykendall, 1996 and 2002; in comparison to Anemone et al., 1991 and 1996). More remarkably are derived data on chimpanzee dental eruption showing clear and significant differences between captive and wild animals (Nissen and Riesen, 1964 in conjunction with eruption schedules from Kuykendall et al., 1992 in comparison to Zihlman et al., 2004) (Table 1). This captive/wild distinction has also been demonstrated in baboons (Kahumbu and Eley, 1991 and Phillips-Conroy and Jolly, 1988).

Within the context of human growth and development and life history studies (Hawkes and Paine, 2006), research in many fields, including psychology, anthropology and auxology, have focussed on an understanding of unique features of humans. One such developmental hallmark of humanness appears to be in the much speculated upon frame termed middle childhood (often times labelled as the "juvenile" phase). Eveleth and Tanner (1990) not only described this phase of childhood but in their analyses of dental growth and development proposed that one unique feature of the human dentition, occurring in concert with other developmental changes, is the disjointing of early dental events (including the calcification of the I1s, I2s, and M1s) from a secondary phase of the developing dentition (including the calcification of the Ps and M2s) (Figure 1). This developmental gap may manifest in the human dentition either by a delay in the initial calcification of the second phase teeth or by a slowing down of dentogenic processes (either in enamel or dentin formation or both). This developmental gap in Phase 1 versus Phase 2 teeth is visible in virtually every population studied and is reproduced in one such visual representation in Figure 2. Given the proposed reduction in chronologic years based on recent human dental development standards, in conjunction with the expansion of chronological years of wild chimpanzee dental development, we asked the question: Does this reduction in human dental developmental years serve to blur the distinction between Phase 1 and Phase 2 teeth? Based on our sample of Philadelphia children (Figure 3), this phase developmental shift is still present and extends upwards of 2 to 3 years. Qualification of this phase shift, and easily applied to dental developmental sequences, is the staging gap between any of the Phase 1 versus Phase 2 teeth. We chose to reproduce this between the latest Phase 1 tooth, the I2, in comparison to the middle Phase 2 tooth, the M2. Using the 14 calcification staging of Moorrees (1963), this calcification gap of at least 2 to 3 stages, occurs in both of the Philadelphia subsamples.

Although no such phase gap exists in reproduced information from captive chimpanzee dental development (Figure 4), it is not possible to determine if elongated maturation patterns of wild chimps results in a phase gap since only eruption data is presented in Zihlman. However, reproduced photos and line drawings clearly show that I2s and M2s are erupting in synchrony - a feature that never occurs in humans and has been well documented in captive chimps by Conroy and Mahoney (1991) (Table 2).

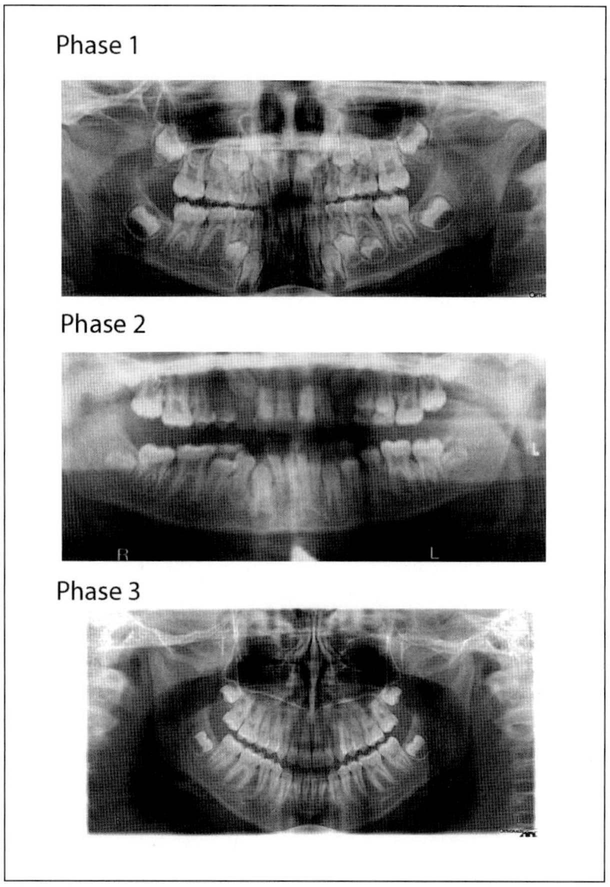

Figure 2. Sample panoramic radiographs of 3 Philadelphia school children in each of the phases of dental maturation.

MANDIBLE	MAXILLA
M1 I1 I2 M2 (P3/P4)	***M1 I1 M2 I2 P4 P3 C***
M1 I1 I2 M2 (P3/P4)	M1 I1 I2 M2 (P3/P4) C
M1 M2 I1 I2 (P3/P4)	M1 I1 I2 (M2/P3/P4)
M1 I1 I2 M2 P4	M1 I1 I2 (M2/P3/P4)
M1 (I1/I2) M2 (P3/P4)	***M1 I1 M2 I2 P3 P4***
M1 I1 I2 M2 P4	***M1 I1 (I2/M2) P3 P4***
M1 I1 I2 (M2/P3/P4)	***M1 M2 I1 (I2/P3/P4)***
M1 (I1/M2) I2	***M1 I1 M2 P4 (I2/P3)***

*Table 2. Various eruption sequences for both the mandible and maxilla of the common chimpanzee. **BOLD** type face indicates situations where the eruption of the M2 actually precedes the eruption of the I2. This array of erupting teeth also characterizes wild chimpanzee populations as summarized by Zihlman (2004). Thus, from eruption data alone, there is no phase distinction between the teeth in either wild or captive chimpanzees.*
Data From: Conroy and Mahoney 1991

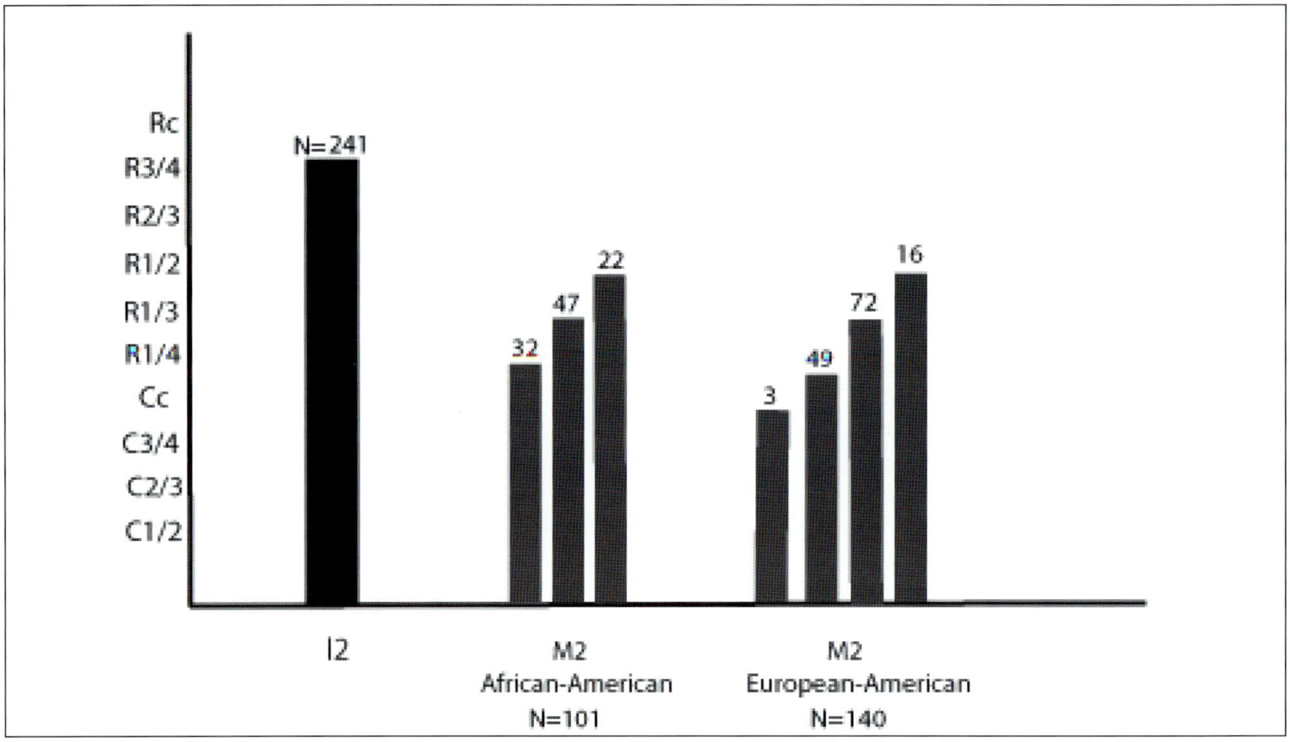

Figure 3. Relationship of the I2 Phase I tooth development to the M2, a Phase 2 tooth. In a sample of 1,245 Philadelphia school children, a total of 241 show the stage of development of the I2 (a first Phase tooth and the last in most cases to develop and erupt in that stage) at R3/4th. Although there are difference in the chronological time frame of development of the teeth between the subsamples, both populations of children show at least a 2 staging delay in development of the Phase 2 tooth - the M2. This delay between the developmental stage of these 2 teeth clearly shows evidence of the Quiescent phase in the developing human dentition. Although there is a clear trend for an earlier maturation of all teeth in comparison to many samples previously published, the retention of the delay between each phase is clearly retained.

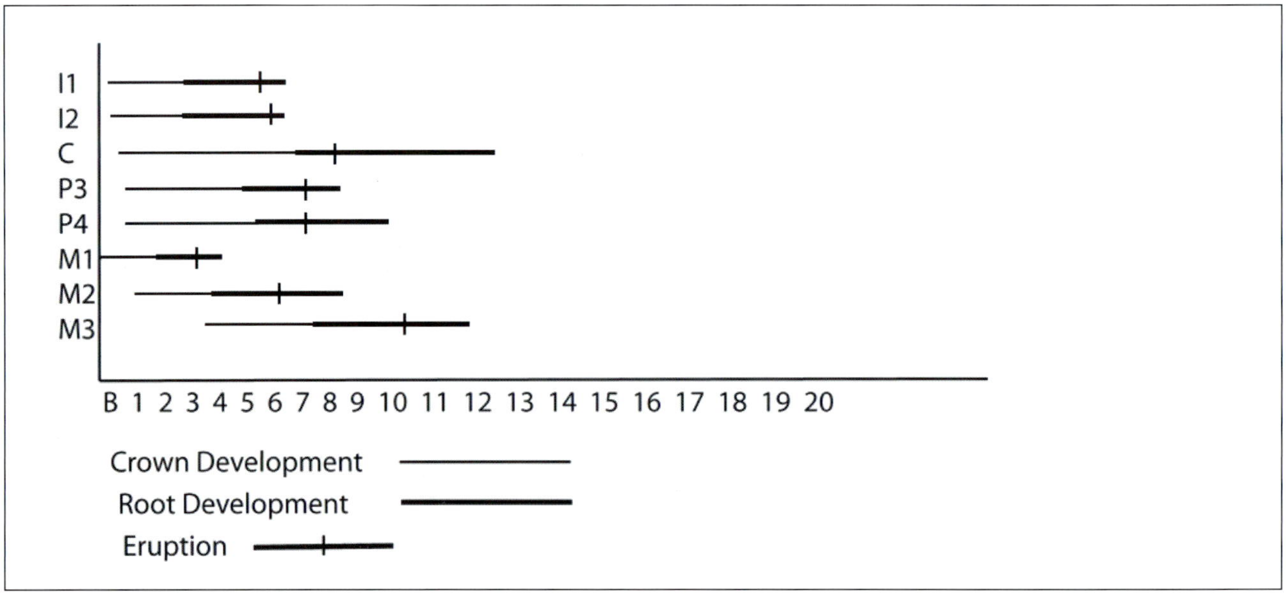

Figure 4. Dental development chart adapted from Smith (1986) and including the data on chimpanzee from Anemone et al. 1991 and 1996. Both captive and wild chimpanzees show a direct overlap in the development of the I2 and the M2. In the wild version of chimps, eruption data indicates that the overall time frame of dental development is shifted to the right along with an assumed eschew of each of the calcification stages. For example, the I2 erupts between 7.4 and 8.6 years; the M2 between 8.2 and 8.4 years (Zihlman et al. 2004).

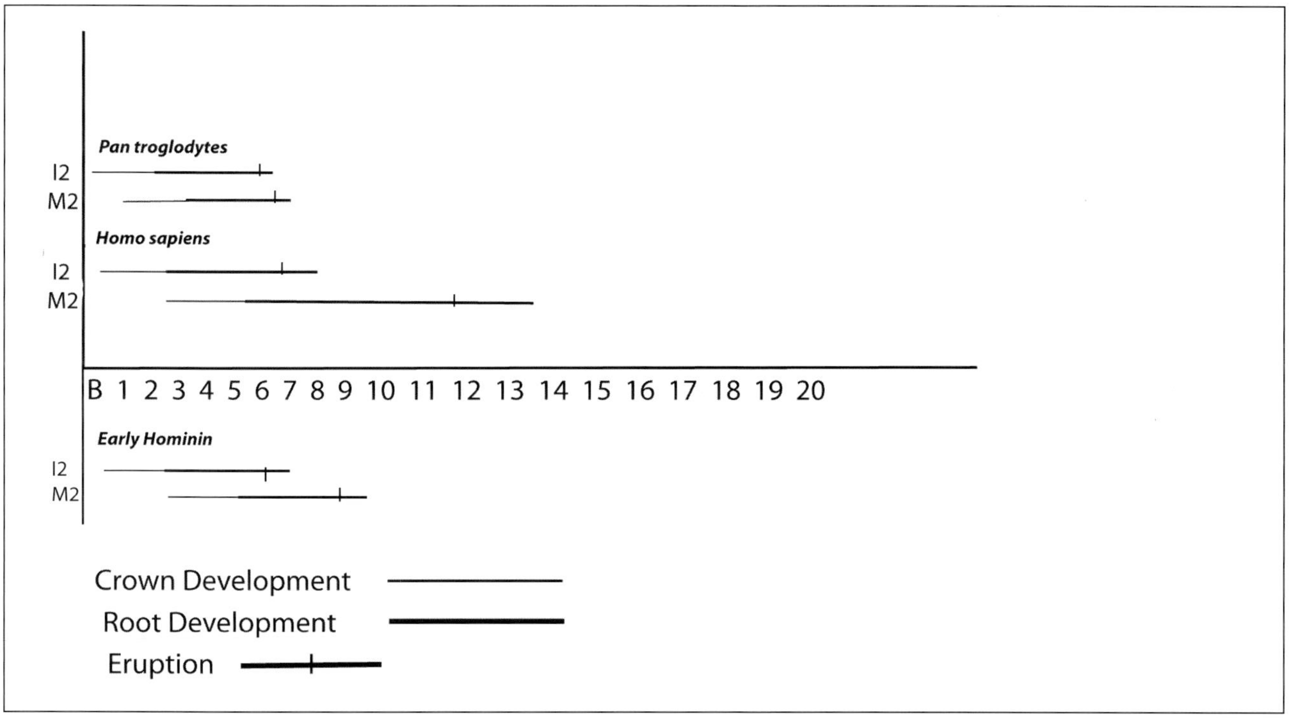

Figure 5. Pan, Homo sapiens and early hominin dental development of the I2/M2 compared. Using a 2.5 year calcification time for both the I2 and M2 crown, and a root timing developed from radiographic studies of chimpanzees, early hominin specimens would show a developmental delay that is the equivalent of the Phase 2 of modern humans. Although the time frame of dental development is not as elongated as it is in Homo sapiens, the initiation of the shift appears relatively early in the evolutionary history of our lineage.

Finally, can Phase 1 and 2 shifts be documented in any fossil forms present in the hominin lineage? Virtually all of the immature fossil hominin specimens where both the developing I2 and M2 are present and imaged, show the identical type of phase delay as shown in all modern humans. This includes members of both the genus Australopithecus and non-modern versions of the Homo encompassing Neandertals. Table 3 presents the data from various published sources. Although different staging techniques were used, and are explained in the figure caption, clearly this pattern of delay is present. Since virtually all hominoid appear to form molar crowns in approximately the same time frame, the earlier hominin comparison is drawn with crown complete achieved in 2.5 years (average of the variation of 2-3 years). Since more limited information is present for the time frame of root development in these extinct forms, the chart is produced using a rapid root development time frame modeled from the genus *Pan* (Figure 5). With these very conservative estimates of both crown and root formation, coupled with at least a 2 stage lag between the I2 and M2, it is clear that even within the early hominins, there is a developmental delay that corresponds to the Quiescent phase.

The Quiescent Period in Human Dental Development

In their world wide survey of variation in human growth, Eveleth and Tanner (1991) describe two active phases in permanent tooth emergence, separated by a period of quiescence (Figure 6). The first active phase lasts one and a half to two years (when we consider the means for populations) M1, I1 and I2 emerge. The second phase lasts slightly longer, M2, C, PM1 and PM2 emerge. The Quiescent phase in between lasts a generally similar time, between two and three years in the male in nearly all populations and between 1.7 and 2.7 years in the female." (Eveleth, P.B. and Tanner, J.M. 1990 Worldwide Variation in Human Growth, 2nd ed. page 159)

Middle Childhood

Developmental psychologists have focused on this period as being a time of crucial importance in the emergence of language based cognitive behaviors as well as the appearance of a greater understanding of, and reliance on, cultural rules.

John Lucy and Suzanne Gaskins (2001:280) have noted that "Regarding the changes in middle childhood, cognitive developmentalists have long recognized this as the period in which the child completes a shift from dependence on more spontaneous, perceptual strategies to reliance on more systematically organized, conceptual ones. In short, the child now enters the world of the adult, which is more heavily guided by systems of shared cultural meaning".

In a review of middle childhood cross culturally,

Hominin		Stage I2	Stage M2
A.robustus	SK 62	**5**	**2 or 3**
A.robustus	SK 63	**5 or 6**	**3**
A.africanus	Taung	**4**	**2**
A.africanus	STS 24	**4 or 5**	**2**
Early *Homo*	KNM-WT 15000	**A1/2**	**R2/3**
Neandertal	Devil's Tower	**Ri**	**C1/4**

Table 3. Stage delay between the I2/M2 in a sample of earlier hominin forms. Australopithecine data from Conroy and Vannier 1991 and based on the stages of dental calcification by Demirjian et al. 1973. Early Homo data on KNM-WT 15000 from Smith 1993, and Devil's Tower Neandertal, from Dean et al. 1986 both based on the staging technique from Moorrees et al. 1963.

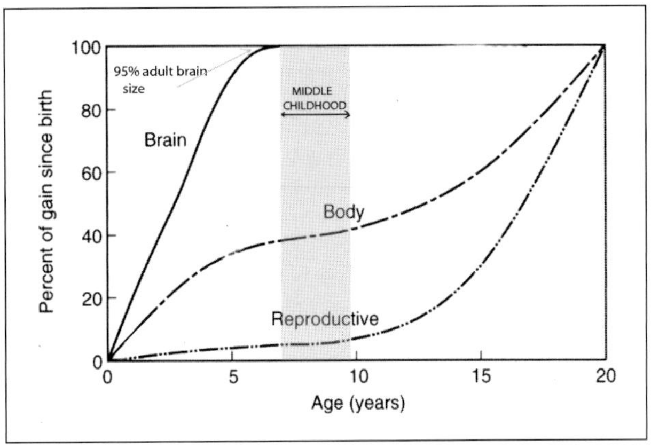

Figure 6. Original Scammon (1930) curves showing the growth of different body tissues (in weight) plotted against chronological age. The Quiescent period corresponds to the age in which brain weight is close to the maximum (95% adult weigh), with body weight gain decelerating and in conjunction with the lengthy attenuation of reproductive organ growth and maturation.

Weisner (1984: 344) notes that "Many cultures also share the belief that between the age of 5 and age 7 children begin to acquire reason or sense, the ability to understand cultural rules and to carry out directions. Rogoff et al. (1975), Super (1981), and J. Whiting and B. Whiting (1960) identified this age period from cross cultural samples, and Nerlove et al. (1974) did so from data from Guatemala.

Sweder (1981) argued that in middle childhood, children begin to acquire self/cultural/moral understandings of their world. He lists a set of ten themes that illustrate this:

1. Personal boundaries: what's me versus what's not me.
2. Sex identity: what's male versus what's female.
3. Maturity: what's grownup versus what's childlike.
4. Cosubstantiality: who is of my kind and thus shares food or blood with me versus who is not of my kind.
5. Ethnicity: what's our way versus what's not our way.
6. Hierarchy: the unequal share of life burdens and benefits.
7. Nature versus culture: what's human versus what's animal-like.
8. Autonomy: independent, dependent or interdependent.
9. The state: what I want to do versus what the group wants me to do.
10. Personal protection: avoiding the war of all against all.

Clearly these conflicts are part of a child's increasing socialization and integration into a society. Further, most reflect factors that represent human cultural phenomena. The difference between the younger, pre-Quiescent children and the older post-Quiescent children is in the way by which the cultural rules are recognized and enabled.

Middle childhood is then the time when the norms governing appropriate behavior within the culture are internalized as part of the development of an integrated social member of the group. These cognitive changes coincide with the maturational time frame brain growth in volume is almost complete and body size dimensions are minimally altered.

Interestingly, in his studies of the evolution of the brain in the human lineage, Holloway emphasized organizational rather than volume metric evolution of the brain in the human lineage. He has emphasized that although early hominin endocasts reveal a brain size in the range of the African apes, details of the position of surface anatomical features, such as the lunate sulcus, suggest that these brains had undergone neurological reorganization. Evidence that even early members of the hominin lineage based on dental developmental studies experienced a middle childhood period of would support Holloway's ideas of reorganization. These data suggest a reevaluation of the ways we view behavior and neurological evolution in early hominin evolution.

Middle Childhood and the Evolution of Human Cognition: Summary

1. Collection of a substantial series of panoramic X-rays of American children aged 4-14 indicates that there are significant changes in the timing and variation of human dental maturation since the last major studies were published in the 1960's and 1970s (Moorrees et al., 1963; Demirjian et al., 1973). This data set, and others that have been collected (i.e. Liversidge 2003, summarized by Guatelli-Steinberg,

2008), indicate there are both changes in the timing of the eruption of many of the permanent teeth as well as an under appreciated level of variation in human dental development that should be considered in reconstructions of earlier hominin development and life history.

2. A recent study by Zihlman and colleagues (2004) has presented data that patterns of dental eruption in a small sample of wild chimpanzees of known age record are significantly retarded compared to dental development in captive animals.

3. While these data sets narrow the timing differences between wild chimpanzee and modern human dental maturation patterns, it may be that when examining and comparing growth and development in humans and chimpanzees, we have failed to appreciate that maturation and growth represent a series of discrete periods. By drawing comparisons across the entire flow of human and chimpanzee maturation and development, shorter episodes in development may have been overlooked.

4. Just such a episode may be present in the dental eruption data that shows consistent differences between chimpanzees and humans in the timing of the eruption of the initial set of permanent teeth (two incisors and first molar) and the second set (canine, both premolars and the second molar). In humans, the time between the eruption of these two permanent teeth sets has been termed the "Quiescent period" (Eveleth and Tanner, 1990); it extends between 1.7 and 3.4 years and occurs at about the same time as a behavioral reordering in children that developmental psychologists term *middle childhood.*

5. It is worth noting that the Quiescent period and middle childhood begin just after the greatest growth of the human brain has been completed.

6. The Quiescent period is clearly marked in the development and eruption patterns of the human dentition but is not present in the dental development of chimpanzees.

7. Immature specimens of hominins of the appropriate dental age, from *Australopithecus* to *Homo sapiens neanderthalensis,* all demonstrate, without exception, the presence of the Quiescent period in their dental development.

8. Numerous publications by Holloway (1969, 1981, 1983a, 1983b, 1996, 2008 and Holloway et al., 2003 and 2004) over the past 40 years have been focused on the evolution of the brain in the human lineage. He has emphasized that although early hominin endocasts reveal a brain size in the range of the African apes, details of the position of surface anatomical features, such as the lunate sulcus, suggest that these brains had undergone neurological reorganization. Evidence that even early members of the hominin lineage experienced a middle childhood period of would support Holloway's (for example, 1983a, 1983b, 1988, 1996) ideas of reorganization. These data suggest a reevaluation of the ways we view behavior and neurological evolution in early hominin evolution.

References

Anemone, R.L, 2002. Dental development and life history in hominid evolution. In: Minugh-Purvis, N. and McNamara, K.J. (Eds.), Human Evolution through Developmental Change. Johns Hopkins University Press, Baltimore, Maryland, pp. 249-280.

Anemone, R.L., Watts, E.S. and Swindler, D.R., 1991. Dental development of known-age chimpanzees, *Pan troglodytes* (Primates, Pongidae). American Journal of Physical Anthropology 86, 229-241.

Anemone, R.L., Mooney, M.P. and Siegel, M.I., 1996. Longitudinal study of dental development in chimpanzees of known chronological age: implications for understanding the age at death of Plio-Pleistocene hominids. American Journal of Physical Anthropology 99, 119-133.

Beynon, A.D., Clayton, C.B., Ramirez Rossi, F.V. and Reid, D.J., 1998. Radiographic and histological methodologies in estimating the chronology of crown development in modern humans and great apes: a review, with some applications for studies on juvenile hominids. Journal of Human Evolution 35, 351-370.

Blankenstein, R., Cleaton-Jones, P.E., Luk, K.M. and Fatti, L.P., 1990. The onset of eruption of the permanent dentition amongst South African black children. Arch. of Oral Biol. 35, 225-228.

Bogin, B., 1999. Patterns of Human Growth, 2nd ed., Cambridge University Press: Cambridge, UK.

Conroy, G.C. and Mahoney, C.J., 1991. Mixed longitudinal study of dental emergence in the chimpanzee, *Pan troglodytes* (Primates, Pongidae). American Journal of Physical Anthropology 86, 243-254.

Conroy, G.C. and Vannier, M.W., 1991. Dental development in South African australopithecines. Part II: Dental stage assessment. American Journal of Physical Anthropology 86, 137-156.

Dean, C., 2000. Progress in understanding hominoid dental development. J. Anat. 197, 77-101.

Dean, C., Stringer, C.B. and Bromage, T.G., 1986. Age at death of the Neanderthal child from Devil's Tower, Gibraltar and the implications for studies of general growth and development in Neanderthals. American Journal of Physical Anthropology 70, 301-309.

De Coster, P.J., Marks, L.A., Martens, L.C., and Huysseune, A., 2009. Dental Agenesis: Genetic and Clinical Perspectives. J Oral Pathol Med 38, 1-17.

Demirjian, A., Goldstein, H. and Tanner, J.M., 1973. A new system of dental age assessment. Hum. Biol. 45, 211-227.

Eveleth, P.B. and Tanner, J.M.,1990. Worldwide Variation in Human Growth, 2nd ed. Cambridge University Press, Cambridge.

Guatelli-Steinberg, D., 2009. Recent studies of dental development in Neandertals: Implications for Neandertal life histories. Evol. Anthr. 18, 9-20.

Harris, E.F. and McKee, J.H., 1990. Tooth mineralization standards for blacks and whites from the Middle Southern United States. J. Foren. Sci. 35, 859-872.

Hawkes, K. and Paine, R.R., eds., 2006. The Evolution of Human Life History. School of American Research Press. Santa Fe: New Mexico.

Holloway, R.L., 1969. Culture: A human domain. Current Anthropol. 10, 395-425.

Holloway, R.L., 1981. Culture, symbols, and human-brain evolution - A synthesis. Dialectical Anthropology 4, 287-303.

Holloway, R.L., 1983a. Cerebral brain endocast pattern of Australopithecus afarensis hominid. Nature 303, 420-422.

Holloway, R.L., 1983b. Human paleontological evidence relevant to language behavior. Hum. Neurobiol. 2, 105-114.

Holloway, R.L., 1988. Some additional morphological and metrical observations on *Pan* brain casts and their relevance to the Taung endocast. American Journal of Physical Anthropology 77, 27-33.

Holloway, R.L., 1996. Toward a synthetic theory of human brain evolution. In: Changeux, J.-P. and Chavaillon, J., (Eds.). Origins of the Human Brain. Clarendon Press: Oxford. pp. 42-54.

Holloway, R.L. 2008. The human brain evolving: A personal retrospective. Ann.Rev.Anthropol. 37, 1-19.

Holloway, R.L., Broadfield, D.C., and Yuan, M.S., 2003. Morphology and histology of chimpanzee primary visual striate cortex indicate that brain reorganization predated brain expansion in early hominid evolution. Anat.Rec.A Discov. Mol. Cell. Evol. Biol. 273, 594-602.

Holloway, R.L., Broadfield, D.C., and Yuan, M.S., 2004. The Human Fossil Record: The Paleoneurological Evidence, Volume 3. Wiley-Liss: Hoboken, NJ.

Kahumbu, P. and Eley, R.M., 1991. Teeth emergence in wild olive baboons in Kenya and formulation of a dental schedule for aging wild baboon populations. American Journal of Primatology 23, 1-9.

Keeley, J., 2002. Life-history evolution in the Miocene and extant apes. In: Minugh-Purvis, N. and McNamara, K.J. (Eds.), Human Evolution through Developmental Change. Johns Hopkins University Press, Baltimore, Maryland, pp. 223-248.

Kuykendall, K.L., 1996. Dental development in chimpanzees (*Pan troglodytes*): the timing of tooth calcification stages. American Journal of Physical Anthropology 99, 135-157.

Kuykendall, K.L., 2002. An assessment of radiographic and histological standards of dental development in chimpanzees. In: Minugh-Purvis, N. and McNamara, K.J. (Eds.), Human Evolution through Developmental Change. Johns Hopkins University Press, Baltimore, Maryland, pp. 281-304.

Kuykendall, K.L., 2003. Reconstructing australopithecine growth and development: What do we think we know? In: Thompson, J.L., Krovitz, G.E. and Nelson, A.J. (Eds.), Patterns of Growth and Development in the Genus *Homo*. Cambridge University Press, Cambridge, pp. 191-218.

Kuykendall, K.L., Mahoney, C.J. and Conroy, G.C., 1992. Probit and survival analysis of tooth emergence ages in a mixed-longitudinal sample of chimpanzees (*Pan troglodytes*). American Journal of Physical Anthropology 89, 379-399.

Liversidge, H.M., 2003. Variation in modern human dental development. In: Thompson, J.L., Krovitz, G.E. and Nelson, A.J. (Eds.), Patterns of Growth and Development in the Genus *Homo*. Cambridge University Press, Cambridge, pp. 73-113.

Liversidge, H.M., 2008. Dental age revisited. In: Irish, J.D. and Nelson, G.C. (Eds.), Technique and Application in Dental Anthropology. Cambridge University Press, Cambridge, pp. 234-252.

Liversidge, H.M., 2008. Timing of human mandibular third molar formation. Annuals of Human Biology 35, 294-321.

Liversidge, H.M. and Speechly, T., 2001. Growth of permanent mandibular teeth of British children aged 4 to 9 years. Ann. Hum. Biol. 28, 256-262.

Liversidge, H.M., Speechly, T. and Hector, M.P., 1999. Dental maturation in British children: Are Demirjian standards applicable. Int. J. Pediatr. Dent. 9, 263-269.

Lucy, J.A. and Gaskins, S., 2001. Grammatical categories and the development of classification preferences: A comparative approach. Bowerman, M. and Levinson, S.C. Language Acquisition and Conceptual Development. pp. 257-283.

Minugh-Purvis, N. and McNamara, K.J., eds. 2002. Human Evolution through Developmental Change. Johns Hopkins University Press: Baltimore, Maryland.

Monge, J.M., Mann, A.E., Stout, A., Roger, J. and Wadenya, R., 2007. Dental calcification stages of the permanent M1 and M2 in U.S. children of African-American and European-American ancestry born in the 1990s. In: Bailey, S.E. and Hublin, J.-J. (Eds.), Dental Perspectives on Human Evolution. Springer: The Netherlands, pp. 263-274.

Moorrees, C.F.A., Fanning, E.A., and Hunt, E.E., 1963. Age variation of formation stages for ten permanent teeth. J. Dent. Res. 42, 1490-1502.

Nadler, G.L., 1998. Earlier dental maturation: Fact or fiction? Angle Orthodont. 68, 535-538.

Nissen, H.W. and Riesen, A.H., 1964. The eruption of the permanent dentition of chimpanzees. American Journal of Physical Anthropololgy 22, 285-294.

Nerlove, S., Roberts, J.M., Klein, R.E., Yarborough, C. and Habicht, J., 1974. Natural indicators of cognitive development: An observational study of rural Guatemalan children. Ethos 2, 265-295.

Olze, A., Schmeling, A., Taniguchi, M., Maeda, H. van Niekerk, P., Wernecke, K.D. and Geserick, G., 2004. Forensic age estimation in living subjects: the ethnic factor in wisdom tooth mineralization. Int. J. Legal Med. 118, 170-173.

Phillips-Conroy, J.E. and Jolly, C.J., 1988. Dental eruption schedules of wild and captive baboons. Am. J. Primatol. 15, 17-29.

Reid, D.J., Schwartz, G.T., Dean, C., and Chandrasekera, M.S., 1998. A histological reconstruction of dental development in the common chimpanzee (*Pan troglodytes*). Journal of Human Evolution 35, 427-228.

Robson, S.L, and Wood, B., 2008. Hominin life history: reconstruction and evolution. J. Anat. 212, 394-425.

Rousset, M.M., Boualam, N., Delfosse, C. and Roberts, W.E., 2003. Emergence of permanent teeth: Secular trends and variance in a modern sample. J. Dent.Child. 70, 208-214.

Scammon, R.E., 1930. The measurement of the body in childhood. In: Harris, J.A., Jackon, C.M., Paterson, D.G. and Scammon, R.E. (Eds.), The Measurement of Man, University of Minnesota Press: Minneapolis. pp. 173-215.

Simpson, S.W. and Kunos, C.A., 1998. A radiographic study of the development of the human mandibular dentition. Journal of Human Evolution 35, 479-505.

Simpson, S.W., Lovejoy, C.O. and Meindl, R.S., 1990. Hominoid dental maturation. Journal of Human Evolution 19, 285-297.

Skinner, M.M and Wood, B., 2006. The evolution of modern human life history: A paleontological perspective. In: Hawkes, K. and Paine, R.R. (Eds.), The Evolution of Human Life History. School of American Research Advanced Seminar Series: Santa Fe, pp. 331-364.

Smith, B.H., 1986. Dental development in Australopithecus and early Homo. Nature 323, 327-330.

Smith, B.H., 1993. The physiological age of KNM-WT15000. In: Walker, A. and Leakey, R. (Eds.), The Nariokotome Homo erectus Skeleton. Harvard University Press: Cambridge, MA. pp. 195-220.

Smith, T.M., Harvati, K., Olejniczak, A.J., Reid, D.J., Hublin, J.-J., and Panagopoulou, E., 2009. Brief Communication: Dental Development and Enamel Thickness in the Lakonis Neanderthal molar. American Journal of Physical Anthropology 138, 112-118.

Smith, T.M., Reid, D.J., Dean, M.C., Olejniczak, A.J. and Martin, L.B., 2007. Molar development in common chimpanzees (*Pan troglodytes*). Journal of Human Evolution 52, 201-216.

Smith, T.M. and Tafforeau, P., 2008. New Visions of Dental Tissue Research: Tooth Development, Chemistry and Structure. Evolutionary Anthropology 17, 213-226.

Smith, T.M., Harvati, K., Olejniczak, A.J., Reid, D.J., Hublin, J-J., and Pangopoulou, E., 2009. Brief Communication: Dental Development and Enamel Thickness in the Lakonis Neanderthal Molar. American Journal of Physical Anthropology 138, 112-118.

Super, C.M., 1981. Behavioral development in infancy. 1981 Munroe, R., Munroe, L. and Whiting, B. (Eds.) Handbook of Cross-Cultural Development. New York; Garland Press.

Sweder, R.A., and Bourne, E., 1981. Beyond self-constructed knowledge: The study of culture and morality. In: Munroe, A.J. and Whiting, B. (Eds.), Cultural Conceptions of Mental Health and Therapy. Boston: D. Reidel.

Tompkins, R.L., 1996. Relative dental development of Upper Pleistocene hominids compared to human population variation. American Journal of Physical Anthropology 99, 103-118.

Thompson, J.L., Krovitz, G.E. and Nelson, A.J. (Eds.), 2003. Patterns of Growth and Development in the Genus Homo. Cambridge University Press: Cambridge, U.K.

Weisner, T.S., 1984. Ecocultural niches of middle childhood: A cross cultural perspective. In: Collins, W.A. (Ed.), Development during childhood: The years from six to twelve. Washington, DC: National Academy of Science Press. pp. 335-369.

Whiting, J. and Whiting, B., 1960. Contributions of anthropology to the methods of studying child rearing. In: Mussen, P. (Ed.), Handbook of Research Methods in Childhood Development. New York: John Wiley & Sons.

Zihlman, A., Bolter, D. and Boesch, C., 2004. Wild chimpanzee dentition and its implications for assessing life history in immature hominin fossils. Proceedings of the National Academy of Sciences USA 101, 10541-10543.

CHAPTER 16

PERIKYMATA COUNTS IN TWO MODERN HUMAN SAMPLE POPULATIONS

MICHAEL SHENG-TIEN YUAN

ABSTRACT

Many studies based on the perikymata, a dental enamel surface microstructure, have attempted to estimate the age-at-death and crown formation times on human and other fossil specimens. However, due to problematic assumptions, the small sample sizes, the wide range of perikymata count estimates, and the limited portions of dentition explored, considerable controversy has resulted. The collection of baseline information on perikymata counts for various hominid and hominoid dentitions represents an important step toward resolving the controversies.

The goal of this study is to establish a database of modern human perikymata counts of the maxillary third premolars for comparative purposes. The results demonstrated: 1) There is sexual dimorphism, though not statistically significant within the two populations: perikymata counts are higher in males. 2) There are no significant differences between right and left sides. 3) There are no significant differences for sex-combined samples between the two ethnic groups. 4) When data are pooled for the East Asian samples, the mean perikymata count for maxillary third premolars is 150 with a standard deviation of 25. 5) The perikymata counts are significantly correlated with their corresponding crown height. 6) Although the data on perikymata counts follow a normal distribution, the variation is high (coefficient of variation= 16 %). This study disputes the perikymata count application in anthropology and questions the interpretation of the ape affinity of australopithecines in hominid dental evolution.

INTRODUCTION

Teeth and their associated structures including the maxillary and mandibular bones and the masticatory musculature, provide one of the best vehicles for studying hominoid and hominid evolution (Dahlberg, 1971; Hillson, 1986; Aiello & Dean in Chapter 8, 1990; Hillson, 1996; Scott & Turner II, 1997).

Due to its high percentage of inorganic components (by weight: enamel, 95%; dentin, 70%, cementum 61%; vs. bone 45%; and by volume: enamel, 86%; dentin, 45%, cementum 33%; vs. bone 23%; see Schroeder, 1991, p. 73, Fig. 1.34) and the almost irreversible growth and development of the calcification process, the tooth itself, once formed, becomes a fossil-like material that is highly durable through time. During the past two decades, investigations of the morphological structures of the teeth, especially the microanatomy of enamel, have become prominent in studies of both hominoid taxonomy and hominoid ontogeny and phylogeny (see review of Macho & Wood, 1995; Mann et al., 1990a; Winkler & Swinder, 1991).

Many studies based on the microstructure of the dental enamel surface, especially the perikymata, and its internal enamel structures, the cross-striations and the striae of Retzius, have attempted to estimate crown formation time and age-at-death in hominoids (Boyde, 1963; Bromage & Dean, 1985; Dean et al., 1986; Bacon, 1987; Dean, 1987; Dean & Beynon, 1991; Stringer et al., 1990). In these studies, fossil teeth of Australopithecines, early (archaic) *Homo sapiens*, and Neanderthals were compared with those of *Homo sapiens* to reveal differences between taxa in crown maturation times and also to estimate age-at-death. The results of these investigations have stimulated several different views on how to infer patterns of human evolution based on dental microanatomy (Bromage & Dean, 1985; Dean et al., 1993; Mann et al., 1990a, 1990b, 1991).

Mammalian enamel ontogeny and comparative histology have been thoroughly reviewed by Boyde (1971, in Chapter 7, pp. 81-94; 1997) and Moss-Salentijn et al. (1997, in Chapter 1, pp. 5-30). Several widely-used textbooks on anatomy, oral development, and histology provide excellent reviews of human enamel structure. These texts include Avery (1994, see Piesco & Avery in Chapter IV, pp. 228-241), Ten Cate (1998, see Eisenmann in Chapter 10, pp. 197-217 and Chapter 11, pp. 218-235), Aiello & Dean (1990, in Chapter 7, pp. 106-132); Moss-Salentijn & Hendricks-Klyvert (1985, in Chapter 11, pp. 229-254, 1990), and Schroeder (1991, in Chapter 1, pp. 38-67). The physiologic and genetic interactions among enamel matrix proteins, minerals, and various components during the secretary and maturation phases, have also been extensively discussed in Chadwick & Cardew (1997).

In the following sections, several features of human enamel structure will be reviewed in detail. These include: 1) the histologic nature of the perikymata on enamel; 2) the lines of Retzius of enamel; 3) the cross-striations of the enamel prisms; and 4) the Hunter-Schreger bands. Additionally, the chronology of dental growth and development, human crown formation times, and the applications and the controversy surrounding the interpretation of hominid and hominoid dental remains based on studies of circadian and infradian characters will be discussed.

A number of attempts have been made to apply knowledge of the circadian and infradian incremental structures on dental enamel to explore the crown formation time (or crown maturation time), to estimate the age-at-death in developing individuals, and to reconstruct life history from the manifestation and counts of incremental structures. In addition, crown formation time and the rate of dental tissue formation, which were derived from the estimation of counts of incremental structures, may provide evidence for species differences in the hominoid and hominid evolution.

The cross-striations in enamel and perikymata counts on the crown surface are the most commonly used incremental structures to estimate crown formation time and age-at-death. Boyde (1963) first suggested that age could be estimated from prism cross-striation counts in non-living specimens. Bromage & Dean (1985) developed an ageing method based on perikymata counts alone, suggesting that lower permanent incisor crown formation started at 3 months of age, and then at approximately 6 months appositional enamel growth began, so that the first perikymata groove would appear at 9 months of age. They assumed a 7-day repeat interval to derive ages-at-death for uncompleted fossil dental crowns. Stringer et al. (1990) later carried out a study to test the age estimation from enamel layering using known age archeological specimens with the assumed cross-striation counts of suggested 7, 8, and 9-day repeat intervals. The best matching result was concluded by applying an 8-day repeat interval between Retzius lines for the age-at-death estimation. Dean & Beynon (1991) similarly applied the cross-striation counts and perikymata counts to estimate crown formation time and age-at-death in a child from the 18/19th century A.D. in London (Shown in Table 1-5).

Efforts have also been made to determine the perikymata counts of modern humans. Results of studies of modern humans should ideally and hypothetically be useful for inferring time of crown formation and age-at-death for fossil hominids, if accurate adjustments are applied. The results of studies of modern humans to date, however, have been very diverse. In addition, many of these studies do not provide sufficient information on the perikymata counts with which to compare to the posterior dentition. The range of human incisor perikymata counts based on a sample of 12 five-thousand-year-old immature human specimens from the Iranian site of Hasanlu is 75-157 with a mean of 116, a median of 118, and a standard deviation of 25 (Mann et al., 1990b, 1991). On the other hand, Bromage & Dean (1985) obtained perikymata counts ranging from 165-202 counts, with a mean of 188, in a sample of 10 unworn modern human lower incisors. Bacon (1987) also reported a range of 111-179 in a sample of 23 modern human incisors.

The collection of crown formation data during radiographic and histologic studies of dentition in modern humans, gorillas, and chimpanzees and the perikymata-derived crown formation time of the australopithecines, has made comparisons possible between extant and extinct hominoid species. For example, the combined upper and lower incisor crown formation time averages 4.21 years and ranges 3.71-4.71 years in *Homo sapiens* (Shellis, 1984). The investigation of *Pan troglodytes* provided crown formation time of the maxillary central incisor is reported to be 5.47 $\pm$ 0.24 years and that of the mandibular, 4.86 $\pm$ 0.37 years (Chandrasekera et al., 1993). The crown formation time of *Gorilla gorilla* based on the histological examinations revealed 4.0 and 3.6 years for maxillary and mandibular central incisors, respectively (Beynon et al., 1991). The mandibular central incisor formation times in *Australopithecus afarensis* and *Australopithecus africanus* were estimated 3 years and 3 years 1.1 months for specimens LH2 and Sts 24a respectively, utilizing the perikymata-derived crown formation times (Bromage & Dean, 1985).

A list of age-at-death estimate for various hominid fossil specimens using perikymata counts is summarized in **Table 1-5**. This approach has created a totally different interpretation of the fossil record, especially with regard to the affinity of australopithecines. Claiming the ape affinity of australopithecines based on the dental eruption pattern and estimation of age-at-death, perikymata counts played a key role in this ongoing controversy.

Nevertheless, we must note that these estimated results are based on the premises that: 1) all enamel microstructures are tightly correlated with their mutual periodicities; and 2) the dentitions are equally accounted for their developmental timings and sequences across the extant and extinct primate species.

Table 1-5. Perikymata count study on record.

Taxon	Specimen	Site	Teeth used	P.C.	Age-at-death Estimate (yr)	Ref.
A. afarensis	LH2	Laetoli, Tanzania	mand. R. 1st incisor	130	3.25	a
A. africanus	Sts24a	Sterkfontein, S.A.	max. R. 1st incisor	135	3.3	a
P. robustus	SK62	Swarkran, S.A.	mand. R. 1st incisor	57	3.35	a
			mand. L. 2nd incisor	64	3.48	a
P. robustus	SK63	Swarkran, S.A.	mand. R. 1st incisor	86	3.15	a
			mand. 1st incisor	75	3.98	g
			mand. L 2nd incisor	84	3.98	g
			mand. R canine	98	3.98	g
P. boisei	KNM-ER 1477	Koobi Fora, Kenya	mand. 1st incisor	92	2.5-3.0	c
P. boisei	KNM-ER 812	Koobi Fora, Kenya	mand. 1st incisor	86	2.5-3.0	c
P. boisei	KNM-ER 1820	Koobi Fora, Kenya	mand. 1st incisor	82	2.5-3.1	c
P. boisei	OH30	Olduvai, Tanzania	mand. 1st incisor	101	2.7-3.2	c
Early Homo	KNM-ER 820	Koobi Fora, Kenya	mand. L. 2nd incisor	105	5.3 (5.3-6)	a
Neanderthal	Gibraltar child	Devil's Tower, Gitraltar	max. 1st incisor	119	3.1	b
	Krapina 90		mand. R. 2nd incisor	205±10	4.4	e
	Krapina 91*			>100±4	-	e
	Krapina 93*			>107±2	-	e
	Krapina 94*		max. R. 1st incisor	>144±7	-	e
	Krapina 95*			>50	-	e
Homo sapien	2179	Spitalfield, London	max. 1st molar	85		
			max. 1st molar	120		
			mand. 1st incisor			
	197					
			mand. 2nd incisor	224	5.25	d
			mand. 2nd incisor	162		
			(?) canine	184		
			(?) canine	182		
		Hasanlu, Iran	(?) incisor	124±1	2.9	f
		(3000 BC)	(?) incisor	134±4	3.1	f
			(?) incisor	99±5	2.4	f
			max. R. 2nd incisor	128±4	2.9	e, f
			mand. R. 1st incisor	157±12	3.5	e, f
			(?) incisor	90±11	2.2	f
			max. R 1st incisor	75±7	1.9	e, f
			(?) incisor	134±2	3.1	f
			(?) incisor	103±1	2.5	f
			(?) incisor	93±1	2.3	f
		Island Field, USA	(?) incisor	148±7	3.3	f
		(AD 800)	(?) incisor	113+3	2.7	f

This table was based on the listed references. Modified from three sources: 1) Aiello & Dean, 1990, p. 131, Table 7.1; 2) Hillson, 1996, p. 179, Table 6.4.; and 3) Mann et al., 1991, p.180, Table 2.
(*) sign represents incomplete crown; (?) sign represents unknown dental location.
(P.C.): perikymata count; (max.): maxillary; (mand.): mandibular; (R): right; (L): left

Reference:
a. Bromage & Dean (1985) *b. Dean et al. (1986)* *c. Dean (1987a)*
d. Dean & Beynon (1991) *e. Mann et al. (1990b)* *f. Mann et al. (1991)*
g. Dean et al. (1993b)

Although during the last decade researchers have repeatedly tried to assess crown formation times and the age-at-death in hominid fossil teeth, the use of perikymata counts in interpreting the fossil record remains problematic at various levels.

Some assumptions have been challenged, such as the true representation of the circadian rhythm of cross-striations, the correlation between the lines of Retzius and cross-striations especially in its notion of circaseptan rhythm, and the wide variation in the periodicity of perikymata counts within and between individuals (Mann et al., 1990b, 1991; FitzGerald, 1998; Risnes, 1998). In addition, during the cuspal, or appositional stage of enamel formation, lines of Retzius do not reach the tooth surface; only in enamel formed later at the imbricational stage, lines of Retzius are manifested at the surface of the crowns. Therefore, crown formation times may not have been precisely predicted, when the differential appositional enamel formation times were ignored or corrected by the estimation. Moreover, perikymata are variably expressed at the cervix of teeth, which increases the possibility of underestimation for age-at-death and crown formation time, even if the perikymata, the lines of Retzius, and the cross-striations were to be truly and exclusively correlated with their periodicity.

In the investigation of age-at-death, the results vary within the same individual. While Bromage & Dean (1985) estimated the age-at-death of a juvenile *Paranthropus robustus* SK 63 to be 3.15 years old, based on the perikymata counts of a mandibular right central incisor; Dean et al. (1993b) applied the histological cross-striation counts of a mandibular right canine and concluded that 4 years would be a more accurate estimate. The discrepancy between the estimates may result from either the high variability of the tissue studied or the inadequateness of the methodology itself.

A test was carried out by Stringer et al. (1990) to investigate the correlation between samples of known-age and the incremental ageing of the perikymata counts in the Spitalfields collection. Three estimates of age at death were calculated from each incisal perikymata count by using 7-day, 8-day, and 9-day periodicity for perikymata. The assumption of 7-day periodicity and early incisal calcification at about 3 months after birth (Dean et al., 1986) consistently underestimates age at death. The 8-day periodicity and an adjustment of a later initiation of calcification, about 6 months for lower central incisors and 9 months for upper central incisors, gives an agreement between real and estimated ages. The 9-day periodicity gives a poor agreement with the real ages. Though the study seems to provide a good match and evidence for their applicability, it also clearly demonstrates that no one choice of periodicity is likely to accurately reflect those of a whole population of individuals.

The controversy surrounding the use of perikymata counts as the ultimate tools of estimating crown formation times and age-at-death can only be resolved through further investigations.

The wide range of perikymata count estimates and very small sample sizes in previous studies have sparked much controversy surrounding the interpretations and inferences of results derived from perikymata counts. One of the major problems relates to the absence of important baseline information on the biological variation in human perikymata counts. We do not have sufficient comparative data across our own and relevant species. The taxonomic premises of such perikymata counts should be tested first in order to offer a true database for comparisons.

In this project, a scanning electron microscope (SEM) was used to determine the perikymata counts of the third premolars (symbol: P3 and clinically called the first bicuspids) in the maxillary region in two modern human samples collected from Taiwan and Japan.

The purpose of this study was to establish a database of modern human perikymata counts to facilitate comparisons by sex, ethnicity, and species, and to help resolve the perikymata counts controversy. This study was designed to achieve several goals which include the following:

1. to investigate the biological variation in modern human perikymata counts;
2. to determine the degree of sexual dimorphism in human perikymata counts;
3. to examine the ethnic differences of Chinese and Japanese or the regional differences between two isolated islands of Taiwan and Japan in their perikymata counts;
4. to analyze the strength of correlations between the perikymata counts and various dental dimensional parameters.

Methods

The dental samples were collected from two different geographic regions in Asia, which also had different ethnic compositions. One region was the Tainan City of Taiwan, Republic of China and the other was the Nagoya City of Japan. Both are the fourth largest cities in their own country.

Since the study applied ethnicity as one of the variables, the sample collection process excluded those specimens which were obtained from aborigines in Taiwan. Taiwan aborigines consist of nine tribes. They are widely dispersed, though most inhabit the remote mountains of central Taiwan, and are rarely seen in the highly populated coastal plain regions. They are believed to be more closely related to the Malay than to the Chinese both morphologically and genetically. Given that Taiwanese aborigines live quite distant from Tainan City, where the samples were collected, we assumed that the statistical error including aborigines in the study was quite low. The ethnic identification of the samples was carried out by the dentists who collected them. If no screening had taken place during sample collection and the sampling had been random, sampling error would have been maximally 2 %.

The definition of Japanese here does not include Korean-Japanese descendants. The population of Japan was estimated 126.18 million as of July 1999. The ethnic groups in Japan includes 99.4% Japanese and 0.6 % other, consisting mostly of Korean descendants (CIA, 1999). Historically, many Koreans immigrated to Japan. As with the Chinese, Korean descendants usually maintained their own ethnic origin by holding on to their Korean family name.

The Ainu aboriginal Japanese who live in the Northern Japanese island of Hokkaido make up a very small percentage of the total population in Japan. Since they are morphologically distinctive and are distant from where our samples were collected, we can ignore the statistical bias of possibly inclusion of Ainu. Even if the Korean samples were accidentally included and the sampling had been random, there would only have been 0.6% error in statistic probability.

The samples were collected during a 5-year period from 1993 to 1998. Local dental clinics (predominantly orthodontic), of Tainan and Nagoya, assisted the dental sample collections for this project. The collected teeth were generally extracted during the course of orthodontic treatment. Since most of the orthodontic patients were juvenile or adolescent individuals, the age of the individual from whom dental specimens were extracted ranged mostly from 7 to 18 years old in our samples. Dental specimens from adults were excluded during the collection process, since attrition and abrasion would have occurred and been noticeable.

After the teeth were extracted, the samples were first divided into several subgroups according to ethnicity, ("Chinese" and "Japanese" as defined above); sex (male and female); side (right and left), and location (maxillary and mandibular). A total of 728 extracted modern human third premolars were collected: 439 from Taiwan and 289 from Japan.

Most of the specimens collected were well preserved in gross morphology with minimal attrition and abrasion. Specimens with obvious abnormality, such as microdontia, or major surface damage, such as fractured or chipped enamel, were excluded. Since most of the dental specimens were removed from individuals of young age during their ongoing apexogenesis (root apex formation), some of them exhibited incomplete root formation, which is in part characterized by an open apex.

All samples were then screened using a Zeiss dissecting microscope with a magnification of 50X to verify their surface morphology, especially the existence of the perikymata microstructure. Specimens were discarded when they either showed very little presence of perikymata or defective enamel on the buccal surface of the tooth. These criteria further reduced the sample size from 399 to 92 cases for the maxillary third premolars. The final perikymata SEM observations included a total of 92 cases of maxillary third premolars, in which 44 dental specimens came from Taiwan and 48 from Japan. The mandibular third premolar samples were stored away for later SEM investigation, and were not included in this research project.

In this study, three sets of dimensional parameters were directly measured from the dental specimens themselves, epoxy resin dental duplicates, and the dental plaster casts. All the measurements were performed by a Mitutoyo digimatic caliper. Each set of measurements included: 1) buccal crown height, 2) lingual crown height, 3) mesiodistal width, and 4) buccolingual width. For each measurement, three readings were carried out and were averaged to reach a final measurement for the record.

All the parameters were defined as the greatest dimensions during the assessment. Buccal and lingual crown heights were measured from the cervico-cemental junction (CEJ) to the buccal and lingual cusp tips. Buccolingual widths were measured between the most convex points on the buccal and lingual crown surfaces. Mesiodistal dimensions were measured between the mesial and distal contact points.

For SEM data collection Coltène PRESIDENT microSystem light body, a commercially available dental impression system, was utilized to produce replicas. Casts derived from these molds were made with Araldite 502. These casts were then coated and examined using an AMRAY 1850 field emission electron microscope.

The perikymata count observation was performed with the electron microscope operating at an acceleration potential of 2.00 KV with magnifications ranging from 7X to 2000X. In each case, a set of four reference micrographs was taken at the magnification of 7X to record its buccal, mesial, and occlusal views (Fig. 1) and at a magnification of 20X to record the occlusal view of the buccal cusp (Fig. 2). 40X to 50X magnification was used to provide adequate resolution for calculating perikymata counts (Fig. 3). Statistical analysis was carried out to calculate the sample size (N), means (M), standard deviations (SD), coefficient of variation (CV), range, minimum (Min.) and maximum (Max.) values, and other descriptive statistics such as skewness and kurtosis . Paired sample t-test, independent samples t-test, one-way Analysis of Variance (ANOVA), Scheffe's procedure for Post Hoc Comparisons, Pearson's correlation, and curve estimation for regression were also tested. A p value of ≤ 0.05 was considered to be significant.

Results

When all cases were combined, the total sample size was 92. For the total sample, perikymata counts had a mean of 150, a standard deviation of 25, a coefficient of variation of 16%, minimum counts of 107, maximum counts of 209, a range of 102 counts, a median of 148, a mode of 140, kurtosis of - 0.275, and skewness of 0.437.

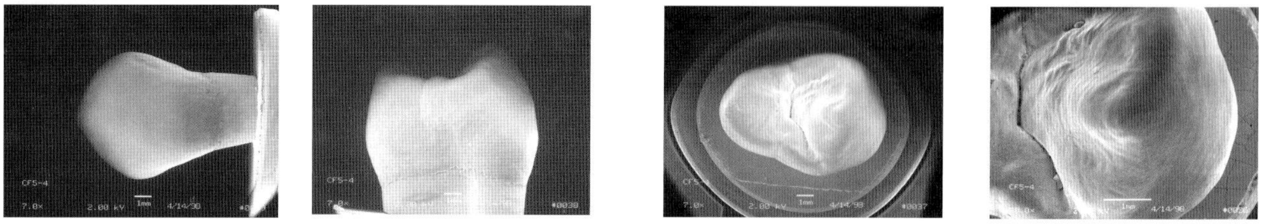

a: buccal view *b: mesial view* *c: occlusal view* *d: buccal cusp*

Fig. 1. A series of micrographs taken as references for the observation of perikymata. The specimen was obtained from a Chinese female. It is a right maxillary third premolar. (a), (b), and (c) are micrographs of buccal, mesial, and occlusal views respectively. (d) is the buccal cusp view. Note that on the (b) mesial view, the mesial groove was clearly seen in the middle of crown, dividing buccal and palatal cusps. (Photo© Michael S. Yuan)

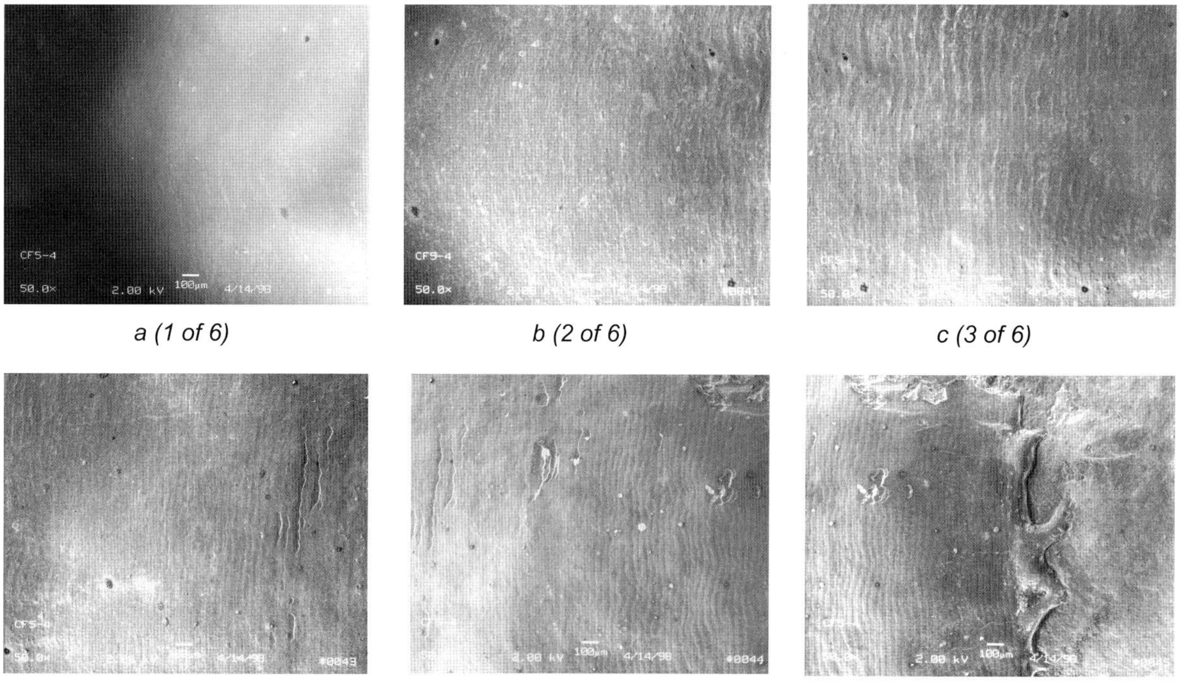

a (1 of 6) *b (2 of 6)* *c (3 of 6)*

d (4 of 6) *e (5 of 6)* *f (6 of 6)*

Fig. 2. An example of a series of continuous micrographs on the buccal surface of a maxillary premolar. The micrographs are in a cusp-cervix sequence. (a): the buccal cuspal region; (b), (c), (d), & (e): regions between cusp and CEJ; and (f): the CEJ cervical region. (Photo© Michael S. Yuan)

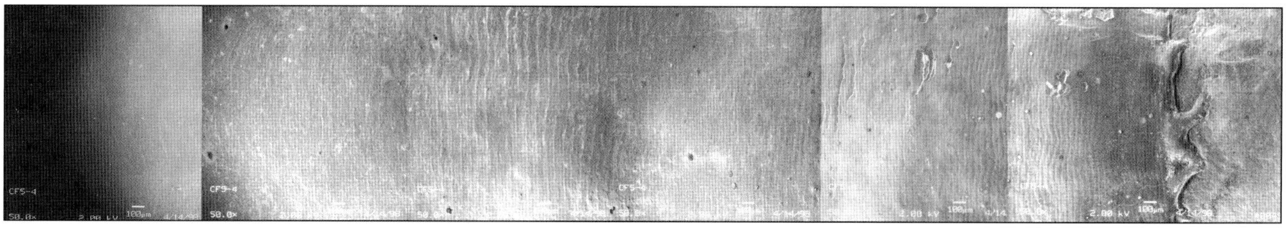

Fig. 3. An example of a collage made from 6 continuous pictures illustrated at Figure 2-12. The collage shows perikymata ridges and grooves on the middle buccal surface of a human maxillary premolar. The cusp is on the left side of the collage, while the CEJ on the right. (Photo© Michael S. Yuan)

Table 1. Perikymata counts of the maxillary third premolar grouped by sex.

Group	N.	Mean	S.D.	C.V.	Range	Min.	Max.
Male	45	153.84	23.58	15.33%	93	111	204
Female	47	146.85	25.02	17.04%	102	107	209
Total	92	150.27	24.45	16.27%	102	107	209

T-test: t=1.378 (p= 0.171).

Table 2. Perikymata counts of the maxillary third premolar grouped by side.

Group	N.	Mean	S.D.	C.V.	Range	Min.	Max.
Right	46	153.87	24.91	16.19%	102	107	209
Left	46	146.67	23.70	16.59%	95	109	204
Total	92	150.27	24.45	16.27%	102	107	209

T- test: t=1.419 (p= 0.159).

Table 3. Perikymata counts of the maxillary third premolar grouped by ethnicity.

Group	N.	Mean	S.D.	C.V.	Range	Min.	Max.
Chinese	44	151.05	23.84	15.78%	100	109	209
Japanese	48	149.56	25.22	16.86%	97	107	204
Total	92	150.27	24.45	16.27%	102	107	209

T test: t=0.289 (p= 0.773).

The distribution is based on a scale of 5 counts per interval for perikymata count distribution. In the primary subgroups, there were no statistical differences in maxillary third premolar perikymata counts between the sexes and between the sides at the significance level of 0.05 (Tables 1 & 2). The variation was relatively high, since the coefficient of variation ranged between 15 to 18%.

Though no statistical differences were found, the mean differences showed that the males had more average perikymata than the females by 7 counts, and the right side had more counts than left side by 6. This suggests that males may require longer time to complete crown formation of the maxillary third premolar. Likewise, the right side may require longer time to complete crown formation as compared to the left.

Comparing ethnicities, there was also no significant difference in perikymata counts of the maxillary third premolar (Table 3). The means of the two subgroups, which were 151 and 150 counts, are practically identical. This result shows that there were no differences between two non-contiguous geographic regions of Taiwan and Japan, where samples were collected.

Further statistical analysis was performed to test the differences among the eight tertiary subgroups. In the tertiary subgroups, the perikymata counts of the maxillary third premolars demonstrated means ranging from 140 to 156 counts, standard deviations ranging from 20 to 30 counts, and coefficients of variation ranging from 20 to 30 %. Once again, there were no significant differences found among any of the tertiary subgroups (Table 4).

The statistical results suggests that given unknown modern East Asian maxillary third premolar specimens, one would not be able to identify their sex, side, and ethnicity by perikymata counts of the enamel microstructures. However, one can distinguish the right and left sides of the maxillary and mandibular third premolars based on their distinctive gross morphology (Kraus et al., 1969; Jordan, 1992). Results comparing buccal crown height, lingual crown height, mesiodistal width, and buccolingual width for all measurement except as follows: 1) there were significant differences between males and females, and right and left sides with regard to lingual crown height, 2) there was a significant difference between males and females with regard to buccolingual width, 3) there was a significant difference between Chinese-male and Japanese-female subgroups with regard to buccolingual width.

Measurements of the perikymata counts and crown dimensions of the maxillary third premolar were tested for correlations (Table 5). Perikymata counts of the maxillary third premolars were significantly correlated with buccal crown height. Such a high correlation makes sense since the perikymata counts were numbered on the buccal crown surface. The Pearson correlation coefficient (r) was calculated 0.475, which resulted in a coefficient of determination (r^2) of 0.23. Perikymata counts were also significantly correlated with the lingual crown height with a correlation coefficient of 0.304 and a r^2 of 0.09. However, perikymata counts were not correlated with the mesiodistal and buccolingual widths in our dental samples.

The dimensional parameters, i.e. the buccal crown height, lingual crown height, mesiodistal width, and buccolingual width, were mutually correlated with one

Table 4. Perikymata counts of the maxillary third premolar grouped by sex, side, and ethnicity (C: Chinese; J: Japanese; M: male; F: female; R: right; L: left).

Group	N.	Mean	S.D.	C.V.	Range	Min.	Max.
C-M-R	11	156.18	19.84	12.70%	62	127	189
C-M-L	12	151.50	26.76	17.66%	91	111	202
C-F-R	10	154.00	25.24	16.39%	91	118	209
C-F-L	11	142.73	23.99	16.81%	77	109	186
J-M-R	11	155.55	25.95	16.68%	68	119	187
J-M-L	11	152.36	23.96	15.73%	87	117	204
J-F-R	14	150.64	29.46	19.56%	97	107	204
J-F-L	12	140.25	20.44	14.57%	57	112	169
Total	92	150.27	24.45	16.27%	102	107	209

ANOVA F test: F= 0.634 (p= 0.727).

Table 5. Pearson correlation analysis of perikymata counts and crown dimensions (N = 92).

Group / r value	Periky. C.	Log Periky.C.	Bu.Cr.Ht.	Li.Cr.Ht.	Me.Di.Wd.	Bu.Li.Wd.
Periky. C.	1.000	0.995*	0.475*	0.304*	0.082	0.027
Log Periky. C.	0.995*	1.000	0.482*	0.318*	0.063	0.024
Buc. Cr. Ht	0.475*	0.482*	1.000	0.631*	0.396*	0.386*
Ling. Cr. Ht	0.304*	0.318*	0.631*	1.000	0.349*	0.391*
Mesiod. Wd.	0.082	0.063	0.395*	0.349*	1.000	0.716*
Bu. Li. Wd.	0.027	0.024	0.386*	0.391*	0.716*	1.000

** Correlation is significant at the 0.01 level (2-tailed).*

Abbreviations: *r: Pearson's correlation coefficient.*

Periky. C.:	*perikymata counts*	*Log Periky.C.:*	*Log 10 base perikymata counts*
Bu. Cr. Ht.:	*buccal crown height.*	*Li. Cr. Ht.:*	*lingual crown height.*
Me. Di. Wd.:	*mesiodistal width.*	*Bu. Li. Wd.:*	*buccolingual width.*

another. The buccal crown height was significantly correlated with the lingual crown height with a correlation coefficient of 0.631 and a coefficient of determination of 0.369. The mesiodistal width was significantly correlated with the buccolingual width with a correlation coefficient of 0.716 and a coefficient of determination of 0.513.

Table 6 shows a correlation table of perikymata counts and its buccal crown dimension subcategorized by sex and ethnicity. All correlations are significant at the 0.01 level except the Chinese group which is significant at 0.05. The results demonstrate a higher correlation in males and in Japanese between perikymata counts and corresponding buccal crown height.

The means of perikymata counts, when rescaled in an increment of 1 mm, showed a consistent positive correlation with buccal crown height in both male and female samples (Table 7), and also in the sex-combined sample (Table 8). In Table 7, the mean counts clearly demonstrated an average of 15 increments corresponding to each mm height increase. A similar pattern is also shown in ethnicity, except that in the 10 mm buccal crown height group, Chinese showed a lower mean, while Japanese showed a much higher mean. This may be attributed to the low sample size.

Negative kurtosis value for most of the perikymata count observations corresponds to the flat, wide shape of the distribution. This large sample (n = 92) satisfies the power of statistical requirement. As noted, there is also an obvious right skewness along perikymata count and crown dimensions as all the results showed positive values.

As we apply correlation and curve estimation equation to explore the regression model between perikymata counts and crown dimensions, we should always bear in mind that such a model only applies to a limited range in reality. Data were transformed into log base 10 to provide a logarithmic approach in correlation with the parameters of crown dimensions. Although there does not exist much difference between the linear and log regression curve estimation between the perikymata counts and crown height, the log approach does fit better and serve as an accurate representation as the counts are one of the products of biological tissues.

Table 6. Pearson correlation analysis of perikymata counts and buccal crown height grouped by sex and ethnicity in a combined sample.

Category	sample size	correlation (R)	R square	p-value	Sig. level
Male	45	0.498	0.248	0.000	$p < 0.01$
Female	47	0.451	0.203	0.001	$p < 0.01$
Chinese	44	0.380	0.144	0.011	$p < 0.05$
Japanese	48	0.547	0.299	0.000	$p < 0.01$
Total	92	0.475	0.226	0.000	$p < 0.01$

Table 7. Mean of perikymata count based on the 1 mm interval of buccal crown height of the maxillary third premolar in East Asians.

Interval	N.	Mean	S.D.	C.V.	Range	Min.	Max.	Kurtosis	Skewness	Median
11 mm	1	180.0	-	-	0	180	180	-	-	-
10 mm	13	167.3	23.4	13.98%	69	140	209	-0.927	0.591	161.0
9 mm	50	152.9	23.9	15.63%	95	109	204	-0.177	0.331	149.5
8 mm	27	137.3	18.9	13.76%	73	107	180	-0.258	0.524	136.0
7 mm	1	119.0	-	-	0	119	119	-	-	-
Total	92	150.3	24.5	16.30%	102	107	209	-0.275	0.437	148.0

Table 8. Mean of male and female perikymata count based on the 1 mm interval of buccal crown height of the maxillary third premolar in East Asians.

Interval	Sex	N.	Mean	S.D.	C.V.	Range	Min.	Max.	Kurtosis	Skewness	Median
11 mm	M	1	180.0	-	-	0	180	180	-	-	-
	F	0	-	-	-	-	-	-	-	-	-
10 mm	M	8	167.0	22.0	13.17%	64	140	204	-0.631	0.470	164.0
	F	5	167.8	28.2	16.80%	67	142	209	-1.019	0.909	152.0
9 mm	M	24	155.7	23.2	14.90%	85	117	202	-0.409	0.554	151.0
	F	26	150.3	24.7	16.43%	95	109	204	0.041	0.233	149.0
8 mm	M	11	141.0	18.9	13.40%	69	111	180	0.790	0.565	140.0
	F	16	134.7	19.2	14.25%	63	107	170	-0.405	0.616	133.5
7 mm	M	1	119.0	-	-	0	119	119	-	-	-
	F	0	-	-	-	-	-	-	-	-	-
Subtotal	M	45	153.8	23.6	15.34%	93	111	204	-0.486	0.433	149.0
	F	47	146.9	25.0	17.02%	102	107	209	0.034	0.526	146.0

Discussion

The use of perikymata counts has been suggested for studies in estimating the age of death as well as the length of crown formation times. While this idea may be appealing, several problems stand in the way of an application of this methodology.

Age of death estimation is not feasible at this time for a number of reasons, some of which are inherent in the histological issues as described below. In addition, the degree of variability is high (maximal difference up to 1.75 years) in the radiographic and histological data on crown developmental times as shown in Table 9 modified from Reid et al. (1998).

The application of perikymata counts in estimating crown formation times requires several assumptions:

1. The full establishment of cross-striations as circadian rhythm;
2. The clear mathematical correlation of the counts of cross-striations to each interval between the line of Retzius, such as 5, 7, 8, 9, 10, 11, more, or irregular periodicities;
3. The correct correlation between lines of Retzius and perikymata.

How certain are we that the circadian rhythm of the cross-striations, the circaseptan rhythm or consistent periodicity of lines of Retzius, and correlations between

Table 9. Human crown initiation-(crown formation)-crown completion times (years).

		Moorrees et al.	Gustafson & Koch	Dean et al.	Reid et al.*
Tooth		(1963)	(1974)	(1993)	(1998)
Max.	I1	-	0.30-(4.20)-4.50	0.32-(3.15)-3.47	0.35-(4.08)-4.43
	I2	-	0.95-(4.15)-5.10	0.69-(3.72)-4.41	1.05-(3.61)-4.66
	C	(3.5)	0.40-(5.80)-6.20	0.38-(4.37)-4.75	0.75-(4.45)-5.20
	P3 b	(3.1-3.4)	1.75-(4.15)-5.90	1.67-(2.85)-4.52	1.85-(3.57)-5.42
	P4 b	(3.1-3.4)	2.15-(4.70)-6.85	2.41-(3.11)-5.52	2.65-(2.95)-5.60
	M1 mb	(2.1)	0.00-(3.10)-3.10	0.00-(2.41)-2.41	0.05-(2.83)-2.78
	M2 mb	(2.8)	2.85-(4.55)-7.40	2.92-(3.13)-6.05	2.80-(3.28)-6.08
	M3 mb	(2.8)	-	-	7.68-(3.27)-10.95
Mand.	I1	-	0.30-(3.90)-4.20	0.32-(3.10)-3.42	0.25-(3.52)-3.77
	I2	-	0.30-(4.20)-4.50	0.69-(3.72)-4.41	0.40-(4.20)-4.60
	C	(3.5)	0.35-(5.75)-6.10	0.38-(4.37)-4.75	0.55-(5.41)-5.96
	P3 b	(3.1-3.4)	1.75-(4.25)-6.00	1.67-(2.85)-4.52	1.85-(3.87)-5.72
	P4 b	(3.1-3.4)	2.25-(4.60)-6.85	2.68-(3.11)-5.79	2.65-(3.46)-6.11
	M1 mb	(2.1)	0.00-(3.00)-3.00	0.00-(2.67)-2.67	0.05-(3.39)-3.34
	M2 mb	(2.8)	2.85-(4.45)-7.30	-	2.90-(3.16)-6.06
	M3 mb	(2.8)	-	6.42-(3.16)-9.58	7.77-(3.09)-10.86

Adopted and modified from Reid et al., 1998, Journal of Human Evolution, p. 474,Table 5.
** Reid et al. (1998) was based on histological data; other results were assessed by radiograph.*
Abbreviations: (Max.): maxillary; (Mand.): mandibular; (b): buccal cusp; (mb): mesiobuccal cusp

lines of Retzius and perikymata are universally true across the extant and extinct primates?

1. Cross-striations

The overall evidence demonstrates that cross-striations characterize a 24-hourly or circadian rhythm (Mirura, 1939, Risnes, 1986, Bromage, 1991). The precise nature of the cross-striations is still unclear at present. The differences in the composition of hydroxy-apatite crystallites and the enamel matrix, as well as the degree of calcification have been proposed to explain the observable fact of cross-striations, but remain to be proven (Semmelink & Nygaard, 1982). While phases of the mineral secretion play the key roles in determining such periodicity, the consistency of the circadian rhythm may have some degree of variation (see Robison et al., 1997, for related issues in the mineral, water, protein, and enzyme distribution throughout the enamel maturation phases).

The difficulty in labeling enamel during its secretion and maturation phases has thus far made it difficult to establish beyond doubt that cross-striations represent a circadian rhythm in enamel formation. Nevertheless, evidence in other tissues such as dentin, bone, and cartilage make it quite reasonable to assume that they are indeed evidence of a circadian rhythm.

2. Relationship between cross-striations and lines of Retzius

Norman & Poole (1974) found in their transmission electron microscopic investigation that lines of Retzius appeared as gaps between rows of enamel crystals. They proposed that the lines of Retzius were a phenomenon of imperfect synchronizations of two or more circadian rhythms and suggested an eight-day periodicity. One example they cited was the study by Lewis & Lobban (1957) which explored the dissociation of diurnal rhythms in human subjects by restraining the subjects to live in abnormal time routines of 21, 24, & 27 hours per day. Surprisingly, they found that while the potassium secretion persisted in a 24-hour rhythm, the other variables adapted to the environmental alterations through time.

The numbers of cross-striations between the lines of Retzius have always been a confusing matter in debates. In the traditional examination of cross-striations, the methodology relies on ground sections of tooth enamel observed under light microscopes. One of the difficulties has been how to reduce the thickness to one or two enamel prisms, which would only be 10 to 15 μm, and still be able to observe the dark-light alternating bands under a light microscope.

Dean & Beynon (1991) recommended and performed their research with 100 μm to achieve a uniform section thickness. Bromage (1991) also applied dental sections of 80 to 100 µm in the enamel labeling study on macaques. Nevertheless, the 100 µm thickness does impose a reading error by overlapping too many layers (about 15) of enamel prisms, therefore possibly leading to inaccurate estimations in cross-striation counts.

FitzGerald (1998) employed 158 anterior teeth of different sex and ethnicity to explore the periodicity between the lines of Retzius. The final dental sample size was 96. He reported a mean of 9.7 and a SD of 1.0 count of cross-striations between lines of Retzius. His effort in trying to dispute the generally believed circaseptan rhythm between lines of Retzius may not be reliable due to his methodology in choosing 100 µm thickness.

Although the reduction of section thickness was managed, Bullion (1987) commented that a poor reading resolution of cross-striations would result, if the enamel sections went under 40 µm in thickness. Therefore, the methodological constraints in tissue preparation and section observation would have introduced some degree of imprecision and inaccuracy in the aforementioned studies.

Moreover, in most of the illustrations in the literature, the counting of cross-striations is on the superficial or outer part of enamel prism and seldom on the deeper or inner part of the enamel prism. This is due to the difficulty in reading the cross-striations in the deeper enamel in which the Hunter-Schreger bands are more obvious. This optical phenomenon is created by the decussating layers of enamel prisms. The exaggerated curvilinear prism path may be the reason why the difficulty in reading the cross-striations occurs.

Moss-Salentijn et al. (1997, p.17) concluded, " In our opinion, all that can be stated at present is that the lines of Retzius in imbricational enamel exhibit an apparent periodicity. The difficulty of establishing the length of the period and the question whether this period is equal among hominids do not permit definitive statements at this time."

3) Correlation between lines of Retzius and perikymata

Perikymata are generally assumed to be the surface or external manifestations of the lines of Retzius of teeth. While this relationship has been well established for coronal and middle thirds of enamel (Kölliker, 1854; Pickerill, 1912; Risnes, 1984), the relationship is not as clearly defined in the cervical enamel. This is particularly troublesome since the distances between the cervical perikymata are progressively smaller, thus involving a relatively large proportion of the count (see Chapter 6, pp. 96-97 for the discussion on the perikymata count variations at the cervical regions).

In addition, the equivalence between the lines of Retzius in imbricational enamel and the perikymata on the enamel surface may not be a universal mammalian pattern. Skobe et al. (1985) have demonstrated that in carnivore enamel the lines of Retzius do not extend to perikymata grooves on the enamel surface.

Several methodological and theoretical assumptions based on the human perikymata counts of the permanent incisors and the radiographic observations on permanent dentition have long been employed as evidence in the application of perikymata counts in anthropological studies to estimate the crown formation time and age at death. We would like to point out again that such attempts have created numerous controversies and misleading results.

First, the currently widely applied assumptions were derived from small sample sizes that only used incisor teeth. The examples of modern perikymata count in human incisors, as was described in Chapter 1, includes two studies on modern humans by Bromage & Dean (1985, mean = 188, range = 165-202, n = 10, mandibular incisors) and Bacon (1987, mean = 145, range = 111-179, n = 23, specimen: incisors). There is one report by Mann et al. (1990b, 1991), using the archeological collection of 3000 B.C. from Hasanlu, Iran and A.D. 800 in Island Field, Delaware, U.S.A. This investigation disputed the aforementioned work (mean = 116, median = 118, SD = 25, range = 75-157, n = 12). In addition to the small sample sizes, none of these studies presented convincing results regarding sex, and ethnicity, nor were the other dental locations, such as canine, premolars, or molars, examined.

Second, Bromage & Dean (1985) verified their results as evidence for the perikymata count applicability in estimating the crown maturation times by the matched overlapping of the perikymata-derived crown maturation times with the radiographically documented crown maturation times. We emphasize, as did Mann et al. (1991), that while exercising these applications we must recognize the fact that since crown maturation times were derived, such an approach has to be considered as hypothetical and requires further proof in testing its accuracy.

Third, while perikymata counts may correspond to the incremental lines of the superficial imbricational enamel, the incremental lines in the appositional enamel can only be estimated, assuming that formation times can be derived from the counts of lines of Retzius or cross-striations. This indirect approach will expand the range of variation or reduce the correlations of the related dental microstructures, thus introducing more errors in estimation.

Fourth, as we examined the crown formation data, we noted that the growth and developmental timings for each tooth differed from one another. Therefore, as was done in this study, the subcomponents of the incisors, canines, premolars, and molars should be treated individually to obtain their respective perikymata counts, instead of pooling them together to form a much larger group for comparisons.

We should also not consider the upper and lower dentitions as essentially mirror images. In much of the literature, researchers tend to pool samples, which may lead to problems in increasing the biological variations and misrepresentation by a larger sample size. While

Table 10. Comparison of perikymata counts in this study vs. the lines of Retzius (LR) counts of Bullion (1987) of the maxillary third premolar.

Group	N.	Mean	S.D.	C.V.	Range	Min.	Max.
Bullion, 1987							
Sleeve LR	4	106.17	8.77	8.26%	20.67	94	114.67
Yuan, 2000							
Perikymata	92	150.27	24.45	16.27%	102	107	209

Significant p= 0.0007 (p < 0.01)

Mann et al. (1990b, 1991) cautioned researchers regarding the misleading conclusion of ape versus human affinity of the australopithecines by applying the perikymata counts to estimate the age of death, it was unfortunate that they grouped all the incisors together and failed to provide a true representation in the variations of each individual incisor. All incisors, whether maxillary or mandibular, central or lateral, have different crown maturation times.

This project, which explores human third premolars in order to establish a data set of perikymata counts, has been the only attempt so far in expanding our knowledge of human biological variations in perikymata counts beyond permanent incisors in the past decade. The sample size of 92 specimens in a single tooth location in permanent dentition, the maxillary third premolar, is also the largest observation ever made. This database will no doubt provide a reliable source for comparisons and further implications.

In this study, not only the sample size was expanded from the above mentioned 12 to 92 specimens, but also perikymata counts were established in a new dental location, maxillary third premolars. In addition, the variations in sex and ethnicity were fully explored. Moreover, the correlation among the perikymata counts and their corresponding crown dimensions were investigated. The results shown in this study not only help us understand the complexity of this phenomenon, but also dispute the present application of perikymata counts until further evidence of the correlation between the enamel microstructures is confirmed.

The perikymata counts of the maxillary third premolar in mixed sex and ethnicity of this East Asian population gives a mean count of 150, a standard deviation of 25, a coefficient of variation of 16%, a median of 148, a mode of 140, a range of 102 with minimum of 107 and maximum of 209.

The results demonstrated that there was neither sex differences between males and females, nor ethnic differences between Chinese and Japanese in perikymata counts. In addition, the results illustrated sexual dimorphism in that males had higher perikymata counts and larger crown dimensions than those of females.

The commonly cited reference in the study of cross-striations and lines of Retzius comes from Bullion (1987). Bullion counted the lines of Retzius of the appositional enamel (referred to as the dome-shaped lines of Retzius) and the imbricational enamel (referred to as the sleeve-shaped lines of Retzius, and is considered as the internal manifestation of perikymata) in a total sample of 48 unworn modern human teeth of 100 μm thickness ground sections (see Bullion, 1987, Chapter 7 for detailed descriptions and the summary of findings in Table 7.1, 7.2, 7.3, & 7.4). The sample mostly came from the Royal Lancaster Infirmary and local dental offices, UK, and partially contributed by Dean.

In Bullion's study, the maxillary third premolar dome-shaped lines of Retzius had average counts in a sample of 4 teeth of 39, 42, 36.33, and 43.33 respectively. The statistical results are n = 4, mean = 40.15, SD = 9.83, CV = 24.47%, range = 36.33-43.33. These reflect high variation.

The maxillary third premolar sleeve-shaped lines of Retzius had average counts in a sample of 4 teeth of 94, 106.67, 114.67, and 109.33 respectively. The statistical results are n = 4, mean = 106.17, SD = 8.77, CV = 8.26%, range = 94-114.67. These represent a very different pattern.

Bullion's results on dental examinations of the maxillary third premolar are reported 106 $\pm$ 9 (n = 4) for the counts of sleeve-shaped lines of Retzius. The result stands in contrast to what we have obtained in the result of perikymata counts in our sample. The 106 $\pm$ 9 counts of sleeve-shaped lines of Retzius demonstrate a major discrepancy to the corresponding counterparts of the 150 $\pm$ 25 (n = 92) perikymata count in this study. We find a significant difference between the two findings ($p < 0.01$) (see Table 10 for summary).

The discrepancy may be accounted for as a result of:

1. Observation errors. As Bullion (1987, p. 141) stated, "...under high magnification, they (lines of Retzius) had the appearance of 'fuzzy brown bands'... they often appeared in a haphazard pattern with none of the regularity observed under low magnification..." This phenomenon should have existed in all of the observations. The resolution needed to improve the detection or identification of these microstructures requires SEM investigation.

2. Small sample size. Bullion only investigated four maxillary third premolars as compared to the 92 specimens examined in this study. The small sam-

ple size may possibly lead to a biased result.

3. Pattern miscorrelation and misrepresentation. If there were no observation errors in Bullion's study, then the correlation between lines of Retzius and their external counterparts, perikymata, would have exhibited an unknown and more complicated pattern than what was previously concluded. As such, this opens a major theoretical gap for the implications of perikymata counts.

The notion of crown formation times and age-at-death estimation based on the perikymata counts should be re-examined.

As discussed above, significant issues remain in the interpretation of the three microstructures in enamel: cross-striations, line of Retzius, and perikymata. As we have seen in the methodology, (1) the estimations and countability of cross-striation counts between lines of Retzius, (2) the constraints in the thickness of section preparation and in the tissue observation, (3) the indefinite role of the curvilinear and decussating structure of the Hunter-Schreger band related enamel prisms, (4) the inconsistency in the deeper enamel layer for the lines of Retzius and cross-striations, these problems render the application of the perikymata counts as truly questionable.

It is obvious that too many unknown factors have contributed and played major and/or minor roles in determining the final outcome.

This study has demonstrated the value of a large data set. Such baseline information will serve not only the relevant anthropological field, but also several areas in biology, with a solid foundation to further examine the kinematics of ameloblasts and their principal product, enamel, during human development and the evolutionary history of hominid lineages.

Investigations of perikymata counts in great apes and monkeys are essential to hominid and hominoid dental anatomy comparisons, fossil record interpretations, and life history reconstructions. Without these data, we will not be able to accurately use the perikymata counts to estimate age in fossil primate species. The next step is to complete the East Asian mandibular third premolar perikymata count study in Chinese and Japanese, since the sample collection is available. The expansion of such a dataset should further include: 1) the comparisons of other teeth in the permanent dentition, and 2) the comparisons of other ethnic groups from different regions, e.g., Africa and Europe.

The controversies about the correlations between perikymata counts, lines of the Retzius, cross-striations, and Hunter-Schreger bands also require immediate attention. The perikymata counts will serve as a fundamental data set as more efforts are made to further explore these correlations.

Conclusion

Perikymata, the incremental lines on the dental crown surface, are of great interest to dental histologists and anthropologists. The use of perikymata counts in estimating the age at death and crown formation times has been a controversial issue in providing evidence for how we interpret the place of australopithecines in hominid evolution.

This study, which collected the largest sample size thus far, examined the perikymata counts at one of the posterior dental locations, the maxillary third premolar, rather than the previously studied incisors in the anterior dentition. A total sample size of 92 maxillary third premolars (P^3), including right and left sides from males (n = 45) and females (n = 47), were investigated for their perikymata counts and crown dimensions in two modern human populations, Taiwanese and Japanese.

The results dispute the utilization of perikymata counts in the estimation of age-at-death and crown formation times. A valuable database is reported here for further comparisons in resolving the controversy surrounding the use of human perikymata counts, based on a large sample of modern *Homo sapiens* teeth.

We conclude that:

1) A review of the relevant literatures on histology, periodicity, research methodology, and the correlations among the enamel microstructures, including cross-striations, lines of Retzius, Hunter Schreger bands, and perikymata, has provided sufficient scientific evidence to necessitate rethinking the assumptions and methodologies underlying many of the fossil interpretations. Clearly, we do not have enough accurate data on the times needed to form crowns. Additionally, there is variability in the number of cross-striations between lines of Retzius. These are sufficient reasons to warrant a moratorium on the use of perikymata for hominid evolution studies.

2) While there is sexual dimorphism in perikymata counts, it is not statistically significant within the two populations. The counts in males are higher than those of females in both Taiwanese and Japanese.

3) There are no differences in perikymata counts between the right and left sides.

4) There is no difference for either males or females between Taiwan and Japan in their perikymata counts.

5) Perikymata counts are significantly correlated to their corresponding buccal crown heights.

6) When data from the two populations were pooled as East Asians (n= 92), the mean perikymata count for maxillary third premolars is 150 with a standard deviation of 25. The perikymata counts in this study are very different from the commonly believed counterpart line counts of Retzius, investigated by Bullion (1987). This significant discrepancy verifies the cautions raised by Mann et al. (1990b, 1991).

7) Although the data on perikymata counts follow a normal distribution, the variation is high (coefficient of variation = 16 %). Such high variation provides

part of the evidence for disputing the perikymata count assumptions.

8) The presumption of applying perikymata counts in estimating age at death and crown formation times should not be encouraged, as this will only compound the unknown issues in dental biology and create misleading results in human evolution.

9) We demonstrated the strength of SEM in this study, just as in other studies in dental anatomy, such as Boyde (1990) and Risnes (1998, 1999). Scanning electron microscopy proves to be one of the most powerful tools in future research of enamel microstructures.

References

Adams, M. & Reeve, J. (1987). Excavations at Christ Church, Spitalfields 1984-86. *Antiquity*, 61: 247-56.

Aiello, L. & Dean, C. (1990). *An Introduction to Human Evolutionary Anatomy*. Chapter 7: The Microanatomy and Development of Teeth. San Diego: Academic Press, pp. 106-132.

Aiello, L. & Dean, C. (1990). *An Introduction to Human Evolutionary Anatomy*. Chapter 8: Hominoid Tooth Morphology. San Diego: Academic Press, pp. 133-159.

Anemone, R.L.; Mooney, M.R.; & Seigel, M.I. (1996). Longitudinal study of dental development in Chimpanzees of known chronological age; Implications for understanding the age of death of Plio-Pleitocene hominids. *American Journal of Physical Anthropology*, 99: 119-133.

Asper von, H. (1916). Uber die Braune Retziusche Parallelstrifung im Schmelz der memschlichen Zahne. *Schweiz Vjschr. Zahnheik*, 26: 275-314.

Avery, J.K. (1994). *Oral Development and Histology* (ed.). 2nd edition. New York: Thieme Medical Publishers, Inc.

Bacon, A.M. (1987). Estimation de l'âge des individus â partir du comptage des perikymaties. In *D.E.A. de Paleontologie*, Universite de Paris VI.

Bevenius, J. & Hultenby, J. (1991). *In vitro* and *in vivo* replication for scanning electron microscopy of the cervical region of human teeth. *Scanning Microscopy*, 5 (3): 731-740.

Beynon, A. (1987). Replication technique for studying microstructure in fossil enamel. *Scanning Microscopy*, 1: 663-669.

Beynon, A.D. & Dean, D.C. (1988). Distinct dental development patterns in early fossil hominids. *Nature*, 335: 509-514.

Beynon, A.D. & Reid, D. J. (1987). Relationships between perikymata counts and crown formation times in the human permanent dentition. *Journal of Dental Research*, 66: 889-890.

Beynon, A.D. (1992). Circaseptan rhythms in enamel development in modern and Plio-Pleistocene hominids. In Smith, P and Tchernoc, E. (ed.) "*Structure, Function, and Evolution of Teeth*". London and Tel Aviv: Freund Publishing House Ltd. pp. 295-310.

Beynon, A.D.; Dean, D.C.; & Reid, D.J. (1991a). Histological study on the chronology of the developing dentition in gorilla and orangutan. *American Journal of Physical Anthropology*, 86: 189-203.

Beynon, A.D.; Dean, D.C.; & Reid, D.J. (1991b). On thick and thin enamel in hominids. *American Journal Physical Anthropology*, 86: 295-309.

Boyde, A. (1963). *Estimation of age at death of young human skeletal remains from incremental lines in the dental enamel*. Third International Meeting in Forensic Immunology, Medicine, Pathology and Toxicology. London

Boyde, A. (1971). Chapter 7: Comparative Histology of Mammalian Teeth. In Dahlberg, A.A., *Dental Morphology and Evolution*. Chicago & London: The University of Chicago Press. pp. 81-94.

Boyde, A. (1975). Scanning electron microscopy of enamel surfaces. *British Medical Bulletin*, 31: 120-124.

Boyde, A. (1976). Amelogenesis and the development of teeth. In "*Scientific Foundations of Dentistry*". edited by Cohen, B. & Kramer, I.R.H. London: Heinemann.

Boyde, A. (1990). Developmental interpretations of dental microstructure. In "*Primate Life History and Evolution*", edited by DeRousseau, C.J. New York: Wiley-Liss. pp. 229-267.

Boyde, A. (1997). Microstructure of enamel. In Chadwick, D.J. & Cardew, G.B. (ed.) "*Dental Enamel*". New York: John Wiley & Sons. pp. 18-31.

Bozzola, J.J. & Russel, L.D. (1998). *Electron Microscopy: Principles and Techniques for Biologists*. 2nd edition. Sudbury, MA: Jones and Bartlett Publishers.

Bromage, T.G. & Dean, M.C. (1985). Re-evaluation of the age at death of immature fossil hominids. *Nature*, 317: 525-28.

Bromage, T.G. (1987). The scanning electron microscopy/replica technique and recent applications to the study of fossil bone. *Scanning Microscopy*, 1: 607-613

Bromage, T.G. (1991). Enamel incremental periodicity in the pig-tailed macaque: a polychrome fluorescent labeling study of dental hard study. *American Journal of Physical Anthropology*, 86: 205-214.

Bullion, S.K. (1987). *Incremental Structures of Enamel and Their Applications to Archeology*. University of Lancaster, Ph.D. dissertation.

Carlson, B.M. (1998). *Human Embryology and Developmental Biology*. 2nd edition. St. Louis: Mosby. pp. 298-303

Chadwick, D.J. & Cardew, G.B. (1997). *Dental Enamel* (ed.). Ciba Foundation Symposium 205. New York: John Wiley & Sons.

Chandrasekera, M.S.; Reid, D.J.; & Beynon, A.D. (1993). Dental chronology in chimpanzee (*Pan troglodytes*). *Journal of Dental Research*, 72: 729 (Abstract)

Colton, T. (1974). *Statistics in Medicine*. Boston: Little & Brown Publishers. pp. 87-92.

Dahlberg, A.A. (1971). *Dental Morphology and Evolution* (ed.) Chicago & London: The University of Chicago Press.

Dawes, C.J. (1994). *Introduction to Biological Electron Microscopy: Theory and Techniques*. Revised edition. Burlington, Vermont: Ladd Research Industries, Inc., Publisher.

Dawson-Saunders, B. & Trapp, R.G. (1990). *Basic and Clinical Biostatistics*. Norwalk, Connecticut: Appleton & Lange. pp. 95-96.

Dean, M.C. & Beynon, A.D. (1991). Histological reconstruction of crown formation times and initial root formation times in a modern human child. *American Journal of Physical Anthropology*, 86: 215-228.

Dean, M.C. (1987). The dental development status of six East African juvenile fossil hominids. *Journal of Human Evolution*, 16: 197-214.

Dean, M.C.; Beynon, A.D.; Reid, D.J.; & Whittaker, D.K. (1993a). A longitudinal study of tooth growth in a single individual based on long and short-period incremental markings in dentine and enamel. *International Journal of Osteoarchaeology*, 3: 249-264.

Dean, M.C.; Beynon, A.D.; Thackeray, J.F.; & Macho, G.A. (1993b). Histological reconstruction of dental development and age at death of a juvenile *Paranthropus robustus* specimen, SK 63, from Swartkrans, South Africa. *American Journal of Physical Anthropology*, 91: 401-419.

Dean, M.C.; Stringer, C.B.; and Bromage, T.G. (1986). A new age at death for the Neanderthal child from Devil's Tower, Gibraltar and the implications for studies of general growth and development in Neanderthals. *American Journal of Physical Anthropology*, 70: 301-309.

Eisenberg, M.J. (1938). A microscopic study of the surface enamel of human teeth. *Anatomical Record,* 71 (2): 221-231.

Eisenmann, D.R. (1998). Chapter 10: Amelogenesis. In Ten Cate, A.R. *Oral Histology: Development, Structure, and Function* (ed.). 5th edition. St. Louis, Baltimore, Toronto: Mosby. pp. 197-217.

Eisenmann, D.R. (1998). Chapter 11: Enamel Structure. In Ten Cate, A.R. *Oral Histology: Development, Structure, and Function* (ed.). 5th edition. St. Louis, Baltimore, Toronto: Mosby. pp. 218-235.

ElNesr, N.M. & Avery, J.K. (1994). Tooth eruption and shedding. In Avery, J.K. (ed.) *Oral Development and Histology*. 2nd edition. New York: Thieme Medical Publishers, Inc. pp. 110-129.

FitzGerald, C.M. (1998). Do enamel microstructures have regular time dependency? Conclusions from the literature and a large-scale study. *Journal of Human Evolution*, 35: 371-386.

Gustafson, G. & Koch, G. (1974). Age estimation up to 16 years of age based on dental development. *Odontologisk Revy*, 25: 297-306.

Hess, A.F.; Lewis, J.F.; & Roman, B. (1932). A radiographic study of calcification of the teeth from birth to adolescence. *Dental Cosmos*, 74: 1053-61

Hillson, S. (1986). *Teeth.* Cambridge: Cambridge University Press.

Hillson, S. (1992). Counting perikymata using a low-power stereomicroscope. *Dental Anthropology Newsletter*, January, p. 12.

Hillson, S. (1996). *Dental Anthropology*. Cambridge: Cambridge University Press.

Hillson, S. (1996). *Dental Anthropology*. Chapter 6: Dental Enamel. Cambridge: Cambridge University Press. pp. 148-181.

Hunter, J. (1778). *The Natural History of the Human Teeth: Explaining their Structure, Use, Formation, Growth and Diseases*. 2nd edition. pp. 1-128. London: J.Johnson.

Jordan, R.E. (1992). *Kraus' Dental Anatomy and Occlusion.* 2nd ed. St. Louis: Mosby Year Book.

Koenigswald, W.v. & Pfretzschner, H.U. (1987). Hunter-Schreger-Bänder im Zahnschmelz von Säugetieren: Anordnung und Prismenverlauf. *Zoomorphology*, 106: 329-338.

Koeningswald, W. v. & Sander, P.M. (1997). *Tooth Enamel Microstructure*. Rotterdam, Brookfield: A.A. Balkema.

Kölliker, A. (1854). *Mikroskopishe Anatomie*. II. Band, 2. Hälfte. Leipzig: Welhelm Engelmann. p. 76.

Kraus, B.S. & Jordan, R.E. (1965). *The Human Dentition before Birth*. Philadelphia: Lea & Febiger.

Kraus, B.S. (1959a). Differential calcification rates in the human primary dentition. *Archives of Oral Biology*, 1: 133-44.

Kraus, B.S. (1959b). Calcification of the human deciduous teeth. *Journal of American Dental Association*, 59: 1128-36.

Kraus, B.S.; Jordan, R.E.; & Abrams, L. (1969). Section 4: The premolars. *Dental Anatomy and Occlusion: A Study of the Masticatory System*. Baltimore: The Williams and Wilkins Company. pp. 48- 74.

Kuykendall K.L. (1996). Dental development in chimpanzees (*Pan troglodytes*): The timing of tooth calcification stages. *American Journal of Physical Anthropology*. 99(1): 135-57.

Line, S.R.P. (2000). Incremental markings of enamel in ectothermal vertebrates. *Archives of Oral Biology*, 45: 363-368.

Lunt, R.C. & Law, D.B. (1974). A review of the chronology of calcification of deciduous teeth. *Journal of the American Dental Association*, 89: 599-606.

Macho, G.A. & Wood, B.A. (1995). The role of time and timing in hominid dental evolution. *Evolutionary Anthropology*, 4 (1): 17-31.

Mann, A.; Lampl, M; & Monge, J (1990a). Patterns of ontogeny in human evolution: evidence from dental development. *Yearbook of Physical Anthropology*, 33: 111-150.

Mann, A.; Monge, J.; and Lampl, M. (1990b). Dental caution. *Nature*, 348: 202.

Mann, A.E.; Monge J.M.; and Lampl, M. (1991). Investigations into the relationship between perikymata counts and crown formation time. *American Journal of Physical Anthropology*, 86: 175-188.

Martin, T. (1997). Incisor enamel microstructure and systematics in rodents. In W.v. Koenigswald & P.M. Martin (ed.) "*Teeth Enamel Microstructure*". Rotterdan: Balkema. pp. 163-175.

Massler, M. & Schour, I. (1946). The appositional life span of the enamel and dentine-formation cells. I. Human deciduous teeth and first permanent molars. *Journal of Dental Research*, 25: 45-150.

McKee, M.D.; Nanci, A; Smith, C.E.; & Warshawsky, H. (1989). Cyclical aspects of enamel maturation and the role of ruffle-ended and smooth-ended maturation ameloblasts. In Fearnhead, R.N. (ed.) "*Tooth Enamel V*". Proceedings of the Fifth International Symposium on the Composition, Properties, and Fundamental Structure of Tooth Enamel and Tissues. Yokahama, Japan: Florence Publishers. pp. 41-50.

Moorrees, C.F.A.; Fanning, E.A.; & Hunt E.E. (1963). Age variation of formation of stages for ten permanent teeth. *Journal of Dental Research*, 42: 1490-1502.

Moss-Salentijn, L. & Kendricks-Klyvert, M. (1985). *Dental and Oral Tissues: Introduction*. 2nd edition. Philadelphia: Lea & Febiger.

Moss-Salentijn, L. & Kendricks-Klyvert, M. (1990). *Dental and Oral Tissues: Introduction*. 3rd edition. Philadelphia: Lea & Febiger.

Moss-Salentijn, L.; Moss, M.L.; & Yuan, M.S. (1997). The ontogeny of the mammalian enamel. In Koeningswald, W. v. and Sander, P.M.: *Tooth Enamel Microstructure*. Chapter 1. Rotterdam, Brookfield: A.A. Balkema. pp. 5-30.

Moyers, R.E. (1988). *Handbook of Orthodontics*. 4th edition. Chicago: Year Book Medical Publishers.

Moyers, R.E.; van der Linden, P.G.M.; & Riolo, M.L. (1976). *Standards of Human Occlusal Development,* Monograph 5, Craniofacial Growth Series. Ann Arbor, Michigan, Center for Human Growth and Development, University of Michigan.

Newman, H.N. & Poole, D.F.G. (1974). Observations with scanning and transmission electron microscopy on the structure of human surface enamel. *Archives of Oral Biology*, 19: 1135-1143.

Nolla, C.M. (1960). The development of the permanent teeth. *Journal of Dentistry for Children*, 27: 254-266.

Ooë, T. & Ogura-Futaki, Y. (1985). Differential termination times of enamel formation on human deciduous anterior-tooth surfaces revealed by tetracycline labeling. *Archives of Oral Biology*, 30 (5): 409-414.

Osborn, J.W. (1965). The nature of the Hunter-Schreger bands in enamel. *Archives of Oral Biology*, 10: 929-933.

Osborn, J.W. (1971). A relationship between the brown striae of Retzius and prism directions in the transverse plane of the human tooth. *Archives of Oral Biology*, 16: 1061-1070.

Osborn, J.W. (1973). Variations in structure and development of enamel. *Oral Sciences Reviews*, 3: 3-83.

Osborn, J.W. (1990). A 3-dimensional model to describe the relation between prism directions, parazones and diazone, and the Hunter-Schreger bands in human tooth enamel. *Archives of Oral Biology*, 35: 869-878.

Pantke, M. (1985). Rasterelektronenmikroskopische Untersuchngen über Retziusstreifen an Milchzähnen. *Zahnärztl. Prax*, 36: 275.

Pedersen, P.O. & Scott, D.B. (1951). Replica studies of he surface of teeth from Alaskan Eskimo, West Greenland natives, and American whites. *Acta Odont. Scan.*, 9: 262

Pfretzschner, H.U. (1994). Biomechanik der Schmelzmikrostruktur in den Backenzähnen von Großsäugern. *Palaeontographica* A, 234: 1-88.

Pickerill, H.P. (1912). *The Prevention of Dental Caries and Oral Sepsis*. London: Baillière, Tindall, and Cox. pp. 52, 69.

Pickerill, H.P. (1913). The structure of the enamel. *Dental Cosmos*, 55: 959-88.

Piesco, N.P. & Avery, J.K. (1994). Chapter 5: Development of teeth: Crown formation. In Avery, J.K. *Oral Development and Histology* (ed.). 2nd edition. New York: Thieme Medical Publishers, Inc. pp. 70-93.

Postek, M.T.; Howard, K.S.; Johnson, A.H.; & McMichael, K.L. (1980). *Scanning Electron Microscopy: A Student's Handbook*. Burlington, Vermont: Ladd Research Industries, Inc., Publisher.

Preiswerk, G. (1895). *Beiträge zur Kenntniss der Schmelzstruktur bei Säugetieren mit besonderer Berücksichtigung der Ungulaten*. University of Basel, Phil. Diss. Basel: C.F. Lendorff. pp. 13-4, 95-6.

Reid, D.J.; Beynon, A.D.; & Ramirez Rozzi, F.V. (1998). Histological reconstruction of dental development in four individuals from a medieval site in Picardie, France. *Journal of Human Evolution*, 35: 463-477.

Retzius, A. (1837). Bemerkungen über den innern Bau der Zähne, mit besonderer Rücksicht auf den im Zahnknochen vorkommenden Röhrenbau. *Arch. Anat. Physiol.*, 486-566.

Risnes, S. (1984). Rationale for consistency in the use of enamel surface terms: perikymata and imbrications. *Scandinavian Journal of Dental Research,* 92: 1-5.

Risnes, S. (1985). Circumferential continuity of perikymata in human dental enamel investigated by scanning electron microscopy. *Scandinavian Journal of Dental Research.* 93: 185-191.

Risnes, S. (1986). Enamel apposition rate and the prism periodicity in human teeth. *Scandinavian Journal of Dental Research*, 94: 394-404.

Risnes, S. (1998). Growth tracks in dental enamel. *Journal of Human Evolution*, 35: 331- 350.

Risnes, S (1999). Mature enamel morphology and forming enamel dynamics: Crystal orientation, crystal continuity, prism diameter, Retzius lines, and prism cross-striations. In Mayhall, J.T. & Heikkinen T (ed.) "*Dental Morphology 1998*". Proceedings of the 11th International Symposium on Dental Morphology, Olulu, Finland, August, 1998. Olulu: Oulu University Press. pp. 312-322.

Robison, C.; Brookes, S.J.; Bonass, W.A.; Shore, R.C.; & Kirkham, J. (1997). Enamel maturation. In Chadwick, D.J. & Cardew, G.B. (ed.) "*Dental Enamel*". New York: John Wiley & Sons. pp. 156-174.

Rosenberg, G.D. & Simmons, D.J. (1980a). Rhythmic dentinogenesis in the rabbit incisor: circadian, ultradian, and infradian periods. *Calcified Tissue International*, 32: 29-44.

Rosenberg, G.D. & Simmons, D.J. (1980b). Rhythmic dentinogenesis in the rabbit incisor: Allometric aspects. *Calcified Tissue International*, 32: 45-53.

Schour, I. & Hoffman, M.M. (1939a). Studies on tooth development. I. The 16μ calcification rhythm in the enamel and dentine from fish to man. *Journal of Dental Research*, 18: 91-102.

Schour, I. & Hoffman, M.M. (1939b). Studies on tooth development. II. The rate of apposition in the enamel and dentine in man and other animals. *Journal of Dental Research*, 18: 161-175.

Schour, I. & Poncher, H.G. (1937). Rate of apposition of enamel and dentine, measured by the effect of acute fluorosis. *Am. J. Dis. Child.*, 54: 757-776.

Schour, I. (1936). The neo-natal line in the enamel and dentine of the human deciduous teeth and first permanent molar. *Journal of American Dental Association*, 23: 1946-55.

Schroeder, H.E. (1991). *Oral Structural Biology: Embryology, Structure, and Function of Normal Hard and Soft Tissues of the Oral Cavity and Temporomandibular Joints.* New York: Thieme Medical Publishers, Inc.

Scott G.R. & Truner II, C.G. (1997). *The Anthropology of Modern Human Teeth: Dental Morphology and its Variation in Recent Human Populations*. New York: Cambridge University Press.

Scott, D.B.; Kaplan, H.; & Wyckoff, R.W.G. (1949). Replica studies of changes in tooth surfaces with age. *Journal of Dental Research*, 28: 31-47.

Scott, J.H. & Symons, N.B.B. (1974). *Introduction to Dental Anatomy*. 7th edition. Edinburgh: Churchill Livingstone.

Scott, J.H. & Symons, N.B.B. (1982). *Introduction to Dental Anatomy*. 9th edition. Edinburgh: Churchill Livingstone.

Shellis, R.P. (1984). Variations in growth of the enamel crown in human teeth and possible relationship between growth and enamel structure. *Archives of Oral Biology*, 29: 607-705.

Simmelink, J.W. & Nygaard, V.K. (1982). Ultrastructure of striations in carious human enamel. *Caries Research*, 16: 179-188.

Skobe, Z.; Prostak, K.; & Stern, D. (1981). Ultrastructure of secretory ameloblasts in a monkey *Macaca mulata*. *Archives of Oral Biology,* 26: 1075-90.

Skobe, Z.; Prostak, K.S.; & Trombly, P.L. (1985). Scanning electron microscope study of cat and dog enamel structure. *Journal of Morphology*, 184: 195-203.

Spears, I.R. & Macho, G.A. (1998). Biomechanical behavior of modern human molars: Implications for interpreting the fossil record. *American Journal of Physical Anthropology*, 106: 467-482.

Spear, I.R.; Noort, R.; Crompton, R.H.; Cardew, G.E.; & Howard, I.C. (1993). The effects of enamel anisotrophy on the distribution of stress in a tooth. *Journal of Dental Research*, 72: 1526-3.

Stringer, C.B.; Dean, M.C.; & Martin, R.D. (1990). A comparative study of cranial and dental development within a recent British sample and among Neanderthals. In "*Primate Life History and Evolution*", edited by de Rousseau, J. New York: Wiley-Liss. pp. 115-152.

SYSTAT software operation manual. *Systat 9: Statistics I*. Chicago: SPSS. p. I-212.

Teaford, M.F. & Oyen, O.J. (1989). Live primates and dental replication: new problems and new techniques. *American Journal of Physical Anthropology*, 80: 73-81.

Ten Cate, A.R. (1989). *Oral Histology: Development, Structure, and Function* (ed.). 3rd edition. St. Louis, Baltimore, Toronto: The C.V. Mosby Company.

Ten Cate, A.R. (1998). *Oral Histology: Development, Structure, and Function* (ed.). 5th edition. St. Louis: Mosby-Year Book, Inc.

Turner, E.P. (1963). Crown development in human deciduous molar teeth. *Archives of Oral Biology*, 8: 523-540.

Wakita, M. & Kobayashi, S. (1983). The three-dimensional structure of Tomes' processes and the development of the microstuctural organization of tooth enamel. In Suga, S.: *Mechanisms of Tooth Enamel Formation*. New York: Quintessenz. p. 165.

Weber, D.F. & Glick, P.L. (1975). Correlative microscopy of enamel prism orientation. *American Journal of Anatomy*, 144: 407-420.

Whittaker, D.K. (1981). Variations in the structure of the surface layer of human dental enamel. *Medical Science*, 9: 247.

Whittaker, D.K. (1982). Structural variations in the surface zone of human tooth enamel observed by scanning electron microscopy. *Archives of Oral Biology,* 27: 383-392.

Wilson, D.F. & Shroff, F.R. (1970). The nature of the striae of Retzius as seen with the optical microscope. *Australian Dental Journal*, 15: 3-24.

Yoshioka, M (1970). Difference in time of crown completion among dental surfaces. *Okajimas Folia Anat. Jap.,* 47: 213-228.

CHAPTER 17

Mosaic Cognitive Evolution: The Case of Imitation Learning

Francys Subiaul

Introduction

Among Ralph Holloway's many contributions to anthropology is thc notion of mosaic brain cvolution; Specifically, the notion that the human brain is more than just a larger primate brain or an expanded rodent brain. Support for this view comes from research on hominid endocasts, the only direct evidence of human brain evolution available, showing that hominid brains underwent a number of organizational changes including prominent changes to visual striate cortex and the parietal lobe (Holloway, 1996) as well as the temporal lobe (Rilling & Seligman, 2002). These changes—produced by evolutionary forces—led to changes in overall size as well as regional changes in volume, hemispheric asymmetry, the distribution of fiber track connections within and between hemispheres, and species-specific variation in neuro-receptor distributions (Holloway, Broadfield, Yuan, 2004).

In 1967 Holloway hypothesized that this pattern of mosaic brain evolution resulted from selection for "complexity management." By complexity management Holloway (1967) referred to a subset of continuous primate behavioral traits "related to the efficiency and fineness of discrimination, and adaptive problem-solving ability, which includes factors such as memory storage [encoding], recall, attention-span, and delay responses" (5). These basic processes contribute to multiple psychological systems as such they represent 'domain-general' cognitive processes. Selection for specific behaviors likely favored a number of neural changes that affected how these domaingeneral processes contributed input to domain-specific mechanisms. That is, mechanisms that solve specific adaptive problems such as theory of mind or the causal properties associated with tool-use. Here I hope to build on some of Holloway's (1967) ideas, specifically, the notion of mosaic brain evolution and the forces that produced such, and further explore Holloway's ideas concerning how selection acted indirectly on the brain through it's selection of the specific actions and behaviors it produced given that behavior is what selection ultimately acts upon (Holloway, 1979; 1981; 1996). As such, mosaic brain evolution is necessarily a reflection of mosaic cognitive evolution at both general (i.e., memory and attention) and specific (i.e., tool-use and language) levels. The notion of mosaic brain evolution contrasts with both domain-general views of intelligence, such as those that propose a conceptual 'g' or general intelligence (Jensen, 2000) or pan selectionist theories such as the Social Intelligence Hypothesis (Jolly, 1966; Humphrey, 1976; Byrne & Whiten, 1990; Whiten & Byrne, 1997) or the Ecological Intelligence Hypothesis (Parker & Gibson, 1977; Parker & McKinney, 1990) and most resembles the view of cognition proposed by Evolutionary Psychology (Buss, 2006; Tooby & Cosmides, 1990).

Here I will focus on the nature and evolution of the imitation faculty; a psychological faculty that has typically been regarded as an all-purpose learning mechanism (Buller, 2006); a type of 'general intelligence,' the product of selection for 'social intelligence.' To the contrary, imitation appears to be a mosaic cognitive faculty whose evolution was not the result of a general selective force favoring social or technical intelligence, but rather its evolution is the product of a confluence of factors some that are 'social' others that are 'ecological' and still others that are 'technical.' These different pressures from these different domains produced different imitation mechanisms, specialized in the imitation of different rules and responses. As a result, our species' seemingly domain-general imitation skill is something of an illusion. It is an illusion because our ability to imitate different types of information results from the operations of many different imitation mechanisms that give the appearance of a 'general purpose' psychological faculty. Such an imitation faculty was likely to be very useful for solving a number of problems. Some of these problems include: (a) the problem of learning dominance relationships, where individual can minimize injury by inferring from observational learning who is likely to be dominant/submissive, (b) the diet problem; learning what is edible and what isn't or the problem of what to eat when, (c) the problem of alliances and cooperation, where individuals can minimize the risks of bad alliances by inferring from observation who is a reliable/unreliable partner, (d) the problem of extractive foraging, where individual can learn from others how to process or acquire protected food products, (e) the problem of social convention, where individuals use others' behaviors to guide where and when they should display species-typical behaviors or behavioral traditions. And, there are certainly others. In each instance, specialized mechanisms in the imitation faculty in coordination with other cognitive faculties grant individuals the flexibility to make rapid inferences about the dispositions of others or the causal structure of actions, bypassing the costs associated with trial and error learning, which in some instances may be lethal (e.g., the diet problem). Some of these instances require 'imitation learning' or novel imitation (when knowledge is first acquired and reproduced) but others only require the copying of species-typical behaviors—familiar imitation (e.g., social conventions)—where previously acquired behaviors (either by imitation or trial and error) are appropriately and adaptively displayed.

The Multiple Imitation Hypothesis' (Subiaul 2007) distinction between different imitation mechanisms may explain many of the similarities and differences reported between human and ape imitation performance. The argument that will be put forth in this essay is that humans and apes share some but do not share all imitation mechanisms. Differences in the number and type of imitation mechanisms available to individual species likely rests on the unique adaptations that resulted from different species-specific problems encountered in the species' environment of evolutionary adaptedness (Buss, 2007; Tooby & Cosmides, 1990) and the consequences of how selection favored different strategies for complexity management in different ape lineages (Holloway, 1967).

The Many Faces of Imitation

The Multiple Imitation Hypothesis

Most view the imitation faculty as a domain- and content-general mechanism that operates across different problem domains and content types, allowing individuals to learn everything from motor rules such as how to use chop sticks, to vocal rules such as *aguacate* ('avocado' in Spanish), to procedural rules such as how to cook your favorite pasta dish. Given what is known about the imitation skills of human children and other primates, it appears that the environment of early hominids favored individuals who were flexible imitators, capable of copying a wide range of behaviors and responses: from using chop sticks, among other tools, to saying *aguacate*, among other novel sounds, to cooking pasta among other procedural rules. However, the representation of auditory stimuli (such as *aguacate*) for the purposes of reproducing that sound must be fundamentally different than the representation of a motor action (such as using chop sticks) for the purposes of copying that action. A general-purpose mechanism capable of performing these different tasks seems unlikely if not improbable. What is more likely is that selection sifted through individuals with varying imitation skills and a unique cognitive-neural imitation profile capable of identifying, representing and copying these different types of information. This process would have produced distinct imitation skills mediated by specific imitation mechanisms dedicated to representing and copying specific types of stimuli. From this it follows that humans are good imitators relative to other primates not because we have an imitation mechanism that primates lack but because our species has evolved a whole suite of distinct imitation mechanisms or 'imitation instincts' that together result in an impressive ability to copy all sorts of responses in a flexible and adaptive fashion.

This view of imitation fundamentally differs from the widely held domain-and content-general view of imitation. The multiple imitation hypothesis proposes that the imitation faculty is similar to other vertical cognitive faculties (Fodor, 1983), such as language, that are modular, specialized and consist of multiple components with discrete functions. However, it's unlikely that the imitation faculty is as encapsulated as Fodor (1983) proposed for visual systems, for example (c.f., Marr, 1982). In this conceptualization, the imitation faculty represents a specialized psychological mechanism with input from a number of domain-general systems like memory and attention as well as domain-specific 'core knowledges' that include 'theory of mind,' 'naïve physics' and 'naïve biology' (Spelke, 2000). Through this kind of domainspecificity, the imitation faculty can copy responses across different domains in a flexible and adaptive fashion.

Like other faculties, the imitation faculty can be divided by its various functions. These functions are best captured by super-ordinate and sub-ordinate imitation mechanisms associated with the processing of specific types of stimuli (e.g., novel, familiar, auditory, motor, social, etc.). The super-ordinate imitation mechanisms include, (a) 'familiar imitation,' or the copying of familiar rules or responses and (b) 'novel imitation,' or the copying of novel rules or responses; often referred to as 'imitation learning,' which is distinguished from 'familiar imitation' in that it requires observational learning. That is, the ability to learn through vicarious (rather than direct) reinforcement (Bandura, 1977). Various researchers have made similar class distinctions, recognizing that different mechanisms likely mediate the learning and copying of a novel behavior(s) and the copying of behaviors that already exist in an individual's repertoire (Byrne & Russon, 1998; Heyes, 2001; Visalberghi & Fragaszy, 2002). However, these investigators have tended to argue that these skills are not related and consequently have tended to give these skills different names, which imply that they exist outside of a dedicated cognitive faculty for imitation. The reason for this being that many of these researchers believe that imitation is a single unitary cognitive process that animals either have or lack entirely (e.g., Tomasello & Call, 1997). In this framework, familiar and novel imitation mechanisms are brought together as part of the same cognitive faculty that mediates the ability to flexibly copy rules or responses across contexts. Moreover, subsumed within those two broad functional concepts are sub-ordinate mechanisms of imitation that specify the type of stimuli that is reproduced by either novel or familiar imitation (i.e., auditory, motor, cognitive).

As has been noted, all the proposed imitation mechanisms are characterized by flexibility *and* specificity. The flexibility requirement means that the behavioral rule that is copied is deliberate or replicable. That is, can be elicited in multiple contexts on multiple occasions; not the result of happenstance or trial and error learning or the product of narrow contextual cues. The specificity requirement emphasizes that individuals must copy a specific 'rule.' The term 'rule' is broadly defined as a response involving more than two steps (e.g., with a distinct 'beginning-middle-end' structure) that are hierarchically organized and structured to achieve a matching response. The requirement that any type of imitation be rule-governed and flexible is necessary in order to differentiate imitation from either perceptual or motivational mechanism that in association with rapid trial-and-error learning may represent an ancestral learning mechanism that predates (and may, perhaps, co-exist) with the imitation faculty, providing critical input to the mechanism mediating familiar imitation, for example. The same is true of narrow species-specific skills such as copying mate preferences that while impressive, learning does not extend beyond a very narrow context (i.e., mating) and is dependent on specific stimuli (i.e., females) (Bshary & Grutter, 2006; Paz y Miño et al., 2004). Nevertheless, such studies provide important evolutionary clues into the origins of the imitation faculty; highlighting for instance, how selection for multiple content-specific observational learning skills could be aggregated by natural selection resulting in an imitation faculty like the one described here.

Super-ordinate mechanisms of imitation: Novel imitation

Part of the confusion in the imitation literature is that 'imitation' has been largely conceptualized as 'novel imitation' or the imitation of novel behaviors. For example, in 1898, Thorndike defined imitation as "learning to do an act from seeing it done" (p. 79). Nearly a half-century later, Thorpe defined imitation more narrowly and in purely behavioral terms: "copying a novel or otherwise improbable act" (p. 122). These definitions are often viewed as synonymous, but they are quite different. One core difference between these two definitions is the requirement that individuals *copy* another's behavior. Copying is, arguably, the essence of imitation. After all, what is imitation if it isn't copying something? Yet, Thorndike's definition doesn't mention or imply copying but rather observational learning. The distinction between observational learning and imitation is critical. It is possible to learn something from another, yet not overtly express the acquired knowledge; for example, learning what *not* to do. In such instances, one can learn from a model without imitating the model. Thorpe's definition, unlike Thorndike's, stresses both (observational) learning and copying. Learning is implied in the criteria that what is copied is 'novel' rather than something that already exists in the observer's behavioral or cognitive repertoire. Despite a number of qualifications and revisions (e.g., Galef, 1988; Tomasello & Call, 1997; Whiten & Ham, 1992), Thorndike (1898; 1911) and Thorpe's (1956) definition of imitation remain influential because of their simplicity and the ease with which they lend themselves to experimentation. Nevertheless, these definitions, which conceptualize imitation as the copying of specific and novel motor responses, have largely ignored an equally important function of the imitation faculty, familiar imitation.

Super-ordinate mechanisms of imitation: familiar imitation

Familiar imitation involves the ability to flexibly and adaptively copy common or recognizable rules/responses that exist within an individual's behavioral repertoire. In the motor domain, everyday actions fall into two distinct and conceptually significant categories: transparent versus opaque. Transparent responses are those responses that are immediately available to the senses such as transitive actions that involve reaching for and interacting with objects and, as a result, may be executed via a visualvisual match (i.e., my hand on an object looks like your hand on an object). However, opaque responses cannot be executed in the same fashion, as they are not available to the senses in the same way as transparent actions. Consider the act of imitating someone scratching their head. What you perceive when you see someone scratch their head is very different from what you perceive when you scratch your own head. The phenomenological experiences are very different. This problem of translating a visual experience into a corresponding proprioceptive response has been termed the "correspondence problem" (Dautenhahn & Nehaniv, 2002).

While to some, the distinction between 'novel' and 'familiar' imitation may be obvious, there is significant debate as to what should count as a 'novel' response. Does 'novel' imply an entirely new behavior? By the most strict of standards this would exclude all species-typical behaviors; a constraint that significantly limits research questions. One way around such a constraint is to require animals to execute a series of familiar behaviors in arrangements that are never (or rarely) observed. This technique—of stringing familiar actions in an arbitrary sequence—has been employed by a number of animal researchers (apes: Whiten, 1998; birds: Nguyen et al., 2005; monkeys: Caldwell & Whiten, 2002) and represents one way of operationalizing 'novelty' in imitation research. Another technique has been to use a tool in novel problem-solving tasks (e.g., Visalberghi & Fragaszy, 1989, 1990, 1995; Whiten & Horner, 2007). Perhaps these studies, more than any others, represent the most strict standards of novelty, as subjects must often learn how to handle the tool and then learn how to use the tool *in relation* to another object. However, this poses a unique problem when comparing human and non-human ape imitation studies that involve tool-use because humans may have unique causal conceptual mechanisms and by extension, species-specific skills pertaining to objects in general and tools in particular that nonhuman primates may lack (Johnson-Frey, 2003; Povinelli, 2000). But there are other ways to operationalize 'novelty' without using tools or specific motor responses. Subiaul and colleagues (2004), for instance, developed a cognitive imitation paradigm, where subjects had to copy novel serial rules independently of copying novel motor actions. All of these tasks require that subjects learn something new in order to be reinforced, and exclude the possibility that subjects already know how to execute the target response. At the same time, such tasks control for the possibility that the ability to execute the motor response interferes with expression of knowledge gained during observation.

Others have tried to operationalize 'novelty' using single and familiar actions on objects (e.g., Apes: Hopper et al., 2008; Monkeys: Bugnyar & Huber, 1997; Voekl & Huber, 2000; 2007). Here, the rationale is that while a behavior such as mouthing is species-typical, mouthing an object in order to open it is novel. The problem is that animals often explore objects using their mouths and certainly use their mouths on objects associated with food. So, while a particular behavior directed toward a specific object may be unique, the actual behavior is not. In this regard, it's more likely that familiar imitation of the familiar action (e.g., mouthing) rather than novel imitation is the primary mechanism underlying the behavioral response in single-action paradigms. Such paradigms also make it difficult to distinguish between various mechanisms of the imitation faculty and the products of perceptual and motivational mechanisms in which, for example, an animal's interaction with an object may direct an observer's attention to that object (stimulus enhancement) or a part of that object (local enhancement), motivating the observer to interact with it (social enhancement). In such instances, these two individual's responses may be very similar, yet the similarities are likely to be the products of stimulus and social enhancement as well as rapid trial-and-error learning, rather than by any mechanisms of the imitation faculty.

Sub-ordinate mechanisms of imitation: cognitive, motor & vocal imitation

In addition to distinguishing between familiar and novel imitation, it is important to distinguish between various sub-ordinate mechanisms that form part of the imitation faculty. These mechanisms involve copying different classes of stimuli, for example, auditory, motor, and cognitive stimuli. The reproduction of these different types of stimuli compromise three particular classes of imitation: vocal imitation (the imitation of vocal/auditory responses), motor imitation (the imitation of motor actions), and cognitive imitation (the imitation of cognitive rules, including rules governing serial order, social conventions and spatial relationships, for example). The distinction between superordinate mechanisms of imitation (e.g., novel v. familiar) and sub-ordinate mechanisms of imitation (e.g., vocal, motor and cognitive) are important because it allows researchers to specify what type of imitation they are capable of. For example, an individual may be able to reproduce *familiar* vocal rules (e.g., words), but may not be able to copy *novel* vocal rules (e.g., novel words). Moreover, individuals may be able to copy novel *cognitive* rules (e.g., serial order), but not novel *motor* rules (e.g., specific action sequence). Some of these dissociations appear to be true in monkeys for instance, which seem unable to copying novel motor rules, but can copy novel cognitive rules (Subiaul et al., 2004; 2007). Interestingly, similar dissociations exist within humans. For example, children with autism, are unable to copy novel motor rules, but can copy familiar motor rules (Williams, Whiten & Singh, 2004). There's also a dissociation in novel imitation performance among individuals with autism that parallels the dissociation in monkeys; in particular a dissociation between novel motor and novel cognitive imitation (Subiaul, Lurie, Romansky, Cantlon, Terrace, 2007).

This framework does not necessarily challenge familiar terms that have become an integral part of the imitation literature such as emulation—where individuals copy the outcomes or 'affordances' of actions—or goal emulation—where individuals copy the 'intended' action of others using idiosyncratic means. Rather, it questions the logic that terms such as emulation are alternatives to imitation or more precisely, that 'emulation' is a mechanism that exists outside the mechanisms of the imitation faculty as described here. Rather, I advance the contrarian's view that terms such as emulation and goal emulation describe the imitation of different types of rules or responses; specifically, copying rules—novel or familiar—about environmental affordances or goals, respectively.

Neurobiology of familiar and novel imitation

Recently, a number of advances have supported the multiple imitation hypothesis (Subiaul, 2007). A functional dissociation between novel and motor imitation is supported by neuropsychological and neurophysiological research. In a series of studies, Rumiati and Tesari (2002, 2003) presented two groups of subjects with two different tasks: one involved copying familiar "meaningful" actions; the other involved copying novel "meaningless" actions. Meaningful (i.e., familiar) actions consisted of common actions such as brushing one's teeth. Meaningless actions (i.e., novel, arbitrary actions)[1] consisted of performing common actions in an arbitrary fashion, for example, a brushing action performed with the right arm extended outwards and the hand held upright. Predictably, subjects copied "meaningful" actions with fewer errors than meaningless actions. Rumiati and Tesari interpreted these results to mean that different systems mediate the imitation of "meaningful" and "meaningless" actions. In their model, the perception of familiar actions are recalled from long-term memory then moved into working-short-term memory in order to generate a matching motor output. The perception of novel "meaningless" actions, however, is processed in working-shortterm memory as there's no memory trace to recall from semantic long-term memory.

Neuroimaging studies conducted by the authors have provided additional support for a dual-processing route. Rumiati, Weiss, Tessari and colleagues (2005), reported that the left inferior temporal gyrus was associated with a significant increase in blood flow when subjects copied meaningful actions. Whereas, greater blood flow to the parieto-occipital junction was associated with copying meaningless actions. When comparing neural activation during the imitation of familiar relative to unfamiliar actions there were differential increases in neural activity in the left inferior temporal gyrus, the left parahippocampal gyrus, and the left angular gyrus, structures associated with longterm memory processes. Whereas, the superior parietal cortex (bilaterally), the right parieto-occipital junction, the right occipital–temporal junction (MT, V5), and the left superior temporal gyrus where differentially active when subjects copied novel actions relative to familiar actions. The primary sensorimotor cortex, the supplementary motor area, and the ventral premotor cortex showed increased neural activity when subjects copied both types of actions (familiar and novel).

There are a number of studies that are consistent with the multiple imitation hypothesis (Buccino et al., 2001; Cochin et al., 1999; Fadiga et al., 1995; Decety & Chaminade, 2005; Heyes, 2002; 2005; Stevens, Fonlupt, Shiffrar & Decety, 2000). For instance, various neurophysiological studies using transcranial magnetic stimulation (TMS), magnetoencephoalography (MEG), positron emission tomography (PET), and functional magnetic resonance imaging (fMRI) have found that when subjects observe an individual executing an action using a specific muscle group, corresponding areas of the observer's motor strip is activated, as if the observer was executing the action themselves rather than passively observing someone else performing the same action (Buccino et al., 2001; Cochin et al., 1999; Fadiga et al., 1995). Consequently, when one sees a conspecific execute actions that are familiar and form a part of one's own motor repertoire, neural regions such as the supplementary motor area (SMA), the premotor cortex, and the superior and inferior parietal cortices—the action preparation system—are activated. This "motor resonance" phenomenon is not triggered by novel actions because they are not present in the motor repertoire of an observer and are yet to be learned. When individuals observe novel actions they have no existing representations of the motor component of these actions. At best, they can call upon related or similar rules or responses. As implicated by the dual-route model (Rumiati & Tessari, 2002; 2003; Tessari & Rumiati, 2004), the match between what is seen and what is ultimately executed must be done online (in working memory) with little or no help from existing cognitive representations of the target action.

The apparent motion paradigm (Shiffrar and Freyd, 1990) has further highlighted the functional and structural differences associated with copying novel as opposed to familiar actions. Using PET technology, Stevens et al (2000) presented participants with a human model engaged in possible (i.e., familiar) and impossible (i.e., novel) biomechanical paths of apparent motion. When the subjects perceived 'possible' paths of human movement, the left primary motor cortex and the parietal lobule in both hemispheres were found to be selectively activated. These areas were not activated when participants observed impossible biomechanical movement paths.

---

[1]. It's important to point out, however, that novel actions could be perceived as meaningful, yet, not exist in the observer's behavioral reportoire. For example, we may observe two American Sign Language (ASL) speakers communicate with one another. Though the actions are novel to us because we are unfamiliar with ASL, the signs are, nevertheless, recognized as being "meaningful." That is, they are recognized by naive observers as having a communicative function. Consequently, individuals may imitate meaningful novel actions differently from meaningless novel actions. Future experiments may wish to more directly assess the role of "meaning" in imitation independently of the familiarity of actions.

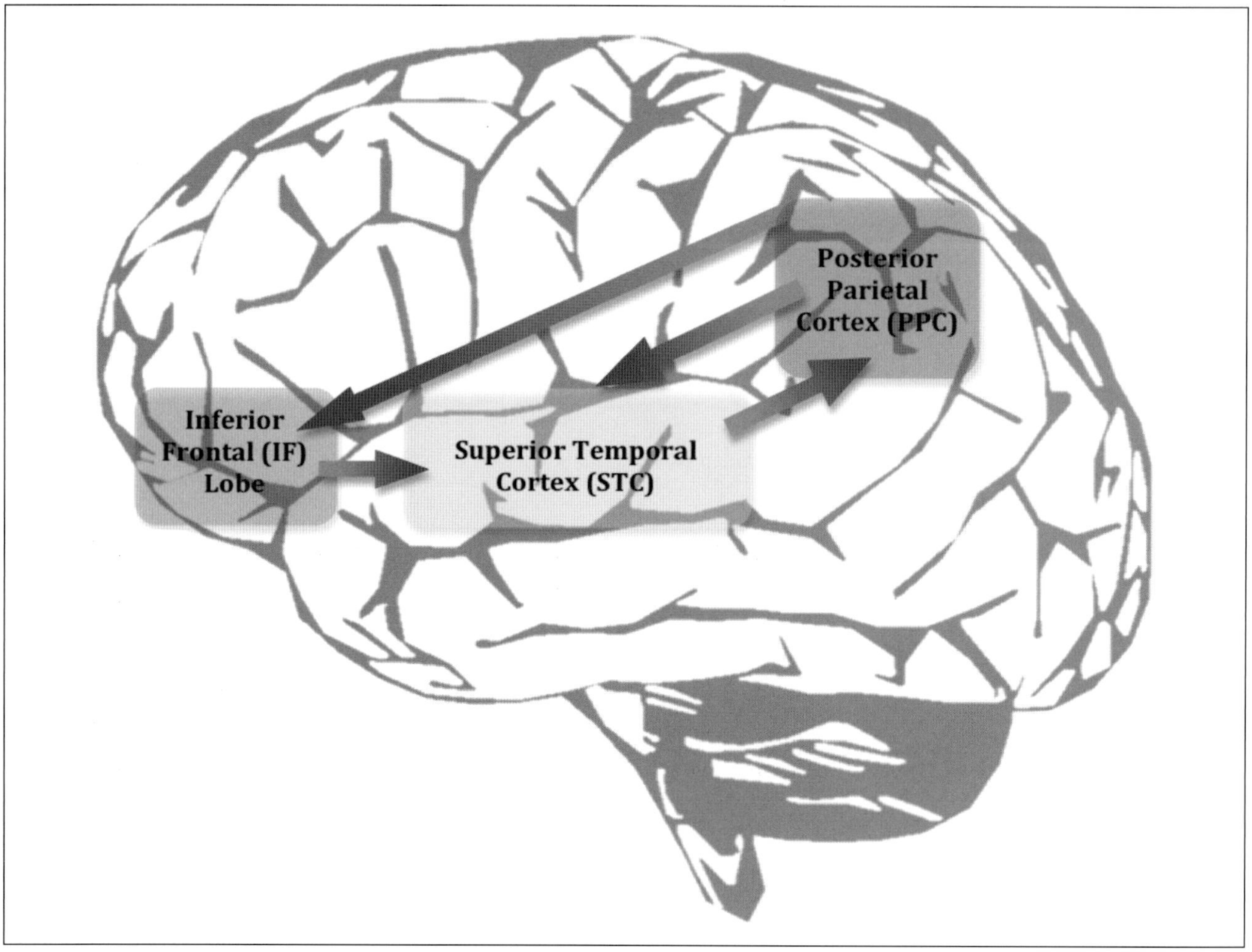

Figure 1. Familiar Motor Imitation Circuit. According to Carr et al. (2003) information flows as follows: (1) the STC codes early visual descriptions of actions and projects these representations to the PPC mirror neurons; (2) the PPC integrates representations of kinesthetic aspects of actions and projects this information to IF mirror neurons; (3) IF codes the outcome or the 'goal' of the target action; (4) IF and PPC send efferent copies of the action plan back to the STC, creating a matching 'resonance' mechanism between visual and motor representations of the same action event; (5) motor execution of imitation is initiated.

The results reported by Rumiati and Tessari as well as those by Shiffrar and Freyd make clear that different neural mechanisms mediate the imitation of novel as opposed to familiar responses. They further demonstrate that the distinction between familiar and novel imitation may best be characterized as a difference between recall and learning. In the case of familiar imitation, individuals recollect past (learned) experiences. Whereas in the case of novel imitation, individuals are encoding novel experiences and knowledge through observation or vicarious learning. In any event, these distinct imitation systems may feed into a more general motor imitation circuit such as that proposed by Carr and colleagues (2003) and summarized in Figure 1.

Additionally, neurobiological studies have demonstrated that observational learning—the core feature of novel imitation—has independent neurobiological circuits. Again, it must be stressed that in the multiple imitation framework, observational learning is not synonymous with imitation, particularly familiar imitation. There are two main differences between observational learning and novel imitation: First, novel imitation requires observational learning, but familiar imitation does not. Second, novel imitation requires observational learning in addition to copying. Observation learning requires only learning, not copying. The rationale here is that one may learn many things from observation (dispositional traits, the worth of things, what *not* to do or how not to behave) but we don't copy all we learn from others.

A number of lesion and single-cell recording studies suggest that observational learning is largely mediated by the right cerebellum. For example, Petrosini and colleagues (1999, 2000; 2007) demonstrated that rats tested in a Morris water maze task learn to locate a hidden platform in a pool one of two ways: by individual, trial-and error learning or by observing an experienced conspecific. To explore the cerebellum's role in this skill, Petrosini and colleagues removed the right hemicerebellum of naïve rats either after they had been given the opportu-

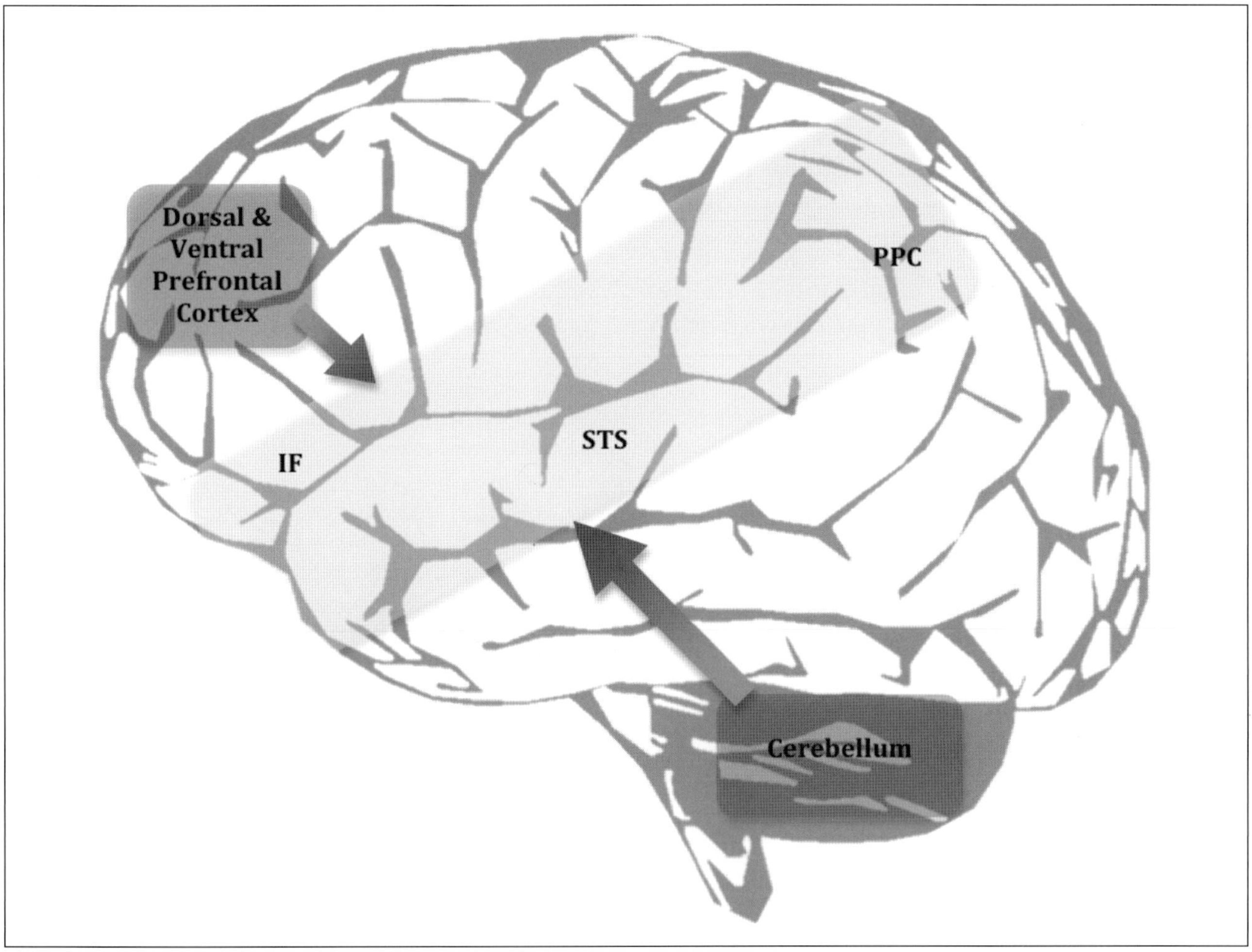

Figure 2. Novel Motor Imitation Circuit. A number of authors have pointed to the left posterior cerebellum as well as the dorsal and ventral prefrontal cortex as critical for (i) the intention to imitate (e.g., Chamindate et al. 2002) and (ii) observational learning (e.g., Petrosini, 2007). Leslie and colleagues (2003) have suggested that these cerebellar and frontal circuits that appear critical for novel motor imitation likely interact with circuits that appear responsible for familiar motor imitation (c.f., Figure 1).

nity to observe expert rats navigate through the pool and settle on a hidden platform (post-observation surgery treatment) or ablated the same part of the cerebellum before naïve subjects had been given the opportunity to observe the expert rat find the hidden platform (pre-observation surgery treatment). Results revealed that rats that received the post-observation surgery treatment learned how to find the hidden platform significantly faster than they would by trial and error. However, rats in the pre-observation surgery treatment failed to learn where the hidden platform was located. As a result, these rats performed randomly, eventually learning where the platform was located by trial-and-error learning.

Though these experiments do not exclude learning by perceptual/motivational mechanisms such as local enhancement, the results reported by Petrosini and colleagues (1999, 2000; 2007) have a number of significant implications. First, the removal of the right hemicerebellum in rats does not extinguish spatial or navigational abilities because all subjects are capable of learning where the hidden platform is located. Moreover, the ablation of this part of the cerebellum did not affect motor movements and/or coordination. Second, the difference between the performance of individuals in the pre- and post-observation surgery treatment demonstrates that the right cerebellum plays a significant role in learning. Third, the cerebellum's potential role in observational learning strongly suggests that a distinct circuit (independent of neural circuits mediating familiar imitation) is at work in social learning tasks in general and novel imitation in particular. This last point is corroborated by at least one other study with human subjects. Grèzes, Costes, and Decety (1998) showed that the left posterior cerebellum is uniquely active when subjects have the intent to imitate a novel response. Results demonstrated that the cerebellum becomes active when subjects are confronted with new rules that must be learned by observation (rather than by trial and error). In humans, as in rats, this cerebellar circuit (Grezes et al., 1998; Petrosini et al., 2000; 2007) appears to be independent of a separate frontal (e.g., BA 6, 9, 10, 46) and parietal (e.g., BA 40 & 7) circuit that have been linked specifically to familiar imitation (Carr et al., 2003; Rizzolatti et al, 2002). Nevertheless, while these cerebellar circuits appear to mediate observational learning, Leslie, Johnson-Frey and Grafton (2003), suggest that information from the left posterior cerebellum as well as the dorsolateral and ventral prefrontal cortex interact with the circuit (i.e., inferior frontal, STS and posterior parietal) associated with familiar motor imitation (Carr et al., 2003) in order to achieve novel motor imitation, for example (c.f., Figure 2).

Neurobiology of cognitive & motor imitation

Theoretically, the brain may imitate in one of two ways: either via a single imitation network involving hippocampal networks for familiar imitation (e.g., Rumiati et al., 2005) and a cortical-straital network for novel imitation or through distinct networks corresponding to the imitation of different types of stimuli such as motor, vocal, cognitive. At present the evidence is mixed. At least one imaging study on the "song system of the human brain" (Brown, Martinez, Hodges, Fox & Parson, 2004) suggests that familiar motor imitation and certain aspects of novel vocal imitation may have overlapping neural structures or be mediated by the same neural systems. While certain aspects of the human song system were unique, such as action in the superior part of the temporal pole (BA 38) others either overlap or are adjacent to the 'mirror neuron system' in the inferior frontal operculum (BA 44) that is known to play a critical role in familiar motor imitation (c.f., Figure 1). However, the overlap in the present study may have been due to sub-vocal rehearsal or the recall of lyrics from songs with a similar melody. Importantly, Brown et al. (2004) report that this system is only active when subjects are actively matching the pitch and rhythm of novel sequences but not when participants are recalling familiar melodies. Another neuroimaging study supports a dissociation between motor and cognitive imitation systems. Chaminade et al. (2002) presented subjects with a model executing one of three different aspects of an event: (a) the complete action arc from start to finish, (b) only the means used to achieve the action, and (c) only the result of the action. Subjects made one of three different responses: (a) passive observation, (b) imitated what was observed, or (c) acted freely. Because the task involved the intentional copying of actions, neural regions associated with higher-order motor representations and sensorimotor transformations in addition to the posterior Superior Temporal Sulcus (STS) were active across conditions. However, different neural regions were active when subjects observed and copied an entire event as opposed to when subjects observed and copied only the means or only the goals of that same event. Specifically, there was significant activation in the cerebellum (bilaterally) and the dorsolateral prefrontal cortex (DLPFC) when subjects copied both the means and the goals of an action. Yet, there was hypo- or no activation in these same regions when subjects copied the entire event. Moreover, despite the fact that some of the same regions were active when copying goals and means, regions of activation within DLPFC were not entirely overlapping. Furthermore, the medial prefrontal cortex was active only when subjects copied the means used to execute the action, whereas the left premotor cortex was active only when subjects copied the goals of the action. The fact that premotor cortex was differentially active in the course of copying goals versus means is of some significance as premotor cortex is associated with "mirror properties" in monkeys and humans (Buccino et al., 2001) and associated with the preparation and execution of goal-directed actions. Chaminade et al. (2002) argue that premotor cortex is only active when subjects copy goals because this is the only condition in which the means of the actions must be inferred from the observation event.

Taken together, these results suggest that the possibility for imitation-specific circuits that correspond to different imitation mechanisms. However, it cannot be overlooked that the studies by Chaminade et al. (2002) investigated goals, means, and action in the context of a motor imitation task rather than a task that involved copying non-motor or cognitive rules (independently of the execution of specific motor actions) as was done by Subiaul and colleagues (2004; 2007), for example. Moreover, this study did not distinguish between copying familiar (familiar imitation) versus unfamiliar (novel imitation) goals and means. So, for example, the system that mediates the copying of novel goals may differ from the system that mediates the copying of familiar goals.

Mosaic Imitation Skills in Apes

The comparative study of imitation: Apes and humans

Certainly, social learning is common in the animal kingdom (Zentall, 2007) and sophisticated local traditions exist in apes (Whiten et al., 1999; van Schaik et al., 2003) and to a lesser degree in monkeys (Panger et al., 2002; Perry et al., 2003). And, as can be seen in Table 1, while there are a number of similarities between human and nonhuman 'cultures' only humans have cultures that build on prior knowledge and accumulate over time (Boyd & Richerson, 1985; Henrich & McElreath, 2003; Subiaul, 2007; Tomasello, 1999; Tomasello, Kruger & Ratner, 1993). Given our species' penchant for cultural learning and the extent to which our survival depends on that learning, there is perhaps no greater question than what underlies such skills. One (arguably) uniquely human skill is the ability to copy a broad range of rules—motor, vocal, cognitive—from a model. Might differences in cultural learning be explained in part by differences in what and how apes and humans imitate?

Table 1. Features of 'Culture.' Below is a list of the characteristics of culture proposed by different authors and their distribution in humans, non-human (NH) apes and monkeys (specifically, capuchin monkeys). The table demonstrates that apes share many features in common and differ from monkeys.

Components of Culture	Humans	NH Apes	Monkeys
Innovation: New behavioral pattern is invented*	+	+	+
Dissemination: Transmitted from individual to individual*	+	+	+
Durability: Pattern persists beyond demonstrator's presence*	+	+	–
Diffusion: Pattern spreads across groups*	+	+	+
Tradition: Pattern endures across generations*	+	+	–
Standardization: Pattern is consistent and stylized*	+	+	~
Species-Valid: Not an artifact of human influence*	+	+	+
Transcendent: Not determined by biophysical environment*	+	+	+
Accumulation: Traditions build over time**	+	–	–
Imitation: Ability to copy novel motor responses‡	+	+	–
Variability: Two or more patterned behaviors in more than one domain§	+	+	+

**Criteria from Krober (1928), ** Tomasello & Call (1997), ‡Galef (1992), Whiten & van Schaik§ (2007), + (present), – (absent), ~ unknown or debatable*

As with the attribution of mental states, there has been a long-lasting controversy over whether or not humans are unique in the ability to learn from others. In fact, Aristotle argued in the *Poetics* that humans are "the most imitative creatures in the world and learn first by imitation." In the past 30 years, interest in imitation learning has experienced a renaissance, particularly as scientists have found that from birth neonate copy the facial expressions of adults (Meltzoff and Moore, 1977) and primatologists have documented various instances of tool traditions in populations of wild chimpanzees (McGrew, 1992; 1994; 2001; Whiten et al., 1999) and orangutans (van Schaik et al., 2003). However, to date only eleven studies have directly compared imitation learning in human and non-human [adult] apes using analogous procedures (Call, Carpenter, Tomasello, 2005; Call and Tomasello, 1995; Herrmann et al., 2007; Horner and Whiten, 2004; 2005; 2007; Horner, Whiten, Flynn & deWaal, 2006; Horowitz, 2003; Nagell et al., 1993; Tomasello, Savage-Rumbaugh, and Kruger, 1993; Whiten, Custance, Gomez et al., 1996). Six of these studies have reported that on an operational task, where a tool or object had to be manipulated in a certain manner to achieve a specific result (or reward), humans reproduce the demonstrator's actions with greater fidelity (i.e., imitation) than did mother-reared apes (Call, Carpenter and Tomasello, 2005; Herrmann et al., 2007; Horner & Whiten, 2007; Call and Tomasello, 1995; Nagel, et al., 1993; Tomasello et al., 1993). The other studies reported both similarities and differences between humans and peer-reared apes when executing specific actions on an object following a demonstration (Horner and Whiten, 2004; 2005; Horner et al., 2006; Whiten et al., 1996). And one, found no differences between the performance of adult humans and other apes (Horowitz, 2003).

Comparing familiar vs. novel motor imitation in primates

Given these results, it is obvious that there's no simple answer to the question, 'Do apes, ape?' How might one explain these seemingly conflicting reports of similarities and differences, particularly if imitation is viewed as one unitary faculty that animals either have or lack entirely? One possibility is that these different studies are measuring different imitation mechanisms. When viewed this way it appears that apes and humans share some imitation mechanisms (hence the similarities in some studies) but do not share all (explaining some of the differences). Using the multiple imitation framework outlined above, studies such as, Horner & Whiten (2004; 2005) and Horner and colleagues (2006) are likely to be tasks of familiar motor imitation, whereas studies such as Horner & Whiten (2007) are tasks of novel motor imitation. Without question, novel motor imitation tasks are harder than familiar motor imitation tasks. What makes novel motor imitation harder is that to be successful the subject must first attend to the relevant information (hand or body part, tool or object), create a new action representation and then match this abstract motor representation with a new action plan. The same is not true for familiar imitation tasks because the observation of a familiar action likely primes that same action in memory (i.e., recognition memory). In this case, the construction of a novel action plan is not necessary as it is recalled from memory.

There are likely to be other differences that contribute to differences in motor imitation performance among apes. Perhaps the most significant has to do with toolknowledge and tool-use. Most studies that require animals to use tools in ways that they do not do naturally in the wild tend to find differences between human and non-human subjects (e.g., Herrmann et al., 2007; Horner & Whiten, 2007). When the imitation task involves using tools in ways that are more 'naturalistic' (i.e., behaviors that typically appear in the wild such as probing with a stick or pushing objects out of the way), more similarities are reported between humans and other apes (e.g., Horner et al., 2006; Hopper et al., 2008). However, there are some studies where apes are required to execute 'familiar' actions—such as pulling or pushing—on unfamiliar objects or in novel experimental circumstances (Call & Tomasello, 1995; Herrmann et al., 2007). These studies, too, tend to report more differences than similarities between humans and other apes. Johnson-Frey (2003; 2004) and Povinelli (2000) have suggested that there may be in some cases subtle and in other cases dramatic differences between humans and other animal's orientation to objects with tool properties. For instances, some of the differences in imitation performance may be due to differences in the "Grasp" and "Manipulation" motor system that are mediated, in part, by circuits in the parietal and frontal lobe. While Johnson-Frey suggests that differences in these two motor systems may be negligible, how these systems interact with conceptual systems mediating causal action likely produces significant species differences, as borne out by a number of comparative studies on chimpanzee tool-use (e.g., Povinelli, 2000). Novel motor imitation likely depends on input from these various systems, without which it cannot operate. The same is likely to be less true for familiar motor imitation, as experience allows individuals to recall existing motor representations and rehearsed motor action plans.

There is some support for the hypothesis that chimpanzees differentially imitate novel versus familiar actions (Myowa-Yamakoshi et al., 1999). Myowa-Yamakoshi and colleagues presented chimpanzees with a number object-based actions that they characterized as general actions (familiar actions on objects that were commonly observed) and non-general actions (relatively novel actions on objects that were not commonly observed). This corresponds roughly to the proposed distinction of familiar versus novel imitation. They applied this scheme to different actions on objects that ranged from copying single but specific actions on objects such as banging the bottom of a bowl, to copying actions that involve directing objects to specific body parts such as putting the bowl on the head, to copying object-object interactions such as putting a ball in a bowl. Results revealed that performance was best for familiar actions and relatively poor for novel actions. Chimpanzees in these studies performed best in the object-object condition and worst in the single-action condition. However, these results are derived from multiple trials and do not represent first trial performance. Unfortunately, no data is presented on 'familiar' versus 'novel' actions in these different conditions. But, Myowa-Yamakoshi and colleagues note that chimpanzees rarely copied any type of action (familiar or novel) on the very first trial. A strong indication that all or any subsequent copying behavior was likely mediated by familiar rather than motor imitation. Yet, given the hypotheses of the multiple imitation framework it's surprising that object-object actions were ultimately easier to reproduce than single actions on objects. There may be two explanations for this result. One possibility is that the objects used in the study constrained or limited the range of object-object responses as compared with the single action on object condition, where many more responses were possible. So, for instance, the object-object action most accurately copied by chimpanzees was the familiar action of putting a ball in a bowl; an object-object interaction with clear causal affordances. Given that the chimpanzees tested in these studies have a lot of experience putting things in bowls, the fact that this action was copied with the highest fidelity shouldn't be surprising even when compared to a relatively simple but arbitrary (and, perhaps, novel) single action like rubbing the bottom of the bowl. A second explanation may have had to do with the fact that when subjects failed to reproduce the action, they received explicit instruction. During the 'Teaching Phase' the demonstrator trained the subject to produce the target action through "verbal and gestural guidance, molding, shaping with verbal praise and food reinforcements, or a combination of these methods" (Myowa-Yamakoshi et al., 1999: 130). One or both of these explanations may explain the difference reported between copying a single action on objects and copying object-object actions.

Recently, a number of studies have focused on a special type of familiar imitation: oral facial imitation. Comparative developmental psychologists have shown no significant differences between a human and a chimpanzee infant's ability to copy the oral-facial expressions of a model. Chimpanzees, like human infants (e.g., Meltzoff and Moore, 1977), reproduce tongue protrusions, lip protrusions, and mouth openings in response to a model displaying the same expression (Myowa-Yamakoshi, Tomonaga, Tanaka, and Matsuzawa, 2004). There are also parallels in the developmental trajectory of oral-facial imitation in both of these species. Myowa-Yamakoshi and colleagues report that after 9 weeks of age, the incidence of oral-facial imitation in chimpanzees slowly disappears. A similar phenomenon has been reported for human infants (Abravanel and Sigafoos,

1984). In short this study found no qualitative differences between human infants and infant chimpanzees in oral-facial imitation. Recently, Ferrari and colleagues (2006) have reported oral-facial imitation in infant rhesus macaques. However, researchers have cast doubt on the notion that matching oral-facial responses is best characterized as imitation (as defined here or elsewhere). First, an extensive review of the literature revealed that only tongue protrusions are matched by human infants (Anisfeld, 1991; 1996; Anisfeld et al., 2001). Second, and perhaps most surprisingly, a number of studies have demonstrated that a moving pen (Jacobson, 1979), blinking light(s) (Jones, 1996) and music (Jones, 2006) are all as likely to elicit tongue protrusions in neonates as is watching a model display the same behavior. However, the study by Ferrari and colleagues on neonatal imitation in macaques is unique in that the experimental design included non-social controls such a spinning disk in addition to the typical social stimuli in such experiments (i.e., mouth opening, tongue protrusions, etc.). Ferrari and colleagues reported that lipsmacking and tongue protrusions occurred significantly more often in response to displays of those same actions than they did to other types of stimuli. However, lipsmacking occurred the most often in response to different types of stimuli, much like tongue protrusions in human infants (Jones, 1996). Ferrari et al. (2006) noting the amount of inter-individual variation and the sensitivity to specific oral-facial movements (e.g., mouth openings and tongue protrusions) in both human and monkey neonatal imitation pointedly caution that "the capacity to respond to the model may not reflect a general imitative skill but rather a sensorimotor sensitivity tuned to specific facial gestures" (p. 1506). At this point it is impossible to say with any certainty whether these results are mediated by a mechanism independent of the imitation faculty.

Taken together, the research reviewed above suggests that the motor imitation skills of primates are a mosaic of many different imitation mechanisms mediating the copying of different types of responses and likely represent adaptive solutions to specific problems. From this it follows that paradigms that conceptualize imitation as one unitary faculty that an individual either has or lacks entirely is problematic. As has been already noted, humans and other primates appear to share some imitation mechanism such as the ability to copy familiar motor actions and even novel cognitive rules (Subiaul, 2007; Subiaul et al., 2004; 2007) explaining the similarities reported by some comparative researchers. However, it's also clear that apes do not possess all the imitation mechanisms of a human 2.5 year old. This conclusion appears to be particularly true for novel motor imitation, a mechanisms that may rely on many higherlevel conceptual mechanisms. The fact that humans possess more imitation mechanisms sensitive to different types of stimuli, rather than a domain- and contentgeneral imitation mechanism, explains our species ability to copy a broad range of behaviors and responses relative to other primates.

The Evolution of the Imitation Faculty

Ecological & technological selection

Any contemplation of the mosaic evolution of the imitation faculty must begin with the question 'What are these different imitation mechanisms for?' How might having a simple imitation faculty consisting of only familiar imitation, for example, be adaptive? How might it increase fitness? Familiar imitation likely solves the problem of where and when to execute species-typical behaviors in appropriate contexts as well as coordinate/affiliative activies. In contrast, novel imitation solves the problems of acquiring information at a low cost. In both cases, imitation reduces the costs (e.g., time, energy) associated with trial-and-error learning. So whereas familiar imitation minimizes the need to learn *where* or *when* to execute familiar responses, in the case of novel motor imitation, it minimizes learning *how* and, perhaps *why* to do a novel action.

As many have noted, these problems are particularly acute in environments that are constantly changing. That environment may be social, it may be physical or it may be both. The more flux, the greater the need to quickly adapt to the new situation and the greater the selection pressures favoring various imitation mechanisms. This view has been supported by mathematical models that have, in effect, demonstrated that the evolution of the imitation faculty is linked to life in ever-changing environments (Boyd & Richardson, 1986; Henrich & McElreath, 2003). An evaluation of animals such as birds and primates who live in variable social and physical environments, suggests that these animals possess social learning skills consistent with at least a basic imitation faculty (Reader & Laland, 2002; Lefebvre et al., 1998). Interestingly, Reader and Laland (2002) have reported that among primates, brain size correlates most significantly with social learning, but also with individual learning ('innovation') and tool-use. In their analysis, social learning, individual learning and tool use are all strongly inter-correlated (Reader & Laland, 2002). Similar data exists for birds (Lefebvre, et al. 1996; Lefebvre, et al., 1998), providing evidence of convergent evolutionary processes.

The above evidence indicates that novel motor imitation is likely to be a derived feature and a characteristic of the hominoid imitation faculty; one that is perhaps intricately linked with tool-use. In this view, the more dependent an organism is on technology or motor learning for subsistence, the more imitation mechanism that animal is likely to possess (Figure 3). The main reason being that the use of technology—tools— requires specialized sensorimotor and inferential mechanisms working in a coordinated fashion to selectively attend to and encode certain types of information that produces a template that serves as the basis for a matching response. Such pressures should be stronger among apes than

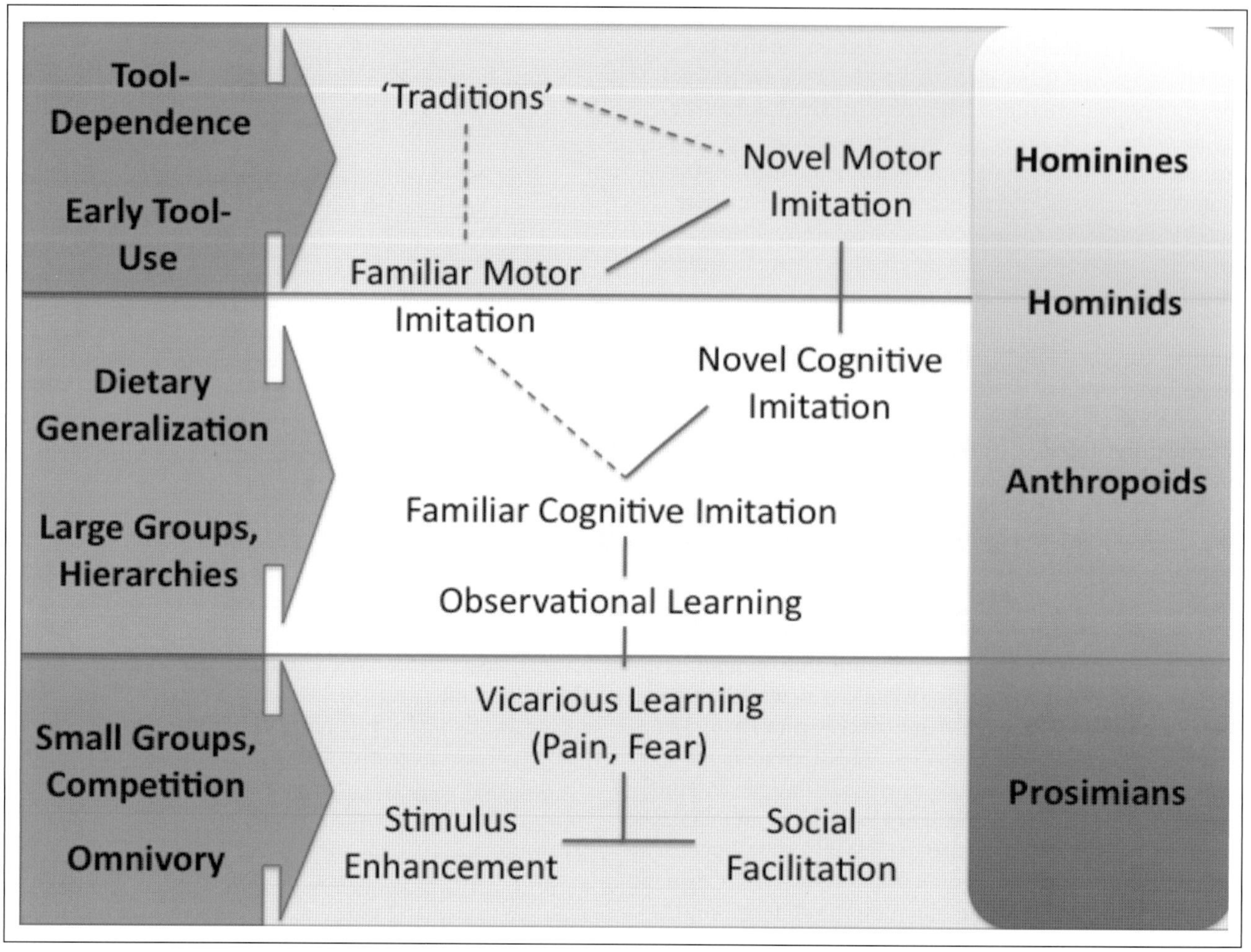

Figure 3. Mosaic Evolution of Imitation. The imitation faculty likely evolved from primitive circuits mediating vicarious learning; specifically, the vicarious learning of fear, disgust and pain. The diagram presents a simplified summary of how certain selective forces and ecological problems may have acted upon these primitive circuits and produced a variety of distinct imitation mechanisms in different primate groups.

monkeys because while monkeys have specialized dentition and digestive systems, apes have somewhat generalized dental anatomy and, with the exception of gorillas, lack specialized digestive systems (Ankel-Simons, 2000). These anatomical differences mean that whereas monkeys are able to enjoy a relatively diverse diet, apes don't have the same luxury. In monkeys, diets range from non-ripe fruits and mature leaves to insects, small animals and gum. Ape anatomy, however, limits dietary options to a narrow range of foods that consist mostly of mature, nonfibrous fruits with high sugar and calorie content (Maier, 1984). As a consequence of these dietary limitations, the great apes occupy a fairly narrow range of ecological habitats, being largely restricted to tropical and woodland forests (Potts, 1998; 2004). Contrast the narrow ecological range of chimpanzees and orangutans to that of macaques that have made a home in the arid lands of Africa as well as the snowy hillside of Japan.

These ecological, morphological and dietary pressures that, among primates, are mostly unique to the great apes, placed a premium on novel behavioral, cognitive, and life history strategies that are critical to fitness (Potts, 2004) and presumably served as a compensatory mechanism for morphological limitations. One such behavioral strategy used to broaden the apes diet is the systematic pursuit of prey in groups—or 'hunting'—(Watts & Mitani, 2002), another has been extractive foraging using tools (Goodall, 1986; Whiten et al., 1999). Yet another, might have been the fission-fussion social organization of chimpanzees and bonobos. Holloway (1967; 1981; 1996) argued that these variables were likely to be "prime interactive agents in human brain evolution" (Holloway, 1996: 97). But I proposed that these behavioral and sociological innovations—hunting, fission-fusion, and tool-use—likely favored an elaboration of the imitation faculty, in particular, the evolution of a robust novel imitation mechanism that was functionally integrated with other domain-specific imitation mechanisms (e.g., motor and cognitive imitation) and by extension shaped neural organization and evolution. Certainly, the novel motor imitation skills of apes are less robust than those known to be present in children as young as 2.5

years of age (Herrmann et al., 2007). These more derived novel motor imitation skills likely date to the first members of the genus *Homo*, where the need and dependence on stone-tool technology and other methods of subsistence including hunting and gathering placed increasing pressures on various mechanisms of the imitation faculty. Some of these elaborations may have included functional connections with other conceptual systems mediating Theory of Mind and causality but also affective systems mediating cooperation and empathy.

Nevertheless, given the ecological circumstances of non-human great apes, an imitation faculty capable of novel motor imitation would immediately increase the fitness of chimpanzees, for example, as it would have provided individuals with the skills to effectively steal the technical knowledge of conspecifics and immediately use that knowledge to supplement their diets. Given the importance of such a skill, it should then be no surprise that apes have elaborate tool-traditions which afford the means to develop and maintain these skills necessary for sustenance (Whiten et al., 1999). Yet, note that traditions as they exist in chimpanzees and orangutans are mostly absent in monkeys (c.f., Table 1). And where they exist, as appears to be the case in capuchin monkeys, they comprise of just 2 or 3 behaviors which lack the diversity and complexity that characterized chimpanzee and orangutan behavioral traditions (Boinski et al., 2003; Panger et al., 2002; Perry et al., 2003). These differences may rest on the fact, in captivity, among chimpanzees such traditions are mediated by motor imitation coupled by a strong tendency to always use the group's preferred technique (see Whiten, 2005 for a review). No comparable evidence exists for capuchin monkeys, or any other monkey species. Perhaps the discontinuity between traditions in monkeys and apes is not surprising, given that monkeys', as a group, are characterized by numerous anatomical specializations that are specifically adapted to their niche, which in no small measure grants them the ability to exploit a wide range of diets and habitats without tools or the need for sophisticated traditions.

Imitation-Brain Co-Evolution

Given the evidence that capuchin, marmoset and rhesus monkeys as well as chimpanzees, orangutans and gorillas share a familiar imitation mechanism, familiar imitation is likely to be the most basic and ancestral feature of the imitation faculty, and the feature that is likely to be present in all animals that possess a faculty of imitation. The models proposed by Boyd & Richardson (1986) and Henrich & McElreath (2003) explain this facet of the imitation faculty best. While it's possible for an animal to possess an imitation faculty that can copy only familiar responses (familiar imitation), it's difficult to imagine an imitation faculty capable of novel imitation, yet incapable of familiar imitation. From this it follows that the evolution of a derived imitation faculty that includes the ability to copy novel responses is premised on mechanisms the mediate familiar imitation. Some of the neurobiological evidence reviewed above provides some insights into how the elaboration of the STS-F5-PS circuit in the macaque brain (c.f., Figure 1), for example, can make at least novel motor imitation possible via the representation of intransitive actions (Rizzolatti, 2005) and input from other neural regions, in particular dorsolater and vertal prefrontal cortex as well the posterior cerebellum (c.f., Figure 2). However, logically, novel motor imitation is premised on novel cognitive imitation. The former seems difficult (if not impossible) without first having the ability to copy novel cognitive rules. But what selection pressures might have driven the elaboration of this faculty? One possibility is the need to develop and acquire more effective extractive foraging techniques; specifically, techniques that require the use of tools. Such selection pressures on observable behaviors certainly affected neural organization and perhaps contributed to mosaic brain evolution (Holloway, 1967; 1996). Perhaps it's no surprise that the regions that Holloway and colleagues have identified as early candidates of reorganization such as parietal, cerebellar and striate cortex, also happen to be areas critical for imitation (c.f., Figures 1 and 2).

Specifically, the evolution of the imitation faculty most certainly involved structural and organizational changes to a number of domain-general and domain-specific neurocognitive circuits: including attentional networks necessary to focus attention on relevant information, memory systems for the purpose of representing, encoding and recalling the target information, as well as changes to the 'reward' and 'empathic' systems, necessary for learning and vicarious reinforcement. For instance, observational learning likely resulted from changes to the 'reward networks' of the brain. Specifically, changes to the left anterior insula, associated with the facial recognition as well as the imitation of 'disgust' (Carr et al., 2003) and along with the anterior cingular cortex, mediating pain empathy (Singer et al., 2004), are most certainly involved in vicarious punishment (c.f., Bandura, 1977). These changes provided individuals with a powerful tool, the power to learn what not to do or what behaviors are most likely to decrease fitness. Such vicariously learned aversions have been reported in many animals including birds, rats and primates (for review see: Olsson & Phelps, 2007). But there must also have been changes to structures that contribute to vicarious positive reinforcement. That is, a mechanisms that promotes fitness-increasing behaviors but through vicarious rather than direct learning. Unfortunately, there's very little to nothing that has been done about vicarious positive reinforcement or the study of positive empathy. A better understanding of the role of the vicarious experience of positive emotions will go a long way to explain vicariously learning; a central component of novel imitation and by extension how pressure to make individuals better novel imitators directed brain evolution and re-organization.

Conclusions

The data summarized above provides compelling evidence that the imitation faculty is mosaic and given its distribution among primates, its evolution and neural organization appears to reflect this fact. Holloway and colleagues have identified a number of neural regions such as cerebellar and parietal cortex that have undergone significant organizational changes. Others, such as Deacon (1997) and Semendeferi et al (2001), have argued for relative expansions of prefrontal regions (see Holloway, 2002 for a critique), structures that have been implied in both familiar and novel imitation. Given our knowledge of tool traditions in contemporary chimpanzee societies and evidence from the cognitive neurosciences identifying frontal, parietal, and cerebellar regions as critical for imitation, it might not be so surprising that these neural regions, central to imitation, appear to have undergone radical changes in the course of human brain evolution.

The mosaic nature of the imitation faculty, consisting of the ability to copy different types of rules and responses including, familiar motor actions (i.e., familiar motor imitation) as well as novel cognitive rules (i.e., novel cognitive imitation), most certainly afforded monkeys the ability to appropriately copy the (familiar) actions of their conspecifics. The evolution of this skill was likely to be a specific adaptation to the pressures of group living, such as pressures associated with managing social hierarchies and group feeding. From this it follows that familiar imitation should be common in most social species where the ability to adaptively copy the familiar behaviors of conspecifics during synchronized activities like foraging, feeding and territory defense would afford important fitness benefits; reviews of social learning in a variety of animals suggests that this is the case (see Zentall, 2006). These cognitive mechanisms mediating familiar motor imitation and novel cognitive imitation as well as observational learning provided the biological raw materials for the evolution of novel motor imitation. Here it's proposed that a combination of sociological, ecological and technological variables favored such a skill. From this it follows that novel imitation should be common in species with generalized anatomies and where technical (or specialized motor) knowledge is critical for survival. Thus, in this view, the elaboration of a critical social cognition skill—imitation—was the product not simply of social factors but physical factors associated with knowledge of tools, motor actions and spatial relations.

References

Abravanel, E., & Sigafoos, A. D. (1984). Exploring the presence of imitation during early infancy. Child Development 55, 381-392.

Ankel-Simons F., 2000. Primate Anatomy: An introduction. Academic Press, London.

Anisfeld, M., 1991. Neonatal imitation: Review. Developmental Review 11, 60-97.

Anisfeld, M., 1996. Only tongue protrusion modeling is matched by neonates. Developmental Review 16, 149-161.

Anisfeld, M., Turkewitz, G., Rose, S.A., Rosenberg, F.R., Sheiber, F.J., Couturier-Fagan, D.A., et al., 2001. No compelling evidence that newborns imitate oral gestures. Infancy 2, 111-122.

Bandura, A., 1977. Self-efficacy: toward a unifying theory of behavioral change. Psychological Review 84, 191-215.

Boinski, S., Quatrone, R.P., Sughrue, K., Selvaggi, L., Henry, M., Stickler, C.M., Rose, L.M., 2003. Do brown capuchins socially learn foraging skills? In: Fragaszy, D., Perry, S., (Eds.), The Biology of Traditions. Cambridge University Press, Cambridge. pp. 365-390.

Boyd, R., Richardson, P.J., 1985. Culture and the evolutionary process. Chicago University Press. Chicago.

Brown, S., Martinez, M.J., Hodges, D.A., Fox, P.T., Parsons, L.M., 2004. The song system of the human brain. Cognitive Brain Research, 20, 363-75.

Bshary, R., Grutter, A.S., 2006. Image scoring and cooperation in a cleaner fish mutualism. Nature 22, 441 (7096), 975-978.

Buccino, G., Binkofski, F., Fink, G.R., Fadiga, L., Fogassi, L., Gallese, V., et al., 2001. Short Communication–Action observation activates premotor and parietal areas in a somatotopic manner: an fMRI study. European Journal of Neuroscience 13, 400-404.

Bugnyar, T., Huber, L., 1997. Push or Pull: an experimental study on imitation in marmosets. Animal Behaviour 54, 817-831.

Buller, D.J., 2006. Adapting Minds: Evolutionary psychology and the persistent quest for human nature. MIT Press, Cambridge.

Buss, D.M., 2006. Strategies of human mating. Psychological Topics 15, 239-260.

Buss, D.M., 2007. The evolution of human mating. Acta Psychologica Sinica 39, 502-512.

Byrne, R., 2005. Social cognition: imitation, imitation, imitation. Current Biology 15, R498-R500.

Byrne, R.W., Russon, A.E., 1998. Learning by Imitation: a hierarchical approach. Behavioral & Brain Sciences 21, 667-721.

Byrne R.W, Whiten A., 1990. Tactical deception in primates: the 1990 database. Primate Report 27, 1-101.

Caldwell, C.A., Whiten, A., 2002. Evolutionary perspectives on imitation: is a comparative psychology of social learning possible? Animal Cognition 5, 193-208.

Call, J., Carpenter, M., Tomasello, M., 2005. Copying results and copying actions in the process of social learning: chimpanzees (*Pan troglodytes*) and human children (*Homo sapiens*). Animal Cognition 8, 151-163

Call, J., Tomasello, M., 1995. Use of social information in the problem solving of orangutans (*Pongo pygmaeus*) and human children (*Homo sapiens*). Journal of Comparative Psychology 109, 308-20.

Carr, L., Iacoboni, M., Dubeau, M.C., Mazziotta, J.C., Lenzi, G.L., 2003. Neural mechanisms of empathy in humans: a relay from neural systems for imitation to limbic areas. Proceedings of the National Academy of Sciences 100, 5497-502.

Chaminade, T., Meltzoff, A.N., Decety, J., 2002. Does the end justify the means? A PET exploration of the mechanisms involved in human imitation. Neuroimage 15, 318-328.

Cochin, S., Barthelemy, C., Roux, S., Martineau, J., 1999. Observation and Execution of Movement: Similarities Demonstrated by Quantified Electroencephalography. European Journal of Neuroscience 11, 1839-1842.

Dautenhahn, K., Nehaniv, C., 2002. Imitation in Animals and Artifacts. MIT Press, Cambridge.

Deacon, T.W., 1997. The Symbolic Species. W.W. Norton, New York.

Decety, J., Chaminade, T., 2005. The neurophysiology of imitation and intersubjectivity. In: Hurley, S., Chater, N., (Eds.), Perspectives on Imitation. From Neuroscience to Social Science-Volume 1: Mechanisms of Imitation and Imitation in Animals. MIT Press, Cambridge. pp.119-140.

Fadiga, L., Fogassi, L., Pavesi, G., Rizzolatti, G., 1995. Motor facilitation during action observation: a magnetic stimulation study. Journal of Neurophysiology 73, 2608-2611.

Ferrari, P.F., Visalberghi, E., Paukner, A., Fogassi, L., Ruggiero A., Suomi, S.J., 2006. Neonatal imitation in Rhesus Macaques. PLoS Biology 4, 501-1508.

Fodor, J., 1983. The Modularity of Mind. MIT Press, Cambridge.

Galef, B.G., Jr., 1988. Imitation in Animals: History, definition, and interpretation of data from the psychological laboratory. In: Galef, Z., (Ed.), Social Learning: Psychological and Biological Perspectives. Lawrence Erlbaum, New Jersey. pp. 3-28.

Goodall, J., 1986. The chimpanzees of Gombe: Patterns of behavior. Belknap Press of Harvard University Press, Cambridge, MA.

Grezes, J., Costes, N., Decety, J., 1998. Top-Down Effect of Strategy on the Perception of Human Biological Motion: A Pet Investigation. Cognitive Neuropsychology 15, 553-582.

Henrich, J., McElreath, R., 2003. The evolution of cultural evolution. Evolutionary Anthropology 12, 123-135.

Herrmann, E., Call, J., Hernandez-Lloreda, M.V., Hare, B., Tomasello, M., 2007. Humans have evolved specialized skills of social cognition: The cultural intelligence hypothesis. Science 317, 1360-1366.

Heyes, C., 2001. Causes and Consequences of Imitation. Trends in Cognitive Science 5, 253-261.

Heyes, C., 2002. Imitation by Association. In: Hurley, S., Chater, N., (Eds.), Perspectives on Imitation. From Neuroscience to Social Science-Volume 1: Mechanisms of Imitation and Imitation in Animals. MIT Press, Cambridge. pp. 157-176.

Heyes, C.M., 2004. Imitation by association. In: Hurley, S., Chater, N., (Eds.), Perspectives on Imitation. From Neuroscience to Social Science-Volume 1: Mechanisms of Imitation and Imitation in Animals. MIT Press, Cambridge. pp. 157-176.

Holloway, R.L., 1967. The evolution of the human brain: Some notes toward a synthesis between neural structure and the evolution of complex behavior. General Systems 12, 3-19.

Holloway, R.L., 1969. Culture: A human domain. Current Anthropology 10, 395-412.

Holloway, R.L., 1979. Brain size, allometry, and reorganization: toward a synthesis. In: Hahn, M.E, Jensen, C., Dudek, B.C., (Eds.), Development and evolution of brain size: behavioral implications Academic Press, New York. pp. 59-88.

Holloway, R.L., 1981. Cultural symbols and brain evolution: a synthesis. Dialectical Anthropology 5, 287-303.

Holloway, R.L., 1996. The evolution of the human brain. In: Lock, A., Charles, P.A., (Eds.), Handbook of Human Symbolic Evolution. Clarendon Press, Oxford.

Holloway, R.L., 2002. Brief communication: how much larger is the relative volume of area 10 of the prefrontal cortex in humans? American Journal of Physical Anthropology, 118, 399-401.

Holloway, R.L., Broadfield, D., Yuan, M., 2004. The Human Fossil Record, Vol. 3: Brain endocasts—The paleoneurological evidence. Wiley-LISS, New York.

Hopper, L., Spiteri, A., Lambeth, S., Schapiro, S., Horner, V., Whiten, A., 2007. Experimental studies of traditions and underlying transmission processes in chimpanzees. Animal Behaviour 73, 1021-1032.

Horner, V., Whiten, A., 2005. Causal knowledge and imitation/emulation switching in chimpanzees (*Pan trogiodytes*) and children (*Homo sapiens*). Animal Cognition 8, 164-181.

Horner, V., Whiten, A., 2007. Learning from others' mistakes? Limits on understanding a trap-tube task by young chimpanzees. Journal of Comparative Psychology.

Horner, V., Whiten, A., de Waal, F., 2006. Faithful replication of foraging techniques along cultural transmission chains by chimpanzees and children. Proceeding of the National Academy of Sciences, 103, 13878-13883.

Horowitz, A.C., 2003. Do humans ape? Or do apes human? Imitation and intention in humans (*Homo sapiens*) and other animals. Journal of Comparative Psychology 117, 325-336.

Humphrey, N.K., 1976. The social function of intellect. In: Bateson, P.P.G., Hinde, R.A., (Eds.), Growing Points in Ethology: Proceedings. Cambridge University Press, London. pp. 303-317.

Jacobson, S., 1979. Matching behavior in the young infant. Child Development 50, 425-430.

Jensen, A.R., 2000. The g factor: psychometrics and biology. Novartis Foundation Symposium 233, 37-47.

Johnson-Frey, S.H., 2003. What's so special about human tool use? Neuron 39, 201-204.

Johnson-Frey, S.H., 2004. The neural bases of complex tool use in humans. Trends in Cognitive Sciences 8, 71-78.

Jolly, A., 1966. Lemur social behavior and primate intelligence. Science, 153, 501-506.

Jones S.S., 1996. Imitation or exploration? Young infants' matching of adults' oral gestures. Child Development 67, 1952-1969.

Jones S.S., 2006. Exploration or imitation? The effect of music on 4-week-old infants' tongue protrusions. Infant Behavior and Development 29, 126-130.

Lefebvre, L., Gaxiola, A., Dawson, S., Timmermans, S., Rozsa, L., Kabai, P., 1998. Feeding innovations and forebrain size in Australasian birds. Behaviour 135, 1077-1097.

Lefebvre, L., Whittle, P., Lascaris, E., Finkelstein, A., 1996. Feeding innovations and forebrain size in birds. Animal Behaviour 53, 549-560.

Maier, W., 1984. Tooth morphology and dietary specialization. Plenum Press, New York.

Marr, D., 1982. Vision: A computational investigation into the human representation and processing of visual information. Freeman, San Francisco.

McGrew, W.C., 1992. Chimpanzee Material Culture. Cambridge University Press, Cambridge.

McGrew, W.C., 1994. Tools Compared: The Material of Culture. In: Wrangham, M., deWaal, F.B.M., Heltne, P., (Eds.), Chimpanzee Cultures. Harvard University Press, Cambridge. pp. 25-40.

McGrew, W.C., 2001. The other faunivory: primate insectivory and early human diet. In: Stanford, C.B., Bunn, H.T., (Eds.), Meat-eating and Human Evolution. Oxford University Press, New York. pp. 160-178.

Meltzoff, A.N., Moore, K.M., 1977. Imitation of Facial and Manual Gestures by Human Neonates. Science 198, 75-78.

Myowa-Yamakoshi, M., Matsuzawa, T., 1999. Factors influencing imitation in manipulatory actions in chimpanzees (*Pan troglodytes*). Journal of Comparative Psychology 113, 128-136.

Myowa-Yamakoshi, M., Tomonaga, M., Tanaka, M., Matsuzawa T., 2004. Imitation in neonatal chimpanzees (*Pan troglodytes*). Dev. Sci. 7, 437-442.

Nagell, K., Olguin, R., Tomasello, M., 1993. Processes of Social Learning in the Tool Use of Chimpanzees (*Pan troglodytes*) and Human Children (*Homo sapiens*). Journal of Comparative Psychology 107, 174-186.

Nguyen, N.H., Klein, E.D., Zentall, T.R., 2005. Imitation of two-action sequences by pigeons. Psychon. Bull. Rev. 12, 514-518.

Panger, M., Perry, S., Rose, L.M., Gros-Louis, J., Vogel, E., Mackinnon, K.C., Baker, M., 2002. Cross-Site Differences in Foraging Behavior of White-Faced Capuchins (*Cebus capucinus*). American Journal of Physical Anthropology 119, 52-66.

Parker, S.T., Gibson, K.R., 1977. Object manipulation, tool use, and sensorimotor intelligence as feeding adaptation in cebus monkeys and great apes. Journal of Human Evolution 6, 623-641.

Parker, S., McKinney, M.L., 1999. Origins of Intelligence: The evolution of cognitive development in monkeys, apes, and humans. Johns Hopkins University Press, Baltimore.

Paz-y-Miño, G., Bond, A.B., Kamil, A.C., Balda, R.P., 2004. Pinyon jays use transitive inference to predict social dominance. Nature 430, 778-781

Perry, S., Panger, M., Rose, L.M., Baker, M., Gros-Louis, J., Jack, K., Mackinnon, K.C., Manson, J., Fedigan, L., Pyle, K., 2003. Traditions in wild white-faced capuchin monkeys. In: Fragaszy, D., Perry, S., (Eds.), The Biology of Traditions. Cambridge University Press, Cambridge. pp. 391-425

Petrosini, L., 2007. "Do what I do" and "do how I do": different components of imitative learning are mediated by different neural structures. Neuroscientist, 13, 335-48.

Petrosini, L., Leggio, M.G., Molinari, M., Neri, P., Graziano, A., Mandolesi, L., 2000. Representation of Actions in Rats: The Role of Cerebellum in Learning Spatial Performances by Observation. Proceedings of the National Academy of Sciences 97, 2320-2325.

Petrosini, L., Leggio, M.G., Neri, P., Graziano, A., Mandolesi, L., Molinari, M., 1999. Cerebellar Contribution to Spatial Event Processing: Characterization of Procedural Learning. Experimental Brain Research 127, 1-11.

Potts, R., 1998. Environmental hypotheses of hominin evolution. American Journal of Physical Anthropology Supplement 27, 93-136.

Potts, R., 2004. Paleoenvironmental basis of cognitive evolution in great apes. American Journal of Primatology 62, 209-228.

Povinelli, D.J., 2000. Folk Physics for Apes: The Chimpanzee's Theory of How the World Works. Oxford University Press, Oxford.

Reader, S.M., Laland, K.N., 2002. Social Intelligence, Innovation, and Enhanced Brain Size in Primates. Proceedings of the National Academy of Sciences U.S.A. 99, 4436-4441.

Rilling, J.K., Seligman, R.A., 2002. A quantitative morphometric comparative analysis of the primate temporal lobe. Journal of Human Evolution 42, 505-533.

Rizzolatti, G., 2005. The mirror neuron system and imitation. MIT Press, Cambridge.

Rizzolatti, G., Fadiga, L., Fogassi, L., Gallese, V., 2002. From mirror neurons to imitation: facts and speculation. In: Metlzoff, Prinz, (Eds.), The Imitative Mind: Development, Evolution, and Brain Bases. Cambridge University Press, Cambridge.

Rumiati, R.I., Bekkering, H., 2003. To imitate or not to imitate? How the brain can do it, that is the question! Brain and Cognition 53, 479-482.

Rumiati, R.I., Tessari, A., 2002. Imitation of novel and well-known actions: the role of short-term memory. Experimental Brain Research 142, 425-433.

Rumiati, R.I., Weiss, P.H., Tessari, A., Assmus, A., Zilles, K., Herzog, H., et al., 2005. Common and Differential Neural Mechanisms Supporting Imitation of Meaningful and Meaningless Actions. Journal of Cognitive Neuroscience 17, 1420-1431.

Semendeferi, K., Armstrong, E., Schleicher, A., Zilles, K., Van Hoesen, G.W., 2001. Prefrontal cortex in humans and apes: a comparative study of area 10. American Journal of Physical Anthropology, 114, 224-41.

Shiffrar, M., Freyd, J.J., 1990. The Apparent Motion of the Human-Body. Psychological Science 1, 257-264.

Singer, T., Seymour, B., O'Doherty, J., Kaube, H., Dolan, R.J., Frith, C.D., 2004. Empathy for pain involves the affective but not sensory components of pain. Science 303, 1157–1162.

Spelke, E., 2000. Core Knowledge. American Psychologist, 55, 1233-1243.

Stevens, J.A., Fonlupt, P., Shiffrar, M., Decety, J., 2000. New Aspects of Motion Perception: Selective Neural Encoding of Apparent Human Movements. Neuroreport. 11, 109-115.

Subiaul, F., Romansky, K., Klein, T., Terrace, H.S., 2007. Cognitive imitation in 2-year old human children (*Homo sapiens*): A comparison with rhesus macaques (*Macaca mulatta*). Animal Cognition. Feb 8 [E-Pub ahead of Print].

Subiaul, F., Cantlon, J., Holloway, R., Terrace, H., 2004. Cognitive Imitation in Rhesus Macaques. Science 305, 407-410.

Subiaul, F., 2007. The imitation faculty in monkeys: evaluating its features, distribution and evolution. Journal of Anthropological Sciences 85, 35-62.

Tessari, A., Rumiati, R.I., 2004. The Strategic Control of Multiple Routes in Imitation of Actions. Journal of Experimental Psychology. Human Perception and Performance 30, 1107-1116.

Thorndike, E.L., 1898. Animal intelligence: An experimental study of the associative processes in animals. Psychol. Rev. Monogr. Suppl. 2, 8.

Thorndike, E.L., 1911. Animal Intelligence. Macmillan, New York, NY.

Thorpe, W.H., 1956. Learning and Instinct in Animals. Methuen, London, UK.

Tomasello, M., 1999 The Cultural Origins of Human Cognition. Harvard University Press, Cambridge, MA.

Tomasello, M., Call, J., 1997. Primate Cognition. Oxford University Press, New York.

Tomasello, M., Kruger, A.C., Ratner, H.H., 1993. Cultural learning. Behavioral and Brain Sciences 16, 495-552.

Tomasello, M., Savage-Rumbaugh, E.S., Kruger, A.C., 1993. Imitative learning of actions on objects by children, chimpanzees, and enculturated chimpanzees. Child Development 64, 1688-1705.

Tooby, J., Cosmides, L., 1990. The past explains the present: emotional adaptations and the structure of ancestral environments. Ethology and Sociobiology 11, 375-424.

van Schaik, C.P., Ancrenaz, M., Borgen, G., Galdikas, B., Knott, C.D., Singleton, I., Suzuki, A., Utami, S.S., Merrill, M., 2003. Orangutan cultures and the evolution of material culture. Science 299, 102-105.

Van Schaik, C.P., Ancrenaz, M., Borgen, G., Galdikas, B., Knott, C.D., Singleton, I., 2003. Orangutan cultures and the evolution of material culture. Science 299, 102-105

Visalberghi, E., Fragaszy, D., 1990. Do monkeys ape? In: Hurley, S., Chater, N., (Eds.), Perspectives on Imitation. MIT PressCambridge.

Visalberghi, E., Fragaszy, D., 2002. Do monkeys ape? Ten Years Later. In: Dautenhahn, K., Nehaniv, C., (Eds.), Imitation in animals and artifacts. MIT Press, Cambridge. pp. 471-479.

Visalberghi, E., Fragaszy, D.M., 1995. The behaviour of capuchin monkeys, Cebus apella, with novel food: the role of social context. Animal Behaviour 49, 1089-1095.

Visalberghi, E., Trinca, L., 1989. Tool Use in Capachin Monkeys: Distinguishing Between Performing and Understanding. Primates 30, 511-521.

Voelkl, B., Huber, L., 2000. True imitation in marmosets. Animal Behaviour 60, 195-202.

Voelkl, B., Huber, L., 2007. Imitation as faithful copying of a novel technique in marmoset monkeys. PLoS ONE 2, e611.

Watts, D., Mitani, J.C., 2002. Hunting behavior of chimpanzees at Ngogo, Kibale National Park, Uganda. International Journal of Primatology 23, 1-28.

Whiten, A., 1998. Imitation of the Sequential Structure of Actions by Chimpanzees (*Pan troglodytes*). Journal of Comparative Psychology 112, 270-281.

Whiten, A., Custance, D.M., 1996. Studies of Imitation in Chimpanzees and Children. In: Heyes, C.M., Galef, B.G., (Eds.), Social Learning in Animals: The Roots of Culture. Academic Press, Inc, San Diego.

Whiten, A., Byrne, W., 1997. Machiavellian Intelligence II: Extensions and Evaluations. Cambridge University Press

Whiten, A., Ham, R., 1992. On the nature and evolution of imitation in the animal kingdom: reappraisal of a century of research. Advances in the Study of Behavior 21, 239-283.

Whiten, A., Horner, V., Litchfield, C.A., Marshall-Pescini, S., 2004. How do apes ape? Learning and Behaviour 32, 36-52.

Horner, V., Whiten, A., 2007. Learning from others' mistakes? limits on understanding a trap-tube task by young chimpanzees (*Pan troglodytes*) and children (*Homo sapiens*). Journal of Comparative Psychology 121, 12-21.

Whiten, A, Goodall, J., McGrew, W.C., Nishida, T., Reynolds, V., Sugiyama, Y., Tuti, C.E., Wrangham, R.W., Boesch, C., 1999. Cultures in chimpanzees. Nature, 399, 682-685.

Whiten, A., Horner, V., Litchfield, C., Marshall-Pescini, S., 2004. How do apes ape? Learning & Behavior 32, 36-52.

Williams, J.H., Whiten, A., Singh, T., 2004. A systematic review of action imitation in autistic spectrum disorder. Journal of Autism Developmental Disorders 34, 285-299.

Zentall, T., 2006. Imitation: Definitions, evidence, and mechanisms. Animal Cognition 9, 335-367.

Zentall, T., 2007. Temporal discrimination learning by pigeons. Behavioural Processes 74, 286-292.

CHAPTER 18

The Foundations of Primate Intelligence and Language Skills

Duane M. Rumbaugh, E. Sue Savage-Rumbaugh, James E. King and Jared P. Taglialatela

It must have been an exciting time when hominids found that they had a level of intellectual operations that clearly was giving them an increased advantage in undertaking the unceasing daily challenges of survival–generally in competition with a variety of animals of the savannah and forest, of getting food and water, avoiding becoming a ready meal for carnivores, dying from exposure, and so on. These challenges were only compounded as they migrated to novel environments in the far corners of Europe and Southeast Asia. Most likely a premium came to be placed on tools. Initially, natural items, such as stout portions of branches, could serve as clubs, and broken cobbles might yield knife-sharp edges for skinning and butchering. But tools are consumable and subject to being lost in less-than-well-coordinated running or left at a site because of distractions from salient distal events.

It is not difficult to conjure the premium afforded by invention and symbolic thought. We have the long-standing view that it was the early evolution of bipedalism in hominids that made it possible for later evolution of general intelligence by selection for brain size, and only secondarily by selection for body size. Prior to the emergence of competent bipedalism, intelligence within the primate order appears to have been a generous corollary of body size. Size always has had its perquisites, providing priority access to the resources that afford not only life but the comforts thereof. The factor limiting size was likely the need for ready nourishment, as well as an environment providing accessible food and water for group living. The larger the quadruped ape, the more food and time to eat were required for life. The encephalization process generously gave the great apes' brains larger than justified on the basis of their body size.

The encephalization of the great apes generally makes them superior to the lesser apes and to the monkeys in formal tests of learning (Rumbaugh & Washburn, 2003). Though excellent learners, even the larger monkeys and baboons lack the readiness to become rational learners in comparision to the apes–that is to learn the overarching principles that differentiate classes of visual discrimination problems. Relational learning includes the ability to learn a general rule that defines a correct response for an entire class of problems containing an unlimited number of exemplars, what Harry Harlow (1949) referred to as "learning sets." Monkeys can achieve that capability, but they require far more experience and training than does an ape reared in a similar environment.

Similarly, the larger monkey species can, with extended experience, become quite proficient at transferring learning and consequently benefit from these increased amounts of learning (e.g., knowledge about how to do a task); by contrast the performance of the smaller primate species with smaller brains might become increasingly compromised with additional training on tasks prior to transfer-of-learning tests (Beran, Gibson, & Rumbaugh, 1999). In other words, if one increases the amount of training even slightly prior to transfer-of-training tests, the apes and other large primates do substantially better, though the small primates do worse. We interpret this to mean that whereas all primates' initial learning of discrimination tasks is basically associative, they have a capability for advancement to relational learning—learning of overarching principles to expedite both learning and the transfer of learning to different situations. This advancement to relational learning is a positive function of the species' brain volume.

The Good Fortune of Having the Apes

Although the early hominids are no longer with us, we are fortunate to have the great apes and dozens of feral populations of monkeys and prosimians around the tropical belt of the planet. Rumbaugh and Washburn (2003) have made a recent analysis of experiments that focus upon the intelligence of primates within a comparative framework; the reader is referred to the relevant studies in their book to better understand the bases for the rest of this chapter.

Because of the close genetic relationship between modern humans and the great apes (Gagneux & Varki, 2001), we believe that much can be learned about the origins of intelligence and language through research with the great apes both in the field and laboratory. Furthermore, and more specific to this volume, such research can help us better understand the emergence of intelligence and complex communication processes in the hominids, and ultimately the essence of our own learning and behavior.

Prior to the recent cognitive revolution, psychology was heavily influenced by radical and methodological behaviorism for the majority of the 20th century (Amsel,1989). For this reason alone it is timely that we reexamine the processes of learning and behavior and, in particular, the presumed role of reinforcement in animals' adaptation to their environments. As we do so, we will reconsider how organisms should be viewed and how the processes of learning and behavior that embrace their root sources, from instincts to conditioning, cognition, intelligence and culture, feed into adaptation and behavior. Rumbaugh, King, Beran, Washburn, and Gould (2007) recently have offered a theory of learning and behavior based on salience, not on reinforcement as it is conventionally defined. We will review the basic principles of that theory after we attend more specifically to the various complex learning and problem-solving skills of the great apes, as well as of their capacity to understand symbolism and certain dimensions of language. The capabilities that we will consider entail the emergence of new behaviors and skills that are well beyond behaviors that would be predicted from a behavioristic interpretation of an animal's specific training and/or reinforcement history. For that reason, we term these new behaviors as *emergents*.

A reinterpretation of how organisms should be viewed and of the processes of learning and behavior that embrace their root sources, from instincts to conditioning, cognition, intelligence and culture, has been recently reported by Rumbaugh, King, Beran, Washburn, and Gould (2007). We will review the basic principles of that theory after we attend more specifically to the various complex learning and problem-solving skills of the great apes, as well as of their capacity to understand symbolism and certain dimensions of language. The capabilities that we will consider entail the emergence of new behaviors and skills that are well beyond behaviors that would be predicted from a behavioristic interpretation of an animal's specific training and/or reinforcement history. For that reason, we term these new behaviors as *emergents*.

How Is It That Apes Can Learn Language and Make Tools?

Among the host of delightfully puzzling questions driving the field of primatology today, as it seeks for an ever-objective definition of the ape mind, are those that ask, basically, "In apparent contradiction to the constraints based on the conventional principles of learning and behavior that have been dominant for the past 75 years, why do apes in particular exhibit emergents that take form as creative problem-solving abilities rather than relatively fixed behaviors in response to specific stimuli? What are their parameters? How are these abilities acquired? How do some apes come to learn the semantic meanings of word symbols, to use them in novel social communication, and even to comprehend human language and its elemental syntax? And how do some of them become sufficiently proficient from only observational learning to make tools and start fires based only on observational learning?"

We clearly need a new and comprehensive framework of learning and behavior that embraces unlearned (i.e., instinctive) behaviors constrained by genetics and the remarkable behaviors brought about through conditioning procedures, yet a framework that also provides for creative and inventive behaviors that emerge from time to time, though without a specific history of training that could account for them. These emergent behaviors come as surprises and are seen as something well beyond the domain of reinforcement, of highly specified training procedures (Rumbaugh et al., 2007; Rumbaugh, Savage-Rumbaugh, & Washburn, 1996; Rumbaugh, Washburn, & Hillix, 1996).

Emergents have their roots in unlearned behavior systems as well as in the respondents and operants of conditioned behavior. Yet, they are something identifiably different that strongly suggests high plasticity and intelligence as foundations. Just how emergents are generated by the normal operations of the brain from the experiences that all of life offers is at present imperfectly understood, to say the least. When asked how emergents are formed, one eminent neuroscientist replied, "God only knows." To come to understand the parameters of emergents at any level will take decades of research at all levels, but to understand them better along the way will be reward sufficient to the task. In the meanwhile, primatologists can make valuable contributions in defining the antecedent and subject parameters of emergents and the impacts of emergents upon subsequent behavior, including learning and all other basic processes.

Lana

Let us take a few select examples from our own laboratory research. First, Lana, a female chimpanzee, was taught dozens of word-lexigrams (geometric patterns glossed as words) by basic operant techniques (Rumbaugh, 1977). Specifically, she learned how to organize them into stock sentences required by the computer to operate specific devices that would vend for her a variety of foods, drinks, music, slides, movies, a view out-of-doors, human companionship, grooming play, and so on.

But well beyond that, she was the one, not us, to initiate conversations by using a keyboard in order to get things that she could not otherwise access and to ask for the names of things. On occasion she would direct caregivers' attention to malfunctioning systems. She accurately differentiated sentence stems that correctly began sentences, (that she then would complete to obtain various rewards), from sentence stems that were in error and, hence, were erased as having no value to her. These are only a few of the highly significant behavioral extensions manifested by Lana that had literally no prior history of reward or reinforcement to account for their emergence.

Sherman and Austin

Second, the chimpanzees Sherman and Austin (Savage-Rumbaugh, 1986) demonstrated their capacity for understanding the semantic meanings of the word symbols (i.e., lexigrams) with which they worked each day. Initially they learned to sort lexigrams for three specific foods and three specific tools, drawn from a larger vocabulary of dozens of symbols. After further training, they were able to label just the lexigrams for these foods and tools used in initial training with two new lexigrams, one glossed "food" and the other "tool." Thus, they categorized the lexigrams for the three foods and three tools used in training with two new lexigrams that served to categorize all examples of food or all examples of tools symbolically. In the final test, they made only a single error between them in sorting 16 other food and tool lexigrams for a variety of foods and tools that had been reserved for the final test.

In brief, they were very precise in labeling these test lexigrams for foods or tools appropriately with the corresponding general food or tool lexigrams even though the test lexigrams for specific foods and tools had never been previously associated with the general food and tool lexigrams. We conclude that their labeling skills in controlled test must have reflected semantic foundations for their lexigrams. How would they have been able to categorize the test lexigrams so accurately if the lexigrams lacked meaning for them? In other words, the food and tool lexigrams had a general meaning for Sherman and Austin that transcended association with only a few specific lexigrams for particular foods and tools. We hold that the meaningfulness of these test symbols enabled covert representations of their physical referents and that it was those representations that, in turn, enabled Sherman's and Austin's remarkable labeling of them at the time of testing.

In sum, Sherman and Austin demonstrated that, for them, word-lexigrams could acquire symbolic meaning, which is absolutely fundamental to language. Their very limited training with only three food and three tool exemplars led to a generalized competence with 16 other lexigrams in a final test.

Kanzi

Third, Kanzi, a male bonobo, came to comprehend human speech, including both the meanings of individual words and their use in novel sentences of request (Savage-Rumbaugh et al., 1993). He acquired these abilities without any specific training to those ends. Instead, Kanzi was raised in a social environment in which language was used by others in a natural context, similar to that experienced by human children. Indeed, we had thought that no ape had the capacity to acquire these skills. In brief, his brain somehow took the experiences of his daily life and gave them structure and function in new vectors, demonstrating new emergent competencies not at all natural to his species.

Rhesus Monkeys

Fourth, two rhesus monkeys (Rumbaugh & Washburn, 2003) rapidly came to discern which of either member of pairs of numerals, from 0 through 5, was the one designated by experimenters to net the larger number of pellets on any given trial. The trials were massed and the monkeys were not food deprived, so the premium of their receiving, say, four rather than three small food pellets on a given trial seems trivial. Their training was then extended to include numerals 6, 7, 8, and 9. In final test, with seven possible pairings of the entire set of real numerals (i.e., 1-9) reserved for this test, they made only two errors.

In other words, they were able to conclude, on the basis of prior experience with the other numerals that had been used in training, which of two numerals encountered for the first time as a pair in final test would net the larger number of pellets. They did not do this because they were required to do so or even specifically trained to do so. Rather, their remarkably extended competence reflects operations by their brains that took the vast array of other relevant experiences and somehow organized them so as to declare the probable "better" choice of numerals on each novel test trial.

These are only a few of dozens of examples of emergents to which we could refer, yet they are sufficient to define our wonderful quandary: Out of their specific and relatively limited rearing and training histories, *how do primates come to manifest a variety of new abilities and even new competencies heretofore unanticipated and unforeseen—that is, emergents?*

This question cannot be pursued to a satisfactory conclusion within the limits of this chapter. Nonetheless, we shall attempt to define the bases for the questions and, finally, to point to a new perspective of learning and behavior that is more in keeping with our current understanding of behavior of all animal forms than is traditional reinforcement theory. All of this is done to the end that the reader have an enlightened perspective of how the salient events of a challenging social and physical environment fostered larger and larger brains, higher levels of intelligence, and the foundations for creative technologies to emerge. It was likely from this path that a socially complex culture as we know it finally took form. And culture, too, served to provide increasingly stimulating environments within which the hominid infants benefited in their intellectual stimulation and development.

Reinforcement Reconsidered

Although we fully agree that the immediate correlates and consequences of behavior are fundamental to the learning process and acknowledge that the concept of reinforcement has played a very significant role in the history of learning and behavior, it is time that we markedly revise our definitions of the term *reinforcement*, if not abandon it altogether. The reasons for so doing are as follows:

> The term carries excess meaning in that it encourages the beliefs that reinforcers actually strengthen associations between stimuli and specific behaviors and that all behaviors have reinforcement histories.
>
> Its definition has been inherently circular.
>
> It emphasizes behavior in its relation to specific stimuli inordinately and does not encourage consideration of whatever the subject might bring as a sentient and knowledgeable agent-of-action to the determination of its behavior.
>
> It inordinately emphasizes fixedness in behaviors and detracts from our likelihood of observing emergent behaviors–those that are creative, new, efficient, and insightful.
>
> By abandoning the term *reinforcement* and using *reward* in its stead, we continue to acknowledge the importance of consequences of behavior yet recognize that what we have called "reinforcement" is really a resource of value to our subjects and that, in essence, it is equivalent to "pay for work done."

The view that we are proposing here is as follows: As they adapt to their changing environments, organisms are fundamentally foragers for information. However, the foraging goes far beyond the usual sense of the word. Animals are constantly seeking optimal environments and, most importantly for our theory, relevant information that yields needed resources bountifully and in relatively safe contexts. The search for relevant information and subsequent use of that information in creative and imaginative ways form the basis for emergent behaviors. In contrast, organisms are not entities that have their behaviors comprehensively shaped and reinforced by the consequences of behavior.

A New Perspective

From a variety of perspectives, perhaps no other construct has survived the past century with greater impact than has reinforcement. At the risk of being too simplistic, we would like to say that all perspectives and definitions of reinforcement assume that if reinforcement occurs soon after the occurrence of a behavior, reinforcement can serve to strengthen the probability that this behavior will reoccur, given a repeat of the situation in which it appeared or was elicited. To us, the effect of reinforcement has always implied a certain degree of fixedness, a predictability, a robot-like predictability, that, at face value, are antithetical to creativity, invention, and intelligence generally for which apes are known. In its most basic traditional definition, reinforcement is posited as a theoretical process that strengthens an association, a stimulus, and a response. But now that we have solid evidence of complex cognitive skills and potentials in animals (including of course, humans), the concept of reinforcement is no longer very appropriate except in the context of simple, predictable responses. On the other hand, reinforcements can be important resources for the organism. We propose that the organisms learn predominantly about the resource values of reinforcements.

We are not asserting that the contingencies or aftermaths of behavior have no effect. Instead, we are arguing that the concept of reinforcement should be supplanted with other terms. We suggest the term *outcome*, with *reward* standing for an appetitive outcome, *resource* meaning an outcome garnered by foraging or taking, and *punishment* standing for an aversive outcome.

So, What Is Being Reinforced?

Historical tradition maintains that the manifestation of learning is expected to be the specific behavior that is being reinforced. Now, to illustrate the difficulties encountered in viewing reinforcement of responses as the determinant of both what is learned and what behavior is to be expected, let us consider a complex video-formatted task in which a rhesus monkey was assiduously trained by traditional operant techniques to capture an erratically moving target by using its foot to control a joystick that moved a cursor. Although the monkey was trained to use its foot exclusively (i.e., it was never permitted to use its hand in training), results made it clear that what was learned was something far more comprehensive.

After the monkey mastered the foot task, it was given its first opportunity to use either its hand or foot. It tried to use a foot not at all! Rather, it used its hand exclusively! Even more impressive was the fact that performance was better with a hand than it had ever been with the monkey's foot.

Though reinforced exclusively in training to use its foot, the monkey clearly had learned much, much more: It accrued an apparently comprehensive understanding about the task, namely the relationship between the movement of the joystick and the movement of the cursor on the video screen! Such comprehension in the rhesus is not to be accounted for by the reinforcement of motor responses but rather by the integrative processes of its brain. (See Rumbaugh & Washburn, 2003, for further details.)

Thus, while reinforcement doctrine would lead us to expect that the conditioning of response modes through extensive reinforcement histories will engender responses that are predictable and relatively stereotyped, reinforcements or rewards do not necessarily have such a limited effects! Stated most simply, though an experimenter might condition a specific response to be learned, the subject actually learns not only the relationships between responses but also how to access resources and avoid risks. In terms of the foraging metaphor used earlier, the contingency between stimulus and reward is simply interesting information that is "foraged" in order to be applied within a wider context.

In other words, the brain takes what it obtains from experience and then runs with it, metaphorically, to form new behaviors and new skills. If such were the case with the early hominids, it is clear how they become the dominant force in the world.

Rumbaugh, King, Beran, Washburn, and Gould (2007) have posited that the design of brains serves to bias the selective perception of events in accord with their salience—natural or acquired—and, also, to organize or interrelate them in accord with the ecological resources and needs of the subject so that it adapts and survives. Major principles of that theory now will be considered in relation to the advanced learning and cognition of primates. Because salience of stimuli and events, produced either externally or internally, are basic to the theory, we shall summarize our perspective of salience and its posited role in learning and behavior.

A Salience Theory of Learning and Behavior

What is salience? We begin with the assumption that consciousness is not necessarily a requisite to perceiving or responding to the salience of events—be they individual units or coupled by parameters outlined below. Notwithstanding, it seems reasonable to assume that organisms attend to stimulus events on the basis of their perceived priority. *Although this might be totally true, to avoid stopping with what is fundamentally a circular definition of salience, let us identify the attributes of objects and events that will declare them as salient.* Salience might be natural, or it might be acquired.

Natural Salience

There are several stimulus events that are inherently salient by reason of the species' genomes. They include ones for which salience is *natural*, such as the following: (a) natural sign-stimuli that are relatively species-specific; (b) intense stimuli (e.g., energies with high decibels, intense illumination and/or pressure levels) that threaten to exceed the sensory thresholds of a given species; (c) biologically predicated need states, as for moisture, nutrients, and an ambient temperature range that varies widely across life forms; (d) novel stimuli; and (e) perceptual integrative/organizing principles as originally defined by Gestalt psychology that serve to group and to otherwise enhance the prospects for an organized percept rather than a random field of stimulation.

These sources of salience are not necessarily dependent upon experience, though they might well be sensitive to requisite stimulation within certain levels of maturation (e.g., within critical age levels). In addition, all unconditional stimuli (of Pavlovian or classical conditioning) that elicit reflexes are inherently salient.

Acquired Salience

Other sources of salience are *accrued*, not necessarily through traditional learning processes but because of what we view as a natural, near-universal principle: Units (stimuli, events, and/or behaviors) that occur reliably in about the same time and space reliably tend to couple, to mix. This is true of most liquids, fumes, and even metals. It clearly is the case in the production of colors and odors. The celebrated neuroscientist Gerald Edelman (2006) has observed that neurons and neural circuits that fire together get wired together] Generally speaking, this mixture is of high probability as a general natural law or principle, echoing the work of D. O. Hebb (1949) and others.

Principles of the Framework

Thus, the first principle of our framework holds that learning is based on the reliable temporal or spatial contiguity of events. Units that co-occur reliably become at least metaphorically coupled or even blended to form an *amalgam*. There are two important corollaries of this principle:

1. In their coupling so as to form an amalgam, the units will mutually share their saliences and their response-eliciting characteristics. Thus, each amalgam will have unique characteristics above and beyond those of the units that have entered into its formation.

2. The merger of two or more units into an amalgam reflects the relative strengths of the individual units.

Consequently, some co-occurring units are coupled naturally because each unit has substantial strength. Thus, the units of lightening and thunder are readily coupled because each unit has substantial strength. It is not necessary that such units in their coupling have inherent sequential organization (e.g., though lightening always precedes thunder, other coupled units might co-occur in any order). That said, if one of two or more co-occurring units differs substantially in its strength from the others, the coupling is more likely if the weaker one(s) precede the stronger ones—as in classical conditioning. Whether units that are relatively weak in strength become coupled if they sequentially follow the units of substantial strength will be determined by the degree to which the salience that inheres in the stronger units obscures or masks the salience that inheres in the weaker units.

Thus we posit that an amalgam of stimulus events will reflect their shared response-eliciting properties as some positive function of the vigor of the responses produced by each stimulus and the relative strengths of their responses when they co-occur in time and/or space. Thus, in classical conditioning, both the conditional stimulus and unconditional stimulus mutually share their response-eliciting properties. It is only because the unconditional stimulus is the stronger of the two that its response-eliciting properties are more strongly manifested with the presentation of the conditional stimulus rather than vice versa. *In other words, the high-strength unconditional stimulus will cause a large change in the response to the conditional stimulus, whereas the low-strength conditional stimulus will have little or essentially no discernable effect on the response to the unconditional stimulus. Nevertheless, there is good reason to hold that the unconditional stimulus accrues an approximation of the relatively minimal response-eliciting property that inheres in the conditional stimulus* (Domjan, 2003).

A second principle of our framework is that species' brains are uniquely designed to process coupled stimulus events, to somehow file and process them to form emergent behaviors and emergent capacities that service the species' adaptation in both familiar and novel challenges. Organisms *detect* coupled events for which their neural systems have been attuned; that is, animals recognize reliable and predictable patterns that might be basic to their adaptation. Either the patterns of events are "out there" in the natural word, or they are reliable consequences of behaviors. They are the regularities if not the invariants of experience across time. The more complex the pattern, the more complex the cognitive system (and the brain) must be to recognize it in detail; notwithstanding, animals learn by detecting predictable temporal or spatial relations if they are extant among co-occurring units that are basic to their adaptation.

Brain Business

We have posited that species' brains or neural systems are attuned to attend to what is basic for their survival and reproduction. As its fundamental model of operation, we posit that the brain produces streams of amalgams as defined above. The brain also functions, perhaps continuously, to relate and interrelate the amalgams into systems that reflect their similarities and their relationships. Metaphorically, we use the term *templates* to label those systems of organized amalgams. We view templates as being either essentially natural or arbitrary. Natural ones are those that reflect the basic adaptation modes and significant processes required of the species. Natural templates receive amalgams that have such fundamental significance to a species' adaptation that rapid learning and adaptation are to be expected. Arbitrary templates are those that are entailed in everything else, such as the complexities of acquiring insights to other-than-natural challenges and of acquiring rules, forming strategies, mastering language, composing music, and inventing. The formation of arbitrary templates might require substantial periods of time, if not years, of experiencing classes of generalized experiences.

In their formation and operation, templates assimilate amalgams that are closely related or similar. When an already existing template cannot accommodate a stream of amalgams being formed as a result of a novel or unexpected pattern of stimulus events, then the assimilation process may adapt by forming a new template to accommodate the novel amalgams. We believe that the tension resulting from the effort to assimilate novel combinations amalgams into new templates may stimulate the formation of emergents as new options that might afford effective and energy-saving adaptations. The flexibility of the template formation process means that the emergent behaviors are emancipated from the constraints of traditional stimulus–response or response–reward mechanisms.

Reinforcement Redefined

The reader will note that reinforcement, according to our frame of reference, does not serve any specific role. Reinforcements obviously have major effects upon behavior due to their strength and response-eliciting properties, either of which might be of natural or acquired origin. In reliable and contiguous association with other stimuli, it shares both its salience and its response-eliciting properties with other current stimuli and behaviors to form amalgams—brain business. Thus, in classical conditioning the unconditional stimulus has natural salience and shares its response-eliciting properties with other stimuli that are contiguous with it, specifically the conditional stimulus. Across trials, the conditional stimulus and the unconditional stimulus form a stream of highly similar amalgams, all sharing a conditional stimulus–unconditioned stimulus temporal contiguity. Hence the conditional stimulus comes to function as though it were the unconditional stimulus, and conditioning is said to have occurred. Since the conditional stimulus is selected by the experimenter because it is weak, nonsalient, and does not elicit strong responses, the unconditional stimulus by itself subsequently shows little readily observable influence after formation of the amalgam between the conditional and unconditional stimulus. Nevertheless, we suspect that subtle influences of a conditional stimulus upon an unconditional stimulus can be detected by appropriate methodology.

We make the following points to clarify further the preceding argument made above. We hold that a process similar to either sensory preconditioning or autoshaping likely prevails in both respondent and operant conditioning situations. Sensory preconditioning enhances the salience of basically neutral stimuli simply by pairing them together temporally with more salient stimuli. In sensory conditioning paradigms, neither of the neutral stimuli would be regarded as an unconditioned stimulus. The less salient of the previously neutral stimuli gains in salience and become functionally equivalent to the more salient stimulus in its role. Thus, if one member of a pair is an unconditional stimulus, then quite likely the other less salient one will assume some of the properties of the unconditional stimulus despite the fact that it originally served as a conditional stimulus in a conditioning procedure.

The phenomena of autoshaping occurs when a neutral stimulus such as a light and a traditional reinforcer such as food are temporally paired. The food presentation is predicted only by the light and is independent of any response that the subject makes (Brown & Jenkins, 1968). After autoshaping, the subject makes responses to the neutral stimulus (e.g. pecking) that were previously make only to the reinforcer. The topography of the conditioned response of pigeons acquired therein (e.g. pecking the light) provides strong support for the frame of reference here advanced—that the functional role of the "reinforcer" is shared with (i.e., becomes elicited by) a visual target or a discriminative stimulus temporally associated with the reinforcer. If the "reinforcer" is grain, the bird pecks at the target as though if it were food; if it is water, the bird pecks as though it were drinking water.

Similarly, pigs described by Breland and Breland (1961) readily learned to pick up wooden nickels and deposit them in a piggy bank for food reward. Across time, however, the nickel-directed depositing behavior became disrupted as the pigs came to root and toss the nickel as though it were food. Thus, the reward came to share its response-eliciting properties with the nickel and resulted in the pigs manifesting their learned rooting and tossing even though it resulted in the absence of food reward.

From our frame of reference, a conditional response is a manifestation of the partial functional equivalence of the conditional stimulus and the unconditional stimulus. The response, once conditioned, never completely duplicates the response elicited by the unconditional stimulus because each unit of an amalgam retains in part its own salience and its own response-eliciting characteristics. Functionally, both "reinforcers" and rewards constitute resources relevant to the organism because of its biological and acquired needs. In conditioning contexts, the organism learns about resources that it can obtain and about how to obtain them. Contingent upon the species of subject and its neural system, the conditioning experiences will be processed to the end that the organism is likely to learn primarily about relationships among the units of the task and how to get the valued resource based on those relationships.

Thus we recommend use of the term *reward* instead of *reinforcer* due to the discredited assumption that a reinforcer directly strengthens a specific response or behavior. Rewards play a much more general role in learning and the directions of behavior than traditional rewards. Rewards give the organism a reason to care and learn about the predictable patterning of stimuli and events that we are constantly experiencing.

Early Environment and its Significance

Among the several sterling contributions made by primatologists is the uncontested principle that conditions and experiences present during early development have long-lasting sculpting effects upon the intelligence, emotions, interests, personalities, and morphology of organisms. In the area of primate behavioral development, Mason (2002) saw the emergence of new behaviors and capabilities as a concept that is fundamental to the understanding of behavioral development and that requires new descriptive categories and measurement. The comprehension of human speech and of the meanings of various word-lexigrams by apes without formal training is a prime example.

The Significance of Recent Language Research With Apes

Earlier assumptions comparing the language ability of apes with human standards of speech, especially in phrase and sentence construction, incorrectly led researchers to conclude that apes do not and cannot have language (Rumbaugh & Savage-Rumbaugh, 1994). Extending this logic has brought the equally incorrect converse implication that language is a uniquely human attribute. Contemporary research in ape language, however, has unequivocally demonstrated the capability of apes to acquire the meaning of symbols and to use those symbols with results that demonstrate the fundamental properties of human language.

Specifically, intensive research across the last 30 years has documented that apes can do the following: (a) learn and use symbols to represent objects or events that are not present. This capability is referred to as *displacement* and is a necessary foundation of *semantics*; (b) use learned symbols among themselves and/or with humans to solve problems by exchanging information; (c) readily organize their learned symbols into conceptual categories (e.g., foods, tools, people); (d) acquire language optimally through daily experiences garnered during infancy, not through formal training; (e) comply with basic rules of grammar and comprehend novel sentences that they hear, sentences that have their meanings syntactically embedded; and (f) understand and respond appropriately to sentences that have not be encountered before (Savage-Rumbaugh et al. 1993). This capability, referred to as *generativity* in language comprehension, is probably the most fundamental of all human language capabilities (Corballis, 1992).

In order for apes to display language capabilities, their understanding of those symbols must have become decontextualized. This capability is necessary in order surmount a classic linguistic puzzle identified by the linguist Quine (1960) as the Gavagai problem. The problem arises when a linguist tries to understand the meaning of words spoken by natives in a language totally different from that of the linguist. If, for example, a native points to an elephant and says a word in the native language, the linguist does not know if the word refers to elephants in general, the name of that particular elephant, a particular body part, a mammal, a quadruped, and so one. Quine speculated that if the native language were sufficiently different from that of the linguist, learning the new language might be virtually impossible. The Gavagai problem led Premack (1986) to a pessimistic view toward the possibility of apes ever mastering a language comparable to human language.

Clearly, because of the Gavagai problem, language experience that is based only on exposing an ape to repeated pairings of a symbol with the same particular exemplar will not produce a full understanding of the symbol's meaning. Instead, the ape should be exposed to the symbol in a wide variety of contexts, just as human children are exposed to the word in different types of linguistic, physical and social settings. In other words, the symbol or word must be experienced in a different setting, that is, it must be decontextualized. This approach was followed in the language acquisition of the bonobo Kanzi (Savage-Rumbaugh et al., 1993). Kanzi experienced both lexigrams and spoken English words in many situations similar to those experienced by a human child during the language-formative years. The later evidence that Kanzi could understand these spoken English words when used in novel sentences was compelling evidence that his prior language-related experiences produced a decontextualized understanding of spoken English words.

King, Rumbaugh, and Savage-Rumbaugh (1999) later noted the similarities between the decontextualized understanding of language-related symbols and the understanding of general personality dimensions including extraversion, conscientiousness, and neuroticism. These dimensions are readily imputed to other individuals as a result of either observing or learning about the responses of an individual in a diverse set of circumstances. Therefore, a personality judgment about someone's degree of extraversion could be viewed as a decontextualized concept extracted from multiple past occurrences of a person's behaving in an extraverted or introverted manner.

The emergence of symbol meaning as well as perception of personality dimensions are both results of inferences about an underlying concept as a result of experiencing multiple instances of multiple exemplars in multiple contexts. Because of the similar logical structure of decontextualized symbols and decontextualized personality constructs, King et al. (1999) suggested that the origins of language in hominids coincided with early intense sociality and increased language use-centered discussion related to personalities of others (see Dunbar, 1996).

The relationship of the exemplars to the symbol meaning or personality perception is far more complex than is the extraction of a set of common elements and lies beyond the scope of this chapter. Yet, symbol meanings and personality perceptions are highly salient parts of our lives. If we return to the previously noted interpretation of salient stimuli as being based on organized aggregations of amalgams into templates, it is clear that the templates for linguistic symbols or personality traits are not a simple sum of all information in the exemplars. Instead, a complex inferential process leading from exemplars to template occurs.

Consequently, on this foundation, current research has accomplished the following broader objectives:

> It has elucidated the evolutionary and ontogenetic roots of language.
>
> It has provided training materials and techniques that greatly benefit children who have language deficiencies because of developmental disabilities.
>
> It has revealed that the basics of language competence probably comprise the abilities (a) to use symbols to represent objects not necessarily present in time or space and (b) to use learned symbols to communicate information that cannot be exchanged via the unlearned modes of communication.
>
> In addition, research into the commonalities of primate language has confirmed similarities among the great apes and humans in the following areas.

Early Environment and the Importance of Logic Structures

Although we know that early environmental stimulation can have generally facilitating effects upon development, research involving apes has confirmed that it is the *logic structure* (recurring patterns of communication, language use, music, and movement) of the early environment that defines the specific dimensions and interests of cognition and competence. A corollary of this important principle is that the specific effects of the logic structure are quite probably related to brain size and complexity. In particular, we can say that early environment is probably much more critical to the cognitive development of children and apes than it is to monkeys and prosimians.

Principles of Continuity

As a result of the research with bonobos, and in particular with Kanzi, the noted comparative psychologist Michael Domjan (2003) concluded that continuities between animals and humans reach far beyond the mere biological: "The language sophistication of Kanzi proves that many important linguistic skills are not uniquely human attributes. Thus, these findings vindicate Darwin's belief, stated in Chapter 6 of the *Origin of Species (1859),* that *Natura non facit saltum* [Nature does not move by leaps but through continuous gradations]" (p. 384).

Neuroanatomical Continuities

One salient feature of how the human brain processes language is the lateralization of this function to one hemisphere. According to the classical model popularized in the second half of the nineteenth century, two cerebral cortical areas larger in the left hemisphere are most commonly associated with language functions. Broca's area is a productive region that encodes vocal signals into meaningful words and sentences; Wernicke's area, a receptive region, processes and integrates auditory sensory information. In other words, Broca's area functions primarily in the planning and execution of speech, whereas Wernicke's area functions to make sense of the speech that a listener perceives.

Human-like neuroanatomical asymmetries have been identified in the posterior temporal lobe and inferior frontal regions in the left hemisphere of the chimpanzee brain, regions considered homologous to Broca's and Wernicke's areas, respectively. Furthermore, asymmetry of the chimpanzee's inferior frontal gyrus, the location of the Broca's area homologue, has been associated with hand use during gestural communication (Taglialatela, et al 2006). These results suggest that both biological and behavioral continuities exist between the communicative systems of the great apes and humans.

Summary

Study of the primate order is very revealing about major trends of evolution to humans. We suspect that the emergence of bipedalism was but one of several major stepping stones, yet with bipedalism came the opportunity for further elaboration of manual dexterity and invention of tools. Intelligence likely was uniquely advanced by selection for dexterity and perceptions of relationships in learning processes. As the processes of learning advanced beyond basic associative problems into realms of learning of relationships there was a tremendous advance that promulgated what we call emergents and have contrasted with the outcome of basic conditioning procedures. Elaboration of the brain, both in size and emphases in organization, facilitated the construction of cultural trends, systems, and institutions.

All of what we know portrays humans as projections of dimensions and of continuities with other forms of primates, not as the creature so apart from the natural world that we are "uniquely unique in being defineable as the totally unique product of nature that we might otherwise want to be."

Notes

Portions of this article were presented by Duane M. Rumbaugh in his Distinguished Primatologist Award Address at the August 2006 annual meeting of the American Society of Primatologists. Preparation of this article and much of the research referenced in it were supported by the National Institute of Child Health and Human Development grants HD-06016, HD-38051, and recently by HD-060563 to David Washburn of Georgia State University's Language Research Center.

References

Amsel, A. (1989). *Behaviorism, neobehaviorism, and cognitivism in learning theory: Historical and contemporary perspectives.* Hillsdale, NJ: Erlbaum.

Beran, M. J., Gibson, K. R., & Rumbaugh, D. M. (1999). Predicting hominid intelligence from brain size. In M. C. Corballis & S. E. G. Lea (Eds.), *The descent of mind: Psychological perspectives on hominid evolution.* New York: Oxford University Press.

Breland, K. & Breland, M. (1961). *The misbehavior of organisms. American Psychologist, 16,* 581-584.

Brown, P. L., & Jenkins, H. M. (1968). Auto-shaping of the pigeon's key-peck. *Journal of the Experimental Analysis of Behavior, 11*, 1-8.

Corballis, M. C. (1992). On the evolution of language and generativity. *Cognition, 44*, 197-226.

Darwin, C. (1859). *The origin of species by natural selection.* London: John Murray.

Domjan, M. (2003). *The essentials of conditioning and learning* (3rd ed.). Pacific Grove, CA: Brooks/Cole.

Dunbar, R. I. M. (1996). *Grooming, gossip, and the evolution of language.* Cambridge MA: Harvard University Press.

Edelman, G. M. (2006). *Second nature: Brain science and human knowledge*. New Haven, CT: Yale University Press.

Gagneux, P., & Varki, A. (2001). Genetic differences between humans and great apes. *Molecular Phylogenetics and Evolution, 18,* 2-13.

Harlow, H. F. (1949). The formation of learning sets. *Psychological Review, 56,* 51-65.

Hebb, D. O. (1949). *The organization of behavior: a neuropsychological theory*. New York: Wiley.

King, J. E., Rumbaugh, D. M., & Savage-Rumbaugh, E. S. (1999). Perception of personality traits and semantic learning in evolving hominids. In M. Corbalis & E. G. Lea (Eds.), *The descent of mind: Psychological perspectives on hominid evolution.* (pp. 98-115). New York: Oxford University Press.

Mason, W. A. (2002). The natural history of primate development: An organismic perspective. In D. J. Lewkowicz and R. Lickliter (Eds.), *Conceptions of development: Lessons from the laboratory* (pp. 105–134). New York: Psychology Press.

Premack, D. (1986). *Gavagai! Or the future history of the animal language controversy*. Cambridge MA: MIT Press.

Quine, W. V. (1960). *Word and object.* Cambridge MA: MIT Press.

Rumbaugh, D.M. (Ed.). (1977) *Language learning by a chimpanzee: The Lana project.* New York: Academic Press.

Rumbaugh, D. M., King, J. L., Beran, M. J., Washburn, D. A., & Gould, K. (2007). A salience theory of learning and behavior. *International Journal of Primatology, 28,* 973-996.

Rumbaugh, D. M., & Savage-Rumbaugh, E. S. (1994). Language in comparative perspective. In N. K. Mackintosh (Ed.), *Animal learning and cognition* (pp. 307-333). San Diego CA: Academic Press.

Rumbaugh, D. M., Savage-Rumbaugh, E. S., & Washburn, D. A. (1996). Toward a new outlook on primate learning and behavior: Complex learning and emergent processes in comparative perspective. *Japanese Psychological Research, 38*, 113-125.

Rumbaugh, D. M., & Washburn, D. (2003). *Intelligence of apes and otherrational beings*. New Haven, CT: Yale University Press.

Rumbaugh, D. M., Washburn, D. A., & Hillix, W. A. (1996). Respondents, operants, and emergents: Toward an integrated perspective on behavior. In K. Pribram & J. King (Eds.), *Learning as a self-organizing process* (pp. 57-73). Hillsdale, NJ: Erlbaum.

Savage Rumbaugh, E. S. (1986). *Ape language: From conditioned response to symbol.* New York: Columbia University Press.

Savage-Rumbaugh, E. S., Murphy, J., Sevcick, R. A., Brakke, K. E., Williams, S. L., & Rumbaugh, D. M. (1993). Language comprehension in ape and child. *Monographs of the Society for Research in Child Development, 58,* 1-242.

Taglialatela JP, Cantalupo C, Hopkins WD (2006) Gesture handedness predicts asymmetry in the chimpanzee inferior frontal gyrus. *NeuroReport,* 17(9): 923-927.

CHAPTER 19

Hominin Brain Reorganization, Technological Change, and Cognitive Complexity

Nicholas Toth and Kathy Schick

Abstract

This chapter will discuss the relationships between hominin brain evolution (encephalization, reorganization) and the prehistoric archaeological record, most notably prehistoric technological material culture and behavioral patterns, to assess the cognitive capabilities evident within different grades of hominins through time. Seven time intervals, spanning from 3.25 million years ago to the present, are sampled to examine major changes in hominin evolution, technology, and behavior over that period of time. Holloway *et al.*'s (2004) proposed three major stages of hominin brain evolution will be discussed in this context. We will argue that each of these three stages appear to be correlated with important changes in material culture and behavior as well. Hypotheses that attempt to explain the causes of hominin encephalization and brain reorganization are discussed.

Key Words

Brain evolution, evolution of human cognition, Palaeolithic archaeology, stone tools

Introduction

"The earliest phases of hominid existence are particularly open to speculative embroidery. But when all is said and done, it remains the stone tool industries or traditions that can inform us most about hominid cognitive abilities. This does not mean that we disregard archaeological contexts such as the faunal remains, home bases, the evidence (or lack of it) for fire, importation over long distances of stones used in making tools, de-fleshing carcasses, or even cannibalism. Holloway (1967, 1969, 1981) suggested that stone tool making and language might have had similar cognitive underpinnings, particularly if the stone tools showed clear evidence of standardization of form from elements (e.g. cobbles) that had very different initial shapes."

Ralph Holloway, Douglas Broadfield, and Michael Yuan, 2004. *The Human Fossil Record, Volume Three: Brain Endocasts - The Paleoneurological Evidence.* Hoboken: John Wiley & Sons. Pp. 288-289.

"If brains were lard, Jethro couldn't grease a skillet."

Jed Clampett discussing his hapless nephew in the 1960's sitcom *The Beverly Hillbillies.*

The hominin fossil record, built up over the past 150 years, has shown us many of the major trends in human evolution pertaining to anatomy and functional morphology. Studies of extant non-mammalian and mammalian (of particular interest, primate) species have shown us relationships between brain anatomy and behavioral, perceptual, and problem-solving capabilities. Ralph Holloway has been at the forefront of human brain evolution studies for over four decades. We have personally known him for three of those decades, and have valued his scholarship, collegiality, and friendship. This chapter is inspired by the corpus of work that Ralph and his colleagues have done in paleoneurology, and the impact this work has had in human origins research.

The human palaeontological record and the prehistoric archaeological record are the two major lines of evidence that shed light on the evolution of human cognitive abilities. Brain endocasts from fossil hominin crania can yield important information regarding brain size, possible brain to body relationships, and brain structure and organization. The archaeological record can yield important information regarding technological patterns, foresight and planning, skill, cognitive capabilities, and dexterity. In this chapter we will review the human palaeontological and archaeological record, sampling time intervals of 500,000 years to explore the relationships between hominin brain reorganization, technological change, and cognitive complexity. We will examine the major stages of brain evolution forwarded by Holloway *et al.* (2004) and see whether these appear to be roughly contemporaneous with behavioral punctuations in the prehistoric archaeological record.

As palaeoanthropologists and experimental archaeologists focusing on Palaeolithic archaeology, we each have over 35 years of experience in stone-knapping and other forms of material culture over a wide range of human technologies through time. This experience hopefully gives us enhanced insights into the level of skill and planning that is required to produce a given artifact form or set of artifacts, as well as into what types of prehistoric artifacts may have been intentionally produced or simply represent by-products of manufacture or use. We also have extensive experience in experimentally using stone tools for a host of activities, including animal butchery for meat consumption, bone-breaking for marrow and brain tissue processing, nut-cracking, wood-working, hide-scraping, etc. These experiments have provided us a wealth of experience to inform insights into the subtle relationships between tool form and tool function.

In past publications we have discussed aspects of the relationships between prehistoric material culture and hominin brain evolution and cognition, including Schick and Toth, 1993, 2009; Toth, 1985a, 1985b, 1990;Toth and Schick, 1993, 2006; Toth, Schick, and Semaw, 2006; Stout *et al.*, 2000, 2006, 2009, 2010. We refer the reader to a number of other publications involving hominin brain evolution and the archaeological record, for example Allman, 2000; Ambrose, 2001; Bar Yosef, 2002; Coolidge and Wynn, 2009; Deacon, 1997; Falk, 1987; Gibson, 1986; Gowlett, 1996; Gibson and Ingold, 1993; Holloway, 1967, 1969, 1981; Isaac, 1986; Lindly and Clark, 1990; McBrearty and Brooks, 2000; McGrew, 1992; Mellars and Gibson, 1996; Mellars et al., 2007; Mithen, 1996; Noble and Davidson, 1996; Parker and Gibson, 1979; Parker and McKinney, 1999; Pelegrin, 2005; Renfrew and Scarre, 1998; Renfrew *et al.* 2009; Roux and Bril, 2005; Schoenemann, 2006; Stout 2002, 2005a, 2005b, 2006; Washburn 1959, 1960; Wynn 1989; and Wynn and McGrew 1989. For overviews of the Palaeolithic archaeological record, see Ambrose, 2001; Delson *et al.*, 2000; Klein, 2009; Oakley, 1976; Schick and Toth, 1993; Toth and Schick, 2007.

This chapter will use the conventional designations for geological periods as opposed of the highly controversial restructuring of the Plio-Pleistocene boundary: here we will show dates for the Pliocene (ca. 5.3 to 1.8 million years ago); Early Pleistocene (ca. 1.8 million to 780,000 years ago); Middle Pleistocene (ca. 780,000-125,000 year ago); Late Pleistocene (ca. 125,000 years ago to 11,000 years ago); Holocene (11,000 years ago to the present.) The dating of hominin fossils, notably those found on the surface in contexts without applicable radiometric dating techniques, is sometimes ambiguous, so we have been as cautious as possible in assigning dates to these important fossils. The earliest dates for *Homo erectus* are somewhat controversial, but here we will use a date of approximately 1.75 million years.

There is much debate regarding the taxonomic status of early hominin fossil as well as evolutionary ancestor-descendant relationships between taxa. In this chapter we will take a conservative position, for example sometimes grouping the taxa *Homo rudolfensis* and *Homo habilis* into "early *Homo.*" We group the proposed taxa *Homo ergaster, Homo georgicus,* and *Homo antecessor* into *Homo erectus*. We also group most non-*erectus* hominin fossils between 750,000 and 250,000 years ago into *Homo heidelbergensis* (what some researchers prefer to call "archaic *Homo sapiens."* We will assign the Neandertals to their own taxon, *Homo neandertalensis*, although some anthropologists would include them in our own species. We will also group most non-Neandertal fossils of the last 250,000 years (excluding relict *Homo erectus* fossils of East Asia and the enigmatic Flores fossils) into *Homo sapiens* (anatomically modern or near-modern humans). This includes all of the African fossils from North Africa and Sub-Saharan Africa in this time period. In this chapter we will also use the mean cranial capacities of hominin taxa and estimated anthropocentric encephalization quotient or EQ (modern human =1.00; modern chimpanzees = 0.34) reported by Holloway *et al.* (2004).

While it is appreciated that at any stage of human evolution, we can only be sure we are seeing *minimal*, but not necessarily *maximal* expressions of cognitive abilities in the form of the material culture and behavioral patterns of prehistoric hominins (for example the relatively simple, traditional material culture of the protohistoric Tasmanians is the product of one *Homo sapiens* society with modern cognitive and language abilities, but clearly not manifesting such complex abilities in their tools and technology relative to contemporary agricultural and industrialized societies), this is nonetheless the most important and reliable source of information that we can recover in the archaeological record. We can also search for the most complex and exceptional forms of material culture at a particular stage of human evolution to see evidence for the most advanced level of cognitive abilities and skill manifested at that time.

Stages of Hominin Brain Evolution (Holloway et al., 2004)

Holloway *et al.* (2004, pp. 289-291) has proposed three major stages of hominin brain reorganization (here we will also add a "Stage 0" to denote a hypothetical ape-like last common African ape/human ancestor):

STAGE 0: Last common ancestor of African chimpanzees/bonobos and hominins (ca. 7-8 million years ago). Ape-like features of brain organization (hypothetical) might include:

a. An ape-like, anterior position of the lunate sulcus indicating more primary visual cortex than seen in hominins
b. Less posterior association cortex than seen in hominins
c. An overall African ape-like (gorilla-, chimpanzee-, bonobo-like) size (ca. 350-450 cc) and ape-like organization of the brain

STAGE 1: Earlier australopithecine grade (e.g. *Australopithecus afarensis* and *africanus*, by ca. 3.5 million years ago). Neurological and cognitive changes at this stage include:

a. Reduction of primary visual cortex (as seen in a more posterior position of the lunate sulcus)
b. Relative increase in posterior association cortex (a human-like pattern)
c. A reorganization of the brain before any major expansion in overall brain size
d. The beginnings of a development in cerebral asymmetries (beyond that seen in modern apes?)
e. By inference, the possibility of more foresight and memory as compared to modern apes

STAGE 2: Early Homo grade (e.g. *Homo rudolfensis, Homo habilis,* early *Homo ergaster/erectus,* by ca. 1.9 million years ago). Neurological and cognitive changes at this stage include:

a. An overall increase in brain volume and encephalization quotient
b. Clear-cut and modern human-like brain asymmetries
c. A prominent Broca's cap region
d. By inference, more strongly developed language capabilities and language behavior
e. By inference, increased postnatal development and learning
f. By inference, social leaning in tool-making, hunting, collecting, scavenging, and reproductive strategies

STAGE 3: *Homo heidelbergensis/neandertalensis/sapiens* grade (by ca. 500,000 years ago to present). Neurological and cognitive changes at this stage include:

a. An overall increase in brain size and encephalization quotient
b. Refinement in hemispherical asymmetries and specializations for visuospatial, verbal, and sociality skills
c. By inference, growing elaboration of cultural skills based on language
e. By inference, arbitrary symbol systems
f. By inference, feedback between behavioral complexity (including stone technology) and brain enlargement

The authors point out that there is little structural or brain size difference evident (based on endocasts) between *Homo heidelbergesis, neandertalensis,* and *sapiens.*

Theories of Encephalization and Hominin Cognitive Evolution

There is a general appreciation that, in primate evolution, larger brains and larger brain/body size ratios are correlated with higher cognitive skills. Popular theories of *why* hominins became encephalized and more cognitively complex has been vigorously debated. One complication in this debate is that many hypotheses have been difficult or impossible to test in the prehistoric record, at least at our present state of knowledge and methodological sophistication. To date, there does not seem to be one overarching theory that has been championed by palaeoanthropologists and palaeoneurologists. Various theories offered to help explain the profound encephalization observed in the course of hominin evolution have included the extracted food hypothesis (Gibson, 1986, 2002), the predation hypothesis (Shipman and Walker, 1989), the social brain hypothesis (Dunbar 1992, 1993, 2003), the expensive tissue hypothesis (Aiello and Wheeler, 1995), the maternal energy hypothesis (Martin, 1996, this volume), and the symbolic hypothesis (Deacon, 1997).

Rationale for this Study and The Time Intervals Sampled Here

We have decided to sample human evolutionary time in half-million year intervals (with the exception of the last time period sampled being the last 250,000 years) in order to see robust changes in hominin brain evolution and material culture as manifested in the prehistoric record. The intervals that we have selected, in our opinion, best show the emergence of new hominin taxa and, potentially, new patterns of hominin neurological reorganization. These neurological changes will be correlated with changes in the material culture and known behavior of these hominins. We have started our first time interval (Time Interval One) to include a phase which pre-dates any definite archaeological record. Time Interval Two includes the emergence of the first definitive flaked stone tools. For each time interval, we will normally only cover the *new* technological and behavioral traits that emerge for the first time in the prehistoric record; it can be assumed that traits that emerged in previous time intervals continue on in more recent times.

Our time intervals and the most encephalized hominins are:

Time Interval One: (3.25-2.75 Ma) *Australopithecus afarensis (*Holloway *et al.* Stage 1)

Time Interval Two: (2.75-2.25 Ma) *Australopithecus garhi/africanus* (Holloway *et al.* Stage 1)

....

Time Interval Three: (2.25-1.75 Ma) early *Homo (habilis/rudolfensis)* (Holloway et. al. Stage 2)

Time Interval Four: (1.75-1.25 Ma) early *Homo erectus* (Holloway *et al.* Stage 2)

Time Interval Five: (1.25-0.75 Ma) later *Homo erectus* (Holloway *et al.* Stage 2)

....

Time Interval Six: (0.75-0.25 Ma) *Homo heidelbergensis* (Holloway *et al.* Stage 3)

Time Interval Seven: (0.25 Ma-present) *Homo neandertalensis/H. sapiens* (Holloway *et al.* Stage 3)

An Examination of the Prehistoric Archaeological Record

Please note: For further general information and detailed references for the sites, fossils, taxa, and archaeological localities discussed here, the reader is referred to encyclopedic references in the field such as Klein (2009) and Delson et al. (2000). References included here will center on specific information and arguments made regarding the subjects discussed below.

Time Interval One: 3.25 Million to 2.75 Million Years Ago (Middle Pliocene).

Overview: This is the time period of the small-brained bipedal hominin *Australopithecus afarensis*. There is no definitive archaeology associated with this hominin form, although there are footprints of three individuals in a volcanic ash deposit at Laetoli, Tanzania at 3.5 million years. (The Stage 1 brain reorganization may have happened at an earlier, *Ardipithecus*-grade hominin, possibly including fossils assigned to *Sahelanthropus* and Orrorin dating to between 4.5 and 6 million years ago, but this has yet to be demonstrated).

Most encephalized hominin: *Australopithecus afarensis.*

Holloway et al. brain evolution stage: Stage 1.

Key hominin cranial fossil sites: Hadar, Ethiopia.

Other fossil sites: Laetoli, Tanzania; Maka (Middle Awash), Ethiopia.

Average cranial capacity: 445 cc (Holloway *et al.*, 2004). For comparison, the mean for modern gorillas is 500 cc, and the mean for chimpanzees is 405 cc.

Estimated encephalization quotient (Homocentric): 0.43. For comparison, the mean EQ for gorillas is 0.24, and the mean for chimpanzees is 0.34.

Technological stage: unknown.

Discussion of the Archaeological Record

There is no definitive evidence of hominin modified stones, bones, or other materials during this time period. Based on own knowledge of chimpanzee material culture and cultural traits (McGrew, 1992; Whiten *et al.* 1999), we can speculate on a range of possible types of tool-use and other types of cultural phenomenon, but without hard prehistoric evidence it still remains a matter of conjecture. In a study carried out by the authors (Toth and Schick, 2009d; Whiten *et al.* 2009), the relationship between the number of shared cultural traits in wild chimpanzees versus distance between study areas was carried out. At the subspecies level (but not the species level) a strong and statistically significant correlation was found, with study areas in closer proximity having more shared cultural traits than study areas further apart. At approximately 700 kilometers there was a drop-off of less than half the number of maximum shared traits (from a maximum of eight to less than four traits), which was used as a model for possible patterns of shared cultures among early hominin Early Stone Age archaeological sites. Archaeological localities closer than 700 kilometers (about 450 miles) to each other at a given time would theoretically share more cultural traits than sites more distant from each other.

It should be noted that a recent claim has been made for stone tool use and bone modification at approximately 3.4 million years (McPherron *et al.*, 2010). This is based on two surface mammal bones at Dikika, Ethiopia with alleged chop-marks and cut-marks. Since these bones were found on the surface, their provenience cannot be ascertained with confidence, and a number of researchers have voiced skepticism regarding the cut-mark evidence, e.g. Dominguez-Rodrigo *et al.*, 2010 and Shipman, 2010. There have been four decades of research in fossiliferous deposits between 2.6 and 3.5 million years ago, and not one stone artifact or cut-marked bone has ever been discovered. Until such time as clear-cut modified stone artifacts or non-controversial tool-modified bones are found in a well-dated, stratified context, such claims must remain unsubstantiated.

Time Interval Two: 2.75 to 2.25 Million Years Ago (Later Pliocene)

Overview: This time interval documents the emergence of the first identifiable stone tools (Oldowan Industrial Complex) and new hominin taxa: in East Africa, the emergence of *Australopithecus garhi* and the megadont *Australopithecus (Paranthropus) aethiopicus*; in South Africa, the emergence of *Australopithecus africa-*

nus. There does not appear to be a significant increase in brain size or EQ from the previous *Australopithecus afarensis* time period. Surface occurrences of cut-marked bones are known also from this time period, notably from the Gona and the Middle Awash of Ethiopia.

Most encephalized hominins: *Australopithecus garhi, Australopithecus africanus, Australopithecus (Paranthropus) aethiopicus.*

Holloway et al. brain evolution stage: Stage 1.

Key hominin cranial fossil localities: Bouri, Middle Awash, Ethiopia (*A. garhi*); Sterkfontein, Makapansgat, and Taung, South Africa (*A. africanus*); West Turkana, Kenya [*A. (P.) aethiopicus*].

Average cranial capacity:

Australopithecus garhi: 450 cc (one specimen).

Australopithecus africanus: 461 cc.

Australopithecus (Paranthropus) aethiopicus: 431 cc (one specimen).

Estimated encephalization quotient (Homocentric): 0.46 (*A. africanus*).

Technological stage: early Oldowan.

Key archaeological sites: Gona, Hadar, and Omo, Ethiopia; Lokalalei (West Turkana), Kenya.

Discussion of the Archaeological Record

It is during this time interval that the first clear evidence of hominin material culture is found. The archaeological sites from this time are all on the African continent. At Gona in the Afar Rift region of Ethiopia, several sites (EG-10, EG-12, OGS-6, OGS-7, DAN 1, DAS 7) are dated to between 2.6 and 2.5 million years ago. Several sites are dated to approximately 2.3 million years ago: sites AL-666 and AL-894 at Hadar, Ethiopia; sites Omo 71, Omo 84, Omo 57, Omo 123, FtJj1, FtJi2, and FtJi5 in the Omo Valley, Ethiopia; and sites Lokalalei 1 and 2c at West Turkana, Kenya. Stone artifacts have also been found in situ at the Pliocene site of Ain Boucherit in Algeria, and biostratigraphy may suggest a similar date (M. Sahnouni, pers. comm.) The archaeological record is characterized by:

- A **simple Oldowan technology**, normally with river cobbles or chunks knapped to produce sharp flakes and fragments. **Unifacial and bifacial chopper cores** are most common, with polyhedrons, heavy-duty scrapers, and discoids also present. Experimentation has shown that these core forms can be produced as a by-product of flake production, although some of the larger, sharper cores could have been used for wood-working or other activities. Summaries of the Oldowan include Schick and Toth, 2006 and Semaw, 2006.
- **Battered percussors** are found in the form of river cobbles that exhibit battering and small-scale flaking on discrete cortical surfaces, indicating their use as a hammerstones.
- **Flakes and fragments** struck from cores, normally **unmodified** (not retouched).
- **Later stages of cobble reduction** (Toth *et al.*, 1985b, 2006) are typical at many sites, suggesting transport of partially-flaked cores and further reduction at the excavated sites.
- Even in this early period there seems to be some indication of **selection for higher-quality raw materials** and **some transport of stone** from their sources (mainly river gravels flowing out of volcanic highlands and quartz-rich outcrops).
- Probable **cut-marks, chop-marks, and hammerstone striations** on animal bones, indicating processing with stone tools for meat and marrow/brains. Surface bones from Gona and bones from the Hata Beds in the Middle Awash of Ethiopia show such features, but confirmation of these features from in situ, excavated, stratified sites (as we will see in the next time interval) will be important in the future.

Experiments with bonobos (Toth *et al.*, 1993, 2006; Schick *et al.*, 1999) have shown that these modern apes (with average cranial capacities just slightly smaller than *Australopithecus garhi*) can master the basic principles of percussive stone fracture of unmodified Gona volcanic cobbles, although the resultant products appear to show less knapping skill than that of the early Oldowan hominins 2.6 million years ago at Gona, Ethiopia. In particular, the bonobo cores are less reduced and show more battering on core edges from unsuccessful hammerstone blows. The evidence suggests that the bonobos are striking the cores with their hammerstones at a significantly lower velocity than the Gona hominins, so that the major differences in skill in stone knapping may be more biomechanical than cognitive.

It would appear that, at our present state of knowledge, the earliest known stone tool-makers were relatively small-brained, bipedal australopithecines that exhibit no evidence of marked encephalization or evidence of a significant tooth reduction. Nonetheless, the earliest flaked stone artifacts suggest a gradual technological and adaptive shift towards the human condition. Whether even earlier Palaeolithic archaeological sites, possibly from Time Interval One (3.25-2.75 million years ago), will be found still remains uncertain.

Time Interval Three: 2.25 to 1.75 Million Years Ago (Late Pliocene/Early Pleistocene)

Overview: This is the beginning of Holloway *et al.*'s brain evolution Stage 2. There is clear evidence of encephalization and probable rise in EQ of these forms, probably important reorganization in the hominin brain leading to more profound hemispheric asymmetries (petalias), and possibly preferential right-handedness in these tool-making populations (Toth, 1985a).

Most encephalized hominins: *Homo habilis/rudolfensis.*

Holloway et al. brain evolution stage: Stage 2.

Other hominins: *Australopithecus (Paranthropus) boisei/robustus.*

Key hominin cranial fossil localities: Koobi Fora (East Turkana), Kenya; Olduvai Gorge (Bed I), Tanzania.

Average cranial capacity:

Homo rudolfensis: 788 cc.

Homo habilis: 610 cc.

"Early *Homo*": 698 cc.

Estimated encephalization quotient:

Homo rudolfensis: 0.66.

Homo habilis: 0.62.

"Early *Homo*": ca 0.64.

Key archaeological sites: Fejej, Ethiopia; East Turkana (KBS Member sites) and Kanjera, Kenya; Bed 1 Olduvai Gorge (Bed I), Tanzania; Ain Hanech and El-Kherba, Algeria; Sterkfontein, and Kromdraai, South Africa.

Technological stage: Oldowan and Developed Oldowan.

Discussion of the Archaeological Record

A larger number of sites date to this time interval compared to the previous time interval. This suggests that flaked stone technologies were becoming more widespread and there was more habitual use of flaked and battered stone technology in these hominin populations. All of the sites in this time period would be grouped into the Oldowan Industrial Complex.

- Oldowan **archaeological sites become more common** in this time interval, all found on the African continent.

- For the first time, **retouched forms,** usually made on flakes or flake fragments (such as "**scrapers**" and "**awls**") become common at some sites after two million years ago. Such retouch might be done to resharpen an edge. Experiments have shown that such retouched denticulated ("toothed") flake edges make very good, long-lasting butchery knives (Toth and Schick, 2009c, in press). Retouch may also be done to shape or strengthen an edge for a specific activity, such as scraping wood or hide, or to remove irregularities or spurs along edges to make the tools more ergonomic in the hand. Based on the fact that it does not appear that hominins prior to 2.0 million years were retouching edges to any great extent, and that our experimental program with bonobos has not shown them to retouch flake edges (even when given a flake whose edge has been ground down to make cutting impossible), it is likely that such lithic retouch of flake edges may require more complex cognitive abilities. For the first time, hominins appear to be *intentionally* shaping stone, albeit in a simple way.

- Highly battered **spheroids and subspheroids** in quartz and lava become more common in this time interval. Experiments have shown that these forms are probably hammerstones, some of which were used for several hours of knapping, either because the spheroid/subspheroid was carried around or because sites were revisited on a regular basis and the hammers re-used (Schick and Toth, 1994; Toth and Schick, 2009c). Using a quartz hammerstone for four hours can produce thousands of flaked stone artifacts (fragments, flakes, cores), so that this time could have been spread over weeks or months.

- **Cut-marks** on animal bones are present at a number of sites, especially at the FLK Zinj site in Bed I of Olduvai Gorge show that early hominins were **exploiting a wide range of large mammals**. Whether these animals were obtained through more passive scavenging (e.g remains of carnivores, or carcasses obtained from streams during migrations), confrontational scavenging, or active hunting is hotly debated by zooarchaeologists.

- Although most raw materials were obtained within a few kilometers of archaeological sites, some sites suggest **transport of some rock for longer distances**, on the order of ten to twenty kilometers. This is far beyond the range of transport of food or tools by chimpanzees today.

Time Interval Four: 1.75 to 1.25 Million Years Ago (Early Pleistocene)

Overview: This time interval documents the emergence of *Homo erectus* and the disappearance of other forms of early *Homo*. Early *Homo erectus* (sometimes called *Homo ergaster*) had a significantly larger brain than earlier and other contemporary hominin forms and had body proportions more like modern humans. The first evidence of hominins outside of Africa is found early in this time interval. The earliest Acheulean handaxe and cleaver industries are also found in this time interval.

Most encephalized hominins: earlier *Homo erectus (ergaster/georgicus).*

Holloway et al. brain evolution stage: Stage 2.

Other hominins: *Homo habilis* (?relict populations); *Australopithecus (Paranthropus) boisei/robustus.*

Key hominin cranial fossil localities: Gona, Ethiopia; Koobi Fora (East Turkana), Nariokotome (West Turkana), Kenya; Olduvai Gorge (Bed II), Tanzania; Nyambusosi, Uganda; Swartkrans, South Africa; 'Ubeidiya, Israel; Dmanisi, Republic of Georgia; Sangiran, Java.

Average cranial capacity: ca. 800 cc.

Estimated encephalization quotient: 0.58.

Key archaeological sites: Konso Gardula, Melka Kunture, Ethiopia; East Turkana (Okote Member sites) and Chesowanja, Kenya. Olduvai Gorge (Bed II) and Peninj, Tanzania; Nyambusosi, Uganda; Swartkrans, South Africa; Dmanisi, Republic of Georgia; 'Ubeidiya, Israel.

Technological stage: early Acheulean (and simpler industries).

Discussion of the Archaeological Record

This time interval includes typical Oldowan sites and so-called "Developed Oldowan" sites (with more retouched forms and battered spheroids and subspheroids), and the emergence of the first Acheulean sites.

- Early Acheulean forms include crude **handaxes, cleavers, and picks** made on **large flakes** struck from boulder-cores (usually lava, quartz, or quartzite) or made on **large cobbles**. Much of the rest of the stone technology associated with early Acheulean sites is very similar to Oldowan and Developed Oldowan assemblages.

- This time period witnesses the first evidence of hominins out of Africa and **dispersal into Eurasia** at sites such as Dmanisi in the Republic of Georgia, 'Ubeidiya in Israel, and Sangiran in Java.

- Although there is evidence for the presence of **fire** at some sites (e.g. Swartkrans Cave in South Africa and the FxJj20 site complex at Koobi Fora, Kenya), it cannot be demonstrated that hominins were the agents of manufacture or use of fire, and these burnt bones and/or thermally fractured stone artifacts could be the result of natural fires sweeping across the landscape. Whether early hominins at this stage could have maintained naturally occurring fires for a certain time period cannot be demonstrated at our present state of knowledge.

Time Interval Five: 1.25 to 0.75 Million Years Ago (Later Early Pleistocene)

Overview: It is during this time interval that the robust australopithecines appear to go extinct, leaving *Homo erectus* as the sole hominin (in all its regional variations in Africa and Eurasia). Average cranial capacity appears to go up in this time interval (ca. 950 cc compared to 800 cc in the previous time interval). In much of Africa and the Near East there is a continuation of the Acheulean handaxe/cleaver industries, but Oldowan-like Mode 1 industries (sometimes called "Tayacian") are common in East Asia and Europe. This technological dichotomy is sometimes called the "Movius Line" after Harvard professor Hallum Movius who was one of the first to note this (Movius, 1948; Schick, 1994).

Most encephalized hominin: later *Homo erectus*.

Holloway et al. brain evolution stage: Stage 2.

Other hominins: possibly *Austalopithecus (Paranthropus) boisei/robustus*.

Key hominin cranial fossil localities: Buia, Eritrea; Daka (Middle Awash), Ethiopia; Sambungmacan and Trinil, Java; Lantian, China; Atapuerca Gran Dolina (TD6), Spain; Ceprano, Italy.

Average cranial capacity: ca. 950 cc.

Estimated encephalization quotient: 0.68.

Technological stage: middle Acheulean (and simpler industries).

Key archaeological sites: Buia, Eritrea; Daka (Middle Awash), Ethiopia; Olorgesailie and Kariandusi, Kenya; Olduvai Gorge (Beds 3 and 4 and Masek Beds); Ternifine (Tighenif), Algeria; Gesher Benot Ya'aqov, Israel; Orce and Atapuerca Gran Dolina, Spain; Nihewan Basin, China.

Discussion of the Archaeological Record

- During this time interval we see **better-made handaxes and cleavers**, more symmetrical and more extensively flaked. Examples of typical Acheulean forms come from sites in the Buia area of Eritrea; sites in the Daka member of the Middle Awash, Ethiopia; Olorgesailie, Kenya; Ternifine, Algeria; and Gesher Benot Ya'aqov in Israel. Hard-hammer percussion seems to be the technological norm in biface (handaxe and cleaver) production, often made on large flakes struck from boulder-cores. By this technological stage there are recurrences of very similar forms that suggest to us the emergence of **style** and a concept of a **"mental template"** or the idea of a distinct artifact form in the mind of the toolmaker (Deetz, 1967). Such templates become even more standardized in the subsequent later Acheulean, Middle Palaeolithic, and Upper Palaeolithic, to be discussed in later time intervals.

- There is evidence of the use of **fire** at Gesher Benot Ya'aqov in Israel (Goren-Inbar et al., 2004), but most sites of this time period do not show the use of fire in the form of hearth features or concentrations of burnt bones or stones. Fire-making is a very complex technology (most modern humans, even if they know the basic principles, cannot easily produce fire), so this may be evidence of the maintenance of natural brushfires rather than production. Consistent use (and presumably knowledge of manufacture) of fire does not appear until the subsequent time interval.

- Gesher Benot Ya'aqov in Israel also bears evidence of early **fish and crustacean (crab) exploitation** on the edge of an ancient lake in the Jordan Rift

Valley (Alperson-Afil et al., 2009). Such exploitation will only become common in the last time scale (last 250,000 years).

- There is evidence for **nut-cracking** in the form of broken nuts (five species) and pitted anvils at Gesher Benot Ya'aqov in Israel, but since wild chimpanzees do similar activities, this is probably not a significant cognitive milestone (Goren-Inbar et al., 2002).

Time Interval Six: 750,000 to 250,000 Years Ago (Earlier Middle Pleistocene)

Overview: This time interval documents the emergence of the larger-brained *Homo heidelbergensis* (sometimes called "archaic *Homo sapiens*"). It also documents the development of finely made Acheulean handaxes and cleavers and a gradual shift to flake tool industries, some with prepared core technologies, of the Middle Stone Age/Middle Palaeolithic. The earliest wooden spears are known from this time as well as the first possible evidence of ritualistic behavior.

Holloway et al. brain evolution stage: The beginning of Stage 3. Hominins in Africa, Europe, and western Asia show encephalization (about 300 cc larger than later *Homo erectus*).

Most encephalized hominins: *Homo heidelbergensis*.

Holloway et al. brain evolution stage: Stage 3.

Other hominins: *Homo erectus* (e.g. Asia, Java).

Key hominin cranial fossil localities: Bodo (Middle Awash) and Gona, Ethiopia; Ndutu, Tanzania; Kabwe (Broken Hill), Zambia; Saldanha, South Africa; Zuttiyeh and Tabun (lower levels), Israel; Swanscombe, England; Atapuerca Sima de Los Huesos, Spain; Arago, France; Altamura, Italy; Steinheim, Germany; Petralona, Greece; Narmada, India; Dali, Zhoukoudian and Yunxian, China.

Average cranial capacity: 1260 cc.

Estimated encephalization quotient: 0.81.

Technological stage: later Acheulean, early Middle Stone Age/Middle Palaeolithic (and simpler technologies).

Key archaeological sites: Bodo, Ethiopia; Olorgesailie, and Kapthurin, Kenya; Isimila, Tanzania; Kalambo Falls, Zambia; Elandsfontein and Montagu Cave, South Africa; Tihodaine, Algeria; Boxgrove, Hoxne, Clacton, and Swanscombe (England); Torralba, Ambrona, and Atapuerca Sima de Los Huesos (Spain); Arago Cave, Lazaret Cave, and Terra Amata (France), Isernia, Italy; Bilzingsleben and Schöningen (Germany); Vertesszollos, Hungary; Zhoukoudian, China.

Discussion of the Archaeological Record

A number of technological advances are observed in the archaeological record during this time interval. These include much more refined forms of artifacts, more formal tool forms, new and more elaborate techniques for tool production, new categories of tools in evidence at some sites, indirect evidence of improved hunting technology, and possible evidence of symbolic behavior, including the use of ocher pigments.

- **Refined handaxes and cleavers** (some made by a soft-hammer technique with careful platform preparation, as discussed below) become common in this time interval. Microwear analysis of later Acheulean handaxes from some well-preserved sites, e.g. Hoxne (Keeley, 1980) and Boxgrove (Mitchell, 1995; Pitts and Roberts, 1997; Roberts and Parfitt, 1999), indicate wear-patterns consistent with **animal butchery**.

- **Soft hammers** of antler, bone, or ivory or softer stone were evidently used at many localities to produce finely flaked stone artifacts beginning around 500,000 years ago. Use of such soft hammers is evident in the pattern of flaking observable in stone tools, and in some instances such soft hammers themselves have been found, e.g. antler and bone percussors from Boxgrove, England (Pitts and Roberts, 1997; Roberts and Parfitt, 1999). Use of such soft hammers allowed for thinner flakes and more controlled shaping of stone tools than did the use of harder stone hammers.

- **Platform preparation** on the edges of cores and bifaces becomes common in this time interval, especially at later Acheulean sites. Such striking platform preparation consists of steepening edges before flake removal, which normally produces flakes (e.g. Levallois flakes and points, handaxe thinning flakes) with "facetted" striking platforms (platforms showing multiple flake scars). Such careful preparation begins around 500,000 years ago.

- We think that **stylistic norms** become more prevalent and more clearly defined in later Acheulean times. Recurrent shapes suggest that hominin toolmakers had more formal "mental templates" than earlier hominins had, although not as standardized as later hominins. They also appear to have had more consistent control and skill in stone flaking. For example, at the later Acheulean site of Hargufia A4 in the Middle Awash, perhaps 300,000 years old, small lava handaxes made on flakes are remarkably similar in size, shape, and symmetry (de Heinzelin et al., 2000).

- **Prepared cores** appear sporadically in the latter part of this time interval, but become more common in the succeeding intervals.

- **Wooden spears** are seen at such well-preserved sites as Schöningen in Germany (ca. 400,000 years old) (Thieme, 1997, 1998, 2005) and the broken spear tip from Clacton in England (ca. 300,000) (Oakley et al., 1977). Carefully sharpened and shaped wooden spears suggest that they were part of hunting paraphernalia, either as hand-held stabbing weapons or as thrown projectiles.

- **Possible big-game hunting** has also been suggested at some sites such as the Acheulean site of Boxgrove in England (ca. 500,000 years ago). The remains from several rhinoceros and horse skeletons bear butchery marks from stone tools (Pitts and Roberts, 1997; Roberts and Parfitt, 1999).
- Micro-wear analysis on retouched flake scrapers from sites of this period (e.g. Clacton, Hoxne) (Keeley, 1980) indicate that a number of these tools were used for **hide-scraping**, suggesting that cured hides could have been used for such items as blankets, simple garments, thongs for stitching or tying things together, or containers.
- **Ground pigment** pieces from sites such as Twin Rivers, Zambia, are believed to be about 300,000 years old (Barham, 2002). These facetted pieces of hematite may have been ground to produce a powder to decorate an object (or body).
- Possible **ritualistic or funerary behavior** may be seen at the Atapuerca locality, Sima de Los Huesos (ca. 400,000 years ago), where the remains of approximately thirty individuals appear to have been disposed of down a forty-foot shaft in a cave (Arsuaga and Martinez, 2004). Found with this incredibly dense concentration of hominin bones was one well-made quartzite red handaxe, which may have been intentionally put there by the hominins. Such behavior is not seen again until the last 100,000 years.
- **Abstract decoration** may be seen in a geometric, evenly-spaced fan-shaped set of cut-marks on a fragment of elephant tibia from the site of Bilzingsleben in eastern Germany, estimated to be between 280,000 and 400,000 years ago (Mania and Mania, 2005). This is an unusual and anomalous occurrence, and such design will not be seen again until the last 100,000 years.
- At the Acheulean site of Berekhat Ram in the Golan Heights, dating to about 400,000 years ago a small lava pebble with three apparent linear grooves has been interpreted by some (but not all) palaeanthropologists as an enhancement to create a crude representation of a human figure (Goren-Inbar and Peltz, 1995). Convincing representations of human figures are only seen in the early Upper Palaeolithic of Europe in the last 40,000 years.

Time Interval Seven: 250,000 Years Ago to Present (Later Middle Pleistocene, Late Pleistocene, Holocene)

Overview: Hominin average brain size increases by about 200 cc over *Homo heidelbergensis* in the previous time interval. During this time, the Neandertals *(Homo neandertalensis*) emerged and flourished in cold-adapted conditions in Europe and parts of western Asia, going extinct ca. 30,000 years ago. Anatomically modern humans, *Homo sapiens*, emerged in Africa over 150,000 years ago and gradually spread to all inhabitable parts of the globe, including the Near East by about 100,000 years ago, much of Eurasia by 40,000 years ago, Sahul (the landmass when Australia, New Guinea, and Tasmania were connected at low sea levels) by 40,000 years ago and the Americas by 12,000 years ago.

Most encephalized hominins: *Homo neandertalensis and Homo sapiens.*

Holloway et al. brain evolution stage: Stage 3.

Other hominin taxa: possibly *Homo erectus* (East Asia and Java); *Homo floresiensis*(?)

Key hominin cranial fossil localities:

Neandertal: Devil's Tower and Forbes Quarry, Gibraltar; La Ferrassie, La Chapelle-aux-Saints, La Quina, Le Moustier and Saint-Cesaire, France; Feldhofer Grotto, Germany; Spy and Engis, Belgium; Krapina and Vindija, Croatia; Moldova, Ukraine; Grotta Guattari (Monte Circeo) and Saccopastore, Italy; Amud and Tabun, Israel; Shanidar, Iraq; Teshik-Tash, Uzbekistan.

Modern or near-modern human: Herto (Middle Awash) and Omo Kibbish, Ethiopia; Ngaloba, Tanzania; Florisbad, Border Cave, Fish Hoek (Skildergat), and Tuinplaas, South Africa; Singa, Sudan; Eyasi, Tanzania; Jebel Irhoud and Dar es Soltane, Morocco; Jebel Qafzeh and Skhul, Israel; Cro-Magnon, Solutre, Chancellade, Abri Pataud, and Combe Capelle, France; Vogelherd, Germany; Grimaldi Caves, Italy; Mladec, Zlaty Kun, Pavlov, Predmosti, Brno, Pavlov, and Dolni Vestonice, Czech Republic; Bacho Kiro, Bulgaria; Zhoukoudian Upper Cave, China; Lake Mungo and Kow Swamp, Australia.

Average cranial capacity:

Homo neandertalensis: 1427 cc.

Homo sapiens: 1496 cc (Pleistocene); today's humans: 1335 cc.

Estimated encephalization quotient:

Homo neandertalensis: 0.99.

Homo sapiens: 0.90 (this e.q. disparity relative to today's humans [1.00] may be due to overestimates of body size in the Pleistocene sample.)

Technological stage: very late Acheulean, Middle Palaeolithic/Middle Stone Age, Late Palaeolithic (Upper Palaeolithic, Later Stone Age, Palaeoindian, etc.)

Key archaeological sites:

1. Neandertal archaeological sites: La Ferrassie, Le Moustier, La Quina, Combe-Grenal, Pech de l'Aze, Arcy-sur-Cure, and Saint-Cesaire, France; La Cotte de St. Brelade, Jersey; Zafarraya, Spain; Tabun Cave, Amud Cave, Kebara Cave, and possibly Quneitra, Israel; Tata and Szeleta Cave, Hungary; Krapina, Croatia.

2. Modern or near modern human archaeological sites: Pietersburg, Klasies River Cave, Die Kelders Cave, Blombas Cave, Howieson's Poort, Apollo-11 Cave; Skildergat, Nelson Bay Cave, Border Cave, Eland Bay Cave, Pinnacle Point (Mossel Bay), and Rose Cottage Cave, South Africa; Mumba, Tanzania; Ishango, Zaire; Dar-es-Soltane, Morocco; Haua Fteah, Libya; Skhul, Qafzeh, and Boker Tachtit, Israel; Ksar 'Akil, Lenanon; Chauvet, Laugerie Haute, Abri Pautaud, Solutre, La Madeleine, Mas d'Azil, Enlene, Pincevent, and Lascaux Cave, France; Parpallo, Castillo, Altamira, Cueva Morin, and El Juyo, Spain; Vogelherd, Hohlenstein-Stadel, and Gonnersdorf, Germany; Dolni Vestonice and Pavlov, Czech Republic; Mezhirich and Menzin, Ukraine; Istallosko, Hungary; Kostienki and Sunghir, Russia; Mal'ta, Siberia; Zhoukoudien Upper Cave (China); Lake Mungo, Australia; Monte Verde, Chile; Blackwater Draw, United States.

Other contemporary hominins: *Homo erectus, Homo soloensis, Homo floresiensis*

Discussion of the Archaeological Record

A number of technological and behavioral changes emerge in the earlier part of this time interval associated with Neandertal populations. Sites bearing evidence of early modern humans early in this time period, contemporary with the Neandertal sites, also tend to show patterns similar to those observed at Neandertal sites. Late in this time interval, when Neandertal populations were declining and disappearing, many new behavioral and technological advances associated with fully modern humans emerge in the archaeological record.

Neandertal technological and behavioral traits include:

- **Retouched flake tools** predominate in the Neandertal Middle Palaeolithic (Mousterian) tool kit, notably side scrapers, denticulated scrapers, backed knives, and points.
- **Points**: Some stone point forms (Levallois points, retouched unifacial Mousterian points, bifacial "Blattspitzen" points) could have been hafted onto wooden shafts, for thrusting or thrown spears. Some of these points have thinned bases, possibly to facilitate this hafting. Unifacial points made on thin flakes are first known from the Middle Palaeolithic of Europe and the Near East.
- **Intentional burials** are known from Neandertal times (e.g. La Ferrassie, Le Moustier, and La Chapelle in France; Shanidar, Iraq; and Kebara in Israel) but usually without any clear grave goods.
- **Prepared cores**: Prepared core forms, most notably Levallois tortoise-cores and Levallois point cores are found in many Middle Stone Age/Middle Palaeolithic assemblages. Neandertals continue using prepared core methods for removing flakes of a predetermined shape. Most notable are disc-shaped Levallois tortoise-cores as well as Levallois point cores for removing triangular flakes. Such cores require careful preparation of the core topographical surface and also careful preparation of the striking platform in order to successfully remove the target flakes. This platform preparation usually included faceting (removing small flakes from the striking platform to steepen and strengthen an edge before flake removal) and carefully shaping the striking platform to isolate one area of high topography to strike with a percussor (giving the edge as shape that the French call the "chapeau de gendarme," showing the profile of a 19th century policeman's Napoleonic hat). These are quite sophisticated cognitive operations, requiring a good sense of three-dimensional geometry as well as the mechanics of stone fracture.
- There appears to be **more variability** in Neandertal lithic assemblages compared to earlier time periods. Archaeologists in Western Europe, for example, have identified a number of Middle Palaeolithic variants, including the Typical Mousterian, Denticulate Mousterian, Quina Mousterian, Ferrassie Mousterian, and Mousterian of Acheulean Tradition A and B. Explanations for these variants have included cultural, functional, and chronological ones. It can also be argued that there are **more tool types** than is seen in earlier time periods.
- **The Chatelperronian**: Late Neandertals in France, around 32,000 years ago, are associated at several sites with blade technologies, backed blades or points, and ornamentation in bone and ivory. This so-called Chatelperronian industry is contemporary with anatomically modern humans in France and Upper Palaeolithic (Aurignacian) blade technology. At present there is a lively debate regarding the nature of the Chatelperronian: Did it develop out of local Middle Palaeolithic tradition, was it adopted by contact with Aurignacian peoples, or is it the product of admixture of materials from Middle and overlying Upper Palaeolithic strata?

Anatomically modern human technological and behavioral traits:

Anatomically modern humans, in the first part of this time interval, show little or no major technological or behavioral differences with contemporary Neandertal populations. But as time goes on, much more complex technological and behavioral innovations can be seen. Major behavioral and technological innovations observed in sites associated with anatomically modern humans during the latter phases of the Pleistocene include:

- **Elaborate burials**: The first burials with elaborate grave goods are found associated with anatomically modern humans. Elaborate burials may include red ochre, stone or bone tools, as well as high-status items, possibly denoting social rank, at some sites. Instances include the very early burial at Skhul in Israel of a modern human male cradling a wild pig mandible approximately 90,000 years ago, and the very elaborate burial of several individuals at the 28,000 year-old site of Sunghir in Russia, including two juveniles and one 60-year old man, with thousands of small beads that appear to have been sewn onto clothing, the two juveniles flanked by long mammoth tusks, and many other special tools and materials accompanying the grave.

- **Blades**: Although sporadic blade industries are found in earlier times, systematic blade production based upon specially-prepared prismatic cores becomes prevalent in the Upper Palaeolithic/Later Stone Age as well as some Middle Stone Age industries of Subsaharan Africa (e.g. Howieson's Poort). Blades, which are flakes at least twice as long as they are wide, can be produced by hard hammer percussion, soft hammer percussion, and punch (indirect percussion) methods. Blades were made into a range of forms such as **end scrapers**, **burins**, **backed knives**, **awls**, and **points**.

- **Personal adornment**: The earliest probable beads are perforated shells from South Africa, North Africa, and the Levant. These include shells from Blombos Cave in South Africa, dated to approximately 80,000 years ago (Henshilwood et al., 2004), specimens from North Africa dated to approximately 82,000 years ago (Bouzouggar et al., 2007), and ones from Israel dated by chemical means to between 100,000 and 125,000 years ago (Vanhaeren et al., 2006). Interestingly, this may have been a short-lived or sporadic tradition, and the next probable beads are found in Kenya at Enkapune Ya Muto at about 40,000 years ago (Ambrose, 1998) and in the Aurignacian of Europe and the Near East between 40,000 and 35,000 years ago.

- **Abstract decoration**: A clear geometric designed was inscribed with a stone tool on a piece of ochre at Blombos Cave in South Africa, dated to between 75,000 and 100,000 years ago (Henshilwood et al., 2009). Abstract designs are well-known in the Upper Palaeolithic art traditions in cave paintings, engravings and mobilary art of Europe, and later around the world. Some of these designs may symbolically represent specific words or concepts, although this is very difficult to verify archaeologically. Although the precise meaning of this abstract art is unknown, it has been argued that much of it could be entopic hallucinatory images seen during shamanistic trances (Lewis-Williams, 2002). Other repetitive abstract designs might stand for specific words or concepts and be a form of incipient writing. Yet others classified as abstract may, in fact, be representational, perhaps illustrating traps, huts or tents, or geographical features.

- **Very refined stone tools**: By the later Upper Palaeolithic of Europe, incredibly skilled stone tools were being made, such as the bifacial Solutrean leaf points. Heat treatment was also reportedly used on silcretes at the site of Pinnacle Point in South Africa about 72,000 years ago (Brown et al., 2009). Possible evidence of pressure flaking has been argued for some bifacial points at the site of Blombos in South Africa from approximately 80,000 years ago (Moure and Henshilwood, 2010), but in our judgement the flaking of these pieces appears to be indistinguishable from delicate soft hammer retouch.

- **Bone, antler, and ivory tools**: Bone tools (excepting soft hammers for knapping) are very rare prior to the Upper Palaeolithic/Later Stone Age. Beginning about 40,000 years ago, there is a major technological shift to other materials that can be shaped into a range of forms that might not be possible in stone, notably needles, spear throwers, and barbed harpoons. These materials would have been worked with stone tools such as burins and scrapers. Other artifacts include points, perforated batons, and pressure flakers.

- **Ground stone tools**: Archaeological evidence from Australia indicates that hunter-gatherers there began to manufacture ground stone axes beginning possibly about 35,000 years ago (Australian Archaeology, December 2010; Morwood and Trezise, 1989). Ground stone tools such as axes are normally not as sharp as a flaked stone tool, but ground edges can stay functional for longer periods of time before resharpening (re-grinding) is required. Ground axes are a substantial investment in time (requiring hours or even days to produce) and are normally hafted to a handle with some form of binding material (hide, sinew, vine, vegetable cordage, adhesive mastic, etc.) Such ground stone tools become especially common in the last 10,000 years with the rise of agricultural communities around the world as forests were cleared to plant crops.

- **Representational art**: The first clear evidence of representational art is seen in the form of early Upper Palaeolithic (Aurignacian) starting after 40,000 years ago, including animal and human sculptures in ivory as seen at Vogelherd (>30,000 years ago) and Hohle Fels (35,000 years ago) in Germany (Conard, 2009), and cave paintings such as at Chauvet Cave (32,000 years ago) in southwest France (Clottes, 2001). Interestingly, the first representational art in the prehistoric record is remarkably skilled and well-executed; unlike other aspects of technology, there is no evidence of a long period of simplicity that predates this expressive competence. The earliest representational art known in Africa comes from Apollo-11 Rockshelter in Namibia in southern Africa (Bednarik, 2003), dating to about 26,000 years ago.
- **Religion:** Although the roots of human religion very likely pre-date anatomically modern humans, many scholars point to the rich symbolic content of Upper Palaeolithic art and elaborate mortuary practices to argue that there must have been well-established religious and ritualistic behavior as well as a belief in an afterlife and a spirit world (Dickson, 1990). Some cave art authorities (e.g. Clottes and Lewis-Williams, 1998; Lewis-Williams, 2002) have suggested that the best interpretation of Upper Palaeolithic religion was a trance-induced shamanistic one. In one study of Upper Palaeolithic art caves in the Pyrenees (Reznikoff and Dauvois, 1988), it was found that high concentrations of art in caves coincided with areas of these caves with unusual acoustic properties (resonance and echoes), and suggested that these painted areas were also areas of chanting or singing during rituals.
- **Musical instruments:** Although there have been claims of musical instruments from the Middle Palaeolithic, these have not been substantiated on closer scrutiny. The earliest definitive musical instruments are in the form of bone flutes and whistles from the Aurignacian stage of the early Upper Palaeolithic, ca. 35,000 to 40,000 years ago (Conard et al., 2009).
- **Notational tallies:** During the Upper Palaeolithic, there are a number of marked bones, antlers, and stones, beginning around 32,000 years ago, which suggest that these objects may have been recording devices for counting or documenting natural phenomena (e.g., days, lunar months, numbers of people or animals, etc.) (Marshack, 1991). Notable examples include the Aurignacian bone plaques from Abri Lartet and Blanchard in the French Dordogne region.
- **"Supernatural" imagery**: Some of the iconography in Upper Palaeolithic art appears to represent entities which are not found in the natural world, and may represent some mythological creatures. This includes the ivory sculpture of a "lion-man" from Hohlenstein Stadel in Germany dated to 32,000 years ago (Wynn et al., 2009), the "Unicorn" (actually a two-horned fantastic animal) from Lascaux Cave from the French Perigord dated to about 15,000 years ago and the so-called antlered "Sorcerer" from Trois Freres Cave (Leroi-Gourhan, 1967) in the French Pyrenees also dated to approximately 15,000 years ago.
- **Tanged points**: The Aterian of North Africa has tanged points which may date as early as 100,000 years ago. Such tangs must denote hafting, presumably for a spear. Tanged points also emerge in the Solutrean of Europe about 25,000 years ago.
- **Spear Throwers**: The first mechanical devices for propelling spears are found in the Upper Palaeolithic of Europe going back to about 20,000 years ago. These hooked sticks (known in the Aztec world as *atlatls*) gave a hunters arm a longer lever with which to impart more speed (and distance and/or penetration power) to a spear.
- **Geometric microliths**: Small retouched forms, often made on blades or bladelets and in geometric shapes (trapezoids, crescents, etc.), could have been used as elements of composite tools such as arrows and knives. These forms emerge especially in the later Upper Palaeolithic and become widespread in the Mesolithic of early Holocene Eurasia and in many Later Stone Age industries in Africa.
- **Ceramic figurines**: At Dolni Vestonice in the Czech Republic, fired clay figurines of animals and humans are dated to approximately 26,000 years ago (Vandiver et al., 1989). These are the earliest known ceramic technologies, pre-dating pottery by some 14,000 years.
- **Mythology and folklore motifs**: It is likely that true folk traditions were passed on from generation to generation by the Upper Palaeolithic. One such folklore motif (a story or joke) may be manifested in the form of a sculpture of a deer or ibex, peering back at its rear, perhaps in the act of either defecating or giving birth. On the end of the protruding object is a bird, which forms the hooked end of the spear-thrower. Several sites in the French Pyrenees (and one site over 100 km north of the sites in the Pyrenees) have yielded such sculptures engraved on antler spear-throwers, including very complete specimens from the Magdalenian site of Mas d'Azil in the French Pyrenees dating to about 16,000 years ago and from the nearby site of Bedeilhac (Bahn, 1982). This recurring motif almost certainly represents a specific story maintained by oral tradition, presumably over a wide area in this

region.

- **Natural history:** Many of the representational animal images in painting, engraving, and sculpture in the Upper Palaeolithic show a remarkable degree of anatomical and behavioral detail. One specimen in particular is exceptionally fascinating. At the French Magdalenian site of Mas d'Azil, an antler carving (possibly part of a spear thrower) shows three horse heads: a small one (possibly a young individual), a larger head (probably an adult), and another large 'flayed' head showing the skull of a horse (Leroi-Gourhan, 1967). This piece could represent the life history of a horse.

- **Architecture**: Although there may be a few structures associated with European Middle Palaeolithic (presumably Neandertal) hominins, most recognizable hut or tent structures are known from the Upper Palaeolithic, such as those from Mezhirich and Molodova (21,000 years ago) in the Ukraine, Dolni Vestonice and Pavlov (27,000 years ago) in the Czech Republic, and Pincevent (12,000 years ago) in France (Vasil'ev et al., 2003).

- **Weaving**: Impressions on fired clay at Dolni Vestonice and Pavlov in the Czech Republic indicate that by 26,000 these occupants were weaving plant materials into baskets, mats, or textiles (Soffer et al., 1998). Before the advent of pottery this could have been a very important technology for making containers to carry foods, material culture, or, if lined with a waterproof material such as pitch, the storage or transport of water.

- **Needles/Sewing** The first bone needles appear about 25,000 years ago in the Upper Palaeolithic, suggesting sophisticated sewn apparel. Upper Palaeolithic figurines from Siberia show humans with parka-like outfits, denoting their adaptation to these severe environments.

- **Grinding stones**: Early mortars and pestles that were used to grind wild cereals are found in early agricultural sites in many parts of the world. Very early instances of such grinding stones have been reported from a Palaeolithic site in Italy, where evidence is claimed for the grinding of starch grains into flour 25,000 years ago (Aranguren *et al.*, 2007), and from an early Holocene site in China, where evidence for the grinding of acorns has been suggested from 11,000 years ago (Liu *et al.*, 2010). Such grinding stones became much more common with the rise of agricultural communities around the world in the last 10,000 years.

- **Hearths/boiling**: Clearly-made hearth structures, sometimes delineated by a circle of stones, are common in the Upper Palaeolithic (e.g., see Movius, 1966 and Leroi-Gourhan and Brézillon, 1983). Large quantities of fire-cracked rocks have suggested to some prehistorians that Upper Palaeolithic people were dropping hot stones into water (ethnographically, this could be done in a greased hide put over a depression in the ground to hold water) to boil the water for cooking (e.g. at El Miron Cave in Cantabrian Spain about 15,500 years ago) (Nakazawa et al., 2009). One advantage of boiling over roasting is that all of the nutrition of a food (e.g. fats, prized by hunter-gatherers) can be retained in a broth rather than drip away into a fire.

- **Lamps**: The earliest stone lamps are known from the Upper Palaeolithic of Europe starting approximately 40,000 years ago, pecked or carved out of a variety of softer stones (Beaune, 1987). These lamps probably used an animal fat as a fuel and a wick of moss or some other vegetable material. It is likely that torches were also used by this period, but no direct evidence has yet been found for these.

- **Pottery vessels**: The first fired **clay pots** are known from late Pleistocene and early Holocene hunter-gatherer populations in eastern Asia including sites in Japan (known as the early Jomon culture and possibly in eastern China and Russia, beginning at least 13,000 years ago (Kuzmin, 2006; Rice, 1999)). Such ceramic vessels could be storage containers or cooking pots (or both). During Holocene times pottery would be independently invented I the Near East, East Asia, and the Americas.

- **The emergence of "ethnicity"**: There is a general appreciation that beginning about 40,000 years ago, hominin material culture becomes much more variable in time and space. Stone tool types and art styles can become very particular and geographically and temporally diagnostic (for example, Aurignacian split-based bone points and carinated end scrapers, Gravettian backed points, Solutrean leaf points, Magdalenian sagaie and harpoons, and parrot-beaked burins) and Azilian painted pebbles.)

- **Longer-distance trade**: The Upper Palaeolithic indicates that trade reciprocity networks were greater than in earlier times, particularly between 25,000 and 15,000 years ago, with raw materials such as high-quality flint sometimes moving more than 200 kilometers, and exotic materials such as sea shells or Baltic amber also moving appreciable distances (Mellars, 1996).

- **The peopling of New Worlds: Australia, the Americas, and Siberia**: In the latter part of this Time Interval, human populations made significant incursions into new areas of the earth not previously inhabited by humans. Early occupation of Australia by modern humans began between 40,000 and 60,000 years ago during a glacial period when sea levels were lower, but still requiring the crossing of at least 60 miles of open sea between the Sunda land mass of southeastern Asia and the Sahel land mass that included Australia, New Guinea, Tasmania, and nearby islands. Siberia was occupied by Upper Palaeolithic times, as early as 40,000 years ago in southern Siberia, and with substantial sites throughout much of northeastern Asia established between 30,000 and 20,000 years ago. This region is widely thought to be the staging ground for human immigrations into Beringia, the land bridge between Siberia and Alaska exposed during the glacial maximum and whose steppe-like grassland likely supported large herds of herbivores, and then for the ultimate spread of these human populations into North America as the ice sheets began to recede. Such migrations would indicate populations armed with the appropriate tools and technology to cope successfully with challenging environments and crossings.

- **Broader-spectrum economies and the rise of farming communities**: Towards the end of the Pleistocene there is evidence, around the world, of intensification of foraging patterns among many hunter-gatherer groups. In many areas, a wider range of food items were exploited, including shellfish (mussels, oysters, crabs, lobster), fish (e.g. salmon), cereals (wheat, barley, millet, sorghum, rice, maize), etc. These so-called "broad spectrum economies" would, at the end of the Ice Age and the beginning of the Holocene, lead to the first farming communities. Intensification of certain types of hunting prey (e.g. sheep and goats) would lead to the domestication of certain animal species as well. Farming would ultimately allow much larger, sedentary communities which would lay the foundations for the first complex state societies or "civilizations."

Summary and Conclusion

Table 1 summarizes the major trends for each time interval discussed in this chapter. Overall, a gradual progression of technological sophistication is observed relative to Holloway *et al.*'s three stages of hominin brain evolution.

During Holloway *et al.*'s Stage 1 of hominin brain evolution, the first evidence of hominin stone tools emerges at 2.6 mya during the time of *Australopithecus garhi*. Fragmentary jaws and teeth may suggest that early *Homo* goes back to 2.3 million years ago (e.g. Hadar jaw AL-666), a few hundred thousand years after the first appearance of stone tools, but without crania and endocasts, it is not clear whether these non-paranthropine forms (i.e., not showing the dental features of robust australopithecines) show any evidence of encephalization or brain reorganization.

Holloway *et al.* Stage 2 (early *Homo*) is seen in the prehistoric record about 700,000 years after the first recognizable stone tools. Many palaeoanthropologists believe that this neurological evolution may be the consequence of expanding diet breadth and higher diet quality supported by the use of stone tools and allowing for the larger brain evident in evolving *Homo* (*Homo habilis* and *Homo rudolfensis*) during this time. During Holloway et al.'s Stage 2, Oldowan sites become more prevalent and widespread on the African landscape, and we see the emergence of Acheulean tools, with large handaxes and cleavers shaped from large flakes or cobbles. The spread of hominins and tool cultures over much of the southern to middle latitudes of Eurasia is also evident during Stage 2 of hominin brain evolution.

During Holloway et al.'s Stage 3 of brain evolution, significant technological and adaptive advances are observed in the archaeological record. During the early part of this phase, associated with *Homo heidelbergensis*, these include the development of refined Later Acheulean tools, commonly including extremely symmetrical and finely fashioned handaxes and cleavers, and often showing very intricate, controlled flaking and use of careful platform preparation and of soft hammer percussion in their production. Additional advances observed in the archaeological record and associated with *Homo heidelbergensis* include the use of wooden spears and apparent evidence of some controlled use of fire (though its incidence was still rare and may not have involved skilled production of fire), as well as possible emergence of early ritual or symbolic behavior. Changes associated with Neandertals add more complex (hafted) tools, apparent habitual use of fire, use of personal adornment, and burial of the dead to the adaptive and behavioral repertoire.

Such trends continued in earnest among the modern human populations as they grew and spread, eventually involving the emergence of flourishing art traditions, addition of other materials (bone, antler, ivory) to the toolmaking systems, appearance of needles and sewing, development of habitual architecture in various forms, and evidence for long-distance transport and trade of materials. By the time of the Upper Palaeolithic, regional patterns emerge in the archaeological record which would appear to indicate geographically distinct clusters of traditions, perhaps indications of 'ethnicity' mirroring that observed among human groups in recent and modern times. The emergence and evolution of complex symbolic behavior during this stage of brain evolution is evidenced in highly endowed burials, often prolific use of ornaments, and the proliferation of artistic traditions (including sometimes elaborate decoration of utilitarian tools). The emergence of such regional patterning and the evidence for complex symbolic behaviors may indicate evolution of complex language systems and abilities during this stage of brain evolution.

Thus, there are important changes in hominin behaviors indicated in the archaeological record correlated

Table 1.

Time Interval	Age	Hominin form	cc	EQ	Holloway et al. Stage	Technology, behavior, other
ONE	3.25-2.75 Ma	*Australopithecus afarensis*	445	(0.43)	Stage 1	(no established evidence)
TWO	2.75-2.25 Ma	*Australopithecus garhi*	450	(.45?)		Oldowan artifacts Early sites in Africa
THREE	2.25-1.75 Ma	*Homo habilis* *Homo rudolfensis* Early *Homo* (grouped)	610 788 698	(.62) (.66) (~.64)	Stage 2	Cut-marked bone Oldowan artifacts
FOUR	1.75-1.25 Ma	Early *Homo erectus*	800	(0.58)	Stage 2	Early Acheulean (& Oldowan) Africa & Eurasia
FIVE	1.25-0.75 Ma	*Homo erectus*	950	(0.68)	Stage 2	Acheulean
SIX	0.75-0.25	*Homo heidelbergensis*	1260	(0.81)	Stage 3	Later Acheulean Soft hammers Spears Some control of fire Ritual?
SEVEN	0.25-0.03 Ma	*Homo neandertalensis*	1427	(0.99)	Stage 3	Middle Palaeolithic Habitual control of fire Hafting (composite tools) Burial Decoration (late)
SEVEN	0.20 Ma – present	*Homo sapiens*	1496	(0.90)	Stage 3	Middle & Upper Palaeolithic Blade tools Habitual control of fire Habitual shellfish exploitation Decoration Iconic art Musical instruments Bone, antler, ivory tools More complex composite tools Habitual architecture Longer-distance transport Accelerating elaboration of technology Last 10,000 years: food production, complex societies

with the progressive stages of hominin brain evolution proposed by Holloway et al. (2004) based upon their study of hominin fossil and endocranial evidence. These changes involve appearance and evolution of new technologies, adaptive behavioral shifts indicated by the new technologies, ultimate spread and adaptation to very new environments, and, eventually, emergence of a full-blown symbolic dimension among human ancestors.

In addition to major changes from one stage to the next in technologies and behaviors, there are also some notable changes over time that can be observed within a single stage in terms of technology, behavior and adaptation. This is particularly during Holloway et al.'s Stage 2 and Stage 3 of hominin brain evolution. Stage 2 involves early *Homo* and *Homo erectus* and archaeological evidence that chronicles significant behavioral changes, including the addition of Acheulean technology to the Oldowan stone tool repertoire, and the spread of hominins into Eurasia and consequent adaptation to new environments. During Stage 3, which involves the hominin forms *Homo heidelbergensis*, *Homo neandertalensis*, and finally *Homo sapiens,* archaeological evidence shows technological transitions from Later Acheulean, to Middle Palaeolithic/Middle Stone Age, and then Upper (or Late) Palaeolithic/Later Stone Age technological periods of the Pleistocene, involving increasing complexity and sophistication of the tool-kit and of overt symbolic dimensions, and, ultimately, cultural developments supporting complex societies and profound, accelerating technological innovations of the Holocene.

Such behavioral transitions within a single stage of hominin brain evolution could be due to various factors, including possible neurological changes and potential cultural dynamics. Possible neurologically-based changes could include reorganization of neural pathways in the brain that might support more complex behaviors and conceptualization or enable increased capacity for more complex communication and or language. In the case of language, other biological complements of such neurological changes, also with a presumed genetic basis, could include structural/musculoskeletal changes in the vocal tract. Cultural factors which might support such profound behavioral changes within a single stage of brain evolution could involve increased contact and sharing among individuals and groups and enhanced cultural means of storing information so as to increase the cultural repertoire of knowledge and, potentially, gradually but dramatically increase the rate of cultural change. It is possible that both major types of factors played a role in cultural elaboration and innovations during Stages 2 and 3 of hominin brain evolution, with the cultural aspects likely playing an increasingly important role over time.

References

Aiello L., and Wheeler, P., 1995. The expensive tissue hypothesis: the brain and the digestive system in humans and primate evolution. Current Anthropology 36, 199-221.

Allman, J., 2000. Evolving Brains. New York: Scientific American Library.

Alperson-Afil, N., Sharon, G., Kislev, M., Melamed, Y., Zohar, I., Ashkenazi, S., Rabinovich, R., Biton, R., Werker, E., Hartman, G., Feibel, C., Goren-Inbar, N., 2009. Spatial organization of hominin activities at Gesher Benot Ya'aqov, Israel. Science 326, 1677-1680.

Ambrose, S., 2001. Palaeolithic technology and human evolution. Science 291, 1748-1753.

Ambrose, S., 1998. Chronology of the Later Stone Age and food production in East Africa. Journal of Archaeological Science 25, 377-392.

Aranguren, B., Becattini, R., Mariotti Lippi, M., Revedin, A., 2007. Grinding flour in Upper Palaeolithic Europe (25,000 years bp). Antiquity 81, 845-855.

Arsuaga, J., Martinez, I., 2004. Atapuerca y la Evolución Humana. Barcelona: Fundació Caixa Catalunya.

Bahn, P., 1982. Inter-site and inter-regional links during the Upper Palaeolithic: the Pyrenean evidence. Oxford Journal of Archaeology 1. 247-268.

Barham, L., 2002. Systematic pigment use in the Middle Pleistocene of south-central Africa. Current Anthropology 43, 181-190.

Bar-Yosef, O., 2002. The Upper Palaeolithic revolution. Annual Review of Anthropology 31, 363-393.

Beaune, S., 1987. Palaeolithic lamps and their specialization: A hypothesis. Current Anthropology 28, 569-577.

Bednarik, R., 2003. The earliest evidence of Paleoart. Rock Art Research 20, 95.

Bouzouggar, A., Barton, N., Vanhaeren, M., d'Errico, F., Collcutt, S., Higham, T., Hodge, E., Parfitt, S., Rhodes, E., Schwenninger, J-L., Stringer, C., Turner, E., Ward, S., Moutmir, A., Stambouli, A., 2007. 82,000-year-old shell beads from North Africa and implications for the origins of modern human behavior. Preceedings of the National Academy of Sciences 104, 9964-9969.

Brown, K., Marean, C., Herries, A., Jacobs, Z., Tribolo, C., Braun, D., Roberts, D., Meyer, M., Bernatchez, J., 2009. Fire as an engineering tool of early modern humans. Science 325, 859-862.

Clottes, J. (Ed.), 2001. La Grotte Chauvet: L'Art des Origines. Paris: Seuil.

Clotte, J., Lewis-Williams, D., 1998. The Shamans of Prehistory: Trance and Magic in the Painted Caves. New York: Abrams.

Conard, N., 2009. A female figurine from the basal Aurignacian of Hohle Fels Cave in southwestern Germany. Nature 459, 248-252.

Conard, N., Malina, M., Munzel, S., 2009. New flutes document the earliest musical tradition in southwestern Germany. Nature 460, 737-740.

Coolidge, F., Wynn, T., 2009. The Rise of *Homo sapiens*: The Evolution of Modern Thinking. Malden, MA: Wiley-Blackwell.

Deacon, T., 1997. The Symbolic Species: The Co-Evolution of Language and the Brain. New York: Norton.

de Heinzelin, J., Clark, J.D., Schick, K., Gilbert, W.H., 2000. The Acheulean and the Plio-Pleistocene Deposits of the Middle Awash Valley, Ethiopia. Tervuren: Musée Royal de l'Afrique Central, Volume 104.

Delson, E., Tattersal, I., Van Couvering, J., Brooks, A., (Eds), 2000. Encyclopedia of Human Evolution and Prehistory, second edition. New York: Garland.

Deetz, J., 1967. Invitation to Archaeology. Garden City: Natural History Press.

Dickson, D., 1990. The Dawn of Belief: Religion in the Upper Palaeolithic of Southwestern Europe. Tucson: University of Arizona Press.

Domínguez-Rodrigo, M., Pickering, T., Bunn H., 2010. Configurational approach to identifying the earliest hominin butchers. Proceedings of the National Academy of Sciences 107.

Dunbar, R. 1992. Neocortex size as a constraint on group size in primates. Journal of Human Evolution 20, 469-493.

Dunbar, R. 1993. Coevolution of neocortical size, group size, and language in humans. Behavioral and Brain Sciences 16, 681-735.

Dunbar, R., 2003. Evolution of the social brain. Science 302, 1160-1161.

Falk, D., 1987. Hominid paleoneurology. Annual Review of Anthropology 16, 13-30.

Gibson, K., 1986. Cognition, brain size, and the extraction of embedded food resources. In: Else, J., Lee, P. (Eds.), Primate Ontogeny, Cognition and Social Behavior. Cambridge: Cambridge University Press, pp. 93-103.

Gibson, K., 2002. Evolution of human intelligence: the role of brain size and mental construction. Brain Behavior and Evolution 59, 10-20.

Gibson, K., Ingold, T., (Eds.), 1993. Tools, Language and Cognition in Human Evolution. Cambridge University Press.

Goren-Inbar, N, Peltz, S., 1995. Additional remarks on the Berekhat Ram figure. *Rock Art Research* 12, 131-132

Goren-Inbar, N., Sharon, G., Melamed, Y., Kislev, M., 2002. Nuts, nut cracking, and pitted stones at Gesher Benot Ya'aqov, Israel. Proceedings of the National Academy of Sciences 99, 2455-2460.

Goren-Inbar, N., Alperson, N., Kislev, M., Simchoni, O., Melamed, Y., Ben-Nun, A., Werker, E., 2004. Evidence of hominin control of fire at Gesher Benot Ya`aqov, Israel. Science 304, 725-727.

Gowlett, J., 1996. Mental abilities of early *Homo*: Elements of constraint and choice in the rule system. In: Mellars, P. and Gibson. K. (Eds.), Modeling the Early Human Mind. Cambridge: McDonald Institute for Archaeological Research, pp. 191-215.

Henshilwood, C., d'Errico, F., Vanhaeren, M., Van Niekerk, K., Jacobs, Z., 2004. Middle Stone Age shell beads from South Africa. Science 304, 404.

Henshilwood, C., d'Errico, F., Watts, I., 2009. Engraved ochres from the Middle Stone Age levels at Blombos Cave, South Africa. Journal of Human Evolution 57, 27-47.

Holloway, R.L., 1967. The evolution of the human brain: Some notes towards a synthesis between neural structure and the evolution of complex behavior. International Journal of General Systems 12, 3-19.

Holloway, R.L., 1969. Some questions on parameters of neural evolution in primates. In: Petras, J. and C. Noback (Eds.), Comparative and Evolutionary Aspects of the Vertebrate Central Nervous System. Annals of the New York Academy of Sciences 16, 332-340.

Holloway, R.L., 1981. Culture, symbols, and human brain evolution: A synthesis. Dialectical Anthropology 5, 287-303.

Holloway, R., Broadfield, D., and Yuan, M., 2004. The Human Fossil Record. Volume Three: Brain Endocasts: The Palaeoneurological Evidence. Hoboken: John Wiley & Sons, Inc.

Isaac, G., 1986. Foundation stones: early artifacts as indicators of activities and abilities. In: Bailey, G. and Callow, P. (Eds.), Stone Age Prehistory: Studies in Honor of Charles McBurney. London: Cambridge University Press, pp. 221-241.

Jerison, H., 1973. Evolution of the Brain and Intelligence. New York: Academic Press.

Keeley, L., 1980. Experimental Determination of Stone Tool Uses: A Microwear Analysis. Chicago: University of Chicago Press.

Klein, R.G., 2009. The Human Career: Human Biological and Cultural Origins. Third edition. Chicago: University of Chicago Press.

Kuzmin, Y., 2006. Chronology of the earliest pottery in East Asia: progress and pitfalls. Antiquity 80, 362-371.

Leroi-Gourhan, A., 1967. Treasures of Prehistoric Art. New York: Abrams.

Leroi-Gourhan, A., Brézillon, M., 1983. Fouilles de Pincevent: Essai d'Analyse Ethnographique d'un Habitat Magdalénien. Paris: Éditions du Centre Nationale de la Recherche Scientifique.

Lewis-Williams, D., 2002. The Mind in the Cave: Consciousness and the Origins of Art. London: Thames and Hudson.

Lindly J., Clark, G., 1990. Symbolism and modern human origins. Current Anthropology 31, 233-261.

Liu, L., Field, J., Fullagar, R., Bestel, S., Chen, X., Ma, X., 2010. What did grinding stones grind? New light on Early Neolithic subsistence economy in the Middle Yellow River Valley, China. Antiquity 84, 816-833.

Martin, R., 1996. Scaling the mammalian brain: the maternal energy hypothesis. News in Psychological Sciences 11, 149-156.

Marshack, A., 1991. The Roots of Civilization. Mount Kisco, New York: Moyer Bell Limited.

McBrearty, S., Brooks, A. The revolution that wasn't: a new interpretation of the origin of modern human behavior. Journal of Human Evolution 39, 453-463.

McGrew, W., 1992. Chimpanzee Material Culture: Implications for Human Evolution. Cambridge: Cambridge University Press.

McPherron, S., Alemseged, Z., Marean, C., Wynn, J., Reed, D., Geraads, D., Bobe, R., Bearat, H., 2010. Evidence for stone-tool-assisted consumption of animal tissue before 3.39 million years ago at Dikika, Ethiopia. Nature 466, 857-860.

Mellars, P., 1996. The Neandertal Legacy: An Archaeological Perspective from Western Europe. Princeton: Princeton

University Press.

Mellars, P., and Gibson, K. (Eds.), 1996. Modeling the Early Human Mind. Cambridge: McDonald Institute for Archaeological Research.

Mellars, P., Boyle, K., Bar-Yosef, O., Stringer, C. (Eds.), 2007. Rethinking the Human Revolution. Cambridge: McDonald Institute for Archaeological Research.

Mitchell, J., 1995. Studying biface utilization at Boxgrove: Roe deer butchery with replica handaxes. Lithics 16, 64-69.

Mithen, S., 1996. The Prehistory of the Mind: The Cognitive Origins of Art, Religion, and Science. London:

Morwood, M., Trezise, P., 1989. Edge-ground axes in Pleistocene Greater Australia: new evidence from S.E. Cape York Peninsula. Queensland Archaeological Research 6, 77-90.

Mourre V., Villa, P., and Henshilwood, C., 2010. Early use of pressure flaking on lithic artifacts at Blombos Cave, South Africa. Science 330, 659-662.

Movius, H., 1949. The Lower Palaeolithic cultures of southern and eastern Asia. Transactions of the American Philosophical Society 38, pt. 4, 329-420.

Movius, H., 1966. The Hearths of the Upper Périgordian and Aurignacian Horizons at the Abri Pataud, Les Eyzies (Dordogne), and Their Possible Significance. American Anthropologist 68, 296-325.

Nakazawa, Y., Straus, L., Gonzalez-Morales, M., Solana, D., Saiz, J., 2009. On stone-boiling technology in the Upper Paleolithic: behavioral implications from an Early Magdalenian hearth in El Mirón Cave, Cantabria, Spain. Journal of Archaeological Science 36, 684-693.

Oakley, K.P, 1976. Man the Tool-Maker. Seventh Edition. Chicago: University of Chicago Press.

Oakley, K., Andrews, P., Keeley, L., Clark, J.D., 1977. A reappraisal of the Clacton spear point. Proceedings of the Prehistoric Society 43, 13-30.

Mania, D. Mania, U., 2005. The natural and socio-cultural environment of Homo erectus at Bilzingsleben, Germany. In: (Gamble, C. Porr, M., Eds.), The Hominid Individual in context: Archaeological Investigations of Lower and Middle Palaeolithic Landscapes, Locales, and Artefacts. London: Routledge, pp. 98-114.

Noble, W., Davidson, I., 1996. Human Evolution, Language and Mind. Cambridge: Cambridge University Press.

Parker, S., Gibson, K., 1979. A developmental model for the evolution of language and intelligence in human evolution. Behavioral and Brain Sciences 2, 367-408.

Parker, S., McKinney, P. (Eds.), 1999. Origins of Intelligence: The Evolution of Cognitive Development in Monkeys, Apes, and Humans. Baltimore: Johns Hopkins University Press.

Pelegrin, J., 2005. Remarks about archaeological techniques and methods of knapping: elements of a cognitive approach to stone tools. In: Roux, V., Bril, B (Eds.), Stone Knapping: The Necessary Conditions for Uniquely Human Behavior. Cambridge: McDonald Institute for Archaeological Research, pp. 23-34.

Pitts, M., Roberts, M., 1997. Fairweather Eden: Life Half a Million Years Ago As Revealed at the Excavations at Boxgrove. London: Century Books, Ltd.

Renfrew, C. and Scarre, C. (Eds), 1998. Cognition and Material Culture: The Archaeology of Symbolic Storage. Cambridge: McDonald Institute for Archaeological Research.

Renfrew, C., Frith, C., Malafouris, L. (Eds), 2009. The Sapient Mind: Archaeology Meets Neuroscience. Cambridge: Cambridge University Press.

Reznikoff, I., Dauvois, M., 1988. La dimension sonare des grottes ornées. Bulletin de la Société Préhistoriques Francaise 85, 238-246.

Rice, P., 1999. On the origins of pottery. Journal of Archaeological Method and Theory 6, 1-54.

Roberts, M., Parfitt, S., 1999. Boxgrove: A Middle Pleistocene Hominid Site at Eartham Quarry, Boxgrove, West Sussex. London: English Heritage.

Roux, V., Bril, B (Eds), 2005. Stone Knapping: The Necessary Conditions for Uniquely Human Behavior. Cambridge: McDonald Institute for Archaeological Research.

Schick, K. 1994. The Movius Line reconsidered: perspectives on the earlier Palaeolithic of Eastern Asia. In: Corruchini, R., Ciochon, R. (Eds) Integrative Paths to the Past: Palaeoanthropological Advances in Honor of F. Clark Howell. Englewood Cliffs, New Jersey: Prentice-Hall, pp. 569-596.

Schick, K., Toth, N., 1993. Making Silent Stones Speak: Human Evolution and the Dawn of Technology. New York: Simon & Schuster.

Schick, K., Toth, N. 1994. Early Stone Age technology in Africa: a review and case study into the nature and function of spheroids and subspheroids. In: Corruchini, R., Ciochon, R. (Eds), Integrative Paths to the Past: Palaeoanthropological Advances in Honor of F. Clark Howell. Englewood Cliffs, New Jersey: Prentice-Hall, pp. 429-449.

Schick, K., Toth, N., 2006. An overview of the Oldowan Industrial Complex: the sites and the nature of their evidence. In: Toth, N., Schick, K. (Eds.), The Oldowan: Case Studies into the Earliest Stone Age. Gosport, Indiana: Stone Age Institute Press, pp. 3-42.

Schick, K., and Toth, N., 2009. The Cutting Edge: New Approaches to the Archaeology of Human Origins. Gosport, Indiana: Stone Age Institute Press.

Schick, K.D., Toth, N., Garufi, G., Savage-Rumbaugh, E.S., Rumbaugh, D., Sevcik, R., 1999. Continuing Investigations into the Stone Tool-making and Tool-using Capabilities of a Bonobo (*Pan paniscus*). Journal of Archaeological Science 26(7), 821-832.

Schoenemann, P.T. 2006. Evolution in the size and functional areas of the brain. Annual Review of Anthropology 35, 379-406.

Semaw, S., 2006. The oldest stone artifacts from Gona (2.6-2.5 Ma), Afar, Ethiopia: Implications for understanding the earliest stages of stone knapping. In: Toth, N., Schick, K. (Eds.), The Oldowan: Case Studies into the Earliest Stone Age. Gosport, Indiana: Stone Age Institute Press, pp. 43-75.

Shipman, P., 2010. The Cutting Edge: Can stone-tool marks on fossils be distinguished from tooth marks? American Scientist 462.

Shipman, P., Walker, A. 1989. The costs of becoming a predator. Journal of Human Evolution 18, 373-392.

Soffer, O., Adovasio, J., Hyland, D., Klima, B., Svoboda, J., 1998. Perishable technologies and the genesis of the eastern Gravettian. Anthropologie 36, 43-68.

Stout, D., 2002. Skill and cognition in stone tool production: a case study from Irian Jaya. Current Anthropology 45, 693-722.

Stout, D., 2005a. Neural foundations of perception and action in stone knapping. In: Roux, V., Bril, B (Eds) Stone Knapping: The Necessary Conditions for a Uniquely Hominin Behavior. Cambridge: McDonald Institute for Archaeological Research, pp. 273-286.

Stout, D., 2005b. The social and cultural context of stone-knapping skill acquisition. In: Roux, V., Bril, B. (Eds) Stone Knapping: The Necessary Conditions for a Uniquely Hominin Behavior. Cambridge: McDonald Institute for Archaeological Research, pp. 273-286.

Stout, D., 2006. Oldowan toolmaking and hominin brain evolution: Theory and research using positron emission tomography (PET). In: Toth, N., Schick, K., (Eds) The Oldowan: Case Studies into the Earliest Stone Age. Gosport, Indiana: Stone Age Institute Press, pp. 267-305.

Stout, D., Toth, N., Schick, K., Stout, J., Hutchins, G., 2000. Stone tool-making and brain activation: positron emission tomography (PET) Studies. Journal of Archaeological Science 27, 1215-1223.

Stout, D., Toth, N., Schick, K., 2006. Comparing the neural foundations of Oldowan and Acheulean toolmaking: a pilot study using positron emission tomography (PET). In: Toth, N. Schick, K. (Eds.) The Oldowan: Case Studies into the Earliest Stone Age. Gosport, Indiana: Stone Age Institute Press.

Stout, D., Toth, N., Schick, K. Chaminade, T., 2009. Neural correlates of Early Stone Age toolmaking: technology, language and cognition in human evolution. In Renfrew, C., Frith, C., and Malafouris, L. (Eds.) The Sapient Mind: Archaeology Meets Neuroscience. Cambridge: Cambridge University Press, pp. 1-19.

Stout, D., Toth, N., Schick, K., 2010. Understanding Oldowan knapping skill: an experimental study of skill acquisition in modern humans. In: (Schick, K., Toth, N., Eds.), The Cutting Edge: New Approaches to the Archaeology of Human Origins. Gosport, Indiana: Stone Age Institute Press, pp. 247-265.

Thieme, H., 1997. Lower Palaeolithic hunting spears from Germany. Nature 385, 807-810.

Thieme, H., 1998. The oldest spears in the world: Lower Palaeolithic hunting weapons from Schöningen, Germany. In: (Carbonell, E., Bermudez de Castro, J., Arsuaga, J., Rodriguez, X., Eds.), The First Europeans: Recent Discoveries and Current Debate. Burgos: Caja de Burgos, pp. 171-193.

Thieme, H. 2005. The Lower Palaeolithic art of hunting: the case of Schoeningen 13 II-4, Lower Saxony, Germany. In: (Gamble, C. , Porr, M. Eds.), The Hominid Individual in Context: Archaeological Investigations of Lower and Middle Palaeolithic Landscapes, Locales and Artefacts. London: Routledge. pp. 115-132.

Toth, N., 1985a. Archaeological evidence for preferential right-handedness in the Lower and Middle Palaeolithic, and its possible implications. Journal of Human Evolution 14, 607-614.

Toth, N., 1985b. The Oldowan reassessed: a close look at early stone artifacts. Journal of Archaeological Science 12, 101-120.

Toth, N., 1990. The prehistoric roots of a human concept of symmetry. Symmetry: Culture and Science 1, 257-281.

Toth, N., Schick, K., 1993. Early stone industries and inferences regarding language and cognition. In: (Gibson, K., and Ingold, T., Eds.), Tools, Language and Cognition in Human Evolution. Cambridge: Cambridge University Press, pp. 346-362.

Toth, N., Schick, K. (Eds), 2006. The Oldowan: Case Studies into the Earliest Stone Age. Gosport, Indiana: Stone Age Institute Press.

Toth, N., Schick, K., 2007. Overview of Palaeolithic archaeology. In: Henke, W., and Tattersal, I. (Eds). Handbook of Palaeoanthropology: Volume 3. New York: Springer, pp. 1943-1963.

Toth, N., Schick, K., 2009a. Early hominids. In: (Cunliffe, B., Goshden, C., and Joyce, R. ,Eds.). The Oxford Handbook of Archaeology. Oxford: Oxford University Press.

Toth, N., Schick, K., (Eds), 2009b. The Cutting Edge: New Approaches to the Archaeology of Human Origins. Gosport, Indiana: Stone Age Institute Press.

Toth, N., Schick, K., 2009c. The Importance of Actualistic Studies in Early Stone Age Research: Some Personal Reflections. In (Schick, K. and Toth, N., Eds,), The Cutting Edge: New Approaches to the Archaeology of Human Origins. Gosport, Indiana: Stone Age Institute Press, pp. 267-344.

Toth, N., Schick, K., 2009d. The Oldowan: the tool making of early hominins and chimpanzees compared. Annual Reviews of Anthropology 38, 289-305.

Toth, N., Schick, K., in press. Why the Acheulean? Experimental studies of the manufacture and function of Acheulean tools. Proceedings, Colloque internationale: Les Culture a Bifaces du Pleistocene Inferieur et Moyen dans la Mode.

Toth, N., Schick, K.D., Savage-Rumbaugh, E.S., Sevcik, R.A., Rumbaugh, D.M., 1993. Pan the Tool-Maker: Investigations into the Stone Tool-Making and Tool-Using Capabilities of a Bonobo (*Pan paniscus*). Journal of Archaeological Science 20(1), 81-91.

Toth, N., Schick, K., Semaw, S., 2006. A comparative study of the stone tool-making skills of Pan, Australopithecus, and Homo sapiens. In: Toth, N., Schick, K. (Eds), The Oldowan: Case Studies into the Earliest Stone Age. Gosport, Indiana: Stone Age Institute Press.

Vandiver, P., Soffer, O., Klima, B., Svoboda, J, 1989. The origins of ceramic technology at Dolni Věstonice, Czechoslovakia. Science 246, 1002-1008.

Vanhaeren, M., d'Errico, F., Stringer, C., James, S., Todd, J., Mienis, H., 2006. Middle Paleolithic shell beads in Israel and Algeria. Science 312, 1785-1788.

Vasil'ev, S., Soffer, O., Kozlowski, J., 2003. Perceived Landscapes and Built Environments: The Cultural Geography of Late Paleolithic. Oxford: Archaeopress.

Washburn, S., 1959. Speculations on the interrelations of the history of tools and biological evolution. In (Spuhler, Ed.) The Evolution of Man's Capacity for Culture. Detroit: Wayne State University Press.

Washburn, S., 1960. Tools and human evolution. Scientific American 203(9), 63-75.

Whiten, A., Goodall, J., McGrew, W., Nishida, T., Reynolds, V., Sugiama, Y., Tutin, C., Wrangham, R., Boesch, C. 1999. Culture in chimpanzees. Nature 399, 682-685.

Whiten, A., Schick, K., Toth, N., 2009. The evolution and cultural transmission of percussive technology: integrating evidence from palaeoanthropology and primatology. Journal of Human Evolution 57, 420-435.

Wynn, T., 1989. The Evolution of Spatial Competence. Urbana: University of Illinois Press.

Wynn, T., McGrew, W., 1989. An ape's view of the Oldowan. Man 24, 383-398.

Wynn, T., Coolidge, F., Bright, M., 2009. Hohlenstein-Stadel and the evolution of human conceptual thought. Cambridge Archaeological Journal 19, 73-84.